A Century of Nobel Prizes Recipients

Chemistry, Physics and Medicine

A Century of
Nobel Prizes Recipients
Chemistry, Physics and Medicine
EDITED BY
Francis Leroy

Translated from:
Dans la Serie: Les Prix Nobel de Science

Dictionnaire Encyclopédique des Prix Nobel de Médecine
Francis Leroy
ISBN: 2-87294-003-X
Dépôt légal: Octobre 1997

Dictionnaire Encyclopédique des Prix Nobel de Chimie
Claude Ronneau et Francis Leroy
ISBN: 2-87294-004-9
Dépôt légal: Octobre 1997

Dictionnaire Encyclopédique des Prix Nobel de Physique
Guy Demortier
ISBN: 2-87294-005-7
Dépôt légal: Octobre 1998

Library of Congress Cataloging-in-Publication Data
A catalog record for this book is available from the Library of Congress.

ISBN: 0-8247-0876-8

This book is printed on acid-free paper.

Headquarters
Marcel Dekker, Inc.
270 Madison Avenue, New York, NY 10016
tel: 212-696-9000; fax: 212-685-4540

Eastern Hemisphere Distribution
Marcel Dekker AG
Hutgasse 4, Postfach 812, CH-4001 Basel, Switzerland
tel: 41-61-260-6300; fax: 41-61-260-6333

World Wide Web
http://www.dekker.com

The publisher offers discounts on this book when ordered in bulk quantities. For more information, write to Special Sales/Professional Marketing at the headquarters address above.

Current printing (last digit):
10 9 8 7 6 5 4 3 2 1

PRINTED IN SPAIN

Preface

There can be no doubt that the prestige associated with the Nobel Prize has been a driving force in the extraordinary advances in science throughout the course of the 20th century. The Nobel Foundation has given awards to no fewer than 450 scientists, men and women whose stories have assimilated themselves into the public consciousness along with their discoveries and inventions. In the beginning, the three "hard sciences"-chemistry, physics, and medicine-were clearly separated, but were later found to overlap frequently, physics often being the source of concepts and equipment that enabled progress in chemistry and medicine. Was not the first Nobel winner in physics, Wilhelm Röntgen, with his discovery of X rays, already on the brink of major therapeutic applications? And have not the radioactive elements that won the Nobel Prize in physics for Irene and Frederic Joliot-Curie been used mainly in chemistry, especially cellular chemistry, allowing us to follow the successive steps of fundamental life processes such as photosynthesis or the Krebs cycle? The attribution of the Nobel Prize also brings to light a social effect that cannot be neglected: the advent of teamwork. It has become rare indeed for the prize to be awarded to one individual. On the contrary, it is increasingly common for two or even three researchers to be cited each year, regardless of the discipline. This corresponds to the interdisciplinary developments among the three aforementioned fields as well as the passion provoked by the extraordinary power of investigation of the scientific method.

The humanistic value of scientific research is tightly linked to man's use of its applications. History has amply demonstrated that the latter can bring a better state of being to society and the public health, for example, the telecommunications industry. We have been great beneficiaries of scientific research and without a doubt it will contribute further in the future. But progress also allowed for the increased destructive power of weapons. Moreover, it enabled us to take advantage of certain discoveries to create applications-for example, the application of genetic ingenuity to human cloning, as well as to modern electronic diffusion-that pose real ethical

If the Committee charged with awarding the Nobel Prize recognizes fundamental research first of all, it is therefore important that it weigh all its authority with states and institutions to promote the adoption of control systems to prevent the work and research of numerous scientists, in particular those implicated in fundamental research, from being used in ways counter to the benefit of mankind.

One hundred years of knowledge acquired since the beginning of the 20th century have essentially turned our understanding of the universe and of life on its ear. What we know today was unthinkable before; it should be the same in the future. The coming decades will probably further increase our perception of the nature of matter and the forms it takes, and the Nobel Foundation will have to distinguish among the multitude of new discoveries those which will best contribute to our heritage as a whole.

Acknowledgments

This work could not have been accomplished without the contributions of the draftsmen and artists whose drawings illustrate the workings of equipment or reproduce images obtained by electron microscopy or scanning microscopy. Several allegorical works pay homage to the works of certain laureates and add a truly artistic dimension to the work. I address my thanks most particularly to Francis Delarivière, Luc Chaufoureau, Philippe Rampelberg, Simon Dupuis, Kamenica Nedzad and Philippe Devallée.

The translation from the French was done by Nathalie Jockmans (reviewed by Guy Demortier) (for the laureates in physics), by Kathleen Broman, Robert Crichton and Jean-Marie André (for the laureates in chemistry) and by Kathleen Broman, Richard Epstein, Rebecca Petrush and János Freuling (for the laureates in medicine). I would especially like to express my gratitude and appreciation to my friend Anna Pavlovna Finkelstein, a young Russian student who graciously presented me with the translation of several excerpts from the original texts in chemistry and medicine.

My sincerest thanks also go to the laureates of the Nobel Prize who kindly contributed the information that enabled me to complete the texts relating to their works. In the field of medicine, this was principally Werner Arber, Francis Crick, James D. Watson, Christian de Duve, Jean Dausset, Paul Greengard, Roger Guillemin, Lee Hartwell, Andrew Huxley, Godfrey Hounsfield, Eric Kandell, Arthur Kornberg, Joseph Murray, Paul Nurse, Timothy Hunt, Erwin Neher, Richard Roberts, Bert Sakmann, Harold Varmus and Torsten N. Wiesel. In the field of chemistry, I thank Sidney Altman, Derek Barton (document submitted by Paul Potier), Paul Boyer, Donald J. Cram, Paul Crutzen, Robert F. Curl, Johann Deisenhofer, Manfred Eigen, Walter Gilbert, Herbert Hauptman, Jerome Karle, Harold Kroto, Yuan Tseh Lee, William Lipscomb, Kary Mullis, George A. Olah, Max F. Perutz, John A. Pople,

In the background is an artist's drawing of a striated muscle cell. This work contains eight original illustrations of the same size.

Frederick Sanger, Richard Smalley and Ilya Prigogine. The University of Toulouse gave me a biographical text on Paul Sabatier. Several texts on recent laureates were written by Jean-Marie André, professor of chemistry at the Facultés Universitaires Notre-Dame de la Paix in Namur, Belgium. Nearly one hundred Nobel laureates, in physics as well as chemistry and medicine, authorized me to reproduce their photographs and I express to them my sincere gratitude.

I am much obliged to my daughters, Claire and Nathalie, for allowing me to reproduce the image oftheir magnificent blue eyes to illustrate the text in reference to Paul Karrer, Nobel Prize recipient in 1937 for his work on carotenoids.

I also thank Marcel Dekker, Inc., who agreed to publish this work. I offer them in return a work in which I gladly invested a significant part of my time and means. I am also grateful to Barbara Mathieu, production editor, for her wonderful help during all the stages of development of this project. Mrs. Julie Chytrowski is not to be forgotten, who played an irreplaceable role as an intermediary between Marcel Dekker, Inc., and myself. Last but not least, I would like to thank Professor Guy Demortier, author of the biographies of the Nobel laureates in physics, as well as Professor Claude Ronneau, author of the biographics of a number of the Nobel laureates in chemistry.

Guy Demortier would particularly like to thank Mrs. Chantal Honhon for her continuing assistance in organizing the presentation of the biographies of the winners of the Nobel prize for physics, and Yvon Morciaux for his skill in creating and designing color illustrations related to atomic, nuclear and particle physics. He also thanks his Belgian colleagues (A. Lucas, P. Thiry, Ph. Lambin, J-P. Vigneron, G. Terwagne, Chr. Iserantant, F. Durant, J. Wouters, J. Bernaerts, P. Rudolph, I. Marenne and C. Goffaux) who contributed pertinent information and / or useful documents to illustrate the physics part of the book.

Francis Leroy

About the authors

Claude RONNEAU teaches inorganic chemistry at the Catholic University of Louvain, Belgium and is the head of the Laboratory of Nuclear and Inorganic Chemistry of this institution. He is an air pollution specialist. His objectives are to describe as precisely as possible, by means of high performance spectroscopic methods, the physico-chemical characteristics of particles emitted by overheated nuclear fuel as a consequence of a reactor accident.

Guy DEMORTIER is Ph.D in physics of the University of Louvain-Belgium (1963). Initially involved in fundamental physics in nuclear reactions , he moved to the University of Namur-Belgium in 1966, took part there in the creation of the LARN (Laboratory for Analysis by Nuclear Reactions,whose main activity is the modification and the characterization of materials using ion beams) in 1969 and is director of this laboratory since 1989. His present research concerns the study of the migration of fluorine in the human tooth enamel(in vivo), non-destructive analysis of archaeological materials (soldering procedures of ancient gold artefacts, depletion gilding of amerindian jewelry items, mode of construction of the Egyptian pyramids, numismatics, paintings...). He was president of the Belgian Physical Society (1991-1993) and is since 2000 President of the Center for Positron Tomography of Namur.

Francis LEROY is owner of Biocosmos Editions, Thuin, Belgium. The author or coauthor of professional publications, he received his first B.S. degree in zoology from the University of Leuven, Belgium, and a second B.S. degree in molecular biology from the University of Brussels, Belgium.

Alfred Nobel (1833 - 1896)

A chemical engineer, the man who would become one of humanity's greatest benefactors, came from a family containing several distinguished scientists. His maternal grandfather discovered the existence of lymph nodes. His father, Emmanuel Nobel, was the self-taught father of several inventions as well, including the submarine torpedo. He experimented with one in Russia, and the authorities of that country awarded him some 25,000 rubles to develop his finding. Child of a family originally of eleven, Alfred Nobel owed his education as a chemist to his father. He himself became known for his invention of a detonator that allowed the ignition of nitroglycerine. This substance, derived from the action of nitric acid on glycerine, showed itself to be a much more powerful explosive than gunpowder. One of the main problems was neutralizing it against shocks because, due to its great instability, nitroglycerine explodes at the smallest shock, hence the difficulty in transporting it.

Alfred Nobel successively perfected diverse procedures to stabilize the product. One of the last consisted of mixing it with a proportion of kieselghur, an inert, porous substance made up of the carapaces of ciliated protozoa. The domestication of the explosive, primarily used to loosen large masses of earth and stone, marked the coming of an impressive series of great works in civil engineering. Let us take, for example, the creation of the Panama and Corinthian canals, and the tunnel of St. Gothard. The use of dynamite-the name given to the mix of nitroglycerine and kieselghur, improved by Alfred Nobel during a stay in Paris-became so widespread that toward the end of the century his invention was the product of some 80 factories in numerous countries

He had become very wealthy while leading a nomadic lifestyle, and rather than leave his fortune to his heirs (whose ability to put it to good use he doubted), Nobel thought to use it in a manner that seemed to him to be the most profitable to humanity: to create a prize that would reward the most commendable research in physics, chemistry and medicine. To this end he created the Nobel Foundation, which also included a prize for the authors of the most remarkable works of literature as well as for those who strove to promote world peace.

Contrary to popular belief, the creation of the foundation was not a gesture by Nobel to exonerate himself for having contributed to the rise of an industry that served war just as much as peace. In reality, a committed pacifist, he saw putting the destructive power of his invention at the disposal of countries as a way to dissuade those countries from using it out of the fear of having the same weapon turned against them. Nobel can, therefore, in some way be considered the instigator of a form of dissuasion comparable to the nuclear dissuasion some 70 years later.

After World War II, more than half a century after the death of the industrial benefactor, the Nobel Foundation created a prize for economics. The work of the foundation as desired by Nobel was not an easy process, as there were political and judicial ripples which accompanied his project. Having named no direct successor at his death, he attracted the criticism of his brothers, who had created the Bakou Petrol Company in Russia and who coveted the enormous capital that had accumulated thanks to the sale of dynamite, and of his Swedish countrymen, who saw themselves dispossessed of what they considered a national fortune. In the same way, France wanted to reclaim what she considered her due, justified by the presence of Alfred Nobel on her soil for many years. However, toward the end of the 19th century, the project became an institution and, beginning in 1901, the first five prizes were conferred in the presence of the Swedish monarchs. Alfred Nobel died in San Remo, Italy, on the 18th of December 1896, leaving an estimated thirty-three 33 million Kronors (about twenty-five million dollars in 2002) to his foundation. Fiscally exempt since 1946, this institution guarantees a substantial sum of prize money to each laureate.

Table of contents

The Nobel Laureates in Chemistry (1901 - 2001)

Claude Ronneau
Francis Leroy

The Nobel Laureates in Chemistry

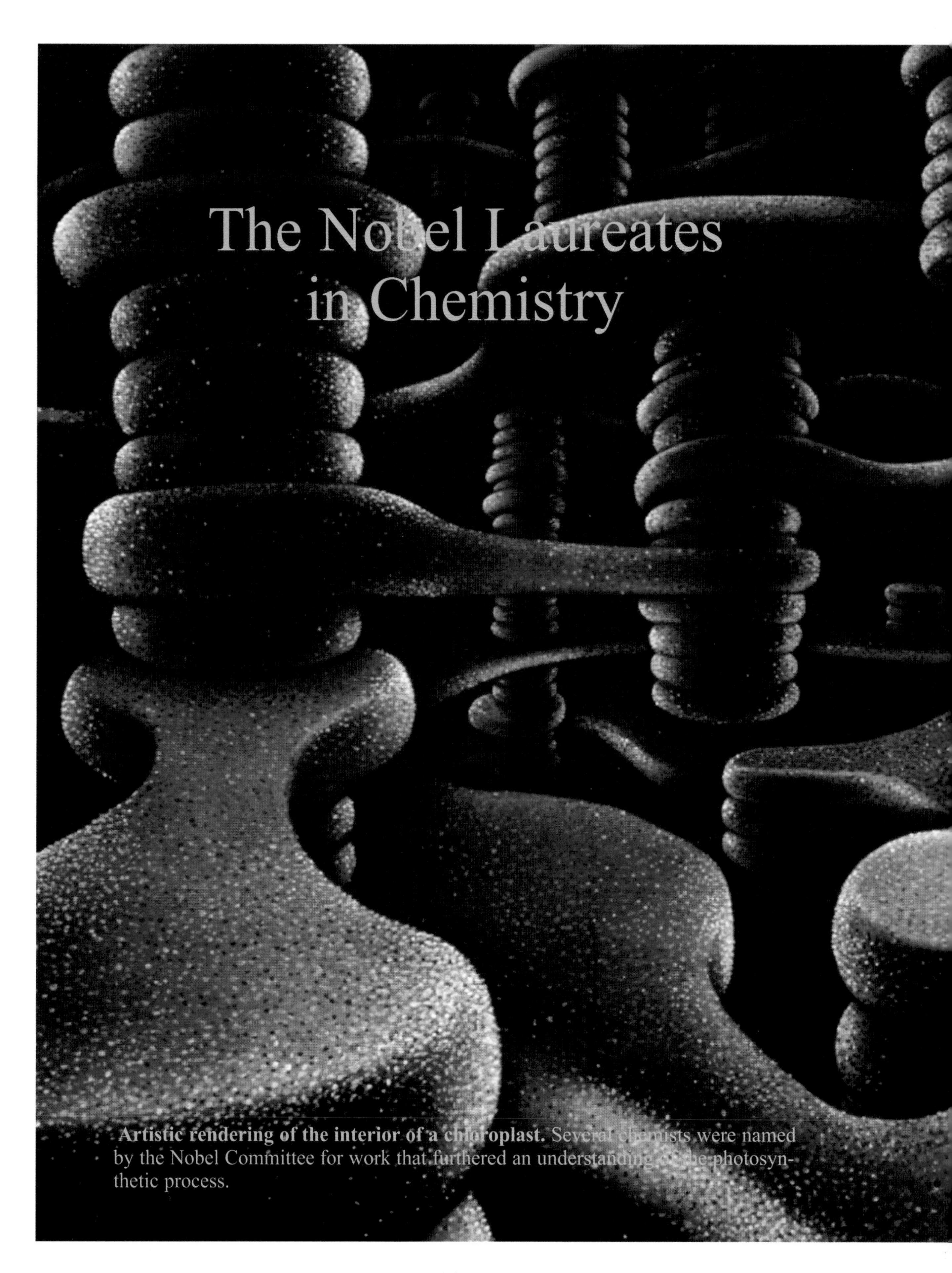

Artistic rendering of the interior of a chloroplast. Several chemists were named by the Nobel Committee for work that furthered an understanding of the photosynthetic process.

Introduction

Chemistry in the 20th century, having inherited the advances of the two previous centuries, was first distinguished by the synthesis of several compounds that are vital to industry, such as nitric acid, ammonia, and methanol. Then came the chemistry of phosphates and synthetic fertilizers, which appeared at the end of World War I and were followed by plastics and the first products of therapeutic chemistry. As progress was made towards understanding the behaviour of atoms and molecules, the distinction between mineral and organic chemistry was replaced with a new classification taking the interactions between these disciplines into account. Molecular and supramolecular chemistry, interface chemistry, and materials chemistry are the new names of modern chemistry.

These considerable advances are due mainly to progress in instrumentation and the analytical methods used in chemistry and physics. Among these methods, it is worth mentioning those used today to determine in a very short time the structure of biological molecules, even complex substances such as proteins and nucleic acids. Major discoveries have stemmed from the use of instruments such as particle accelerators and synchrotrons, or methods such as spectroscopy and nuclear magnetic resonance.

This part of the book gives a panoramic view of the evolution of chemistry from 1850 to our time.

Louis Pasteur The French microbiologist postulated that nature has a chiral asymmetry.

1901

Hoff, Jacobus Henricus van 't (August 30, 1852, Rotterdam, The Netherlands - March 1, 1911, Berlin, Germany). Physical chemist. Nobel Prize for Chemistry for work on rates of reaction, chemical equilibrium, and osmotic pressure. This student trained in mathematics was to have an outstanding destiny, receiving his scientific education from renowned scientists such as Friedrich August Kekulé, discoverer of the structure of benzene. van 't Hoff can be considered as the founder of physical chemistry. At the age of 22 he was one of the great physico-chemists of his time. He demonstrated his precocity in an article where he related the optical activity of certain molecules to their structure. Such studies led him to postulate that the chemical bonds of a carbon atom linked to four different partners adopt an asymmetrical, tetrahedral geometry. The French microbiologist Louis Pasteur had previously suspected such asymmetry at the molecular level when he observed macroscopic symmetry defects in crystals formed by optically active compounds. The young Dutchman's breakthrough (he was then only 26 years old) led to his being appointed Professor of Chemistry in Amsterdam. Later observations on certain oxidation reactions led him to demonstrate how reagent concentrations and temperature influence the reaction rate. He also showed how the temperature affects the equilibrium of a reaction, expressing this relationship in an equation that bears his name. In addition to this discovery of prime importance, he noted a similarity between the behaviours of gases and solutions. This led him to formulate the famous equation of state for perfect gases, familiar today to students worldwide. This equation :$PV = nRT$ is applicable in solution to phenomena involving osmotic pressure. This observation led to a theory of osmotic pressure

Jacobus van ‘t Hoff

and provided a basis for interpreting the lowering of vapour pressure and freezing temperature observed in solution. Jacobus van 't Hoff's brilliant career reached completion in Berlin where he held the post of professor and was appointed a member of the Prussian Academy of Sciences.

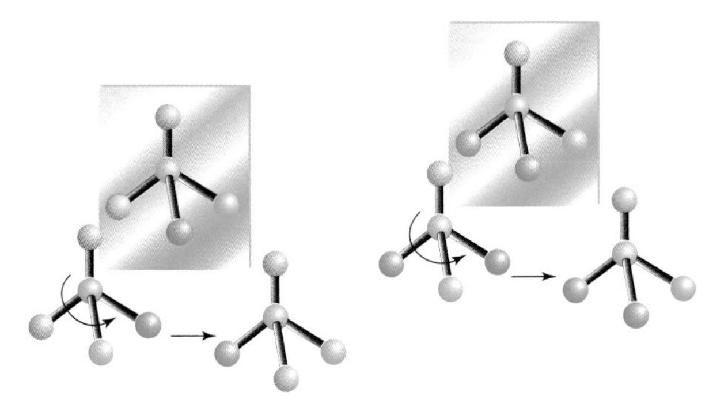

Molecular asymmetry. The figure shows that contrary to a compound with symmetric carbon atoms (at right), a compound with asymmetric carbon (at left) cannot be superimposed on a mirror image of the original.

1902

Fischer, Emil Hermann (Euskirchen, Prussia, October 9, 1852 - Berlin, Germany, July 15, 1919). German chemist. Son of a businessman. Nobel Prize for Chemistry for his investigations of the sugar and purine groups of substances. The training of this scientist, whose name was to resound in the world of chemistry, took place first under Kekulé in Bonn and then, between 1872 and 1881, under the famous **Adolf von Baeyer** in Strasbourg and Munich. He then taught at the Universities of Erlangen and Würzburg before moving to the University of Berlin, a renowned post in Germany at that time. His work with von Baeyer led him to focus on dyes derived from triphenylmethane. This is when he discovered phenylhydrazine, a substance he would later use to synthesise sugars. Contrary to an opinion that prevailed among chemists at the time, Fischer was convinced - and here lies his merit - that any organic compound could be synthesised in the laboratory from simpler molecules. The set of steps involved in this process would be called "total synthesis". This scientist's work took an unusual turn in 1898, with the total synthesis of purine, a molecule which gives rise to a whole family of compounds deriving one from another, urea being one of them. In 1903 he succeeded in synthesizing veronal, a soporific whose name recalls Verona, the Italian town. Yet Fischer's truly monumental contribution to chemistry was the fruit of work conducted between 1884 and 1894, on the structure of sugars. Not only did he describe their structure with outstanding accuracy, but he gave a stereochemical explanation of isomerism. The consistency of this work makes it a reference in the scientific world in general. And this is not all! A little later, between 1899 and 1908, he became a pioneer in the systematic description and study of enzymes and proteins, large molecules about which little was known at the time. We owe to Fischer a precise description of the peptide bond which links amino acids in proteins. It is remarkable that Fischer was able to cover so systematically such a wide variety of substances intervening in cell structures and functions. Today, the unique character of his contribution to our understanding of living organisms is acknowledged by all. He is also famous for his role in developing the German chemical industry. Like other chemists of his time, he made possible the industrial production of strategic substances, such as nitrates and artificial fertilisers, of which his country was deprived because of the wartime blockade. Fischer was also convinced of the irreplaceable role of basic research in the development of science. This philosophy is doubtless behind the extraordinary emulation produced by the creation of the famous "Kaiser Wilhelm Institutes", later to be called the "Max Planck Institutes". These European halls of excellence in pure research have developed outstandingly over the years.

Emil Fischer

CHO
IO — C — H
CH2OH

CHO
H — C — OH
CH2OH

Fischer projections show that enantiomers molecules (here a sugar) are nonsuperimposable mirror images of each other.

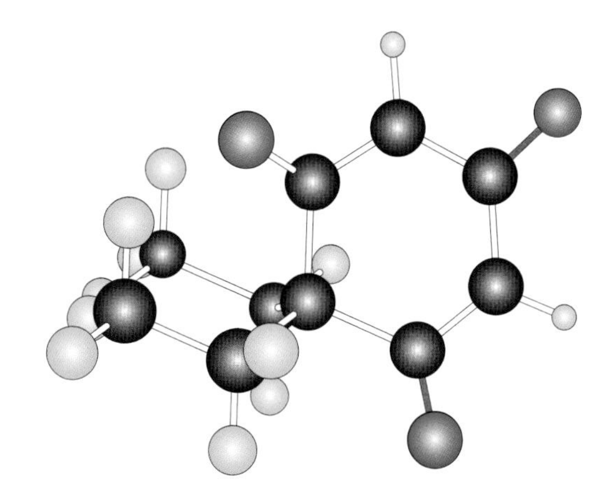

Veronal. This substance acts as a long - acting depressant of the central nervous system.

1903

Arrhénius, Svante August (Wyk-Uppsala, Sweden, February 19, 1859 - Stockholm, Sweden, October 2, 1927). Swedish physical chemist. Son of a land surveyor employed by the University of Uppsala. Most well-known for his studies on electrolyte conductivity, he is considered as one of the founders of physical chemistry. The physicists and chemists of his time were interested in the finding that certain salts dissolved in water could conduct an electrical current. They wondered why these same salts when dry or in pure water could not. Arrhénius hypothesised that these salts, and other substances such as acids and bases, could dissociate in aqueous solution. Yet his doctoral thesis on the subject, produced in 1884 under the title "Research on the Galvanic Conductivity of Electrolytes", was not accepted by the University of Stockholm, and Svante Arrhénius was not offered a job until **Wilhelm Ostwald**, a German physico-chemist who had appreciated the results Arrhénius had communicated to him, intervened in his favour. Arrhénius established the relationship between the strength of acids and bases and their degree of ionisation. He further showed that each of the ions resulting from the dissociation of an electrolyte had the same effect as molecules on the osmotic and cryoscopic properties of solvents. Arrhénius's scientific curiosity was wide-ranging. In the 1890s he published a study showing that the carbon dioxide present in the atmosphere can trap infrared radiation emitted by the earth's surface. He proposed that this effect should cause the atmospheric temperature to rise above that predicted by black-body theory. Today, this observation feeds concern over the thermal stability of our environment, given the rising levels of so-called "greenhouse gases". Svante Arrhénius is also well known for his hypothesis regarding the origin of life. He defended the theory of an extraterrestrial origin of life, believing that the whole universe was filled with seeds of life, perhaps bacteria propelled by light pressure, and coined the term "panspermia".He also formulated the

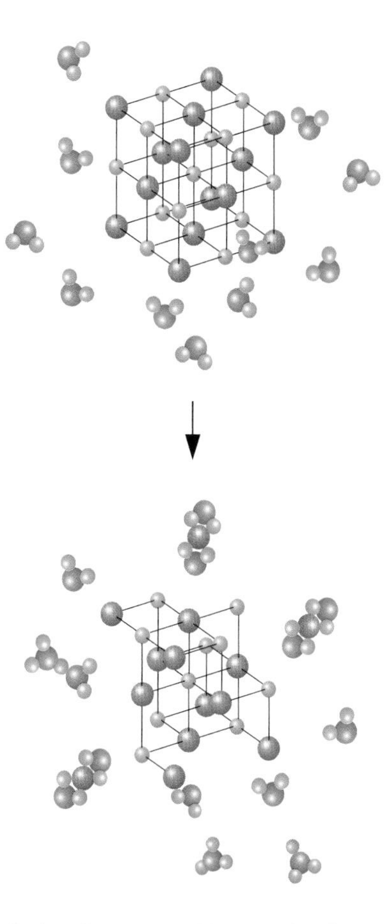

Salt dissociation in water. Water molecules are miniature dipols, with the positive pole near the hydrogens and the negative pole near the oxygen. When a salt crystal is dropped into water, the water molecules accumulate around the outside of the sodium and chloride ions, surrounding them with oppositely charged ends of the water molecules. Thus insulated from the attractiveness of other salt molecules, the ions separate, and the whole crystal gradually dissolves.

hypothesis that a collision between two "cold" stars should lead to formation of a nebula, which in turn would generate stars. This hypothesis has not been confirmed by modern cosmology. Arrhénius was educated at the Cathedral School in Uppsala and studied physics and chemistry at the university of the same town before moving to Stockholm and working on electrolysis under Erik Edlund. During his scientific career he met several of the most famous scientists of his time, among them **Hans Euler-Chelpin**, **Jacobus Van 't Hoff** and Ludwig Boltzmann. Arrhénius wrote many popular scientific books for the general public.

1904

Ramsay, William (Glasgow, Scotland, October 2, 1852 - Hazlemere, England, July 23, 1916). Scottish chemist. Nobel Prize for discovery of the noble gases. William Ramsay discovered five inert gases whose existence was unsuspected prior to his work: the so-called 'noble' gases. This discovery added a column to Mendeliev's periodic table. It also opened the way to a more accurate description of the atom, the chemical inertia of the "noble" gases being attributable to the fact that their outer electron layer is full. This concept of stability linked to a complete electron layer also led Gilbert N. Lewis to propose, in 1916, his model of the covalent bond. Ramsay's research on the inert gasses began in 1892, when attempting to interpret Cavendish's 1785 discovery that when all oxygen and nitrogen is eliminated from a sample of air, there remains a small

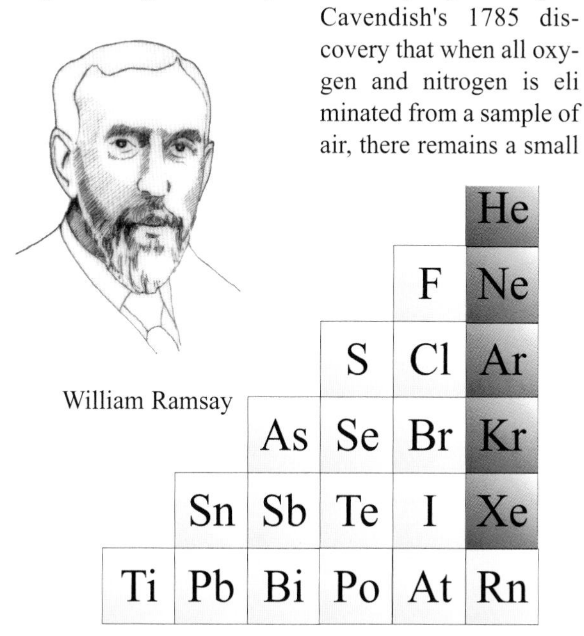

Part of the Table of Mendeleev emphasizing the elements (Noble gases) discovered by William Ramsay (in red).

volume of gas. The remaining gas had not been identified. Ramsay used metallic magnesium to eliminate the oxygen and nitrogen chemically. He performed a spectroscopic analysis of the non - consumed gas and discovered in its spectrum the rays of an unknown element. He called this element argon, meaning "inert". A determination of its atomic weight led him to place it between chlorine and potassium in the periodic table, and thus to postulate the existence of a new group of elements. Analysing in similar fashion the gas produced by a radioactive ore, he discovered rays observed as early as 1868 in the sun. They corresponded to helium, from the Greek "helios", meaning sun. He further identified krypton, neon, and xenon, meaning respectively "hidden", "new", and "strange", and isolated these gases by distil- ling 15 L of liquid argon. With **Frederick Soddy**, he discovered that helium is formed through the radioactive decomposition of radium, an element isolated shortly before then by **Marie Curie**. Ramsay was knighted in 1902. A very clever student, William Ramsey began his doctoral thesis at the University of Tübingen, Germany, when he was only 19. Back in Glasgow, he became an assistant at Anderson College, which was to become the Royal Technical College. He was appointed professor of chemistry at the University College of Bristol in 1880. Seven years later he moved to London and taught chemistry at the University College.

1905

Baeyer, Johann Friedrich Adolf von (Berlin, Prussia, October 31, 1835 - Stanberg, Germany, August 20, 1917). German chemist. Son of a staff officer. Nobel Prize for Chemistry in recognition of the services in the advancement of organic chemistry and the chemical industry, through his work on organic dyes and hydroaromatic compounds. In 1856, Johann von Baeyer was studying chemistry with Bunsen in Heidelberg. Then he taught in Berlin and Strasbourg and succeeded Liebig at the Uni-versity of Munich as a professor of chemistry in 1875. There he was influenced by Kekulé, who first proposed a model for the structure of benzene. From 1864 onward he continued the research begun by the latter on uric acid. In 1871 he synthesised phenolthalein and fluorescein, two substances that were to find wide applications in the dye industry. von Baeyer also discovered phenolformaldehyde resin, a product that would be industrialised thirty years later by the Belgian Baekeland under the name Bakelite, the first commercial plastic in the world. In 1883 he synthesised indigo, one of the most important natural dyes, and proposed a structure for this compound that turned out to be roughly correct. In 1887 the work of the chemist Heumann allowed BASF to proceed with its industrial synthesis. After centuries of costly production from the indigo plant, indigo dye thus became a common industrial product. In collaboration with his famous student **Emil Fischer**, von Baeyer spent more than 20 years studying a group of synthetic dyes related to the phtaleins. In the course of these investigations, he understood that the colouring properties of these substances are linked to the nature of the chemical bonds in each of their molecules. Another important field of research concerned the benzene ring and von Baeyer proposed a model slightly different from the Kekulé model.

Adolf von Baeyer

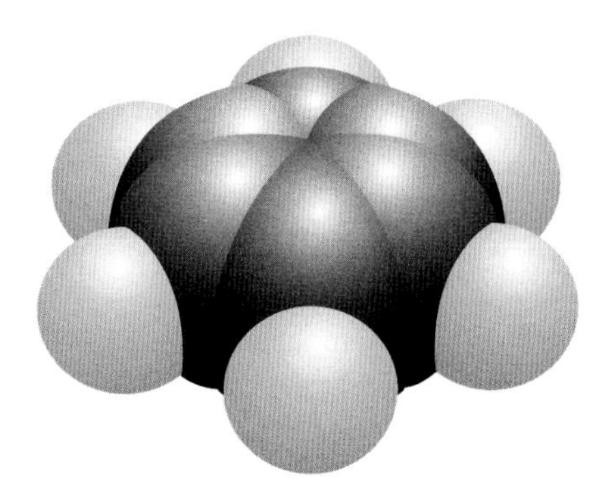

Benzene. A space-filling molecular model. The flat, hexagonal ring of six carbon atoms is called the benzene ring. This ring occurs also in proteins and many drugs. This model was first proposed by A. Kekulé, a German chemist. Bayer proposed a variant of this molecule.

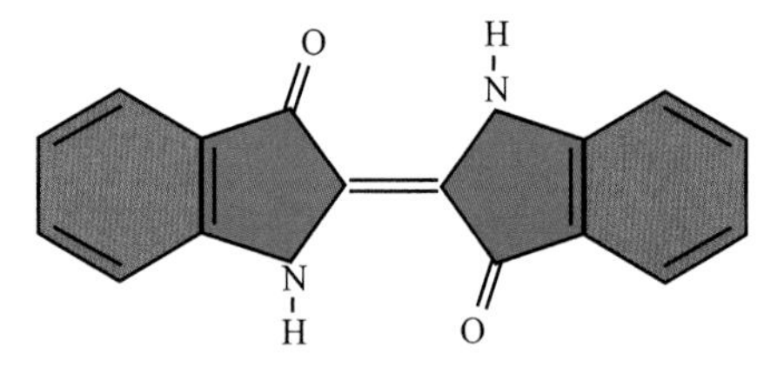

Phenolphtaleine. This indicator dye is colourless in acids and deep red in bases.

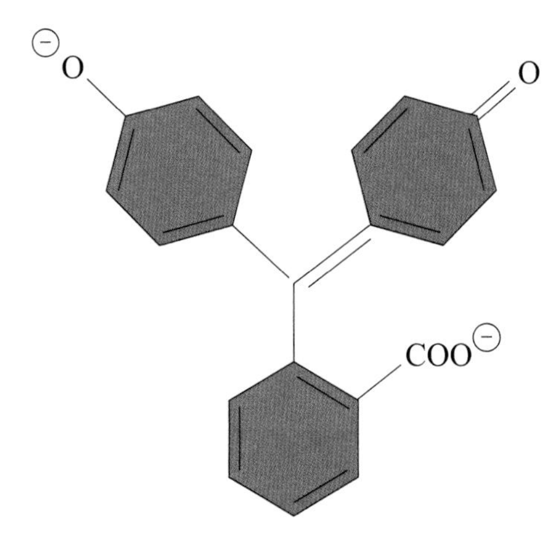

Indigo. This dye was obtained around 1900 from plants of the genera Indigofera and its chemical structure was announced by Adolf von Baeyer in 1883.

In the field of cyclic molecules, his merit was to develop the concept of tension in bonds between carbon atoms. This concept explains why six-atom cycles, being more stable, are more accessible and more numerous. In 1885 von Baeyer was knighted by King Louis II of Bavaria.

1906

Moissan, Henri (Paris, France, September 28, 1852 - Paris, February 20, 1907). French chemist. Son of a lawyer. Nobel Prize for Chemistry for the isolation of the element fluorine and the development of the Moissan electric furnace. Henri Moissan is best known for his success in isolating the very reactive element fluorine and for his invention of an electric arc furnace. Moissan studied chemistry at the "Ecole de Pharmacie de Paris", France. As far back as 1886, he became professor of toxicology and then, in 1889, professor of inorganic chemistry in the same school. In 1900, he was appointed professor of inorganic chemistry at the Sorbonne. In 1884 he undertook to study fluorine compounds. The element fluorine, ninth in the periodic table, is the most powerful oxidant of all the halogens. This makes it extremely reactive and explains why all attempts to prepare it in its elemental state had failed. In 1885, he succeeded in preparing the reactive gas fluorine by electrolyzing a solution of potassium fluoride in fluoridric acid. The isolation of fluorine allowed him to study its properties. Moreover, he could explain how fluorine reacted with other elements. In 1892, Moissan developed an electric-arc oven operating under vacuum. Thanks to the absence of air and to the temperatures reached, he was able to explore a range of substances then reputed to be unmeltable because they oxidized so readily. He thus succeeded in preparing nitrides, borides, and metal carbides, extremely hard compounds that imposed themselves in industry as abrasives, for instance. This furnace also allowed him to isolate metals such as niobium and molybdene. This scientist was also known for his attempts to manufacture diamonds. Greatly impressed by the presence of tiny fragments of this form of carbon in meteorites, he imagined that if the conditions believed to reign in space could be reproduced, it would be possible to synthesise an artificial diamond from carbon. In 1893 he announced he had been successful in making diamonds by placing a mixture of carbon and iron in its furnace at very high temperature and then subjecting this mixture to great pressure by sudden cooling in water. Although he claimed that he had succeeded, the scientific community never endorsed his results. On the other hand the use of his electric arc furnace allowed him to sucessfully devise a commercially profitable method of producing acetylene.

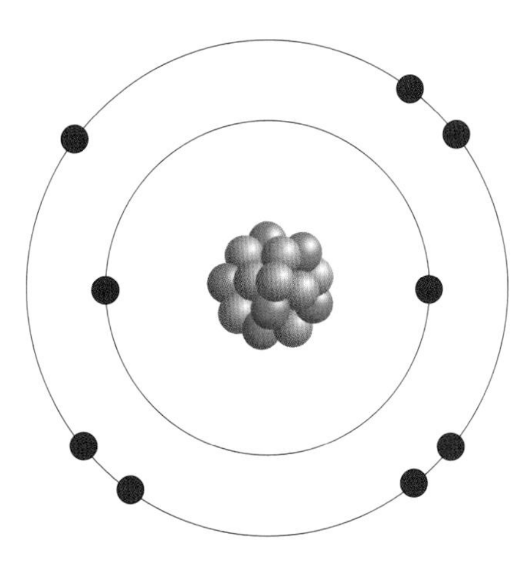

Fluorine atom. This is a relatively scarce nonmetallic element, but one that combines with other substances even more vigorously than oxygen does. Fluorine was isolated by Moissan by electrolysis of anydrous hydrogen fluoride.

1907

Buchner, Eduard (Munich, Bavaria, May 20, 1860 - Focsani, Roumania, August 1917). Son of a physician. Nobel Prize for his discovery of cell-free alcoholic fermentation of sucrose. After beginning studies in chemistry at the Technical University of Munich and, under **Adolf von Baeyer**, at the Bavarian Academy of Sciences, Eduard Buchner studied the fermentation of sugar to alcohol at the Institute of Plant Physiology. His first research was published in 1886. In 1897, Buchner showed that fermentation does not need to take place in the yeast cell itself, but in the presence of enzymes contained in these cells. He isolated one of these enzymes from barm and named it "zymase" which means "ferment" in greek. He then observed that zymase belonged to a very large class of natural substances. We are also indebted to him for the extraction of invertase and lactase from lactic ferment. It is noteworthy that as early as 1980 the French chemist Berthelot had the premonition that such "soluble ferments" might exist. This led him into conflict with Pasteur, convinced that the intervention of a living cell was required in fermentation phenomena. In 1897, two years after Pasteur's death, Buchner disproved the vitalist theory and introduced a new vision of living cells, showing that what had been called 'vital energy' actually reflected the existence of chemical reactions whose mechanisms scientists were beginning to elucidate. After having studied at the University of Munich, he taught there for some years and the held posts at the universities of Kiel and Tubingen. In 1898, Buchner became professor of general chemistry at Berlin's Agricultural College. He later conducted research at the universities of Breslau and Wurzburg. Buchner was killed during World War I (1914-1918).

Eduard Buchner

Yeast cell. Eduard Buchner demonstrated that the enzymes secreted by yeast cells are also able to drive sugar fermentation (i.e. to break them down to carbon dioxide and alcohol) if they are extracted from these cells.

1908

Rutherford, Ernest (Spring Grove, New Zealand, August 30 1871 - Cambridge, England, October 19, 1937). New Zealander physical chemist. Nobel Prize for Chemistry for his work on radioactivity. Ernest Rutherford was educated at the Trinity College, Cambridge, under **Joseph Thomson** (1906 Nobel Prize for Physics) who was Head of the Cavendish Laboratory. He was asked in 1896 to study the properties of X-rays, discovered the year before by **Wilhelm Conrad Röntgen** (1901 Nobel Prize for Physics). Rutherford thus acquired some celebrity. Yet it's another discovery, that of radioactivity by Becquerel in 1896, that established his scientific fame. In 1898, he was appointed Professor at McGill University in Montreal, where a top - level laboratory was placed at his disposal for

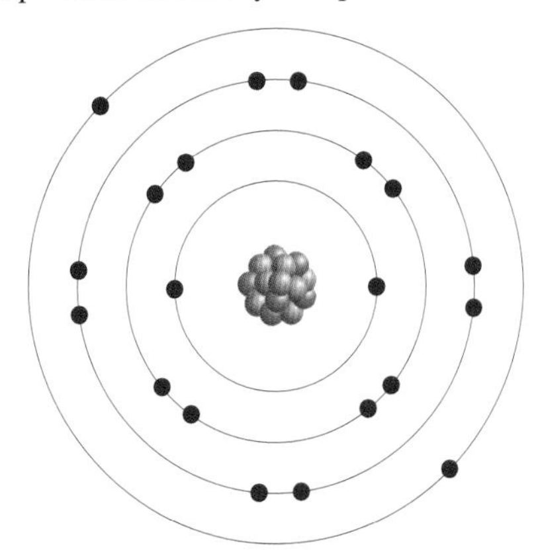

Atomic structure. Atoms are made up of a very small but heavy central nucleus with a positive charge, surrounded by a negatively charged cloud of electrons.

studying radioactive phenomena. In 1907 he was appointed in Manchester. In 1899 he discovered an emanation of thorium: isotope 220Rn of the gas radon, and measured the radioactive period of this isotope. In 1901, in collaboration with **Frederick Soddy**, he formulated the laws of radioactive derivation. It was to radioactivity that he attributed the thermal energy produced by the earth, energy which much later would explain the plate tectonics of the earth's crust. At the University of Manchester he was surrounded by young scientists such as Ernest Marsden, **Niels Bohr**, Henry Moseley and Hans Geiger (inventor of the radioactivity counter that bears his name). This was a team of exceptional scientists whose scientific production was to be rich indeed. The names of Rutherford and Marsden are linked to a major discovery: that of the atomic nucleus. By bombarding the atoms of a very thin sheet of gold with alpha particles, they observed that most of the radiation crossed the sheet wi- thout deviating; only a fraction of the particles were deviated from their trajectories, or even "bounced" back. The interpretation was that the mass and positive charge of the atom are concentrated in a minute volume at the centre of the atomic edifice. The dimensions of this nucleus are some ten thousand to a hundred thousand times smaller than those of the atom. The electrons, which determine the dimensions and chemical properties of the atom, revolve around this nucleus. The discovery of this atomic structure, proposed in 1911, opened the way towards a model developed by Niels Bohr in 1913. It is also worth mentioning that in 1919 Ernst Rutherford accomplished the transmutation of the nitrogen atom to oxygen, again by bombarding it with alpha particles. Using nuclear techniques, he thus made the old utopic dream of the alchemists come true. The same year, he was appointed Head of the Cavendish Laboratory.

1909

Ostwald, Friedrich Wilhelm (Riga, Lithuania, September 2, 1853 - Grossbothen, Germany, April 4, 1932). Lithuanian chemist. Nobel Prize for Chemistry for his work on catalysis, chemical equilibrium, and reaction velocities. Son of an artisan. Very early on, Wilhelm Ostwald showed deep interest in many chemical phenomena that had not yet been explained. For this reason he is considered as one of the founders of physical chemistry. This branch of chemistry explains chemical phenomena on the basis of physical arguments. Ostwald developed this discipline at a time when chemistry was devoted almost entirely to the study of organic molecules. His doctorate, received in 1878 from the University of Dorpat, Estonia, was devoted to the study of light refraction by aqueous solutions. He extended the concept of chemical affinity (studied before him by J. Tomsen and Marcelin Berthelot) on the basis of the heat released in the course of a chemical reaction. Among the first works done by Ostwald was the study of the catalytic action of acids and bases on certain chemical reactions. This work led him to devise an exact definition of the role of catalysers, in keeping with the role proposed by Jacob Berzelius in 1936. He was appointed professor of chemistry at the Riga Polytechni-cal University.There he studied chemical reaction rates and thus laid the foundation of chemical kinetics by quantifying the reactivity of evolving chemical systems. It was in Riga that he became familiar with Arrhénius's ion dissociation theory and realised its importance, despite the scepticism of the scientists of the time. He thus engaged in active research in the field of electrochemistry, introducing the notion of affinity. Electric conductivity measurements led him to formulate the dilution law stating that a weak electrolyte should be completely dissociated at infinite dilution. The concept of the strength of acids and bases derives from this. Among the scientists who supported him were **Svante Arrhénius**, **Jacobus van 't Hoff** and **Walter Nernst**. In 1887 he was offered a position at the University of Leipzig and thus became the first foreigner in Germany to be appointed Professor of Physical Chemistry. It was also in 1887 that he founded with van 't Hoff the "Zeitschrift für Physikalische Chemie", which was soon to become one of the top - ranking journals of physical chemistry worldwide. In 1900, with the help of Eberhard, he elaborated a method of synthesizing nitric acid (via nitric oxyde) by oxidizing ammonia with platinum acting as a catalyst.Combined with the synthesis of ammonia developed in 1913 by **Fritz Haber** and **Karl Bosch**, this technique soon showed great economic importance as a means of producing nitric acid, and thus nitrates, basic fertilisers for agriculture. The prestige of this positive scientist is not tarnished by the fact that he was the last chemist to accept the atomic theory, formulated a hundred years earlier by Dalton!

Friedrich Ostwald

1910

Wallach, Otto (Könisgsberg, Prussia, March 27,1847 - Göttingen, Germany, February 26, 1931). German chemist. Son of a politician. Nobel Prize for Chemistry for analyzing fragrant essential oils and identifying the compounds known as terpenes. Otto Wallach received his doctorate in Chemistry from the University of Göttingen in 1869; his thesis concerned the isomers of toluene. Isomers are chemicals having the same chemical composition but different structures and properties. This is the type of work that was to lead to the importance of the isomer notion in chemistry, and especially biochemistry, where the stereospecificity concept provides precious information on enzyme function. From 1869 to 1896, he was Kekulé's assistant at the University of Bonn. Appointed Professor at this same university, he taught pharmacy there and, in 1881, identified the terpenes. These are natural odorants with a geometry based on that of isoprene, a basic structure with five carbon atoms. Terpenes played an important role in the development of the fragrance and food industries. Examples include rose oil, peppermint, menthol, and camphor. Otto Wallach can also be viewed as one of the pioneers of the extraordinary development of the pharmaceutical industry and of the fragrance and aroma industries.

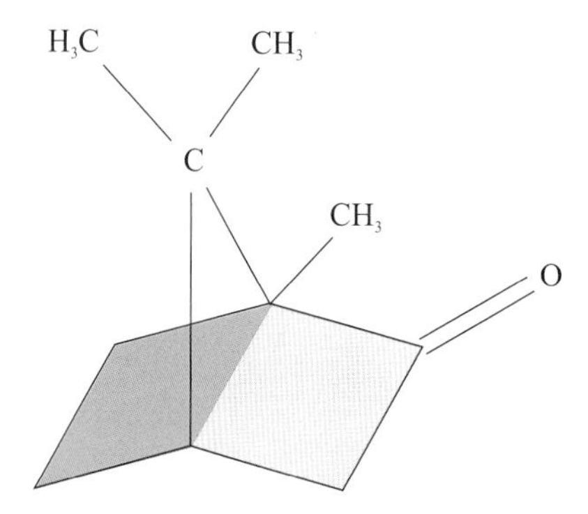

Camphor. This substance is extracted from the camphor tree, Cinnamomum camphora, common in China and Japan. It has been used for many centuries as a component of incense but also as a medicinal.

Otto Wallach

Peppermint This plant is frequently used to relieve stomach and bowel spasms and promote the expulsion of flatus.

1911

Curie-Sklodowska, Marie (Varsovia, Poland, November 7, 1867 - Sancellemoz, France, July 4, 1934). Polish-born French physicist. Nobel Prize for Chemistry for the discovery of radium and polonium. She was also awarded the Nobel Prize for Physics, with **Henri Becquerel** and her husband, **Pierre Curie**. Marie Sklodowska left her family at the age of 17 to become a governess in a wealthy family. Her aim was to help her older sister, who had registered as a student at the Paris School of Medicine. In 1891 it was Marie's turn to go to Paris to study at the Sorbonne. At the time she was one of very few women students at that university. Her difficult financial situation forced her to live very frugally, to the point of endangering her health - already quite poor. Her extraordinary perseverance was rewarded in 1894, when she received her bachelor's degree in physics. The same year, she met **Pierre Curie** who headed the "Ecole de Physique et de Chimie Industrielle de Paris". He encouraged her to undertake research on radioactivity, just discovered by Becquerel in 1896. This discovery was a revolution in science because it posed the problem of the origin of radiative energy, which seemed to contradict the law of the conservation of energy. In 1895 she married P. Curie and began doctoral research aiming to determine the origin of the energy responsible for radioactivity. She focused on pitchblende, an ore extracted from mines in Bohemia and the basis of Becquerel's discovery. Actually, this ore had long been exploited for its uranium, in which people were interested because it yielded a yellow oxide used to colour glass. Having noticed that certain waste products derived from

				He
	N	O	F	Ne
	P	S	Cl	Ar
	As	Se	Br	Kr
	Sb	Te	I	Xe
	Bi	Po	At	Rn

H		
Li	Be	
Na	Mg	
K	Ca	Sc
Rb	Sr	Y
Cs	Ba	Lu
Fr	Ra	Ac – Li

Marie Curie

Part of the Table of Mendeleev. The position of polonium and radium, two elements of the Table of Mendeleev discovered by Marie Curie, are identified by a blue background.

the ore were more radioactive than the ore itself, Marie Curie undertook the gigantic task of selective extraction, in order to concentrate the substance hidden in the ore. Starting with over a ton of pitchblende waste, she successively isolated two radioactive elements: first polonium, then radium. Ending up with 0.1 gram of radium chloride, she showed in 1902 that radium was in fact a new element. This was the start of her scientific reputation. The same year she was appointed as a teacher at the "Ecole Normale Supérieure de Paris". During the year 1903, Marie defended her thesis based on her work on radioactive substances other than uranium and thorium, entitled " Research on Radioactive Substances ". One year later, her husband was appointed professor of physics at the Sorbonne, Paris. Following the death of Pierre Curie, - he was knocked down and killed by a horse-drawn cart in 1906, - Marie Curie replaced him as professor of physics at the Sorbonne, becoming the first woman to teach there. She would eventually be appointed as a tenured professor in 1908. Later yet she succeeded in isolating radium under its metallic form. We also owe to Marie Curie the discovery of thorium, the same year as the German Schmidt. She founded the Institute of Radium in 1906 in order to promote research on radioactive substances. In the meantime the therapeutic properties of radium in the treatment of tumors were discovered. From this stemmed some intense industrial activity aimed at producing ever-greater quantities of this element. Marie Curie always refused to patent her discovery, which was to save many lives. The treatment of cancers by radiotherapy has now evolved, and radium, too dangerous, has been replaced by more specific irradiation techniques that target the diseased tissues more accurately while preserving the surrounding healthy tissues. Until recently, the Curie was the activity unit of a radioactive substance. It equals 3.7.1010 disintegrations per second, the activity produced by 1 gram of radium. The element 96 of the periodic table was named curium in honor of this scientist. The element 84, polonium, was also named by Marie Curie in honor of her mother's native country, Poland. Marie Curie died from pernicious anemia caused by overexposure to radiation at the Sancellemoz sanatorium in Haute-Savoie in 1934.

1912

Grignard, Victor (Cherbourg, France, May 6, 1871 - Lyon, France, December 13, 1935). French chemist. Nobel Prize for chemistry along with **Paul Sabatier** for his development of the Grignard reaction. Son of a boat builder, Victor Grignard did his doctoral research under Professor Barbier, interested in the

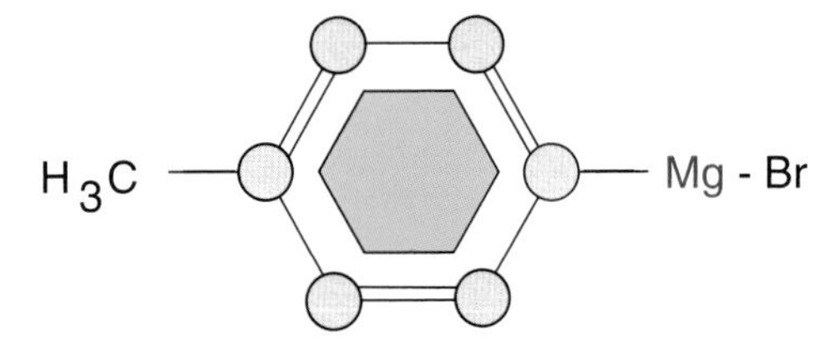

Structure of an organo - magnesium complex. This type of organo - metallic complex is characterized by a carbon - metal bond which has properties intermediate between covalent and electrovalent. The most interesting organo - magnesium complexes are also known under the name of "Grignard reagents".

binding of organic molecules to various metals. Grignard's research topic was the reaction between organic radicals and the metal magnesium. Compounds of this type were already known, but some of them, such as organosodium and organoposassium compounds, were too unstable, to the point of being dangerous to handle. Others, such as organomercury compounds, were on the contrary too inert to be of any use. Working in anhydrous ether, Grignard developed an organomagnesium compound particularly suited for the chemical synthesis of many organic substances. His thesis was published in Lyon in 1901, and in 1908, he was appointed Professor of the University of Lyon. Grignard was elected Member of the Academy of Science in 1926. His monumental "Traité de chimie organique", 20 volumes, was published in 1950.

Sabatier, Paul (Carcassonne, France, November 5, 1854 - Toulouse, France, August 14, 1941). French organic chemist. Nobel Prize for Chemistry with **Victor Grignard** for research in catalytic organic synthesis, and particularly for discovering the use of nickel as a catalyst in hydrogenation. Paul Sabatier received the aggregation degree in physical science in 1877 and obtained his doctorate in 1880. In 1884, he was named to the chair of chemistry at Toulouse, when he was thirty, the minimum age for the position, Sabatier's intial research were organic studies within the thermodynamical tradition of Berthelo's laboratory. The Mond's preparation of nickel carbonyl instigated him to study gaseous molecules which might behave analogously to carbon monoxyde: In 1892, he succeeded in fixing nitrogen peroxyde on copper, cobalt, nickel and iron. One year later he repeated the experience of Henri Moissan and Moreu with unsaturated hydrocarbons and reduced nickel:he found that ethylene and acetylene were hydrogenated. With his student, J.-B. Senderens, he demonstrated the generality of this method to the hydrogenation of non-saturated and aromatic compounds, ketones, aldehydes, phenol, nitriles, nitrites, etc. In contrast to previous theories, Sabatier postulated that, during catalysis, a temporary, unstable intermediary between the catalyst and one of the reactants forms on the surface of the catalyst. He predicted the reversibility of the reaction:a catalyst of hydrogenation will be equally one of dehydrogenation. Paul Sabatier was a very reserved man. Elected Professor of chemistry at Toulouse in 1894, he was ever faithful to this town and turned down many offers of attractive positions in Paris. In 1913, he became the first scientist elected to one of six chairs newly created by the Academy for provincial members.

Paul Sabatier

1913

Werner, Alfred (Mulhouse, France, December 12, 1866 - Zurich, Switzerland, November 15, 1919). Swiss chemist born in France. Son of a foreman. Nobel Prize for Chemistry for research into the structure of coordination compounds. Werner graduated from the University of Zurich where he earned his Ph.D. in 1890. After a working stay in the laboratory of Berthelot in Paris he returned to Zurich and was named professor there. His work greatly influenced the development of chemical bond theory, particularly as regards the so - called "complex salts" in which true chemical reactions occur. Chemical bond theory had already been applied with success to ionic bonds and covalent systems in organic chemistry, but it seemed not to apply to complexes, where a metal could display different valences. In 1893, 26 - year - old Werner proposed that metal ions, and particularly ions of the transition-group metals, could form true covalent bonds with certain molecules rich in free electron pairs, such as water, ammonia, halides, cyanides, etc. These molecules, called

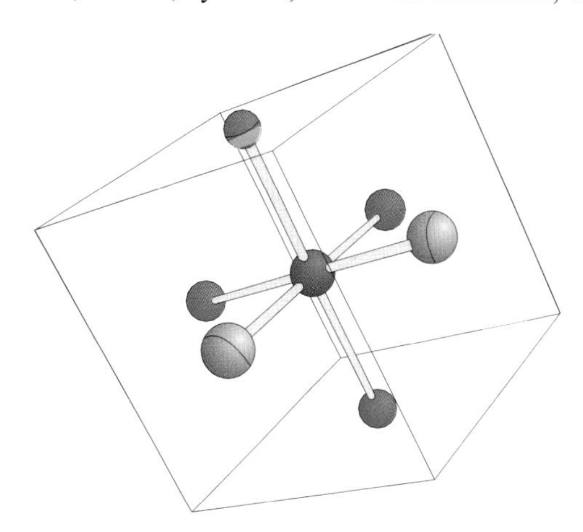

A coordination complex. A structure formed by "injection" into the empty orbitals of a metal, of free electronic pairs supplied by the ligands (H_2O, Cl^-, NH_3, ...). In the scheme presented here, six ligands are bound to a central metal ion to form an octahedral edifice.

ligands, share one or more electron pairs with the metal ions by "injecting" them into the free atomic orbital of the ions and thus creating a strong covalent bond. This mechanism was not explained until long after Werner's death. Such combinations can form complex ions and associate with other anions or cations to form salts. Werner demonstra- ted convincingly that these compounds could no longer be viewed as "double salts". In his experiments performed with complex salts such as cobalt-amine chlorides, he showed the distinction between ligands and ions by precipitating the chloride ions with silver nitrate and showing that the ligands remained tightly bound to the metal. In addition, Werner suggested that coordination complexes adopt a structure with maximal symmetry, the metal ion being at the centre of the complex. An octahedral geometry was thus proposed for a central atom bound to six ligands. This hypothesis was based on the existence of isomers, the number of which was determined by the geometry of the complex. Werner demonstrated this long before it could be confirmed by X - ray diffraction. This scientist is also to be credited for discovering and separating optically active stereoisomers.

1914

Richards, Theodore William (Germantown, Pennsylvania, January 31, 1868 - Cambridge, Massachusetts, April 2, 1928). American chemist. Son of a painter. Nobel Prize for Chemistry for determination of the atomic weights of a number of elements. Theodore W. Richards studied chemistry at Harvard and earned his doctorate in 1888. He was then 20 years old. The first American to receive the Nobel Prize, he was very skilful in chemical analysis. Devoting himself to the precise determination of atomic weights, he found - contrary to the current belief - that they were not whole-number multiples of the mass of the hydrogen atom. For oxygen, for instance, he determined 15.869 and not 16. The relative lack of precision of previous analyses, conducted mostly by the Belgian scientist Stas, had led to the belief that the different elements were assemblages of hydrogen atoms. Richards was to disprove this hypothesis, proposed by Proust but already questioned by Stas. It is now known that the most significant deviations of atomic weights from whole numbers usually reflect the fact that elements are mixtures of isotopes. The atomic weight of an element is the weighted average of the atomic masses of the isotopes composing it. And the masses of individual isotopes are not whole numbers either. This reflects the fact that the energy that binds the nucleons results from a slight loss of mass, in accordance with Einstein's relationship between mass and energy. Richards also made a significant contribution to the study of radioactive disintegration, showing that the atomic mass of the lead produced radioactively from plutonium is different from that of ordinary lead.

1915

Willstätter, Richard (Karlsruhe, Germany, August 13, 1872 - Locarno, Switzerland, August 3, 1942). German chemist. Son of a textile merchant. Nobel Prize for Chemistry for his researches on plant pigments, especially chlorophylls. Richard Willstätter was one of the major figures of German chemistry at the beginning of the 20th century. He studied at the University of Munich and showed keen interest in alkaloid chemistry. His doctoral thesis was on the properties of cocaine. On the advice of **Adolf von Baeyer**, his illustrious professor, he accepted a position at the Polytechnic Institute of Zurich. Then, encouraged by **Emil Fischer** who visited him personally, he went to the Kaiser Wilhelm Institute in Berlin, where he became engrossed in the study of pigments that give flowers and fruit their colour. His renown as a chemist was such that he finally became Baeyer's successor to the Chair of Chemistry in Munich. He quit this job, however, at the first sign of anti - Semitic actions in 1924, remained in Germany as long as he did not feel his life was threatened, and finally emigrated to Switzerland in 1939, abandoning his precious library and many works of art. This great chemist died on the banks of Lake Maggiore in 1942. We owe to him the synthesis of cocaine and better knowledge of atropin, the alkaloid which inhibits the action of acetylcholine. It was around 1912 that he began his famous work on the pigments involved in photosynthesis. Thanks to the chromatographic technique developed by the Russian botanist Mikail Semyonovitch Tsvett, he showed that there actually exist two types of chlorophyll, called chlorophyll a and chlorophyll b. He also managed to manipulate their chemical structure and to reveal the presence of a magnesium atom in each. Other work concerned the yellow and blue

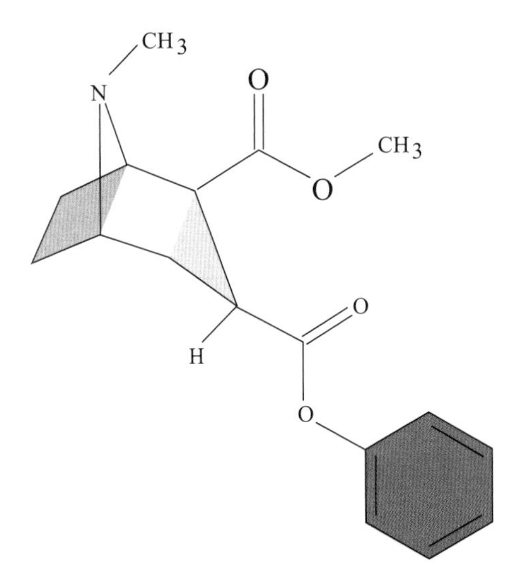

Cocaine. This alkaloid is extracted from the leaves of the coca plant (*Erythroxylon coca*) and has been used by Andean Indians for hundreds of years. Cocaine acts as a cerebral stimulant and narcotic.

pigments of the photosynthetic complex, respectively carotene and anthocyanin. His theory on the non-protein nature of enzymes first met with some success, but was later disproved by the resounding discovery of the American chemist **John Northrop**, who demonstrated in 1930 that urease is indeed a protein.

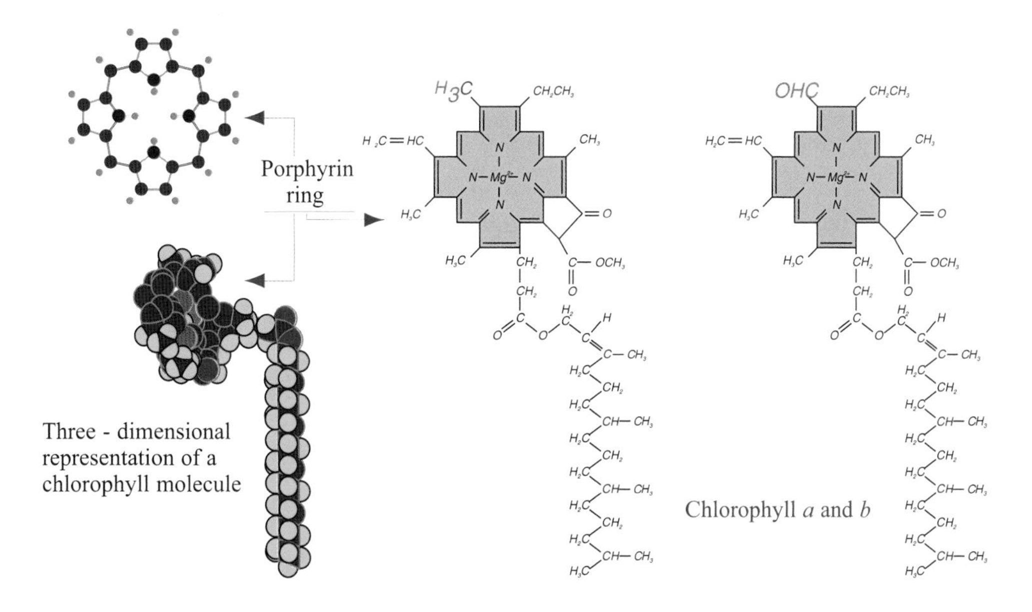

Chlorophyll. Chlorophyll is a magnesium-porphyrin compound with a long hydrocarbon tail. The porphyrin ring is related to the heme group of hemoglobin and cytochromes. The free electrons of the metal atom and in the porphyrin ring makes chlorophylls good absorbers of visible light, in particular blue and red wavelengths.

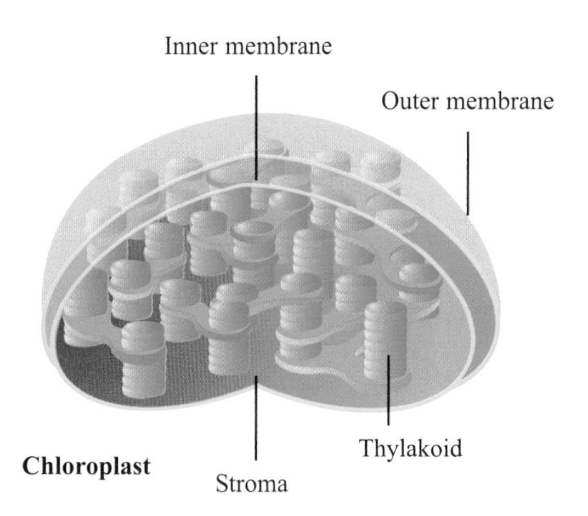

The chloroplast. Their thylakoid membranes are organized into paired folds that extend throughout the stroma. The chlorophyll molecules as well as the carotenoids are located almost exclusively in these lamellae.

Richard Willstätter

1916 to 1917 Not awarded

1918

Haber, Fritz (Breslau, Prussia, December 9, 1868 - Basel, Switzerland, January 29, 1934). German chemist. Nobel Prize for Chemistry for his development of a method of synthesizing ammonia. Son of a dye and pigment importer, Fritz Haber was probably one of the most unsettling and controversial scientists of his century, because he devoted as much energy to devising tactics for dumping suffocating gases onto the front during the First World War as to developing a technique for synthesising ammonia for fertiliser production. Although endowed with an exceptional intelligence and extraordinary working power, he was not appointed Professor in Karlsruhe until 1908 and then in Berlin in 1911. In 1906, with the help of **Karl Bosch** and Alwin Mittasch, he undertook to synthesise ammonia industrially from nitrogen and hydrogen by the reaction: $N_2 + 3H_2 = 2NH_3$. This is an example of an exothermic equilibrium reaction, which should thus ideally be carried out at low temperature. Yet the kinetic conditions at low temperature are unfavourable if the tempe rature is too low, the reaction is literally 'frozen'. This makes it necessary to carry out the reaction at 400°C, but at this temperature the equilibrium is not in favour of ammonia formation. To compensate for this handicap, it is necessary to work at very high pressure, since ammonia formation entails a decrease in volume. This constraint posed major technological problems. To scale up his invention for industrial use, Haber called upon the company BASF, the 'Badische Anilin und Soda Fabrik'. This is where Karl Bosch intervened, developing an industrial reactor capable of working at pressures above 200 atmospheres. On the other hand, to work at a temperature as low as possible without the process becoming too slow, Mittasch developed an iron-based catalyser enabling the reaction to take place at a reasonable rate. The process became operational in 1913, creating a revolution in industrial chemistry. The Nobel Prize awarded to Haber for this process was a consecration of integrated research combining theoretical chemistry with chemical engineering and leading to production at minimal cost of a substance of great importance to humanity. Ammonia synthesis was indeed important because it opened the way to production of nitric acid by catalytic oxidation of ammonia on platinum - a process invented by **Wilhelm Ostwald** in 1907 - and hence to the production of nitrates and fertilisers essential to increasing agricultural yields. Unfortunately, every medal has its other side and the nitrates so useful in feeding populations could also be used to produce explosives. This enabled Germany, during World War One, to exploit the Haber-Bosch process for military purposes, while its weapons industry was deprived of nitrates from Chile as a result of the blockade imposed by the British Navy. It was unsettling to see what extraordinary energy Haber deployed during the First World War to develop offensive techniques based on the use of suffocating gases. In truth, scientists on both sides used their science to serve war. Nevertheless, the attribution of the Nobel Prize in 1918 elicited a violent outcry among American, French, and English scientists. Contradictory in the extreme, Haber devoted himself after the conflict to developing insecticides for agricultural use. The company he founded for this purpose developed a preparation that was to enter history under the name "Zyklon B", infamous because of the use of this preparation in the gas chambers of World War Two. Haber was never to witness the disastrous effect of this product on human beings. In collaboration with Born, he developed a thermodynamic method for analysing the ionic bond. The work led to the concept of lattice enthalpy, a parameter used to quantify the cohesion of ionic systems and to characterise their macroscopic properties such as hardness, melting temperature, etc. In 1933, deprived of his official responsibilities by the Nazi Party because of his Jewish origins, Haber found refuge in Cambridge, England. He died of a heart attack in Basel in 1934, on his way to Palestine where Chaïm Weitzmann, future president of the State of Israel and himself a chemist, had offered him a position as Head of the Physical Chemistry Department of the Sieff Research Institute in Rehovot.

Fritz Haber

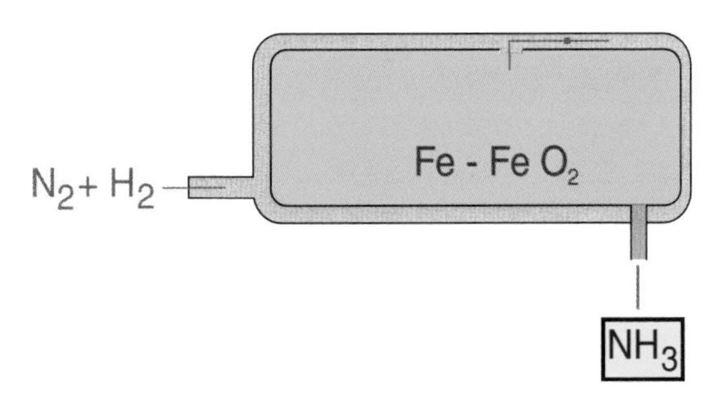

The synthesis of ammonia by the Haber-Bosch process. A representation of the famous process for nitrogen fixation developed by the German chemists. It results in the incorporation of nitrogen in substances such as ammonia. By oxidation to nitrate, the nitrogen can be used by plants to make nitrogen - containing substances, notably proteins.

1919 Not awarded

1920

Nernst, Walther (Briesen, Prussia, 1864 - Muskau, Germany, November 18, 1941). German physical chemist. Son of a judge. Nobel Prize for Chemistry in recognition of his work in thermochemistry. Walter Nernst defended his doctoral thesis at the University of Würzburg in 1887. This thesis dealt with the influence of magnetism and temperature on electrical conductivity. The same year he was appointed assistant to Professor **Wilhelm Ostwald** in Leipzig. Nernst proposed a mechanism explaining how ionic compounds disolve in water through the action of water molecules that surround the ions, isolate them from one another, and cause them to disperse. He thus explained a phenomenon about which scientists had remained skeptical when **Svante Arrhénius** first proposed it. From this mechanism he deduced a law relating the electromotive force of a galvanic cell to its standard potential E and to the concentrations of the ions extended to electrochemical systems the free enthalpy concept of Gibbs. It made possible an extraordinary development of the thermodynamics of oxidation-reduction reactions. Nernst was appointed professor at the University of Göttingen in 1894 and founded the Institute of Physics and Electrochemistry. There he directed research on the chemistry of solutions. He also developed methods for measuring the pH of solutions,

Walther Nernst

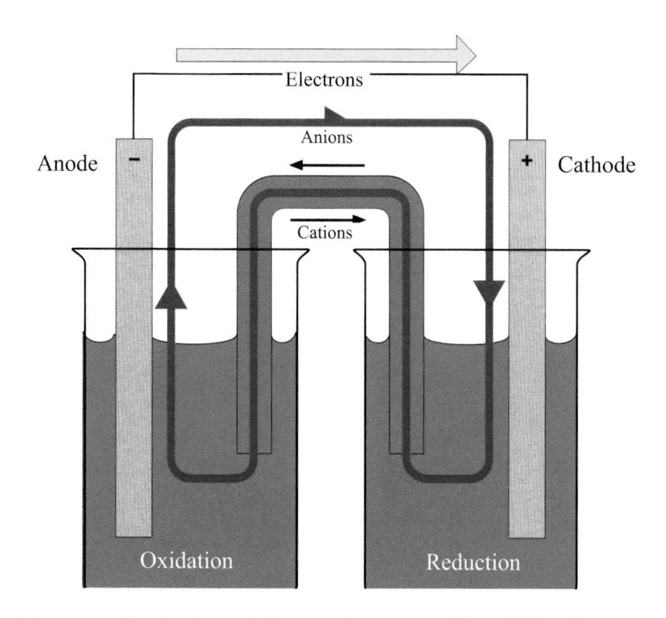

Diagram of a battery. A battery is an assembly which allows chemical energy to be transformed into electrical energy. In the left hand side of the diagram (anode), a metal (zinc, for example) oxidises and dissolves, releasing electrons into the external circuit. These electrons pass to the right-hand compartment (cathode), where they reduce the cation of a more noble metal (copper, for example).

defined the solubility product of a salt and, in 1903, formulated the principle of action of buffer solutions. He was offered the Chair of Physical Chemistry at the University of Berlin in 1905. Changing his orientation, he then focused on the thermodynamics of chemical reactions at very low temperature, showing in 1906 that specific heats and dilation coefficients approach zero when the temperature approaches the absolute zero. This led him to formulate the third principle of thermodynamics, according to which it is impossible to reach absolute zero temperature. It is worth mentioning that in 1997 the Nobel Prize of Physics rewarded research that had led to reaching a temperature of about three billionths of a degree Kelvin! In 1922, Nernst was appointed President of the "Physikalische-Technische Reichanstalt." He left this institution in 1924 and headed the Institute of Experimental Physics until his retirement in 1934, caused by his firm opposition to Nazism. It can be said that Svante Arrhénius, Wilhelm Ostwald and Walter Nernst were the founders of physical chemistry, the science that studies the physical laws which govern chemical phenomena.

1921

Soddy, Frederick (East-bourne, England, September 2 1877 - Brighton, England, September 22, 1956). Son of a wheat merchant. Nobel Prize for Chemistry for investigating radioactive substances and for elaborating the theory of isotopes. Frederick Soddy registered at Oxford University in 1896 and remained there for two years after receiving his degree in chemistry (in 1898). In 1900 he was appointed Assistant at McGill University in Montreal, where he worked with **Ernest Rutherford** on radioactive decay. There he took part in an important discovery: Rutherford's team showed that radioactive decay changes the chemical nature of the isotope emitting the radiation. This meant that the phenomenon affected the nucleus of the atom, which Rutherford was to characterise a few years later. Upon his return to England in 1903, Soddy was appointed assistant to Ramsey in London. Then came the demonstration that helium is released when the radium atom decays. This corroborated the discovery made in Montreal. He was appointed Lecturer in Chemical Physics at the University of Glasgow in 1904. After teaching chemistry at the University of Aberdeen he moved to Oxford. The main focus of Soddy's research was radioactivity. In 1902 he explained the relationship between a radioactive isotope and the "progeny" deriving from it upon emission of alpha and beta particles or gamma radiation. He also defined the notion of radioactive isotopy, showing that three "elements" with different radioactive properties were in fact three different isotopes of radium. His discoveries did not stop there. He noticed that the product formed when an isotope emits an alpha particle has the atomic number of the initial isotope minus two. He had thus discovered the phenomenon of transmutation resulting from the emission of charged particles from the atomic nucleus. This was of course a purely nuclear phenomenon....and to think alchemists had devoted so many sterile efforts to achieving it! In 1914, Soddy left Glasgow and was appointed Chairman of Chemistry at the University of Aberdeen. At this point he made another remarkable discovery: studying the atomic mass of lead extracted from a bed in Ceylon, he observed that it differed significantly from the accepted value of the time. Without comprehending the meaning of this finding, he had in fact discovered that the element lead comprises several isotopes that vary in proportion according to the age of the bed. This opened the way to explaining the radioactive decay pathway leading from uranium to lead. It was in 1936 that Frederick Soddy retired from this exceptionally prolific professional life.

Frederick Soddy

1922

Aston, Frederick (Harborne, England, September 1, 1877 - Cambridge, England, November 20, 1945). British chemist. Son of a farmer. Nobel Prize for Chemistry for his development of the mass spectrograph. This instrument allows, with unequalled precision, the study of the mass of the isotopes which compose the elements of the periodic table. Frederick Aston thus elucidated the famous problem raised by **William Richards**, Nobel Prize winner in Chemistry in 1914, who had measured atomic masses significantly different from whole numbers. Aston showed

Frederick Aston

that in reality the atomic mass of an element is an average of the masses of the various isotopes of which it is composed. In 1893, he personally built the equipment which allowed him to conceive experiments carried out on various electric phenomena. This particular talent earned him a scholarship at Birmingham University in 1903. Then he joined **Joseph Thomson**, the inventor of the mass spectrograph, at the University of Cambridge. This incomparable research tool derives in fact from electric gas discharge tubes, still called Crookes tubes, which had made it possible for the British physicist to discover the ratio of the charge-to-mass of the electron in 1897. In 1910, Aston and Thomson used it to study neon and discovered therein the existence of isotopes whose masses are, at first glance, whole numbers. These results persuaded Aston to perfect the mass spectrograph. Thanks to his technical improvements, he observed that the isotopic masses are, in reality, slightly lower than the whole numbers that they should reach. This is the question of "mass defect", which Albert Einstein had shown corresponds to a release of energy. Of course, it concerns energy released during the union of the particles which constitute the atomic nucleus. It thus proved that the higher the released energy, and thus the more significant the mass defect, the more the formed core is stable. This was a most significant discovery for nuclear physics.

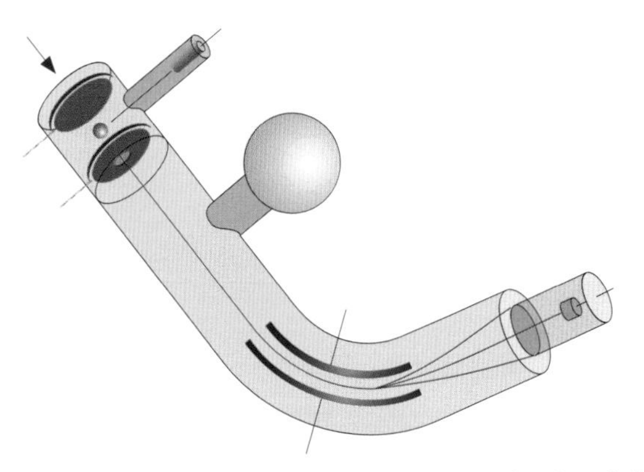

Mass spectrograph. This instrument was developed by Frederick Aston. It operates on the principle that moving ions may be deflected by electric and magnetic fields. The mass spectrograph permits the separation and the identification of isotopes of chemical elements and can determine their masses with high accuracy. It is also used for the analysis of inorganic and organic molecules in impurities or to determine structural formulas of complex substances.

1923

Pregl, Fritz (Laibach, Austria, September 3, 1869 - Graz, December 3, 1930). Austrian chemist. Nobel Prize for Chemistry for developing techniques in the microanalysis of organic compounds. Son of a treasurer. Pregl achieved his medical studies at the University of Graz in 1893 and was appointed assistant in physiological chemistry in the same institution in 1899. He headed the Department of Chemistry in Insbrück from 1910 to 1913 but he then returned to Graz and headed the Institute of Medicinal Chemistry After a short working stay with **Wilhelm Ostwald** and **Emil Fischer** in Germany, Pregl conducted research on bile acids and urine, two products of human secretion. Influenced by Fischers's work, he focused particularly on proteins, but encountered an obstacle: traditional chemical analyses required much greater quantities of reagents than could be isolated from a sample of a physiological fluid such as bile. To overcome this obstacle, Pregl undertook to develop new microanalytical techniques. He thus oriented his career towards analytical chemistry. First, with the help of Kuhlman (a manufacturer of chemistry instruments), he developed a microbalance with a sensitivity of 0.001 mg. He also miniaturised combustion systems for organic samples, developed during the previous century

Fritz Pregl

by Justus Von Liebig. He finally reached a point where he could perform precise analyses on 1- to 2-mg samples. This revolutionised biochemical research. Fritz Pregl then used this approach to study bile components, enzymes, serum, and trace substances whose detection is sometimes required in forensic medicine.

1924 Not awarded

1925

Zsigmondy, Richard (Vienna, Austria, April 1, 1865 - Göttingen, Germany, September 29, 1929). Austrian chemist. Son of a dentist. Nobel Prize for Chemistry for research on colloids. After graduating in organic chemistry from the University of Munich and working in Berlin and Graz, Richard Szigmondy headed the Institute for Inorganic Chemistry at the University of Göttingen. His first centre of interest was the chemistry of glass colour, which is sometimes caused by very fine particles scattered through the bulk of the glass. This research led him into the field of colloids, substances formed by ultramicroscopic particles suspended in a liquid such as water. Such particles escaped ordinary microscopic observation and were therefore not well known. To better study this peculiar organisation of matter, he invented in 1903 with Siedentopf, a physicist working for the Zeiss Company, an ultramicroscope in which light was beamed through the colloidal particle suspension perpendicularly to the direction of observation. Thanks to the light-scattering (Tyndall) effect of the suspended colloid particles, it was possible to count them and to observe their Brownian motion, even though they could not be seen clearly and appeared only as spots of light.

1926

Svedberg, Theodor (Fleräng, Sweden, August 30, 1884 - Örebro, Sweden, February 25, 1971). Swedish chemist. Son of an engineer. Nobel Prize for Chemistry for his studies in the chemistry of colloids and for his invention of the ultracentrifuge. His doctoral thesis, presented at the University of Uppsala in 1907, was about colloids. He created these by producing an electric spark in a liquid. He thus obtained fairly pure colloidal suspensions in which he studied Brownian motion. Appointed lecturer, he continued this work, which led him to establish firmly the existence of molecules. He thus confirmed Dalton's hypothesis, proposed 120 years earlier, which some scientists as highly regarded as **Wilhelm Ostwald** still contested. The observation of Brownian motion demonstrated the existence of collisions between the colloid particles and the molecules of the surrounding liquid. Thanks to Zsigmondy's microscope, Svedberg observed that colloid particles obey the laws of classical physics. He was able to determine their diameter by measuring their sedimentation rate. This rate is so low that he had to develop ultracentrifugation, a technique accelerating the sedimentation of the finest particles. Developed in collaboration with Nichols at the University of Wisconsin, this technique was used to reach rotational velocities of some 30,000 rotations per minute and to create a centrifugal acceleration several thousand times greater than the earth's gravitational acceleration. In 1935, their ultracentrifuge managed to produce accelerations up to 750,000 times that produced by the earth's gravity. Svedberg was then able to study proteins such as haemoglobin and albumin, thus revealing the potential of a technique that was to impose itself in biology and lead to basic results in physics. The 1926 Nobel Prize in Physics was indeed awarded to **Jean-Baptiste Perrin** (1926 Nobel Prize for Physics) for his work on colloids, and specifically for his determination of Avogadro's number based on Svedberg's work.

Theodor Svedberg

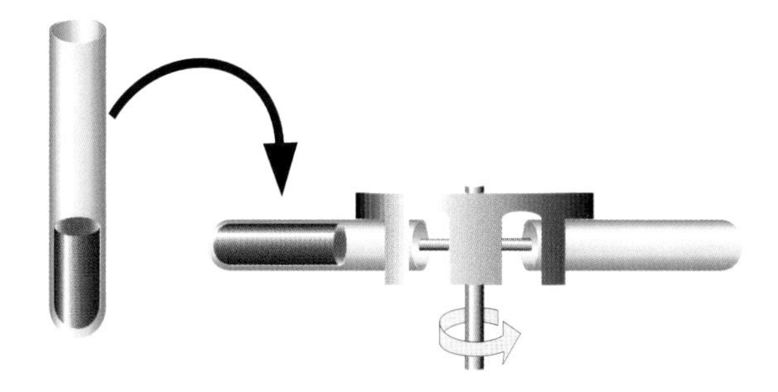

Ultracentrifuge. Since its initial development by T. Svedberg, this apparatus has been widely used to mesure the purity of cell components and to estimate their molecular sizes on the basis of sedimentation rate.

1927

Wieland, Heinrich (Pforzheim, Germany, June 4, 1877 - Munich, August 5, 1957). German chemist. Son of a chemical pharmacist. Nobel Prize for Chemistry for his determination of the molecular structure of bile acids. Considered to be one of the greatest organic chemists of his century, Heinrich Wieland is known principally for his work on biological oxidations and bile acids. In the oxidation field he notably showed that, contrary to what people thought at the start of the 20th century, oxidation proces-

Heinrich Wieland

ses do not usually result from the direct action of oxygen on substrates but from their dehydrogenation. Wieland showed, for example, that substances like methylene blue can oxidise a molecule by removing hydrogen, without any intervention of oxygen. His work on bile acids, for which he received the Nobel Prize, proved particularly

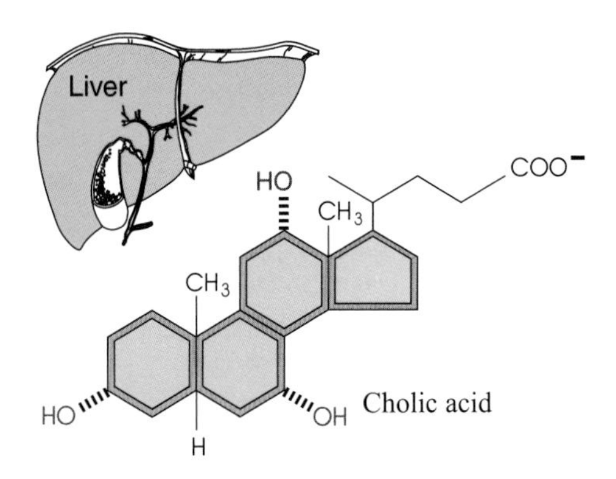

Liver and production of bile acids. Bile acids are critical for digestion and absorption of fats and fat - soluble vitamins in the small intestine.

arduous. Produced in the liver, these acids are stored in the gall bladder and then carried into the intestine where they play an important role in the digestion of fats. Wieland exploited the advances in this field of **Adolf Windaus**, who had shown that these acids are related to cholesterol and had determined their structure. Later it appeared that the cholesterol nucleus constitutes the basic structure of many sex hormones (progesterone, testosterone), adrenal hormones (cortisone, hydroxycortisone), and molecules such as digitalin, used as a heart stimulant. We are also indebted to Heinrich Wieland for his contribution to the structures of morphine, lobeline, and strychnine. The latter substance, quite complex, is an alkaloid extracted from the seed of a plant called *Nux vomica*, which grows wild in India. It is used as a stimulant of the central nervous system, with its effect being to increase sensory perceptions such as taste, smell, or vision.

1928

Windaus, Adolf (Berlin, Germany, December 25, 1876 - Göttingen, June 9, 1959). German chemist. Son of artisans. Nobel Prize for Chemistry for research on sterols, notably vitamin D, that play important biological roles.

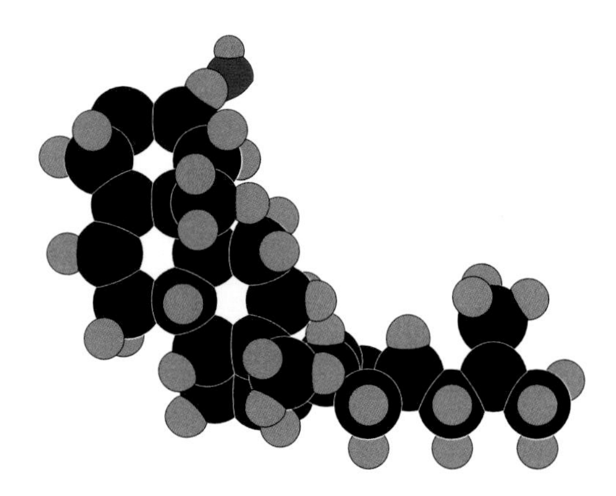

Cholesterol. It is an essential component of biological membranes. It is also a raw material for substances like steroid hormones, bile salts as well as some insect hormones like ecdysone. The molecule was extensively studied by **Konrad E. Bloch** who showed that acetate was a precursor. (See Windaus).

After studying the heart stimulant digitalin and making it the subject of his doctoral thesis, this ex-student of Emil Fischer devoted himself almost exclusively to the chemistry of cholesterol. His work progressed in close association with that of **Heinrich Wieland** and led to the determination of the structure of the sterol nucleus in 1932, some 30 years after he began his research in this field. Windaus was also active in another research area:

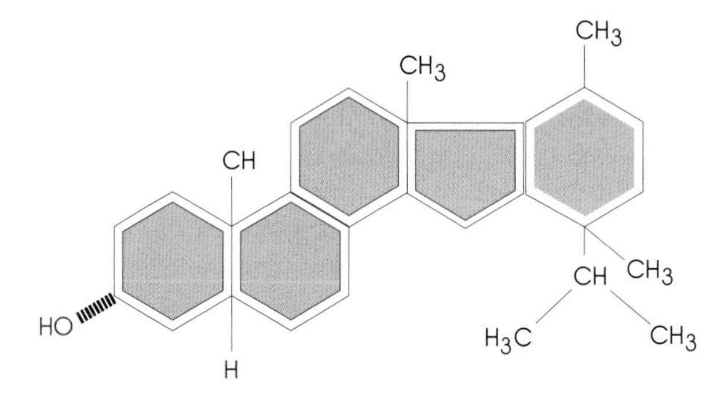

Vitamin D, sunlight and skin. Vitamin D is produced in the skin by the action of sunlight on its precursor molecule, 7-dehydrocholesterol.

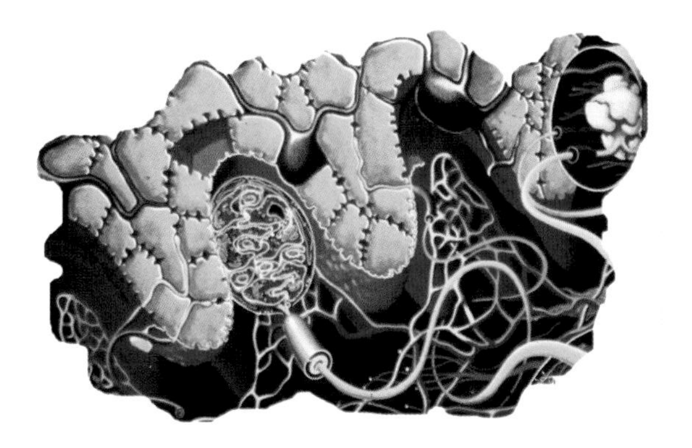

Human skin produces vitamin D when exposed to sunlight. People also obtain the nutrient by consuming vitamin supplements, fish oil, and breakfast cereals and milk fortified with vitamin D.

the study of vitamin D. The physiologist **Alfred Hess** (1946 Nobel Prize for Medecine) had suggested to him that this vitamin might be related to cholesterol. Cooperation between Windaus and Hess led to confirming this relationship while revealing the existence of several forms of vitamin D. Whereas ergosterol or vitamin D_2 is found in plants, vitamin D_3 or cholecalciferol, involved in calcium metabolism, is synthesised by the cells of the epidermis by activation of dehydrocholesterol in response to ultraviolet radiation. This chemist's career was also enriched by the discovery of histamine, a substance responsible for the inflammatory response, and by the elucidation of its structure. Meanwhile Adolf Windaus had become Head of the General Chemistry Laboratory of the University of Göttingen and one of his students had been **Adolf von Butenandt**, who had demonstrated the sterol nature of several sex hormones and notably progesterone.

1929

Harden, Arthur (Manchester, England, October 12, 1865 - Bourne End, England, June 17, 1940). English biochemist. Nobel Prize for Chemistry, along with **Hans von Euler - Chelpin**, for work on the fermentation of sugar. The young Arthur Harden, the only son in a family of 9 children, was first welcomed at the Owens College of the University of Manchester, where he became a chemistry

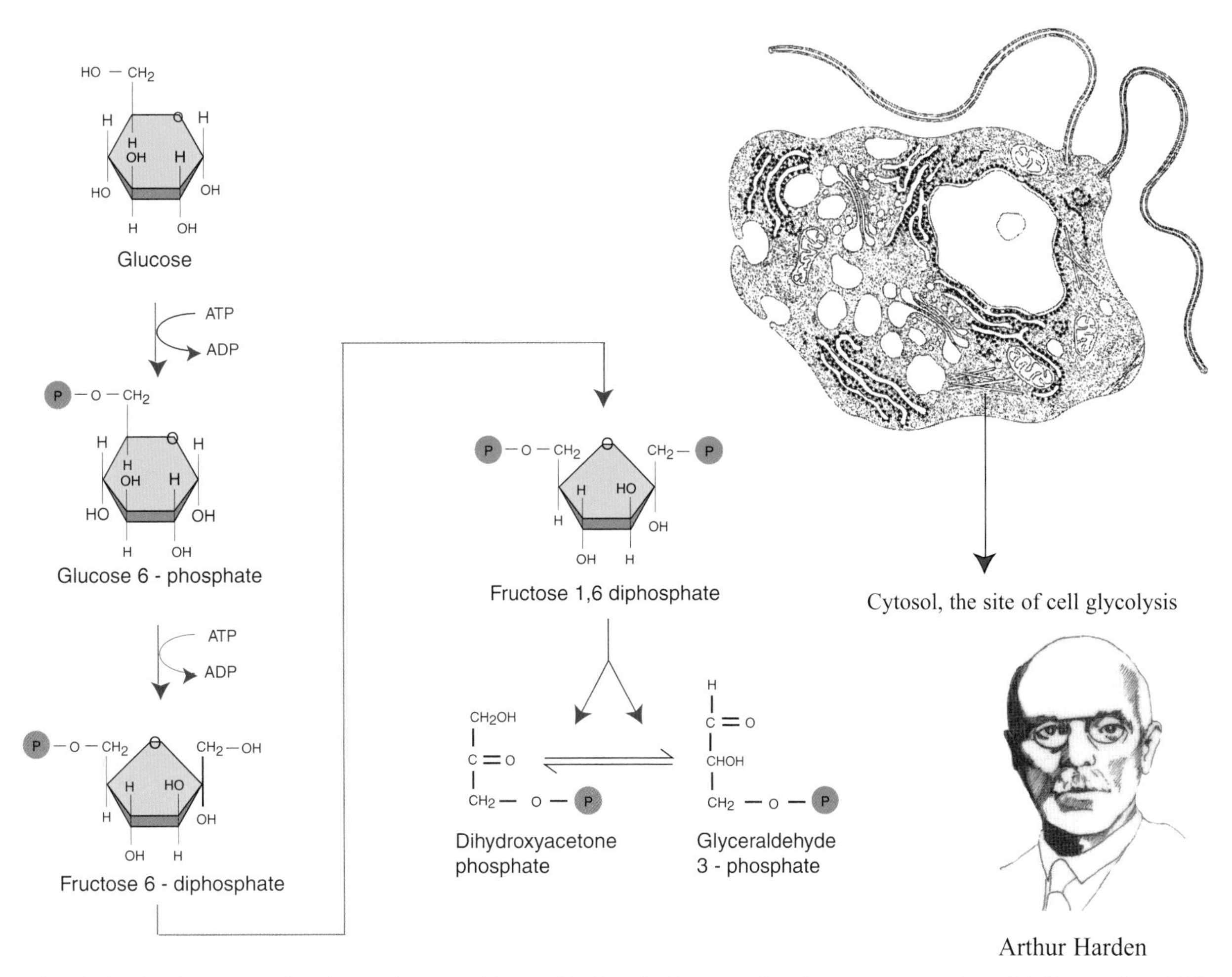

Cytosol, the site of cell glycolysis

Arthur Harden

Glycolysis, the first steps. As shown, the sugar (glucose) is first doubly phosphorylated and two moles of ATP are utilized. The six-carbon is then spit into two three-carbon units. The major oxidation of the substrate is catalyzed by glyceraldehyde 3-phosphate dehydrogenase.

enthusiast. A visit to the University of Erlangen in Germany enabled him to extend his knowledge in this field, notably under **Ernst Otto Fischer**. Yet back in England, he did not orient his career towards research but towards the history of science. At this point he wrote a very important book on the work of Dalton, the English chemist who developed atomic theory. Then Eduard Buchner's discovery that alcoholic fermentation can occur without the intervention of living cells is probably what made him want to plunge wholeheartedly into research in biochemistry. One of the main discoveries of Harden was to demonstrate that a yeast enzyme involved in alcoholic fermentation is composed of two parts, the enzyme itself and a small non-protein molecule which was later called coenzyme. The chemical nature of this coenzyme was studied principally by von Euler Chelpin, and it soon became clear that vitamins, substances whose discovery had begun with **Christiaan Eijkman**, are indispensable to life because they constitute an important part of many coenzymes. It was later discovered that minerals such as copper, zinc, and other atoms, also required in trace amounts, contribute like vitamins to the function of coenzymes. Another main discovery of Harden was to show, with William Young, the involvement of phosphate groups in the biochemistry of organisms and in particular during fermentation. The importance of these chemical groups was later underlined by **Carl** and **Theresa Gerty Cori**, who elucidated some steps of the glycolytic patway. Arthur Harden was knighted in 1936.

Euler-Chelpin, Hans Von (Augsbourg, Germany, February 15, 1873 - Stockholm, Sweden, November 6, 1964). Swedish biochemist. Nobel Prize for Chemistry shared with **Arthur Harden** for work on the role of enzymes in the fermentation of sugar and for his discovery of nucleotide adenine nicotinamide. Few men have had such illustrious scientists to supervise their studies. Hans von Euler-Chelpin, whose father was a cavalry captain and whose mother was related to the Swiss mathematician Euler, first wanted to paint. Yet the chemical phenomena linked to pigmentation led him into science, and particularly physical chemistry. This is when he worked under masters of great stature, such as **Max Planck**, **Otto Warburg**, **Emil Fischer**, **Walter Nernst**, **Svante Arrhenius**, **Eduard Buchner**, and **Jacobus van 't Hoff**, all of them future Nobel Prize laureates. He chose to study fermentation and this led him to add major contributions to the earlier work of Harden. This chemist had shown that enzymes contain a nonprotein part, a coenzyme. Working out the structure of one of them, called nicotinamide adenine dinucleotide, he proved that it was made up from a nucleotide similar to that found in nucleic acid. One of his sons, Ulf Svante von Euler, was awarded the Nobel Prize for Medicine, for his work on neurotransmitters.

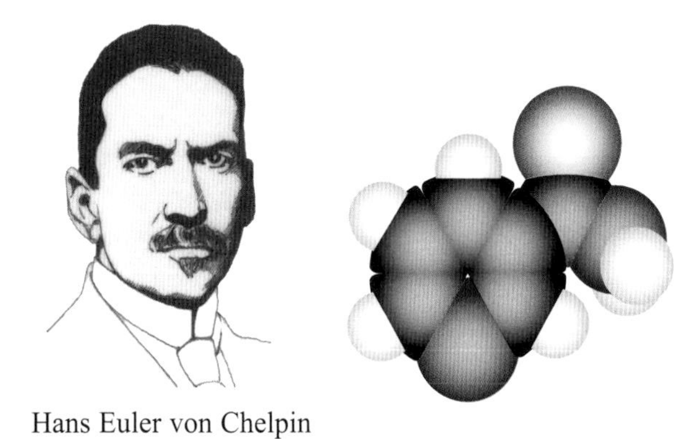

Hans Euler von Chelpin

Nicotinamide. This molecule is a member of the B complex of vitamins, nutritionally equivalent to nicotinic acid. It occurs widely in living organisms.

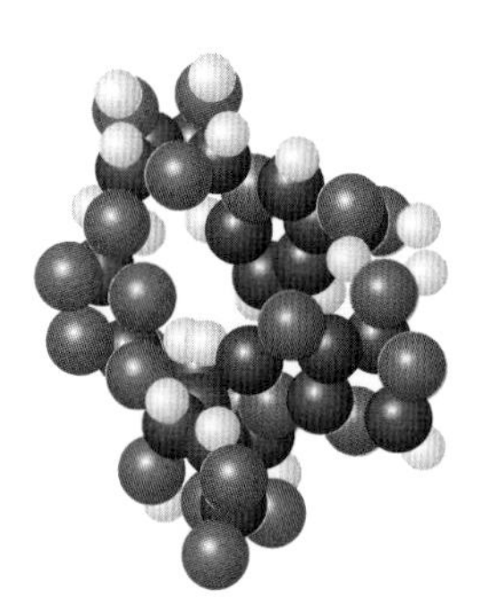

Three-dimensional structure of nicotinamide adenine dinucleotide, or NAD. Discovered by the Swedish chemist Hans von Euler - Chelpin, the main function of this coenzyme is to carry electrons in numerous cell oxidoreduction reactions. Like flavine adenine dinucleotide (FAD), NAD is a major electron acceptor in the oxidation of fuel molecules. NAD is used primarily for the generation of ATP. NADPH is a major electron donnor in reductive biosynthesis.

1930

Fischer, Hans (Höchst Am Main, Germany, July 27, 1881 - Munich, March 31, 1945). German biochemist who was awarded the Nobel Prize for Chemistry for research into the constitution of hemin, the red blood pigment, biliburin and chlorophyll, the green pigment in plants. The first major success was the synthesis of porphyrins, substances consisting of four pyrrole rings linked together by methine bridges. Nearly all porphyric pigments in nature arise from the presence of different substituents on carbons 1 through 8. In 1929, Hans Fischer succeeded in synthesizing the heme group, the porphyrin bound to hemoglobin and determined its structure. The function of hemoglo-

Hans Fischer

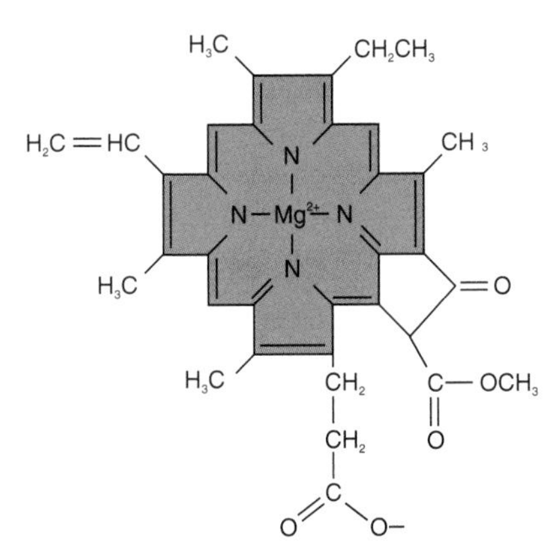

The porphyrin ring of chlorophyll. It contains a magnesium atom.

bin is to carry oxygen in the blood. Later, he showed that the heme group contained an iron atom. At the reception following the award of the Nobel Prize, he stressed the similarity between the porphyric nucleus of chlorophyll and the haem of haemoglobin. He also showed that the metal atom present in the molecule of chlorophyll is magnesium and not iron. Hans Fischer and his collaborators synthesized more than 130 different porphyrins. The end

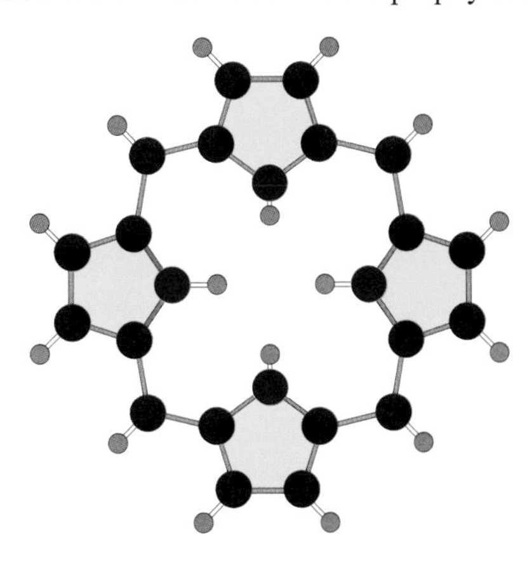

Structure of a porphyrine. This type of molecule consists of four pyrrole nuclei linked together through methene bridge. This structure is widely distributed among naturally occuring substances, especially electron transport molecules such as chlorophylls, hemoglobins and cytochromes.

of his life was marked by Allied bombardments having completely destroyed his laboratory during World War Two and his tragic suicide.

1931

Bergius, Friedrich (Goldschmieden, Prussia, October 11, 1884 - Buenos Aires, Argentina, March 30, 1949). German chemist. Son of the manager of a local chemical factory. Nobel Prize for Chemistry along with **Carl Bosch** in recognition of their contribution to the invention and development of chemical high pressure methods. Many current processes used to refine petroleum fractions are still based on these methods. After receiving his Ph.D. from the University of Breslau in 1907, Friedich Bergius worked with Nernst in Berlin and with Haber in Karlsruhe in order to develop the synthesis of ammonia. As early as 1909 he began to test high - pressure techniques and to develop reactors that were sufficiently resistant. In 1913,

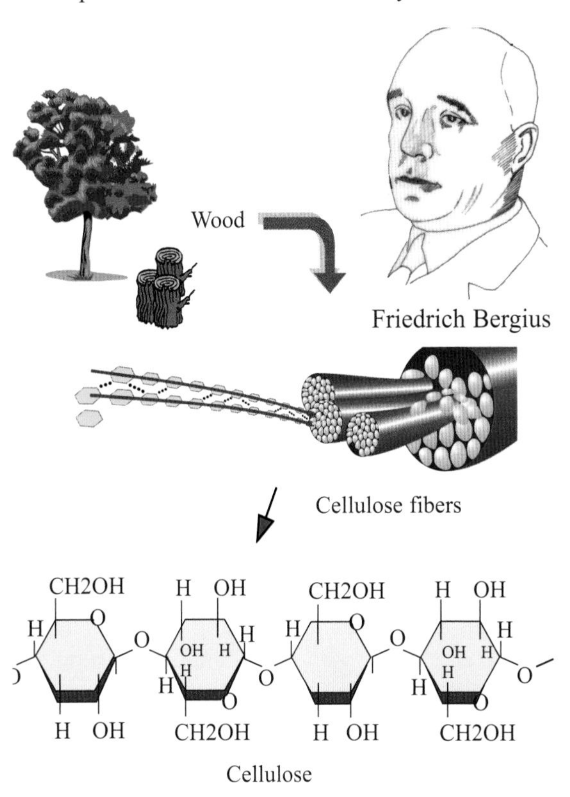

Saccharification procedure for treatment of cellulose. Elaborated by Bergius it converts cellulose, a sugar polymer, into a variety of fermentable compounds such as glucose and fructose.

he was granted a patent for the so - called "Bergius process" for synthesising liquid hydrocarbons from coal and hydrogen. After this patent was granted while he was working at the Badische Anilin und Soda Fabrik (BASF), he conti- nued his research there in order to improve the yield of the process, which was not very cost-effective. The Bergius process was later applied industrially by I.G Farben industrie. Between 1940 and 1945, hydrogenation of coal enabled Germany to maintain the fuel supply needed for its war effort. Bergius also produced sugars by acid hydrolysis of cellulose, a process called saccharification. Here again the technique proved important to wartime Germany, cut off from its normal food supplies. Between 1946 and 1949, Bergius worked as an expert at the Ministry of Industry of the Argentine government. As reserves of liquid hydrocarbons diminish, we can expect the Bergius process to have another heyday. When this happens we will have to become aware of the economic importance of hydrogen, and use methods for synthesising it that offer satisfactory energy yields.

Bosch, Karl (Cologne, Prussia, August 27, 1874 - Heidelberg, Germany, April 26, 1940). German industrial chemist Nobel Prize for Chemistry along with Friedich Bergius for the Haber - Bosch process for devising chemical high - pressure methods in ammonia synthesis. Karl Bosch was trained in metallurgy and mechanical engineering at the Technical University of Charlottenburg. Having become interested in phenomena occurring during industrial syntheses, he entered in 1896, the University of Leipzig, where he presented a doctoral thesis on carbon compounds. Bosch then joined the Badische Anilinund Soda-Fabrik (BASF) in 1899. In 1925 he was manager of I.G. Farbenindustrie, re-sulting from the merger of BASF with six German companies that specialised in dye production. Close collaboration with Fritz Haber was to lead to the development of a process for synthesising ammonia directly from nitrogen and hydrogen. Haber had sold his patent to BASF in 1909 and Bosch made it his task to scale up the process for industrial use. He had to solve several problems to achieve this: separating the hydrogen from the carbon monoxide present in the "water gas" resulting from the action of water on coal at high temperature, isolating nitrogen from air by distillation, developing a cheaper catalyser than the osmium used by Haber, and finally building a reactor capable of withstanding the particularly harsh process conditions of high temperature:500°C; high pressure:200 atmospheres. No such installations existed in industry at the time. Bosch designed a reactor with a double wall and circulated the cold hydrogen-nitrogen mixture inside the wall, at high pressure, before injecting it into the reactor. The inner wall made of soft steel was resistant to hydrogen at high temperature whereas the outer wall, made of carbon-rich steel, could withstand high pressures. In 1911, industrial production of ammonia began at BASF. Again under Bosch's responsibility, the company undertook in 1925 to study the so-called Bergius process for producing liquid hydrocarbons by the reaction of hydrogen with coal dust. Although this method could not compete with petroleum derivatives, it did sustain Germany's war effort between 1940 and 1945. In addition to extending the study of catalytic reactions applied to industrial conditions, Karl Bosch also became interested in photochemistry and polymers. He openly opposed the Nazi regime. He was appointed President of the Kaiser Wilhelm Institute in Berlin in 1937.

Karl Bosch

1932

Langmuir, Irving (Brooklyn, New York, January 31, 1881 - Falmouth, Massachusetts, August 16, 1957). American chemist. Son of a clerk. American physical chemist. Nobel Prize for Chemistry for studies of molecular films on solid and liquid surfaces that opened new fields in colloid research and biochemistry. After graduating as a metallurgical engineer from Columbia, Irving Langmuir worked for some years in Göttingen under Walter Nernst. In 1909 he headed a research laboratory at the General Electric Company, in Schenectady, until 1950. He was given considerable resources for conducting long-term research on tungsten filaments in incandescent lamps. This is a concrete example of industrial development leading, thanks to tenacious scientific genius, to discoveries in basic science. In 1916, Langmuir applied for a patent covering the invention of a lamp with a long life and a higher yield than anything produced before. The problem of incandescent lamps, although very practical, touched upon very unusual physico - chemistry involving interactions between a metal surface, residual gases pre-

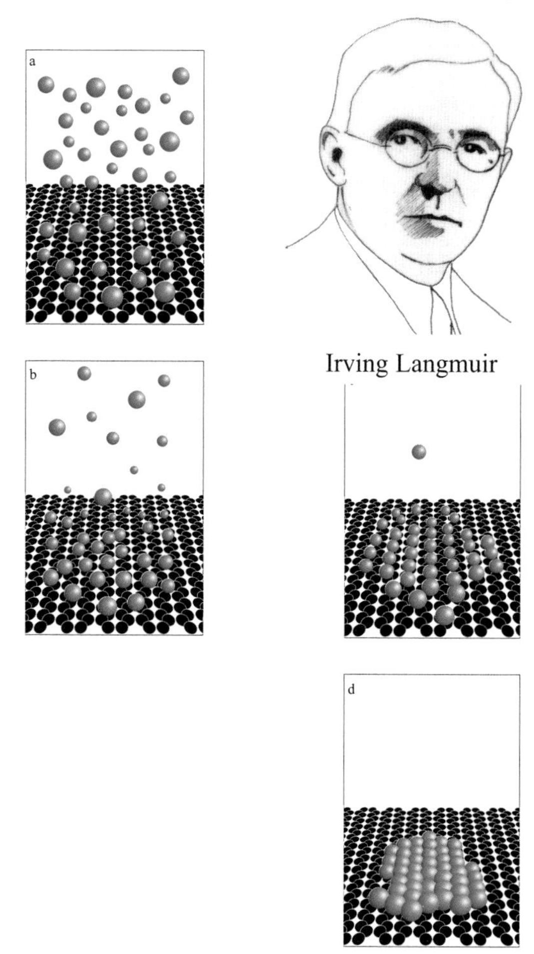

Irving Langmuir

Adsorption of gas molecules by a metal.

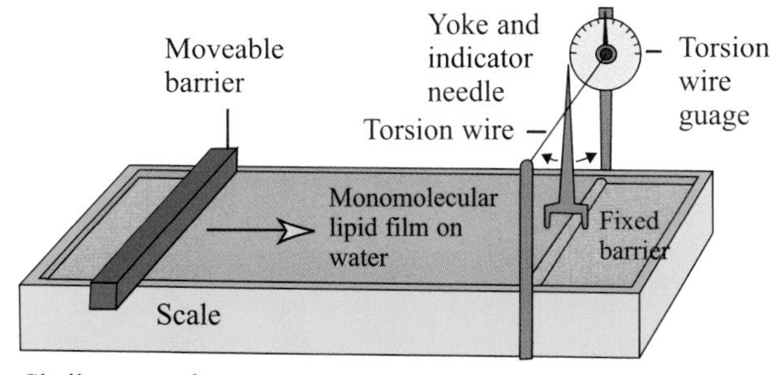

Langmuir trough. This apparatus measures the dependence of the surface force exerted by a film of a substance floating on a liquid with which it is immiscible, on the area of the film.

sent in the light bulb, and the glass surface. Langmuir was then led to study the adsorption of gaseous molecules on solid surfaces in function of the temperature and pressure. This was the first description of the thermodynamic phenomena occurring at a gas - solid interface. These studies were to have a deep influence on the description and elucidation of the mechanisms of heterogeneous catalysis, which involve reactions between gas molecules activated by a solid catalyst. He thus developed an equation known as the "Langmuir isotherm", describing the parameters that determine the chemisorption of molecules to the surface of a solid. The scientists most familiar with this equation are probably those who study surface phenomena. The scientific legacy of Irving Langmuir remains the basis of many industrial syntheses using heterogeneous catalysis (developed first by Nobel laureates such as Haber, Bosch, and Bergius). His research also led to significant discoveries concerning electron emission from the surface of heated metals. These discoveries were to have well-known applications in the production of electronic tubes, the epitome of this development being the television screen.

1933 Not awarded

1934

Urey, Harold Clayton (Walkerton, Indiana, April 29, 1893 - La Jolla, California, January 5, 1981). Son of a school-master. Nobel Prize for Chemistry for his discovery of the heavy form of hydrogen known as deuterium. Harold Urey registered at Montana State University in 1914, obtaining his bachelor's degree in zoology in 1917. At the end of World War One, he returned to Montana to teach chemistry, but in 1921 he resumed his studies at Berkeley under the famous American chemist Gilbert N. Lewis. After a postdoctoral stay at the University of Copenhagen under the direction of **Niels Bohr**, he returned to the States in 1925. Until 1929, he worked at the Johns Hopkins University, Baltimore, but left this establishment and taught chemistry at Columbia in 1934. Since the discovery of isotopes by **Frederick Soddy**, Urey hoped to demonstrate the existence of a heavy isotope of hydrogen. He attempted the feat by distilling four litres of liquid hydrogen, from which he isolated a residue that did turn out to be a heavy form of hydrogen: deuterium, a hydrogen isotope with two times the atomic mass of ordinary hydrogen. A logical consequence of this was his discovery in 1931 of heavy water, which contains an atom of oxygen and two atoms of deuterium and thus has a molecular mass of 20 (instead of 18 as for normal water). Heavy water melts at 3.8°C and boils at 101.4 °C. Because deuterium, unlike hydrogen, absorbs few neutrons, heavy water constitutes an excellent, if costly, moderator for nuclear reactors. Its production was the object of intense rivalry during World War Two: within ten years of its discovery it had acquired capital strategic importance in the development of plutonium-producing nuclear reactors. Later, the laboratory headed by Urey discovered isotopes of nitrogen, oxygen, and sulfur. Credit is also due to this tenacious scientist for the observation that different isotopes of an element, although chemically identical, can react chemically at slightly different rates. In the framework of the "Manhattan Project" aiming to produce the atomic bombs that were to be dropped on Hiroshima and Nagasaki, Urey was in charge of enriching uranium in its fissile isotope ^{235}U. The bomb used to destroy Hiroshima contained 64 kg of this isotope. This enrichment was achieved by gaseous diffusion of uranium hexafluoride through porous walls. Urey strongly opposed the dropping of the nuclear bombs on Japan and later campaigned in favour of limiting the extension of nuclear energy production. At the end of World War II he joined the Enrico Fermi Institute for Nuclear Studies, Chicago, and held a post as a professor in 1952. Later he became

Harold Urey

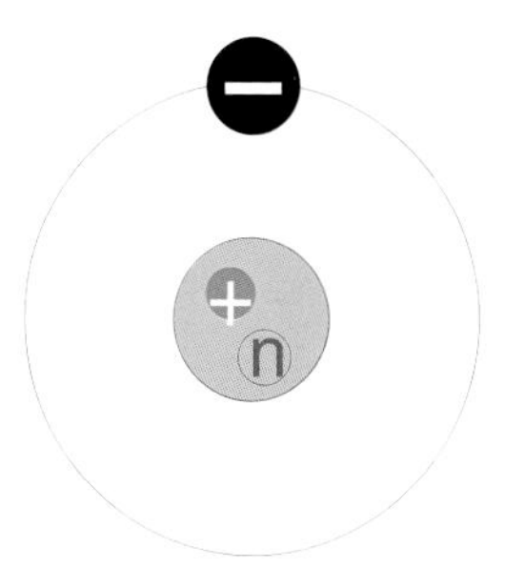

Structure of a deuterium atom.

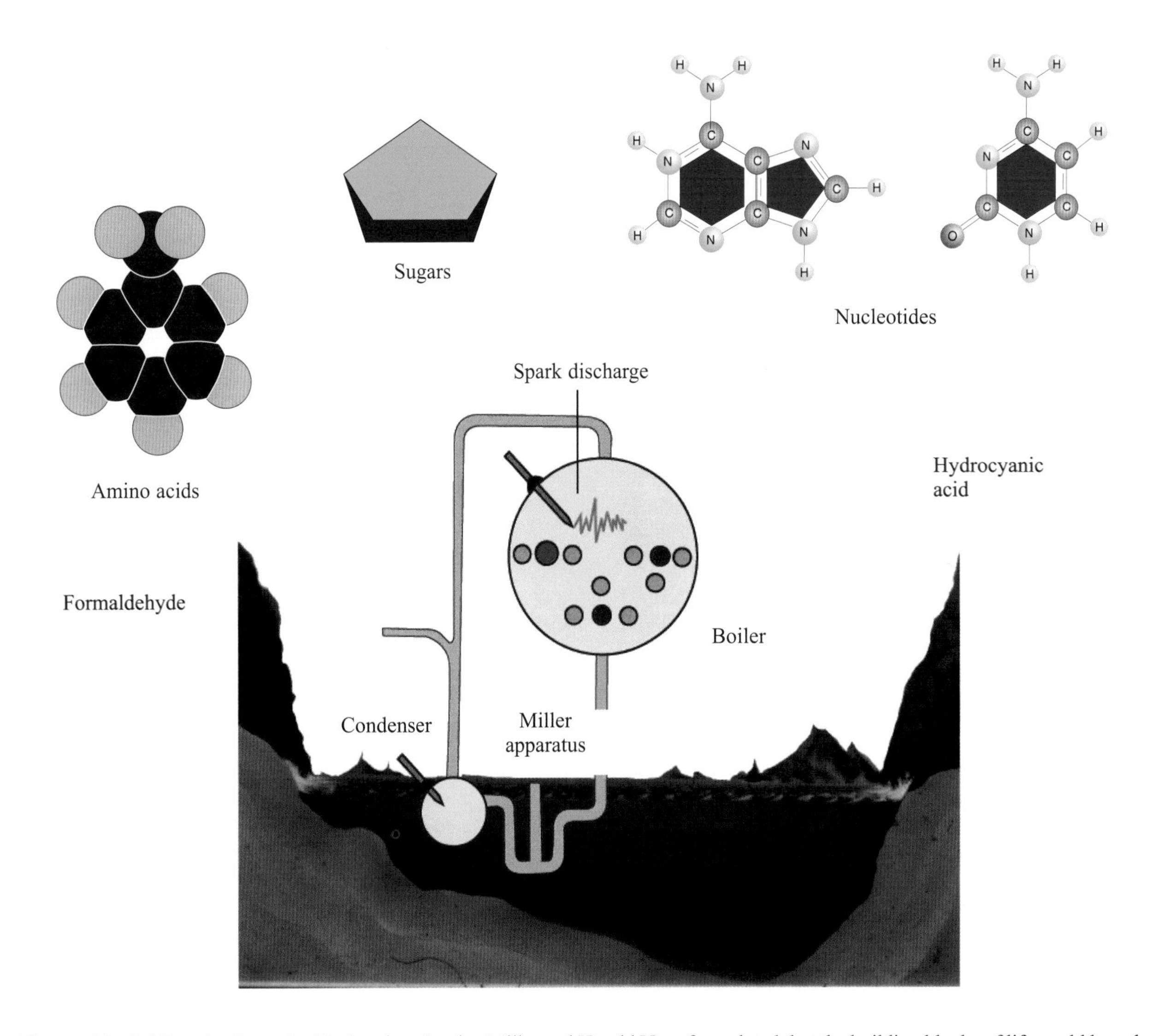

The world of Miller. As shown in this drawing, Stanley Miller and Harold Urey formulated that the building blocks of life could have been formed spontaneously in a reducing atmosphere. **Stanley Miller apparatus**. It was designed by the graduate student of Harold Urey for the circulation of methane, ammonia, water vapor and hydrogen. Water was boiled in the flask at lower left. Energy for chemical reactions came from electric discharges. Among the products of the chemical reactions are several amino acids and several organic compounds.

interested in the study of chemical phenomena in the universe. He notably focused on the origin of the earth's atmosphere and on the atmospheric chemical reactions that might have led to the synthesis of the first biological molecules from which life originated. In 1951 Stanley Miller, one of his graduate students, set up an apparatus into which he introduced such molecules and submitted it to electrical discharges and confirmed Urey's hypothesis. In 1958 Harold Urey became a professor at the University of California, San Diego, and stayed there until his retirement.

1935

Joliot-Curie, Jean Frédéric (Paris, France, March 12, 1900 - Paris, August 14,1958). Physical chemist. Nobel Prize for Chemistry for his discovery of new radioactive elements prepared artificially. Frederic Joliot graduated in physical engineering at the École de Physique et de Chimie Industrielle de la Ville de Paris in 1920, which was then directed by Paul Langevin. Some time later he was appointed as Marie Curie's assistant. There he met **Irène Curie** and married her in 1926. His doctoral thesis was on the electrochemistry of polonium, a radioactive element discovered by **Marie Curie** in 1898. This gave him the opportunity to make major improvements to "Wilson's chamber", an instrument sho-wing the trajectory of ionising radiation. After having received his Ph.D. in science in 1930 he was appointed lecturer at the Paris Faculty of

Harold Urey, Stanley Miller and the origin of life. This image symbolizes the possible passage from the inert world to the highly organized world characteristic of living beings.

Sciences. Irène Curie and Frédéric Joliot collaborated on research on the structure of atom. Their most significant discovery was artificial radioactivity, from which humanity was to benefit from 1934 onward. They bombarded several elements such as aluminum and magnesium with alpha particles. Much to their surprise they succeeded in that manner in forming radioactive isotopes of nitrogen, phosphorus, silicon, and aluminum. These isotopes are unstable and disintegrate rapidly, unlike the natural primordial isotopes whose very long half-lives have enabled them to subsist since their formation in stars. The Joliot-Curie's discovery was the first step in extraordinary developments that are continuing today. In 1937, Joliot became Professor at the Collège de France, where he directed the construction of the first cyclotron in Western Europe. Developed by the American **Ernest Lawrence** (1939 Nobel Prize for Physics) in 1930, a cyclotron is a particle accelerator. It is the instrument of choice for producing radioisotopes, so it is easy to understand Joliot's interest in this technique. The discovery in 1939 of the fission of uranium by the German scientists Fritz Strassmann and **Otto Hahn** triggered, among nuclear physicists, frenetic research on this type of reaction, the energy-producing potential of which soon became obvious. In Rome, **Enrico Fermi** discovered the moderating power of heavy water on the neutrons produced in nuclear fission. These neutrons, slowed down by a series of collisions with the hydrogen atoms of water, are excellent propagators of the fission chain. The prospect of a nuclear reactor appeared increasingly feasible. With Von Halban, the French physicist **Jean-Baptiste Perrin** and Lew Kowarski, Joliot studied the fission chain reaction and developed a plan for a nuclear reactor using uranium and heavy water. World War Two put an abrupt end to this research of capital importance: upon the arrival of the German troops in France, Frédéric Joliot carried off to England his equipment and scientific documents. During the war, he took an active part in the resistance movement, and in 1942, he joined the communist party. At the request of General de Gaulle, Frédéric established the Commissariat à l'énergie atomique (CEA). He was the first high commissioner for atomic energy. In 1948, ZOE, the first French nuclear reactor, was commissioned by the CEA. In 1950, Frédéric was dismissed from the CEA. On March 17, 1956, Irène Juliot-Curie died from leukemia and Frédéric replaced Irène as director of the Curie Laboratory (Radium Institute), and was appointed professor at the Paris Faculty of Science.

Jean Frédéric Joliot-Curie

Joliot-Curie, Irène (Paris, France, September 12, 1897 - Paris, March 17, 1956). French chemist. Daughter of **Marie** and **Pierre Curie**. Physical chemist. Nobel Prize for Chemistry for her discovery of new radioactive elements prepared artificially. The same year the Nobel Prize was also awarded to her husband, **Jean-Frédéric Joliot**. Furthermore, she was the daughter of Pierre and Marie Curie, both of them also Nobel laureates. In 1925, Irène Curie defended a thesis on rays emitted by polonium. Appointed lecturer at the Paris Faculty of Science in 1932, she became Pro-fessor there in 1937. From 1936, she headed the Institute of Radium. Her marriage with Frédéric Joliot in 1926 led to intense collaborative research, first on natural radioactivity, then on artificial radioactivity (which she discovered with her husband in 1934). This scientific couple discovered a way to induce radioactivity by means of alpha particles from polonium. Since then, many other kinds of radiation have been used to create radionuclides, so that now we have a vast panoply of substances displaying a wide variety of radiative properties. For instance, these elements were widely used in radiotherapy and medical imaging. From 1938 onward, the Joliots launched into the study of the effects of neutron beams on matter. The neutron had been discovered by the British physicist **James Chadwick** in 1932; it is a neutral particle capable of penetrating into many atomic nuclei and making them radioactive. The action of neutrons on heavy nuclei was to lead in 1938 to the discovery of the fission of uranium. Irène Joliot-Curie's career was prestigious : after the war, she helped her husband create the Atomic Energy Commission, which she headed from 1950 to 1956, the year of her death. She was also one of the founders of the Nuclear Physics Centre in Orsay. Frédéric Joliot continued her activities after 1956.

Irène Joliot-Curie

1936

Debye, Petrus Josephus Wilhelmus (Maastricht, The Netherlands, March 28, 1884 - Ithaca, New York, November 2, 1966). Dutch physical chemist. Nobel Prize for Chemistry for investigations of dipole moments, X - rays, and light scattering in gases. Peter Debye was educated at the Technische Hochschule, Aachen, where he gained a degree in electrical technology in 1906. The same year he held post as a theoretical physicist in Munich, where he met Arnold Sommerfeld. He received his Ph.D. in

Josephus Debye

physics from the university of this town in 1908. In 1911 he was offered the Chair of Theoretical Physics at the University of Zurich, which **Albert Einstein** had held just a short time before him. It is probably in Zurich that Debye began to reflect on the physical properties of molecules. He was to focus on these until 1950, and his work was to lead to linking their structure to their dipole moment, now measured in units called "debyes". After a short stay at the University of Utrecht he joined the University of Göttingen where he studied X - ray diffraction, a technique invented some time before by **Max von Laue** and **William Henry Bragg**. In 1920 Debye returned to Zurich and became interested in the problem of electrolyte dissociation. He suspected the presence of ions in certain solids (called ionic solids for this reason), although **Svante**

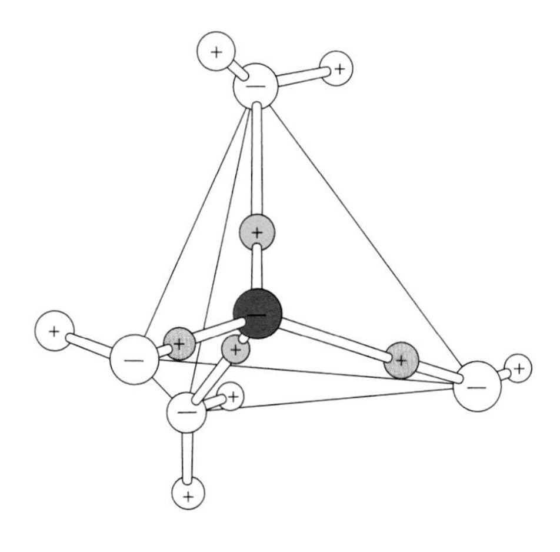

Tetrahedral representation of a water molecule and its dipole moment. The geometry of the molecule is due to the presence of two free electron pairs (not involved in a bond), constituting a total of four electronic sites which, by mutual repulsion, are arranged in a tetrahedron.

Arrhénius had only seen a dissociation of molecules when an electrolyte was dissolved. According to Debye, the reason why ions don't behave ideally in solution is that the movements of each ion are disturbed by those of the ions of the opposite sign that surround it. With the help of Hückel, he elaborated a mathematical description of ionic behaviour in solution. This untiring researcher developed in 1923 a mathematical expression explaining the Compton effect produced by X - rays. Debye conducted research at the University of Zurich from 1927 to 1934 before joining the University of Berlin where he worked until the end of his career. Ordered to opt for the German nationality, he left his adoptive country in 1940 to teach

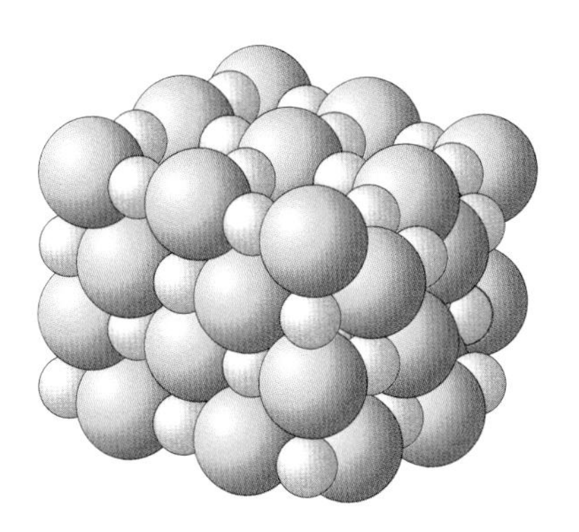

A sodium chloride crystal. The sodium ions and chloride anions are held to their position by ionic forces.

chemistry at Cornell University (Ithaca, New York). There he became interested in macromolecules and developed a method for determining their molecular weight from the way they scatter light when in solution. Peter Debye became a naturalised American citizen in 1946.

1937

Haworth, Walter (Chorley, England, March 19, 1883 - Birmingham, March 19, 1950) . British chemist. Shared the Nobel Prize for Chemistry with **Paul Karrer** for his work in determining the chemical structures of carbohydrates and vitamin C. Walter Haworth studied chemistry at the University of Manchester and worked under Perkin on terpenes, a class of lipids. After graduating in 1906, he became William Henry Perkin's assistant. In 1909, he went to Göttingen and met Otto Wallach, who was also engaged in research on terpenes. In 1912, he obtained his Ph.D. in chemistry and went back to Manchester and worked on carbohydrates. He taught at the University of Durham from 1920 to 1925, then at the University of Birmingham until 1948 when he retired. His first research

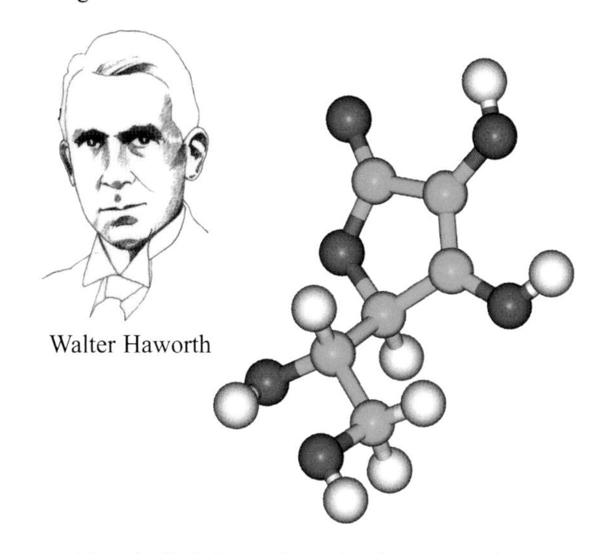

Walter Haworth

Vitamin C. A three - dimensional representation.

focused on monosaccharides, (i.e. simple sugars). He notably determined the cyclic structure of glucose, with five carbon atoms and one oxygen atom. Haworth showed that the carbon atoms in cyclic sugars are linked by oxy-

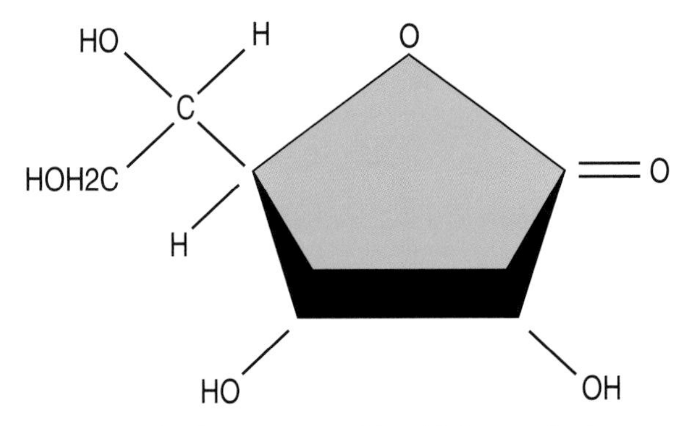

Vitamin C. Also known as ascorbic acid. It may play important roles in the brain and in central nervous system tissue.

gen. He focused his attention to more complex sugars like lactose and fructose, two disaccharides. One of his main works was the determination of the structure of vitamin C. This work was accomplished in 1933. He also succeeded in synthesizing it. Vitamin C had just been discovered by the Hungarian biochemist Albert Szent-Györgyi, who already suspected that its structure must resemble that of hexuronic acid, abundant in oranges and cabbages. Because of the action of vitamin C against scurvy (scorbutus), Haworth proposed to call it ascorbic acid. Walter Haworth was knighted in 1948.

Karrer, Paul (Moscow, Russia, April 21, 1889 - Zurich, Switzerland, June 18, 1971). Swiss chemist. Nobel Prize for Chemistry with **Norman Haworth** for investigating the constitution of carotenoids, flavins, and vitamins A and B2. Son of a dentist. In 1908 Paul Karrer joined the University of Zurich where he studied under the direction of **Alfred Werner** and where he gained his Ph.D. in 1911. His main interests then focused on cobalt complexes and arsenical compounds. In 1912, the German bacteriologist **Paul Ehrlich** invited him to hold a post at the "Georg Speyer Haus", a research institute based in Frankfort. After Ehrlich's death in 1915, Karrer focused on the chemistry of plant products. In 1918 he went to Zurich, where he succeeded Werner (deceased in 1919). He became Professor within the University's Department of Chemistry, then Rector of the university from 1950 to 1953. His first work was on anthocyanins, pigments that give flowers their blue and red hues. He studied the struc-

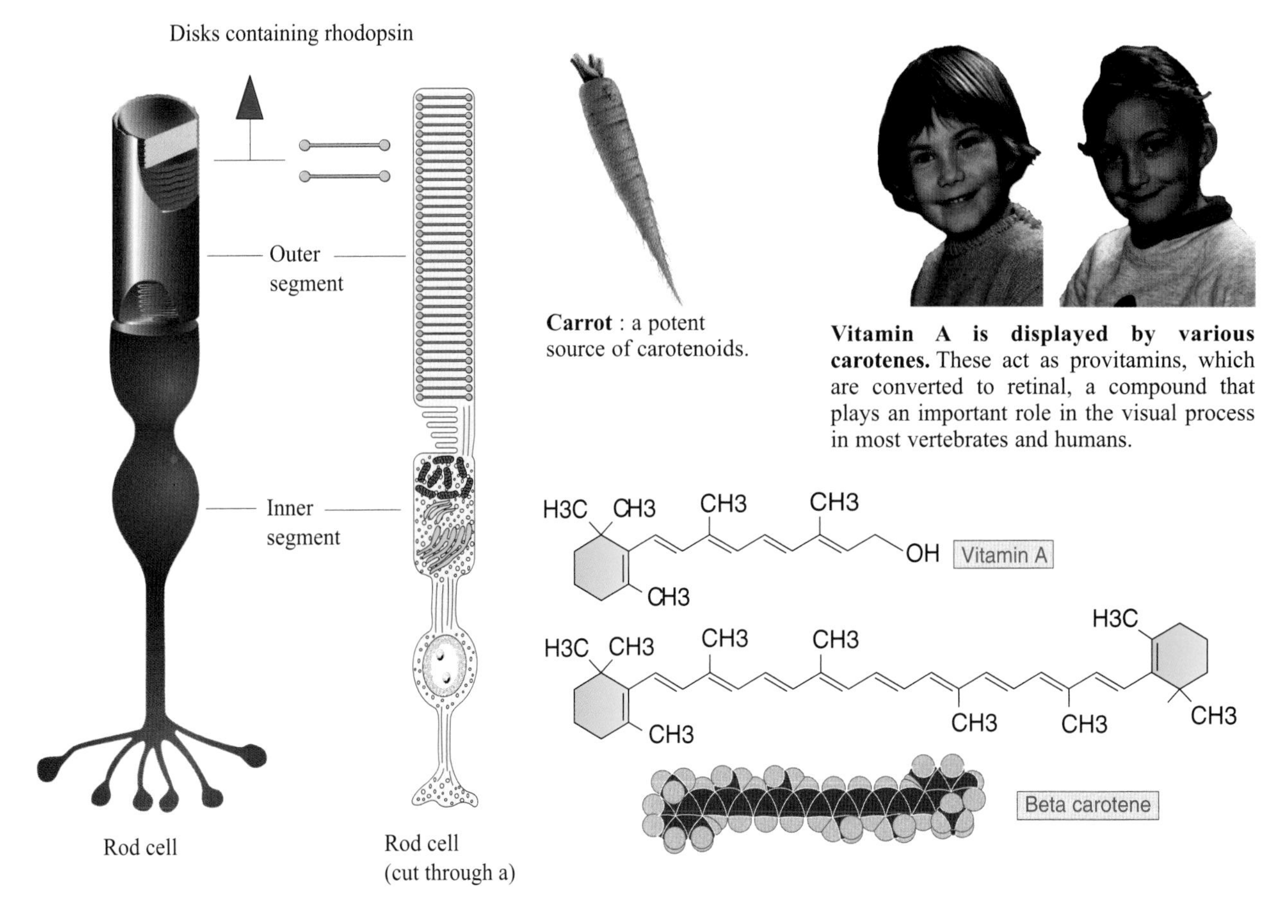

Carrot : a potent source of carotenoids.

Vitamin A is displayed by various carotenes. These act as provitamins, which are converted to retinal, a compound that plays an important role in the visual process in most vertebrates and humans.

Vitamin A or retinol (visual pigment in retina cells) is displayed by various carotenes acting as provitamins. These compounds are then converted to retinal in the small intestine in most vertebrates.

ture of these pigments, using enzymes that split molecules into simpler compounds. Then focusing on caro-tenoids (yellow-orange plant pigments), he determined the structure of beta-carotene, isolated previously by **Robert Kuhn**. This work led him to understand that vitamin A is synthesised in the organism by splitting of carotene, now called "provitamin A", to be found in carrots and several other plants. In 1931, Karrer succeeded in describing the structure of vitamine A. Continued research on vitamin A over the following decade led to its synthesis. He then proceeded to study and synthesise riboflavin (another B vitamin), followed by vitamins E and K. Another success was the elucidation of the structure and the function of the coenzyme nicotinamide adenine dinucleotide, NAD, involved in electron transfer in cells.

Paul Karrer

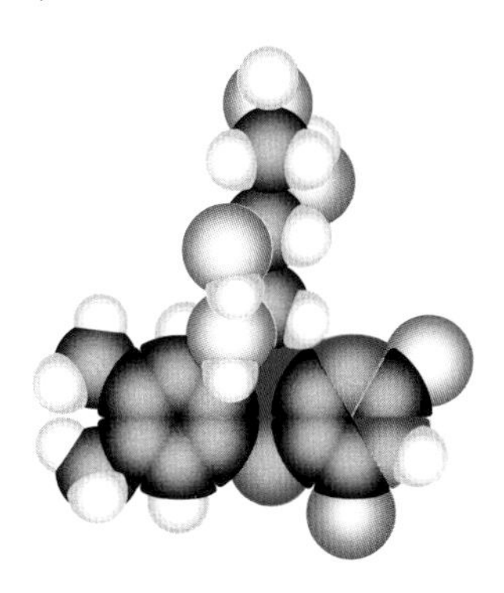

Riboflavine. It is the precursor of flavine adenine mononucleotide and flavine adenine dinucleotide, two important coenzymes that serve as electron carriers.

1938

Kuhn, Richard (Vienna, Austria-Hungary, December 3, 1900 - Vienna, Austria, July 31, 1967). German physician and biochemist. Son of an engineer. Nobel Prize for Chemistry for work on carotenoids and vitamins. Richard Kuhn, like **Paul Karrer**, was interested in studying plant pigments and molecules such as vitamins. He succeeded in synthesizing vitamin A, vitamin B_6, a pyridoxin, in a highly purified form. To obtain vitamin B_6, he used about 14,000 gallons (19,000 litres) of skim milk. He also succeeded in isolating vitamin B_2. Like many other German scientists of the between-wars period, this scientist was a victim of nazism. In 1938 he was refused permission to accept the Nobel Prize awarded to him, because the Nazi government forbade giving any credence to the prize. He had to wait until 1945 before he was allowed to receive the prize. Richard Kuhn graduated from the University of Munich in 1922. He became professor at the University of Heidelberg and was later made director of the Kaiser Wilhelm Institute for Medical Research in the same city.

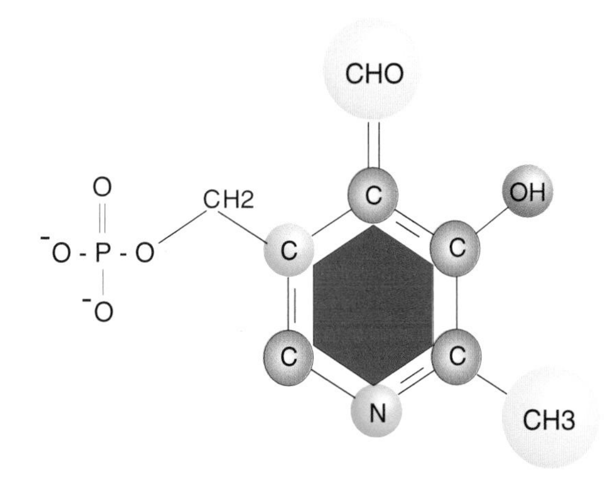

Vitamin B_6. Its biologically active form is pyridoxal -5 -phosphate, a coenzyme that participates in the catalysis of many reactions involving amino acids.

1939

Butenandt, Adolph Frederick Johann (Bremerhaven-Lehe, Germany, March 24, 1903 - Munich, Germany, January 18, 1995). German chemist. Son of a trademan. Nobel Prize for Chemistry, with **Leopold Ruzicka**, for pioneering work on sex hormones. Adolf Butenandt obtained his Ph.D. in 1927 under **Adolf Windaus** at the University of Göttingen. In 1929, he isolated the first sexual hormone, oestrone, also called follicular hormone, from the urine of pregnant women. Made within the pharmaceutical firm Schering, this discovery was also made independently at around the same time by the American **Edward Doisy**. In 1931, Butenandt married his assistant Erika von Ziegner, who had developed the bioassays for this discovery. The same year he isolated 15 milligrams of the hormone androsterone from 3960 gallons of human urine. Butenandt was also the first to crystallize an insect hormone, ecdysone, and he showed that it was derived from cholesterol. Butenandt then directed

Adolf Butenandt

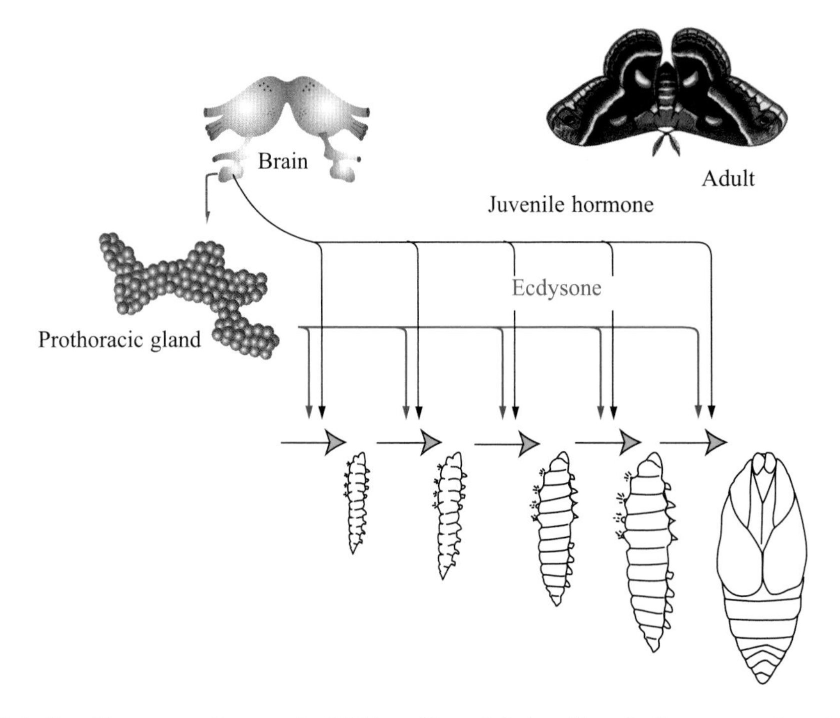

Metamorphosis in the silkworm moth cecropia. Molting of larva is induced by a brain hormone that stimulates the prothoracic gland. This gland then secretes ecdysone, which stimulates development. Ecdysone is balanced by juvenile hormone secreted by the *corpus allatum*. In the pupal state the level of juvenile hormone drops. A shift in the balance between ecdysone and juvenile hormone allows pupa to develop into adults. The structure of ecdysone was elucidated by Adolf Butenandt.

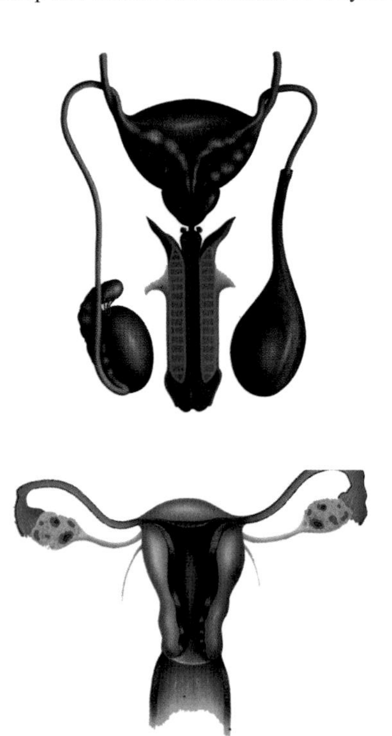

Female and male reproductive systems. They produce hormones, respectively oestrogens and androgens. The androgens and oestrogens are responsible for the characteristic changes of male and female adolescence.

the famous Max Planck Biochemistry Institute in Tübingen, transferred later to Munich, and in 1960 was appointed to succeed **Otto Hahn** at the Head of the Max Planck Society. In 1939, he was awarded the Nobel Prize but he was forbidden to accept it by the Nazi government.

Ruzicka, Léopold (Vukovar, Croatia, September 13, 1887 - Zurich, Switzerland, September 26, 1976). Nobel Prize for Chemistry, along with **Adolf Butenandt** for his

Leopold Ruzicka

Musk production. This name regroups a variety of substances secreted by a specialized organ of some members of the deer family (Cervidae). With its pleasant and penetrating odour,it is used in perfumery. Plant musk also exists, extracted from the mauve mallow.

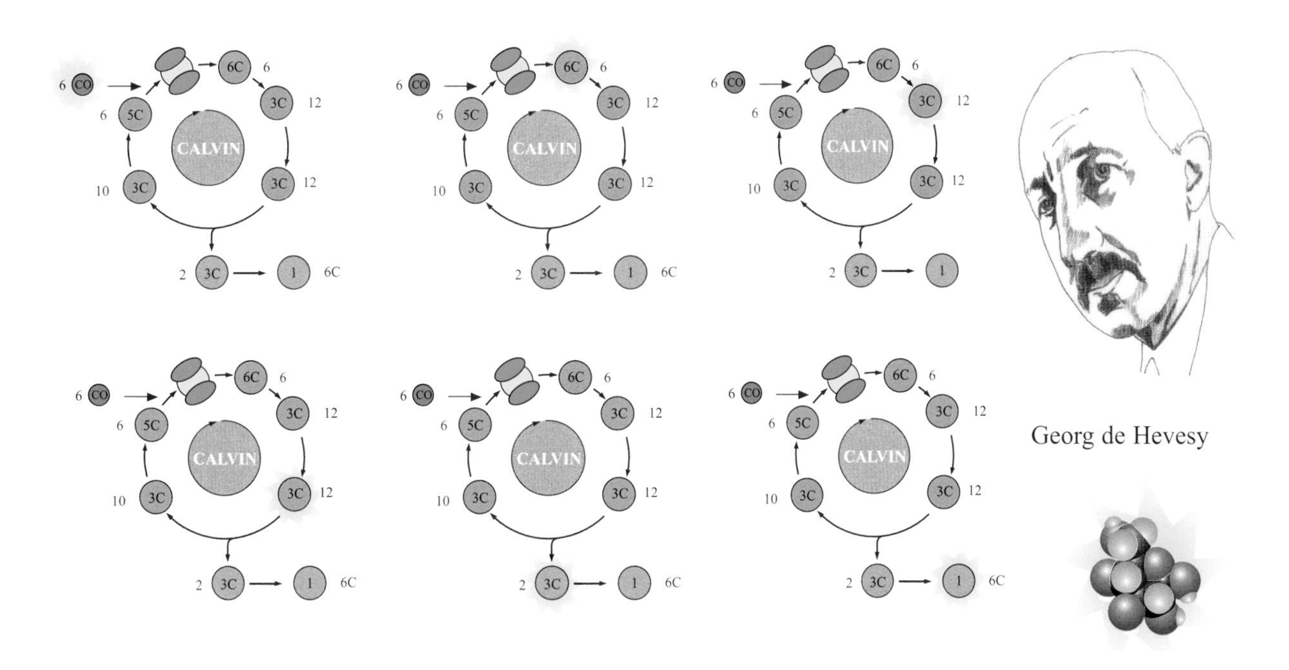

Radioactive labeling. Injection of a radioactive tracer (for example carbon - 14), into the Calvin cycle and its conversion to carbohydrates such as glyceraldehyde can be observed by radiation detection methods.

work on terpenes and sex hormones. Son of a cooper. Léopold Ruzicka graduated in chemistry from the Karlsruhe Institute of Technology, in Germany and, in 1929, he was appointed professor of organic chemistry at the Federal Institute of Technology at Zurich, Switzerland. Ruzicka is known for his research on natural fragrance molecules (such as those of musk and civet) and on terpenes. He was notably the first to synthesise pentacyclic triterpenes, but it's for his work on sex hormones that he received the Nobel Prize, jointly with Butenandt. Leopold Ruzicka succeded in synthesizing androsterone, an hormone of which 15 milligrams had been obtained in a pure form by von Butenandt from urine.

1940 to 1942 Not awarded

1943

de Hevesy, Georg (Budapest, Austria - Hungary, July 5, 1885 - Freiburg im Breisgau, Germany, August 1, 1966). Hungarian chemist. Son of an industrialist. Nobel Prize for Chemistry for development of isotopic tracer techniques that greatly advanced understanding of the chemical nature of life processes. This scientist had a complex career. He worked successively at the Univer-sities of Budapest and Freiburg, at the Polytechnic School of Zurich, in Karlsruhe under **Fritz Haber**, and in Manchester where he joined a prestigious team including figureheads such as **Niels Bohr**, Henry Moseley, and Hans Geiger. Manchester is also where Frederick Soddy's discoveries had contributed to a better understanding of the consequences of radioactive phenomena on the transmutation of elements. In 1913, Hevesy returned to Hungary. Appointed Lecturer at the University of Budapest, he became full Professor after the war. He then focused once again on exploiting radioactive phenomena, but this time he imagined "tracing" the behaviour of molecules in biological systems so as to monitor the transformations of various active compounds. He thus studied the absorption of radioactive lead by plant roots. In 1923, Hevesy discovered hafnium. In 1926, Hevesy taught chemistry at the University of Freibourg but, in 1934, because of the rise of nazism, he left Germany for Denmark. Still using his tracer method, he used heavy water to study water transfer in biological systems. The discovery of artificial radioactivity by **Frédéric Joliot** in 1934 opened prospects for using radioisotopes of elements having greater biological significance. For example, he now had 32S and 32P; later, he would use ^{45}Ca, ^{42}K, ^{24}Na, ^{38}Cl, and ^{14}C. Our current chemical knowledge of the metabolism of living organisms is based on such research. In 1944 Hevesy was appointed Professor at the Institute of Organic Chemistry in Stockholm and he took the Swedish nationality in 1945. His subsequent research concerned the separation of isotopes, i.e. atoms with identical chemical properties but slightly different masses. He obtained 41K, which constitutes 6.7% of natural potassium. Another of his scientific initiatives was to use nuclear activation by neutrons to analyse and quantify certain elements with extreme sensitivity. In 1936, shortly after James Chadwick's discovery of the neutron, de Hevesy published results of research that involved irradiating a sample of yttrium to measure its dysprosium content. This was the first example of neutron activation analysis.

1944

Hahn, Otto (Frankfurt Am Main, West Germany, March 8, 1879 - Göttingen, July 28, 1968). Son of a businessman. German chemist. Nobel Prize for Chemistry for the discovery of nuclear fission.

Otto Hahn was educated at the University of Marburg where he studied chemistry. He obtained his Doctor of Science degree after presenting a thesis in organic chemistry. In 1901, he joined the laboratory of William Ramsay, at the University College of London to study radioactive elements.

Otto Hahn

Then, joining **Ernest Rutherford** in Montreal in 1905, he discovered radioactinium, a thorium isotope. In 1912, he held a post at the Kaiser Wilhelm Institut of Berlin, which he directed in 1928. Also in 1905, Hahn discovered radiothorium (thorium - 228). In 1907, after beginning research at the Berlin Institute of Chemistry, he announced the discovery of mesothorium (radium - 228). Then he began a collaboration that was to prove very fruitful with Lise Meitner, an Austrian physicist who had been Max Planck assistant Berlin. Their work, which focused on radioisotopes emitting beta radiation, led in 1917 to the discovery of protactinium. Yet Hahn's fame is due to his discovery of nuclear fission in 1938. Right after **James Chadwick** identified the neutron, physicists and chemists launched enthusiastically into creating isotopes by bombarding various elements. The neutron, a neutral particle, has an interesting property: it is readily absorbed by many atomic nuclei. Hahn and Strassmann wanted to create a transuranian element - element 93 - by bombarding uranium, the heaviest element naturally present in the earth's crust, with neutrons. Yet instead of synthesising the expected transuranian element, they detected one that was lighter than uranium: radioactive barium. Meitner, meanwhile, had to escape Nazi Germany and flee to Sweden. There she proposed the hypothesis that instead of adding weight to uranium, Hahn and Strassmann had caused it to break and form two lighter fragments, by a process we now call "fission". Hahn and his coworker managed to find the second fragment, krypton. The sum of the atomic numbers of krypton and barium - 36 and 56 - equals 92, the atomic number of uranium. It was soon observed that other fragments could form through fission. Fission of the nucleus of uranium 235, called fissile uranium, is actually a random phenomenon yielding a heavy fragment, with a mass ranging from 130 to 145, and a lighter fragment, with a mass between 90 and 105. Hahn and Strassmann thus opened the way to an extraordinary research effort that was to lead, in 1942, to the development of the first nuclear reactor by the Italian physicist **Enrico Fermi** (1938 Nobel Prize for Physics). Sadly, this discovery also made possible the production of the atomic bombs that were to destroy Hiroshima and Nagasaki. After the war, Hahn was appointed President of the Kaiser Wilhelm Society, and devoted himself to re-launching scientific research in Germany. All the while he was very active in the international nuclear arms control movement.

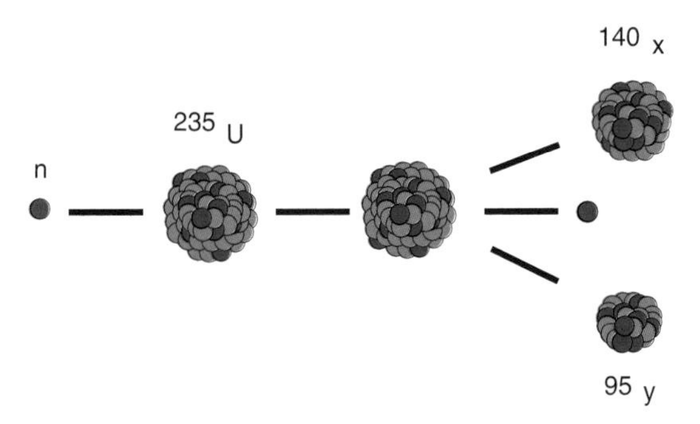

Nuclear fission. This phenomenon is produced by neutron bombardment of uranium235. A slow neutron captured by U-235 results in the highly unstable compound nucleus U-236, which ruptures into a heavy fragment and a light fragment. Such an event liberates a great deal of energy.

1945

Virtanen, Arturi Ilmari (Helsinki, Finland, January 15, 1895 - Helsinki, November 11, 1973). Finnish biochemist. Nobel Prize for Chemistry for investigations directed toward improving the production and storage of protein-rich green fodder. Arturi Virtanen studied biology, chemistry, and physics at the University of Helsinki, where he obtained his bachelor's degree in science in 1916. After a short stay as a chemist at the Central Industrial Laboratory of Helsinki, he returned to the university and received his doctorate in 1919. During a stay in Zurich in 1920, he undertook postdoctoral studies in physical chemistry. He then went to Stockholm where he focused on bacteriology and worked in enzymology with **Hans Euler von Chelpin**. His renown in the dairy product field led to his being appointed Professor of biochemistry in 1939. There he studied acid fermentations and the enzymes involved. Virtanen then became interested in leguminous plants, which can assimilate nitrogen from the air and incorporate it into biological molecules. This property makes these plants particularly interesting as organisms contributing to soil fertilisation. Another research focus was forage used to feed livestock and its bacterial degradation during storage. This phenomenon reduces the food value of the forage and decreases milk production. Virtanen discovered that it was sufficient to add a dilute solution of mineral acids to inhibit bacterial action and maintain the essential nutritional qualities of the forage. This procedure was first applied in Finland in 1929. In 1948 Arturi Virtanen was rewarded for his research with the title of President of the Finnish Academy of Science and the Arts.

1946

Northrop, John Howard (Yonkers, New York, July 5, 1891 - Wickenburge, Arizona, May 27, 1987). American biochemist. Son of a zoologist. Nobel Prize for Chemistry, along with **James Sumner** and **Wendell Stanley** for successfully purifying and crystallizing certain enzymes, specially ribonuclease. In 1913 Northrop received the Master of Arts degree from the University of Columbia, having specialised in biochemistry. He gained his Ph.D. in sciences. The same year he was an assistant at the Rockefeller Institute for Medical Research, New York, a post offered by Loeb. His career was interrupted during World War I. During this period he conducted research on fermentation processes in order to produce acetone and ethylic alcohol. He was to be appointed to a similar position in 1941-1945, with the task of studying combat gases.

John H. Northrop

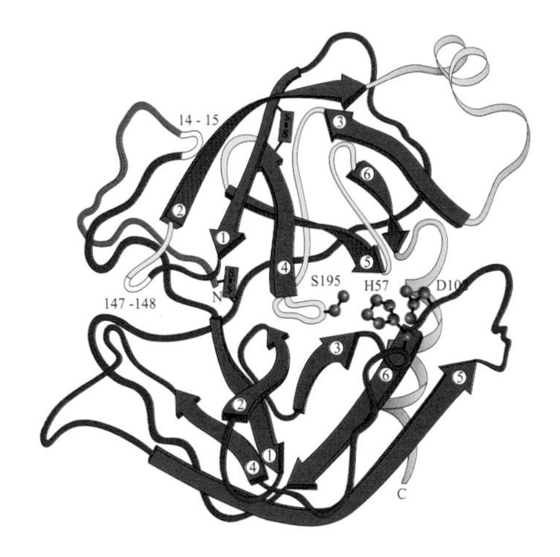

Pepsin. A ribbon model of the proteolytic enzyme. The path of the ribbons describes the fold of the polypeptide chain which makes up the protein.

In 1919 he began to study the digestive enzymes pepsin and trypsin, which he obtained in the crystalline state and which he proved to have a protein nature, thus confirming Sumner's views. Following Loeb's death he joined the Department of Animal Pathology of the Rockefeller Institute at Princeton and studied bacteriophages, the viruses that infect bacteria. There he succeeded in isolating and purifying a viral nucleoprotein from a culture of infected bacteria, *Staphylococcus aureus*, and demonstrated that this substance formed the essential part of viruses. In 1941 he succeeded in isolating and crystallizing the antibody elaborated by the immune system against diphteria. In 1949 he held a post as a teacher at the University of California, Berkeley, and carried on his research on bacteriophages. This researcher was also very interested by the origin of the viruses.

Stanley, Wendell Meredith (Ridgeville, Indiana, August 16, 1904 - Salamanca, Spain, June 15, 1971). Spanish American biochemist. Son of a publisher. Nobel Prize for Chemistry, along with **John H. Northrop** and **James B. Sumner** for his work in the purification and crystallization of viruses. Wendell Stanley obtained his Ph.D. in chemistry from the University of Illinois in 1929. There, as an assistant, he undertook research on biphenyl and on a drug potentially useful in treating leprosy. In 1930 he did a post-doctorate at the University of Munich under **Heinrich Wieland**. Back in the United States 1931, he was appointed research assistant at the Rockefeller Institute in New York, where he studied potassium ion transfer from seawater into plant cells. In 1932, he joined the Department of Vegetal Pathology at the Rockefeller Institute, Princeton. He became interested in viruses, about which little was known at the time. Focusing particularly on the tobacco mosaic virus, he concentrated, isolated, and purified it until he finally obtained it in the crystalline state, noting that crystallisation did not in the slightest affect its virulence. The discovery that a living organism could be crystallised while remaining infectious was quite revolutionary. After this success, two British scientists showed that this same virus consists of 94% protein and 6% nucleic acids, and a German team revealed by electronic microscopy that the virus has a rod-like structure. During World War Two Stanley took part in research aimed at fighting the viral diseases threatening soldiers on the front. Upon isolating the influenza virus and inactivating it with formaldehyde, he observed that a virus inacti-

Wendell M. Stanley

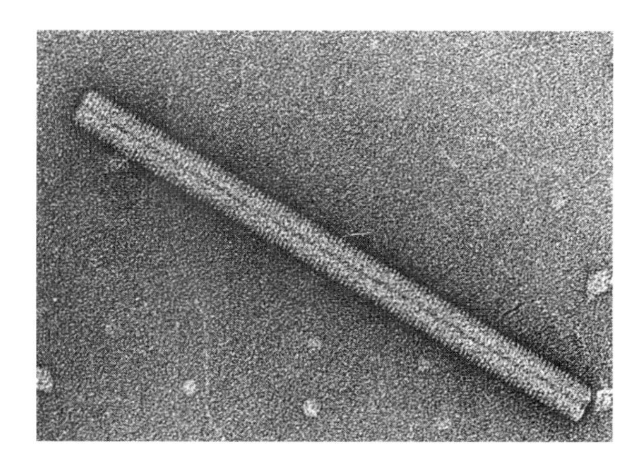

Electron micrograph of a tobacco mosaic virus. The illustration shows the helical array of protein subunits around a single-stranded RNA molecule.

vated in this way could still trigger antibody production in man. He thus developed during this period a flu vaccine that was subsequently used by the U.S. Army. From 1946 untill his death he directed the Virus Research Laboratory at the University of California. There, he demonstrated the protein nature of the virus envelope and showed that the virus biological activity was essentially due to its ribonucleic acid. Finally, in 1960, Wendell Stanley determined the complete amino acid sequence of the tobacco mosaic virus's coat protein.

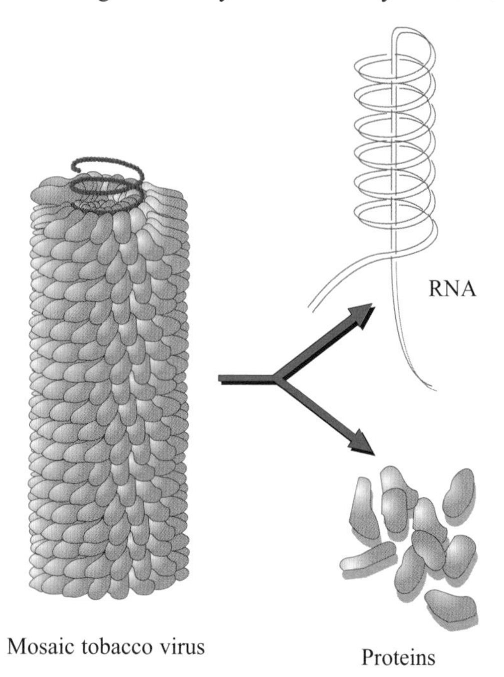

Tobacco mosaic virus. One of the best studied viruses and the first to be prepared in crystallized form. Within the protein sheath is a twisted RNA composed of 6400 nucleotides. The protein sheath or capsid consists of 2130 identical protein subunits.

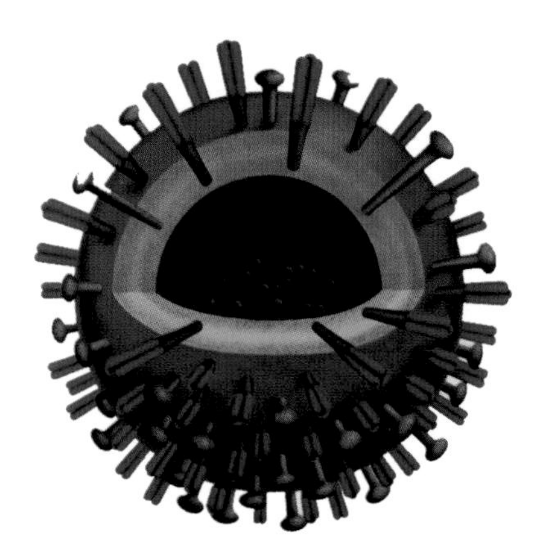

Influenza virus. The transmembrane glycoproteins form the spikes seen in electron micrographs. One of these glycoproteins enables the virus to enter and infect susceptible cells.

Sumner, James Batcheller (Canton, Massachusetts, November 19, 1887 - Buffalo, New York, August 12, 1955). Son of a cotton manufacturer. Nobel Prize for Chemistry, along with **John Northrop** and **Wendell Stanley** for successfully purifying and crystallizing certain enzymes, specially ribonuclease. When he was only seventeen, he lost an arm in a shooting accident. He compensated for this handicap by working hard. Sumner was educated at Harvard, where he obtained his Ph.D. in 1914. In the same year he took up an appointment at the Cornell Medical School where, in 1929, he became professor of biochemistry. As he was studying urease, the enzyme catalysing the decomposition of urea to ammonia and car-

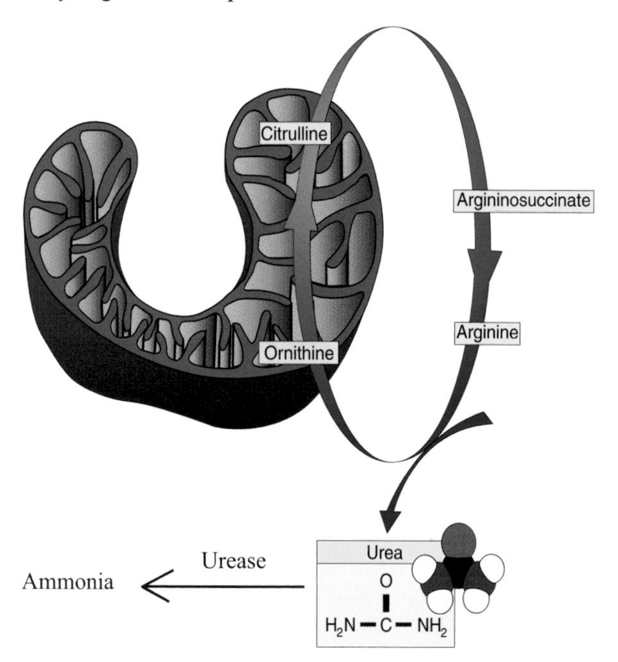

Urea cycle. Also called ornithine - citrulline cycle. By this process carbon dioxide and ammonia are combined to form urea. This cycle is the pathway by which ammonia, the waste product of protein metabolism, is made less toxic to cells and prepared for excretion. This cycle was discovered by **Hans Adolf Krebs**. Some steps occur in the mitochondria.

bon dioxide, Sumner discovered that it was actually a protein. Having accidentally left a urease preparation on a table overnight on a cold night, he noted the appearance of crystals. This meant that a protein from a living organism could crystallise. This was 1926, and his discovery was a major step forward. Crystallisation of a molecule opens the way to determining its structure by X - ray diffraction, and hence to elucidating its mechanism of action. Since Sumner's preparations were not entirely pure, the protein nature of enzymes was not generally accepted. Conclusive proof of the protein nature of enzymes was documented by the American chemist John Northrop in 1935.

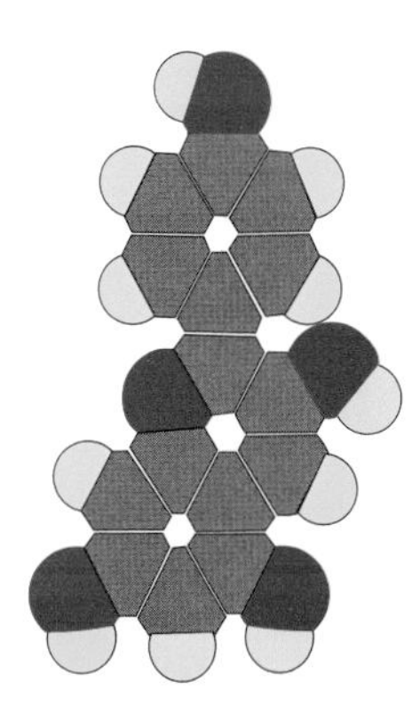

Anthocyanin. This type of molecule is responsible for most of the scarlet, mauve, purple and blue colouring in higher plants, especially of flowers.

1947

Robinson, Robert (Bufford, England, September 13, 1886 - Oxford, February 8, 1975). British chemist. Son of an inventor. Nobel Prize for Chemistry for his research on a wide range of organic compounds, notably alkaloids. Robert Robinson studied chemistry under the direction of Perkin at the University of Manchester and obtained his

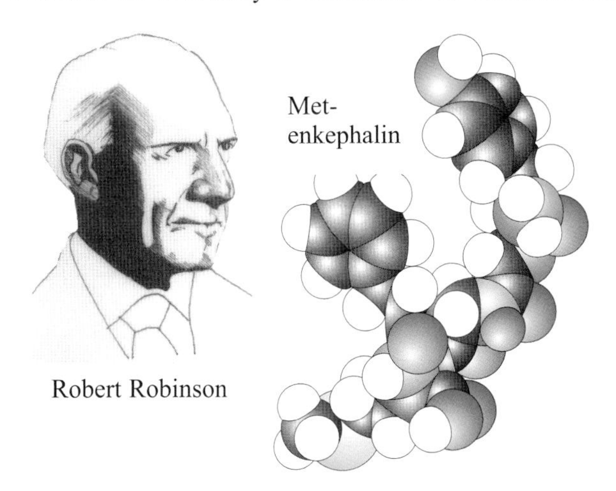

Morphine. The most important opium alkaloid. It exerts its pharmacological effects by mimicking molecules such as methionine enkephalin that are normally present in the bodies of vertebrates.

Ph.D. in 1910. From 1912 until 1915, he taught chemistry at the University of Sidney, Australia. Back in England, he held the Chairs of Organic Chemistry in Liverpool in 1915, Manchester in 1927, London in 1927, and Oxford in 1930. Oxford is where he completed his career in 1955. Robinson was mainly interested by natural dyes such as brazilin and hematoxylin and their study allowed the synthesis of floral pigments, especially anthocyanins and flavones. But his most important research concerned alkaloids. These highly complex natural substances containing nitrogen and a combination of carbon cycles have very important biochemical effects. Many alkaloids are violent poisons. In 1925, Robinson's studies of the chemical reactions that form alkaloids in plants led him to discover the structure of morphine, he also succeeded in synthesizing it, and in 1946, strychnine. He is also credited for the synthesis of papaverine, tropinone and other narcotics. His career was also marked by the study of penicillin, which he succeeded in synthesizing.

1948

Tiselius, Arne Wilhelm Kaurin (Stokholm, Sweden, August 10, 1902 - Uppsala, Sweden, October 29, 1971). Swedish chemist. Son of an insurance broker. Nobel Prize for Chemistry in 1948 for his work on electrophoresis. Like **Svante Arrhénius**, Tiselius was from the University of Uppsala, Sweden, where he studied under **Theodor Svedberg**. His name is associated with electrophoresis, a method for separating charged chemical substances. Tiselius's doctoral thesis was on the electrophoresis of proteins. He visited Princeton from 1934 to 1935, working under Taylor. The first Chair of Biochemistry in Sweden was inaugurated for him in 1938. Electrophoresis is a technique by which ions or charged particles, including macromolecular substances, move differentially through a fluid under the influence of an electric field. The support fluid often includes porous paper. This technique was applied by Tiselius to analyse the composition of blood serum. Owing to this method, Tiselius was able to confirm the existence of different groups of proteins including albumins, globulins and antibodies. From 1946 to 1950, Arne Tiselius was Chairman of the Swedish Natural Science Research

Arne Tiselius

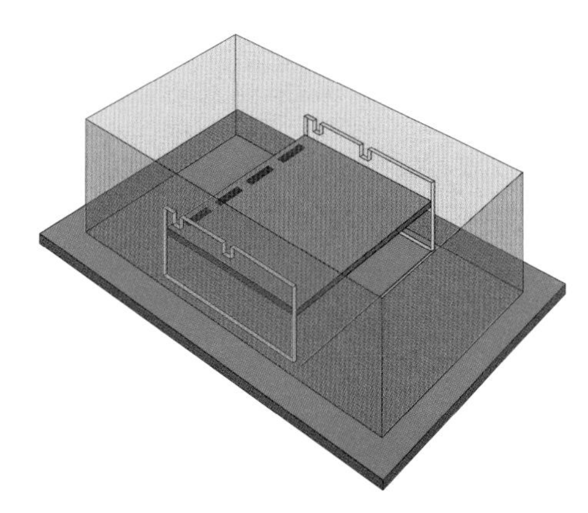

Electrophoresis. This technique has been widely used in separating proteins, nucleic acids and in the preparative fractionation of heterogenous populations of dispersed cells or other types of macroscopic particles.

Council. He was then successively made Vice President (1947 - 1960) and President (1960 - 1964) of the Nobel Foundation. Tiselius also provided a clear classification of chromatographic methods.

1949

Giauque, William Francis (Niagara Falls, Canada, May 12, 1895 - Berkeley, California, March 28, 1982). Canadian-born American physical chemist. Nobel Prize for Chemistry for his studies of the properties of matter at temperatures close to absolute zero. Son of a station-master. Having received his Ph.D. in chemistry from the University of Berkeley in 1922, William Giauque was named professor in 1934 and worked at Berkeley until his retirement in 1962. There he became greatly interested in the properties of materials as well as in the thermodynamics of gases at very low temperatures. He focused particularly on the third law of thermodynamics formulated in 1906 by **Walter Nernst**:the entropy of a substance approaches zero when its temperature approaches absolute zero (0° Kelvin). In 1924, independently of **Peter Debye**, he developed a process of achieving very low temperatures using a technique called adiabatic demagnetization. He thus managed to bring the temperature down to 0.004°K, breaking the record of 1 °K reached by the Dutchman **Kammerlingh-Onnes** between 1900 and 1910. It is worth noting that the 1997 Nobel Prize crowned research that had led to reaching about 3 billionths of a degree Kelvin! In 1929, Giauque and Johnston observed that the spectrum of oxygen also displayed, in addition to a series of intense rays, two series of less visible rays. He attributed these to two heavy isotopes of oxygen, ^{17}O and ^{18}O, then unknown. Three years later he discovered deuterium (^{2}H).Deuterium is a component of "heavy water", which was to acquire strategic importance during World War Two, in research conducted principally in Germany on the use of nuclear energy.

1950

Alder, Kurt (Königshütte, Prussia, July 10, 1902 - Cologne, Germany, June 20, 1958). German chemist. Shared the Nobel Prize for Chemistry along with **Otto Diels** for their development of the Diels-Alder reaction. Alder received his degrees in chemistry from the University of Berlin. He then moved to the Christian Albrecht University at Kiel, where he joined Diels and where both of them conducted research in organic chemistry. From his collaboration with this scientist emerged a manufacturing process for dienes, called "Diels-Alder synthesis". This reaction opened the way to synthesising various substances of prime importance in biochemistry. Among these substances were reserpine, morphine and steroids. This collaborative work also led to synthesising commercial products such as the insecticides dieldrin and aldrin, thus named in honour of these two researchers. Alder also discovered the polymerisation of butene and

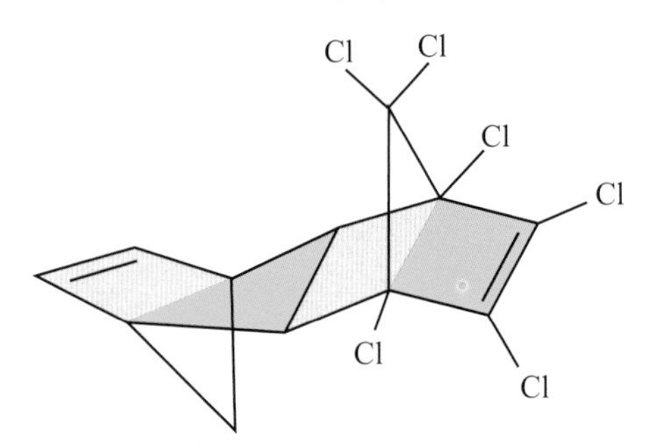

Aldrin. An excellent insecticide obtained by the Diels-Alder synthesis. This organochlorine compound is now banned on account of its toxicity.

butadiene to BUNA. The German authorities used this process abundantly during World War II to make up for the lack of natural rubber. The collaboration between Alder and Diels stopped in 1936, when Alder was appointed Head of the Research Laboratory of Bayer Werke in Leverkusen. However, he left this society in 1940 and headed the Experimental Chemistry Department at the University of Cologne until his death.

Diels, Paul Otto (Hamburg, Germany, January 23, 1876 - Kiel, March 4, 1954). German organic chemist. Nobel Prize for Chemistry with **Kurt Alder** for their joint work in developing a method of preparing cyclic organic compounds. Son of a professor of litterature. Paul Otto Diels first attracted the attention of the scientific world by inventing a dehydrogenation process involving selenium which, applied to cholesterol, enabled him to deduce its basic structure. This structure is based on that of a tetracyclic compound, 3-methyl-1,2 cyclopentenophenanthrene, or "Diels's hydrocarbon". Many natural substances turned out to share this basic architecture, particularly steroids, bile acids, and the active components of digitalis, used as heart stimulants. Yet the work for which Diels and Alder were awarded the Nobel Prize was the

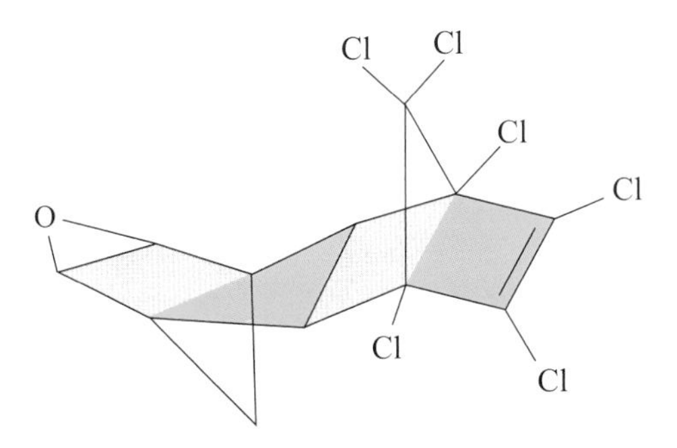

Dieldrin. Like aldrin, this compound can also damage the ner - vous system.

elaboration of the reaction which bears their name, the so-called Diels-Alder reaction. The Diels-Alder reaction is essentially a cycloaddition reaction in which an alkene adds to a 1,3-diene to form a 6-membered ring. Because it can be carried out at an ordinary temperature under gentle conditions and without catalysers or condensing agents, the Diels-Alder reaction has proved to be one of the most useful reactions in chemical synthesis. It made possible the creation of bridged structures such as norcamphor and camphor, the latter being a respiratory stimulant. It was there that he elaborated several synthesis schematics thanks to his "retrosynthetic method," which adapts perfectly to computers. Among other possibilities it also allowed the creation of bridged structures such as camphor and norcamphor. Thanks to the Diels-Alder reaction it became possible to manufacture artificial rubber, the most important of all elastomers. During World War II, hundreds of thousands of tons of synthetic rubber were produced in factories controlled by Germans. The loss of two of his children killed on the Russian front in World War Two, the bombardment of his home, the death of his wife in 1945, and a painful arthritis tragically marked the end of this scientist's career.

1951

McMillan, Edwin Mattison (Redondo Beach, California, September 18, 1907- El Cerrito, California, September 7, 1991). American nuclear physicist. Son of a physician. Nobel Prize for Chemistry with **Glenn T. Seaborg** for his discovery of element 93, neptunium. McMillan received his Ph.D. from the University of Princeton. In 1932 he rejoined the group of **Ernest Lawrence** (1939 Nobel Prize for Physics) - Lawrence was the inventor of the cyclotron - at the Berkeley Radiation Laboratory, University of California, and stayed there for 40 years. In 1940, in collaboration with Philip Abelson, he bombarded uranium with particles (accelerated by the cyclotron of Berkley) and caused its transmutation into neptunium. The two scientists had thus created the first transuranian element, an artificial element coming after uranium in the periodic table. This transmutation had already been attempted by the German physicists **Otto Hahn** and Fritz Strassmann, and had led in 1938 to the discovery of nuclear fission. The very first artificial radioactive element, technetium, had been detected by Perrier and **Emilio Segré** (1959 Nobel Prize for Physics) in 1937, among the fission products of uranium. Many studies subsequently focused on synthesising heavier and heavier transuranian elements. McMillan and **Seaborg** isolated plutonium (atomic number 94) in 1941.All of the transuranians up to Z = 104 were thus produced in a cyclotron by bombarding lighter

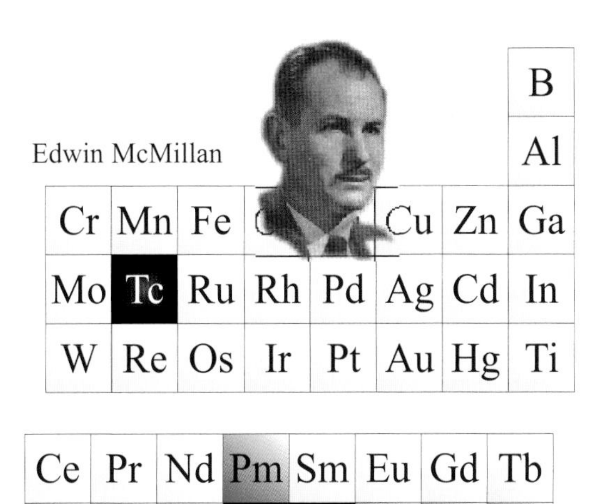

Part of the Table of Mendeleev. emphasizing neptunium and plutonium, two transuranian elements synthesized by Edwin McMillan by bombardment of uranium.

transuranians. It is worth noting that quite a few heavy elements were isolated by treating spent nuclear fuel. They derived from uranium - 238 through a succession of neutron capture steps involving increasingly heavy nuclei. This is the case of plutonium, today extracted in such high quantities from spent nuclear fuel that it is itself used, mixed with uranium - 235 (the mixture is called "MOX"), as a nuclear reactor fuel. Fissile isotopes of plutonium are also recovered in the name of peace in the process of dismantling nuclear weapons. McMillan is further credited for the discovery of spallation, a phenomenon in which a nucleus splits into several fragments. This happens when the particles that cause the nucleus to split have a very high energy. Edwin McMillan retired in 1973. In 1990 he received the prestigious National Medal of Science.

Seaborg, Glenn Theodore (Ishpming, Michigan, April 12, 1912 - Lafayette, California, February 25, 1999). Son of a stage-setter. Nobel Prize for Chemistry, along with **Edwin McMillan**, for his work on isolating and identifying transuranians.

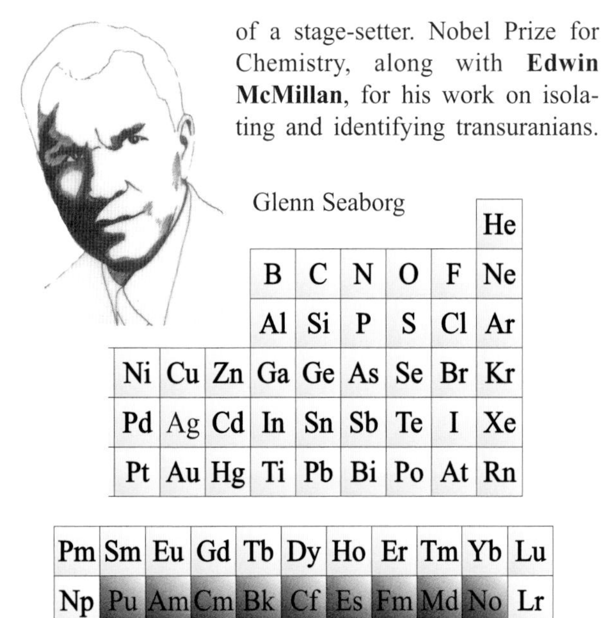

Elements discovered by Glenn Seaborg. Otto Hahn's discovery of nuclear fission and the completion of the first fission reactor by **Enrico Fermi** provided the researchers with an intense source of slow neutrons which, absorbed by uranium nuclei, created new elements with masses greater than that of uranium. Thus Seaborg was one of the first to isolate and characterize a good number of transuranians not found in nature.

Glenn Seaborg entered the University of California at Los Angeles in 1929. At Berkeley he began work on his doctorate with the great American chemist Gilbert N. Lewis and with the collaboration of **Ernest Lawrence** (1939 Nobel Prize for Physics). Named a doctor of science in 1937,

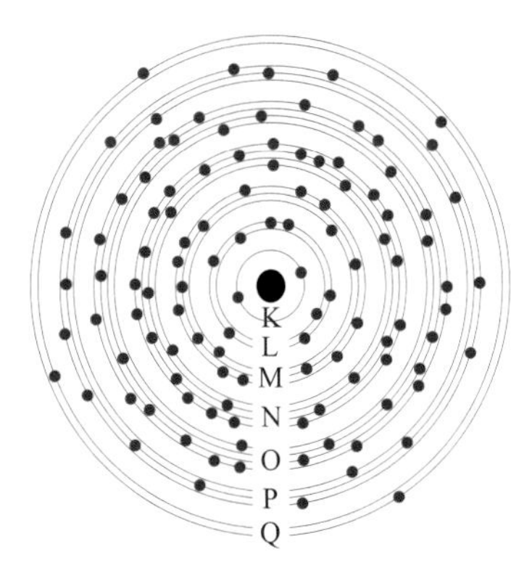

Plutonium atom. Artificial chemical element produced by the nuclear chemist Glenn T. Seaborg. The first industrial production of plutonium was undertaken in uranium reactors, and Seaborg had the primary responsibility for isolating plutonium from the reaction products. Isotope 239 is fissile and may be used in nuclear reactors.

Seaborg specialized in radiochemistry. In 1941, in collaboration with Edwin McMillan, he isolated plutonium. Later, in Chicago, he developed the sequence of chemical procedures required to isolate plutonium from the fission products present in spent nuclear fuel. These separation techniques made it possible to prepare the plutonium that was to be used in the bomb dropped on Nagasaki in 1945. This bomb, the second to have been used against civilian populations, contained 64 kilograms of plutonium, a fissile substance produced in the plutonium-generating reactors built at Hanford under the Manhattan Project. In 1945 he isolated americium and curium, again by reprocessing nuclear fuel extracted from uranium reactors. He then returned to Berkeley in 1946. From 1948 to 1959 came the discoveries of berkelium, element 97, and californium, element 98, followed later by einsteinium, fermium, mendelevium, and nobelium. Seaborg was probably, with Ramsay, the chemist who added the largest number of elements to the periodic table, despite the fact that all of the elements he discovered are radioactive and subsist only for a short time after their synthesis in nuclear reactors or particle accelerators. Seaborg participated actively in the crusade against nuclear weapony. In 1961, he was a member of the American delegation to sign the treaty limiting nuclear testing.

1952

Martin, Archer John Porter (London, England, March 1, 1910 -). British biochemist. Son of a physician. Nobel Prize for Chemistry, with **Richard Laurence Millington Synge**, for the development of paper partition chromatography. Archer Martin was educated at the University of Cambridge where he studied chemistry, physics, mathematics, and mineralogy. Following a meeting with Professor John S.H.Haldane, he became interested in biochemistry. Earning a Bachelor of Sciences in 1932, he began a research career in physical chemistry. In the early 1930s, he collaborated with Synge to develop chromatographic separation techniques. This technique had already been used by the German chemist **Richard Wilstätter** to separate plant pigments, but the latter had failed in his attempts to separate more complex molecules. Martin and Synge managed to separate carotene constituents on a column filled with starch, cellulose, or silicagel. From 1938 to 1946, Martin worked as a Ph.D. in the wool industry, where he developed a technique for separating and analysing amino acids extracted from proteins. From 1946 to 1948, he worked at Boot Pure Drugs Research Company where again, thanks to paper chromatography, he managed to separate and identify small peptides in complex mixtures. Continuing his work with Synge at Trinity College in Cambridge, he perfected methods of chromatographic analysis. In 1952 he headed the Physical Chemistry Department of the National Institue for Medical Research at Mars Hill in London. There he studied sugars and fatty acids.He also adopted gas

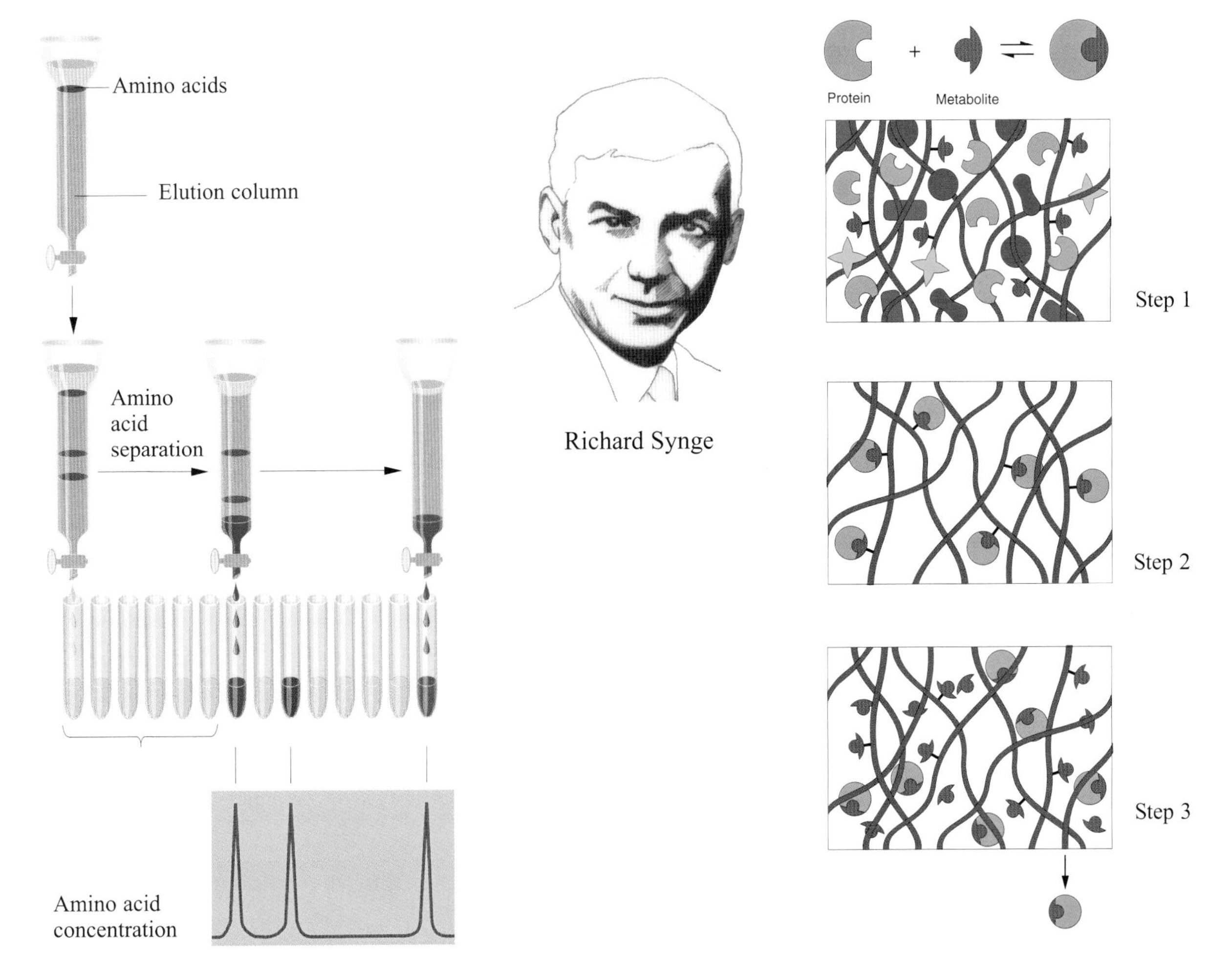

An example of chromatographic analysis, ion exchange chromatography. A sample containing a mixture of proteins is deposited on the summit of a column filled with a gel matrix carrying positive or negative charges. As elution proceeds, the different proteins, which have different electrical charges, migrate as a function of their affinity for the ionic matrix and appear as different coloured bands. The different fractions are subsequently collected and analysed.

Affinity chromatography:three steps. This technique takes advantage of the high affinity of many proteins for specific chemical groups. **Step 1** - a metabolite is fixed to an insoluble matrix such as agarose. **Step 2** - a specific protein from a cell extract binds to this metabolite. **Step 3** - complexes formed by metabolites bound to specific proteins are released from the column.

chromatography, which can be viewed as an extension of paper chromatography to gaseous samples.

Synge, Richard Laurence Millington (Liverpool, England, October 28, 1914 - Liverpool, August 18, 1994). British biochemist. Son of a stockbroker. Nobel Prize for Chemistry shared with **Archer, J.P.Martin** for their development of partition chromatography. Having studied at Trinity College (University of Cambridge), Richard Synge received his Ph.D. in 1941 and began a collaboration with Martin on methods for isolating vitamin A. Together they attempted to separate molecules using the adsorption chromatography technique developed by Mikhail Semyonovitch Tsvett, a Russian botanist who had succeeded in separating plant pigments by this means. This process, however, was limited by the nature of the adsorbing material. Another existing method for separating molecules was based on counter-current extraction; it consisted in using two non-miscible liquids and separating various substances according to their affinities for the two liquids. Synge and Martin managed to combine the two techniques. They placed a drop of the sample mixture to be resolved at one end of a piece of cellulose paper (used as an adsorbant) and allowed that end to soak in a mixture of solvents, for instance water and alcohol. Having chosen the sample mixture so that one of its constituents would have a greater affinity for the cellulose adsorbant, they allowed the mixture to migrate through the strip of paper. In such a set - up, the migration speed of each constituent depends on its affinity for the adsorbant and for the two liquids. When the separation is finished, the different fractions appear on the paper as spots that can be revealed with specific dyes or under ultraviolet light. Synge thus determined the structure of a simple protein, gramicidin S, by identifying its constitutive amino acids. Synge and

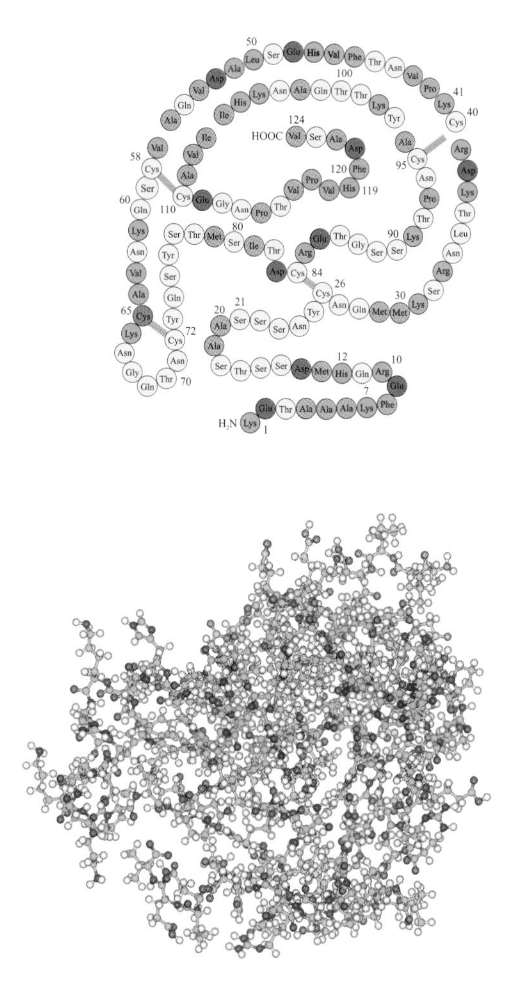

Primary structure (above) **and three-dimensional structure** (below) **of ribonuclease**. The primary structure is the amino acid sequence of the protein. After being synthesized, protein chains fold into their characteristic three - dimensional conformation.

Martin then considerably improved the separating power of this approach by developing a two-dimensional chromatography technique in which the solvents migrated at right angles with respect to each other. Partition chromatography has become a necessary tool for separating the molecules of living cells, and particularly for se parating amino acids or proteins. This technique allowed molecules labeled with radioactives dyes to be easily identified. This type of analysis was abundantly used by **Melvin Calvin** to study photosynthesis and by **Walter Gilbert** and **Frederick Sanger** in their DNA sequencing research. Richard Synge also had the opportunity to work in 1946 with **Arne Tiselius** at the University of Uppsala, where he applied electrophoresis to separation problems.

1953

Staudinger, Hermann (Worms, Germany, March 23, 1881 - Freibourg im Breisgau, September 8, 1965). German chemist. Son of a professor of philosophy. Nobel Prize for Chemistry for demonstrating that polymers are long - chain molecules. Staudinger graduated from the University of Halle under Krebs, a botanist. There he studied chemistry in depth so as to acquire the base he needed to understand biological phenomena. In reality, it was to chemistry that he ended up devoting all his intellectual energy. His doctoral thesis, presented in Halle in 1903, concerned the malonic esters of unsaturated compounds. In 1905 while he was assistant to Professor Thiele in Strasbourg, he discovered ketenes, a family of extremely reactive and toxic gasses, the simplest of which is $CH_2 = C = O$. He then focused on the chemical properties of these compounds. He was named professor at the Federal Polytechnic Institute in Zurich in 1912 and conducted research there on ketenes and other highly reactive substances. In 1926 he accepted a position at the University of Freiburg, as Head of the Laboratory of Chemistry, and remained there until his emeritus in 1951. In 1924 he proposed a hypothesis on the nature of rubber molecules. Observing that the size of these molecules varies considerably and is quite temperature-dependent, he proposed that they could be chains of interlinked smaller molecules, and that the chains formed could be very long. Staudinger's colleagues responded to this hypothesis with skepticism or even fierce opposition. He nevertheless succeeded in proving, by means of meticulously elaborate experiments, that rubber does consist of macromolecules, not aggregates. In these extremely long molecules (called polymers), a single motif (monomer) is repeated indefinitely. During his research he demonstrated the relationship between the degree of polymerization of macromolecules and their viscosity:this relationship is known as the

Hermann Staudinger

Rubber. Staudinger established the polymeric nature of this substance.

law of Staudinger. This law is still used to measure the degree of polymerisation of macromolecules. It took many years before the views of the German chemist were accepted by the scientific community. In particular, the ultracentrifugation technique developed by **Theodor Svedberg** came to the rescue of his theory.

1954

Pauling, Linus Carl (Portland, Oregon, February 28, 1901 - Big Sur, California, August 19, 1994). American chemist. Son of a pharmacist. Nobel Prize for Chemistry for applying quantum mechanics to the study of molecular structure and for his work on proteins. His most celebra - ted book, entitled The Nature of the Chemical Bond, and the Structure of Molecules and Crystals was probably the most influential chemistry textbook of the early- to mid-20th century. By tackling the study of the molecules of living organisms, and particularly the high-molecular-weight molecules which are proteins, Linus Pauling became a true pioneer. Towards the end of his long life, he lived 92 years, he was convinced that the study of the intimate structure of these substances would also lead to understanding the links between chemistry and the living world.

Linus Pauling. At left, an alpha helix.

In the years following World War II, the scientific community knew practically nothing about the architecture and extraordinary complexity of proteins, even though their existence had been known since the late 19th century. His early research having yielded inconclusive results, Pauling gave up studying these molecules directly and focused on amino acids, the building-blocks of proteins. While elucidating the structures of several amino acids, he examined how they assemble in protein molecules. This research led him to postulate with **Elias Corey** in 1952 the existence of alpha-helices and folded beta-sheets. These simple, elegant configurations are now recognised in most proteins. The discovery of these structures through model

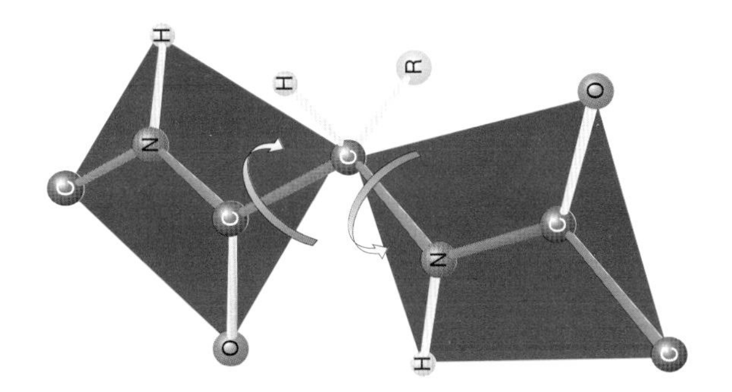

Peptide bond. It is a linkage between two amino acids. A peptide bond involves bonding between a carbon atom and nitrogen atom on one side, and the same carbon atom and oxygen on the other.

building ranks as one of the landmarks of structural biochemistry. Millions of different proteins exist in nature, each of them having specific properties. Pauling worked on a variety of problems in chemistry and in biology. Pauling also promoted the therapeutic value of vitamin C and claimed that the daily intake of high doses of this substance would improve the health in reducing the risk of heart disease. This hypothesis, however, led to conclusions that were contradicted by many scientists. For his cam-

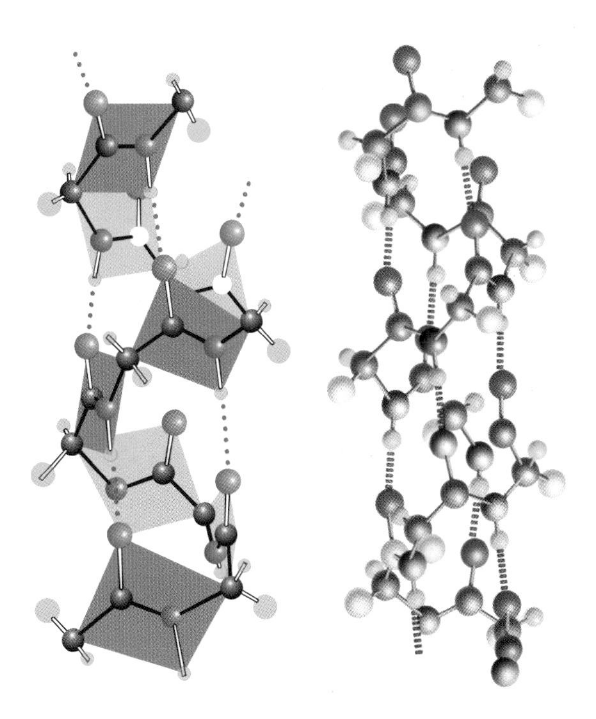

Alpha - helix. One of the standard forms of polypeptide described by Carl L. Pauling. The helix is shaped by hydrogen bonds that extend from one turn to the turns above and below. The alpha - helix is mainly found in fibrous proteins of hair, wool, skin and also in many enzymes. At right, a schematic diagram of an alpha helix

paign against the development of nuclear weapons he was awarded the Nobel Prize for Peace in 1963.

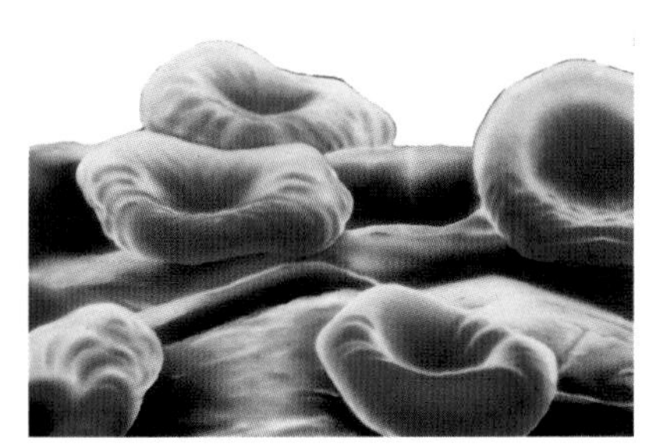

Red blood cells (normal form)

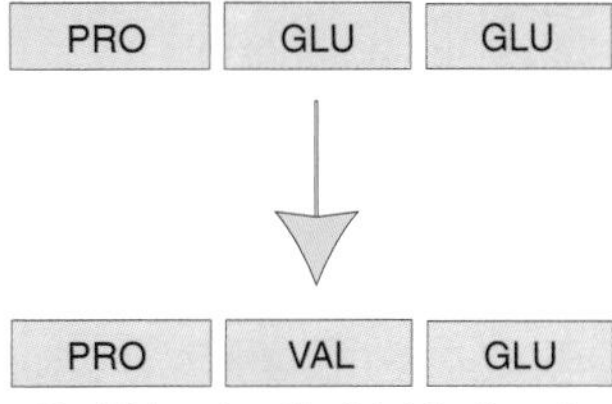

Red blood cells (sickle form)

Sickle-shaped red blood cells. A single mutation in DNA causes glutamic acid, an amino acid in the beta chain, to be replaced by valine, another amino acid. This replacement yields a defective hemoglobin which has an altered affinity for oxygen and which may affect the shape of the red blood cell containing it.

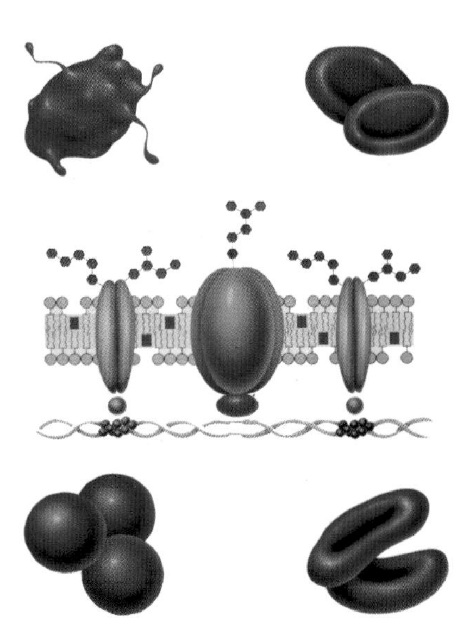

Except the one shown at the bottom right of this figure, the other forms of red blood corpuscles are abnormal, and several of them may be due to defects occurring in the membranes of these cells following diverse mutations involving one of their protein components**.**

Three-dimensional structure of bacteriorhodopsin. This membrane protein, which contains retinal, resembles animal rhodopsin. It is formed by *Halobacterium halobium* and serves as a light-operated proton pump to translocate protons through the bacterial membrane, from the inside to the outside.

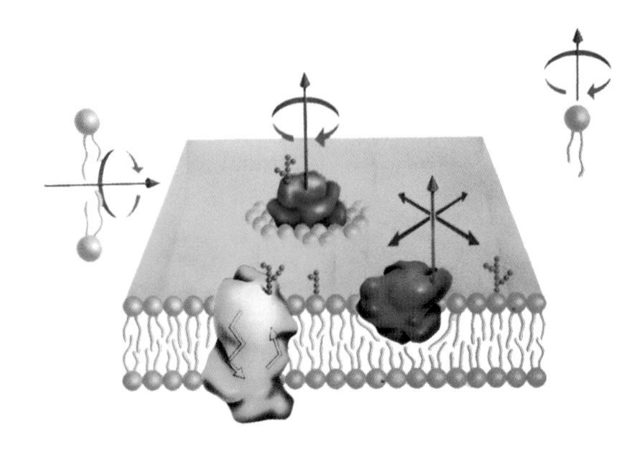

Protein movements in the cell membrane. Individual proteins as well as phospholipids can rotate and move laterally in the lipid bilayer.

Collagen. Its basic unit, tropocollagen, consists of three intertwined polypeptide chains, each of them arranged in a helical fashion.

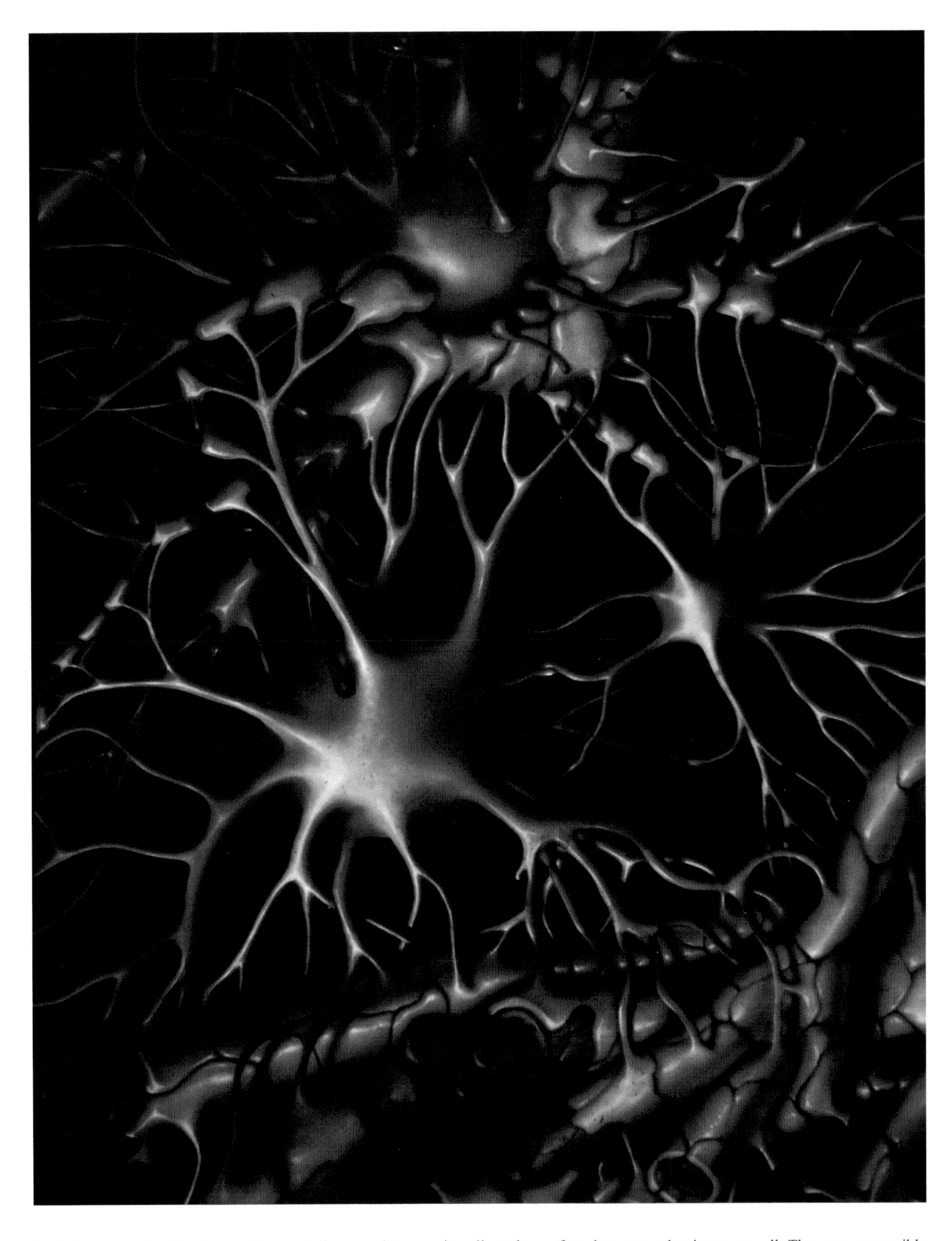

Proteins and cells. Proteins are the most diverse substances in cells and are often the most voluminous as well. They are responsible for most cellular functions, particularly those requiring a high degree of specificity. For example, in the cells of the nervous tissue represented here, proteins command the workings that enable them to take nutrients in the blood vessels that irrigate the brain (glucose in particular) and transfer them to neurons. Glial cells assume other functions and bring other categories of proteins into play.

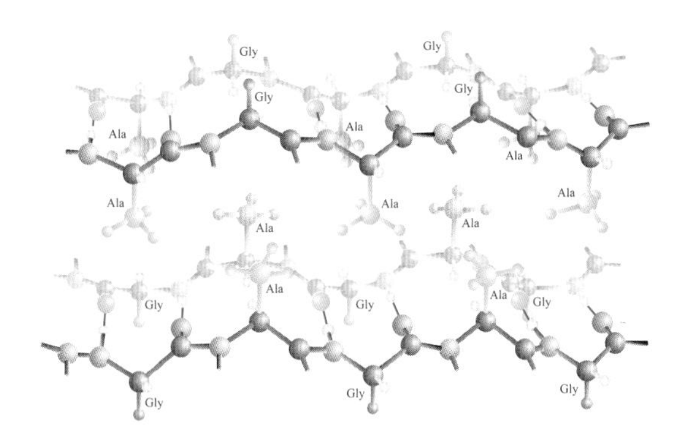

Silk. Silk fibers are based on a basic structure - a beta sheet - which was described by Carl Linus Pauling. Here, the protein chains are almost completely extended and are bound into sheets by hydrogen bonds from one chain to their neighbor. (Redrawn after I. Geiss; Sinauer Associates).

1955

du Vigneaud, Vincent (Chicago, Illinois, May 18, 1901 - White Plains, New York, December 11, 1978). American biochemist. Nobel Prize for Chemistry for the isolation and synthesis of two pituitary hormones. His father was a machine designer and inventor, fascinated by science since his early childhood. Vincent du Vigneaud's studies in chemistry and biochemistry first led him to focus on sulfur - containing amino acids like methionine, homocystine, cysteine, and cystathionine. Focusing on insulin, he began with the assumption that this protein contained sulfur. One of his first major achievements was to isolate cystine from insulin, thus proving the presence of sulfur in this protein. His name is also linked to several discoveries in the field of small peptides of physiological importance. One of them is glutathione, a tripeptide composed of glycine, cysteine and glutamic acid. He managed to synthesise it and to show that its reducing power was sufficient to break the disulfide bridges linking the two polypeptide chains of insulin, thus deactivating the hormone. Another type of research led him to show that the methyl group of methionine can be transferred from one chemical compound to another. This opened the way to the important study of metabolic trans-methylations. In 1938, he headed the Biochemistry Department at Cornell and four years later he succeeded isolating and elucidating the structure of vitamin H (biotin). Another breakthrough was the isolation of oxytocin and vasopressin (antidiuretic hormone), two related small peptides secreted by the neurohypophysis. Oxytocin stimulates uterine contraction during childbirth; vasopressin indirectly affects the concentration of sodium ions by regulating the amount of water in the urine. du Vigneaud isolated and synthesised these hormones, giving medicine two precious substances in sufficient amounts. Although oxytocin, to the scientist's surprise, was found to contain only eight amino acids, it took him no less than ten years to discover its sequence. In 1946, Vincent du Vigneaud and his colleagues also achieved the synthesis of penicillin.

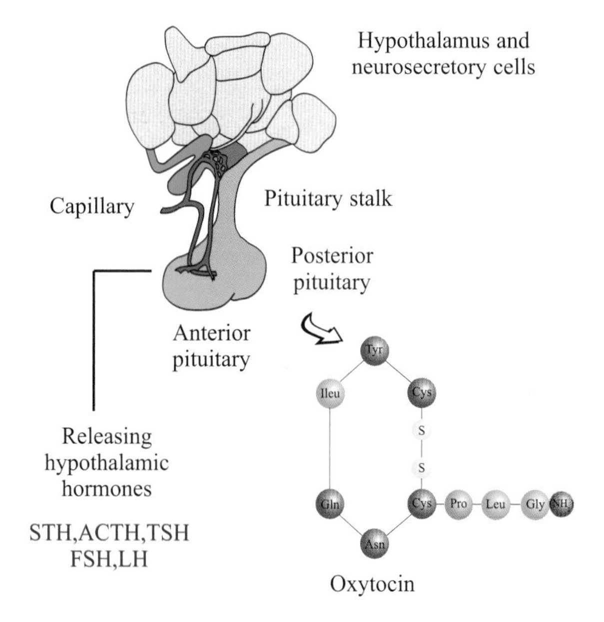

Oxytocin. A hormone secreted by the posterior hypophysis (neurohypophysis). It facilitates the ejection of milk in the mammary gland and stimulates contraction of the uterus.

1956

Hinshelwood, Cyril Norman (London, England, June 19, 1897 - London, October 9, 1967). British chemist. Son of a jurist. Nobel Prize for Chemistry with the Soviet scientist **Nikolaï Semënov** for works on reactions rates and reaction mechanisms. Cyril Hinshelwood began his career as a chemist by working for the Army during World War I. This is probably what aroused his interest in the kinetics of fast chemical reactions. He was educated at the Balliol College, Oxford, and was named professor of chemistry at the University of Oxford in 1937. The 1920s saw the beginning of outstanding developments in the kinetics of chemical reactions, both in Oxford and in Semenov's laboratory in Leningrad. These studies were applied specifically to reactions in the gaseous phase. Hinshelwood first studied the thermal decomposition of vapours of organic molecules. The mechanisms deduced from these studies led him to postulate the intervention of unstable intermediates, called transition species or "acti-

Cyril Hinshelwood

vated complexes". In short, before reacting, a molecule has to reach a high level of excitement. And the kinetic theory of gases assumes that this activation is caused by collisions with other molecules. This model (including an

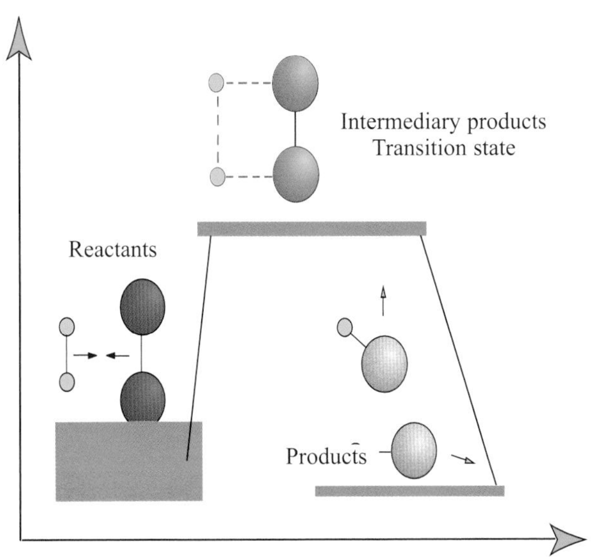

Activated complex and transition state. The latter is the state in which the two partners of a chemical reaction are in an active state. Represented at the top of the figure, it has a geometrical and electronic structure which favors the reaction towards the formation of the products.

intermediate state in a reaction process) was to prove very fruitful and triggered the renewal of chemical kinetics. In 1927 Hinshelwood began studying how hydrogen and oxygen combine to form water. Again using a model involving activation by collision, he determined that such a mixture should become explosive only within a certain pressure range. Accepting the views of Bodenstein and **Walter Nernst** on the propagation of the reaction between hydrogen and chlorine (published respectively in 1894 and 1916), and in accordance with Semenov's conclusions, he adopted the hypothesis of a mechanism governed by progressive accumulation of reactive species within the explosive mixture. This is the so - called "free-radical branched-chain reaction" mechanism, a "free radical" being a species possessing an unpaired electron. A similar propagation mechanism intervenes in the amplification of fission in nuclear explosives. The rest of Hinshelwood's career was devoted to the study of bacterial development, in which he used the existing models of chemical kinetics as tools. Modern biochemistry owes much to his pioneering works. Cyril Norman Hinshelwood was not only a scientist but a polyglot as well. As such, he was appointed as President of the Modern Language Association. He was knighted in 1948.

Semënov, Nicolaï Nicolaevitch (Saratov, Russia, April 6, 1896 - Moscow, USSR, September 25, 1986). Soviet physical chemist. Nobel Prize for Chemistry with **Cyril Hinshelwood** for research in chemical kinetics. Nicolaï Semënov was the first Soviet citizen living in Russia to receive the Nobel Prize for Chemistry. In 1917, he graduated from the University of Petrograd (formely St. Petersburg, and later Leningrad). At the age of 20 he published a study on the collisions between electrons and molecules. In 1920, after spending some time at the University of Tomsk, he returned to the Institute of Physics and Technology of Petrograd. In 1928, he was put in charge of the Mathematics and Physics Department of the Polytechnic Institute of Leningrad. Named Director of the Institute of Physical Chemistry of the Academy of Science of the USSR in 1931, he held this position untill 1944, before directing the Department of Chemical Kinetics at the University of Moscow. From the beginning, Semënov's research was mainly oriented towards chemical kinetics in the gaseous phase. It is noteworthy that this scientist and Hinshelwood conducted their research practically independently for nearly 25 years, but reached similar conclusions. This is why they are so closely associated in people's minds in connection with work having led to the improvement of the internal combustion engine, to the development of rocket engines and high-performance explosives, and conversely to the development of fire prevention and fire - fighting techniques. These last-mentioned developments stemmed from Semënov's observation that the reactor walls and certain impurities present in explosive mixtures exerted an inhibitory effect on the free - radical reactions he was studying. The interpretation of these observations is that these surfaces and impurities are able to "trap" the free radicals that propagate the combustion. Semënov thus described the function of these "quenchers" which were to play such an important role in fire prevention and in controlling firedamp explosions in coal mines.

Nicolaï Semënov

1957

Todd, Alexander Robertus (Glasgow, Scotland, October 2, 1907 - January 10, 1997). Scottish biochemist. Nobel Prize for Chemistry for research on the structure and synthesis of nucleotides, nucleosides, and nucleotide coenzymes. Alexander Todd was born into a humble family. His interest in nucleotides stemmed from prior studies on the B vitamins thiamine, riboflavin, and niacin, which contain nucleotides. The chemical structures of the nucleic acids were elucidated by the early 1950s, largely through the efforts of Todd. These substances are also

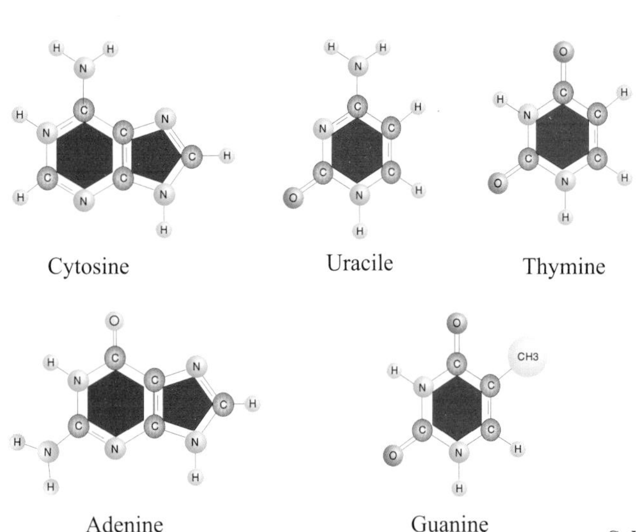

Alexander Todd

Schematic structure of the five nucleic bases that enter the composition of DNA and RNA.

present in the nucleic acids RNA and DNA. Todd's work led to a better understanding of the composition of genes. They provided a few crude landmarks that guided the elucidation of the DNA structure and helped **James Watson**

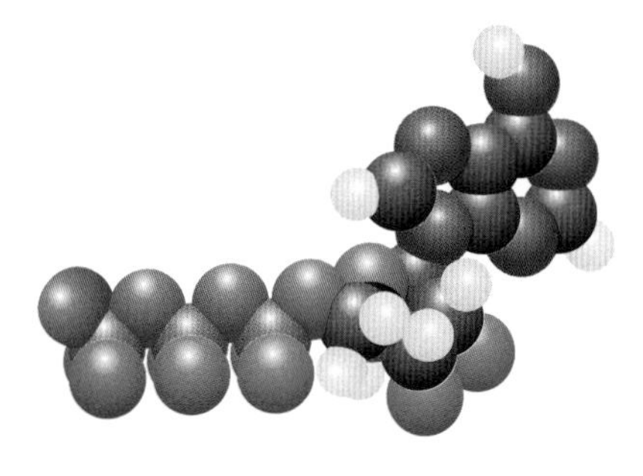

Adenosine triphosphate. A three - dimensional representation.

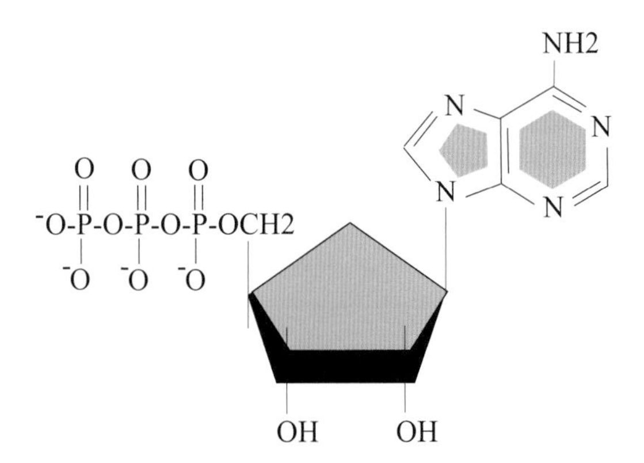

Adenosine triphosphate. The molecule is composed of adenine, an organic base, linked covalently to ribose, a five carbon sugar, to form adenosine. Adenosine in turn is bound to three phosphates to make ATP (phosphodiester bonds). The energy released by hydrolysis of one or two of the phosphodiester bonds is required by many biolo - gical processes such as muscle contraction and membrane exchanges.

and **Francis Crick** in their builiding of the double helical model of DNA. It is also to Todd that we owe our know - ledge of the structures of coenzyme flavin adenine dinucleotide (FAD), adenosine diphosphate (ADP), and adenosine triphosphate (ATP). Alexander Todd was the son-in-law of **Henri Dale**. He was knighted in 1954.

1958

Sanger, Frederick (Rendcombe, England, August 13, 1918 -). English biochemist who was twice the recipient of the Nobel Prize for Chemistry. Son of a physician. Frederick Sanger received the Nobel Prize twice. The first time was in chemistry in 1958 for methods of amino acid

Frederick Sanger

sequencing; the second time was in 1980 with **Walter Gilbert** and **Paul Berg** for a method of nucleic acid sequencing. Around 1943, upon graduating from

Cambridge University, he took up a position in that institution's famous molecular biology laboratory, where great minds in biochemistry such as **Francis Crick**, **James Watson**, **Max Perutz** and others were already conducting the research that would allow molecular biology to be born. Sanger's plan was to determine not

Amino acid sequence of insulin. Insulin consists of two polypeptide chains, one of 21 and the other of 30 amino acids. These chains are joined by two disulfide bonds.

only the amino acid composition of a protein, but also the sequence of amino acids in the protein chain. Since the development of paper chromatography by **Archer Martin** and **Richard Synge**, it had been possible to determine the proportion of each amino acid in a given protein. Sanger chose to focus on insulin, which had been isolated by **Frederick Banting** one quarter of a century earlier. After eight years of hard work, he managed to reconstitute the exact sequence of this small hormone containing some 50 amino acids and consisting, in fact, of two small chains linked by disulfide bonds. With today's methods it takes no more than a few days to sequence a protein of that size, but at the time it was a genuine feat. Sanger's demonstration - this was in 1953 - that a protein has a precise but non-repetitive sequence was proof that the sequence information required to produce such molecules must be encrypted in a genetic code. Frederick Sanger's work would be of great usefulness in the studies that would lead to the discovery of this code.

1959

Heyrovsky, Jaroslav (Prague, Bohemia, Austro-Hungary, December 20, 1890 - Prague, Czechoslovakia, March 27, 1967). Czech chemist. Nobel Prize for Chemistry for his discovery and development of polarography. Son of a jurist. Polarography is a microanalytical method allowing scientists to determine, with a high precision, the concentration of an electroyte without modifying its concentration. Jaroslav Heyrovsky was educated at the Charles University of Prague and at the University College, London, where he earned his M.S. degree in 1913. It was in this city that he became interested in electrochemistry, the discipline in which he conducted his doctoral research. During World War I he served in a hospital in Prague, but he continued his research and in 1918 he presented his thesis on

Jaroslav Heyrovsky

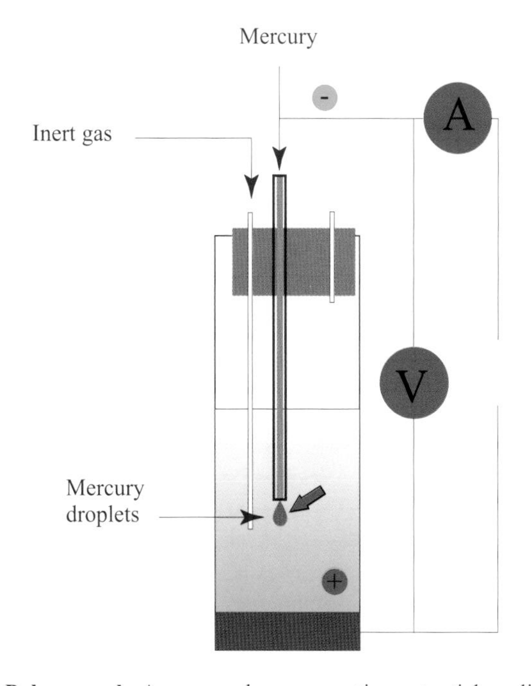

Polarograph. A more and more negative potential applied to a mercury drop electrode which progressively captures the different cations present in the solution. A voltage - amperage curve is obtained, from which the nature and concentrations of the cations in the solution can be deduced.

the electroaffinity of aluminium at Charles University. His career as a scientist and as a teacher began in 1919. He directed the Department of Chemistry of Charles University from 1922. Appointed as professor and director of the Institute of Physical Chemistry in 1924 he held this post until 1950 before heading the Polarography Institute at the Czechoslovakian Academy of Science until 1963. During Heyrovsky's first years of research in electrochemistry in London under Donnan, the principle of polarography took shape. Polarography is the successive microelectrolysis of various ions at different characteristic voltages. The intensity of the current produced by each ion is proportional to its concentration in the solution, and the very small amount of matter deposited does not significantly alter this concentration. The first operating polarograph was ready in 1924. For many years thereafter, Heyrovsky strove to perfect the process; he managed to accelerate it so as to allow determinations within a fraction of a second. Heyrovsky's polarograph was one of the first instruments to be suitable for routine automated analysis of substances. On the other hand this research made possible significant progress in the development of selective electrodes for various measurements and monitoring in fields as diverse as medicine, biochemistry, and industry.

1960

Libby, Willard Frank (Grand Valley, Colorado, December 17, 1908 - Los Angeles, California, September 8, 1980). American chemist. Son of a farmer. Nobel Prize for Chemistry for his method to use carbon-14 for age determination in archeology, geology, geophysics and other branches of sciences. Williard Libby was educated at the University of California and gained his Ph.D. in chemistry in 1933. During World War II, he was associated with the Manhattan Project and worked on the separation of uranium isotopes in view of developing the atomic bomb. In 1945 he held a post as a teacher in chemistry at the University of Chicago first, then at the University of California, Los Angeles. Libby owes his celebrity to research having a much more pacific finality than the research conducted during the war. Let's recall a few facts. Cosmic rays were discovered by American scientists as early as 1939. On the other hand, Lawrence's cyclotron in Berkeley had made it possible to synthesise carbon-14. On the basis of these experiments, Libby hypothesised that carbon-14 could be formed at very high altitude through the action of cosmic rays on atmospheric gases. The carbon-14 formed would associate with the oxygen of air to yield $^{14}CO2$, which would diffuse to lower altitudes and take part in chlorophyllian syntheses. The carbon-14 produced naturally by cosmic rays would thus be gradually integrated into the food chain and end up in the tissues of all living organisms on the planet. As long as an organism was alive, its content in these radionuclides would remain constant, as a steady state would result from mutually compensating absorption through food and excretion. After the death of the organism, absorption would stop and the radionuclide should simply decrease according to its radioactive half - life. Libby's idea was to count the radioactivity of carbon-14 in a sample of any non-living biological material to determine the date at which the organism that produced the material died, and thus the approximate period at which the material was used to create an object. A test was performed on a piece of wood from an Egyptian tomb of well - known age. The result was conclusive: the age determined by counting the sample's radioactivity corresponded almost perfectly with the archaeological age. The carbon-14 dating method was greeted with enthusiasm. It stimulated extraordinary progress in the timing of events in human history and also in paleobotany, recent geology, etc. Carbon-14 dating made a significant contribution in many fields. The technique is suited for samples ranging in age from a few decades to some 40,000 years, spanning a broad panorama of recent human history. The only limit to its application concerns the required sample size - the sample must contain at least 1g of carbon, and it is not acceptable to mutilate rare or precious objects for dating purposes -. A recent improvement of this dating technique consists in using a mass spectrometer instead of a radioactivity counter to assay carbon-14 in the sample. This approach requires sacrificing only a few milligrams of organic matter. This made it possible, for instance, to date the Turin Shroud, now estimated to have been made around the start of the 14th century. Libby also exploited the release of tritium into the atmosphere caused by nuclear tests to study water cycles in the United States. From 1955, Libby also served on the U.S. Atomic Energy Commission but he retired in 1959. The same year he directed the Institute of Geophysics, Los Angeles, and he continued to develop techniques of radioactive dating.

Williard Libby

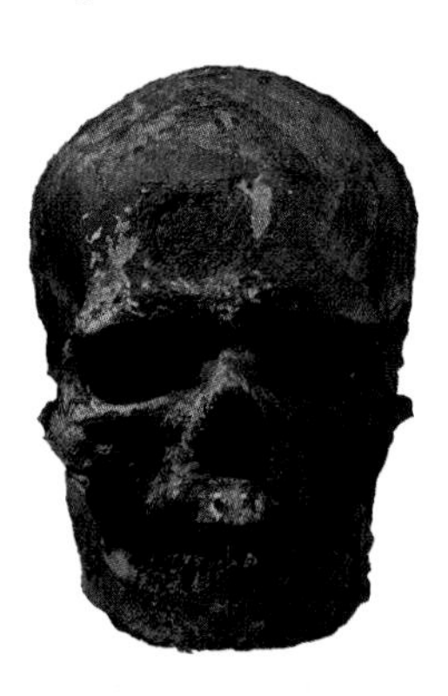

The dating technique discovered thanks to the carbon -14 isotope made possible an extraordinary advance in the chronology of the events of human history.

1961

Calvin, Melvin (St. Paul, Minnesota, April 8, 1911 - Berkeley, California, January 8, 1997). American biochemist. Nobel Prize for Chemistry for his discovery of the chemical pathways of photosynthesis. Born of working - class parents having emigrated from Russia. His initial studies, at the Michigan College of Mines and then in Manchester, England, led him gradually into chemical engineering. This explains why the essence of his scientific career is best described in terms of the practical aspects of chemistry and their interactions with biochemistry. His work focused particularly on coordination chemistry. At the University of Manchester, he worked under the direction of John Polanyi and became interested in metalloporphyrins, especially chlorophylls and the role of these substances in photosynthesis. In 1938 he went back to the United States and joined the University of Berkeley, California. This is where most of his work was conducted; it focused on elucidating the basic mechanism of photosynthesis. It is noteworthy that for Calvin, who had trained as a physical chemist and whose only know - ledge of biology was what he had learned from a course in palaeontology, it was first necessary to study biochemistry. He applied himself to this task for several years. Photosynthesis is the biochemical mechanism by which a plant absorbs carbon dioxide from the air or water and uses solar energy to convert it to sugars, while relea- sing oxygen. This biological process can be viewed as one of the most important ones in the biosphere, because the survival of most living organisms depends on it. In Calvin's time it was already known that the above-mentioned conditions were sufficient for photosynthesis, but it was also suspected

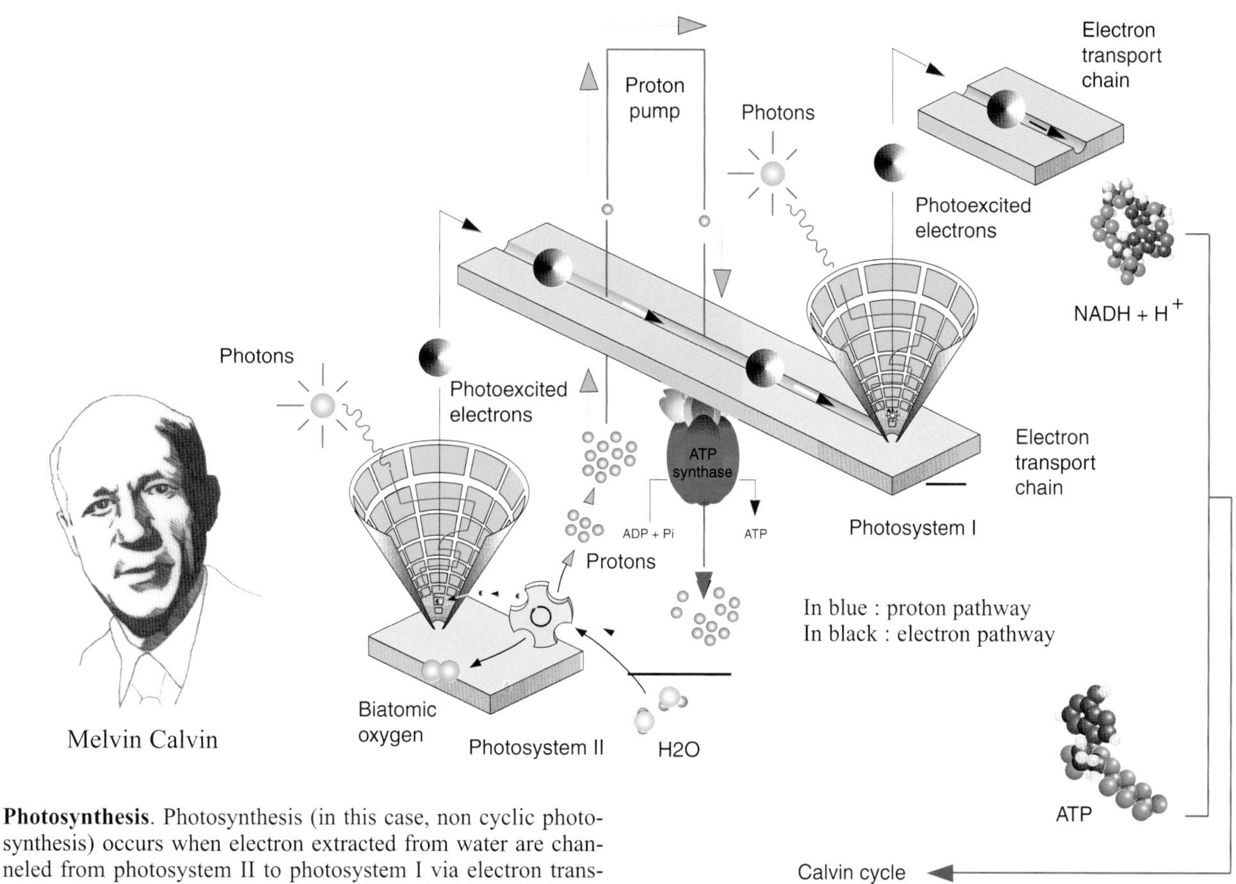

Melvin Calvin

Photosynthesis. Photosynthesis (in this case, non cyclic photosynthesis) occurs when electron extracted from water are channeled from photosystem II to photosystem I via electron transporters. This process is coupled to a mechanism that allows the pumping of protons, the synthesis of ATP and the production of sugars (see Paul Boyer).

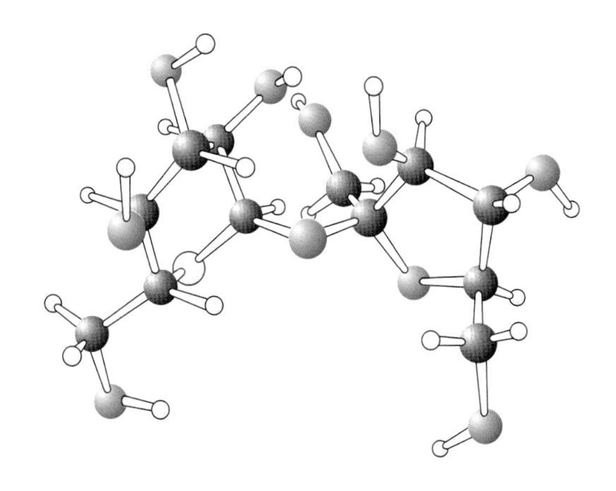

Saccharose, a spatial model

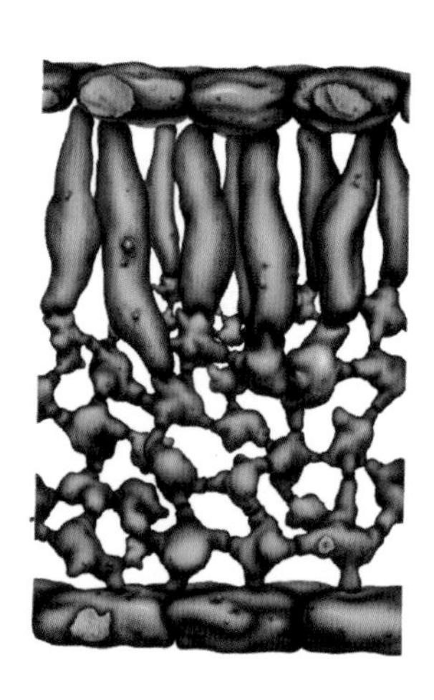

A cut through a leaf. Note the mesophyll cells lying between the upper and the lower epidermis. They are the major sites of photosynthesis in green plants.

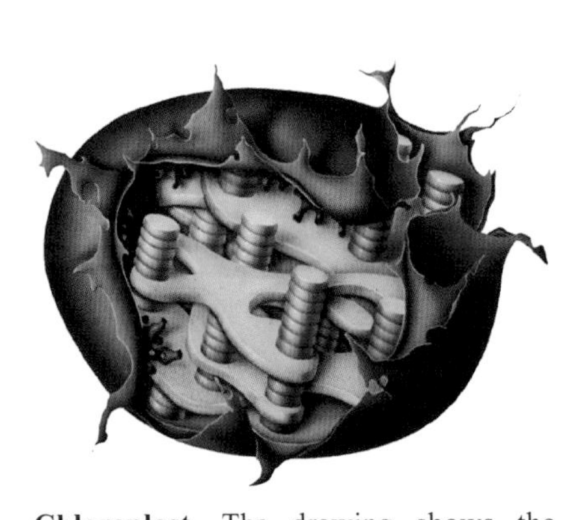

Chloroplast. The drawing shows the chloroplast membrane and the thylakoid membrane running lengthwise through the liquid matrix, the stroma. While the enzymes and pigment molecules for the light reactions of photosynthesis are bound to the thylakoids, the enzymes for the dark reactions (the Calvin cycle) are free - floating in the stroma.

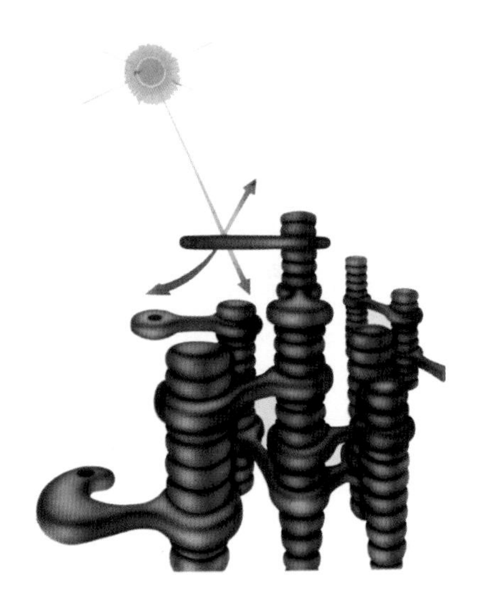

Three - dimensional representation of thylakoid membranes. The chlorophyll light reactions take place in these membranes whereas the Calvin cycle (production of sugars) takes place in solution inside the thylakoid discs.

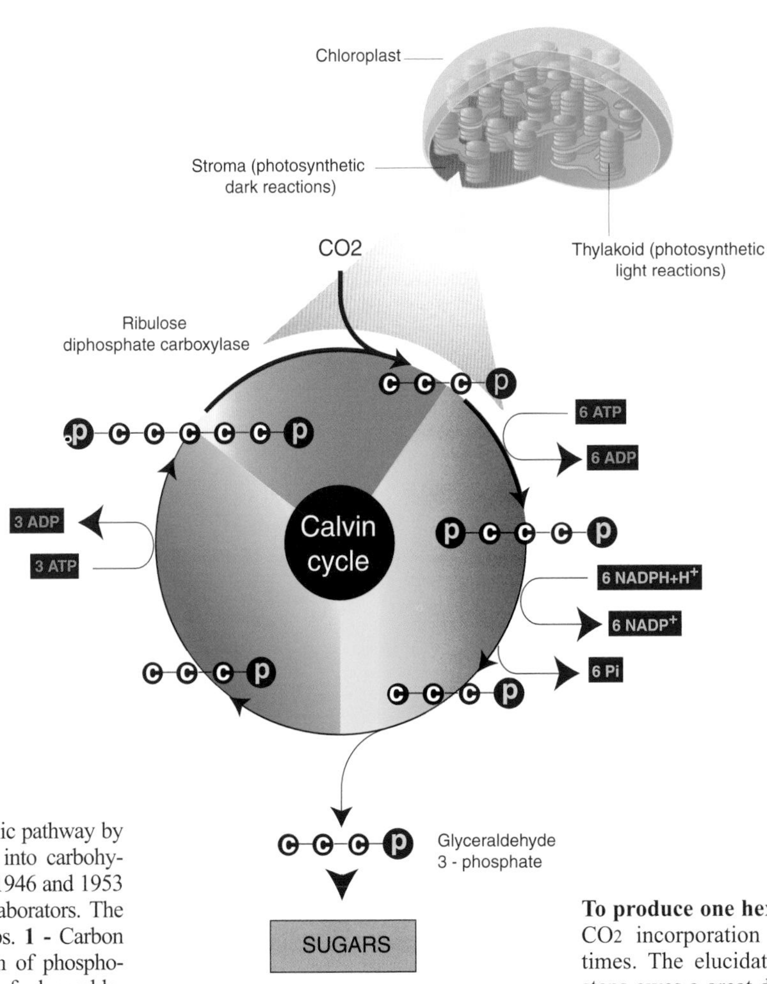

The Calvin cycle. The metabolic pathway by which plants incorporate CO_2 into carbohydrates was elucidated between 1946 and 1953 by Melvin Calvin and his collaborators. The cycle is divided into three steps. **1** - Carbon dioxide fixation; **2** - Reduction of phosphoglyceric acid and production of glyceraldehyde 3 - phosphate (sugars); **3** - regeneration of ribulose from glyceraldehyde 3 - phosphate.

To produce one hexose (six carbons), the CO_2 incorporation must be repeatd six times. The elucidation of these different steps owes a great deal to the radiocarbon labeling techniques used by Melvin Calvin and developed by **Georges de Hevesy**.

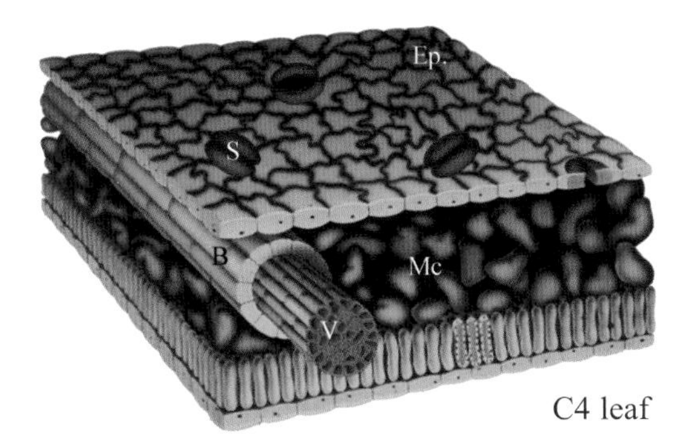

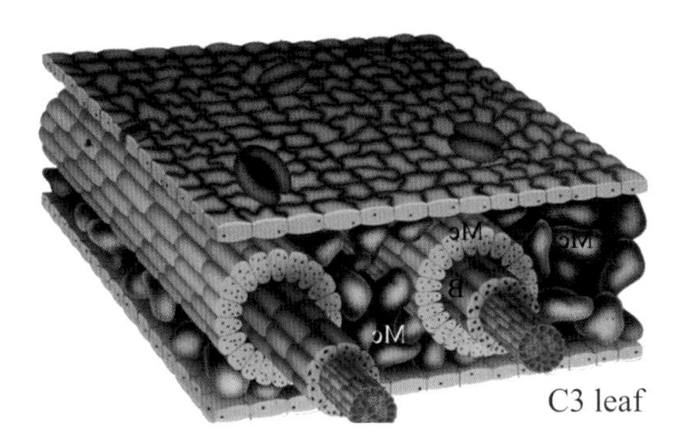

Carbon-dioxide fixation in plants. In C4 leafs, carbon dioxide is initially fixed in mesophyll cells and then transported to bundle - sheath cells. C4 plants are characteristically found in arid environments where they have evolved means to circumvent photorespiration, a process that occurs concomitantly with photosynthesis (especially in C3 plants) whereby oxygen is consumed and carbon dioxide is evolved. (See Calvin).

that incorporation of the mineral carbon into sugars was a multiple-step and probably very complex process. Yet this process remained a mystery. By allowing the algae *Chlorella* to absorb radioactive carbon dioxide during very short times, he succeeded in identifying a series of compounds formed during the course of photosynthesis. He then showed that these compounds belonged to a cycle of reactions, today known as the Calvin cycle. This cycle incorporates carbon dioxide from air into the sugars produced by plants. The end of this great scientist's career was marked by many experiments aimed at determining how a prebiotic atmosphere simulated in a laboratory could produce the characteristic ingredients of life. A much-debated question of the time - the 1960s - was whether Earth's first atmosphere was a reducing or an oxidising atmosphere. This period was also marked by the research of Miller, student of the chemist **Harold Urey** and famous for his simulations of prebiotic syntheses, and by the work of Cecil Ponnemperuma, student of Melvin Calvin interested in questions surrounding the origin of life.

1962

Kendrew, John Cowdered (Oxford, England, March 24, 1917 - Cambridge, England, August 23, 1997). British biochemist. Son of a climatologist. Nobel Prize for Chemistry, shared with **Max Perutz**, for determination of the structure of the muscle protein myoglobin. John Kendrew graduated from Trinity College, Cambridge, where he received his Ph.D. in 1949. Using X - ray diffraction methods, he succeeded in determining with a very high resolution the three-dimensional structure of myoglobin of sperm-whale. About ten years were needed to accomplish this task. Founder of the "Journal of Molecular Biology", Kendrew promoted the building of EMBL, European Molecular Biology Laboratory, in Heidelberg, Germany, which he directed until 1982. John Kendrew was knighted in 1963. He and Max Perutz were the founders of the Medical Research

John Kendrew

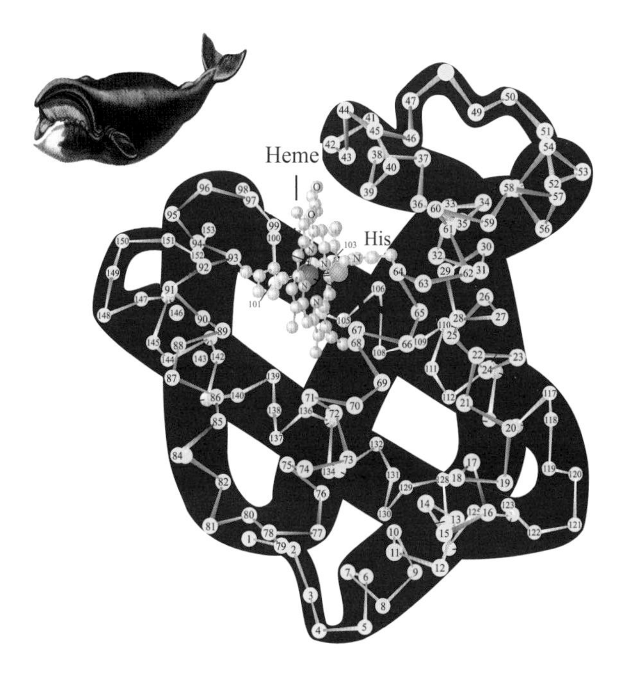

Three-dimensional structure of myoglobin. The structure of this protein was elucidated through X - ray diffraction analysis by John Kendrew. This oxygen - carrying hemoprotein occurs in muscles. A ferroheme prosthetic group (in green) is bound to a single polypeptide chain of 153 amino acid residues. Myoglobin stores oxygen in muscle cells and delivers it to the electron transport chain when the supply becomes limited. (Redrawn from I. Geiss).

Council for Molecular Biology at Cambridge. John Cowdered Kendrew was the head of the Molecular Research Council, Cambridge, England.

Perutz, Max Ferdinand (Vienna, Austria, May 19, 1914 - Cambridge, England, February 6, 2002). Austrian - born British biochemist. Nobel Prize for Chemistry, shared with **John Kendrew**, for his X - ray diffraction analysis of the structure of hemoglobin. Son of a textile manufacturer. Perutz studied chemistry at the University of Vienna and took his Ph.D. in X - ray crystallography at the University of Cambridge, England. In 1947 he became

Max Perutz

Head of the newly-founded Medical Research Council Unit for Molecular Biology. In 1962, he was appointed Chairman of the Laboratory of Molecular Biology, built by the Medical Research Council for him and his colleagues. Perutz began his scientific work on the structure of proteins in 1937. In 1953 he discovered a method of solving their structure. This led to the solution of the structure of myoglobin by Kendrew and that of haemoglobin by Perutz and his colleagues. The Medical Research Council Unit for Molecular Biology was the place where the structure of DNA was discovered by **James Watson** and **Francis Crick** in 1953 and the triplet nature of the

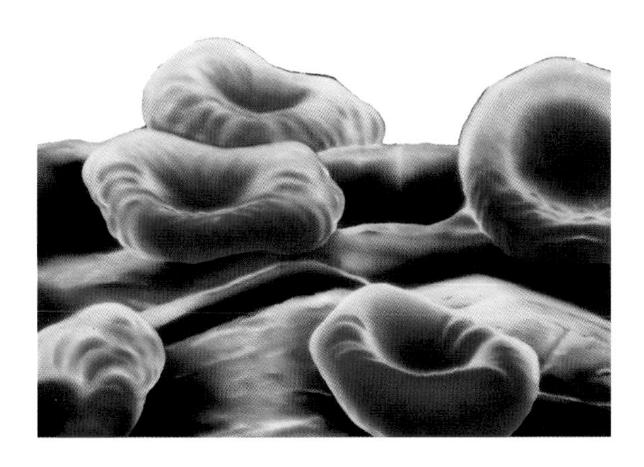

Red blood cells. A scanning electron micrograph (colored).

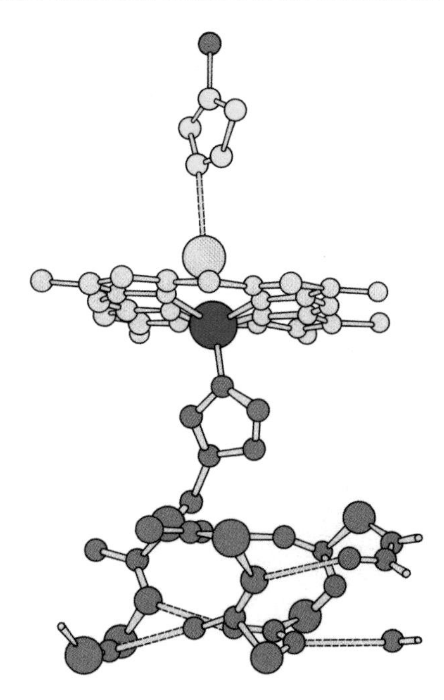

The hema group of the hemoglobin molecule. A tridimensional drawing.

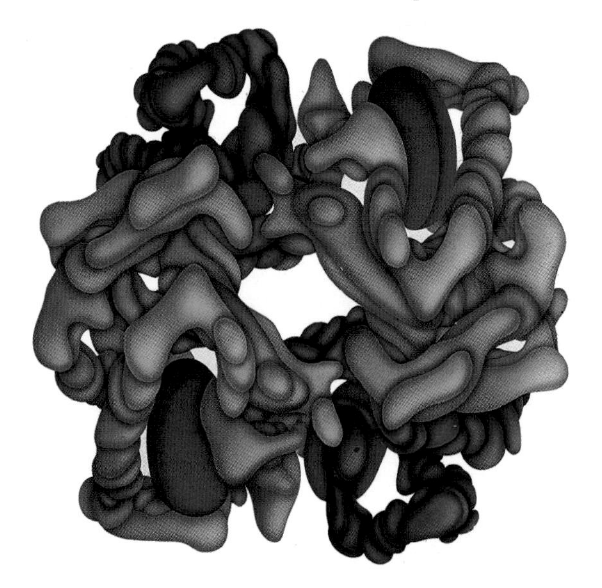

Three-dimensional representation of the hemoglobin molecule. The four chains of the hemoglobin molecule each are folded and packed together into a compact unit. The four heme pockets are exposed on the outside of the molecule and are available for binding four oxygen molecules. The four chains cooperate to permit the transport of molecular oxygen in the blood.

Red blood corpuscles, white globules, and platelets. Red blood corpuscles are by far the most numerous of cells. There are approximately five million per mm3 of blood. Each corpuscle encloses 300 million molecules of hemoglobin, and one molecule of hemoglobin is made up of several thousand atoms.

genetic code was found in 1962. Many other developments in the new science of Molecular Biology originated there. At the new laboratory, Max Perutz worked out the molecular mechanism of hemoglobin, the molecular pathology of inherited haemoglobin diseases, the stereochemistry of drug binding and other topics. In 1994 he suggested a molecular mechanism for Huntington disease which subsequent research has confirmed.

1963

Natta, Giulio (Imperia, Italy, February 16, 1903 - Bergame, May 1, 1979). Italian chemist. Nobel Prize for Chemistry, along with **Karl Ziegler**, for the development of high polymers. Son of a judge. He was the first Italian chemist to receive this distinction. Natta earned his degrees in chemical engineering at the Polytechnical Institute of Milano in 1924. He taught chemistry at the University of Pavia from 1933 to 1935. After moving to the University of Roma for two years he also taught chemistry in Torino from 1937 to 1938. Returning to Milano

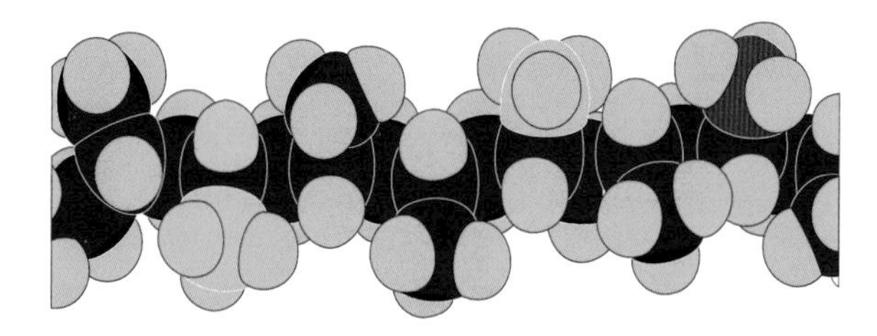

Polypropylene. This substance was discovered in 1954 by the Italian chemist Giulio Natta. Polypropylene is a thermoplastic resin built up by the polymerization of propylene.

in 1939 he headed the Industrial Chemistry Department of the Roma Polytechnic Institute. There he devoted himself to the study of polymer chemistry. In 1953, Ziegler was

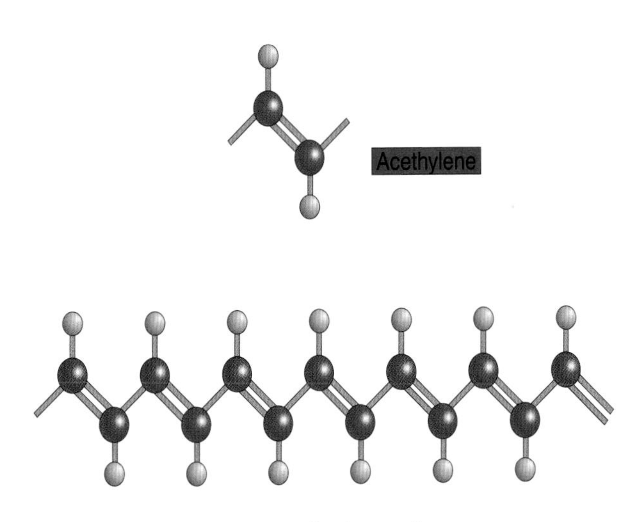

Structure of polyacetylene

developing catalysers for polymerising ethylene to polyethylene. Their action was to produce linear polymer chains, thus considerably improving product quality. Natta succeeded in using this catalyst to polymerize propene and synthesize polypropylene. Thanks to crystallographic determinations by X - ray diffraction, he showed that these high-quality polymers chains - syndiotactic and atactic polymers - are characterised by a very regular arrangement of the lateral methyl groups of the polypropylene. The term "stereospecific polymers" was coined to describe these polymers having high commercial value because they are very resistant to heating and expansion.

Ziegler, Karl Waldemar (Helsa, Germany, November 26, 1898 - Mülheim, August 12, 1973). German chemist. Nobel Prize for Chemistry with **Giulio Natta** for research that greatly improved the quality of plastics. Karl Ziegler received his Ph.D. from the University of Marburg in 1923. Then he taught at Frankfort, Heidelberg and Halle before heading the Kaiser Wilhelm Institut in Mülheim for research on coal. From 1923 to 1943, he studied organomagnesium and organoaluminium derivatives. Lithium and aluminium derivatives proved particularly useful in organic synthesis. Ziegler managed to assemble long chains of aluminium derivatives from which he produced

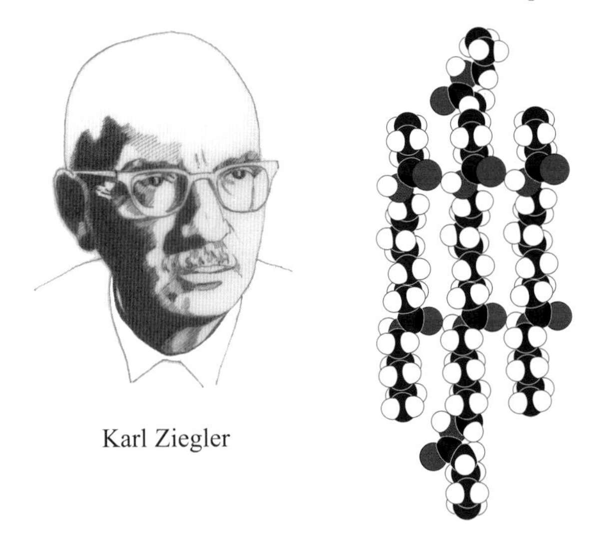

Karl Ziegler

Nylon. A common form of nylon results from a condensation reaction between hexamethylenediamine and adipic acid, a dicarboxylic acid. Because of their high melting points, their insolubility, and their resistance to fracture, they are useful in the manufacture of a number of products, for example in the textile industry.

alcohols useful in the composition of detergents. In 1953 he discovered that ethylene (chemical formula: CH2=CH2), the simplest of the olefins, readily polymerised in a hydrocarbon solution under the catalytic effect of an organoaluminium and titanium tetrachloride. The polymer produced was a linear chain, contrary to a previously synthesised branched product which was soft and whose properties were uninteresting. The polyethylene synthesised by Ziegler's team was stiff and suitable for spinning. This was naturally both a major research discovery and one with important practical applications. Guilio Natta then extended the process, making polypropylene from propylene, $CH_2{=}CH_2{-}CH_3$. He showed that the role of the catalyser used by Ziegler was to control the geometry of the macromolecules, and in particular, the regular distribution of lateral methyl groups.

1964

Hodgkin, Dorothy Crowfoot (Cairo, Egypt, May 12, 1910 - Shipston-on-Tour, England, July 29, 1994). English chemist. Daughter of an archaeologist. Nobel Prize for Chemistry for determination of the structure of vitamin B_{12}. After her studies in chemistry at Oxford, Dorothy Hodgkin went to Cambridge where she worked with John Desmond Bernal on the determination of the structure of molecules using X - ray diffraction techniques. Then, in 1934, she returned to Oxford where she held a post as a Research Professor of the Royal Society and researched sterols. Because of their folded structure, the study of proteins was particularly hard. Nevertheless, this structure is crucial for their biological activity. So it was an important task to elucidate it. After the work of **George Minot** and **William Murphy**, who shared the Nobel Prize for Medicine in 1934, it appeared that pernicious anaemia was caused by a genetic incapacity to synthesise a factor intrinsically associated with vitamin B_{12}, stored in the liver but synthesised by microorganisms. In 1957, Hodgkin determined the structure of this vitamin and showed the presence of a cobalt ion in the porphyrin nucleus. Other resounding successes were the elucidation of the three-dimensional structure of penicillin in 1947 and the determination of that of insulin in 1969. Appointed Fellow of the Royal Society in 1947, Hodgkin did not become full Professor at Oxford until 1957, when her prestige finally overcame the latent sexism of the time.

Dorothy Hodgkin

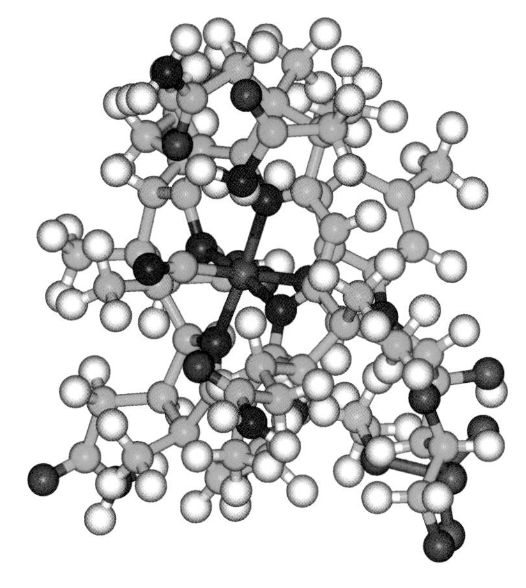

Vitamin B12. A vitamin with a complex structure. It functions as a coenzyme in nucleic acid metabolism. Its deficiency is responsible for pernicious anaemia.

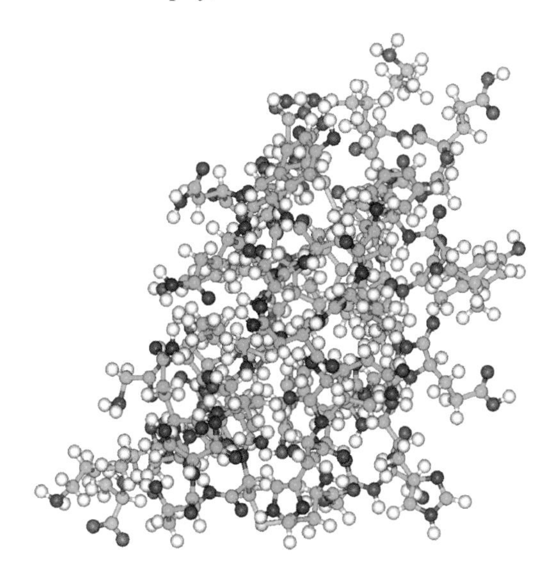

Insulin molecule This protein decreases blood glucose levels by increasing the uptake of glucose into cells and converting it to glycogen, especially in the liver. Insulin also regulates fat metabolism. It was first isolated by **Frederick Banting**. The spatial structure of this protein was determined by Dorothy Hodgkin.

1965

Woodward, Robert Burns (Boston, Masachusetts, April 6, 1917 - Cambridge, Massachusetts, July 18, 1979). American chemist. Nobel Prize for Chemistry for his fundamental contribution to organic synthesis. Robert. Woodward is rightly considered as the most prestigious synthetic chemist of the 20th century. Even very young, he displayed an amazing mastery of this discipline. Having obtained his Ph.D. at the age of 20, he pursued his career at Harvard where he was named professor in 1950. His fame arose from the total synthesis of molecules important in biology and in pharmacology. A total synthesis implies

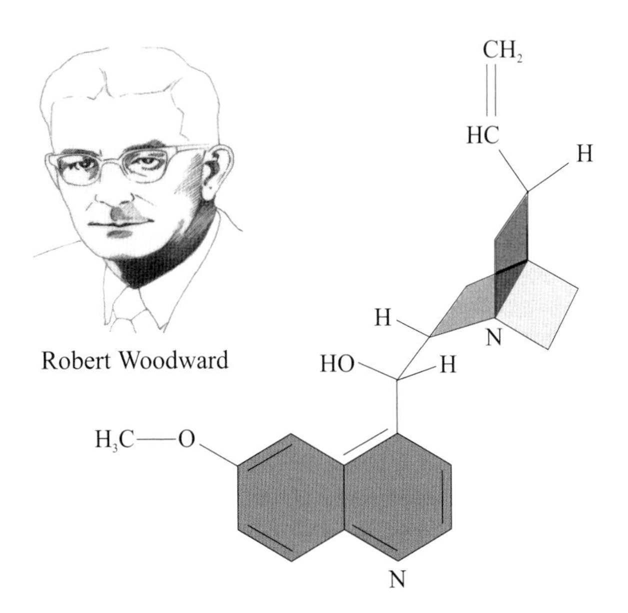

Robert Woodward

Quinine. A bitter-tasting alkaloid extracted from cinchona bark. Quinine is used chiefly for the prevention and the treatment of malaria.

the production of a given molecule from simple starting materials. To accomplish this, Woodward put into play the entire arsenal of the most evolved techniques and theories of his day: quantum mechanics, stereochemistry, state-of-the-art analytical methods, etc., all within the framework of highly rigorous research programs. In collaboration with **Roald Hoffmann**, from Cornell University, he established the rules which bear their names and which allow a construction of organic molecules in the correct configuration. During the course of his brilliant career, Woodward synthesized numerous important molecules such as quinine (used in the treatment of malaria), cholesterol, cortisone, strychnine, colchicine, reserpine (a substance used for the treatment of psychiatric disorders), chlorophyll as well as tetracyclin (an antibiotic). In 1965 he synthesized cephalosporin C and, in 1971, vitamin B_{12}. Such total syntheses also proved the structures of these molecules. In principle, all the chemical know-how required for the elaboration of practically any natural mo- lecule as well as to produce artificial molecules displaying new, unexpected properties now became available. It should be mentioned, however, that the total synthesis of such complex molecules is very expensive: their production on an industrial scale can be prohibitive so that pre- ference is often given to natural substances which can be extracted from plants or animals.

1966

Mulliken, Robert Sanderson (Newburyport, Masachusetts, June 7, 1896 - Arlington, Virginia, October 31, 1986). American chemist. Son of a chemist. Nobel Prize for Chemistry for "fundamental work concerning chemical bonds and the electronic structure of molecules." Mulliken graduated from the Massachusetts Institute of Technology, Boston, and from the University of Chicago where he defended his doctoral thesis in 1921. In 1922 he was the first to imagine separating isotopes by centrifu- ging gaseous forms. This technique has aroused fresh interest as a means of en-riching natural uranium in its isotope ^{235}U, because it is less energy-consuming than gas diffusion. Mulliken then focused on molecular spectra and the application of quantum mechanics to decipher the electronic sta-tes of molecules. With Hund he developed the molecular orbital theory in order to explain the covalent chemical bond. Instead of representing atomic orbitals as systems that do not change during the formation of chemical bonds, he proposed that molecular orbitals are formed through linear combination of the orbitals of the atoms that are uniting. He thus developed a mathematical description of the bond and extended the atomic orbital model to the molecule. According to his model, which remains the basis of current developments in quantum chemistry, the electrons in a molecule can "feel" the field of all the nuclei in the molecule and their behaviour can be described as a combination, also called hybridisation, of

Robert Mulliken

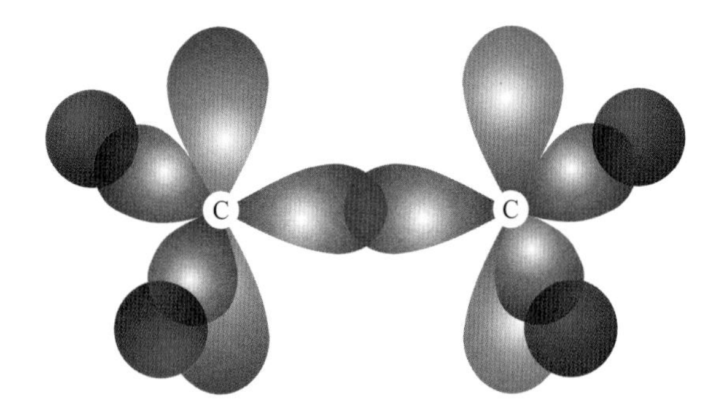

Atomic orbitals in ethylene. These orbitals are arranged as they would be for combination into four C - H single bonds, one C-C bond of σ symmetry, and one c - c bond of π symmetry. This bond model predicts a H - C - H bond angle of 120°, close to the actual value of 117°.

the original atomic orbitals. Mulliken also focused on the concept of electronegativity, i.e. the ability of an atom within a molecule to attract the electrons of the structure. He showed that the electronegativity parameter can be calculated by a very simple formula: $EN = 1/2\ (E + I)$, where I is the atom's ionisation energy and E is its electron affinity. During World War II, Mulliken took part in the Plutonium Project.

1967

Eigen, Manfred (Bochum, Germany, May 9, 1927 -). German chemist. Son of a musician. Nobel Prize for Chemistry with **Ronald George Wreyford Norrish** and **George Porter** for work on extremely rapid chemical reactions. Manfred Eigen invented a so - called "relaxation technique" for studying chemical systems at equilibrium by disturbing the equilibrium with pulses of high - frequency sound waves. Thanks to this method he was able to monitor reactions on a microsecond time scale. This method has been widely used the study of biochemical reactions catalyzed by enzymes. It also proved to be valuable in the studies of problems concerning the genetic code. His scientific interest focused almost exclusively on problems concerning evolution. In 1971, he published a pioneering paper dealing with the concepts of "self - organization of matter" and "evolution of biological macromolecules". These concepts have since become classical concepts in the field of evolution. Initially his thoughts on molecular evolution, and thus the development of life on earth, centered around self - replicating molecules that existed about 3.8 billion years ago. From the very beginning these molecules must have possessed such diverse structures that they were subject to the process of natural selection. Most recently his interest has focused on the technological utilization of ideas concerning evolution. By employing the so - called evolution machines that utilize the principles of biological evolution, new compounds can

Manfred Eigen

become optimally adapted for particular functions. In the late 1960s, Rudolf Rigler and Manfred Eigen conceived a new analytical method called FCS, for fluorescence correlation spectrometry, for detecting single molecules. FCS is now currently employed for studying molecular evolution. Eigen has named this new research field "evolutionary biotechnology". In 1993, in collaboration with co-workers and colleagues, he initiated the foundation of the firm EVOTEC Biosystem GmbH in Hamburg. This firm is investigating technological applications of evolutionary biotechnology. This includes pharmacoscreening, molecular diagnostics (e.g. of viruses) as well as the evolutionary optimization of agents. After obtaining a Ph.D. in physics, Manfred Eigen began his scientific career in 1951 at the Institute for Physikalische Chemie at the University of Göttingen. In 1958 he was elected a scientific member of the Max-Planck-Gesellschaft, now called Max-Planck-Institute for Biophysikalische Chemie. He became Head of the Department of Chemical Kinetics in 1962, and director at the institute, in 1964.

Norrish, Ronald George Wreyford (Cambridge, England, November 9, 1897 - Cambridge, June 7, 1978). Nobel Prize for Chemistry, along with **Manfred Eigen** and **George Porter** for his studies of very fast chemical reactions. Ronald Norrish earned his Ph.D. in chemistry at the University of Cambridge in 1924, under E.K.

Ronald Norrish

Rideal, who stimulated his interest in studying the kinetics of photochemical reactions. It was at Cambridge that he spent all his career. Observations on diverse molecules led him to discover that light can decompose ketones, aldehydes, and other molecules and that among the decomposition products there are "free radicals", that is unstable chemical species characterised by the presence of unpaired electrons. These facts led him to study chain reactions, and notably the reaction of hydrogen with chlorine. The chemists **Norman Hinshelwood** and **Nikolaï Semënov** had already begun to study this field. Norrish's work led him quite naturally to focus on branched - chain combustion reactions propagated by free radicals, as imagined by the Russian scientist. Norrish and his collaborators discovered that formaldehyde was a necessary intermediary in combustion mechanisms. This toxic molecule can be produced by incomplete combustion of hydrocarbons. It is one of the undesirable substances detected in automobile engine exhaust. Being photosensi-

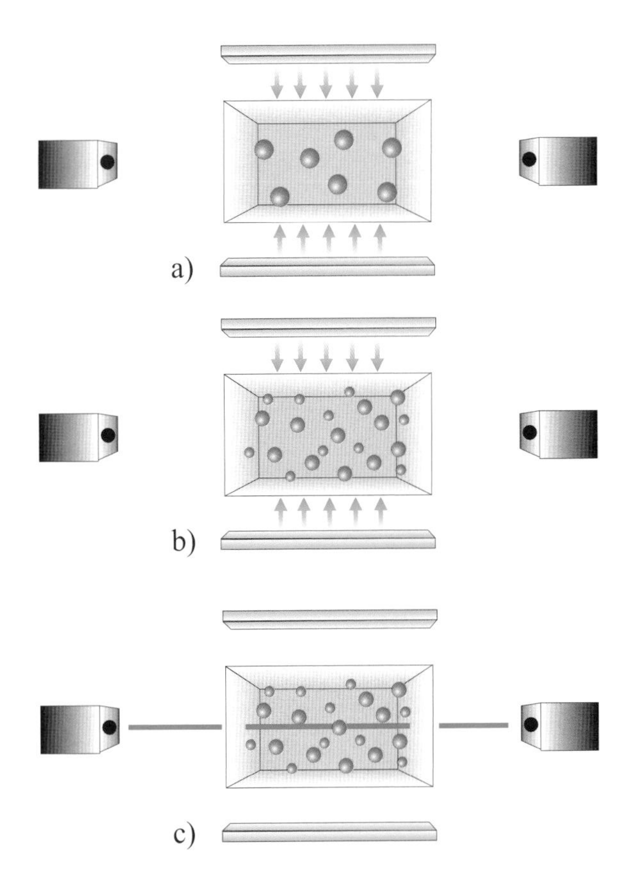

Flash photolysis. A reaction mixture is submitted to a brief, but intense luminous excitation. **(a)**. The transitory species produced by this excitation **(b)** are analysed by spectroscopy of a light beam which is transmitted across the mixture a few fractions of a second after the excitatory "flash" (**c**).

tive, some of these substances undergo photolysis and thus participate in the phenomenon of "photochemical smog". His experiments on photolysis are what brought Norrish international fame and the Nobel Prize. After the war, and with the help of George Porter, he elaborated the so-called "flash photolysis" technique which allowed them to study the mechanism of very fast reactions. The principle was to subject a photosensitive reaction mixture to a brief but intense flash of light. This operation creates unstable species (radicals) characterized by a very short lifetime. Immediately after the flash, within micro - to milliseconds, the chemical species present in the gas are analysed spectrometrically by means of a second, low - intensity flash directed perpendicularly to the excitatory flash. The intensity of the second flash is measured at certain characteristic wavelengths so as to determine the amount of light absorbed by the reaction species present in the excited gas.

Porter, George (Stainforth, England, October 6, 1920 -). British chemist. Nobel Prize for Chemistry, along with **Manfred Eigen** and **Ronal Norrish** for development of a technique for studying very fast chemical reactions. Son of a railroad employee. George Porter entered the University of Leeds in 1938 and studied chemistry as well as electronics and electromagnetic waves. After World War Two, in which he served as a radar operator, he began a doctorate at the University of Cambridge under Norrish, who had begun to develop flash photolysis. He earned his Ph.D. in science in 1949. Thanks to very brief light flashes, Norrish and Porter were able to study reactions occurring in a few millionths of a second. They detected free radicals, i.e. highly reactive, short-lived chemical species

George Porter

essential to the propagation of extremely fast reactions such as flames, explosions, etc. Flash photolysis was to contribute considerably to the elucidation of the detailed mechanisms of these reactions, of great importance in energy production. Porter was appointed professor at the University of Sheffield and in 1963 he headed the Department of Chemistry. It was also at the University of Sheffield that he used flash photolysis for studying the interactions between oxygen and hemoglobin and the mechanisms in which chlorophyll is involved.

1968

Onsager, Lars (Oslo, Norway, November 27, 1903 - Coral Gables, Florida, October 5, 1976). Norwegian - born American chemist. Nobel Prize for Chemistry for development of a general theory of irreversible chemical processes. Son of a barrister - at - law. Onsager earned his degree as a chemical engineer at the Norwegian Technical School of Trondheim in 1925. Very early in his career as a chemist and even during his studies, he became interested in Arrhenius's hypothesis of electrolytic dissociation into ions in aqueous solution. Hückel and the Dutch chemist **Peter Debye** had just improved this theory, suggesting that ions pre-exist in the solid before it is dissolved. They had deduced an equation that

Lars Onsager

fit almost exactly the behaviour of ions in solution. Onsager proposed an equation that markedly improved the mathematical description of electrolytic conduction. Impressed by this success, Debye offered the young Norwegian a position in his Zurich laboratory. Onsager was to occupy this position for two years before emigrating to the USA in 1928. He first taught at Johns Hopkins, then at Brown University and finally at Yale. It was at Brown University that he developed the rationale that led him to lay the foundations of irreversible process thermodynamics. As he was trying to generalise his earlier research on the movement of ions in solution, he developed his "reciprocal relations". In 1931 he formulated a more general law applicable to non-equilibrium situations, whereas classic thermodynamics deals solely with phenomena at equilibrium. At first, the reciprocal relations did not arouse much interest among physical chemists. Not until after the war were they recognised as a basis for elucidating mechanisms governing processes such as electro-osmosis and thermodiffusion, and also various biolo- gical phenomena. Onsager can be considered as the author of the fourth law of thermodynamics. Onsager's scientific production reflects the energy he put into his scientific thought. His training as a theoretical chemist and his capacity to develop mathematical models to describe the phenomena he explored opened to him an impressively wide field of investigation, spanning subjects as diverse as phase transitions in solids, low-temperature physics, the diamagnetic properties of metals, biophysics, and radiation chemistry.

1969

Barton, Sir Derek Harold Richard (Gravesend, England, September 8, 1918 - College Station, Texas, March 16, 1998.) British chemist. Son of a carpenter. Nobel Prize for Chemistry with **Odd Hassel** for his work in the field of conformational analysis. Derek Barton was educated at the

Derek H. Barton

Imperial College, London. Till 1978 he held post as a professor of organic chemistry at Birbeek College. The same year he headed the "Institut de Chimie des Substances Naturelles", Gif-sur-Yvette, France. Finally, in 1986, he took a distinguished professorship at Texas Agricultural and Mechanical

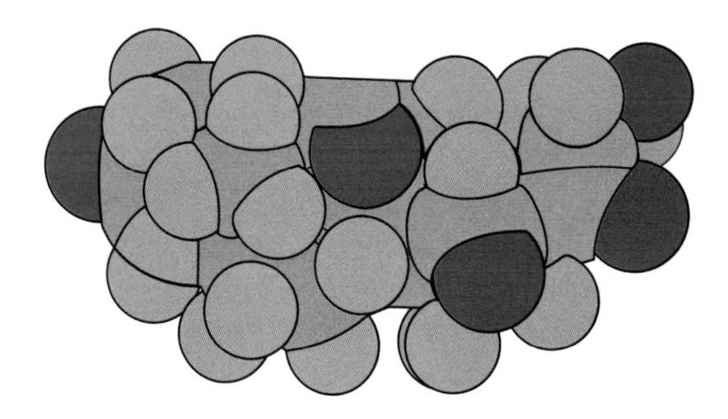

Aldosterone. One of the hormones secreted by the adrenal cortex. Isolated by **Tadeus Reichstein**, its structure was determined by Derek Barton. Its main action is to increase reabsorption of sodium ions by the distal renal tubules and thus to regulate water and electrolyte metabolism.

University, Texas. Barton extended his research into the structural analysis of complex molecules of natural origin: terpenes, steroids and alkaloids. The results he obtained constitute an impressive quantity of work. He then proceeded to the study of biogenesis and the biosynthesis of secondary metabolites. He left aside the indole alkaloids on which many others worked, including, in particular, another giant in chemistry **Robert Robinson**. Instead, he worked on other "sensitive" alkaloids:morphine and its derivatives, colchicine and its derivatives (or precursors). His great flair combined with his encyclopaedic knowledge of chemistry, allowed him to succeed quickly where others before him had strayed or failed. The work he performed on the biosynthesis of alkaloids remains one of the historic steps in the development of modern chemistry. Based on the oxidative reactions of complex phenolic systems, the biogenic hypothesis which Derek Barton put forward at that time were almost all confirmed either at once by himself or much later by others. Conformational analysis, structural analysis, and the biosynthesis of complex natural products therefore represent the major part of his work. Professor at Imperial College (1978), to which he had returned after his time first at Birkbeck College, London, and then in the Department of Chemistry at the University of Glasgow, he decided to return and to take up a new challenge, that is to direct the "Institut de Chimie des Substances Naturelles du CNRS", Gif-sur-Yvette, at the invitation of the CNRS. During the eight years spent there he had already extensively studied the radical reaction which he helped to make known and to tame. Derek Barton also accomplished another gigantic task:that of publications of all kinds involving organic chemistry and its applications in biological chemistry, therapeutic chemistry and biology. In 1984, he took a very interesting offer by the State of Texas at Texas Agricultural and Mechanical University at College Station.

Hassel, Odd (Oslo, Norway, May 17, 1897 - Oslo, Norway, May 11, 1981). Norwegian physical chemist. Nobel Prize for Chemistry, with **Derek H. R. Barton**, for his work in establishing conformational analysis. Son of a gynacologist. In 1920, Odd Hassel joined the University of Berlin and gained his Ph.D. the same year. During his stay in this city, he became familiar with the technique of X - ray diffraction applied to the analysis of the crystallographic structure of solids. In 1925 he returned to Oslo and was appointed Associate Professor. In 1930 he began the study of the tridimensional structure of organic molecules and, in particular, of cyclic organic molecules. X - ray diffraction as he had used it in Berlin was applicable only to the analysis of solids, but in 1938 he received an electron diffractometer and could begin to study gases and liquids. Hassel then demonstrated what he had suspected fifty

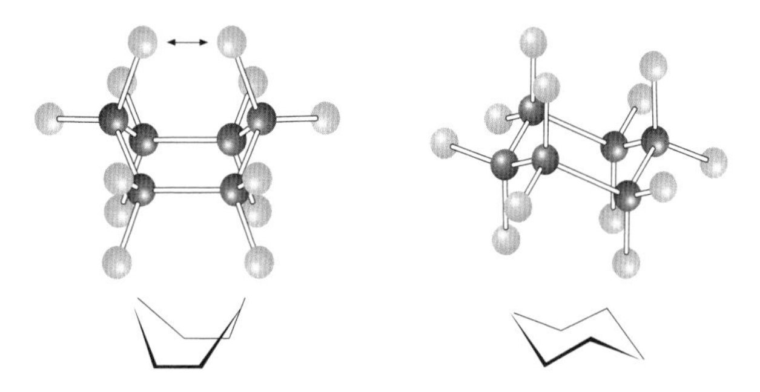

Chair and boat conformations of pyranose rings. These common names designate two distinct spatial conformations for compounds that have at least a saturated ring with six carbon atoms. Chair and boat conformations are interconvertible. The chair form is energetically the most favorable.

years earlier: that cyclohexane can exist in two forms, the "boat form" and the "chair form". This discovery allowed him to explain the properties of molecules derived from cyclohexane. He thus associated himself with the work of Barton on the three - dimensional structure of molecules. In 1943 after the university was closed by the German Army, he was sent to a concentration camp near Oslo. This did not stop him from continuing his studies. He spent the post-war years studying charge transfer compounds and perfecting their description. In 1950 he devoted himself to the study of halogenated organic compounds. Odd Hassel wrote a book entitled " Crystal Chemistry", a classical work about X - ray diffraction, published in 1934.

1970

Leloir, Luis Frederico (Paris, France, September 6, 1906 - Buenos Aires, Argentina, December 2, 1987). Argentine biochemist. Nobel Prize for Chemistry for his investigations of the processes by which carbohydrates are converted into energy in the body. Son of a ground landlord. Despite using poor equipment, Luis Leloir managed to surprise his European and American colleagues. He showed that the synthesis of glycogen in vivo does not follow exactly the way described by Carl Cori and his wife, Theresa Cori. These resear-rchers had discovered the mechanism of glycogen breakdown to lactic acid in vitro and the mechanism by which lactic acid is converted into glucose. They had concluded that these events also take place in vivo. Leloir showed that this pattern was not correct in vivo and announced the discovery of a new enzyme, uridine triphosphate, UTP , which reacts with glucose 1 - phosphate to form a sugar - nucleotide called uridine diphosphate glucose, UDPG. In the presence of a specific enzyme and a chain of sugars used to prime the reaction, uridine diphosphate glucose can add glucose to the growing chain while regenerating uridine diphosphate. The latter is then converted to UTP by ATP hydrolysis. It later became clear that this is indeed how glycogen is synthesised in vivo. Luis Leloir graduated from the University of Buenos Aires in 1932. There, he started his scientific career under Bernardo Houssay. He also conducted research at the Laboratory of Biochemistry of Cambridge University in England. This laboratory was then directed by Frederick Gowland Hopkins.

Luis Leloir

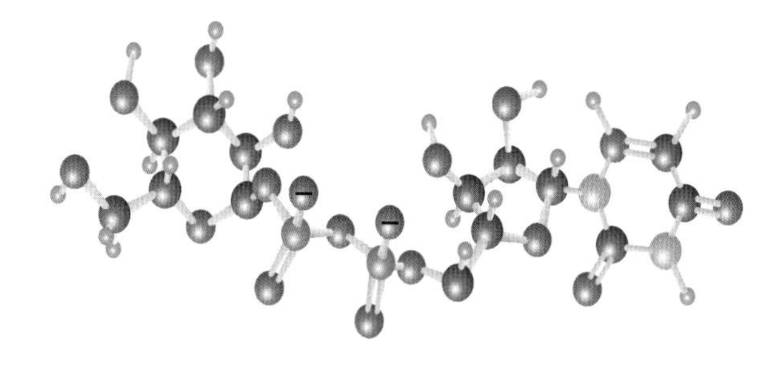

UDP-glucose. Sugar nucleotides like UDP - glucose serve as carriers for introducing glucose into polymers such as starch, glycogen or cellulose and more complex macromolecules. Their role as sugar carriers was established by **Luis Frederico Leloir**.

1971

Herzberg, Gerhard (Hamburg, Germany, December 25, 1904 - 1999). Canadian physicist. Nobel Prize for Chemistry for his work in determining the electronic structure and geometry of molecules. Herzberg received his Ph.D. from the Technological Institute of Darmstadt in 1928. He was interested in questions concerning interactions between electromagnetic rays and matter. After a working stay at Göttingen and Bristol, he returned to Darmstadt in 1930. Because of his Jewish origins he was forced to leave Germany in 1933. After working at the University of Saskatchewan he was granted a Canadian citizenship in 1945. From 1945 to 1948 he conducted research at the Yerkes Observatory, United States, before returning to Ottawa, Canada, where he founded a laboratory of spectroscopy for the National Research Council. Since its beginning, spectroscopy was a key tool in the study of matter, in particular for the study of atoms. It allowed **Niels Bohr** to elaborate his atomic model in 1913. With incontestable success, Herzberg applied this technique to the study of two - atom molecules like nitrogen, oxygen, and hydrogen, whose spectra are much more complex than those of atoms. He succeeded in determining the energy le -vels of electrons in molecules. Using new experimental techniques, he also examined the properties of free radicals, (i.e. short-lived intermediates in certain chemical reactions). During his stay in Yerkes, he was the first to identify certain free radicals in interstellar gases.

1972

Anfinsen, Christian Boehmer (Monessen, Penn-sylvania, March 26, 1916 - Randallstown, Maryland, May 14, 1995). American biochemist. Received the Nobel Prize for Chemistry, along with **Stanford Moore** and **William H. Stein**, for research on the molecular structure and biological functions of proteins. A great music-lover, Anfinsen also showed interest in chemistry. In the course of his career he held positions at several prestigious institutions. Among them, the prestigious Harvard Medical School, the Nobel Medical Institute of Stockholm, the National Institutes of Health, Bethesda, the Weismann Institute of Sciences, Israël, and the Johns Hopkins University. One of the first questions that Anfinsen tackled was how a protein synthesised in a linear form, and particularly an enzyme, can rapidly adopt its correct spatial configuration enabling it to function. This question was pertinent because, with all

Christian B. Anfinsen

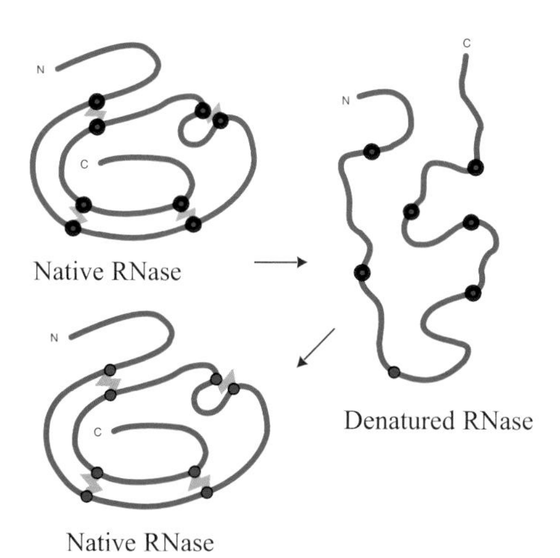

Protein denaturation and renaturation. Christian Anfinsen and his collaborators clearly showed that if the disulfide bonds of pancreatic ribonuclease are reduced to form sulfhydril groups in the presence of urea, the enzyme is denatured and the enzymatic activity is lost. However, the removal of the urea and addition of mercaptoethanol allow the reformation of the native enzyme.

the bonds amino acids can form with each other, an amino acid chain should in principle be able to adopt multiple folds, most being non-functional. It was to solve this problem that around the 1950s he focused on ribonuclease, a 124 - amino-acid enzyme capable of degrading RNA chains. Stanford Moore and William Stein sequenced this enzyme in 1960. By subjecting ribonuclease to treatments that cleaved the disulfide bonds by reducing them to sulfhydryl (SH) groups, he obtained several forms of the protein in which the enzymatic activity of the protein was completely lost. But Anfinsen made the observation that this activity could be recovered if the denatured protein was incubated under native conditions that allowed the polypeptide chain to spontaneously refold and disulfide bonds to reform. Verified on many other substances, these results showed that the primary sequence of a protein determines its functional configuration. Other work also marked Anfinsen's career. Let us cite in parti- cular the works that led to the discovery of interferon and its characterization. The role played by this substance in the immune defense against viruses is well known.

Moore, Stanford (Chicago, Illinois, September, 4 1913 - New York City, New York, August 23, 1982). Nobel Prize for Chemistry, along with **Christian Boehmer Anfinsen** and **William Howard Stein** for their research on the molecular structures of proteins, especially ribonuclease. Stanford Moore studied chemistry at the Vanderbilt University, Tennessee but received his Ph.D. at the University of Wisconsin in 1938. He then joined Max Bergmann laboratory at the Rockefeller Institute for Medical Research, New York. Moore's main interest was the structure of proteins. With Stein, he began a long and highly fruitful collaboration. From 1942 to 1945, he took part in American military research on "mustard gas". In 1945, he returned to the Rockefeller Institute where the two researchers resumed their collaboration. With Stein, he elaborated new techniques of chromatography in order to analyse amino acids and peptides obtained by the hydrolysis of proteins. They also improved column chromatography by filling the columns with an ion-exchange resin, much more selective in separating different fractions. In 1949, they succeeded in separating the amino acids of blood and urine. In 1950, Moore taught chemistry at the University of Brussels. He created there a laboratory of amino acid analysis. As "Visiting Professor" in Cambridge, he worked with **Frederick Sanger** in 1951 to determine the amino acid sequence of insulin. He and Stein developed an automatic amino-acid analyser and, in 1959, they succeeded in determining the amino acid sequence of an enzyme, ribonuclease. In 1959, Stanford Moore returned to Rockefeller University where he resumed his studies on deoxyribonuclease, the enzyme that degrades DNA.

Stanford Moore

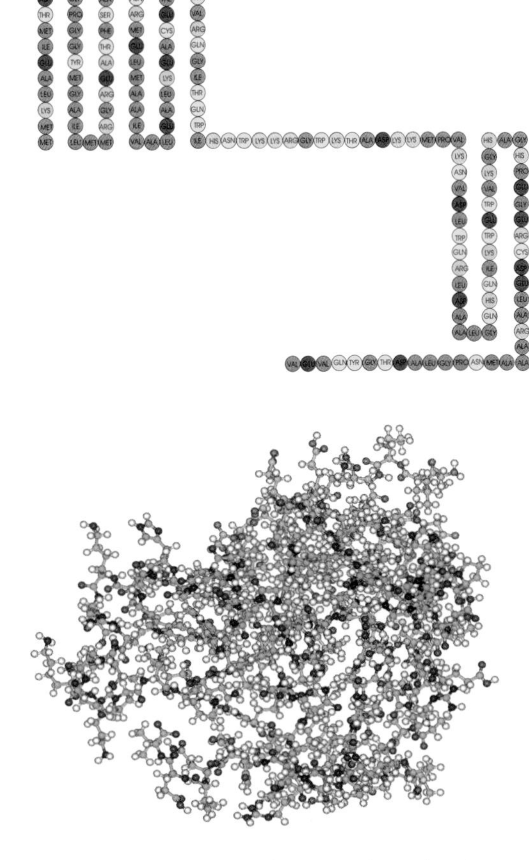

Primary structure (above) and **three - dimensional structure** (below) **of ribonuclease**. The primary structure is the amino acid sequence of the protein. After being synthesized, protein chains fold into their characteristic three - dimensional conformation.

Stein, William Howard (New York City, New York, June 25, 1911 - New York City, February 2, 1980). American biochemist. Son of a tradesman. Nobel Prize for Chemistry, along with **Stanford Moore** and **Christian . Anfinsen** for their studies of the composition and functioning of ribonuclease. William Stein graduated from Harvard in 1933 and specialised in inorganic chemistry, then completed his training at the University of Columbia, New York, where he received his Ph.D. in 1938. His thesis concerned the amino acids composing elastin, a protein component of veins and arteries. This research marked the start of a career devoted to studying the chemical function of proteins. At this point he set out to improve techniques for purifying amino acids. In 1939 he was joined by Moore, with whom he began a long collaboration. They first focused on the analysis of glycine and leucine. After an interruption during World War Two, these two researchers set to work on techniques such as partition chromatography, column chromatography, and ion exchange. With Moore, Stein automated chromatography, notably by inventing a sample collector. Their work turned chromatography into an irreplaceable, high-performance tool used wordlwide in amino acid research. In 1969, Stein contracted a disease that made him tetraplegic.

William Stein

1973

Fischer, Ernst Otto (Solln, Germany, November 10, 1918 -).German chemist. Son of a professor of physics. Shared the Nobel Prize for Chemistry with **Geoffrey Wilkinson** for his identification of a completely new way in which metals and organic substances can combine. Ernst Otto Fischer studied at the University of Technology of his hometown, received his diploma in chemistry in 1949 and three years later obtained the degree of Dr.rer.nat., became private lecturer in 1954 and was appointed associated professor of Inorganic Chemistry at the University of Munich in 1957. From 1964 until his retirement he was a full professor and holder of the chair of Inorganic Chemistry at the Munich University of Technology. His scientific work focused on an area hith-

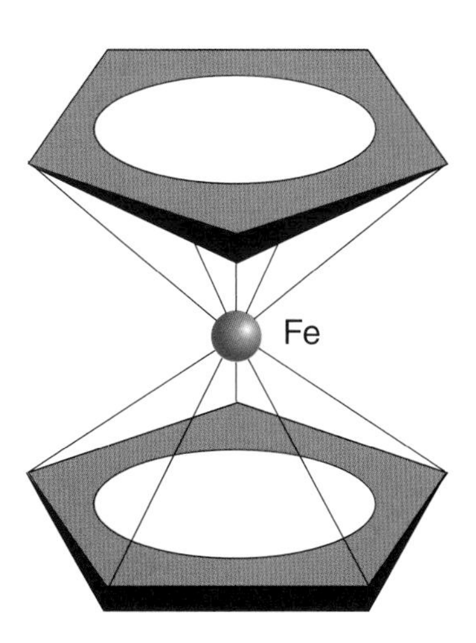

Ferrocene. The best known of the so - called sandwich - compounds synthesized by Otto Fischer. Here, the metal atom is iron and the formula is $(C_5H_5)_2Fe$.

erto considered rather exotic in character:metallo-organic compounds. It was mainly thanks to the studies undertaken by **Otto Fischer** and his British colleague Geoffrey Wilkinson that these substances were wrested from oblivion and gained considerable importance fairly rapidly. Independently, and more or less by coincidence, Fischer discovered that many organo-metallic substances consist of complex molecules in which the central atom, (e.g. iron), is surrounded by slice or sandwich - like hydrocarbon rings. The "symmetric "coordinative" bond to the iron atom is nowadays called π - complex. Ernst Otto Fischer has also been a renowned university lecturer, who has been awarded four honorary doctorates.

Wilkinson, Geoffrey (Todmorden, England, July 14, 1921 - London, England, September 26, 1996). British chemist. Nobel Prize for Chemistry with **Ernst Otto Fischer** for their independent work in organometallic chemistry. Geoffrey Willkinson graduated in chemistry from the Imperial College in London and earned his Ph.D. in 1946 after a working stay in Canada. He then conducted research at the Massachusetts Institute of Technology, Boston, and at Harvard before rejoining the Imperial College where he was appointed Director of the Institute of Organic Chemistry. His research focused first on the study of complexes having a metal-hydrogen bond. He used such compounds to develop a catalyser for hydrogenating double bonds between carbon atoms (present in olefins). This catalyser, which bears Wilkinson's name, is also used industrially to synthesise aldehydes from alkenes in the presence of hydrogen and carbon monoxide. For ferrocene, discovered in 1951, Wilkinson proposed a structure in which the iron atom is sandwiched between two cyclopentadiene cycles. Bonds are formed between the iron atom, which has empty orbitals, and all the carbon atoms of the two pentadiene cycles, via the electrons of the *pi* bonds.

Geoffrey Wilkinson

1974

Flory, Paul J. (Sterling, Illinois, June 19, 1910 - Big Sur, California, September 9,1985). American physical chemist. Son of a clergyman. Nobel Prize for Chemistry for his investigations of synthetic and natural macromolecules. In 1934 Paul Flory gained his Ph.D. in chemistry at Ohio State University. After having worked in the industry, du Pont de Nemours, Standard Oil, GoodYear Tire Company, he joined Cornell University and then Stanford University. His research aimed at understanding the properties of high polymers. Flory established relationships between the physical properties of polymers and their chemical structure. He also studied processes for synthesising these molecules, leading to the development of methods for industrial-scale production. Nylon and synthetic rubber production notably benefited from his work. Some of his books, namely "Principles of Polymer Chemistry" and "Statistical Mechanics of Chain Molecules" were best-sellers in the field of chemistry.

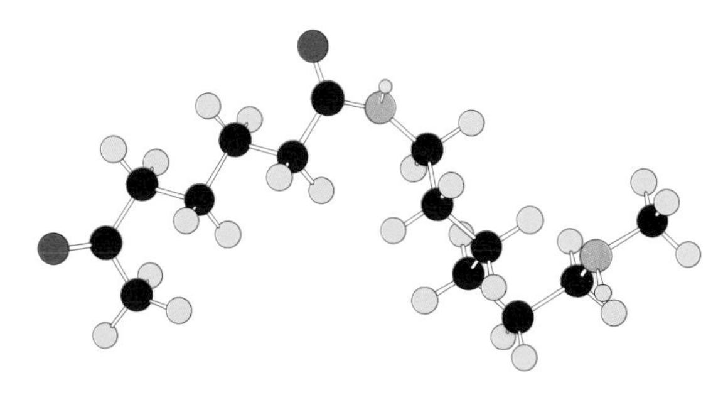

Structure of nylon. This class of substances are long-chain polyamides. They are widely used in industry, for example for the manufacture of laboratory apparatus or as an alternative to nitrocellulose required in hybridization techniques.

1975

Cornforth, Sir John Warcup (Sidney, Australia, November 7, 1917-). Australian chemist. Nobel Prize for Chemistry shared with **Vladimir Prelog** for his work on the stereochemistry of enzyme - catalyzed reactions.

Working in London between 1946 and 1962, he developed the techniques required for the study of the stereospecificity of enzymes. Another major contribution was the development of a technique initially elaborated by **Georg de Hevesy** and which allowed the use of hydrogen radioactive isotopes to trace the different steps of a metabolic pathway in cells. With this method he succeeded in elucidating the 12 steps leading to the synthesis of squalene, a cholesterol precursor widely distributed in nature. This molecule had been previously identified by **Konrad Bloch** during the 1950s. Some years before these achievements he had been involved in the synthesis and the description of a variety of natural substances such as penicillin in collaboration with **Robert Robinson**. Handicapped by early deafness, Cornforth could thank his wife, who also had a Ph.D. in science, for helping him in his work. She even signed with him a major paper on squalene. As the deafness became more severe, this help became all the more precious. It was essential to maintaining contact with his fellow scientists. In 1972, Cornforth was appointed director of the Milstead Laboratory of Chemical Enzymology for Shell Research. In 1974, he became a member of the Royal Society and was knighted by the Queen of England in 1977.

John Cornforth

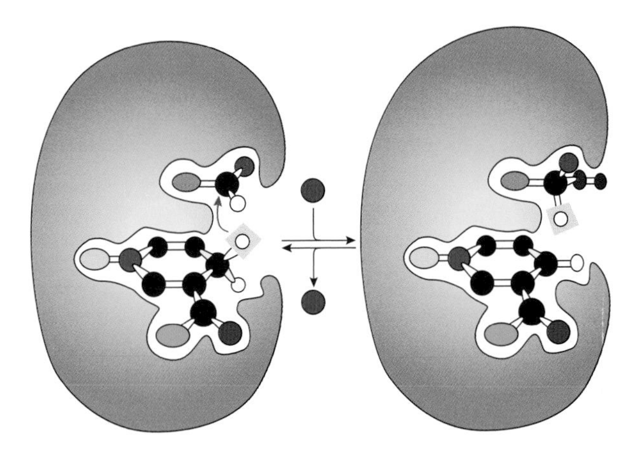

An example of stereospecificity, the stereospecificity of hydride transfer in dehydrogenases as a consequence of the asymmetric nature of the active site.

Prelog, Vladimir (Sarajevo, Austria-Hungary, July 23, 1906 - Zurich, Switzerland, January 7, 1998). Swiss chemist. Son of an historian. Nobel Prize for Chemistry with **John W. Cornforth** for his work on the stereochemistry of organic molecules and reactions. Vladimir Prelog graduated in chemistry from the Technological Institute of Prague and earned his Ph.D. in sciences in 1929 under Votocek. This scientist had introduced him to the problem of the structure of organic molecules. Prelog was offered a teaching position at the University of Zagreb in 1935, but the meager resources at his disposal forced him to take a second job in a pharmaceutical company, where he was initiated to the synthesis of organic molecules. In 1941, thanks to **Leopold Ruzicka**, Prelog settled down with his family in Zurich, Switzerland and was appointed professor at the Federal Polytechnic Institute in 1945. His work was then subsidised by the CIBA Corporation, a giant of the pharmaceutical industry. He was the fifth Director of the Polytechnic School to receive a Nobel Prize. After the war he travelled extensively, forming collaborations with scientists in many countries (notably **Derek Barton** and **Robert Woodward**). He focused for a while on natural alkaloids, then on enzymes and antibiotics. This research led to the synthesis of rifamycin, an antibiotic used in the treatment of tuberculosis and leprosy. In 1960 he became a member of the Board of Directors of CIBA. Owing to the extraordinary development of the spectroscopical techniques Prelog devoted himself to the determination of the spatial structures of essential biochemical molecules. He noticed that slight variations in spatial orientation generated major differences in chemical properties. The chemistry of such molecules does not merely reflect their elemental composition, but also their organisation and geometry. This means that their three - dimensional architecture must be taken into account in attempts to synthesise them. Prelog realised that it was necessary to describe these molecules in terms of all their properties, including their 3D spatial structure. In collaboration with Cahn and Ingold, he elaborated a stereochemical nomenclature called the CIP system, which is now used universally to describe complex molecules such as those playing an essential role in life phenomena.

1976

Lipscomb, William N. (Cleveland, Ohio, December 9, 1919 -). American chemist. Nobel Prize for Chemistry for his research on the structure and bonding of boron compounds and the general nature of chemical bonding. William Lipscomb graduated from the University of Kentucky in 1942 and joined the California Institute of Technology where he gained his Ph.D. in 1946 under the direction of Linus Pauling. He taught at the University of

William Lipscomb

Minnesota from 1946 to 1959 and achieved his career at the University of Harvard until his retirement in 1990. Boranes, which were artificial compounds, had been first synthesized by the German chemist Alfred Stock in the beginning of the twentieth century. Stock first produced pure boranes (B_2H_6, B_4H_{10}...etc). However, the structures and chemical bonding remained unknown until the early 1940s when it was shown that B_2H_6 had two BHB units, which were described as two three-center BHB bonds. The structure of B_4H_{10}, B_5H_{11}, B_6H_{10} and many new hybrides were elucidated by the X - ray diffraction method in Lipscomb's laboratory. The extension in 1953 to three-center BBB bonds in tetrahedral B_4L_4 and other compounds led to rules for the different types of two-center and three-center bonds in these molecules. These structures and bonding descriptions took into account the one-electron deficiency of B relative to C, and explained the compact polyhedral geometry for boranes. Transformation of molecular orbitals to localized orbitals by the procedures of S.F.Boys, or of C.Edmiston and K. Ruedenberg, gave objective and persuasive support to the main features of these valence structures. Even expectations of excess electron density in the center of appropriate B3 triangles (and not always along the BB distances) was supported, and later proved in X - ray diffraction studies of $B_{10}H_{14}$ and in the B_{12} icosahedron in alpha-A_1B_{12}. Lipscomb's predictions from theoretical bonding descriptions in this new area have contributed to the very large increase in this new chemistry. More recently,

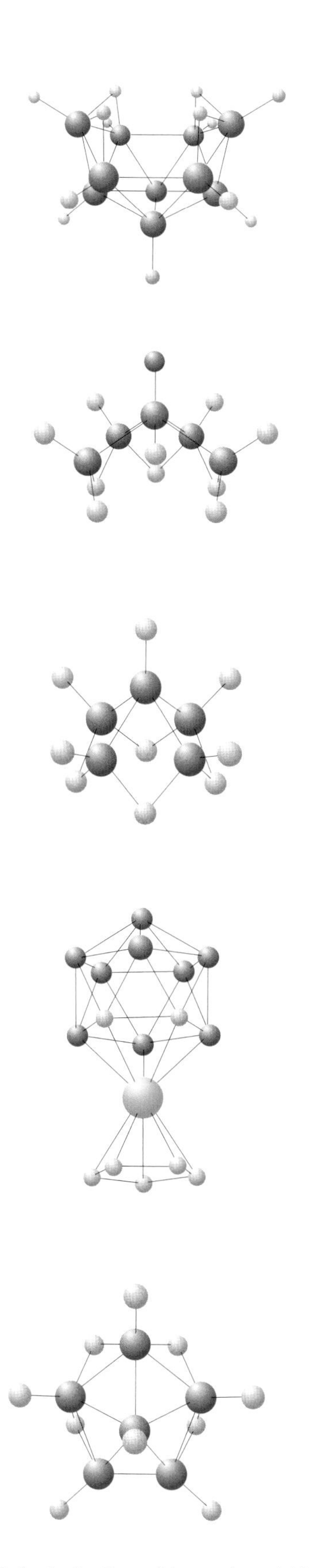

Boranes and their derivatives. Lipscomb and his associates were able to map the molecular structures of numerous boranes and their derivatives. Boranes are compounds of boron and hydrogen.

William Lipscomb has been interested in the structural analysis of proteins and enzymes.

1977

Prigogine, Ilya (Moscow, Russia, January 25, 1917 -). Russian-born Belgian physical chemist. Nobel Prize for Chemistry for contributions to nonequilibrium thermody-

Ilya Prigogine

namics. Ilya Prigogine is the son of a chemical engineer, owner of a company established in Moscow. Faced with the threat of the confiscation of their goods, the family left Russia when the young Ilya was four years old. After a short stay in Lithuania the Prigogine family found themselves in Berlin, at a time when the economic situation was deteriorating sharply. An increasingly virulent anti-Semitism pushed the family once again towards exile and they finally arrived in Brussels in 1929. The young Ilya showed a great interest in various human sciences; an excellent pianist, at one point he considered taking up an artistic career. Eventually, it was chemistry which appealed to him and so he entered the Free University of Brussels where, in 1941, he received his doctorate in chemistry with a thesis entitled:A thermodynamic study of irreversible phenomena. This thesis was carried out under the supervision of De Donder, one of the first scientists to have tackled the problem of non-equilibrium systems and who had developed a mathematical tool to study this very complex aspect of the evolution of matter. Prigogine succeeded De Donder in 1947 as ordinary professor at the Free University of Brussels. He surrounded himself with a multidisciplinary team in order to proceed with studies on thermodynamics. This discipline formulates that, in a closed system, that is without the possibility of exchanging matter and energy with the outside world, disorder can only increase with time. The result of such a formulation is that the universe had to evolve to a maximum disorder characterised by a standardised temperature, a state known as "thermal death". Initially, Prigogine tackled the study of systems not far from equilibrium, in which tiny changes can lead towards a new state of stability which is reached with production of a minimum of entropy. Later, he showed that, far from equilibrium conditions, a system can become unstable and give birth spontaneously to new organised structures. Called "dissipative structures", these systems have an order which results from self-organisation. During their evolution, states of instability or light fluctuations are created, which can grow to give birth to a new structure, the phenomenon being governed by the chance. However, the development of this model remained theoretical until 1965, the year in which a chemical reaction discovered fourteen years earlier by two Russian chemists, Biélousov and Jabotinsky, was taken into account. It concerned a cyclic reaction during which the conditions described by Prigogine's model were developed. Thereafter, many chemical reactions could be better understood thanks to this theory. The evolution of living beings in this theory gave an explanation of the mechanisms by which increasingly complex living forms can be built. Ilya Prigogine has been an emeritus professor of the Free University of Brussels since 1985. He presented the essence of his ideas in two books which have deeply influenced modern thought: The New Alliance published in 1980 and Order Out of Chaos: Man's New Dialogue with Nature, published in 1984 in collaboration with Isabelle Stengers. Prigogine is an honorary foreign member of the United States' Academy of Sciences and of the Russian Academy of Sciences.

1978

Mitchell, Peter Dennis (Mitcham, England, September 29, 1920 - Glynn, April 10, 1992). British chemist. Nobel Prize for Chemistry for clarifying how adenosine diphosphate is converted into the energy - carrying compound adenosine triphosphate in the mitochondria of living cells.

Peter D. Mitchell

Peter Mitchell joined the University of Cambridge in 1939 and specialized in biochemistry. He gained his Ph.D. in 1951 and taught biochemistry from 1951 until 1955. He then left Cambridge and developed a laboratory at the University of Edimburgh. In 1963 he retired in an ownership called "Glynn House", Cornwall, and founded the Glynn

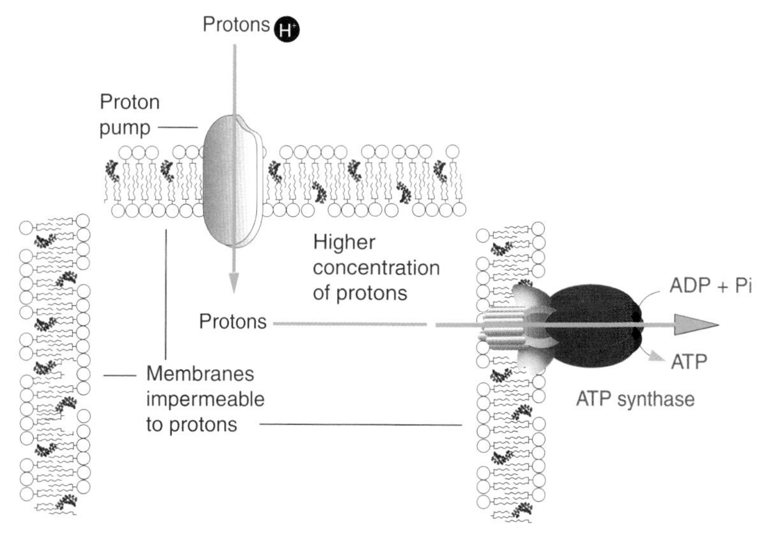

Bacterial photosynthesis. The schematic drawing emphasizes the role of the proton pump in the process of synthesizing ATP.

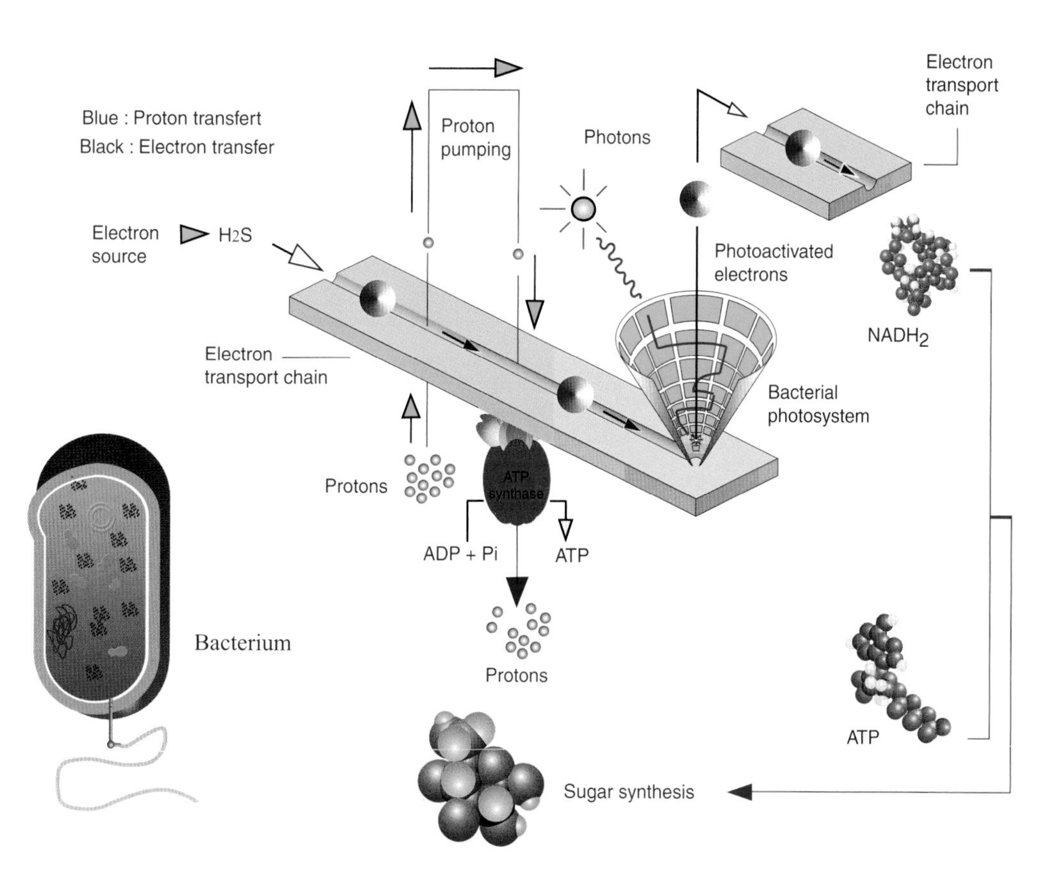

Proton pump and ATP-synthase. A transfer of protons, carried out against a concentration gradient, itself created by a transfer of electrons, establishes their increased concentration on one side of a biological membrane. Their reflux through a proton pump, carried out however by an ATP - synthase, enables this enzyme to use the energy so released to synthesize the phosphodiester linkage of ATP. The scheme presents an electron transport chain in plants and in aerobic photosynthetic bacteria, both of which are molecular oxygen - producing organisms.

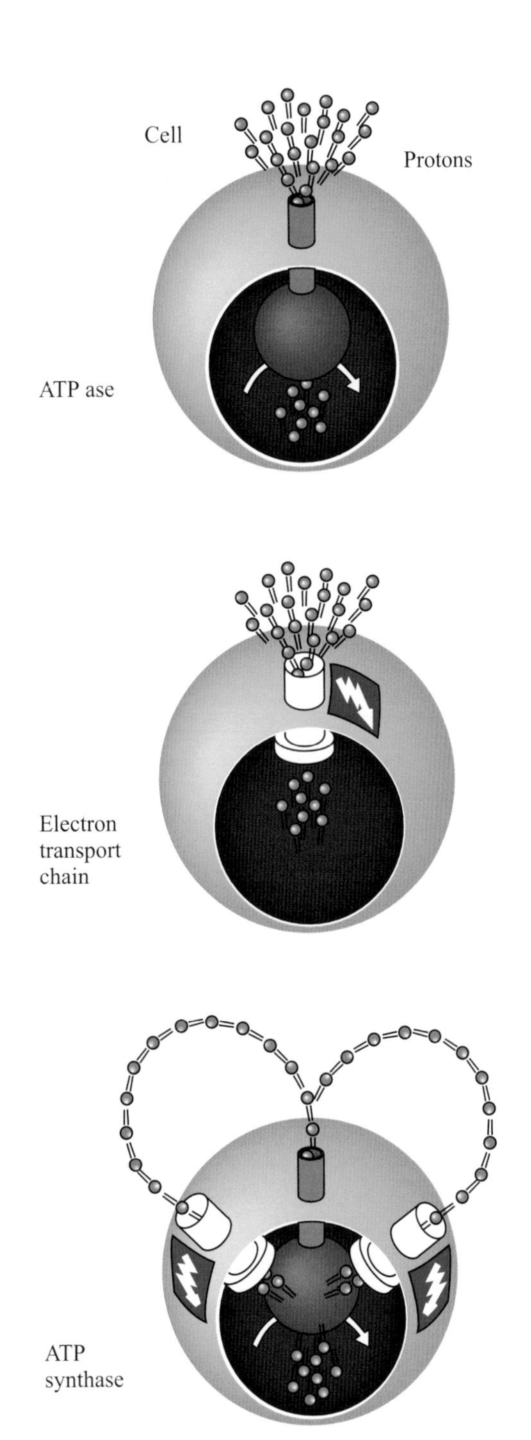

Evolution of ATP-synthases. Following the continuous acidification of the environment by fermentation products, it was necessary to prevent the cells from acidification. **Step 1**. A proton pump, which used energy from ATP hydrolysis, evolves to pump protons out of the cells. **Step 2**. Proton pumping evolves so that the energy required is furnished by an electron transport chain, as in present day organisms. **Steps 3**. The favorable gradient created by proton pumping allows protons to flow in the direction of the cytosol through ATP - driven proton pumps, thereby functioning in reverse so that they can make ATP. The primitive ATPase was thus converted into an ATP synthase.

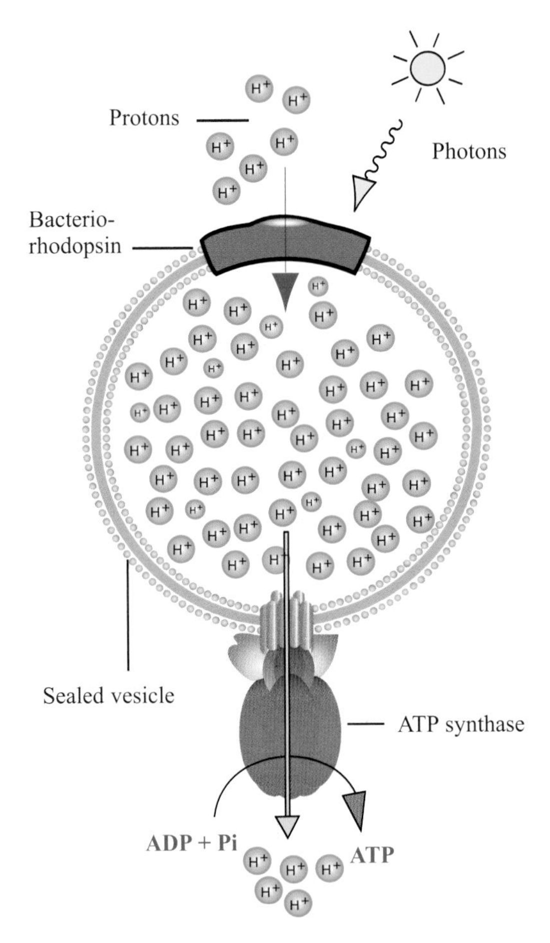

Reconstituted vesicles containing ATP synthase. Upon illumination, bacteriorhodopsin pumps protons into the vesicle and the resulting proton gradient is sufficient to drive ATP synthesis by the ATP synthase.

Research Laboratories in collaboration with Moyle. To explain how cells synthesize ATP Peter Mitchell proposed a new theory called "chemiosmotic theory". In this theory, he predicted that inner mitochondrial membrane can establish a powerful proton gradient by actively transporting protons against a concentration gradient from the mitochondrial matrix to the outside. This gradient is caused by the expenditure of energy during electron transport along the mitochondrial electron transport chain. The flow back of these electrons through the inner membrane causes temporary and reversible changes in the conformation of ATP - synthases, large protein complexes involved in the phosphorylation of ATP. This mechanism also occurs in chloroplasts. This discovery allowed understanding of the mechanism by which cells obtain energy from food or electromagnetic radiations and convert this energy in order to do work and maintain themselves.

1979

Brown, Herbert Charles (London, England, May 22, 1912 -). American chemist. Nobel Prize for Chemistry along with **Georg Wittig** for pioneering work with inorganic and organic boron compounds. Herbert Brown's parents, bearing the name Brovarnik, emigrated to the United States in 1914 and anglicized their name. Self taught and a passionate worker, the young Brown studied diboron chemistry under Schlesinger's guidance. An important first step in his scientific carer was the synthesis of sodium borohydride, a reagent often used in reduction processes in organic chemistry.Working and teaching at the University of Perdue, Indiana, he found a simple way of preparing diborane, B_2H_6. By reacting the compound diborane with alkenes, he succeeded in producing a new group of substances called organoboranes. These have the property of temporarily holding large molecules together before creating a carbon-carbon bond between them.

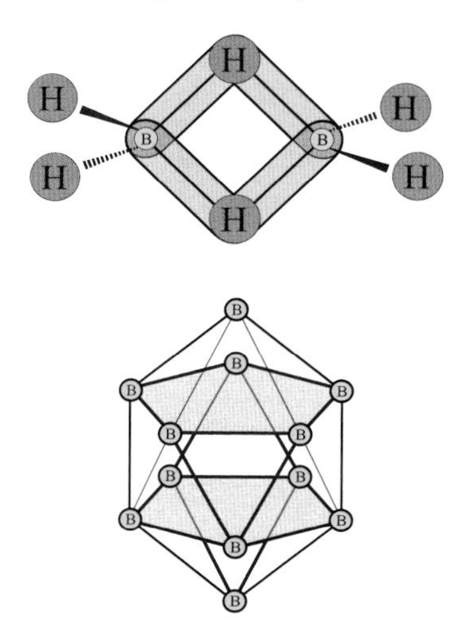

Diborane molecule. Composed of two boron atoms and six hydrogen atoms, this substance has, in reality, a complex structure in which a bridge between the two boron atoms is ensured by a combination of the orbitals hydrogen and the hybrid orbitals of boron.

Wittig, Georg (Berlin, Germany, June 16, 1897 - Heidelberg, August 27, 1987). German chemist. Son of a professor of fine art. Nobel Prize for Chemistry, with **Herbert Brown**, for his studies of organic phosphorus compounds. "Wittig's reaction" creates a double bond between two carbon atoms. Very important in organic synthesis, this reaction is routinely used to prepare high - molecular - weight compounds such as steroid, vitamin D derivatives, and pesticides (see Herbert Brown). Georg Wittig was educated at the University of Marburg in 1923 and earned his Ph.D. there in 1926. He successively taught and conducted research at the universities of Brauns-chweig, Freiburg and Tübingen. In 1956, he entered the Chemical Institute of Heidelberg, where he became emeritus in 1965.

1980

Berg, Paul (New York City, New York, June 30, 1926 -). American biochemist. Nobel Prize for Chemistry, along with **Walter Gilbert** and **Frederik Sanger**, for developing methods to map the structure and function of DNA. Paul Berg received his B.S. from Pennsylvania State University and his Ph.D. in biochemistry from Case Western Reserve University in Cleveland. After serving on the faculty of Washington University in St. Louis, he went to Stanford in 1959 where he was executive Head of the Department of Biochemistry from 1969 to 1974. It is generally believed that both accomplishments led to the explosive advances in molecular genetics and the emergence of the biotechnology industry. Berg's experiments involved introducing into the mammalian SV40 virus genome three genes from *Escherichia coli* responsible for the microbe's ability to metabolize galactose. The method of recombination relied on the ability to synthesize short homopolymeric extensions on the two linear DNA molecules to be joined: poly dA on each end of one of the molecules and poly dT on each end of the other molecule. When the two modified DNAs are mixed they join by forming short stretches of dA:dT double strands. Small gaps are filled in using appropriate enzymes

Paul Berg

and the two DNAs are covalently linked to produce a closed double stranded circular DNA containing

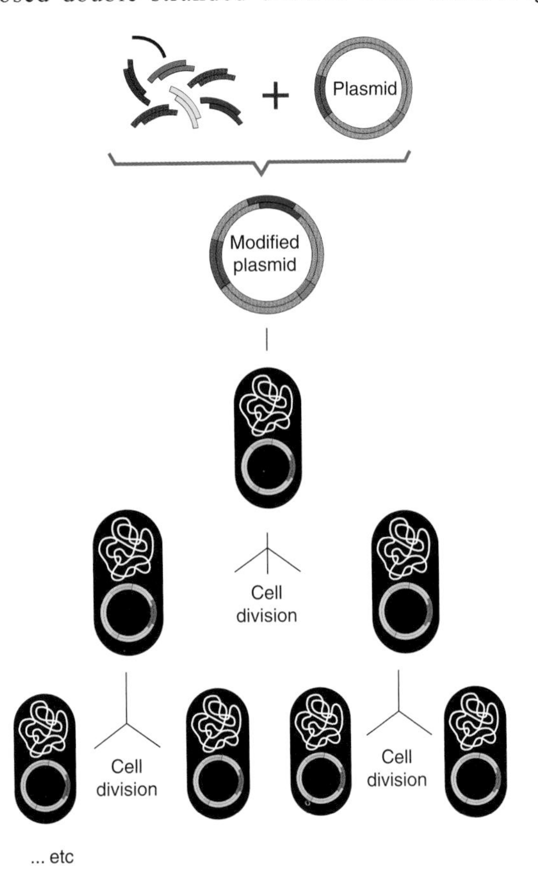

One of the numerous examples of genetic engineering. DNA fragments coming from a mammal or from man, one of which one wishes to obtain in large amounts, are introduced into a bacterial plasmid. After reintroduction in a bacteria, the plasmid vectors reproduce and amplify the DNA fragments.

SV40 virus genes and bacterial genes. Subsequently, simpler means were discovered for making "cohesive" ends on DNA and using these to join any two DNAs. This paved the way for manufacturing a large variety of recombinant DNA molecules, some being used to produce mammalian proteins (i.e. human insulin and growth hormone) in bacteria and yeast.

Walter Gilbert

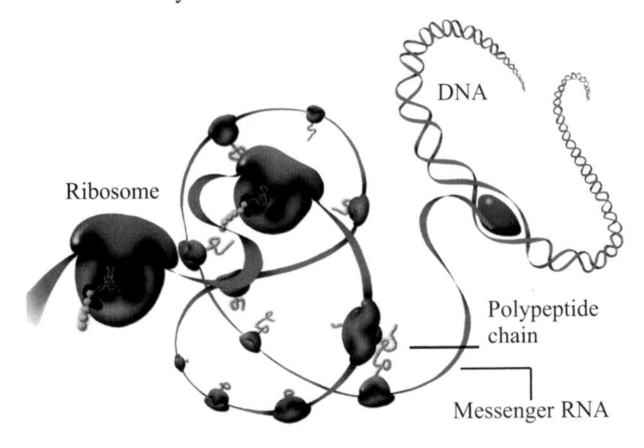

Messenger RNA and proteins. Since Jacob and Monod's 1961 hypothesis, it has been realized that RNA acts as an obligatory intermediate that directs the synthesis of proteins.

Gilbert, Walter (Boston, Massachusetts, March 21, 1932 -) American biophysicist. Nobel Prize for Chemistry, shared with Paul Berg and **Frederick Sanger**, for his development of a method for sequencing nucleic acids. The complex career of this scientist, son of a Harvard economist and trained as a mathematical physicist, he obtained his Ph.D. in physics in 1957, under the direction of Abdus Salam, (1979 Nobel Prize for Physics), followed three major lines:his collaboration with **James Watson** (co-discoverer with **Francis Crick** of the double - helix structure of DNA) and his own research aiming to isolate messenger RNA, the discovery of the repressor of the lactose operon, and the development of a DNA sequencing method. Walter Gilbert's research on messenger RNA aimed at understanding the mechanisms by which a ribosome reads the messenger. He was one of the first researchers to realise that the ribosome literally "runs" along the messenger as it translates the message into a chain of amino acids. Regarding the identification of the putative repressor of the lactose operon in the bacterium E. coli, **Jacques Monod** and **François Jacob** working at the Pasteur Institute in Paris had hypothesised a few years earlier that there must exist a substance capable of controlling this operon and repressing it if necessary. The search for this repressor was long and difficult, as this molecule is present in only about ten copies or even less per bacterial cell. An ideal quantity for detecting the molecule would have been a few hundred million copies! Not only did Gilbert succeed in isolating it but he localized its binding site into the operon. Walter Gilbert showed a great interest in the evolution of the intron / exon strutures of genes. Are the introns ancient structures, used to assemble the first genes four billion years ago or are they more recent acquisitions, used for exon shuffling in recently evolved proteins? His work ranges from theoretical estimates for the size of the universe of exons to theoretical arguments that the introns are very old and that exons are related to folding subunits of proteins. He is studying the assembly of protein shapes from molecular substructures. Walter Gilbert has held professorships at Harvard University in the Department of Physics, Biophysics, Biochemistry, and Biology, and since 1985, in Molecular and Cell Biology. He presently holds the Carl M. Loeb University professorship.

Sanger, Frederick (see year 1958)

1981

Hoffmann, Roald (Zloczow, Poland, July 18, 1937-). Polish-born American chemist. Nobel Prize for Chemistry, with **Kenishi Fukui**, for their independent investigations of the mechanisms of chemical reactions. Son of an engineer. Having survived the Nazi occupation, in 1946 he left Poland and finally arrived in the United States, 1949, at the age of eleven. Roald Hoffman began graduate study

Roald Hoffmann

at Harvard University in 1958. There, through the influence of **Elias James Corey**, he became interested in structural and mechanistic problems in organic molecules. At the end of his stay at Harvard he began his collaboration with **Robert Burns Woodward** on the theory of concerted reactions. In 1965, he joined the Department of Chemistry at Cornell University where he is now professor of chemistry. He is also Frank H.T. Rhodes Professor of Humane Letters. Hoffmann's research interests are in the electronic structure of stable and unstable molecules, and of transition states in reactions. He applies a variety of quantum chemical computational methods as well as qualitative arguments to problems of structure and reactivity of both organic and inorganic molecules of medium size to extended systems in one, two, and three dimensions. Hoffmann and his collaborators have also been exploring the structure and reactivity of inorganic and organometallic molecules. Approximate molecular orbital calculations and symmetry-based arguments have been applied by his group to explore the basic structural features of every kind of inorganic molecule, from complexes of small diatomics to clusters containing several transition metal atoms. Hoffmann also writes popular articles on science including a column for American Scientist. In addition he has another career as a poet. In the 1980s he joined an informal poetry group at Cornell. He has been publishing poetry since 1984. The Methamict State, was pu- blished by the University of Central Florida Press in 1987. The same publisher issued a second collection, Gaps and Verges, in 1990; his more recent collection, Memory Effects, 1999, was published by the Calhun Press of Columbia College, Chicago.

Fukui, Kenishi (Nara, Japan, October 4, 1918 - Kyoto, January 9, 1998). Japanese chemist. Son of a tradesman. Shared the Nobel Prize for Chemistry with **Roald Hoffmann** for their independent investigations of the

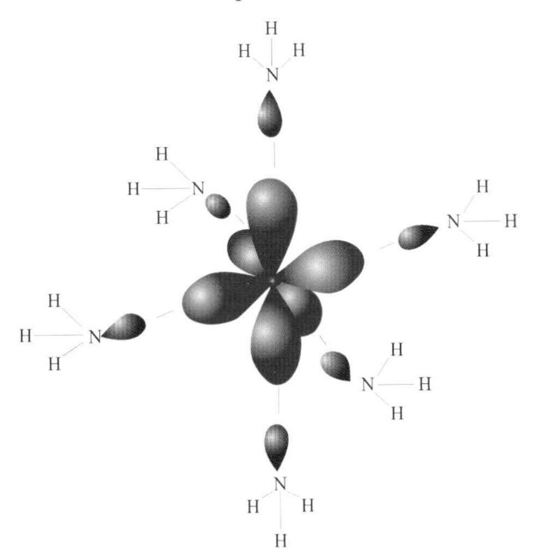

A schematic diagram of orbitals. Each of them is characterized by a particular energy value and may be occupied either by one electron or by two electrons of opposite spins, or it may be unoccupied.

mechanisms of chemical reactions. His father played a determining role by convincing him to study chemistry. During World War II, he devoted himself to conducting research on synthetic fuels. Rejoining the University of

Kyoto in 1945, he earned his Ph.D. in applied chemistry in 1948 and began to teach physical chemistry in 1951. He focused on the theory of chemical reactions and on quantum chemistry. This led him to conduct a theoretical study of electron orbitals in molecular reactions. In 1952, he developed the "frontier orbital" approximation method. Frontier orbitals are where the outermost electrons of mo- lecules are located. These electrons are the most directly active in chemical reactions. Fukui showed that a chemical reaction normally involves a reaction between the highest occupied molecular orbital - the so - called HO orbital - of one compound and the lowest unoccupied orbital - the so - called BV orbital - of the other. Mathematical description of Fukui's theory is so complex that it can hardly be clear for chemists. Hoffmann and Robert Woodward, two researchers who worked independently of Fukui, arrived at the same conclusions but by an easier way. Kenishi Fukui was the first Japanese recipient of the Nobel Prize. He was appointed President of the Japanese Society for Chemistry in 1983 and 1984.

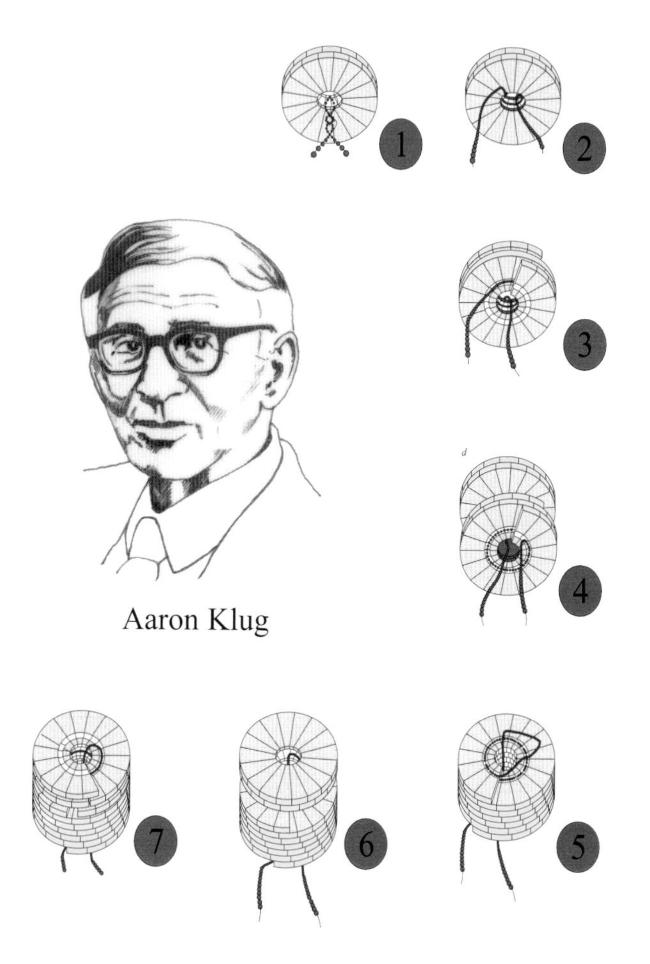

Aaron Klug

Assembly of the tobacco mosaic virus. An important consequence of Klug's studies on this particle was that it resulted from a natural assembly of nucleic acid (RNA in this case) and proteins.

1982

Klug, Aaron (Lithuania, August 11, 1926 -). British chemist. Nobel Prize for Chemistry for his investigations of the three-dimensional structure of viruses and other particles that are combinations of nucleic acids and proteins, and for the development of crystallographic electron microscopy. When he was only three years old Aaron Klug moved with his parents from Lithuania to South Africa. In 1942 he entered the University of Witwatersrand. There in 1945 he gained his Ph.D. and studied crystallography by X-ray diffraction. In 1949 he joined the Cavendish Laboratory at the University of Cambridge, England and, under the direction of **Max Perutz** and **John Kendrew** he studied the structure of biological molecules. Yet it's his work on the structure of steel that earned him the title of Doctor of Science in 1952. Nevertheless, Klug wanted to remain in biochemistry, and in 1953 he was given the opportunity to work at the Birbeck College in London under Rosalind Franklin, a pioneer of crystallography. Franklin oriented Klug towards the structure of viruses. Franklin's death in 1958 led to the appointment of Klug as Head of the Virus Structure Research Group of Birbeck College. In 1962, he went back to the Cavendish Laboratory. Aaron Klug became its Director in 1986. There he developed the crystallographic technique that consists in carrying out X - ray diffraction on a photographic image produced by electron microscopy. This method required a particularly complex mathematical analysis. With the help of protein crystallography, he studied the structure of viruses, cell walls, subcellular particles and he showed that the assembly of viral envelopes was a spontaneous process. This process allows the creation of helical viruses, in which the coat protein subunits associate to form helical tubes or to create spherical viruses, in which coat protein subunits associae to form polyhedral shells. Klug's renown stems mostly from his crystallographic studies on the three - dimensional conformation of protein - nucleic acid complexes like those found in the tobacco mosaic virus.

1983

Taube, Henry (Neudorf, Canada, September 30, 1915 -). Canadian-born America chemist. Son of a farmer. Nobel Prize for Chemistry for his extensive research into the properties and reactions of dissolved inorganic substances, particularly oxidation-reduction processes involving the ions of metallic elements. Taube's parents were Ukrainians

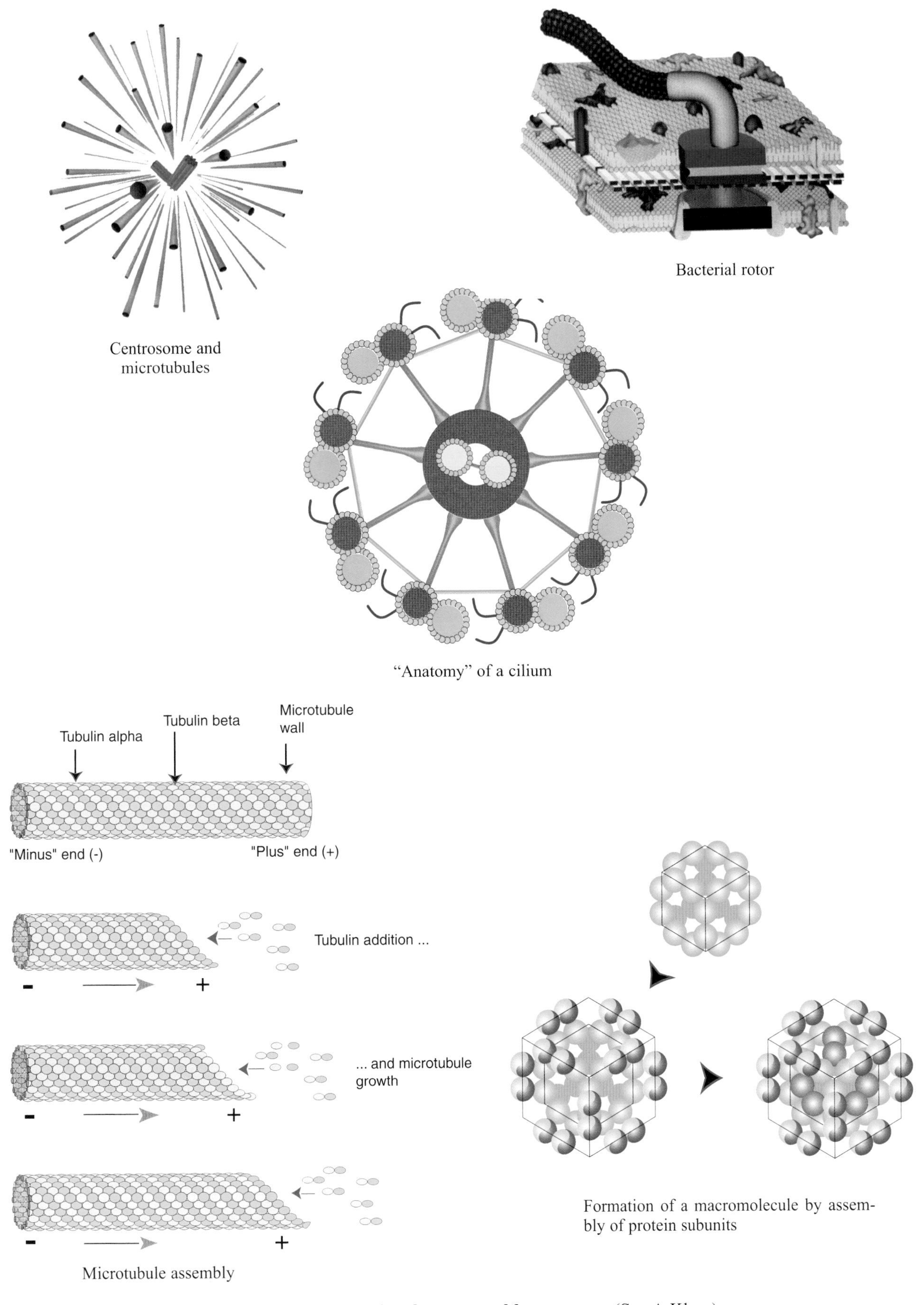

Cell structures resulting from assembly processes.(See A.Klug.)

who fled the Tsarist regime in 1911. He studied at the University of Saskatchewan, Saskatoon and the University of California, Berkeley, where he received his Ph.D. in science in 1940. In 1946 at the University of Chicago, he began to study the kinetics of coordination chemistry. He realised that the mechanisms of the substitution reactions he had already studied could be extended to inorganic complexes. In 1952 he wrote a first paper on the influence of the electronic structure of transition metals on the ligand substitution rate. He thus laid the foundations of the theory of complex reactivity in organic chemistry. From 1956 to 1959, Taube headed the Department of Chemistry in Chicago. In 1961 Stanford University offered him the chance to escape his administrative tasks and to devote himself fully to research. He then studied osmium and ruthenium, two metals characterized by an uncommon electronic structure.He noticed that variations occurred during electric discharges and in the geometry of molecules during electronic transfer. This meant that what was viewed as a simple transfer of electrons actually involved intermediates that could be characterised. Henry Taube thus explained the rate differences observed with different ions. His observations triggered fresh interest in inorganic chemistry.

1984

Merrifield, Robert Bruce (Fort Worth, Texas, July 15, 1921 -). American biochemist. 1984 Nobel Prize for Chemistry for his development of a simple and ingenious method for synthesizing chains of amino acids in any predetermined order. Robert Bruce Merrifield studied chemistry at the University of California, Los Angeles and gained his Ph.D. in 1949. His first research concerned the synthesis of a complex amino acid, dihydroxyphenylalanine, which plays an essential role in the transmission of the nerve influx. The same aminoacid was applied in the treatment of Parkinson disease. The same year he joined the Rockefeller Institute for Medical Research, New York, where he held a post as a teacher in 1984. His research mainly concerned the proteins, the fundamental substances of living organisms. He tried to assemble amino acids into peptide sequen-ces from two to a few dozen amino acids long. The synthetic process in use at the time, elaborated by **Emil Fischer** at the beginning of the 20th century, was long and laborious and required many operations. Merrifield's innovation in this field was to elaborate a technique which allowed automatic synthesis of oligopeptides. This ingenious method involved binding the first amino acid of a desired sequence to a solid medium such as a resin, and adding amino acids to the chain one by one in a given order. Once completed, the chain was detached from its support. Automated peptide synthesis greatly facilitated research on peptides, which are particularly important in physiology and pharmacology. It also reduced the cost of synthetic peptides. This made possible the industrial synthesis of enzymes and hormones (e.g. insulin), and to achieve an unequalled degree of purity. In 1965, Merrifield succeeded in automatizing the process and, in 1969, he announced the total synthesis of ribonuclease, an enzyme. Although his technique has become a basic one in genetics and biochemistry, Robert Merrifield always refused to patent his invention.

Robert Merrifield

1985

Hauptman, Herbert Aaron (New York City, New York, February 1917-). American mathematician and crystallographer. 1985 Nobel Prize for Chemistry, along with **Jerôme Karle** , for developing mathematical methods for deducing the molecular structure of chemical compounds from the patterns formed by X - ray diffraction of their crystals. Son of a printer. Herbert Hauptman is a world renowned mathematician who pioneered and developed a mathematical method that has changed the whole field of chemistry and opened a new area of research in determination of molecular structures of crystallized materials. Today, Dr. Hauptman's direct methods, which he has continued to improve and refine, are routinely used to solve complicated structures. Hauptman's work is concerned with the development of methods for determining molecular structures, that is the arrangement of the atoms in molecules, using the technique of X - ray diffraction. The work is important because it relates molecular structure

Herbert Hauptman

with biological activity and therefore permits a better understanding of life processes. In this way one can devise better methods for the diagnosis and treatment of disease. Before Hauptman and Karle developed their techniques, about two years were necessary to deduce the structure of a simple biological molecule; now, using powerful computers to perform the complex calculations needed, one could do it in about two days. Hauptman left the Laboratory for Naval Research in 1970 and moved to the Laboratory of Physics at the State University of Buffalo in 1970. He simultaneously held a post as a professor of biophysics at the State University of New York in the same city. From 1972 until 1985 he was a Vice President of the Medical Foundation of Buffalo and then President in 1985.

Karle, Jérôme (Brooklyn, New York, June 18, 1918 -). American chemist and crystallographer. 1985 Nobel Prize for Chemistry, along with **Herbert A. Hauptman**, for his development of mathematical methods for deducing the molecular structure of chemical compounds from the patterns formed by X - ray diffraction of their crystals. Son of

Jerôme Karle

a house painter. The reasearch that resulted in the award of a Nobel Prize concerned the problem of determining the arrangement of the atoms in a crystal by use of the X-ray scattering technique. In this technique, a narrow beam of X - ray is made to strike a tiny crystal whose atomic arrangements are of interest. For particular angles of incidence of the X - ray beam on the crystal, achieved by making suitable rotations of the crystal, the incident beam is scattered and recorded for location in space and its intensity. In the course of an experiment, many hundreds or thousands of scattered beams are recorded. These are called diffraction patterns; every substance has its own diffraction pattern. It was thought that, in general, the diffraction patterns did not contain enough information to permit the structure, i.e., the arrangement of the atoms to be deter-

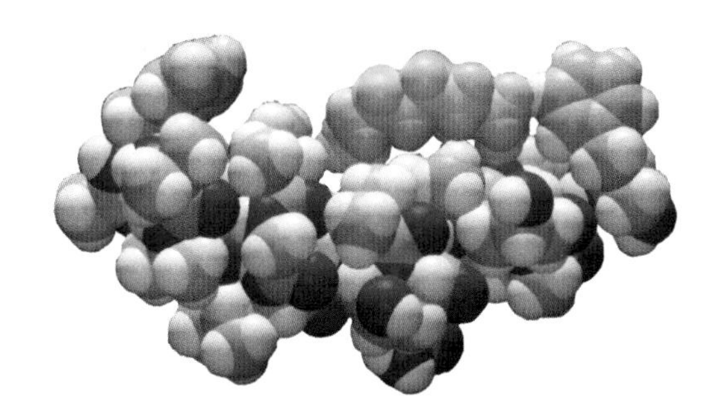

Structure of antiamoebin. This is a diagram of the antibiotic called antiamoebin. Antiamoebin acts by being attached with the membranes that surround living cells and altering the transport of ions into those cells. The attachment occurs between the non-polar part of the antiamoebin molecule and the non-polar cell membranes. The other polar part of the antiamoebin molecule combines with similar parts of two other molecules to form a polar channel for the transport of ions. The figure shows a non-polar octanol molecule (in yellow) attached to the non-polar part of antiamoebin in imitation of the action of a cell membrane. This structure was solved by Isabella Karle who used procedures that she and Jerôme Karle developed almost 40 years ago.

mined. This view was not correct and from the efforts of Jerome Karle and those of collaborators, Herbert Hauptman and Isabella Karle, a general method for determining crystal structures ensued. Major applications of the structural information have occured, for example, in mineralogy, synthetic organic chemistry, biochemistry and the design of new drugs in the pharmaceutical industry. Karle first studied chemistry and biology, but also mathematics and physics at City College, New York. In 1937 he joined the University of Harvard to improve his knowledge in biology. Between 1939 and 1940, he held a post at the New York State Department of Health, Albany, before beginning research in chemistry at the University of Michigan. From 1943 to 1944, he worked on the Manhattan Project at the University of Chicago and earned his Ph.D. in physical chemistry in 1944 with research on electron diffraction in gases. With his wife Isabella, he moved in 1946 to the Naval Research Laboratory, Washington D.C., where he met Hauptman again, becoming chief scientist in the laboratory for researches on the structure of matter in 1968.

1986

Herschbach, Dudley Robert (San Jose, California, June 18, 1932 -). American chemist. Son of a stock - breeder. Nobel Prize for Chemistry, with **Yuan Tse. Lee** and **John C. Polanyi** for his pioneering use of molecular beams to analyze chemical reactions. After studying mathematics at Stanford University, Herschbach entered Harvard University, where he opted to research the dynamics of chemical reactions. There he carried out his doctoral work on the spectroscopy of microwaves emitted by molecules in a methanol solution. He received his doctorate in physical chemistry in 1958, and then he began his career as a researcher and teacher at Berkeley and later at Harvard, where he also chaired the chemistry department beginning

Dudley Herschbach

in 1963. Mechanically, research on the dynamics of chemical reactions had not progressed very far since the start of the century. Chemists had been content to determine reaction rates and establish their energy balance. Very little had been done to describe the elementary processes occurring in chemical reactions. Herschbach then had the idea to borrow from physics a method for studying colli-

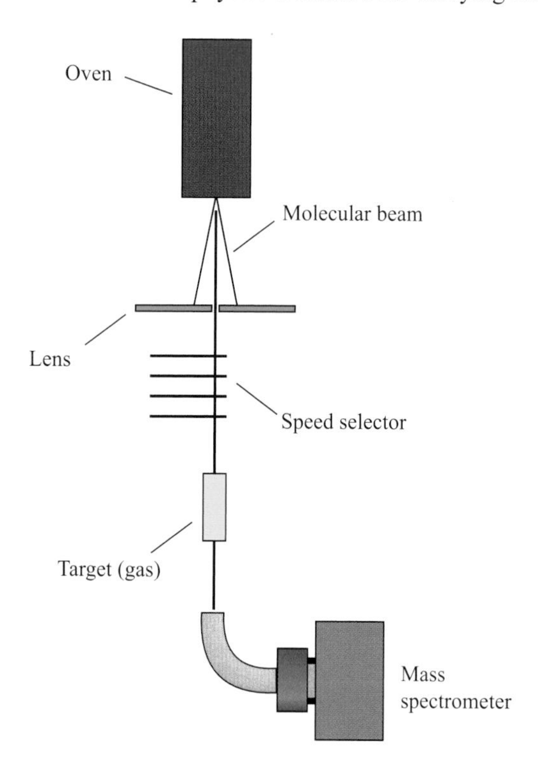

Apparatus designed to produce excited molecular states

sions between molecules. At low pressure he crossed two supersonic molecular beams, one containing alkyl halide molecules and the other containing atoms of an alkali metal, causing the atoms and molecules to collide violently. He used a tungsten wire heated by an electric current to capture the reaction products, and measured how this caused the electric current in the wire to vary. In this way he was able to study the dynamics of the collision reactions. In 1967, Herschbach associated with Lee, who expanded the analytical range of this technique by using a mass spectrograph to detect the species formed in collisions. This enabled Herschbach to develop models that explained precisely the mechanisms of molecular excitation. This excitation enables molecules to pass an energy barrier and brings them into an intermediate state leading to product formation.

Lee, Yuan Tseh (Hsinchu, Taiwan, November 19, 1936 -). Taiwanese-American chemist. Nobel Prize for Chemistry, along with **Dudley R. Herschbach** and **John C. Polanyi** for his role in the development of chemical-reaction dynamics. Son of an artist. Yuan Tseh Lee received his B.S. degree from the National Taiwan University in 1959. After finishing his M.S. degree at Tsinghua University, he pursued his Ph.D. thesis research at the University of California at Berkeley. In 1965 he began to conduct reactive scattering experiments in ion-molecule reaction as a post-doctoral fellow in Mahan's laboratory.

Yuan-Tseh Lee

In 1967, Y.Lee joined Dudley Herschbach's group at Harvard as a research fellow where they took molecular beam experimentation beyond the alkali age. Yuan Lee developed an instrument for studying the elementary processes characterising the collisions that take place in crossing molecular beams. He installed a mass spectrometer to detect the species formed in the molecular beams. Although physicists were already familiar with this method, Herschbach and Lee were the first scientists to apply it to the study of chemical reactions. They were able to determine the structure and chemical behaviour of highly reactive polyatomic radicals and unusual transient species. This technique was one of the most significant breakthroughs in the field of chemical reaction dynamics. After being appointed assistant professor at the University of Chicago in 1968, he rapidly made his laboratory the North American capital of molecular beam study. Yuan Lee returned to Berkeley as a full professor in 1974 and significantly expanded his research to include, in addition to crossed molecuar beams, studies on reaction dynamics, investigations of various primary photochemical processes, and the spectroscopy of ionic and molecular clusters.

In 1994, he retired from his position of University Professor and Principal Investigator for the Lawrence Berkeley Laboratory at the University of California at Berkeley and assumed the position of the President of Academia Sinica in Taiwan.

Polanyi, John Charles (Berlin, Germany, January 23, 1929 -). German chemist born in Hungary. 1986 Nobel Prize for Chemistry, along with **Yuan Tse. Lee** and

John Polanyi

Dudley Herschbach for his contribution to the field of chemical - reaction dynamics. Born in Berlin but educated in England, Polyani began his studies in 1946 at the

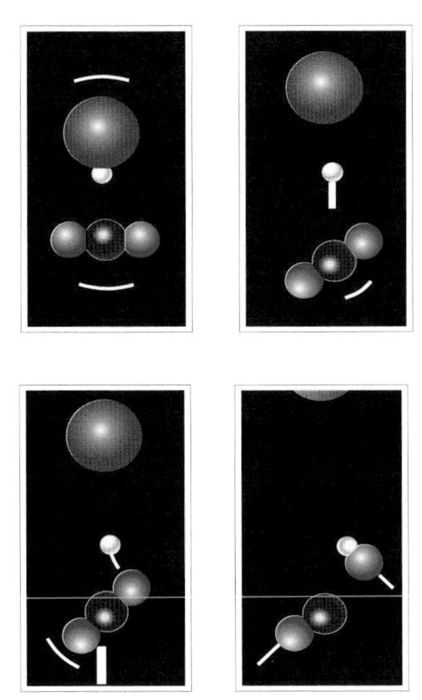

Ultra - fast chemical reactions. Thanks to Norrish and Porter's perfection of this method, today we can successfully determine the chain of events that occurs in the millionth of a billionth of a second following a collision between, for example, a molecule of hydrogen iodide and a molecule of carbon dioxide.

University of Manchester where his father, also a chemist, had obtained a teaching position. The latter's research had focused on molecular bases in chemical reactions. Thus Polyani began his own research along with one of his father's students on the theory of chemical kinetics that describes reactions as the result of collisions between molecules. Armed with a doctorate of science in 1952, he traveled to Toronto to work on a post - doctorate at the Canadian Research Council. It was in the framework of his research in Canada that he attempted to determine whether the reaction rate was in agreement with transition-state theory, which his father had helped to elaborate. Unfortunately, the empirical data concerning this transition state was insufficient to support the theory. Polanyi then spent a few months in the laboratory headed by Gerhard Herzberg and became familiar with the rotational and vibrational spectra of iodine molecules. In 1954 he was invited to do post - doctorate research at Princeton University. This is where he was informed of the existence of a phenomenon called chemiluminescence, which appears in certain chemical reactions. Polanyi concluded from this phenomenon that it was possible to determine the excitation energy characterising product formation by such reactions. In 1956 he returned to the University of Toronto to teach chemistry and earned the title of professor in 1974. He continued the study of chemical luminescence and led projects that were the origin of the chemical laser effect. Such a laser was installed at Berkeley by Pimentel.0

Cram, Donald James (Chester, Vermont, April 22, 1919 - June 17, 2001). American chemist. Nobel Prize for Chemistry along with **Charles J. Pedersen** and **Jean-**

Donald Cram

Marie Lehn for his creation of molecules that mimic the chemical behaviour of molecules found in living systems. Donald Cram developed an interest in chemistry at high

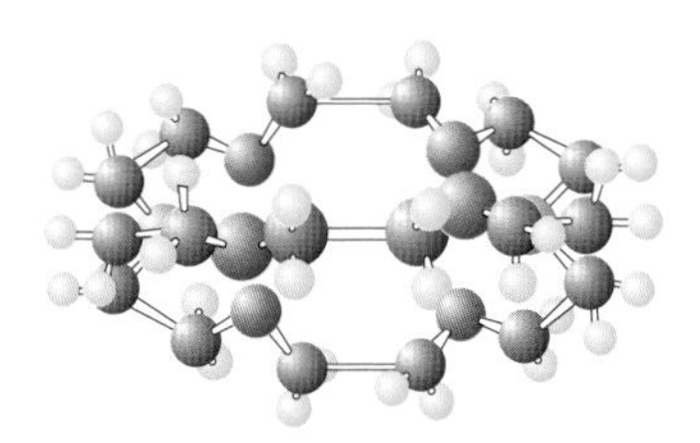

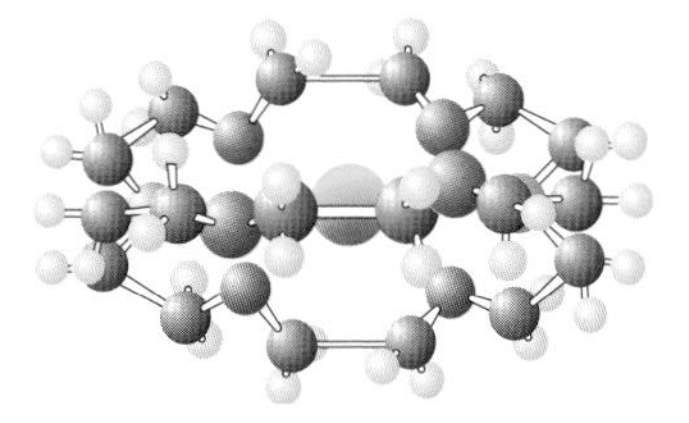

Crown ether. This loose, flexible ring of carbon atoms is punctuated at regular intervals with oxygen atoms. This kind of two-dimensional organic compound is able to bind various ions at the center of the crown.

school. He received a B.S. Degree from Rollins College in 1941, an M.S. from the University of Nebraska in 1942, and a Ph.D. from Harvard University in 1947. At Rollins College, a chemistry professor told him that he would

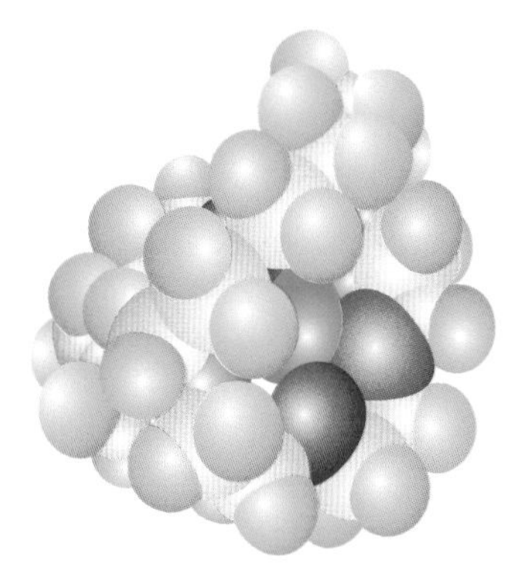

Cryptand. A three - dimensional representation. An ion (in green) is trapped inside the structure.

make a good industrial researcher, but probably not a good academic one. Thereupon, Cram resolved to pursue an academic career in chemistry! While still in his thirties, Cram was already a household name in chemistry. There were two main reasons for this early rise to fame. One was his rule of asymmetric induction, dating from 1952, and the other was his epochal textbook of "Organic Chemistry" written with George Hammond and published in 1959. It introduced students to a completely new way of learning the subject according to reaction types and following mechanistic principles. For those of us studying organic chemistry in the 1960s, "Cram and Hammond" was a real breath of fresh air. Cram was an exceptionally talented stereochemist who forged new ground in molecular organic chemistry However, it was his bold entry, in his fifties, equipped with little more than his beloved bag of space-filling molecular models, into the unknown area of intermolecular conformational analysis, for which he will be remembered above all else. With the realization that synthesis does not begin and end with the making and breaking of covalent bonds and his pioneering foray into the noncovalent synthesis of host - guest complexes, he set the agenda for the dawning of a new age of chemistry as a science to be performed on the nanoscale level. A Professor at UCLA for 54 years, Donald Cram died of cancer on June 17, 2001 at age 82.

Jean-Marie Lehn

Lehn, Jean-Marie (Rosheim, France, September 30, 1939 -). French chemist. Son of a baker. Nobel Prize for Chemistry, with **Charles J. Pedersen** and **Donald J. Cram**, for his contribution to the laboratory synthesis of molecules that mimic the vital chemical functions of molecules in living organisms. In the mid - 1960s Jean-Marie Lehn developed an interest in the process occuring in the

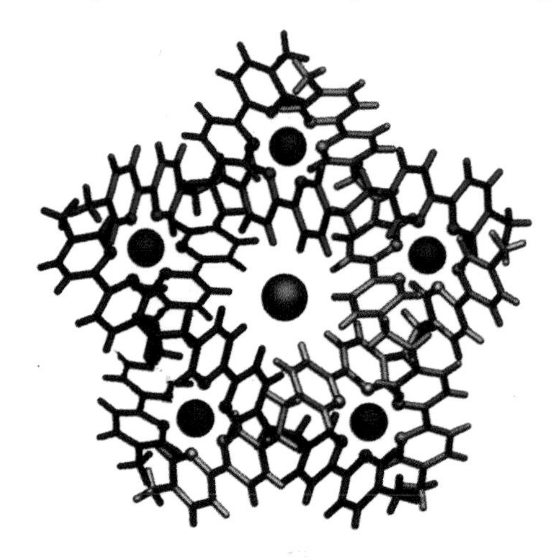

Cryptand. Such an organic compound has two or more rings linked together by bridgehead atoms to form a three - dimensional cagelike molecular structure. The inner space is sufficient to accomodate various ions and form a cryptate.

nervous system, in particular in the fact that the electrical events in nerve cells rest on changes in the distribution of sodium and potassium ions across the membrane. Around the same period several research groups has established that some macrocyclic antibiotics (nonactin, valinomycin) have complexing ability toward alkali cations. Thus, when Charles Pedersen reported the cation binding properties of crown ether - 1 Lehn rapidly realized that the recognition of a spherical species (of alkali cation) would be much more efficient if the ligand had a spherical cavity instead the circular surrounding of the cation observed with the crown ether in 2. This idea led to the synthesis of the first "cryptand" - 3 which forms a very stable potassium complex - 4 named "cryptate". The first synthesis was followed by the modification of the cavity size by altering the lenght of the bridging chains. This led to a series of cryptands exhibiting high selectivity of complexation toward almost all cations of groups I and II. Cryptands incorporating other donor atoms (like sulfur or nitrogen) in place of oxygen were also synthesized and studied. Later Lehn developed systems able to bind efficiently a tetrahedral species, the ammonium cation, and its substituted derivatives. A further step in complexity was accomplished by the design of ligands able to complex more than one cation or substrates bearing several functional groups capable to interact with the binding sites of the receptor. These polytopic receptors can form for example multinuclear metal complexes or bind linear diammonium substrates. The next step was the synthesis of ligands bearing anion-binding-sites (ammonium or guanidinium units) opening the area of anion coordination chemistry. In the course of his investigations Lehn realized that the appropriate design of a receptor able to bind a specific substrate was based on the precise control of (weak) intermolecular forces and he proposed the terms "supramolecular chemistry" for this new domain. He defined "supramolecular chemistry as chemistry beyond the molecule, bearing on the organized entities of higher complexity that result from the association of two or more chemical species held together by intermolecular forces". The field has been extended tremendously, mainly by the studies on the self -assembly process and the design of programmed molecular systems. Lehn graduated in chemistry from the University of Strasbourg and earned his Ph.D. in 1963. In 1970, he was appointed professor of chemistry at Louis Pasteur University in Strasbourg. In 1979, he was also named professor at the Collège de France in Paris.

Pedersen, Charles John (Pusan, Korea, October 3, 1904 - Salem, New Jersey, October 26, 1989). American chemist born in Norway. Nobel Prize for Chemistry, along with **Donald Cram** and **Jean-Marie Lehn** for his synthesis of the crown ethers. Charles Pedersen was the son of a Norwegian navigator who emigrated to the United States in 1922. He first studied chemical engineering at Dayton University before graduating in organic chemistry at the Massachusetts Institute of Technology (MIT), Boston, in 1927. The same year he began a research career at du Pont de Nemours, where for 42 years he was allowed complete freedom to orient his research programme. Pedersen never received a Ph.D. and never had an academic career. His research at du Pont de Nemours focused on macromolecules and polymerisation catalysts. In 1960, as he was trying to develop such a catalyst, an experiment yielded a strange substance that he had not aimed to produce in the scope of his research. Attempting to understand what had happened, Pedersen discovered that he had synthesised a cyclic polymer, in fact a crown polyether; he further showed that this molecule could trap metal ions within its ring and form stable complexes, notably with alkali metals. These new chemical structures are synthetic macrocycles, the so - called crown ethers, and their behaviour is somewhat analogous to that of enzymes and other biological molecules. This unexpected discovery opened new orientations in biochemical research, because the macrocycles described by Charles Pedersen have counterparts in biological systems such as ion pumps or molecules that transport ions across cell membranes. The crown ethers opened the way to the synthesis of specific drugs.

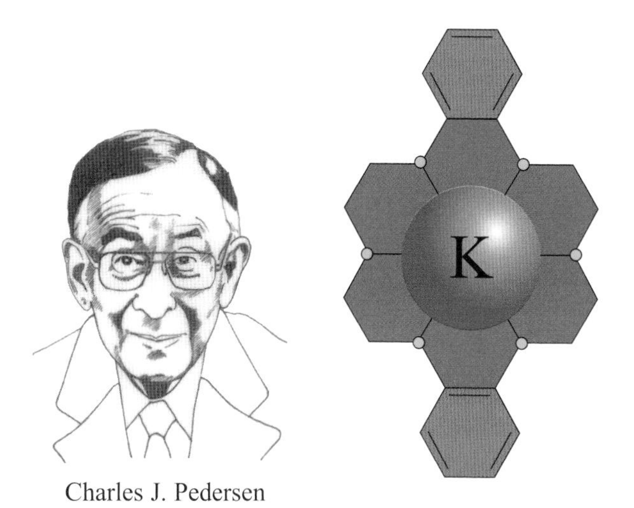

Charles J. Pedersen

Crown ether. The original crown ether discovered by Charles Pedersen has six oxygen atoms exposed along the inside wall of the ring. When atoms of certain metalic elements such as sodium or potassium pass through the center of the ring, they attach themselves to the exposed oxygen atoms and fit like a key in a lock.

1988

Deisenhofer, Johann (Zusamaltheim, Germany, September 30, 1943 -). German biochemist. Son of a

Johann Deisenhofer

farmer. Nobel Prize for Chemistry, along with **Hartmut Michel** and **Robert Huber**, for their determination of the structure of Rhodomonas photosynthetic centre. After studying physics at the Technical University Munich, Johann Deisenhofer switched to X - ray crystallography of

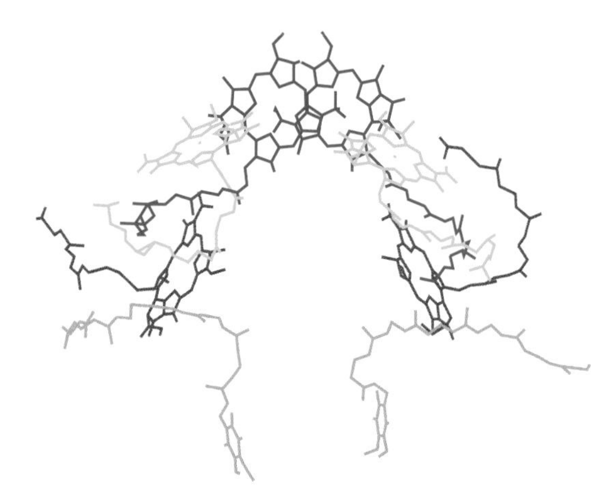

The steps of electron transport during bacterial photosynthesis. The electron excited by luminous photons is channeled from a special pair of chlorophyll (in red) and finally reaches a first (bottom right) and then a second quinone acceptor (bottom left).

proteins at the Max - Planck - Institute for Biochemistry in Martinsried. After receiving a Ph.D. degree in experimental physics from the Technical University Munich in 1974, he continued at the Max - Planck - Institute as a postdoc and staff scientist, working mainly on the X - ray structure analysis of immunoglobins. In 1982 he joined Hartmut Michel in the structure determination of the photosynthetic reaction center of *Rhodopseudomonas viridis*. This substance was a photosynthetic reaction center through

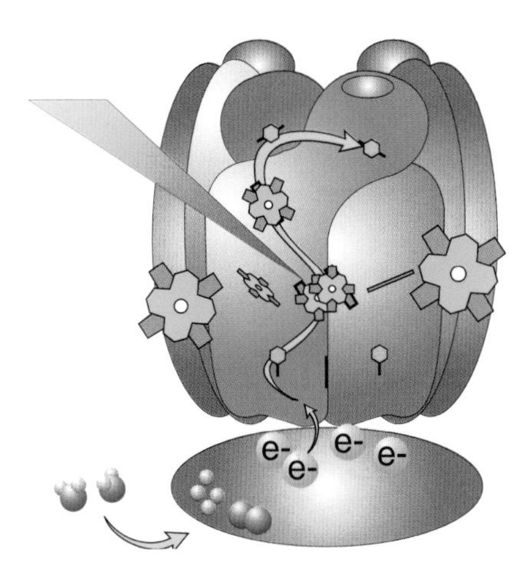

A photosynthetic reaction center. Note that the prosthetic groups of the bacteriochlorophyll and the bacteriophytin molecules are spatially situated so that rapid electron transfer from P870 to QB is facilitated.

which electrons acquired energy for triggering photosynthesis in the bacterium. Between 1982 and 1985, the X-ray crystallography of this photosynthetic reaction center allowed them to determine exactly the atomic structure of the protein complex. Scientists believe that once its detailed structure and electron transfer mechanism have been elucidated, it will be possible to create artificial photosynthetic reaction centres. In 1988 Deisenhofer became Professor of Biochemistry, and Investigator in the Howard Hughes Medical Institute at the University of Texas Southwestern Medical Center at Dallas, where he is now Regental Professor and holds the Virginia and Edward Linthicum Distinguished Chair in Biomedical Science. His main scientific interests include the determination of 3D structures of membrane proteins and of proteins involved in electron transfer and energy transduction.

Huber, Robert (Munich, Germany, February 20, 1937-). German biochemist. Son of an engineer. Nobel Prize for Chemistry shared with his co-worker of many years, **Johann Deisenhofer** together with **Hartmut Michel**, jointly received the Nobel Prize for "the determination of the three - dimensional structure of a photosynthetic reaction center". Robert Huber studied at the Technische Hochschule in Munich. After his graduation in 1963 he worked as lecturer and researcher at this university until 1972. Subsequently he was offered the chair of the "Structural Research II"

Robert Huber

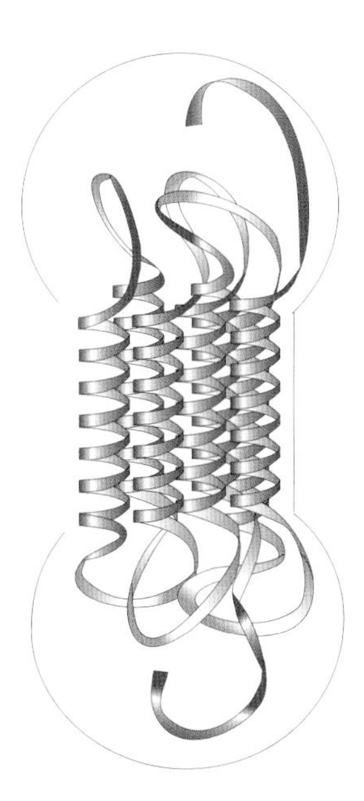

Three - dimensional structure of bacteriorhodopsin. This membrane protein, which contains retinal, resembles animal rhodopsin. It is formed by *Halobacterium halobium* and serves as a light - operated proton pump to translocate protons through the bacterial membrane, from the inside to the outside.

department of the Max-Planck-Institutefor Biochemistry, which had just moved to a new building at Martinsried just outside Munich. This department mainly deals with the structural analysis of biological macromolecules and the development of advanced methods for the X - ray analysis of protein complexes. In 1976, Huber was appointed Associate Professor at the University of Technology. Three years earlier their fundamental findings on the central photosynthesis in purple bacteria of the genus *Rhodopseudomonas* had been published by an international journal, Nature. They were the first researchers who succeeded in elucidating the spatial structure of a protein complex molecule integrated into the cell membrane down to the atomic level. This constituted a pioneering contribution to new insights into life processes, as representatives of an institute which is said "to have produced more protein structures than any other crystallographic laboratory in the world" (**Max Perutz**, Nobel laureate 1962). Huber received many awards and distinctions and is member of German and foreign academies of science.

Michel, Hartmut (Ludwigsburg, Germany, July 18, 1948 -). Nobel Prize for Chemistry, along with **Johann Deisenhofer** and **Robert Huber**, for their determination of the structure of Rhodomonas photosynthetic centre. German biochemist. Son of a joiner. Hartmut Michel attended the Friedrich-Schiller-Gymnasium in his native city and soon aroused attention because of his keen interest in science, physics, chemistry and, especially biology, but also history. After graduation he initially planed to study geology. After military service, however, he decided in favor of biochemistry which, at that time, could be studied only in Tübingen. Later he went to Munich and, finally, to

Hartmut Michel

Würzburg, where he obtained his Ph.D. in 1977. Two years later, he switched to the Max Planck Institute of Biochemistry in Martinsreid near Munich, to the Department of Membrane Biochemistry, to work with Professor Osterhelt who supervised his Ph.D. thesis. In 1986, he obtained his degree at the Ludwig-Maximilians-Universität in Munich. One year later, he changed to the Max Planck Institute for Biophysics in Frankfort where he was appointed Head of the Molecular Membrane Biology Department on September 1, 1987. In his work at the Martinsreid Membrane Biochemistry Department in 1982, he had succeeded in isolating and crystallyzing the photosynthetic reaction center in purple bacteria. Together with Deisenhofer from Huber's Structural Research Department he was now able to use X - ray diffraction to determine the structure of this complicated protein complex. The three scientists published their findings in "Nature" in December 1985.

1989

Altman, Sidney (Montreal, Canada, May 7, 1939 -). Canadian-American molecular biologist. Son of a grocer. 1989 Nobel Prize for Chemistry, along with **Thomas R. Cech**, for his discoveries concerning catalytic properties of RNA. Sidney Altman has been concerned with the way genetic information is expressed in our cells from a very early point in his research career. His best known work is a study of an enzyme, ribonuclease P, a catalyst inside cells that speeds up biochemical reactions, that is composed of two subunits. One of these is made up of protein, as are most enzymes, and the other is made up, surpri-

Sidney Altman

singly, of RNA. In fact, the RNA component turns out to be the catalyst, itself. The latter observation was completely novel, as previously all biological catalysts were

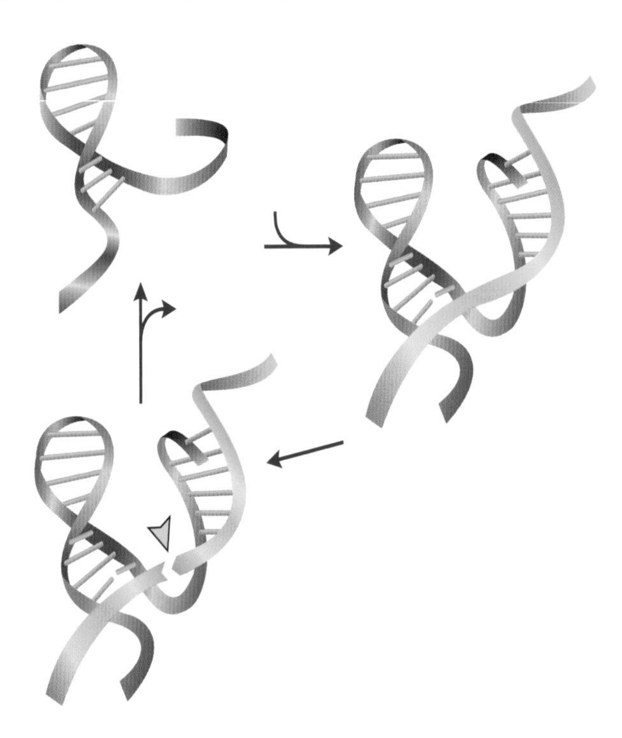

Ribozyme. This RNA molecule with catalytic activity was first discovered by Sidney Altman in 1981 in Rnase P and by Thomas Cech in 1982 in splicing reactions of ribosomal RNA from *Tetrahymena.*

thought to be made of protein. It altered thinking about the nature of biological catalysts and stimulated speculation about some of the important steps in the origins of life on earth. More than one hundred catalytic RNAs are known today. As an outgrowth of studies on the essential features of substrates for Rnase P, a general method was developed to inactivate the expression of any targeted gene in any organism. To date, the efficacy of this method has been demonstrated in laboratory cultures of *Escherichia coli* and human tissue culture cells. For example, the drug resistant phenotype of *E. coli* can be reversed using this technology and influenza virus infection of mouse cells can be completely blocked. Although Altman first studied physics - he received his Ph.D. in physics from the Massachusetts Institute of Technology in 1960 - he became interested in molecular biology and earned a Ph.D. in biophysics at the University of Colorado in 1967. His early research concerned the replication of phage T4,

Thomas Cech

a virus involved in bacterial infection. He pursued his academic career at the University of Harvard and then at the Medical Research Council in Cambridge, England, where he collaborated with **Sidney Brenner** and **Francis Crick**, the co-discoverer of the double helix structure of DNA.

Ribozyme. This RNA enzyme is a kind of molecular scissors that cut RNA. Its discovery by Cech in the 1980s was a major discovery since until then proteins were thought to be the only substances capable of behaving as enzymes.

Altmann currently teaches molecular biology at Yale University.

Cech, Thomas R. (Chicago, Illinois, December 8, 1947 -). American biochemist and molecular biologist. Son of a physician. Nobel Prize for Chemistry, along with **Sidney Altman** for their discoveries concerning RNA. In 1982 Tom Cech and his research group announced that an RNA molecule from *Tetrahymena*, a single - celled pond organism, cut and rejoined chemical bonds in the complete absence of proteins. Thus RNA was not restricted to being a passive carrier of genetic information, but could have an active role in cellular metabolism. This discovery of self-splicing RNA provided the first exception to the long - held belief that biological reactions are always catalyzed by proteins. In addition, it has been heralded as providing a new, plausible scenario for the origin of life; because RNA can be both an information - carrying molecule and a catalyst, perhaps the first self-reproducing system consisted of RNA alone. Only years later was it recognized that RNA catalysts, or "ribozymes," might provide a new class of highly speci fic pharmaceutical agents, able to cleave and thereby inactivate viral RNAs or other RNAs involved in disease.

1990

Corey, Elias James (Methuen, Massachusetts, July 12, 1928 -). Nobel Prize for Chemistry for his development of the theory and methodology of organic synthesis. A precocious youth, Corey entered MIT in Boston at the age

Elias Corey

of 16. There he began research on organic synthesis and, more precisely, penicillin synthesis. In 1951 he earned his doctorate in chemistry and was offered a teaching assistantship at the University of Illinois at Champaign-Urbana. In 1955 he was made a full professor of chemistry. He took a one-year sabbatical in 1957, which led him to Harvard University where he collaborated with **Robert Woodward**. He also spent some time in Europe. At this time, organic synthesis was still a matter of know-how based on intuition more than on a systematic approach. Corey undertook to rationalise the process to make it more reproducible. His research focused on the logic of chemical syntheses and he proposed elaborating a synthesis diagram beginning with the desired end - product rather than with the reagents. This meant examining all access routes to the end product and the resulting diagram resembled a 'tree' with branches describing the possible access routes. This eliminated the trial - and - error approach which had been the rule up to then. Harvard University offered him a position as a professor in 1959. There he continued his research on logic in the elaboration of his synthesis schematics, thanks to his so-called "retrosynthetic method", which is easily adaptable to computers among other things, this allowed the creation of bridged structures such as norcamphor and camphor, the latter being a stimulant to respiratory functions. In 1968 he succeeded in perfecting the synthesis of five prostaglandins, important hormones that play a part in many vital processes such as reproduction, blood pressure regulation, blood clotting, etc. The synthesis ofprostaglandins is of great importance in the world of medicine since it has rendered these substances available in significant quantities and at a much better cost. In 1988, Corey successfully synthesized ginkgolide B, a molecule used in treating asthma and circulatory problems. Corey's work has had a tremendous impact on organic synthesis procedures. Moreover he personally oversaw the education of many chemists. The importance of his contribution to the field of chemistry has been distinguished by numerous awards: he is a doctor *honoris causa* at the universities of Oxford and Chicago, a member of the American Academy of Arts and Science as well as the American Association for the Advancement of Science. In 1989, he published a book that sums up his life's work as a scientist: The Logic of Chemical Synthesis.

1991

Ernst , Richard Robert (Winterthur, Switzerland, August 14, 1933 -). Swiss researcher. Nobel Prize for Chemistry for his development of techniques for high-resolution nuclear magnetic resonance (NMR) spectroscopy. Richard Ernst studied chemistry at Polytechnic Federal Institute, Zurich, and finished with a doctorate in physical chemistry under the direction of Hans Primas and Hans-Heinrich Guenthard. His subject was NMR instrumentation and theoretical aspects of NMR with stochactic excitation. In 1963, he joined Varian Associates in Palo Alto,

California, where he developed together with Dr. Weston A. Anderson Fourier transform NMR which revolutionized the field. He also continued to work on stochastic decoupling techniques and he did some of the first on-line

Richard Ernst

computer experiments in NMR. In 1968, he returned to ETH Zurich where he became Full Professor of Physical Chemistry. He continued to develop novel techniques in the context of Fourier transform spectroscopy together with his research group. From 1972 onwards, he became very active in the development of two - and multi - dimensional NMR, stimulated by conceptual work by Jean Jeener (University of Brussels). A close collaboration with Kurt Wuethrich and his research group was essential for the development of techniques for the determination of biomolecular structures in solution. Richard R. Ernst has tuned nuclear magnetic resonnance into a tool of great value and importance in chemistry, biology and clinical medicine. He introduced parallel data acquisition by means of Fourier transform techniques and dramatically enhanced the information flux. Today, NMR is truly indispensable for determining the shape, the mobility, and the interaction of biologically relevant molecules and for understanding biological processes. Magnetic resonnance imaging (MRI) allows revealing insights into the human body for diagnosing diseases and even for understanding the functioning of the human brain.

1992

Marcus, Rudolph A. (Montreal, Canada, July 21, 1923 -). Canadian-born American chemist. Nobel Prize for Chemistry for his work on the theory of electron-transfer reactions in chemical systems. Rudolh Marcus received his doctorate in sciences at McGill University in 1946 with the kinetics of chemical reactions as his research topic. Then he conducted experiments on free-radical reactions at the National Research Council of Canada. At the University of North Carolina, in the United States, he developed his theories on reaction rates by taking into account the modifications of energy undergone by a molecular system in relation to the changes of the structure which it undergoes during the reaction. The rate of a chemical reaction was thus connected to the duration of the state of transition, itself dependent on the properties of the molecules engaged in the

Rudolph Marcus

reaction. The Polytechnic Institute of Brooklyn then welcomed him as an associate professor in 1951 and, from 1961, he was director of the Department of Physics and Chemistry of this institution. From that time, Marcus devoted himself to the purely theoretical aspects of the mechanisms of chemical kinetics. In 1964 he joined the Illinois State University and his attention turned increasingly to the dynamics of chemical reactions in relation to electron transfers and, in particular, to those which governed the course of the photosynthetic processes. In 1978, appointed professor of chemistry at the California Institute of Technology in Pasadena, he returned to his earlier work, which consisted in describing the reactivity of a chemical system on the basis of the molecular structure involved. The period spent in Pasadena was extremely productive. There he applied the theory of electron transfer to many fields of chemistry and, in particular, to photosynthesis. It was also there that he perfected a simple method to estimate the change in energy that occurs in electron transfer reactions and he explained the origin of the energy, which animates the phenomenon. Rudolph Marcus became a naturalised American citizen in 1958.

Kary Mullis

1993

Mullis, Kary. (Lenoir, North Carolina, December 28, 1944 -). American biochemist and inventor. Son of a salesman. Shared the Nobel Prize for Chemistry with **Michael Smith** for his invention of the polymerase chain reaction (PCR). In 1972, Kary Mullis earned a Ph.D. degree in biochemistry from the University of California, Berkeley and lectured in biochemistry there until 1973. The same year, he became a postdoctoral fellow in pediatric cardiology at the University of Kansas Medical School, with emphasis in the areas of angiotensin and pulmonary vascular physiology. In 1977 he began two years in postdoctoral work in pharmaceutical chemistry at the University of California, San Francisco. Mullis joined the Cetus Corporation in Emeryville, California, as a DNA chemist in 1979. During his seven years there, he conducted research on oligonucleotide synthesis and invented the polymerase chain reaction. This technique, which Mullis conceptualized in 1983, is hailed as one of the monumental scientific techniques of the twentieth century. A method of amplifying DNA, PCR multiplies a single, microscopic strand of the genetic material billions of times within hours. The process has multiple applications in medicine, genetics, biotechnology, and forensics. It is especially required when rare but specific stretches of DNA are to be analyzed. PCR, which was the theoretical basis for the novel and motion picture Jurassic Park because of its ability to extract DNA from fossils, is in reality the basis of a new scientific discipline, paleobiology. Mullis has authored several major patents. His patented inventions include the PCR technology and UV-sensitive plastic that changes color in response to light. The many publications of

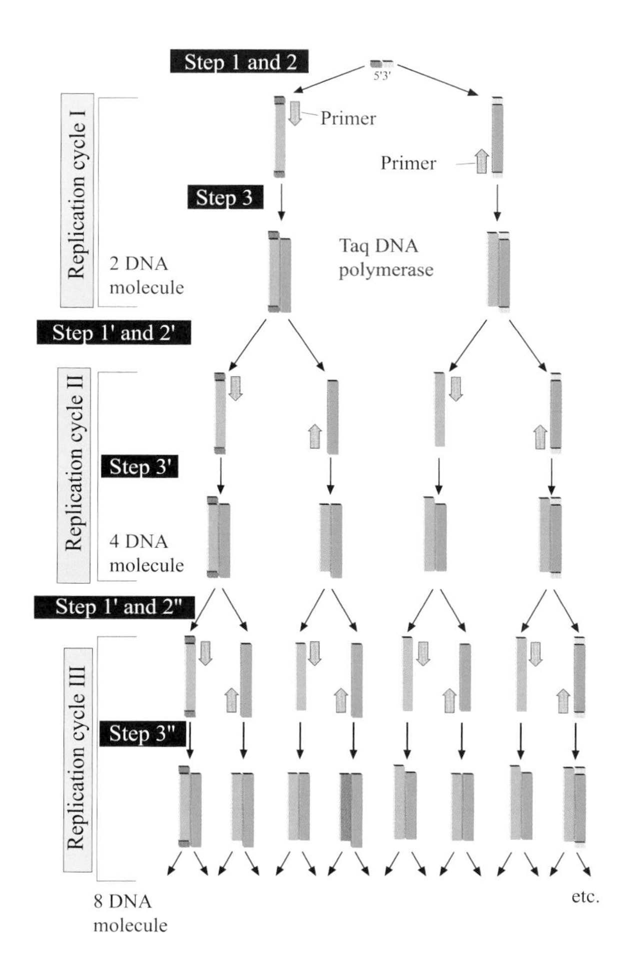

Amplification of DNA by PCR. A method developed by Kary Mullis. Double-stranded DNA is heated in order to separate the two strands and then cooled to allow DNA primers to bind to separate strands. *Taq* polymerase is then used to synthesize new DNA strands from the primers. This process is repeated for multiple cycles, each resulting in a twofold amplification of DNA.

Mullis include "The Cosmological Significance of Time Reversal" (Nature) and "The Unusual Origin of the Polymerase Chain Reaction" (Scientific American). He has also written an autobiographical book entitled "Dancing Naked in the Mind Field" (1998). Kary Mullis is currently working at Burstein Technologies in Irvine, California, where he is vice-president and director of molecular biology. Burstein Technologies is developing a direct bridge between diagnostics and electronics based on optical disc technology.

Smith, Michael (Blackpool, England, April 26, 1932 -). British-born Canadian biochemist. Son of a market - gardener. Nobel Prize for Chemistry, along with **Kary Mullis**, for his development of a technique called oligonucleotide - based site - directed mutagenesis. Michael Smith's contributions in research have been in the field of the syn-

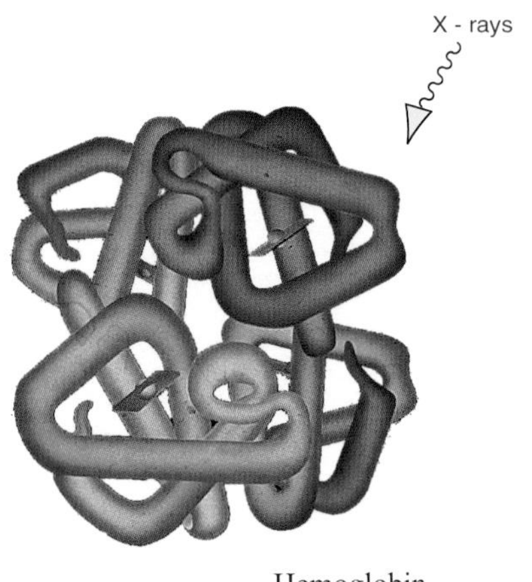

Hemoglobin
(normal form)

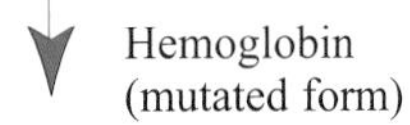

Hemoglobin
(mutated form)

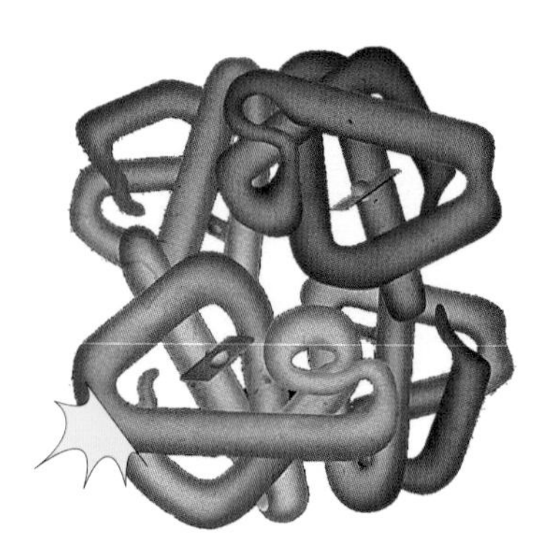

Site-directed mutagenesis. By such a technique, defined mutations can be made in vitro in a cloned DNA. Site-directed mutagenesis enable cloned genes (for example, genes that code for the hemoglobin molecule), to be mutated at will and the functions of these genes to be elucidated by the effects of introducing their mutated forms into an organism.

thetic chemistry of nucleotides and nucleic acids and its applications to studies in biochemistry and genetics. The most important of these studies has been in the use of synthetic DNA fragments, oligonucleotides, as probes to identify specific genes and to identify mutations within these genes as primers in the sequencing of double-stranded deoxyribonucleic molecules and as tools to create specific mutations, site directed mutagenesis, in DNA. The applications of directed mutagenesis are many. It is possible, for instance, to create proteins in which certain known properties are accentuated, for use in diverse fields. Or to attempt to improve agricultural yields by creating a photosynthetic protein with improved carbon-dioxide - absorbing properties. An important part of his career took

Michael Smith

place at the University of British Columbia, where he directed the Department of Biotechnology.

1994

Olah, George Andrew (Budapest, Hungary, May 22, 1927 -). Hungarian-American chemist. Naturalized US citizen in 1971. Nobel Prize for Chemistry for work on carbocations. George Olah was educated in Budapest, Hungary. From 1949 to 1954 he was assistant and then professor of organic chemistry at the Technical University,

George Olah

Budapest. From 1954 to 1956, he was Head of the Department of Organic Chemistry and Associate Scientific Director of Central Research Institute of the Hungarian Academy of Sciences. After emigrating from Hungary during the revolution of 1956 he held a post as a research scientist at the Dow Chemical Company in Ontario, Canada. In 1964 he moved to the United States. A detailed understanding of reactive intermediates is at the very heart and the essence of chemical transformations

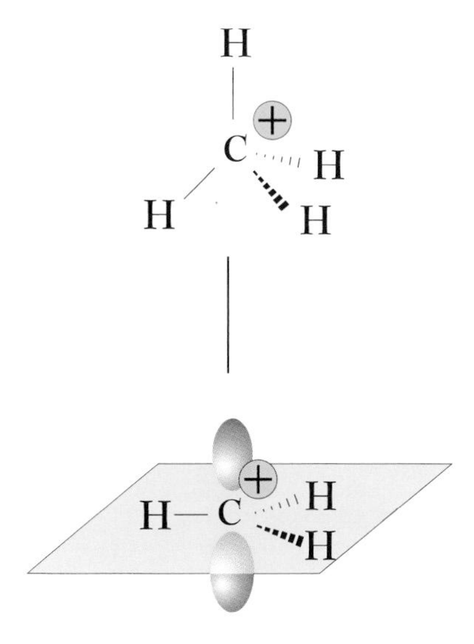

Carbocation. An organic molecule which plays an important role as a reactive intermediate. It can only be isolated and studied in an extremely acidic medium.

and, hence, of modern chemistry. Carbocations and related electron-deficient species represent the most important intermediates in all of organic chemistry. George Olah discovered ways to observe carbocations as persistent, long-lived species, as the Swedish Academy of Sciences wrote "He gave the cations of carbon longer life". His development of superacids, billions of times stronger than sulfuric acid, and superelectrophiles has changed the field of chemistry. The methods and reactions pioneered by Olah are now being applied in organometallic, synthetic organic, bioorganic and industrial chemistry. Of major significance is Olah's realization that carbon in many of its electron deficient species, including carbocations, can simultaneously bound to five, six even seven neighboring atoms or groups (hypercarbon compounds). This extends Kekule's concept of the limiting four valency of carbon to higher coordinate carbon compounds and opened up the field of hypercarbon chemistry. Olah's discoveries in the field of carbocations and novel electrophiles provided the insights into his fundamental discoveries of electrophilic activation of C - H and C - C single bonds and form the basis of the development of new and improved hydrocarbon transformations. These discoveries have major and significant implications in petroleum chemistry including improved production of high octane gasoline. George Olah is now professor and director at Loker Hydrocarbon Research Institute, University of Southern California.

Paul Crutzen

1995

Crutzen, Paul (Amsterdam, The Netherlands, December 3, 1933 -). Dutch chemist. Shared the Nobel Prize for Chemistry with **Mario Molina** and **Franck Sherwood**

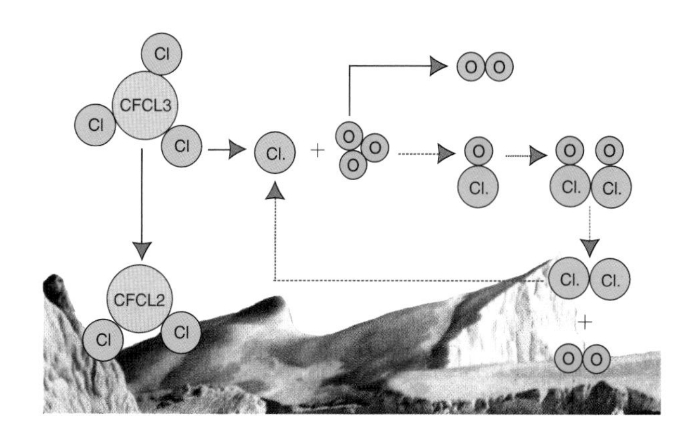

Destruction of stratospheric ozone. Decomposition of chlorofluorocarbons (CFCs) in the stratosphere by UV light gives rise to chlorine atoms. These atoms attack ozone and have a depleting effect on the ozone layer.

Rowland for demonstrating, in 1970, that chemical compounds of nitrogen oxide accelerate the destruction of stratospheric ozone. By explaining the mechanisms that affect the ozone layer, these three researchers contributed to alerting the world to a problem which, if not solved, could have catastrophic consequences. Crutzen obtained his Ph.D. in sciences from the University of Stockholm in 1973. His main discoveries were concerned with the very large works of the highly reactive NO and NO_2 as catalysts in ozone destruction or production. Ozone destruction occurs in the stratosphere, above about 15 km, ozone destruction in the troposphere, below 10 km. Because human activities add significant quantities of NO both to the stratosphere, in the aircraft exhaust, and the troposphere, high temperature fossil fuel combustion and biomass burning, both regions of the atmosphere are in opposite ways. Crutzen's studies in 1970 and 1971 showed for the first time, that mankind was able to affect the global ozone layer.

Mario Molina

Molina, Mario (Mexico City, Mexico, March 19, 1943 -). Mexican-born American chemist. Nobel Prize for Chemistry, along with **Frank Sherwood Rowland** and **Paul Crutzen**, for research in the 1970s concerning the decomposition of the ozonosphere. Molina's work involves the chemistry of the stratospheric ozone layer and its susceptibility to human-made perturbations. Mario Molina graduated in chemical engineering from the University of Mexico. After moving to Europe he joined the University of California, Berkeley and gained his Ph.D. in chemistry in 1972. He then met Rowland and worked with him as a postdoctoral student. The collaboration that began between these two researchers was to mark the beginning of a famous scientist's career. In 1974, together with F.S. Rowland, he predicted that chlorofluorocarbon (CFC) gases that were being used as propellants in spray cans, as refrigerants, and as solvents, would eventually deplete the ozone layer, laying the groundwork for the discovery in 1985 of the "ozone hole" over Antartica. He also proposed and demonstrated in the laboratory a new reaction sequence which accounts for most of the observed ozone depletion in the polar stratosphere, as well as a fundamentally new chemical reaction occuring on the surface of ice cloud particles involving decomposition products from the CFCs. More recently, he has also been involved with the chemistry of air pollution of the lower atmosphere. He is also pursuing interdisciplinary work on tropospheric pollution issues, working with colleagues from many other disciplines on the problem of rapidly growing cities with severe air pollution problems. Molina went to the Massachusetts Institute of Technology (MIT) in 1989 with a joint appointment in the Department of Earth, Atmospheric and Planetary Sciences and the Department of Chemistry and was named MIT Institute Professor in 1997. He has served on the President's Committee of Advisors in Science and Technology, the Secretary of Energy Advisory Board, and on the boards of US-Mexico Foundation of Science and other non-profit environmental organizations. Professor Molina is currently Institute Professor at the Massachusetts Institute of Technology.

Franck S. Rowland

Rowland, Frank Sherwood (Delaware, Ohio, June 28,1927). American chemist. Nobel Prize for Chemistry with **Mario Molina** and **Paul Crutzen** for research on the depletion of the Earth's ozone layer. Doctor of chemistry from the University of Chicago, Rowland first worked under the direction of Libby to perfect the carbon-14 dating method. He then went to Princeton University, and in 1956 he became a professor at the University of Kansas. He went on to the same position at the University of California at Irvine, where he was also Director of the Department of Chemistry. Influenced by the ecological movement, Rowland began in 1972 to explore the potential dangers of chlorofluorocarbons (CFCs), substances being produced at an increasingly rapid rate since the 1940s. Actually, the production of CFCs was doubling nearly every five years and measured around 10 million tons per year. Where it was a question of perfectly inert molecules, at least in the troposphere, the scientific and industrial world

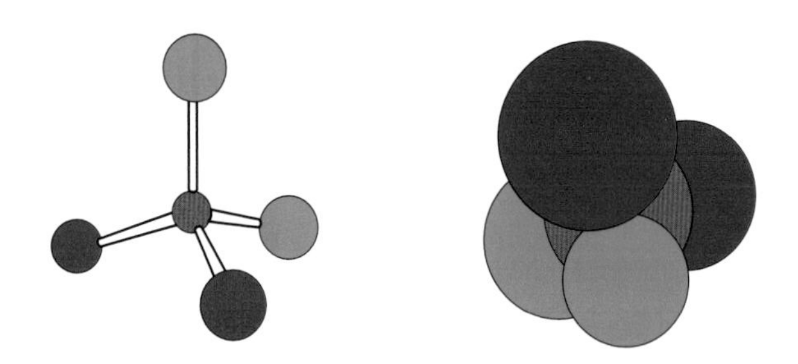

Ozon and chlorofluorocarbon

was convinced that they were perfectly innocuous. Nevertheless, it is precisely this inertness in the lower layers of the atmosphere to which the molecules owe their extended life expectancy, to the point where it is possible for them to migrate toward the stratosphere. Rowland and Molina's demonstration of the harmful nature of CFCs led to the signing of the Montreal Protocol, by which the leading industrial nations attempted to reduce their CFC production by 50% by the year 2000. Rowland continued to study the terrestrial atmosphere on a global scale. Conscious of the significant impact that human activity could have on the chemical stability of the atmosphere, he undertook the task of analyzing the composition of air everywhere industrial development posed a threat to Earth's atmosphere. Sherwood Rowland is the author of over 330 scientific publications in the areas of atmospheric chemistry, radiochemistry, and chemical kinetics. He is an elected member of the U.S. National Academy of Sciences, for which he is currently the Foreign Secretary, and has served as President and chairman of the board of the American Association for the Advancement of Science.

1996

Curl, Robert Floyd (Alice, Texas, August 23, 1933 -). American chemist. Nobel Prize for Chemistry along with **Richard E. Smalley** and **Harold W. Kroto** for discovery of the fullerene. During a brief period in the fall of 1985, Harry Kroto, Rick Smalley, Jim Heath, Sean O'Brien and Robert Curl discovered that carbon vapor condensing under the right conditions forms a new class of spheroidal cage molecules, the most abundant of which is a highly stable, highly symmetric cage isomer of C60 with a bonding pattern resembling the seams on a soccer ball. They dubbed these carbon cage molecules fullerenes in honor of the geodesic domes of R. Buckminster Fuller, which they resemble. It was soon realized that C60 was only the most stable member of a whole class of spheroidal cage carbon molecules named fullerenes. It was found that a metal atom could be introduced inside the C60 cage and that C60 is unusually inert to chemical reagents and highly resistant to photodissociation. As a result of this and subsequent work based upon it, chemists have a new field of organic chemistry to investigate, and materials scientists have a new form of carbon to explore and potentially exploit. Robert Curl received his Ph.D. in

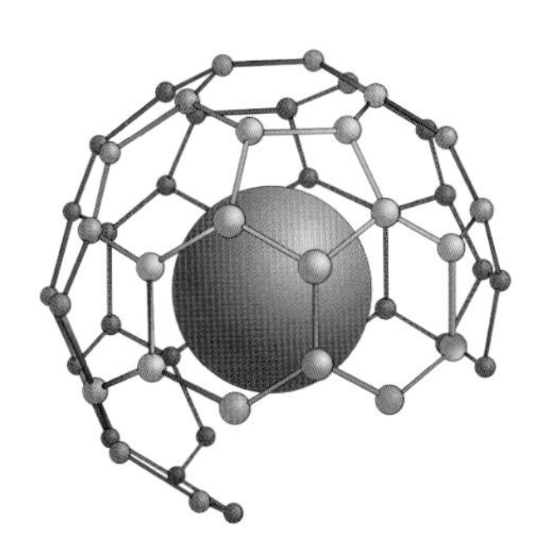

Robert Curl

Robert Curl in his laboratorium

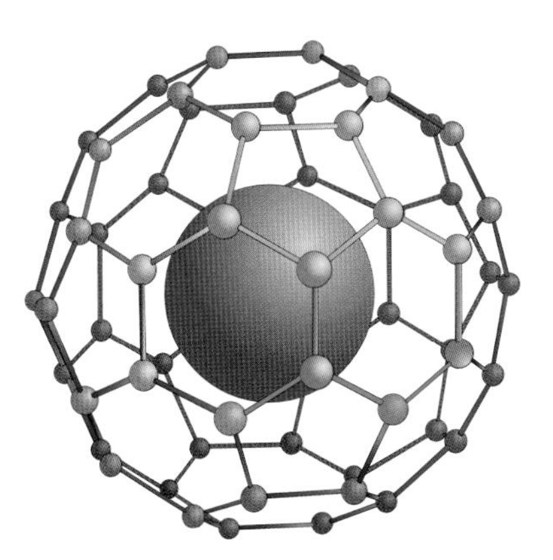

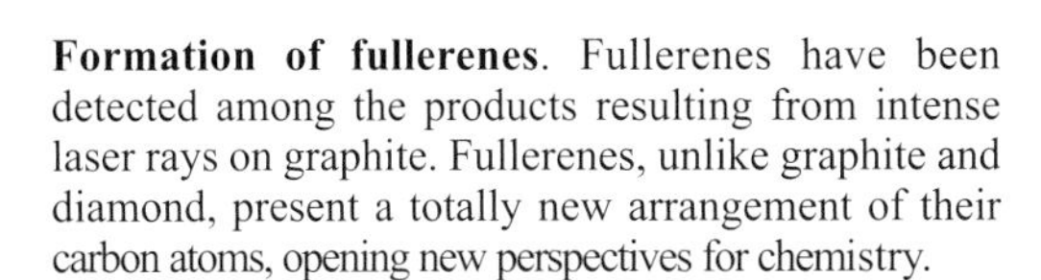

Formation of fullerenes. Fullerenes have been detected among the products resulting from intense laser rays on graphite. Fullerenes, unlike graphite and diamond, present a totally new arrangement of their carbon atoms, opening new perspectives for chemistry.

Galaxy centaurus. Since Urey hypothesis, several scientists have proposed that the building blocks of life arrived from space. The space molecules discovered until now range from water, ammonia or hydrogen cyanide to more complex substances, including chains of up to eight carbon atoms.

chemistry from the University of California in 1957 but most of his scientific career took place at Rice University, Houston, where he held a post as Head of the Biochemistry Department until 1996.

Harold Kroto

Kroto, Harold W. (Wisbech, England, October 7, 1939 -). British chemist. Nobel Prize for Chemistry, along with **Richard E. Smalley** and **Robert F. Curl** for his discovery of the carbon compounds called fullerenes. Research fields conducted by Kroto cover three major topics:**1** - Earlier research focused on the creation of new molecules with multiple bonds between carbon and elements, mainly of the second and third row of the periodic table, which were reluctant to form such a link. These studies showed that many of these previously assumed impossible species could be produced, studied by spectroscopy and used as valuable synthons leading to a wide class of new phosphorus-containing compounds. In particular the spectroscopic studies of molecules with carbon-phosphorus multiple bonds were the pioneering studies that led to the now extremely active fields of phosphoalkene/alkyne chemistry. **2** - Laboratory and radioastronomy studies on long linear carbon chain molecules, the cyanopolyynes, led to the surprising discovery, by radioastronomy, that they existed in interstellar space and also in stars. Since these first observations the carbon chain has become a major area of modern research by molecular spectroscopists and astronomers interested in the chemistry of space. **3** - The revelation (1975-1980) that long chain molecules existing in space could not be explained by the then accepted ideas on interstellar chemistry and it was during attempts to rationalise their abundance that C60 Buckminsterfullerene was discovered. Laboratory experiments at Rice University, which simulated the chemical reactions in the atmospheres of red giant carbon stars, serendipitously revealed the fact that the C60 molecule could self-assemble. This ability to self- assemble has completely changed our perspec tive on the nanoscale behaviour of graphite in particular and sheet materials in general. The molecule was subsequently isolated independently at Sussex and structurally characterized. Present research focuses on Fullerene chemistry and the nanoscale structure of new materials, in particular nanotubes. Some parts of the research have been successful due to their interdisciplinary nature and this has been the result of synergistic collaborations involving primarily:colleagues J.F Nixon, R. Taylor and D.R.M. Walton at Sussex University, T.Oka at NRC, Canada, and R.F. Curl and R.E. Smalley at Rice University, Texas. Harold Kroto started his academic career ay the University of Sussex, Brighton, in 1967, where he became a professor in 1985 and in 1991 he was made a Royal Society Research Professor.

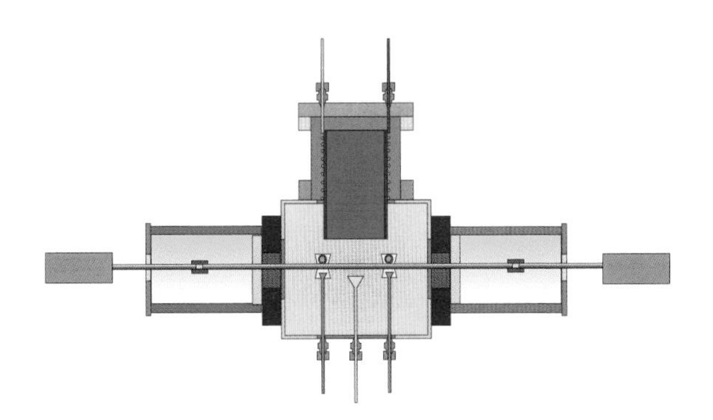

An apparatus elaborated by R. Smalley in order to product fullerenes.

Smalley, Richard E. (Akron, Ohio, June 6, 1943 -). American chemist. Nobel Prize for Chemistry with **Robert F. Curl**, Jr., and **Harold W. Kroto** for his discovery of the fullerenes. His research at Rice University has

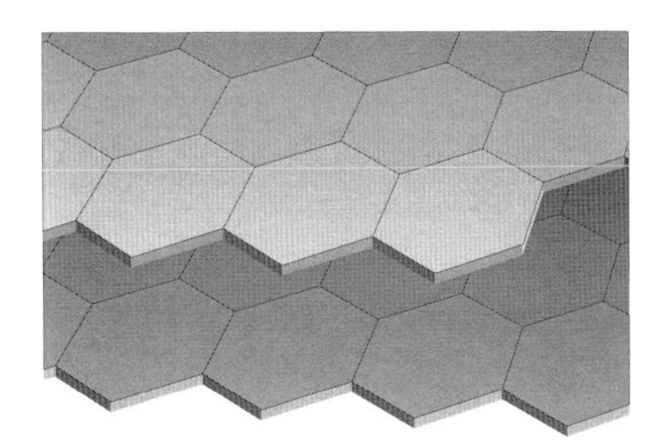

Graphite, a polymorph of the element carbon

Richard Smalley

made pioneering advances in the development of new experimental techniques such as super-cold pulsed beams, ultrasensitive laser detection techniques, a laser - driven source of free radicals, triplets, metals, and both metal and semiconductor cluster beams, and has applied these techniques to a broad range of vital questions in chemical physics. Smalley is widely known for the discovery and characterization of C60, Buckminsterfullerene, a soccer-ball - shaped molecule which, together with other fullerenes such as C70, now constitutes the third elemental form of carbon

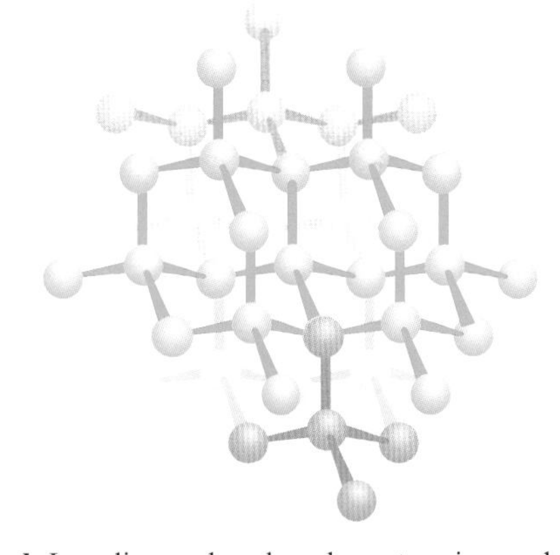

Diamond. In a diamond each carbon atom is covalently bonded to four others, making a tetrahedron. The tetrahedrons are linked together to form the carbon crystals known as a diamond.

after graphite and diamond. His group has also been the first to generate fullerenes with metals trapped on the inside. His current research is focused on the production of continous carbon fibers which are essentially giant single-fullerene molecules. Just a few nanometers in width, but many centimeters in length, these fullerene fibers are expected to be the strongest fibers ever made, 100 times stronger than steel. Professor Smalley received his Ph.D. from Princeton in 1973, with an intervening period in industry research with Shell. During an unusually productive postdoctoral period with L. Wharton and D. Levy at the University of Chicago, he pioneered what has become one of the most powerful techniques in chemical physics:supersonic beam laser spectroscopy. Apart from the Nobel Prize in Chemistry, he is the recipient of numerous awards.

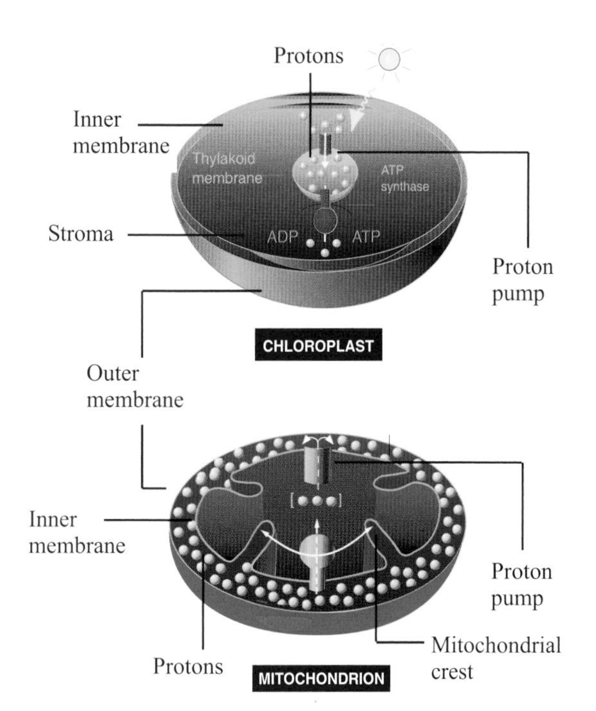

ATP synthase in mitochondria and in chloroplast. The enzyme functions essentially in the same manner, driven by a proton flux. In mitochondria, protons are pumped from the matrix space to the intermembrane space. In chloroplast, protons are translocated from the stromal side of the thylakoid vesicle into the lumen of the organelle. (See Boyer.)

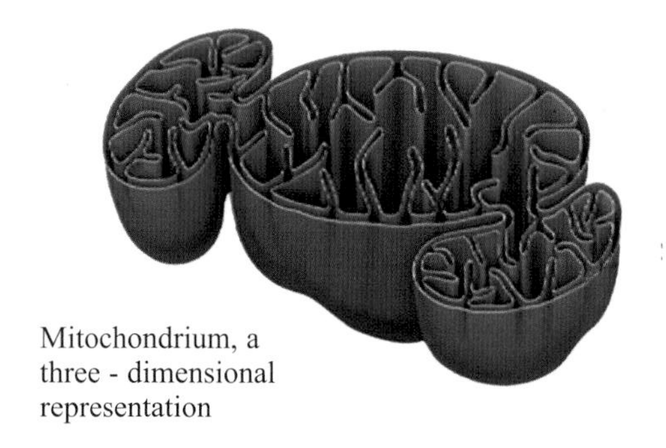

Mitochondrium, a three - dimensional representation

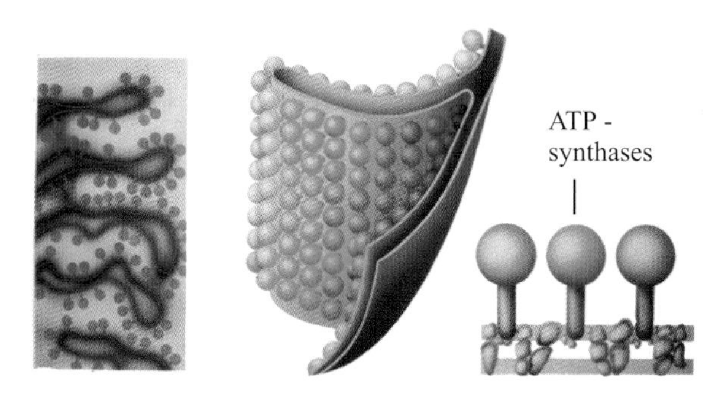

Fine structure of a mitochondrium. As shown, the organelle is bounded by an envelope of two membranes. While the outer is smooth, the inner membrane is convoluted, having infoldings called cristae.Many proteins that function in respiration, including ATP synthase, are built into the inner membrane. (See Boyer.)

1997

Boyer, Paul (Provo, Utah, July 31, 1918 -). American biochemist. Nobel Prize for Chemistry, along with **John E. Walker** and **Jens C. Skou**, for their explanation of the enzymatic process involved in the production of the energy-storage molecule adenosine triphosphate. In living organisms a key enzyme, ATP synthase, serves to capture energy from oxidations or light to make ATP. An active person will make and use nearly a body weight of ATP in a day. Based on isotope exchange and related studies, Paul Boyer proposed a binding change mechanism in which the principal role of energy is to bring about a release of tightly bound ATP. This is accomplished by changes of three catalytic sites such that as one site binds ADP and inorganic phosphate, a second site is catalysing formation of tightly bound ATP, and a third site is releasing ATP. All three sites passed sequentially through the same conformations. The changes were proposed to be driven by a novel rotational catalysis in which proton translocation across the membrane in which the ATP synthase resides

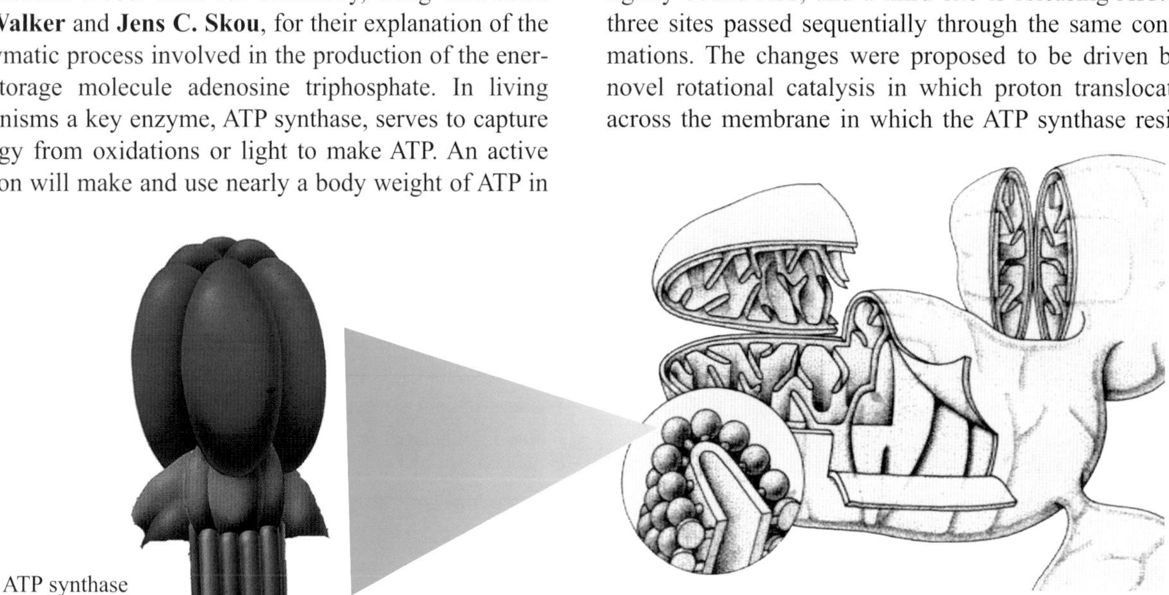

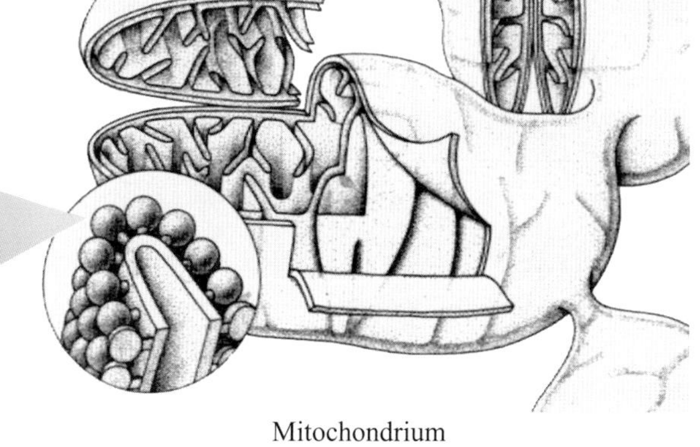

Mitochondrium

Mitochondrium, a schematic drawing The outer membrane has been partially removed in order to show the interior of the organelle. ATP synthase molecules appear like tiny anchored spheres in the inner membrane.

Paul Boyer

causes rotation of single copies of small subunits. This rotational movement brings about the sequential conformational changes of the circularly arranged subunits that form the catalytic sites. Boyer shared the Nobel price with John Walker, of Cambridge University, whose research group achieved a three-dimensional structure of the enzyme that confirmed the principal features of Boyer's suggested mechanism.

Skou, Jens C. (Lemvig, Denmark, October 8, 1918 -). Danish biophysicist. Son of a coal merchant. Nobel Prize for Chemistry, along with **Paul Boyer** and **John E. Walker**, for his discovery of the sodium-potassium-activated adenosine triphosphatase (Na^+-K^+ ATPase). This pump controls the coupled movement of K^+ and Na^+ ions against their electrochemical gradients through the lipid bilayer of the plasma cell membrane. It is embedded in this membrane in such a manner that it has access to both sides. Jens Skou and his collaborators showed the the protein complex has an ATP - binding site and three high-affinity sites for binding Na^+ ions on his cytosolic surface. Two high-affinity binding sites for K^+ ions exist on the outer site of the enzyme. Such pumps also exist in other parts of the cell, for example, the Ca^{2+} ATPase pump of the red blood cells which holds the cytoplasmic Ca^{2+} concentration at a low level. Another Ca^{2+} ATPase pump exists in the sarcoplasmic reticulum. In 1947 Skou got a position at the Institute for Medical Physiology at Aarhus University. In 1963, he was appointed professor and chairman. Jens Skou was appointed as a professor of biophysics in 1977. He is now a member of the Danish Academy of Sciences.

Walker, John E. (Halifax, England, January 7, 1941-). British chemist. Nobel Prize for Chemistry, shared with **Paul Boyer** and **Jens Skou** (who received the Nobel Prize for discovering of Na^+/K^+ adenosine triphosphatase) for their explanation of the enzymatic process that creates adenosine triphosphate. In the early 1980s John Walker was working on mitochondria. He developed an interest in the enzyme complexes that are embedded in the inner membrane of the organelle, which carries out oxidative phosphorylation. Such enzyme complexes also exist in chloroplast and in aerobic or photosynthetic bacteria.

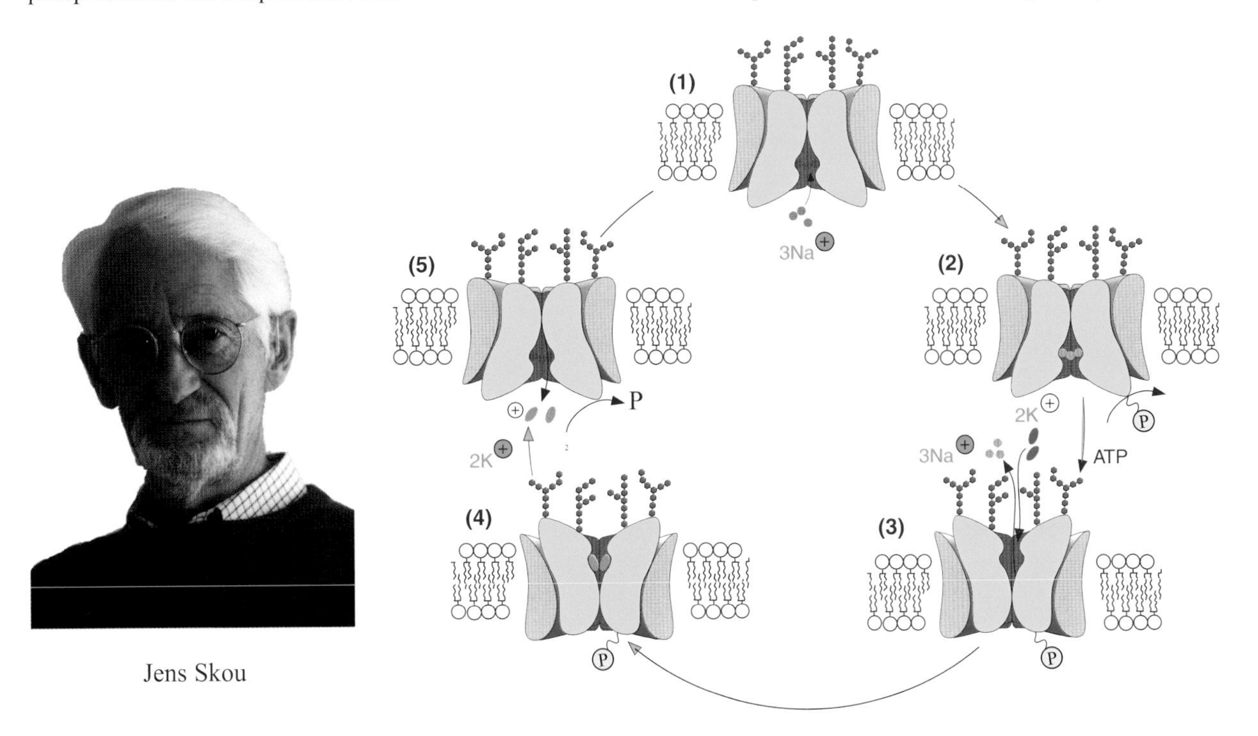

Jens Skou

NA⁺ / K⁺ pump. Also called Na^+ / K^+ ATPase. This ion pump is responsible for maintaining gradients of Na^+ and K^+ ions across the plasma membrane. Transport of these ions against their electrochemical gradient is driven by ATP hydrolysis. During this process, ATP - driven conformational changes occur in the protein complex.

Walker's early work focused on the study of the structure of enzymes such as the cytochrome C oxydase. With his collaborators he showed that some subunits of this electron carrier were encoded by the mitochondria. Later, he began a structural study of the ATP synthase from bovine heart mitochondria. This work eventually resulted in a complete amino acid sequence analysis of the complex and in an atomic resolution structure of the F_1 catalytic domain of the enzyme. Walker received a bachelor's degree in chemistry in 1964. Then he began research on peptide antibiotics with E.P. Abraham in the Sir William Dunn School of Pathology in Oxford. He was awarded the D.Phil. degree in 1969. From 1969 to 1971 he taught at the School of Pharmacy of Wisconsin University and went to Paris in 1971. In 1974, he moved to the Protein and Nucleic Acid Chemistry Division of the Laboratory of Molecular Biology in Cambridge. There, he became senior scientist in 1982 and was finally named Director of the Medical Research Council's Dunn Human Nutrition Unit in 1998. John Walker received a Knighthood in 1999 and was named an honorary member of the British Physical Society in 2000.

1998

Kohn, Walter (Vienna, Austria,1923-) Nobel Prize for Chemistry along with **John A. Pople** for their development of computational methods for use in quantum chemistry. Born into a middle class Jewish family in Vienna, Walter Kohn was forced to emigrate due to the rise of nazism in 1939, first to England and one year later to Canada. He began his armed training in Kent. But, due to a serious illness, he decided to attend the East Grinstead County School in Sussex where he began learning mathematics, physics, and chemistry. In 1940, he left for Quebec City and was transferred among various refugee camps until he passed the McGill University admission examinations in mathematics, physics, and chemistry. The last year of World War II, he was enlisted in the Canadian Infantry Corps. After the war, he continued his Ph.D. at Harvard with Julian Seymour Schwinger (1965 Nobel Prize for Physics) who was already one of the most famous theoretical physicists in the scientific world. His 1948 thesis contains the first formulation of the theorem known today as Kohn's variational principle for scattering. The theorem is of the greatest use in nuclear atomic or molecular problems. Under the influence of John **Hasbrouck van Vleck** (1977 Nobel Prize for Physics), he moved towards the field of solid - state physics and found a position in the Polaroid Laboratories in Cambridge, Massachusetts. During the 1950s, he worked at Copenhagen with Bohr, Princeton with Robert Oppenheimer, and the Carnegie Institute of Technology (currently Carnegie Mellon University). Kohn obtained US citizenship in 1957. He was a professor at the Carnegie Institute of Technology in Pittsburgh from 1959 to 1960 and at the University of California in San Diego from 1960 to 1979. From 1979 to 1984, he was Director of the Institute of Theoretical Physics at the University of California at Santa Barbara, where he is still active. Usually, the calculation of the structure and of the properties of molecules is based on the description of the motion of the various electrons. Those calculations are mathematically very demanding due to the large number of integrals to be calculated and the large size of the systems of equations to be solved. Kohn proved that it is not necessary to consider the individual motion of each electron. It is sufficient to know the electronic density (a concept connected with the average number of electrons) at any point in space. The famous Hohenberg-Kohn theorem was formulated in 1964. It proved that the total energy of a system described by the laws of quantum mechanics is uniquely determined by the spatial distribution of this

Walter Kohn

electronic density. This Hohenberg-Kohn theorem and the subsequent Kohn-Sham theorem made it possible to achieve an effective method of calculation of molecular properties known as DFT (density functional theory). One unsolved question is to establish how energy depends on electron density. The field remains the object of intense research.

Pople, John A. (Burnham-on-Sea, England, October 31, 1925 -). British chemist. Nobel Prize for Chemistry shared with **Walter Kohn** "for their development of computational methods for use in quantum chemistry". John Pople entered Trinity College, Cambridge, UK, in October 1943 to study pure mathematics. After the war, he decided to put his mathematical skills into practice and, after comparing the current's fields of research, he turned toward theoretical chemistry with Sir John Lennard - Jones beginning in 1948. In 1951, he obtained his Ph.D. in Mathematics at Cambridge and became a lecturer at the same institution from 1954 to 1958. In 1952, he planned to simulate the molecular properties of chemical species using the available mathematical models and thus to develop the new field of computational chemistry. His research on p-conjugated hydrocarbons, parallel to the independent work of Pariser and Parr, led in 1953 to what is now called the PPP method (Pariser-Parr-Pople). In 1964, he was appointed Professor of Physical Chemistry at Carnergie-Mellon University where he had been invited a few years earlier by Robert Parr. There he decided to develop his ambitious project of computational chemistry. This project would prove particularly judicious because of the emerging development of increasingly more powerful electronic computers. In the early 1960s, he proposed an improvement of the PPP method, known as CNDO (complete neglect of differential overlap). Contrary to the PPP method, restricted to the p-electrons of conjugated molecules, CNDO is able to handle all the valence electrons of a molecule. In one step characteristic of Pople's scientific approach, this CNDO method would be logically extended to two other more precise techniques: INDO (intermediate neglect of differential overlap) and NDDO (neglect of diatomic differential overlap). All the semi-empirical schemes still used today are directly issued from these three methodologies. At the beginning of the 1970s, Pople and his collaborators developed the first prototype of a series of programs able to treat all the electrons (σ or π, core or valence) of a molecule by non - empirical so - called *ab initio* methods. It is the birth of the series of Gaussian programs. In 1986, Pople became a professor of chemistry at Northwestern University where he is still active. We are indebted to Pople for his coherent project to develop computational methods that allow a precise theoretical study of the electronic structure of molecules. He has popularized the computational techniques of increa- sing complexity by designing the Gaussian series of computers programs and making them easily accessible to researchers. He is coauthor of three books on high nuclear magnetic resonance and molecular orbital theory. In 1962, Pople also published with Walmsley an important paper that contains many of the ingredients of the soliton theory for explaining the properties of conducting polymers. It is interesting to note that the development of conducting polymers in 1977 would lead to the 2000 Nobel Prize for Chemistry being awarded to Alan Heeger, MacDiarrmid, and Hideki Shirakawa. In the 1960s, John Pople was also instrumental in developing with Per-Olov Löwdin, Raymond Daudel, and Robert Parr, among others, the International Academy of Quantum Molecular Science that has largely contributed to the organization of a close community of scientists in the field of quantum chemistry.

John Pople

1999

Zewail, Ahmed H. (Damanhur, Egypt, February 26, 1946 -). Nobel Prize for Chemistry for the pioneering investigation of fundamental chemical reactions, using ultra - short laser flashes, on the time scale on which the reactions actually occur. Ahmed Zewail studied chemistry at the Faculty of Science of the University of Alexandria. He received a grant to the University of Pennsyl-vania and graduated there with a Ph.D. in 1974. After two years at the University of California at Berkeley, he joined the California Institute of Technology (CalTech) where he has occupied the prestigious Linus Pauling Chair of Chemical Physics since 1990. Zewail has both Egyptian and American citizenships. The work that led the Nobel Prize goes back to the end of the 1980s when Zewail conceived a series of experiments which would give rise to the new field of femtochemistry. In particular, his technique makes it possible to view the atoms and the molecules in real time and to see how the chemical bonds are broken or created in a chemical reaction. It allows one to catch images of the reagents in the state of transition which leads to the products of the reaction. The time taken by the atoms in a molecule to perform one vibration is typically 10 femtoseconds (10^{-15} s). It corresponds to a rate of atomic motion of the order of one mile per second. Since these displacements cover a distance of about one Angström (10^{-9} m, one millionth of one millimeter), it is necessary to observe the molecules each femtosecond to see a movie of the reorganization of the atoms during chemical reaction. Zewail's technique may thus be described as the fastest camera in the world. It is obtained by laser flashes of very short duration. The field of femtochemistry has helped to understand why certain chemical reactions take place but others do not. It also explains the dependence of reaction

Ahmed Zewail

rates and yields with temperature. The applications are numerous. They cover fields as varied as the mechanisms of catalysis, the design of molecular electronic components, or the fundamental mechanisms of life processes.

2000

Heeger, Alan J. (Sioux City, Iowa, January 22, 1936-). Alan Heeger went to the University of Nebraska for his undergraduate studies, where he first studied engineering before changing to physics. There he developed his interest in science in general and particularly in mathematics. He graduated from the University of California, Berkeley, and received his Ph.D in condensed matter physics. Then he started an academic career at the University of Pennsylvania where, over the next 20 years, he climbed the ranks to the position of Vice - Provost for research. The main part of the research for which he was awarded the Nobel Prize for Chemistry in 2000 with Alan **MacDiarmid** and **Hideki Shirakawa** was conducted during those years at Penn State. In 1977, they discovered the high electrical conductivity of polyacetylene chains doped by halogens. The disco- very opened a field of research of great of importance for chemists as well as physicists. These new materials are now routinely available for various technological applications in molecular electronics. After this essential discovery Heeger moved, in 1982, to the the University of California at Santa Barbara (UCSB) under pressure from **John Schrieffer**, who won the 1972 Nobel Prize for Physics for the BCS theory (**Bardeen-Cooper**-Schrieffer theory) of supraconductivity. He is a Professor of Physics and Director of the Institute of Polymers and Organic Solids. In 1990, he founded the UNIAX Corporation that would develop the technological applications based on conducting and semiconducting polymers. A physicist, Alan Heeger also developed fundamental physics concepts in order to understand the factors that determine the electronic, optical, and nonlinear optical properties of semiconducting and metallic polymers. With Su and Schrieffer, he proposed the SSH (Su, Schrieffer, Heeger) soliton theory of conducting polymers that is still applied in order to understand the characteristics of one-dimensional organic conductors.

Alan J. Heeger

MacDiarmid, Alan G. (Masterton, New Zealand, 1927-). Nobel Prize for Chemistry, jointly with **Alan Heeger** and **Hideki Shirakawa**, for the discovery of conducting polymers, now commonly named synthetic metals. Alan G. MacDiarmid received his higher education at the University of New Zealand. In 1953, he received a first Ph.D. at the University of Wisconsin before getting a second one at the University of Cambridge, United Kingdom, in 1955. The same year, he moved to the University of Pennsylvania where he began his academic career. Since 1988, he has been Blanchard Professor of Chemistry at the same University. In 1977, he performed the first chemical and electrochemical synthesis of doped polyacetylene,$(CH)_x$, the prototype conducting polymer. He also rediscovered polyaniline, one of the main conducting polymers used industrially. In 1973, he began his work on molecular crystals of $(SN)_x$, a unusual polymer with metallic conductivity (and even supraconductivity, at low temperatures). His interest in conducting organic polymers began in 1975 when Shirakawa showed him the metallic-like reflection of polyacetylene samples synthetized at the Tokyo Institute of Technology. The ensuing collaboration among MacDiarmid, Shirakawa, and Heeger led to the historical discovery of metallic conductivity in an organic polymer. The fact that an organic polymer could be easily doped to a metallic regime opened a completely new and unexpected field in the frontiers of chemistry and physics. The field of "electronic plastics" was born. The study of the interrelationships between the chemistry, the structure, and the electronic properties of semiconducting and metallic organic polymers remains one of the fields of essential concern to many scientists. Technological applications of these materials cover fields as various as rechargeable batteries, electromagnetic shielding, anti-static dissipation, corrosion inhibition, flexible plastic transistors, or electroluminescent polymeric diodes.

Shirakawa Hideki (Tokyo, Japan, 1936-). Japanese chemist. Born in Tokyo in 1936, Hideki Shirakawa received his Ph.D. from the Tokyo Institute of Technology in 1966 where he started his research career. In 1979, he was promoted to Associate Professor at the Institute of Material Science of the University of Tsukuba. He became a full Professor of Chemistry in 1982. Shirakawa explored a new field of the chemistry of polymers. He was awarded the 2000 Nobel Prize for Chemistry, along with **Alan Heeger** and **Alan MacDiarmid** for the discovery of conducting polymers also sometimes called synthetic metals. It was often said that this exceptional discovery resulted from an accident. An 1,000 times over-

Hideki Shirakawa

sized catalyst concentration was added during the synthesis of polyacetylene and resulted in a beautiful silvery substance without metal conductivity. Lecturing in Japan in 1975, MacDiarmid invited Shirakawa to the University of Pennsylvania in Philadelphia as a post - doctoral fellow. During this stay, they came to a conclusion with Heeger that it is possible to introduce mobile charged carriers in polymers by doping polyacetylene by oxidation with halogen vapor. In 1977, Shirakawa, **Heeger**, **MacDiarmid**, and their colleagues published their discoveries in "Synthesis of Electrically Conducting Organic Polymers:Halogen Derivatives of Polyacetylene $(CH)_n$" (The Journal of Chemical Society; Chemical Communications). These exceptional discoveries opened a new field of research in the chemistry and physics of polymers. Since then, the field has grown greatly and has also led to many new applications in the present day. For example, LEDs (light emitting diode), new color screens and batteries. It is probable that in the 21st century this development will proceed further from polymer - based electronics to real molecular-based nano - scale integrated circuits instead of presently used silicon - based electronics.

Noyori, Ryoji (Kobe, Japan, September 3, 1938-) Nobel Prize of Chemistry shared with **William S. Knowles** and K. Barry Sharpless for his work on chirally catalysed hydrogenation reactions". Ryoji Noyori designed and synthesized the chiral disphosphine - binaphtyl currently called BINAP, whose complexes with transition metals are remarkable asymmetric hydrogenation catalysts. Some of them convert an unsaturated carboxylic acid to the anti-inflammatory agent naproxen with a yield of 92% and an enantiomeric purity of 97%. A single one of his catalyst molecules can yield millions of product molecules. His methods have gained very large practical importance in industrial productions of antibiotics and anti - inflammatory pharmaceuticals, agrochemicals, flavors, and fragrances. According the Royal Swedish Academy, the discoveries have had a very large impact on academic research and the development of new drugs and materials. They are now routinely used in many industrial syntheses of drugs and other biologically active compounds. Ryoji Noyori earned his Ph.D. in 1967 at Kyoto University. Since 1972, he has been Professor of Chemistry at Nagoya University and since 2000, Director of the Research Center for Materials Sciences at Nagoya University.

2001

Knowles, William S. (USA, 1917-). 2001 Nobel Prize for Chemistry shared with **Ryoji Noyori** and **K. Barry Sharpless** for his work on chirally catalysed hydrogenetion reactions. In the late 1960s, William Knowles showed that a chiral transition metal-based catalyst could transfer chirality to a non - chiral substrate, resulting in a chiral product. He helped to develop a commercial process that could be used to synthesize individual enantiomers of chiral compounds directly. After having been awarded the Nobel prize, Knowles said that the key message he would like to deli- ver is that, in his opinion, industry does too little exploratory research. His research provides an excellent example of how a modest and inexpensive exploratory effort in industry can produce significant results. His work has changed the face of part of modern medicine in fields like the treatment for Parkinson's disease by L-DOPA industrially synthesized. William S. Knowles earned a bachelor's degree in chemistry at Harvard in 1939 and obtained a Ph.D. in 1942 at Columbia University. He accepted a position at Monsanto Company, St Louis, directly after graduating from Columbia. Knowles retired in 1986.

William S. Knowles

Sharpless, K. Barry (Philadelphia, Pennsylvania, 1941-) 2001 Nobel Prize for Chemistry, shared with **Ryoji Noyori** and **William S. Knowles** for his work on chirally catalysed oxidation reactions. In 1980, Sharpless's group developed transition - metal and tartrate catalysts for the asymmetric epoxidation of allylic alcohols. He discovered that molecular sieves could be used to further improve the efficiency of asymmetric epoxidation. The latter process led to the ton - scale industrial production of chiral glycidols, which are used to synthesize beta - blocker heart medicines, among other products. It is also broadly used in producing medicines against ulcers. His research group is also conducting work on the study of acetylcholinesterase inhibitors as well as on neurogeneration agents for Huntington's Disease and drugs for oncogene-based human cancers. K. Barry Sharpless earned a Ph.D. in chemistry in 1968 from Stanford University. Since 1990, he is W.M. Keck Professor of Chemistry at the Scripps Research Institute, La Jolla.

K. Barry Sharpless

The Nobel Laureates in Physics (1901 - 2001)

Guy Demortier

The Nobel Laureates in Physics

Introduction

The cultured women and men of the beginning of the 21st century cannot omit to bring some interest to the knowledge of the universe. Since the moment when W. Röntgen was awarded the first Nobel Prize for Physics, more than 165 scientists all over the world have received this very great honor. Physics seems to be a difficult matter for most of us, but it is probably the scientific field in which the number of fundamental processes is the lowest. The daunting aspect of physics could probably be found in the fact that it is often associated with technology, or with advanced mathematics.

Our present knowledge of the laws of the universe allows us to understand most of the relations between the four field forces which govern all our physical universe which seemed to be disconnected only half a century ago. Weak nuclear force, electromagnetism and strong nuclear forces are understood in the present standard model. Only the gravitational force is not completely integrated to the other fields of forces in the frame of quantum mechanics. But as it could be understood in the subjects studied by recent Nobel Prize winners, this goal is close to being achieved.

The present part of this encyclopedical review of Nobel Prize winners for Physics from 1901 to 2001 gives a description of the works performed by those brilliant scientists, but also briefly reports on their lives, their personalities and their hobbies.

Thanks to the abundant illustration of what could be depicted without entering complicated theoretical investigation, this book should be accessible to a large audience and is mainly intended to encourage young people to enter faculties of sciences for their advanced studies.

Albert Einstein playing the violin and its famous mass - energy equation.

Wilhelm Röntgen

The ESRF (European Synchrotron Radiation Facility) installed in Grenoble which produces a very high flux of monochromatic X - rays.

1901

Röntgen, Wilhelm Conrad (Lennep, Prussia, March 27, 1845 - Munich, Germany, February 10, 1923). German experimenter physicist. In November 1895, Röntgen discovered that a high voltage inducing current, in a tube of rarefied gas, gave birth to a glow on a screen covered with platinum cyanide. Within a few weeks, he had provided the correct explanation of the phenomenon, in ten separate articles, which he submitted to the medicine and physics society at Würzburg, the university where, two years later, he would occupy the professorship of physics. He did not think of patenting his discovery which was quickly turned to good account in medicine. A radiography of the hand of a surgeon friend and then of his wife's (and of both rings that she wore) made her jump: she directly associated this image of the skeleton with that of death. This fear was justified, since many experimentalists who carried out this research suffered from the harmful effects of these rays, or X-rays as Röntgen himself termed them. During one of the conferences that Röntgen gave on the subject (immediately after his discovery) in front of a gallery of high personalities, one of them had, however, proposed to call them "Roentgen stralhen"; but the name given by their author still prevails today. The discovery of the X - ray occupied the widest scientific community during a good quarter century and their applications were crowned by the attribution of the Nobel Prize in Physics to **Max von Laue**, **William Henry** and **William Lawrence Bragg**, **Charles Barkla**, **Karl Manne Siegbahn**, indirectly to **Albert Einstein** and **Arthur Compton**, but also for having

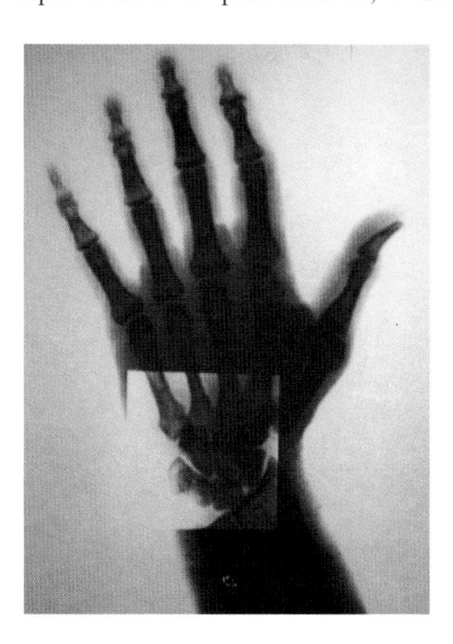

The radiography of the hand of G. Charpak by the conventional method with, in the insert, the improved resolution using the multi-wires detector.

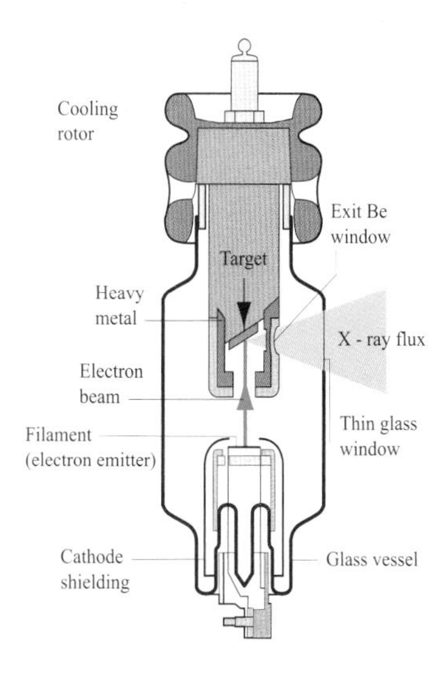

X-ray tube. A modern version.

applied their diffraction to the study of powders to **Peter Debye**, (1936 Nobel Prize for Chemistry). The use of X - rays is part of our modern life: beyond their usefulness in detecting opaque areas in our body, they are still used with an identical aim for the non-destructive control of luggage before it is loaded on aircraft, for the examination of often unique archaeological objects and for the identification of crystalline structures. Large particle accelerators (synchrotrons) produce them in abundance for very specific uses in elemental analysis, crystal structure, and probably soon for medical diagnosis. Born of weaver parents Röntgen spent a part of his chilhood in Holland, his mother's native country. He studied mechanical engineering at the Federal Polytechnic School of Zurich. He graduated from there in 1868 and presented his Ph.D. the following year. He was especially interested in thermodynamics and in the physics of gases. Some time later he worked as an assistant, then as a professor in Strasbourg (then in Germany), Giessen, Würzburg and Munich. In Würzburg, he had as colleague and research companion R. Kohlrausch (whose chair of physics he later occupied) and **Albert Michelson**. He was well - known both for his innovative theoretical interpretations as well as for his ability in the design of equipment for verification of the theories. Röntgen did not have any children but took care of one of his nieces. He liked excursions, climbing and hunting: he was a fine shot despite his difficulty in distinguishing the reddish-brown game in green foliage due to his colour blindness. A widower after 1919, he bemoaned this state of loneliness to his numerous friends. An upright man, full of humour and curious about all that occurred in science, he was strongly self - critical. The Nobel Prize Committee certainly made a happy choice by awarding him the very first Nobel Prize in Physics.

1902

Lorentz, Hendrik Anton (Arnhem, The Netherlands, July 18, 1853 -Haarlem, The Netherlands, February 5, 1928). Lorentz, born in a country which since the Middle Ages has been the source of intense intellectual activity, was one of the greatest figures of all time in theoretical physics and was universally recognised to have inaugurated, during half a century of prodigious activity, a new era in the history of physics. When, in 1886, **Peter Zeeman** (who shared the Nobel Prize with him) highlighted the displacement of the frequencies of light emitted by atoms submitted to a strong magnetic field, it was Lorentz who gave a simple explanation for it, connected with the existence, in matter, of very light particles subject to a force which makes them "oscillate". In 1904, two years after receiving the Nobel Prize, he probably made his most important contribution to theoretical physics, by showing the inaccuracy of the transformations of space and time into systems in relative motions, which were universally accepted at that time and known as Newtonian laws of transformations. Based on calculations showing at first that the length of bodies traveling at speed v relative to an observer was contracted by a factor $\sqrt{1 - v^2/c^2}$ (c, the speed of light) while the time measured by this same observer was increased by the same factor, he demonstrated that these transformations made Maxwell equations invariant (in all inertial reference frames). For this contribution he should have received another Nobel Prize. **Albert Einstein** continued the thinking in this direction. Lorentz

Hendrik A. Lorentz

Lorentz transformations

$$x' = \frac{x - vt}{\sqrt{1 - v^2/c^2}}$$

$$t' = \frac{t - (v/c^2)x}{\sqrt{1 - v^2/c^2}}$$

The Lorentz transformations indicate the relation of lengths and times measured by a fixed observer (x, t) compared to those measured by a mobile observer (x', t'). In these relations *c* is the velocity of light: 299, 792.458 m/sec.

The participants in the Solvay Council of physics of 1921. Principal subject: atoms and electrons. From left to right and seated: **A. Michelson,** P. Weiss, M. Brillouin, E. Solvay, **H. Lorentz, E. Rutherford, R. Millikan, M. Curie.** Second row: M. Knudsen, **J. Perrin,** P. Langevin, **O. Richardson,** J. Larmor, **H. Kamerlingh - Onnes, P. Zeeman,** M. de Broglie. Back row: **W.L. Bragg,** E. Van Aubel, W. Haas, E. Herzen, **G. Barkla,** P. Ehrenfest, **M. Siegbahn,** J. Verschaffelt, L. Brillouin (in bold : the Nobel laureates).

studied physics at the University of Leiden, Netherlands, and gained his Ph.D. in 1875 (he was then only 22 years old). The subject of his thesis "Theory of the reflection and the refraction of light" became integrated into the spirit of the Maxwell electromagnetic theory. Appointed Professor in 1877 (only 24 years old), he ceased his academic career in 1912 to become Director of the Teyler Laboratorium in Haarlem, although he continued to give a weekly seminar in Leiden which was very much appreciated by the students. He was a polyglot and wrote many books in French, Dutch, English and German. He took part as President in the Solvay Councils of Physics in Brussels in 1911, 1913, 1921, 1924, 1927. After the First World War, he was an active supporter of the reinforcement of scientific collaboration in the ex-belligerent countries. In 1926, as thanks for the basic lectures that he generously gave to the students in Medicine, he was elected "Honorary Doctor in Medicine" at the University of Leiden. At his funeral, on February 10, 1928, the oration was given by **Ernest Rutherford** (1908 Nobel Prize for Chemistry). On this occasion, the Dutch telephonic and telegraphic services were interrupted, for 3 minutes, in recognition of his services.

Zeeman, Peter (Zonnemaire, The Netherlands, May 5, 1865 - Amsterdam, The Netherlands, October 9, 1943). Like many physicists at the end of the 19th century, Peter Zeemalasticityn spent the most of his life as a researcher in the study of the close links between optics and electromagnetism. Born the same year as Maxwell (1831 - 1879) published the electromagnetic theory of light, he took part, at the University of Leiden (Holland), together with **Hendrik Lorentz**, in the experimental verification of the concepts of Michael Faraday (1791 - 1867) and Joseph Larmor (1857 - 1942) on the influence of a magnetic field on the frequency of light emitted by atoms. At that time, the existence of electrons as carriers of the elementary electric charge was not known. In 1896, looking at the light of a flame produced in a sodium vapor, Zeeman observed the modification of the light frequency emitted in the presence of a magnetic field and announced that the radiant element was not the atom (as Larmor thought) but an "electron" of this atom whose ratio between the charge and the mass (e/m) he measured. This ratio proved to be identical to that which **Joseph J. Thomson** would determine by a completely different process. The study of the modification of these spectral lines, by an electric field, would earn **Johannes Stark** the 1919 Nobel Prize in Physics. Zeeman did his university studies in Leiden under the supervision of Lorentz and **Heike Kamerligh-Onnes**. A doctor at the age of 28, he defended a thesis on the Kerr effect, discovered in 1875. He then spent six months in Strasbourg before returning to Leiden for some time and then returning Amsterdam in 1897. He became a professor there in 1900 and, in 1908, he succeeded Johannes van der Waals as Head of the Institute of Physics. Finally he took over the direction of the laboratory (which bears his name presently) from 1923 to 1935, the year of his retirement. His first 10 years in Amsterdam were painful, because of less efficient equipment than in Leiden: the cause, the instability of the building due to traffic. An exceptional researcher (he successfully reproduced 's experiments), Zeeman was also an excellent teacher and an inspired organizer of seminars and meetings with students at the end of their second cycle of studies. He took part in the Solvay Councils of Physics in Brussels in 1921 and 1930.

Peter Zeeman

1903

Becquerel, Antoine Henry (Paris, France, December 15, 1852 - Le Croisic, France, August 25, 1908). Henry Becquerel, a French research physicist, worked in 1896 in the favorite research field of the age: the interaction of light with matter in the presence of magnetic fields, following in the steps of Faraday (1791-1867) and Larmor (1857-1942), at the same time as the Dutchmen **Hendrik Lorentz** and **Peter Zeeman**. Becquerel defended his doctoral thesis in this field in Paris in 1888. Introduced, by his thesis, to the phenomena of luminescence in crystals, and having learned of the discovery of X - rays in 1895 by **Wilhelm Röntgen**, he tried to produce very penetrating radiation for the study of crystalline structures. In March 1896, he reported to the Academy of Sciences (of which he became a member in 1899) that uranyl potassium sulphate crystals, known for their luminescent capacity, acted on photographic plates even in total darkness. Then he observed the same phenomenon while testing non-luminescent crystals and in particular pure uranium. He had just discovered radioactivity (the emission of particles) while he was expecting a more profound study on the radiation emission (X - rays). The name radioactivity was given by **Pierre** and **Marie Curie**, winners of the Nobel Prize in physics at the same time as Becquerel. When Becquerel received from the Curies a small quantity of radium bromide, he realized that in the presence of a magnetic field (one of his preferred devices) the emission was spread into three categories: the first two "radiations" deviated in opposite directions (it is known today that they concern particles α positive and ß negative) and a third one moving in a straight line, analogous to X - radiations, namely gamma rays. In 1901 Becquerel reported the evidence of these three types of natural radiation. These dis-

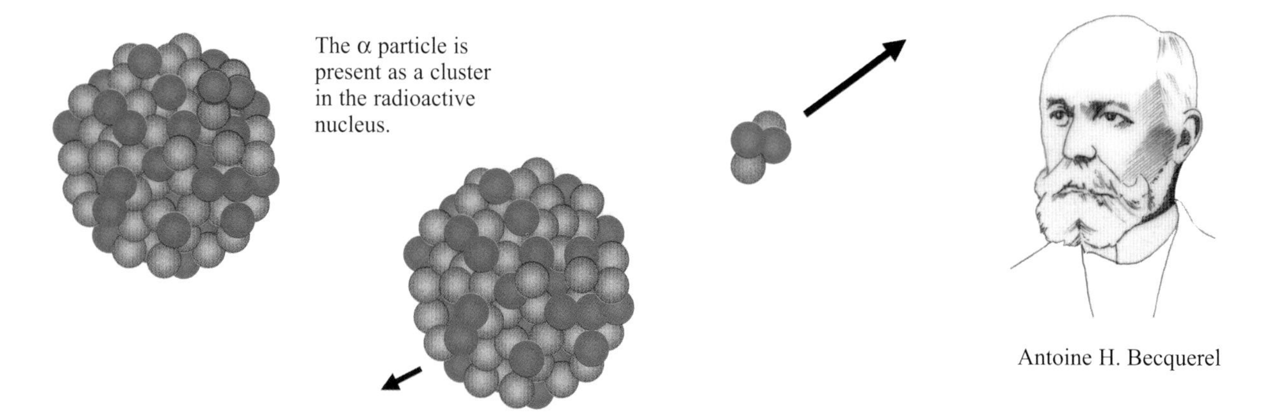

Antoine H. Becquerel

α emission. An α particle, produced by a radioactive nucleus, leaves it by causing the recoil in the opposite direction of the residual nucleus.

coveries, coupled with those of the Curies, opened a significant way for curative medicine. The radioactivity caused by the particle accelerators (see **Ernest Lawrence**, **John Cockcroft** and **Ernest Walton**) and by the nuclear reactors (see **Enrico Fermi**) would open other roads towards fundamental knowledge of the subatomic world and towards applications in isotopic medicine, for the non-destructive analysis and production of nuclear energy, but also, unfortunately, for the manufacture of dreadful atomic bombs. Born into a French family already involved in science (his grandfather, Antoine, and his father Edmond, were, like him, members of the French Academy of Sciences), Becquerel held a chair in physics at the National Academy of Arts and Trades and was head engineer at the National Administration of Bridges and Highways. His son Jean, also a physicist, did research on the optical and magnetic properties of crystals at low temperatures.

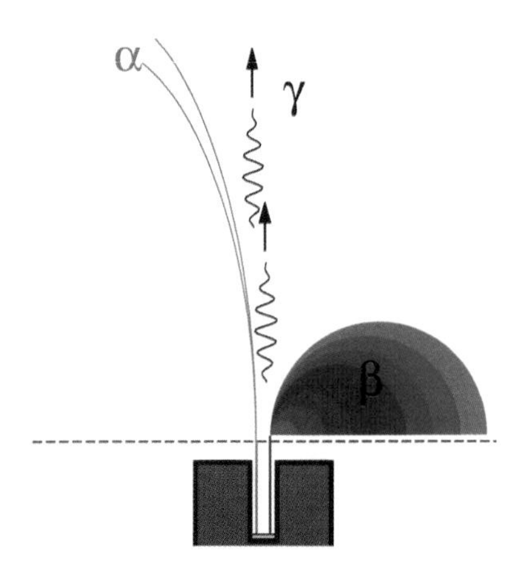

Radioactivity a, ß, γ. A radioactive source (in green) emits α^+ particles (helium nuclei) which are deviated towards the left in two narrow trajectories, $ß^-$ (electrons) deviated towards the right in a widely open cone. The γ rays are not deviated. A magnetic field perpendicular to the sheet and produced by a magnet curves the trajectories in opposite directions for the particles charged with opposite signs.

Curie, Pierre (Paris, France, May 15, 1859- Paris, France, April 19, 1906). Pierre Curie, a French physicist, first studied (with his elder brother Jean-Paul) the phenomenon of piezoelectricity (a property that under high pressure makes electric charges appear on the surface of several crystals: a phenomenon that became profitable with the creation of needles for reading long-playing records). He abandoned this research activity in 1883 when his brother left Paris for a post as Professor in Montpellier. Curie then began research on crystals and magnetism (very fashionable at that time). He was a pioneer in the study of the properties of, first, diamagnetic and paramagnetic materials, then ferromagnetism. He gave his name (Curie point) to the transition temperature between ferro- and paramagnetism. After his marriage to **Marie Sklodowska** in 1895, he started a third career in the chemical separation of new elements: radium in 1898, polonium and actinium in 1899 (in collaboration with one of his students, A. Debierne). At the time he received the Nobel Prize, he officially made known his fears that the fruit of his discoveries could be badly used. The disintegration of radium indeed gave birth to a major release of energy, the harmful effects of which he had studied in mice and pigs. More than 40 years before Hiroshima, he said on the occasion of his speech in Stockholm: "I wonder whether it is useful to let men know the secrets of nature. I am one of those who, like Alfred Nobel, ask whether future discoveries will bring more good than harm to mankind." Pierre was born into a doctor's family, which led an unassuming life. He had a tutor up to the age of 14, his parents believing that this method of education was better suited to his independent and dreamy temperament. He obtained his degree when he was 16 years old and his doctorate when he was 22. Known for his great

Pierre Curie

kindness and modesty, he taught at the Physics and Chemistry School in Paris for 22 years and obtained the chair of physics at the Sorbonne in 1904. On April 19, 1906, crossing the Place Dauphine in heavy rain, he slipped and was killed under the wheels of a cart. Since April 20, 1995, he has rested with his wife in crypt VIII of the Pantheon.

Curie-Sklodowska, Marie (Varsovia, Poland, November 7, 1867-Sallanches, France, July 4, 1934). In addition to the 1903 Nobel Prize for Physics, Marie Curie also received the 1911 Nobel Prize for Chemistry. Invited in 1891 by her sister Bronia and her brother - in - law Casimir Dluski, both doctors, Marie Sklodowska settled down in Paris to continue scientific studies started in Warsaw two years earlier thanks to personal financing from work as a governess. Four years after her arrival in Paris, she obtained two degrees (in physics and in mathematics). Her first thesis work was devoted to the magnetic properties of iron (a fashionable subject during the last years of the 19th century). Then, learning of **Henri Becquerel's**

Marie Curie

discoveries, she chose to re-orient her doctoral thesis towards the study of the radiation emitted by uranium, an element whose property, shown by **Joseph J. Thomson**, was to electrify the ambient air. She received a great quantity of pitchblende, a uranium ore, from the Austrian government. In a laboratory deprived of the most elementary equipment, she identified polonium (400 times more radioactive than uranium) and then radium (900 times more radioactive than uranium). The Curie couple gave the name of radioactivity to these "emanations". In 1902, Marie Curie had isolated 0.1 gram of radium but she ma-

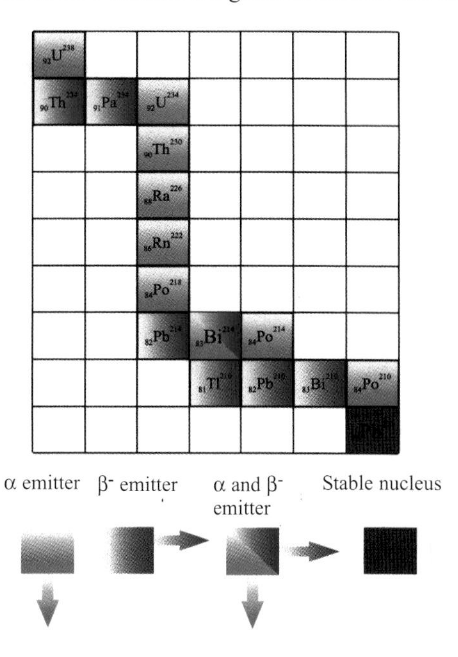

The successive disintegrations of U^{238} producing Ra^{226} to finally lead to the stable nucleus of Pb^{206}. Each step of disintegration involves emission of α^+ particles (red) or β^- (electrons in blue).

naged to produce more than one gram of it in 1910 with the assistance of Andre Debierne who had also taken part in the discovery of actinium. The explanation of the radioactive phenomena would not be completed by the Curie couple but instead by **Ernest Rutherford** and **Frederick Soddy** (1908 and 1921 Nobel Prizes for Chemistry respectively), an explanation that the Curies took time to accept. More optimistic than her husband in the peaceful use of radioactivity and confident about the industrial and medical applications of their discovery, Marie Curie accepted the patronage of a French industrialist, Armet de l'Isle, who financed the construction of a pitchblende treatment plant. The Curies gave up any fortune by not filing for a patent and not requiring royalties. In 1906, Marie Curie succeeded her husband to the general physics chair at the Sorbonne (the first woman in this position). A crowd of interested people followed her first lecture. The discovery of radioactivity brought her international fame. In spite of her reserve about any social gatherings, she was a "star" of the 1900 universal exhibition in Paris. In 1910, at the Brussels Radiology Congress, the curie was unanimously recognized as the unit of radioactivity (the number of disintegrations per second of one gram of radium, thus 3.7 x 10^{10} disintegrations per second). At present, the recognized unit is the Becquerel, which represents one disintegration per second. Curie's international reputation was marked by many events, including not only the creation of research centers in Poland and in France (supported financially by the Lazard brothers and Baron Henri de Rothschild) for the establishment of standards of exposure permitted by the users of radioactive products, but also

The participants in the Solvay Council of Physics of 1911. Principal subject: theory of radiation and quanta. From left to right and seated: **W. Nernst,** M. Brillouin, E. Solvay, **H. Lorentz,** Warburg, **J. Perrin, W. Wien, M. Curie,** R. Poincaré. Standing: V. Goldsmidt, **M. Planck,** H. Rubens, A. Sommerfeld, F. Linderman, **L. de Broglie,** M. Knudsen, F. Hasenohrl, Hostelet, E. Herzen, J. Jeans, **E. Rutherford, H. Kamerlingh-Onnes, A. Einstein,** P. Langevin (in bold: the Nobel laureates).

centers for cancer treatment and diagnostics (radioactive tracers). Many countries issued stamps with pictures of Marie Curie. In 1921, American women offered to the Marie Curie laboratory 1 gram of radium in recognition of the relentless work of this exceptional woman who was elected as head of the radiology service of the Red Cross. Of Jewish origin, she accepted her eviction from a seat at the French Academy of Sciences in 1911, a seat which was allotted to Edouard Branly, a pioneer in the development of the telegraph. Nevertheless, Marie Curie expressed her great attachment to France during the First World War, personally organizing, thanks to donations, a mobile radiology service for the French Army. She took personal responsibility for the formation of specialist doctors. In August 1922, she became one of the twelve founding members of an international committee for intellectual cooperation at the League of Nations. The year she received the Nobel Prize for Chemistry, she was invited to the first Solvay Council of Physics in Brussels and later took part in those of 1913, 1921, 1924, 1927, 1930 and 1933. She was the fifth child of a family of Polish intellectuals (she had three sisters: Sonia, Bronia and Helena and a brother, Joseph). Her father Wladislaw was director of a gymnasium; her mother Bronislawa was administrator of a private school for girls. Widowed in 1906, she took care of the education of her daughter **Irène Curie** (1935 Nobel Prize for Chemistry). In 1911, however she was criticized (by the Parisian press mainly), for rumors of a love affair with Paul Langevin. Curie suffered much from this and was even forced to stay in a convalescent home for several weeks. At the beginning of the 1930s, her health declined quickly. Suffering from tuberculosis (the disease which also carried off her mother) and from severe lesions on the fingers (due to handling radioactive products), she had to undergo, moreover, several eye operations (cataract, in particular). In 1932, she found the strength to go to Warsaw to inaugurate the Marie Sklodowska - Curie Institute of Radium. Practically blind by January 1934, the year her daughter Irène discovered artificial radioactivity, she was to die in the French Alps on July 4, 1934. On April 20, 1995, crypt VIII of the Pantheon was opened for the deposit of the remains of Pierre and Marie Curie.

1904

Strutt, John William (Witham, England, November 12, 1842 - Witham, England June 30, 1919). Third Baron of Rayleigh. Following the scientific ideas of Prout (1785 - 1850), who had already postulated in 1815 that all atomic weights should be multiple integer numbers of hydrogen, John Strutt measured the atomic and molecular masses of the great majority of gases. During his measurements on nitrogen (the major air component) and after having determined with precision the molecular weight of oxygen, he deduced that there had to exist an impurity, in the form of a new element that he called argon, a name from the Greek word meaning "inactive". Strutt served as President of the Aeronautics Scientific Committee during the First World War. **William Ramsay** (1904 Nobel Prize for Chemistry), one of his collaborators, pursued research in the same field and discovered another noble gas: helium. Strutt also worked with Maxwell just after the creation of the Cavendish Laboratory (1874) and served for more than 10 years as a secretary of the Royal Society, London. Son of James Strutt, second Baron of Rayleigh, he studied in Cambridge under E.J. Routh, a mathematician, and received his first degree in astronomy in 1864. He went to the United States in 1868. He served as President of the Scientific Committee of Aeronautics during the First World War. From 1908 until his death he held post as Chancellor of the University of Cambridge.

1905

Lenard, Philippe (Pozsony, Hungaria, but today Bratislava, Slovak Republic, June 7, 1862 - Messelhausen, Germany, May 20,1947). Experimenter physicist. During his long career (almost 60 years), Lenard was interested in many problems connected with the distribution of electrical charges in materials and flames. His main contribution on cathode rays (which would be identified later as an electron beam) earned him the Nobel Prize. He was one of the first to think that atoms were composed of two types of charges, grouped in very tight couplets that he called "dynamides" and, thus, developed the precursor of the atomic model of **Ernest Rutherford** (1908 Nobel Prize for Physics). While studying cathode rays he made known his disillusion at not having discovered X - rays (**Wilhelm Röntgen**, First Nobel Prize for Physics in 1901). In 1902 he highlighted the photoelectric effect (ejection of electrons under the impact of high frequency electromagnetic radiation), but it took **Albert Einstein** to explain its mechanism (Nobel Prize in 1921 for this particular subject). Born in Pozsony, he studied at Budapest, Vienna, Berlin and Heidelberg where he gained his Ph.D. and worked. An ultra nationalist during the First World War, he often claimed that the work of German physicists "was plagiarised" by the British. Then he involved himself deeply in National Socialism and was a passionate defender of Hitler's ideas. An inveterate anti-Semite, he was a fierce opponent of Albert Einstein's ideas throughout his life. In 1929 he published a work "The great men of science" in which he developed a racist ideology and attacked notably **Gustav Hertz's** theoretical work which he called a "Jewish legacy". He left Heidelberg in 1905 for Messelhausen where he died in 1947.

1906

Thomson, Joseph John (Cheetham Hill, Manchester, England, December 18, 1856 - Cambridge, England, August 30, 1940). Joseph Thomson, a talented experimental physicist, used the discovery of X - rays (**Wilhelm Röntgen**, First Nobel Prize for Physics in 1901) as an ideal means to make conductive a gaseous system subject to high voltage. He gave the name of ionization to this phenomenon which was fundamental in the identification of the electron. Foreshadowed in the experiments that earned **Hendrik Lorentz** and **Peter Zeeman** the 1902 Nobel Prize in Physics (already evoked in 1833 in the experiments of Stoney who continued experiments on electrolysis proposed by Faraday), the existence of the electron was definitively confirmed by Thomson by means of the famous cathode - ray tube. From the measurement of e/mv by the electron deflection in a magnetic field and of e/mv^2 by the calculation of energy deposited on the screen of the tube, Thomson concluded that the mass of this particle was almost 2000 times lower than the smallest mass highlighted by Faraday: that of the hydrogen atom. Beginning these experiments again in the Cavendish Laboratory, he concluded that any matter contained these small particles. Entering Owens College at age 14, he learned, under the supervision of Osborne Reynolds, "to think in total independence instead of referring to the bibliography". He received his bachelor's degree from Trinity College of Cambridge when he was 24 years old. Not long afterwards, he was already convinced that energy was present both in the mass and the linear momentum of the particles, thus already perceiving, but without writing down anything, the famous equation of Albert Einstein: $E = mc^2$. He held positions as a teacher at Trinity College and at the Royal Institution of Great Britain. He took part in the Solvay Council of Physics in Brussels in 1913. He was the son of a Scottish bookseller and publisher. He had an easy-going contact with his students who were spread in more than 50 universities throughout the world. Seven of his students were awarded the Nobel Prize. He was buried at Westminster Abbey next to Isaac Newton, Charles Darwin, Lord Kelvin (born William Thomson), William Herschel and Ernest Rutherford (1908 Nobel Prize for Chemistry).

Joseph J. Thomson

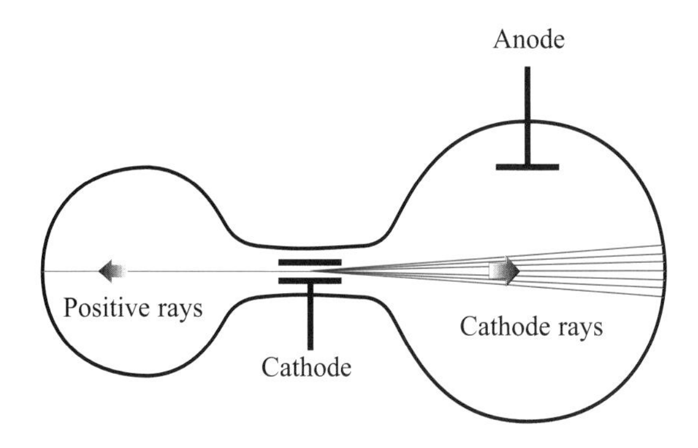

Cathode rays (electrons) and positive rays (ions) produced in a Crookes tube.

1907

Michelson, Albert Abraham (Strelno, Prussia, December 19, 1852 - Pasadena, California, May 9, 1931). In 1887 Albert Michelson, in collaboration with Edward Morley, developed an optical device of extreme precision, able to measure displacement variations of one part in 4 billion, with the intention of measuring whether the orbital velocity of Earth could influence the velocity of light. The result of the experiment proving that this was not the case is at the heart of the verification of the basic hypothesis of **Albert Einstein's** theory of relativity, even if, in 1894, the

Albert Michelson

latter claimed that he did not remember having used the result of the Michelson-Morley experiment to write his famous article on the subject. Contrary to sound, the transport of light does not require the intervention of any medium. Michelson migrated to the United States when he was 2 years old and was then completely educated in USA. Having

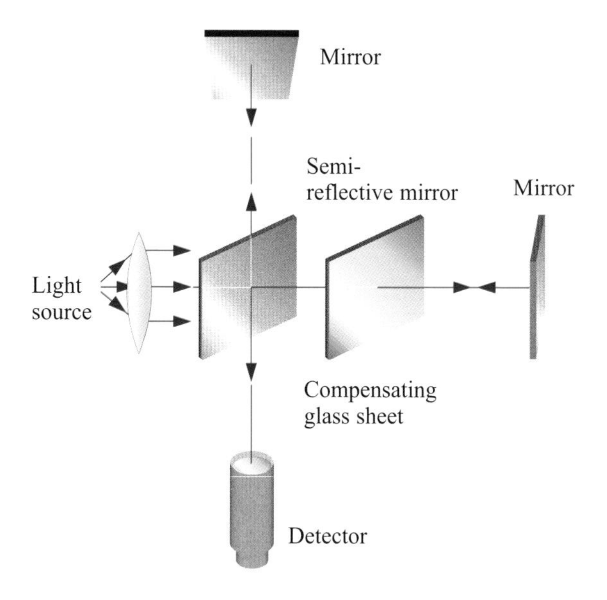

The interferometer. The light beam is initially split and then recombined to give rise to an interference pattern. The apparatus assembled on a rigid frame is able to turn at least 90° around an axis perpendicular to the plane of the drawing. Not observing any change in the figure of interference during this rotation at any hour of the day or night at any period of the year, Michelson and Morley concluded that the speed of light was not altered by any movement of an environment supporting the light waves: an experimental confirmation of the theory of relativity (A. Einstein).

first joined the Navy, from which he resigned in 1881, Michelson became Professor at a school of Applied Physics at Cleveland (Ohio). From 1890 until 1930, he taught at the University of Chicago, where he received the personal help of President Harper as encouragement for his research on high precision devices. That is how he improved the measurements previously carried out by **Peter Zeeman** (1902 Nobel Prize for Physics). He took part in the Solvay Council of Physics in Brussels in 1921. That year, Einstein was absent!

1908

Lippmann, Gabriel (Hollerich, Luxemburg, August 16, 1845 - Atlantic Ocean, July 13, 1921). 1908 Nobel Prize for Physics. Gabriel Lippmann, the inventor of a process of colour photography, explained his method in front of the Swedish Academy of Sciences as follows: "a plate covered with a photosensitive layer is placed in a support containing mercury. During the exposure, the mercury on the sensitive layer produces a mirror. After exposure, the development is carried out in the conventional way and, once dry, the colours appear, visible by reflection and fixed: the phenomenon is induced according to a process of interference in the film itself. Later, he repeated the experiment in public. Born in the Grand Duchy of Luxembourg of French parents, he spent his childhood in Paris. He participated in the translation into German of publications in the "Annals of Physics and Chemistry". In 1873 he stayed in Heidelberg where, in the department of W. Kühne, Professor of physiology, he studied the relation between the surface tension of mercury and the contact voltage with a wire. He also developed a capillary electrometer. Returning to Paris, he became interested in piezoelectricity, like **Pierre** and **Marie Curie**. He taught at the Faculty of Sciences in Paris and did research in astronomy and seismology. He died in 1921 while returning by boat from a visit to the United States and Canada.

1909

Braun, Karl Ferdinand (Fulda, Germany, June 6, 1850 - New York, New York, April 20, 1918). Karl Braun, a wireless communication scientist, shared the Nobel Prize in physics with **Guglielmo Marconi**. He is also the inventor of a particular cathode-ray tube, which is at the origin of many instruments like the TV receiver and various mea- suring devices. Born in the principality of Hesse (Germany), he studied physics in Marburg and in Berlin, where he obtained his doctorate. He then occupied a position of assistant in Würzburg, then as head of curriculum courses in Leipzig while continuing to teach in the secondary studies cycle. He received positions at universities of Strasbourg and Karlsruhe and returned finally to Tübingen where he undertook works in electrostatics and thermodynamics. While in Strasbourg Braun became interested in wireless telegraphy, the field that earned him the Nobel Prize. He showed that transmitters gained power by using a closed capacitor circuit, contrary to the open circuit used by Marconi's transmitters. He also invented directional antennae. The first long distance transmissions that he organized were sent from the Eiffel Tower toward Strasbourg. Summoned to the United States at the end of the First World War, he was unable to return to Strasbourg and died in Brooklyn on April 20, 1918.

Marconi, Guglielmo (Bolognia, Italy, April 25, 1874 - Rome, Italy, July 20, 1937). Guglielmo Marconi, the inventor of wireless telegraphy, made his discovery by intuition. Using the **Gustav Hertz** spark discharge, the aerial of Popov (1859-1906) and the coherer of Branly (1844-1940), he succeeded in transmitting electromagnetic waves over a few hundred meters in 1896. By improving each of the components he was able to send a message from Cornwall to Newfoundland in 1901. As early as 1902 he established a Europe-United States communications link which allowed, after 1904, an information service to be available on all transatlantic steamships. The performance of wireless communication increased with the use of short waves, proposed by **Karl Braun**. Soon, his device provided valuable assistance in the location of ships and for rescue operations. A businessman, he worked mainly from England where he had settled in 1896. As early as 1897 he had convinced the chief engineer of the Post Office (Sir William H. Preece) of the achievements of his invention. In 1897 Queen Victoria was able to communicate from dry land with the Prince of Wales' yacht. From 1921 Marconi used his personal yacht Electra as both a residence and a laboratory.

Guglielmo Marconi

1910

van der Waals, Johannes Diderik (Leiden, The Netherlands, November 23, 1837 - Amsterdam, The Netherlands, March 9, 1923). The 19th century in physics was marked by the detailed study of the properties of gases and of thermodynamics. This science, founded by Sadi Carnot (1796 - 1832), was further developed by Rudolf Clausius (1822 - 1888), James Clerk Maxwell (1831 - 1879) and Ludwig Boltzmann (1844 - 1906). It was on reading Clausius's description of gases as molecules constantly knocking against each other, that Johannes van der Waals understood that there had to be a continuity in the behavior of a gas (or of a vapor) during its transition into the liquid state. This subject-matter, which received an enthusiastic welcome, was developped in his doctoral thesis and presented at the University of Leiden in 1873. This thesis was rapidly translated into German, English and French. The proposed gas-liquid continuity was contained in a general law lin- king pressure (p), volume (V) and temperature (T) of a fluid: $(p + a/V^2)(V - b) = RT$ where a/V^2 has the physical dimension of a pressure and takes into account the short range interactions (due to forces which would be named after van der Waals) and b summarizes the limits of compressibility due to the non-null dimension of atoms or molecules. It was due to this equation (which also contained information on the metastable states) that van der Waals could make the distinction between vapor (liquefiable at constant temperature) and gas (non-liquefiable without a decrease in temperature). This distinction is important when we want to liquefy gases that were improperly termed as "permanent" like air, nitrogen and the rare gases. van der Waals taught in Deventer and in The Hague even before the presentation of his Ph.D., at the age of 36 years old (he had to wait to be exempted from taking a test in Greek and Latin to graduate!). In 1877 he was appointed Professor at the University of Amsterdam where, alone, he carried out the teaching of physics for 20 years. He exerted a profound influence on scientific research, first in The Netherlands (and in particular on the work that brought **Heike Kamerlingh - Onnes** the Nobel Prize), but also on the work of all those who used the properties of fluids like **Charles Wilson**. A small reserved man, he led a very simple life.

Johannes van der Waals

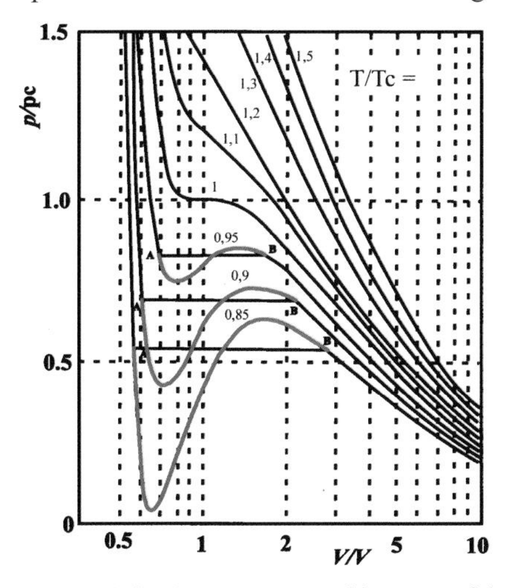

The van der Waals isotherms expressed in terms of the reduced variables p/p_c, V/V_c and T/T_c. The red lines show the partially accessible theoretical values for metastable states, the continuous lines the experimental values.

1911

Wien, Wilhelm Carl Werner Otto Fritz Franz (Gaffken, Eastern Prussia, January 13, 1864 - Munich,

Germany, August 30, 1928). Both a theorist and experimentalist, this brilliant physicist is particularly known for two general laws governing the radiation emitted by bodies in relation to their temperature. All bodies emit radiations in a wide range of wavelengths (or frequencies). This distribution is called emission spectrum. Wilhelm Wien showed, in 1893, the link between the temperature and the wavelength corresponding to the maximum of the emission. The higher the temperature, the shorter the wavelength of the radiations which appear with greatest abundance. Thus, the measurement of the emission spectrum of solar radiation informs us about the surface temperature of our star. It is close to 6000 K. Wien then applied himself to the determination of the shape of the emission spectrum, focusing only on the role played by the emitting molecules obeying classical mechanics and classical thermodynamics. He obtained a satisfactory agreement with the experimental spectrum by adjusting only two parameters. The perfect agreement of the calculation with the experiment would be later established by **Max Planck**, the final explanation returning to **Albert Einstein** with the introduction of the concept of stimulated emission. It was from Wien's laws, better known as the emission laws of the black body, that recent measurements have permitted the clarification that the continuous spectrum of radiations fixes the average actual temperature of the universe (see **Arno Penzias** and **Robert Wilson**) around 3 K (-270°C). Wien spent his youth on the family farm in Drachenstein.

Wilhelm C. Wien

From his historian mother he received an education marked by culture. He had a French tutor, so the young Wilhelm spoke French even before he was able to write German. At the age of 18, he entered the University of Göttingen to study mathematics and natural sciences. But he decided to stop this type of training after 6 months. Introverted and independent, he then decided to travel. In 1884 he had the opportunity of rejoining the laboratory of von Helmholtz, a thermodynamicist who was the founder and the first Director of the University of Berlin. He defended his doctoral thesis there in 1866. An only son, he went back to the family farm to help his parents. Uncomfortable outside research laboratories, he profited from economic problems which obliged his parents to sell their farm in 1890 to take a position as assistant in Charlottenburg. He also went to Berlin, Aachen and Giessen. In 1900 he took over from **Wilhelm Röntgen**, for the first time, in Würzburg and, for the second time, in 1920 in Munich, where he stayed until his death in 1928. Apart from the laws of the black body, he was also interested in the diffraction of light and in the electricity in gases. He showed that canal rays are atoms bearing a positive charge. A great traveler, he visited Norway, Spain, Italy, England and Greece. In 1913 he gave six remarkable

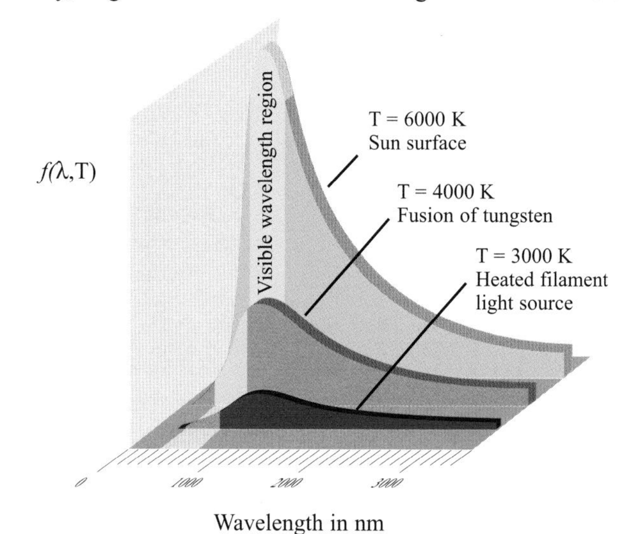

The emission of radiation by the "black body". The spectrum of the radiation emitted by a heated body depends on the temperature of the emitter. The intensity is maximum for λ_M wavelength. The temperature of the emitter is related to λ_M by the formula $\lambda_M T$ = constant. The total intensity in the entire spectrum is proportional to T^4.

lectures at Columbia University and resided for a short time at Yale and Harvard. He took part in the Solvay Councils of Physics in Brussels in 1911 and 1913.

1912

Dalen, Nils Gustaf (Stenstorp, Sweden, November 30, 1869 - Stockholm, Sweden, December 9, 1937). An engineer trained at the Polytechnic School of Zurich, Nils Dalen was the inventor of beacons and luminous acetylene buoys and of various mechanisms allowing control of the duration and frequency of the flashes, their automatic switching on at dusk and switching off at sunrise. These control mechanisms are based on the absorption of radiation, different for dark surfaces compared with the polished surfaces of various metals. In 1977 it was learned that the 1912 Nobel Prize in Physics could have been awarded jointly to Edison and Tesla, following in the path of the discoveries of Guglielmo Marconi and Karl Braun. The prize escaped them as Tesla refused to be honored with Edison because of a financial disagreement. Born into a farming family in the south of Sweden, Dalen followed an industrial career in the field of turbines and compressors but was regarded as the "benefactor of sailors". He was blinded by an explosion in 1912. His brother, a doctor, replaced him at the official awards ceremony of the Nobel Prize.

1913

Kamerlingh-Onnes, Heike (Groningen, The Netherlands, September 21, 1853 - Leiden, The Netherlands, February 21, 1926). An admirer of the work of **Johannes van der Waals** and especially of the power of the equation of state, which predicted the transitions between the liquid phase and the gaseous phase, Heike Kamerlingh - Onnes applied himself to implementing it to reach low temperatures in order to study special properties of matter. Rewarded for these two reasons, he is the first solid-state physicist to receive the Nobel Prize. To reach these low temperatures, energy needs to be extracted from the samples, while transferring heat to the outside. This is accomplished by plunging these samples in a liquefied vapour which evacuates energy while boiling. To liquefy this vapour it is isothermally compressed at a temperature lower than the critical temperature (given by the law of van der Waals). Passing through the intermediate stage of liquid hydrogen and applying the thermodynamic cycles proposed by Carnot (1796 - 1832) and developed by Rudolf Clausius (1822 - 1888), Kamerlingh - Onnes liquefied helium at a temperature considered at that period very close to the zero of Kelvin's scale (4K). Liquid helium is currently the most used product to realize thermostats intended for cryogenic studies. New properties of liquid helium were studied by **Pyotr Kapitza**. Kamerlingh - Onnes dedicated his discovery to his friend van der Waals whose theory guided all his reflections. The device developed by Kamerlingh - Onnes allowed the study of the behavior of the electric resistance of metals and the observation of the superconductor state (zero state of resistance to the passage of a current) for lead and mercury. The ingeniousness of Kamerlingh - Onnes earned him the flattering nickname of "gentleman of the absolute zero". Superconductivity earned other Nobel Prizes in Physics, much later, for Leon Cooper, **John Schrieffer** and **John Bardeen**, in 1972, for their theoretical explanation of the phenomenon, and for **Karl Müller** and **Georg Bednorz** in 1987, for their proposals of new structures allowing the superconductor state to be reached at more easily accessible temperatures. The particular behavior of liquid helium was explained by **Lev Landau**. Kamerlingh - Onnes did his university studies in Heidelberg after having received a gold medal at the end of his secondary studies, on the occasion of a competition organised by the University of Utrecht. However, he returned to Groningen in 1879 to present his doctoral thesis on "A new proof of the earth's rotation". In Heidelberg, he had worked with Bunsen and had been accepted as one of the only two students in the private laboratory of Gustav Kirchhoff. He was chosen as Professor of physics and Director of the laboratory of Leiden. He stayed there for 42 years making Leiden the worldwide centre of low-temperature physics. He took part in the Solvay Council in Brussels in 1911, 1913, 1921 and 1924.

Heike Kamerlingh-Onnes

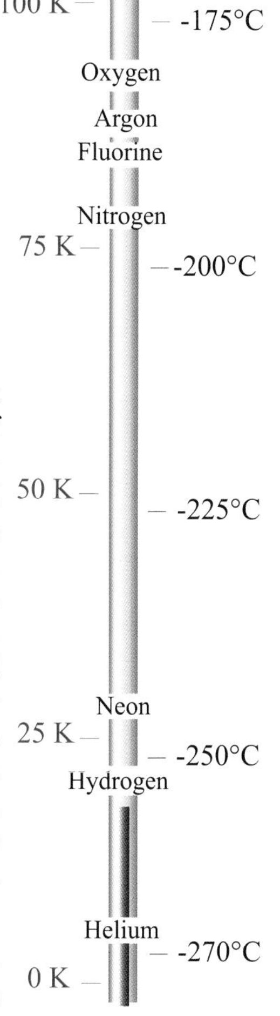

"Symbolic" scale of the liquefaction temperatures of some gas species.

1914

von Laue, Max Theodor Felix (Pfaffendorf, Germany, October 9, 1879 - Berlin, Germany, April 24, 1960). An all - round physicist, Max von Laue received the Nobel Prize in physics, aged 35, for having proved, in the same experiment, that the X - rays discovered by **Wilhelm Röntgen** were (short wavelength) electromagnetic waves and that crystals were geometrically well-organized structures. These two characteristics were predicted by several persons, but it was von Laue who provided the proof by following a way similar to the explanation of visible light diffraction by gratings. von Laue's hypothesis was experimentally demonstrated by W. Friedrich and P. Knipping in 1912. By irradiation of a copper sulfate crystal with X - rays, they obtained the first "Laue's diagram" characterized by a distribution of the diffracted intensities in a series of spots whose fitting and relative intensities are characteristic of crystals. This method is still used today to identify multiple substances and is moreover an outstanding tool for the determination of the structures of synthesis products like medicines. The complete identification of diagrams is presently accomplished by using numerical processing. It should be observed that the 1914 Nobel Prize for Physics but also the 1915 and 1917 Prizes were attributed for discoveries in the field of X - rays (respectively for **William Henry Bragg** and **William Lawrence Bragg** and for **Charles Barkla**). von Laue first graduated from

Max T. von Laue

Strasbourg and defended his doctoral thesis in theoretical physics (optics) in 1903. He stayed at the University of Berlin while working under **Max Planck** until 1909. Early on, von Laue firmly supported **Albert Einstein's** theory of relativity. He voluntarily joined the theoretical physics group of A. Sommerfeld in Munich, and then occupied a teaching position at the University of Zurich in 1912. He was named a professor in Frankfort in 1912. During World War I, he also collaborated with **Wien** in Würzburg in order to improve wireless communications in the German Army. After the war, he used his fame to collect funds intended to boost scientific research. As Chairman of the German Society of Physics, he took a position against Einstein who refused to work for the Nazi regime, but also opposed the admission of **Johannes Stark**, a convinced Hitlerite, to the Academy of Prussia. In return, he was violently attacked (like the other theorists A. Sommerfeld, **Werner Heisenberg** and Albert Einstein) by Stark after 1934, when the latter was elected Head of the Deutsche Forschungsgemeinschaft. During the Second World War, he welcomed the Dutch, Danish and French scientists who were sent to Germany as workers or prisoners. He retired from teaching in 1943 and then lived in Würtenberg-Hohenzollern, mainly looking after the revitalization of several scientific societies. He took part in the Solvay Council of Physics in Brussels in 1913.

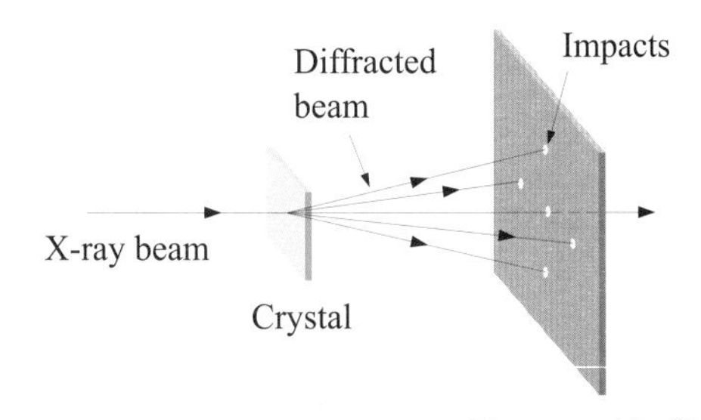

General principle of diffraction of monochromatic X- rays by a single crystal.

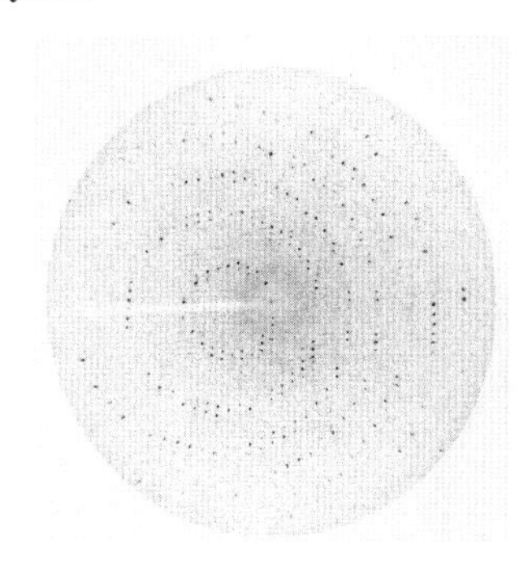

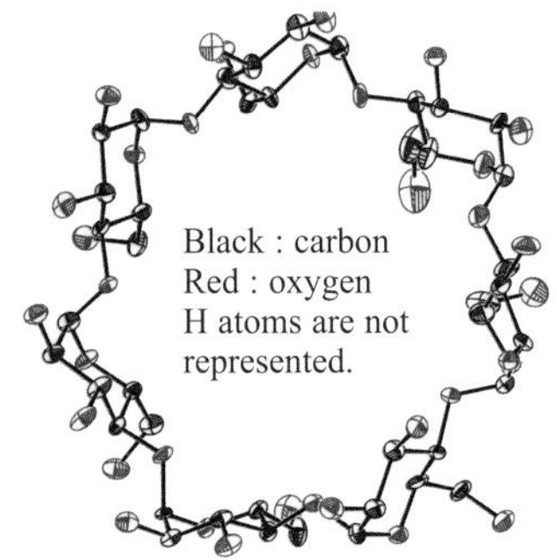

Diffraction pattern of X - ray diffracted by a complex molecule (β - cyclodextrine) (top) and its geometrical distribution (bottom).

1915

Bragg, William Henry (Westwood, England, July 2, 1862 - London, England, March 12, 1942). William Henry Bragg was honored (the same year as his son **William Lawrence**) for crystallographic studies by means of X - rays, just like **Max von Laue**, his predecessor as Nobel Prize winner in physics. He began his career in Australia as Professor of physics and mathematics at the University of Adelaide. Probably trying to attribute a particle beha viour to X - rays, he was only convinced of their wave nature after he became acquainted with von Laue's results. With his son Lawrence, a researcher at the Cavendish Laboratory, he established the formula relating the wavelength (λ) of radiation diffracted to the distance (d) between the adjacent atomic planes and the diffraction angle $n\lambda = 2d \sin \theta$. This formula is currently known as the Bragg equation. After receiving the Nobel Prize, he directed his research towards the construction of X-ray detection devices by using their property to ionize gases. He established various relations between the wavelength, the **Planck** constant, the atomic mass of the emitter and of the absorber. Convinced therefore of the double corpuscular and wavenature of light, he frequently said that he used one from Monday until Wednesday, the other from Thursday until Saturday. After Bragg had taught in Leeds he got the position of a teacher at the University College in London in 1915. During

William H. Bragg

The β - cyclodextrine is a large molecule with extended void which finds applications in cosmetics, textiles, pharmacology and farm-products.

the war he was a governmental official and in 1920 he was granted the title of nobleman. Then he became the Head of the Laboratory of the Royal Institution. He took part in the Solvay Councils in Brussels in 1913, 1924 and 1927. Born on a family farm in Cumberland, he was the son of a former merchant fleet officer. Bragg was a man of the world. Being a brilliant golf player, he occasionally allowed his caddy to assist his elder son Lawrence, thus teaching him to play golf, too. He was a very religious person and avoided any action which might inflict eternal damnation upon him. He showed respect for traditions and proved to be generous, passionate and modest.

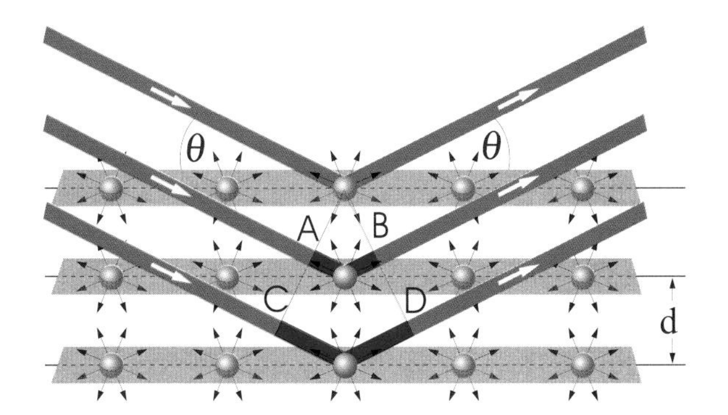

The Bragg diffraction. A beam of X - rays interacting with successive planes of a crystal undergoes a preferential reflection when the distances from A to B and from C to D are respectively equal to once and twice the X - ray wavelength. The images of diffraction allow the determination of the crystal structure of the materials, (see von Laue - 1914).

Bragg, William Lawrence (Adélaïde, Australia, March 31, 1890 - London, England, July 1, 1971). We can read in the preceding biography the reasons for which William Lawrence Bragg received the Nobel Prize in physics jointly with his father. He is the youngest physicist to have received the Nobel Prize in physics. Although they worked in the same field, they belonged to different institutions. Bragg was born in Australia. He entered Adelaide University, where his father was a teacher. When his family returned to England, he entered the Trinity College in Cambridge. In 1914 he started lecturing at Cambridge. Between 1919 and 1937 he taught in Manchester and was the Head of the National Physical Laboratory until 1938. Then he taught general physics at Cambridge University until 1953. Afterwards, until 1966, he presided at the Royal Institution of Great Britain. He took part in the Solvay Councils of Physics in Brussels in 1921, 1927, 1948, 1951, 1954, 1958 and 1961. William Bragg was technical adviser to general headquarters in France during the Second World War. After the war, he organized the International Union of Crystallography of which he was the first President. His research subjects also concerned the structure of silicates, dislocations, phase transitions and polymorphism in metallurgy. He even tackled the difficult problem of the chemistry of proteins. The attribution of a Nobel Prize to a father and son pair in the same year is unique. The **Bohr**, **Curie**, **Siegbahn** and **Thomson** families also received this honor, but for work done at different periods.

William L. Bragg

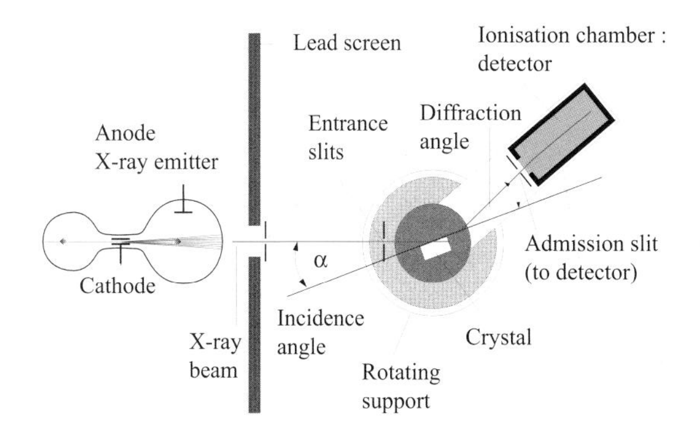

The X - ray spectrometer. The experimental use of Bragg diffraction. Rotation of the crystal at a θ angle and simultaneously the detector at a 2θ angle, the wavelengths of diffracted X - rays are selected. By using several materials as anticathode, W. L. Bragg was able to determine the wavelengths of the X-ray characteristics of the elements of this anticathode.

1916 Not awarded

1917

Barkla, Charles Glover (Widnes, England, June 27, 1877-Edinburgh, Scotland, October 23, 1944). After **Wilhelm Röntgen** in 1901, **Max von Laue** in 1914, and **William Henry Bragg** and **William Lawrence Bragg** in 1915 for their work on X - rays, Charles Barkla (an ingenious experimenter and very skilful in the construction of equipment) received the Nobel Prize for Physics for the discovery of the law relating the energy of characteristic X - rays to the atomic weight of the emitting elements. Working on gas targets, he also showed that the intensity of the emitted X - rays is proportional to the gas density (thus to their atomic number or total number of electrons). He proved moreover that the X - rays are propagated in the form of transverse waves, that the diffusion of X - rays by matter is split into two components, one of an unchanged wavelength, the other of a bigger wavelength, and that several substances emit two series of them, of very different wavelengths, that he named L and K. Taking up those results, Henri Moseley (1885-1915) proposed to use this property to justify the classification of the elements in Dimitri Mendeleev's table. Moseley would certainly have been associated with Barkla for the 1917 Nobel Prize if he had not been killed during the war. The discoveries of Barkla and Moseley form the basis of many high performance analytical applications used today. The excitation modes of the X - rays are multiple: either X - radiations (as in the original process), or electrons (electron microprobe), or heavier particles (protons, helium) produced by accelerators or radioactive sources (PIXE: Particle Induced X - ray Emission). Regarding X - rays as waves, Barkla was criticized (wrongly, it is now known) by William Henry Bragg who claimed that X - rays were particles. Barkla was the son of a chemist. He studied at Liverpool and at King's College, Cambridge. He was Professor at the University of Edinburgh where both his human and teaching qualities were very widely appreciated. He took part in the Solvay Council of Physics in Brussels in 1921.

Charles G. Barkla

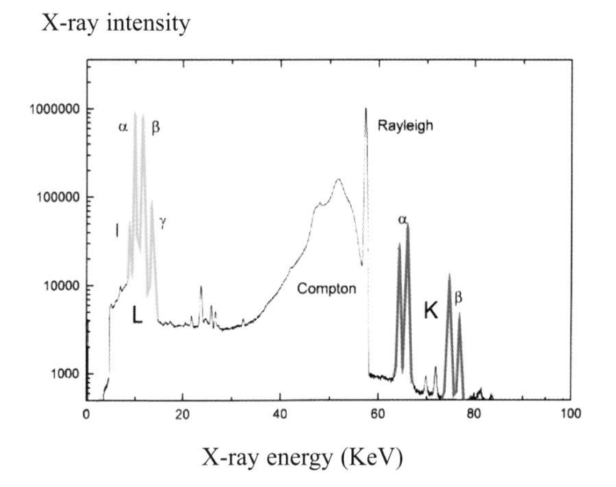

X-ray spectrum, induced by a radioactive source in a sample of pure gold: in yellow the series of L lines; in red the series of K lines; in black a component of the exciting lines elastically scattered (Rayleigh) and the X - rays from inelastic scattering (Compton).

1918

Planck, Max Karl Ernst Ludwig (Kiel, Germany, April 23, 1858 - Göttingen, Germany, October 3, 1947). Max Planck, a theoretical physicist, can be considered as the father of quantum physics. Just like **Albert Einstein**, he had "to be patient" before obtaining the honour of the Nobel Prize because his ideas were not very "classical", disturbing for the science of this period, but very relevant for a revival of physics which took off at the beginning of the 20th century. He received his prize on the occasion of the ceremony in 1919. Since 1896 he had continued **Wilhelm Wien**'s proposals on the distribution of radiations emitted by black bodies (a black body is, contrary to the given name, the body which produces emission of maximum intensity at a given temperature). Planck was inspired by Boltzmann's results on the distribution of the energy of the molecules of a gas and adapted them to the radiation emitted by black bodies considered as an assembly of antennae. He proposed (thus giving rise to quantum physics) that these emissions could take only discrete values of energy, that is, integer multiple of some quantity ε. In 1900 he concluded that the ratio of the energy emitted on the value of its frequency, thus E/ν, must be a constant that he called h. In the units of our MKS system, the value of h is 6.62×10^{-34} joule.second. Guessing the significance of his proposal, he confided to his young son during a walk in the forest: "Today I have made a discovery as important as that made by Newton". He was certainly right. In 1905 Einstein was the first to be interested in Planck's proposal in the study of the photoelectric effect. The explanation given by Einstein was also awarded with the Nobel Prize. Besides his work on the thermodynamics of radiation, Planck determined the charge of the electron within an accuracy of a thousandth. Planck studied at the University of Munich from 1874 to 1877 and prepared his doctoral presentation in Berlin in 1879 under the direction of Hermann von Helmholtz and Gustav Kirchhoff. He taught in Munich and Kiel before taking over for Kirchhoff in Berlin. On the occasion of his 80th birthday, his name was given to a small planet: "Planckiana". In 1911 and 1927 he attended the Solvay Councils of Physics in Brussels, despite the negative opinion that he had given concerning such events. Max Planck was born in Kiel (Prussia) but he lived in Munich from the age of nine. His father was a professor of law at the university in this town. From his first marriage in 1887, he had a son and two daughters and from his second marriage, in 1911, he had a son. He had to suffer a number of family misfortunes: a son was killed at Verdun, his two daughters died in confinement; his last son was executed in 1945 for an attempt to overthrow the Nazi regime. Planck was always proud of his German nationality but refused to subscribe to the idea of an "Aryan physics". A good friend of Einstein, they played chamber music together. Buried in Göttingen, his tomb bears the inscription: $h = 6.62 \times 10^{-27}$ erg.sec [... in CGS units], thus 6.62×10^{-34} J.sec in the international system adopted today.

Max K. Planck

1919

Stark, Johannes (Schickenhof, Germany, April 15,1874 - Traunstein, Germany, June 21, 1957). Attempting, around 1905, to reproduce, by interaction with an electric field, the effect of separation of the spectral lines obtained by **Peter Zeeman** in a magnetic field, Johannes Stark had to make several further attempts before finally succeeding in 1913. In the course of his numerous attempts, he measured the Doppler effect on the canal rays (positive ions produced in a gas tube). It was for these two results as a whole that he received the Nobel Prize. Since 1905 he tried to use this last result to confirm the proposals of **Albert Einstein** (theory of relativity) and of **Max Planck** (quantum physics) but, after 1913, he adopted a completely opposite attitude, probably for political reasons. A passionate supporter of Hitler, he was elected President of the Physikalische-Technische Reichanstalt., in 1933, and of the Deutsche Forschungsgemeinschaft, in 1934. He used this to reinforce his attacks against all the supporters of "modern" physics, in particular **Max von Laue**, Arnold Sommerfeld, **Werner Heisenberg**, Max Planck and Albert Einstein. He only remained on good terms with **Philippe Lenard**, another Hitlerite. Johannes Stark studied in Bayreuth, Regensburg and Munich where he obtained his doctorate in 1897. He then held posts successively in Munich, Göttingen, Hannover, Aachen, Greifswald and Würzburg where he succeeded **Wilhelm Wien**. Deprived of his double scientific presidency in 1936, he retired to Bavaria, his native region.

1920

Guillaume, Charles Edouard (Fleurier, Switzerland, February 1861-Sèvres, France, May 13, 1938). A Nobel prizewinner for his experimental studies on the nickel-steel alloy, Guillaume is the second solid - state physics specialist to receive this distinction after **Heike Kamerlingh-Onnes**. Solid-state physics, although at first sight not very spectacular for students entering university, is actually the field that provides the most jobs for physicists in industry. All of Guillaume's work, produced at the Bureau of Weights and Measures in Sèvres (Paris region) was directed toward research on precision. Thus he was interested in the construction of mercury thermometers, platinum-iridium alloys (alloys in which the metre and the kilogram of reference are preserved), nickel - steel and, in particular, invar (alloy containing 35.6% nickel and having the lowest known expansion coefficient) and elinvar (steel with nickel-chromium whose elasticity remains constant over a wide variation in temperature). One hundred years after the discovery of the properties of invar, this alloy is still used abundantly and almost exclusively today for the storage of cryogenic liquids. It is still used for the masks in high resolution TV tubes, in the high precision construction of wave guides, optical instruments and in particular lasers and telescopes and in all technology requiring the constancy of lengths. Guillaume was a graduate of the Federal Polytechnic School of Zurich. In 1915 he became Director of the Bureau of Weights and Measures.

1921

Einstein, Albert (Ulm, Germany, March 14, 1879 - Princeton, New Jersey, April 18, 1955). Einstein is considered by many people, scientists and non-scientists alike, to be the greatest genius of the 20th century. His Nobel Prize for Physics was awarded for a relatively small piece of his giant work: his explanation of the photoelectric effect, in which electromagnetic radiation induces the emission of electrons from matter, and his contribution to mathematical physics. This constituted one of a series of four papers published by Einstein in 1905. A second was his explanation of Brownian motion (the continuous movement of molecules in a fluid, whose velocity increases with temperature), which he used to determine the size of these molecules. The other two subjects were interrelated: his Special Theory of Relativity, which was based on the fundamental hypothesis that the speed of light as a constant, in every inertial frame and the consequential equation relating mass to energy. To explain this relationship in terms of physical interaction and associated energy exchange, he concluded that mass acts as a reservoir of energy according to the relation $E = mc^2$. In other words, the total energy of an object is a function of its mass and the speed, not of the object but of light. Yet he showed that this mass of a body did not have a constant value but depended on its speed of displacement according to the relationship $m = m_0 (1 - u^2/c^2)^{-1/2}$ where m_0 is the mass of the object at rest, m being the mass of the same object when it is moving at speed of u. This relation, which go-verns all processes of energy exchange, is the starting point for research into nuclear physics and its applications. Although the British physicist **Joseph J. Thomson** had already foreseen this relation, he had never ventured to write it formally. It is certainly this part of Einstein's work that is the most spectacular and innovative. The award of the Nobel Prize to this genius in 1921, 16 years after his first publication, reflects the difficulty Einstein's peers had in accepting his early work, but also the prudence of the Nobel Committee in its choice of laureates. In 1916, Albert Einstein formulated the General Theory of Relativity, showing that Newtonian gravity only represents a special case. This followed his conjecture in 1911 that space is curved in the vicinity of a mass, and that this

Albert Einstein

also leads to the curvature of light. The British astrophysicist Eddington began to confirm some of Einstein's predictions during a total eclipse of the sun in 1919. Finally, in 1953, Einstein attempted to establish a generalized formula for all physical forces, but was not successful. Born of Jewish parents, Einstein lived in Munich from the age of six. His father ran an electrical technology business,

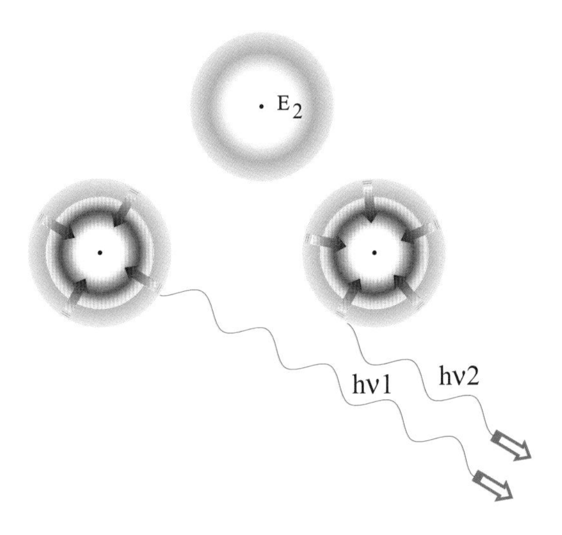

The stimulated emission of the $h\nu_2$ photon by an excited atom is identical to that of the $h\nu_1$ photon emitted by the first atom. The second photon is emitted in phase with the first one.

while his mother was an accomplished pianist. Einstein did not speak his first words until the age of three and has been described by his younger sister, Maja, as an extremely ill-tempered child. Although rebelling against traditional education, his keen interest in numerical manipulation began at a very young age. He would admit later that he did not have a great head for words, preferring conceptual

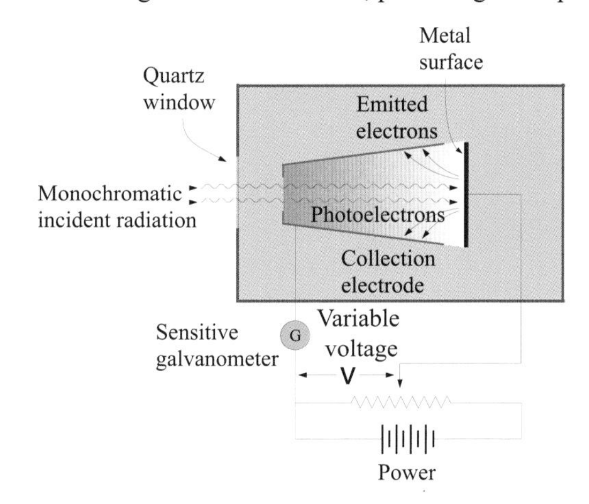

The experimental device highlighting the quantum aspect of the photoelectric effect: the ejection of an electron out of a metal by the interaction of incident radiation. The adjustment of the voltage until the cancellation of the current in the galvanometer (G) allows the measurement of the kinetic energy of the ejected electrons to be made. For each ejected electron one single photon of hν energy has disappeared.

and diagrammatic representations. Albert Einstein finished his secondary education in Aarau, in Switzerland, and had to re-take the entry exams to the Institute of Technology where he graduated in 1900. He presented his thesis in 1905. From 1902 to 1909 he was technical assistant at the Bern Patent Office, before becoming an associate professor at the University of Zurich. In 1913, he

The participants in the Solvay Council of Physics of 1927. Principal subject: electrons and photons. From left to right and seated:**I.Langmuir, M.Planck, M.Curie, H.Lorentz, A.Einstein**, P.Langevin, Ch.Guye, **C.T.R.Wilson**, **O.Richardson, P.Debye**, M.Knudsen, **W.L.Bragg**, H.A.Kramers, **P.Dirac, A.Compton, L. de Broglie, M.Born, N.Bohr,** A.Picard, E.Henriot, P.Ehrenfest, E.Herzen, Th. De Donder, **E.Schrödinger**, E.Verschaffelt, **W.Pauli, W.Heisenberg, R.H.Fowler**, L.Brillouin (in bold: the Nobel laureates).

became Director of the renowned Kaiser Wilhelm Physicalisch Institüt in Berlin, and was also nominated as a Professor at the University of Berlin, a post he occupied from 1914 to 1933. During this time, he was frequently invited to England and to the United States. His uneasiness with Hitler's regime, and with some of his fellow physicists who supported it, encouraged him to accept a post at the Princeton Institute for Advanced Study. He took American citizenship in 1940. His biographer, P. Frank, remarked that the enthusiasm with which the public greeted Einstein on his arrival in New York was one of the historic cultural events of the 20th century. At Princeton, he chose to lead the life of the solitary scientist. His lack of interest in human contact, both socially and with his family, was in stark contrast to his strong sense of justice and social responsibility. Einstein did not play an active part in the development of the Manhattan Project (construction of the atomic bomb) but was asked to put pressure on President Roosevelt to pursue American research in this direction, which explains why he is frequently associated with those who were politically active in this area. Nevertheless, Einstein remained resolutely against all forms of warfare. The Solvay Physics Congress in Brussels was honored four times by his visits: in 1911, 1913, 1927 and 1930, where he helped to clarify quantum theory and quantum mechanics. He was absent from the Council of 1921 because of a trip to the States, and he refused to attend the one in 1924 because of the exclusion of German physicists for political reasons. His opposition to **Niels Bohr**'s interpretation of quantum mechanics had its roots in his 1927 visit to the Solvay Congress, during which Bohr discussed the Principle of Complementarity

From left to right : **Robert Millikan**, Georges Lemaître and **Albert Einstein**

and, in particular, wave - particle duality and the **Werner Heisenberg** uncertainty principle. For Bohr, quantum mechanics was an established theory; for Einstein it was only an approximation of a theory to come. The concept of randomness in quantum mechanics troubled Einstein, who declared "God does not play dice." During the Solvay Physics Congress meeting in 1927, Einstein proposed the 'photon in a box' experiment, as a contradiction to Bohr's point of view. At their next Solvay Congress in 1930, the two physicists returned to the subject, where Einstein's counter-experiments (such as the Einstein - Podolsky - Rosen Paradox) were, on the contrary, interpreted by Bohr as confirmation of quantum theory. Today, almost all physicists side with Bohr. In addition to the Solvay Congress, Einstein had a number of other contacts in Belgium, notably with Georges Lemaitre, one of the fathers of the Big Bang theory, and with the royal family. In 1901, Einstein married Mileva Marec, a mathematician he had met in Zurich. They had two sons: Albert and Edward, both of whom returned to Switzerland in 1913 after their parents' divorce. In 1917, he married his cousin, Elsa. She died in 1936. An ardent defender of the Jewish people, Einstein was a vigorous supporter of the creation of the state of Israel. This stance led to his being proposed as successor to President Chaim Weizmann in 1952, but he refused. Einstein was considered a moral authority, and was often asked for his opinion on the most varied of subjects.

1922

Bohr, Niels Hendrik David (Copenhagen, Denmark, October 7, 1885 - Copenhagen, Denmark, November 18, 1962). It was in 1913 that Niels Bohr provided the basis of what was going to become the new physics: one which provided an explanation of the quantum behavior of the microscopic world. Taking up **Max Planck's** proposal (which showed how atoms absorb and emit energy by quanta of radiation) and linking to it the model of the atom that **Ernest Rutherford** (1908 Nobel Prize for Chemistry) described in the form of a nucleus (gathering the major part of the mass in a very reduced and positively charged volume) surrounded by electrons (carrying together an equivalent negative charge), Bohr established the quantum model of the simplest atom: hydrogen. He postulated in particular that the sole electron of this atom could exist only in a series of well-defined quantum states. The transition towards a state of higher energy (smaller) takes place by absorption (emission) of a quantum of radiation. As long as the electron remains in a constant state of energy, it "does not radiate" any energy, in obvious contradiction with classical Maxwell electromagnetism. This difficulty was to be definitively solved only by **Louis de Broglie** and **Erwin Schrödinger**. Bohr was able to prove with experiments the exactitude of his proposal for a sin-

Niels Bohr

gle hydrogen atom but he clearly stated that all atoms had to have identical behavior and that it is the electrons, which are the less bonded to the structure, that take part in the chemical reactions. **Gerhard Herzberg** (1971 Nobel Prize for Chemistry) and **Wolfgang Pauli** would provide the definitive explanation of this. See the **Albert Einstein** bibliography for the epic discussions that he had with Bohr on quantum mechanics during the 1927 and 1930 Solvay Councils of Physics in Brussels. Niels Bohr also took part in the Solvay Councils in 1933, 1948 and 1961.

He studied physics at the University of Copenhagen where his father taught physiology. In 1911 he received his doctorate and then stayed at Cavendish Laboratory with

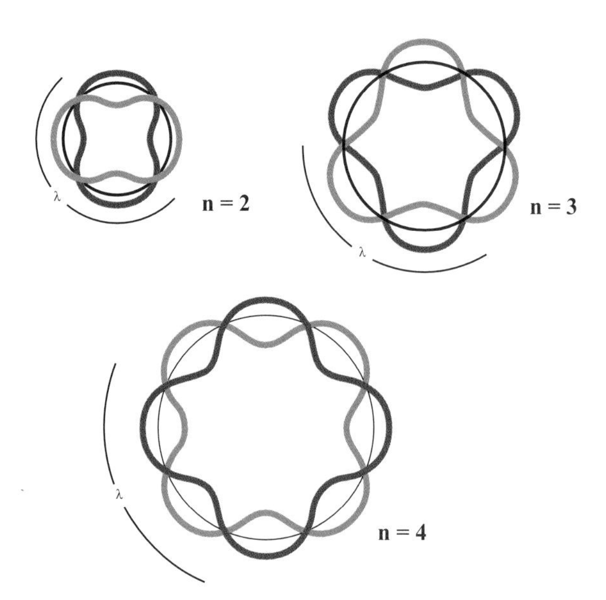

Stationary orbitals. When the wavelength associated with the circular motion of an electron is a submultiple of the length of its (classical) trajectory, the particle undergoing a centripetal acceleration (due to the electric force of the nucleus) does not radiate energy.

Joseph J. Thomson at the head, as well as Manchester University, directed by Ernest Rutherford (1908 Nobel Prize for Chemistry). Bohr's first scientific work was devoted to the study of the surface tension of water droplets in a jet of air, a model which he used later to explain the excitation of atomic nuclei (1937) and the mechanism of uranium fission (1939). Between 1916 and 1962 he taught at the University of Copenhagen. In 1920 Bohr was appointed Head of the Institute of Theoretical Physics and he occupied this post until 1962. In 1943, when Denmark was occupied by Germany, Bohr and his family fled to Sweden and then to the United States. But in 1945 they returned to their home country. Niels Bohr and his son **Aage** worked on the Manhattan Project. It was during a trip to the United States in 1939 that he brought the news that **Otto Hahn** (1944 Nobel Prize for Chemistry) and Lise Meitner, in Germany, had probably succeeded in uranium fission by slow neutrons. In 1994,

The participants in the Solvay Council of Physics of 1930. Principal subject: magnetism. From left to right and seated:Th. De Donder, **P.Zeeman,** P.Weiss, A.Sommerfeld, **M.Curie,** P.Langevin, **A.Einstein, O.Richardson,** B.Cabrera, **N.Bohr,** W.J. de Haas. Standing : E.Herzen, E.Henriot, J.Verschaffelt, Ch.Manneback, A.Cotton, J.Errera, **O.Stern,** A.Picard, W.Gerlach, C.Darwin, **P.Dirac,** H.Bauer, **P.Kapitza,** L.Brillouin, H.A.Kramers, **P.Debye, W.Pauli,** J.Dorfman, **J.van Vleck, E.Fermi, W.Heisenberg** (in bold, the Nobel laureates).

rumours circulated, wrongly, that he had given confidential information, in 1945, on the American nuclear project to the Soviets: in fact the writings and diagrams related to discussions that he had had with **Werner Heisenberg** had been discovered. Being a pacifist, Bohr was also President of the Danish Atomic Energy Commission and in 1955 he took an active part in the first Atom for Peace Conference in Geneva. Bohr was one of the founders of CERN. He also took an interest in biology and launched the project of the Institute of Genetics in Cologne, but failed to accomplish this task and to bring the project to life. In his college years he was a brilliant physicist, and at the same time a superb soccer player. In 1922 Bohr was granted the Nobel Prize in physics. The same year, the Nobel committee presented the Nobel Prize to Albert Einstein, who won it in 1921

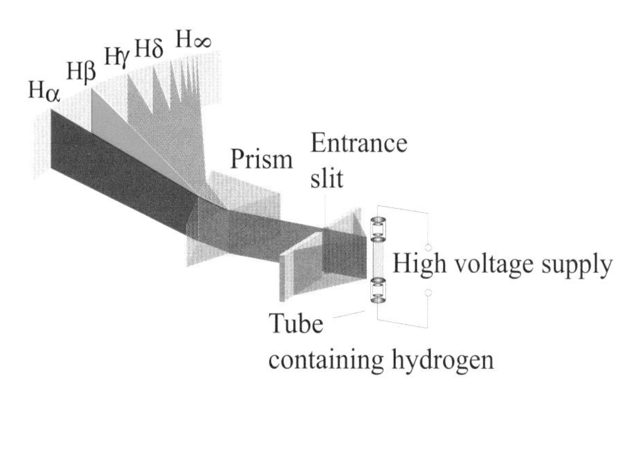

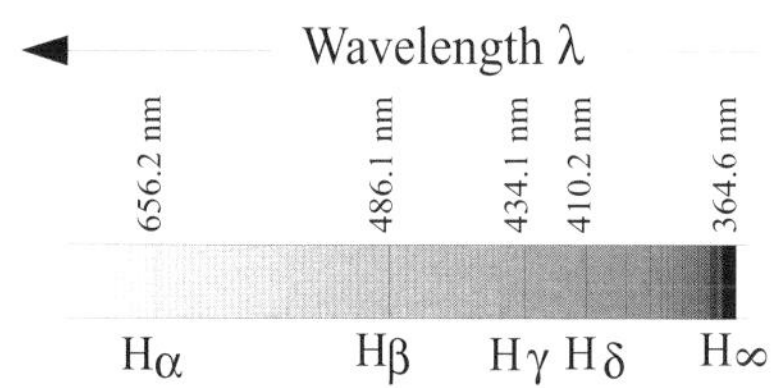

The quantum states of the hydrogen atom: the experimental arrangement (top), the recorded wavelengths (bottom).

but was not able to attend the ceremony in Stockholm in due time.

1923

Millikan, Robert Andrews (Morrison, Illinois, March 22, 1868 - San Marino, California, December 19, 1953). American experimenter physicist. We owe to Robert Millikan the creation of the device which definitively established the electron as carrier of the elementary electric charge of the smallest stable particle. In 1911, pulverising fine droplets of oil (which are charged at the exit of the pipette or by interaction with an X - ray beam), he maintained them in equilibrium in the variable electric field, created between the two plates of a flat capacitor. This equilibrium is reached when the force of gravity on the droplet is counterbalanced by the electrostatic force. From the results, Millikan concluded that these droplets always carried a charge which is an integer multiple of 1.6 x 10^{-19} coulomb, the elementary charge. He also elaborated the work necessary for the extraction of an electron from a metallic plate by photons and, as a result, confirmed **Albert Einstein's** explanation of the photoelectric effect. His other research activities concern, inter alia, cosmic rays. For this purpose he built sophisticated equipment that he took to the bottom of lakes and to the highest mountains of America, and to the Andes in particular. Having passed his baccalaureat at 23 years old, he graduated from the University of Oberlin at age 25 and defended his thesis there two years later. Then he made visits to Göttingen and Berlin. From 1896 until 1921, he taught, brilliantly, at the University of Chicago and moreover gave priority to his teaching activities to the detriment of his research activities. In 1921 he became Director of the emerging California Institute of Technology, an institute that, thanks to him, would become very famous. He took part in the Solvay Council of Physics in Brussels in 1921. Millikan was the son of a Congregational minister. His mother was a teacher and the director of an institute for young girls. **Carl Anderson**, one of his students, recognized that it was Millikan who instilled in him the vocation for research.

Robert A. Millikan

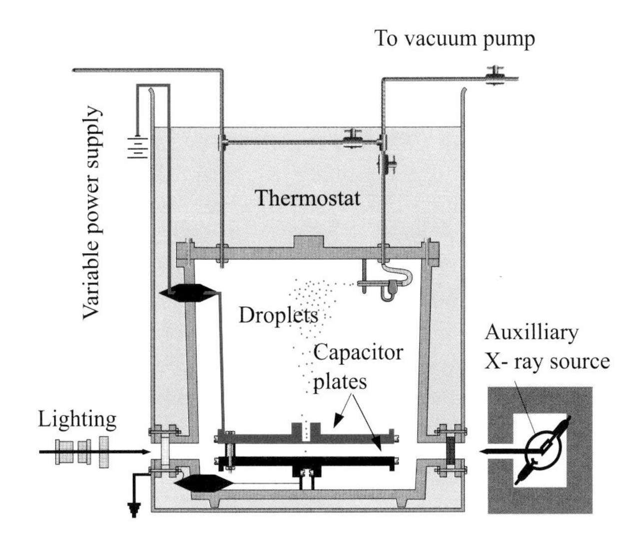

Device for the measurement of the elementary electric charge. Oil droplets vaporized in a vessel are charged by friction when outgoing from the nozzle. They can be maintained immobile by an electric field created in the bottom part. Any modification of this charge by X - ray irradiation requires a readjustment of the electric field which leads to a new balance. From all the measurements, Millikan deduced that the total electrical charge could only be an integer multiple of 1.6 10^{-19} Coulomb. This elementary charge is that of the electron (negative) or the proton (positive).

1924

Siegbahn, Karl Manne Georg (Örebro, Sweden, December 3, 1886 - Stockholm, Sweden, September 25, 1978). Continuing the work of **Charles Barkla**, Karl Manne Siegbahn built an X - ray spectrometry device based on their diffraction by crystals. This spectroscopy had already been developed by **William Henry** and **William Lawrence Bragg** but Siegbahn considerably improved it in order to obtain a precision of 1/100,000 in the measurement of wavelengths. He made a complete study of the spectral lines emitted by the vast majority of the elements of the periodic table, confirmed Barkla's results on the K and L lines of the elements, specified that these lines were multiple, and also identified the lines of the M series of a larger wavelength. This information is still used today for the identification of the elements by X - fluorescence and by X - spectroscopy induced by electrons and by other charged particles. This information also confirms both the wave nature of X - rays and the quantified structure of electronic shells predicted in **Niels Bohr's** atomic model. Karl Manne Siegbahn took part in the Solvay Council of Physics in Brussels in 1921. He entered the University of Lund in 1906 and obtained his doctorate in 1916 after a working stay at Göttingen and

Karl M. Siegbahn

Silver coin struck by a Bishop Prince of Liège (Belgium) during the 14th century A.D.

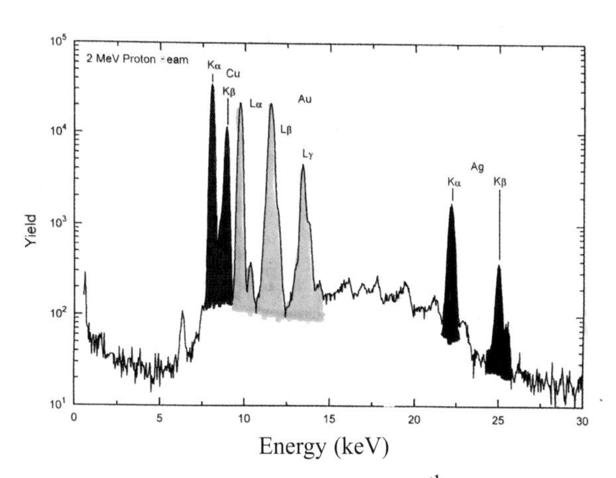

A pendant from the Achemenide period (4th c.B.C.) belonging to the Louvre Museum and studied at LARN by the PIXE technique (particle induced X-ray emission). The X-ray spectrum indicates the presence of copper, gold and silver in the pendant. In red, the copper signals; in yellow, the gold signals; in green, the silver signals.

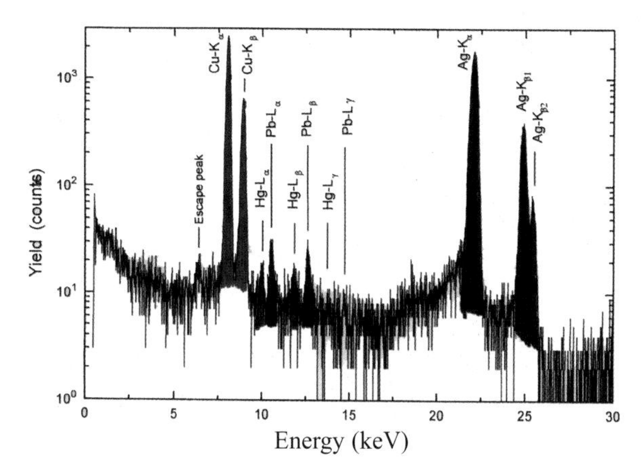

X - ray spectrum induced by PIXE on the silver coin:silver is in green, copper in red, traces of mercury and lead in blue and black respectively.

Munich. From 1914 until 1923 he worked as J.R. Rydberg's assistant and then taught physics in Lund and Upssala, respectively. He finished his scientific career at the Stockholm Academy of Sciences from 1937 until 1964. He headed the Nobel Institute of Physics from 1939 until 1964. His son, **Kai**, was also awarded the Nobel Prize (1981).

1925

Franck, James (Hamburg, Germany, August 26, 1882 - Göttingen, Germany, May 21, 1964). The famous experiment of the Germans James Franck and **Gustav Hertz**, carried out by sending electrons to mercury atoms, established in 1914 the exactitude of the electron structure of the atom proposed by **Niels Bohr** and was verified concurrently by **Karl Manne Siegbahn's** experiments using X - rays. Franck and Hertz showed that the ionization is indeed carried out by a resonant procedure (namely by exchange of an energy quantum being equivalent to an integer multiple of 4.9 eV for mercury). Previously, Franck, together with Condon, had studied the vibration modes of diatomic molecules. He was educated at the Institute of Chemistry in Heidelberg then at the Institute of Physics in Berlin. He obtained his doctorate in Berlin in 1906. He ran the Zweite Physikalische Institüt in Göttingen from 1921, with Max Born as a theorist colleague until 1933. In 1935 he accepted a position in Baltimore and then in Chicago until his retirement in 1949. In 1933 he protested violently against the racial laws of the Hitler regime, which was the reason for his resignation from the position that he held in Göttingen. Although during the First World War he served his native land, during the Second, he took an active role with the Americans in the construction of the atomic bomb, as a weapon of dissuasion, but he protested against the launching of the bomb on a Japanese city. He received a number of medals and decorations, and he had a multitude of friends among his famous physicist colleagues. James Franck received the 1953 Max Planck Medal and he was honoured by the town of Göttingen, which named him an honorary citizen. Regarded as a man of good sense and extremely generous, he always preferred the simple experiment and pure logic to long mathematical demonstrations. He died in his native country in 1964 while visiting in Göttingen.

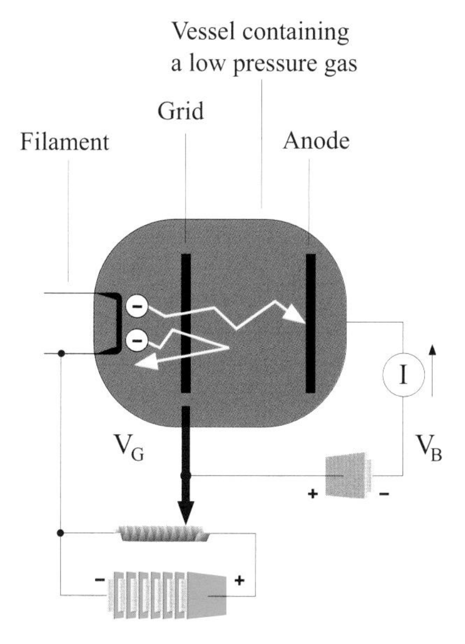

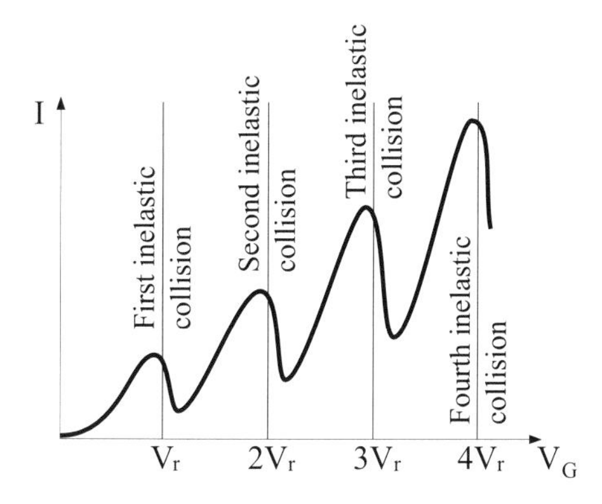

Quantified binding energy of atomic electrons. When the electrons accelerated in an electric field have a kinetic energy just equal to the ionization energy of the gas (maintained at low pressure in the bottle), a sharp decrease of the current is observed. After a first collision where the kinetic energy is reset to zero, the process can be reactivated in the electric field in an identical manner.

Hertz, Gustav Ludwig (Hamburg, Germany, July 22, 1887 - East Berlin, Germany, October 30, 1975). We refer to James Franck's bibliography for the description of the experiment, which earned both of them the Nobel Prize. Hertz studied in Munich then in Berlin where he obtained his first position (with a forced interruption during the First World War). From 1920 until 1925 he worked at the Philips research laboratory in Holland, returned as professor to Halle, then went to Berlin and Charlottenbourg until 1934, at which time he rejoined the management of the Philips and Siemens research laboratories until 1945. After Berlin's fall in 1945, he moved to the Soviet Union where he worked until 1955. The same year he returned to Leipzig and taught physics until his retirement (1961). He then lived in East Berlin for 14 years.

1926

Perrin, Jean-Baptiste (Lille, France, September 30, 1870 - New York, New York, April 17, 1942). Jean-Baptiste Perrin, an experimental physical chemist, received the Nobel Prize for having proved the discontinuous aspect of the distribution of matter in gases and liquids and for the explanation of the sedimentation equilibrium (exponential distribution of the density in a fluid system). He used an ultra microscope to show the exactness of **Albert Einstein's** description of Brownian motion. He had previously worked on cathode rays to show that they consisted of a beam of particles (negatively charged) at a time when a large majority of scientists thought that they consisted of radiation being propagated in the form of waves. He was also the first to try to determine the mass/charge ratio of the electron, work that was completed in 1897 by **Joseph J. Thomson**. Trusting experiment above all, he regularly approached his theoretical colleagues with the sentence: "it is difficult to create a theory which is completely false".

Jean-Baptiste Perrin

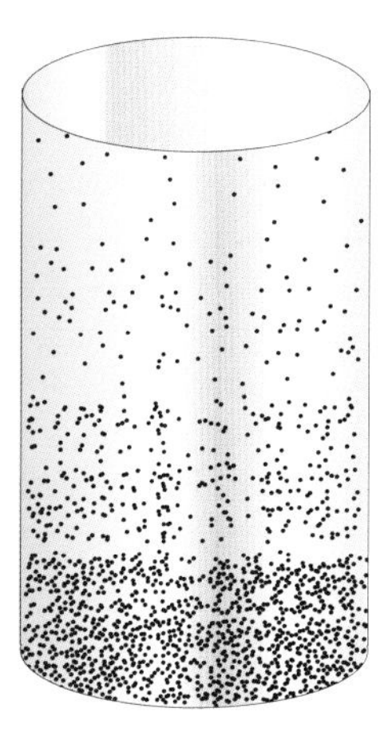

Sedimentation. The exponential distribution of the suspended particles.

Perrin received his Ph.D. from the "Ecole Normale Supérieure" in 1897 and taught there from 1894 until 1897. He took part in the Solvay Councils of Physics in Brussels in 1911 and 1921. From 1910 to 1938 he taught physical chemistry at the Sorbonne in Paris. Perrin also published excellent books dealing with the atomic structure of matter. He served as an officer in the Engineer Corps during the First World War. In the 1930's he was at the heart of the creation of the National Centre for

Scientific Research (CNRS) and the Palace of Discovery (the museum of interactive sciences inaugurated in Paris in 1937). He settled in the United States around 1938 at the time of the rise of Fascism, which he condemned, and he vigorously supported, from abroad, the party created by General de Gaulle. He died in New York in 1942 but his ashes were transported to Paris for burial at the Pantheon on June 20, 1948.

1927

Compton, Arthur Holly (Wooster, Ohio, United States, September 10, 1892-Berkeley, California, March 15, 1962). Arthur Compton, a research physicist, received the Nobel Prize for Physics for the demonstration of the transfer of momentum to a loosely bound electron in matter by X - rays. **Albert Einstein** and **Max Planck** had already shown that the energy of a photon was measured according to the formula E = hν, but it was Compton who demonstrated (in about 1923) that these photons had also a momentum (even if they did not have any mass) whose value is hν/c (c: the speed of light). The transfer of energy and momentum of an individual X - ray to an electron is accompanied by the re-emission of a photon of weaker energy and momentum: this phenomenon is today called the Compton effect. Compton received the Nobel Prize the same year as **Charles Wilson**, who built a useful apparatus which confirmed his discovery.

Arthur H. Compton

Educated initially at Wooster College, he graduated from Princeton where he worked under the supervision of **Owen Richardson** (Master's in 1914, Doctorate in 1916). Then he taught at the University of Manitoba (Canada), worked in Cambridge (U.K.) in 1919 with **Ernest Rutherford** (Nobel Prize in Chemistry in 1908) then returned to the United States in 1920 to work, alternately, at Washington University, St. Louis, and Chicago. He attended the Solvay Council of Physics in Brussels the year he received his Nobel Prize. After his reputation was established by this prize, he took an interest in astrophysics and showed that the cosmic rays were particles (coming from distances further away than the solar system and even than the Milky Way) and not electromagnetic radiations as **Robert Millikan** had vigorously argued. In November 1941 he submitted a report to the National Academy of Sciences of the United States on the possible military applications of atomic energy, a report that was decisive for the establishment of the Manhattan Project, in which he contributed actively to the production of plutonium. After the war, he became famous for his conferences, accessible to wide audiences, on the topic: "science and responsibility." While at school, he was very well known for his skills in football, baseball and basketball.

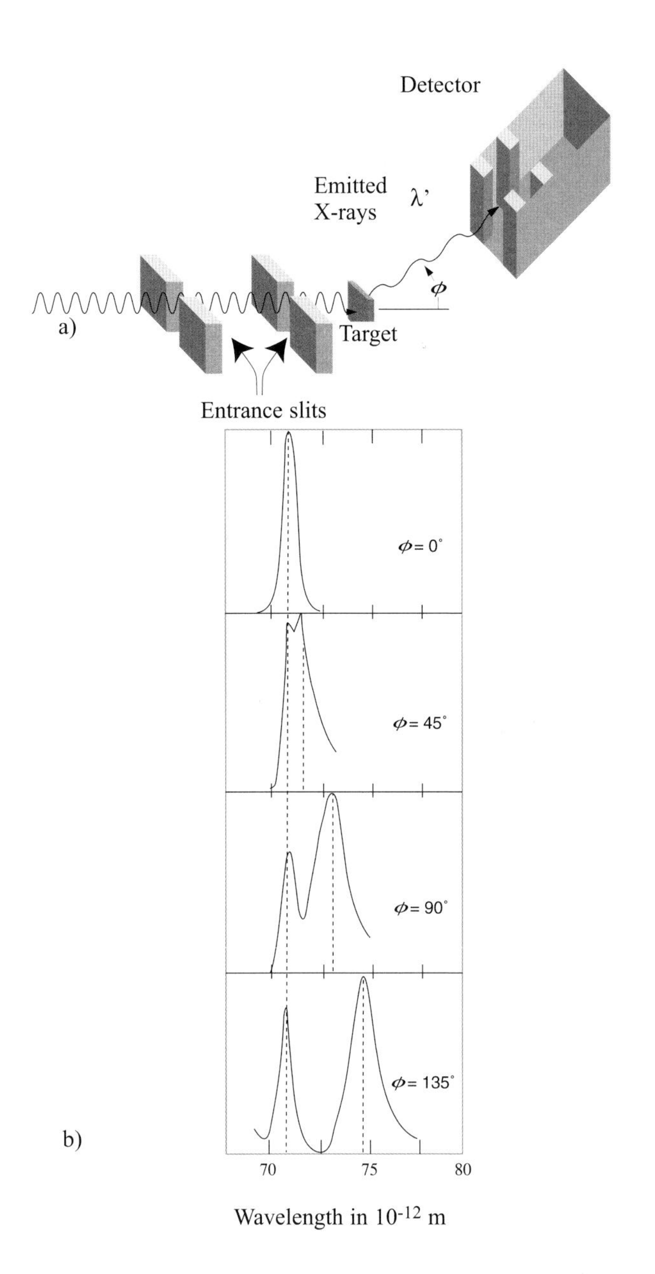

The Compton effect. A quantum of X-radiation interacting with a single (free) electron is scattered by undergoing an increase in its wavelength $\Delta\lambda = h(1 - \cos\phi)/mc$; h: Planck's constant; m: mass of the electron; c: light velocity, h/mc is called the Compton wavelength of a particle of mass m. **a)** device; **b)** progressive displacement of X - ray lines towards larger wavelengths with the increasing angle of emission. The stationary line on the left corresponds to the Rayleigh elastic scattering (no change of wavelength).

Wilson, Charles Thomson Rees (Glencorse, Scotland, February 14, 1869 - Edinburgh, Scotland, November 15, 1959). Using the meta-stable properties of vapor in the vicinity of their condensation conditions, as described by **Johannes van der Waals**' equation of state, Charles Wilson developed, in 1895, a display device of the traces that the charged particles leave while crossing a vapor. Droplets of liquid are indeed formed around the ions formed along the trajectory of a charged particle, if an abrupt expansion is caused in this vapor (contained in an enclosure called a "Wilson cloud chamber"). **Donald Glaser** (in 1960) and **Luis Alvarez** (in 1968) received the Nobel Prize in Physics for the creation and development of an opposite device (this time using the transition from the liquid state to the vapor state) intended for the detection of very high - energy particles. Wilson, did not initially think that his invention would be used in particle physics but for the study of electricity in the atmosphere especially lightning. Born into a farming family, he was the youngest of eight children. He lost his father when he was four years old. At 15 years old, he began to study in Manchester to become a doctor, but his interest in physicochemical phenomena made him change orientation. He studied under **Joseph J. Thomson** and taught physics at the Cavendish Laboratory, then at the University of Cambridge (United Kingdom), until his retirement in 1934. He nevertheless remained active in science for many more years: furthermore, he presented a paper to the Royal Society at the age of 87. He took part in the Solvay Council of Physics in Brussels the year he was awarded the Nobel Prize. A brilliant mind, an exceptional teacher, he was the archetype of the modest and unassuming gentleman.

1928

Richardson, Owen Williams (Dewsbury, England, April 26, 1879 - Alton, England, February 15, 1959). Richardson won the Nobel Prize for his studies on the emission of ions by heated solids. He was the first to propose that these ions came from filaments (he suggested the term thermionic) and not from the residual gas around the wire as was generally thought by his contemporaries. He established a law, which bears his name, that connected the saturation current of these ions to their extraction work and to the temperature of the filament. This law survived the revolution brought about by quantum mechanics of the 1920s. A young prodigy, Richardson started to become interested in sciences from the age of 12 and received numerous prizes during his adolescence. Graduating in physics from Trinity College, Cambridge in 1897 (where he also obtained certificates in chemistry and botany), he worked, until 1906, in the Cavendish Laboratory where he collaborated with **Joseph J. Thomson**, **Ernest Rutherford** (1908 Nobel Prize for Chemistry) and **Charles Wilson**. He was appointed Professor of Physics at the University of Princeton (United States). Arthur Compton was one of his students. He conducted research at King's College until 1944. He took part in the Solvay Councils of Physics in Brussels in 1921, 1924, 1927, 1930, 1933 and 1948. He served in the telecommunications division of the British Army during the First World War and was a specialist in radar, sonar, magnetrons and klystrons during the Second World War. He was Knighted in 1939 for his activities in many learned societies.

Owen W. Richardson

1929

de Broglie, Louis-Victor Pierre Raymond (Prince) (Dieppe, France, August 15, 1892 - Louveciennes, France, March 19, 1987). In 1922, one year before **Arthur Compton's** proposal to attribute a linear momentum hν/c to photons (quanta of radiation), in addition to the energy hn that **Max Planck** had allotted to them, Louis de Broglie took a decisive step toward the unification of the energy and momentum concepts of particles and radiation. Thus he attributed first to electrons and then to all particles a wavelength: $\lambda = h/mv$ (connected very simply to their mass and their speed). This is known as the de Broglie wavelength. In 1927 **Clinton Davisson** (assisted by Germer) and **Georges Thomson** proved the exactitude of de Broglie's views by identifying the electron diffraction by crystals, as the **Braggs** had done (father and son) (1915) by means of X - rays. This experimental proof of the reality of these de Broglie waves was a determining factor in the choice made by the Nobel Committee for the 1929 prize. de Broglie's conversations with **Niels Bohr**, **Erwin Schrödinger** and **Max Born** were most profitable for the progress of modern physics. de Broglie first graduated from the Sorbonne in 1909 in history. Influenced by **Albert Einstein's** articles, he then turned toward theoretical physics and obtained his degree at the Faculty of Sciences of Paris in 1913, his Ph.D. at the Sorbonne (in 1924 only, because of the First World War). The subject of his thesis won him this Nobel prize. He taught at the Sorbonne (Institut Henri Poincaré) from 1928 until 1962

Louis de Broglie

and he published a number of articles. He attended the Solvay Concil of physics in Brussels in 1911, 1913, 1927 and 1933. The scientific career of de Broglie was greatly

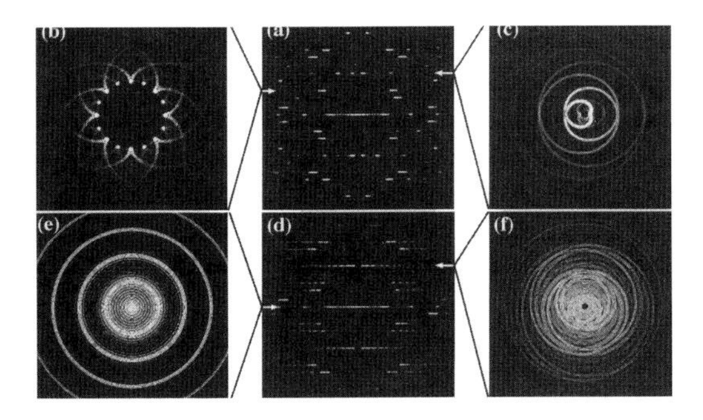

Central part: electron diffraction by a carbon nanotube. Left and right: pattern conversion in the momentum space.

influenced by his brother Maurice, a most notable experimental physicist. With Maurice, he became an adviser to the French Atomic Energy Commissariat in 1945. They supported the peaceful development of nuclear energy but were opposed to its use for military purposes. They are part of an aristocratic family originally from Piedmont, ennobled by Louis XIV in 1740. During World War I, he was attached to the wireless telegraphy section of the French Army and served in the station at the Eiffel Tower.

1930

Raman, Chandrasekhara Venkata (Trichinopoly, today Tiruchirappalli, India, November 7, 1888 - Bangalore, India, November 21, 1970). Chandrasekhara Raman, an experimenter physicist, was the first Indian, and also the first Asian, to win a Nobel Prize in Physics. In 1928 he extended **Arthur Compton's** discovery on the interaction of the X - ray with loosely bound electrons by a similar study on the modification of the wavelength of (visible) light at the time of its interaction with vibrating molecules. This data is now exploited by physical chemists for the identification of various molecular species and extensively used to identify pigments in oil paintings. Raman received the Nobel Prize only two years after his discovery. Born in a family of ancient landholders originating from the country of Madras, he was educated in Vishakapatan where his father taught physics and mathematics. Then he moved to Madras. Because of poor health he was unable to pursue his education overseas. After having received his Nobel Prize (part of which he devoted to the purchase of crystals intended for his research), he became interested in the various crystallographic forms of diamonds. In 1933, he headed the Institute of Sciences, at Bengalore. A national hero after the award of the Nobel Prize, he was a refined man, appreciating music, flowers and physiology.

1931 Not awarded

1932

Heisenberg, Werner Karl (Wurzburg, Germany, December 5, 1901 - Munich, Germany, February 2, 1976). Werner Heisenberg, a theoretical physicist and supporter of **Niels Bohr's** theory, applied it to the mathematical theory of matrices. He explained all the spectral lines of the hydrogen molecule of which he defined the two allotropic forms connected to the (two) orientations of the electrons' spins (the spin is a quantity attached to the spinning movement of a particle) which was introduced by **Paul Dirac**. Describing a molecule in motion on a microscopic scale based on **Louis de Broglie's** idea, he concluded that it was impossible to measure simultaneously two quantities (called complementary) like the x component of momentum mvx and position x with an optimum accuracy. Any attempt to measure more precisely one of the two quantities leads to a loss of precision on the measurement of the complementary quantity. The product of the inaccuracy on the measurement of one, by the inaccuracy on the measurement of the other cannot be lower than $h/2\pi$, h ($6.62\ 10^{-34}$ joule second) being the constant determined by **Max Planck**. This inequality, $\Delta x\ \Delta p_x \geq h/2\pi$ is called the Heisenberg principle of uncer-

Werner Heisenberg

$$2\pi\ \Delta p\ \Delta x = h$$
$$2\pi\ \Delta E\quad \Delta t = h$$
$$2\pi\ \Delta L\ \Delta\varphi = h$$

Heisenberg's principles of uncertainty specify the degree of simultaneous knowledge that we can have of two conjugated data. In these formulas h represents Planck's constant, x the position, p the momentum, E the energy, t the time, L the angular momentum, φ the phase.

tainty. After 1932, having learned of the discovery of the neutron by **James Chadwick**, Heisenberg proposed a model of the atomic nucleus in the form of an association of protons and neutrons, and not of protons and electrons as was envisaged at this time. A consequence of the principle of uncertainty prevents electrons from remaining locked in a volume as tight as that of an atomic nucleus (of which the radius is about 10^{-15} m). Heisenberg studied theoretical physics and presented his thesis in 1923 at Munich in A. Sommerfeld's group (1868 - 1951). For some incomprehensible reason, the latter never received the Nobel

Prize. Heisenberg was then part of the groups of **Max Born** and Niels Bohr in Göttingen and Copenhagen respectively, before returning to Leipzig where he taught from 1927 until 1941. From 1941 until 1945 he was Director of the Max Planck Institute and taught simultaneously at the University of Berlin, finally moving to Munich. He took part in the Solvay Council of Physics in Brussels in 1927, 1930, 1933 and 1961. In February 1943 he gave a series of courses which attracted a great deal of attention at the Trinity College of Dublin on the subject: "What is Life?". There, he asked fundamental questions that a physicist could tackle in this field. What is the physical structure of the molecules during the division of the chromosomes? By what process can we understand this reproduction? How do molecules keep their individuality from generation to generation? How do they succeed in controlling the metabolism of the cells? How do they create the observable organisation in the structure and the operation of very developed organisms? During the Second World War, he inspired enormous fear in the American experts in charge of manufacturing the first atomic bomb: the latter indeed thought that the Germans were making headway in a similar program under Heisenberg's direction. However, it seems that he was more occupied in developing the application of fission for industry than in the manufacture of a bomb. He reproached Chancellor Adenauer about the lack of financial support to finish his project and even refused to represent the Federal German Republic in Geneva (in 1955) at the time of the first international conference on the Peaceful Use of Nuclear Energy. Except for **William Lawrence Bragg** who received the Nobel Prize in Physics aged 25 years old, Heisenberg is, in 1932, the youngest to have received this prize. He was then only 31 years old.

1933

Schrödinger, Erwin (Vienna, Austria, August 12, 1887 - Alpach, Austria, January 4, 1961). Theoretician physicist. The name of Erwin Schrödinger can not be dissociated from that of **Paul Dirac** with whom he shared the prize of 1933. The equation that he developed is still, today, the royal road for the teaching of modern physics. Instead of the model of **Ernest Rutherford** (1908 Nobel Prize for Chemistry), which described the motion of electrons in orbits, like planets around their sun, he substituted another model associating the **Louis de Broglie** "particle waves". This formulation allows the electrons to not lose energy as required by Maxwell's classical equations. This model, which initially did not require probabilistic interpretation like the model of **Niels Bohr**, was more easily acceptable to his contemporaries: **Albert Einstein** in particular. In addition to his work in wave mechanics, he carried out research on the specific heat of solids, on statistical thermodynamics, on colors, on space and time and on science and human behavior. He even wrote poems. After being taught by private tutors, he studied at the University of Vienna and defended his doctoral thesis in 1910. There he held post as **Wilhelm Wien's** assistant before teaching in Stuttgart and Zurich. As early as 1927 he succeeded **Max Planck** at the University of Berlin. In 1933, with the rise of Nazism (of which he was a vehement critic), he decided to accept a position at Oxford. Homesick for his native country, he returned to Graz in 1936 but was obliged to flee to Italy in 1938, owing to the annexation of his native country by Germany. He then took the route to Princeton. With the support of the Irish Prime Minister, E. De Valera, himself a scientist, he moved to the Institute for Advanced Studies in Dublin, which he left in 1956, when he returned definitively to Graz. He took part in the Solvay Councils of Physics in Brussels in 1924, 1927, 1933 and 1948. Not very mindful of worldly practices, his behavior on arrival at the Métropole Hotel of Brussels, where the participants in these Councils were lodged, often caused him difficulty in obtaining a room.

Erwin Schrödinger

Energy

$E\psi$ becomes $i\hbar\dfrac{\delta\psi}{\delta t} = H\psi$;

linear momentum

p^2 becomes : $\left(\dfrac{\hbar}{i}\right)^2 \left(\dfrac{\delta^2\psi}{\delta x^2} + \dfrac{\delta^2\psi}{\delta y^2} + \dfrac{\delta^2\psi}{\delta z^2}\right)$

general equation

with potential energy V

$$H\psi(r) = \left\{-\frac{\hbar^2}{2m}\Delta + V(r)\right\}\psi(r) = E\psi(r)$$

Correspondence of classical and quantum mechanical entities for writing Schrödinger equations.

Dirac, Paul Adrien Maurice (Bristol, England, August 8, 1902 - Tallahassee, Florida, October 20, 1984). A theoretical physicist, Paul Dirac (who shared the Nobel Prize with the Austrian physicist **Erwin Schrödinger**) introduced the relativity concepts developed by **Einstein** into wave mechanics. The latter had shown how mass and energy could be exchanged. With kinetic energy, the linear momentum and mass being linked together in Einstein's equation ($E^2 = m_0^2c^4 + p^2c^2$), Dirac suggested that the extraction of the square root gave rise to two solutions: one containing m_0c^2 (which he associated with a particle).the other containing $-m_0c^2$ (which he associated with an antiparticle). Thus the way was mathematically open toward research on antimatter and in particular the anti - electron (or positron) that **Carl Anderson** discovered in 1932. The Dirac equations also implicitly contained the spin theory, also proposed by **Werner Heisenberg** and postulated by the Dutchmen (originating from Djakarta), Goldsmit (1902 - 1978) and Uhlenbeck (1900 - 1988), to explain the spectra of alkaline atoms (Na, K,). In the 1920's, theorists like **Louis de Broglie**, Paul Dirac,

Paul A. Dirac

The creation of a pair of particles (matter + antimatter) during the interaction, in the presence of matter, of a high energy γ - ray. The energy of the γ - ray is converted into mass (two particles having charges with opposite signs and following two spirals in opposite directions when moving in a perpendicular magnetic field). The two particles have kinetic energy in agreement with Einstein's famous equation: $E = mc^2$. The incident γ - ray is not visible. The path with a spiral shape indicates the progressive decrease of the kinetic energy of the electron and the positron when travelling in the material of the detector.

Werner Heisenberg and Erwin Schrödinger found valuable allies in researchers like Anderson and **James Chadwick**, which led to a complete change in the way of thinking of the physics world. Thus they provided an explanation for all the microscopic phenomena revealed by the discoveries, which earned their authors the vast majority of the first 30 Nobel Prizes in Physics. Dirac had the audacity to propose (and with good reason) the creation of matter and

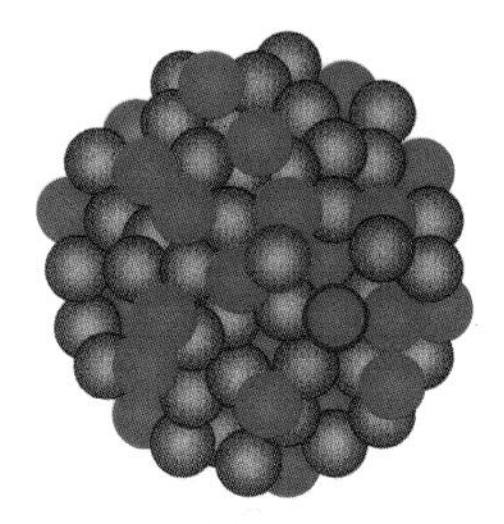

β^+ emitter (before emission)

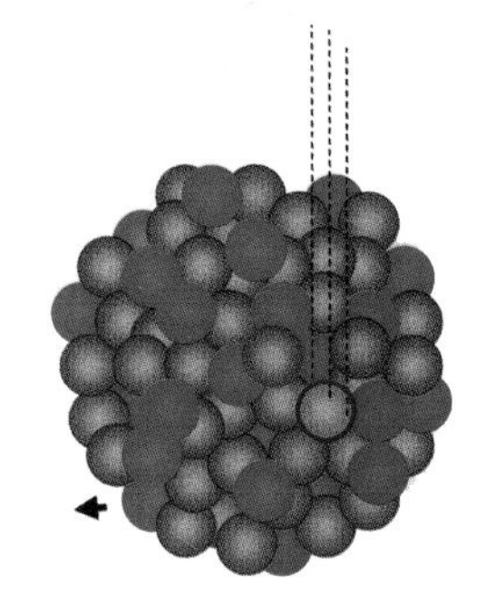

A proton of the nucleus becomes a neutron and a particle with a mass equivalent to that of the electron with a positive charge is emitted. Simultaneously the nucleus emits a neutrino.

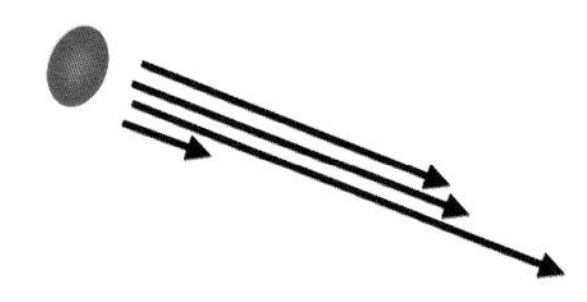

The β^+ emission. In some atomic nuclei a proton can be less stable than if it was transformed into a neutron. This transformation leads to the emission of a positron (positive electron) and a neutrino. See, by comparison, the β^- emission (Fermi).

antimatter from electromagnetic radiation. A visionary genius, he received his first scientific training in Bristol and Cambridge where he defended his thesis in 1926. A great traveller, he regularly visited the universities of Wisconsin, Michigan, Princeton and Miami in the United States, other centers in Japan (where he opened the way for **Yukawa's** work) and in Siberia. His academic career in Cambridge began in 1932 and lasted 47 years. He took part in the Solvay Councils of Physics in Brussels in 1927, 1930, 1933, 1948 and 1961. Paul Dirac was born in Bristol, England. His father was a Swiss citizen. In 1926, he married Margit Wigner, sister of **Eugene Wigner**, 1963 Nobel Prize for Physics. He is among the five youngest physicists of this century to have received the Nobel Prize for Physics.

1934 Not awarded

1935

Chadwick, James (Macclesfield, England, October 20, 1891 - Cambridge, England, July 24, 1974). At the beginning of the 1930s, James Chadwick, a research physicist, exploited a set of data obtained by **Walther Bothe** and the **Joliot-Curie** couple relating to the emission, under the impact of a particle, of a new type of projectiles, themselves able to eject protons from a paraffin film. Resuming these experiments, Chadwick suggested that they were particles, of a mass very close to that of the proton, but carrying no electrical charge. He called this particle the neutron, a component existing in a stable state only inside atomic nuclei. Its half-life was then measured as 11.2 minutes when it is in a "solitary" state. With the discovery of this particle, **Ernest Rutherford's** global model of the atom took its final form: a small size nucleus containing protons and neutrons surrounded, in a (spherical) volume of radius 100,000 times greater, by a number of electrons equal to the number of the protons in the nucleus. These electrons occupy various states of energy described by **Niels Bohr's** model. Having no charge, the neutrons play a major role in the study of nuclear reactions. After the discovery of **Enrico Fermi** and **Otto Hahn** (1944 Nobel Prize for Chemistry) in collaboration with Lise Meitner, the neutrons were used for creating new useful radionuclides which were rapidly used as tracers in medicine and technology, and for creating nuclear reactors for the production of electricity, but also for nuclear bombs of uranium and plutonium. Going to Manchester University in 1908,

James Chadwick

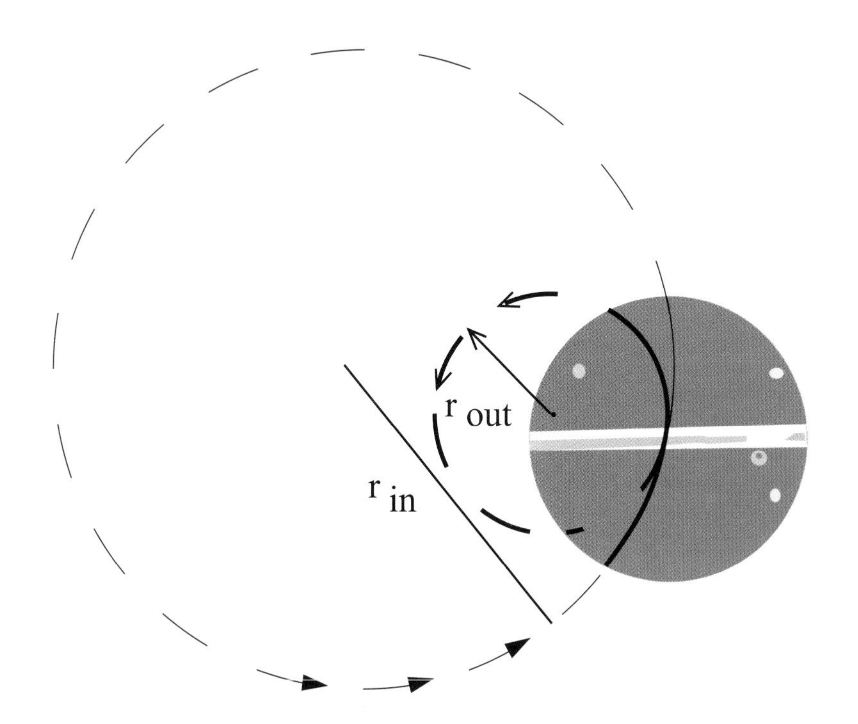

The positron. An historic image indicating the existence of antimatter: a particle with a positive charge follows a trajectory (here upwards) curved in an anti-clockwise direction by a magnetic field. The density of ionization along the trajectory indicates that the mass of the particle is the same as that of the electron. It was produced by the interaction of cosmic radiation in the material below the instrument.

Chadwick obtained his graduate diploma in 1911 in Rutherford's team. Holder of a grant, which obliged him to change laboratories, he continued his research in Berlin under the supervision of Geiger. Forgetting to return to England in time, he was imprisoned for the duration of the First World War but, nevertheless, German physicists supported him during his detention. He went back to Cambridge in 1919 and then became involved in research on nuclear reactions induced by a particles. He was appointed deputy manager at the Cavendish Laboratory in 1923 and occupied the position of Professor at Liverpool University from 1935 until 1948. He supervised, with the greatest integrity, the delegation of British experts involved in the manufacture of the first atomic bomb in the United States, but was worried by the possible disastrous effects of this weapon. After his return to England in 1946 he argued for fundamental research in nuclear physics to be a field reserved exclusively for universities, challenging the Harwell Laboratory's activities in this field. Disappointed by university policy in this matter, he resigned his position in 1958. He was present at the Solvay Council of Physics in Brussels in 1933.

1936

Anderson, Carl David (New York City, New York, September 3, 1905 - San Marino, California, January 11, 1991). An experimental physicist, Carl Anderson discovered, in 1932, the positron in the products of the reaction of cosmic radiation with the Earth's atmosphere. The positron is the anti-particle of the electron. This anti-particle had been postulated by the **Paul Dirac** theory, but this theory had not been well received by the experimenters who considered it confusing. In addition, the identification of the positron confirmed definitively the existence of anti - matter: an essential component in the understanding of the overall structure of the Universe. The association of the positron to astrophysical questions was confirmed by the award, in 1936, of the Nobel Prize for Physics to **Victor Hess**. We owe to Anderson and **Robert Millikan** the construction of detectors able to measure the charge and the mass of the particles of cosmic radiation. As early as 1935 this mechanism also allowed the identification of (furtive) particles for which the mass is intermediate between that of the electron (and therefore of the positron) and of the proton, which were called mesons. Anderson identified the μ meson, which was confused for a time (wrongly) with the π meson, postulated by **Hideki Yukawa** to explain the strong binding between nucleons (protons and neutrons) of the atomic nucleus. The distinction between π and μ would be established in 1947 by **Cecil Powell**. Anderson studied at Cal Tech (created by **Millikan**) in Los Angeles. He obtained his Ph.D. there in 1930 and was appointed professor in 1939. In 1941 he joined a research team of the Defense Department and took part, as an artillery expert, in the battles at the time of the Allied landing in Normandy. As a result, he obtained, after the war, the use of a B-29 for the continuation of his experiments on cosmic rays. He is among the five youngest physicists of the 20th century to have received the Nobel Prize for Physics.

Carl D. Anderson

Hess, Victor Franz (Waldstein, Austria, June 24, 1883 - Mount Vernon, New York, December 17, 1964). Victor Hess, an astrophysicist, was interested in the explanation of ionisation phenomena in the atmosphere. Observing the same effects inside hermetically sealed containers, he attributed its origin to "cosmic rays", the name given by **Robert Millikan**. Using an electroscope at various altitudes (Eiffel Tower, mountain, balloon) he noticed a decrease of ionisation while progressing from the ground to around 600 to 700 m, then a rapid increase. Noticing that the effect was the same at night as well as during the daytime, or during a solar eclipse, he concluded that the origin had to be cosmic. It was not until 1925, with the support of Millikan, that the scientific community accepted the overall interpretation: cosmic rays are mainly composed of particles and not of radiation. **Carl Anderson** (who earned the Nobel Prize the same year as Victor Hess) identified the positron and **Cecil Powell** the π meson in the products of the interaction of these cosmic rays with atmospheric gases. The development of particle accelerators (the cyclotron of **Ernest Lawrence** and the electrostatic accelerator of **John Cockcroft** and **Ernest Walton**) owes a lot to Hess's discoveries. The accelerators were in fact created in order to have available (on earth) these high-energy ions that are found in the cosmos. Their use is now general in numerous techniques of analysis of the elements. Hess studied in Graz and defended his doctoral thesis in 1906 at the University of Vienna. He pursued his research there. From 1919 until 1938 he taught physics at the Universities of Vienna, Graz and Innsbruck. The Nazi occupation forced him to leave the country. He held the post of Professor of Physics at the Federal University of New York until 1956. He was granted American citizenship in 1944. Besides his work on cosmic rays, Hess established the foundations of the dosimetry of radiation: his motivation probably came from the amputation of his thumb necessitated by the effects of the manipulation of radioactive products.

Victor Hess

1937

Davisson, Clinton Joseph (Bloomington, Illinois, October 22, 1881 - Charlotteville, Virginia, February 1, 1958). American experimental physicist. Clinton Davisson shared the Nobel Prize in Physics with **Georges Thomson** for their demonstration of the diffraction of electrons by crystals, experiments which confirmed **Louis de Broglie's** hypothesis relating to the wave property associated with moving particles. Working in the research department of General Electric Company on electron emitter tubes since 1919, Davisson observed, in 1925, (although without having built the experiment to demonstrate it) that the scattering of the electrons by crystals depended on the relative orientation of the crystal compared to the directions of incidence and emergence of the electrons. In 1926, during a visit to England, he was informed of the theory announced by de Broglie. On his return to the United States, and convinced that these particle waves were related to the phenomena which he had already observed, he started again, with Lester Germer, the previous

Clinton J. Davisson

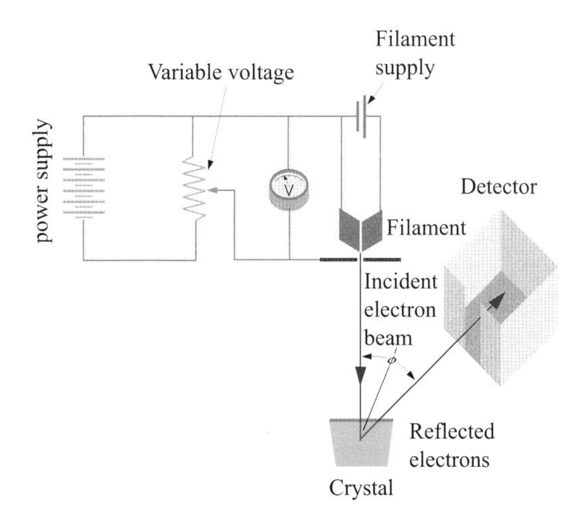

The experimental device used by Davisson to produce a monoenergetic electron beam in order to induce diffraction in a crystal.

experiment using a well-defined energy (and therefore a well - defined wavelength of the incident electron beam). Davisson was born in 1881 into a modest family of artists. After gaining a diploma in his birthplace, he began university studies in 1902 at Chicago, where he was influenced by **Robert Millikan's** teaching. He had to interrupt his studies for financial reasons and sought work.

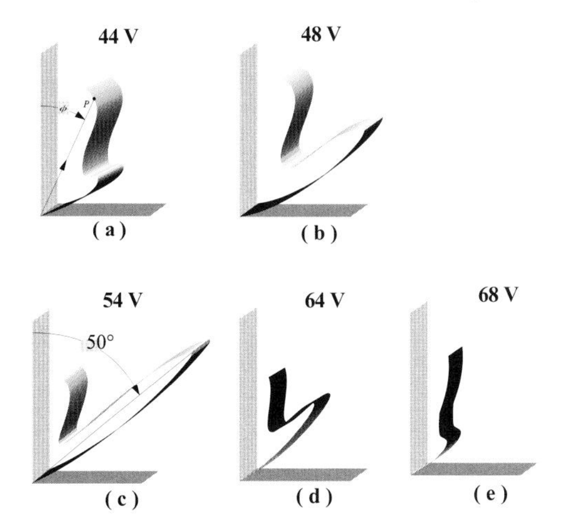

The angular distribution of the diffracted electrons by a crystal structure. Note the largest intensity at an angle of 50° when the accelerating voltage is 54 Volts. This behavior is due to the wave property associated to moving particles as proposed by Louis de Broglie.

However, he obtained a position as assistant at Purdue in 1904 and a position as tutor from 1905 until 1910 at Princeton where he had **Owen Richardson** as professor. He defended his thesis there in 1911 on the emission of positive ions by alkaline salts. From 1911 until 1917 he taught physics at Pittsburgh before taking a job in the Research Laboratory of the Western Electric Company

George P. Thomson

(later Bell Telephone Laboratories). He retired from this company in 1946 and was appointed Professor at the University of Virginia.

Thomson, George Paget (Cambridge, England, May 3 1892 - Cambridge, England, September 10, 1975). British experimenter physicist. In 1927 George Thomson

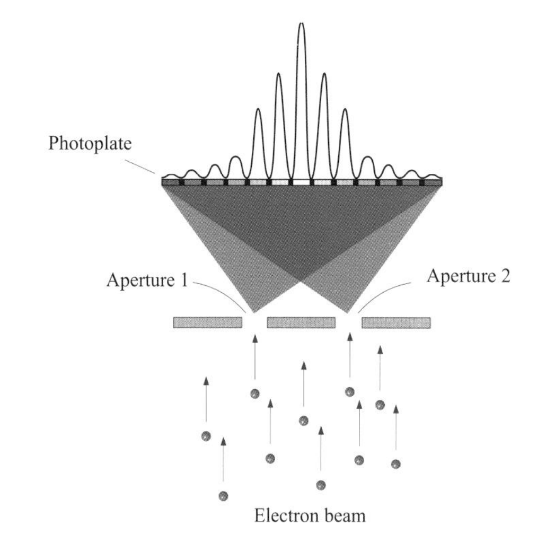

The principle of diffraction of monokinetic electrons by two apertures giving rise to an interference pattern.

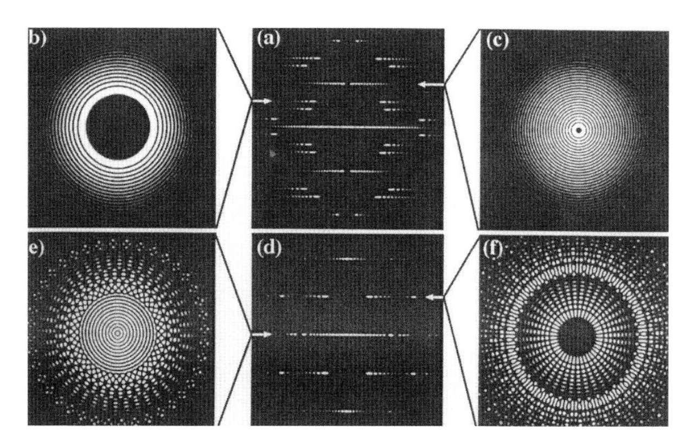

Central part : diffraction of electrons by a carbon nanotube. **Left and right** : pattern conversion in the momentum space.

observed diffraction fringes when an electron beam passed through thin metal sheets. At the same time, and independently of **Clinton Davisson**, he had shown that electrons behaved like waves, as proposed by **Louis de Broglie**. He was also the first to suggest the use of these electron beams to build a microscope. However, the first development (in 1928) of such a microscope, and later improvement giving characteristics close to those available today, were the work of **Ernst Ruska**. This wave property, which permits the electrons to undergo diffraction, is used to probe surfaces, a variant in relation to X - rays which undergo this diffraction in the volume of a crystal (as was shown by **William Henry** and **William Lawrence Bragg**). G. Thomson, the son of **Joseph J. Thomson**, followed a career very similar to that of his father. Both lived for 84 years! Graduating in 1914, he was, during the first war, attached to various British Army Corps in France, England and the United States. He went back to Cambridge in 1919 and taught there until 1927. The same year he was appointed Professor in Aberdeen. There he made the discovery for which he received the Nobel Prize. Moving to the University of London in 1930, he became interested in the reactions induced by slow neutrons, a subject which led him to study the possibility of creating a bomb based on the phenomenon of fission of uranium 235. However, it was **James Chadwick** who became the British coordinator in the American project. Thomson was elected as Head of the British Scientific Office in Canada. He returned to the Air Ministry of England in 1943. He was an ardent defender of the use of nuclear energy for peaceful purposes, in particular in the use of isotopes in biology and medicine. He retired in 1952 and devoted himself to the writing of reflections on science and to the bibliography of his father.

1938

Fermi, Enrico (Rome, Italy, September 29, 1901 - Chicago, Illinois, November 28, 1954). Theoretical and experimental physicist. In 1933 Enrico Fermi had shown that the neutron identified by **James Chadwick** was not stable as a single particle but gave rise to a proton, an electron and an

. The publication of this result, considered too speculative, was refused by the journal Nature. In 1934 he learned that the production of artificial radioisotopes by the bombard-

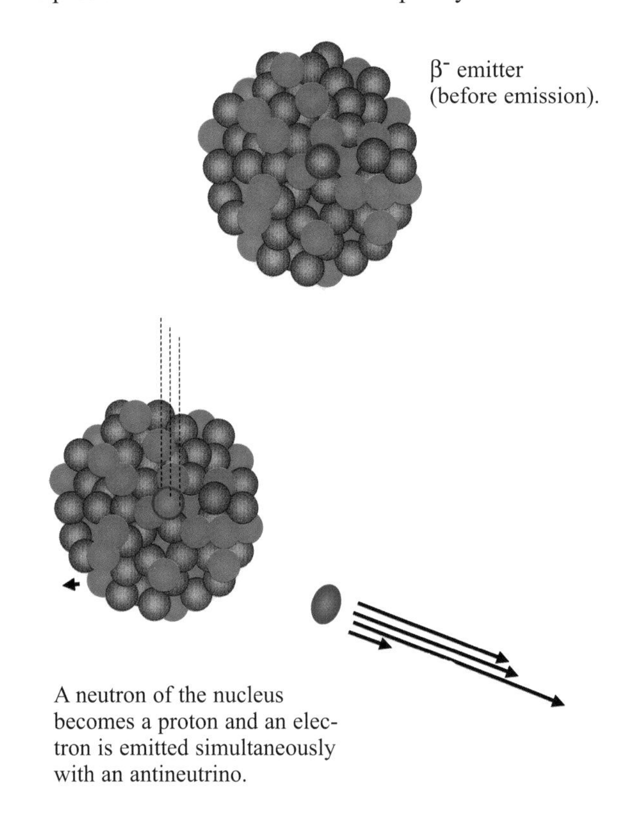

The ß⁻ emission. The principle of the transformation of a neutron into a proton was explained by E. Fermi. Simultaneously to the emission of a β^-, an antineutrino is produced to preserve the total energy and the zero momentum of the initial nucleus. See, by comparison, the β^+ emission (Dirac).

ment of light nuclei with a particles had been carried out in Paris by **Frédéric Joliot** and **Irène Joliot-Curie** (1935 Nobel Prize for Chemistry). Fermi realized then that a transmutation could be more easily induced by neutrons,

Enrico Fermi

non - charged particles. Avoiding the electrostatic repulsion that a particles undergo, the neutrons can transform all the atomic nuclei equally, including the heaviest.

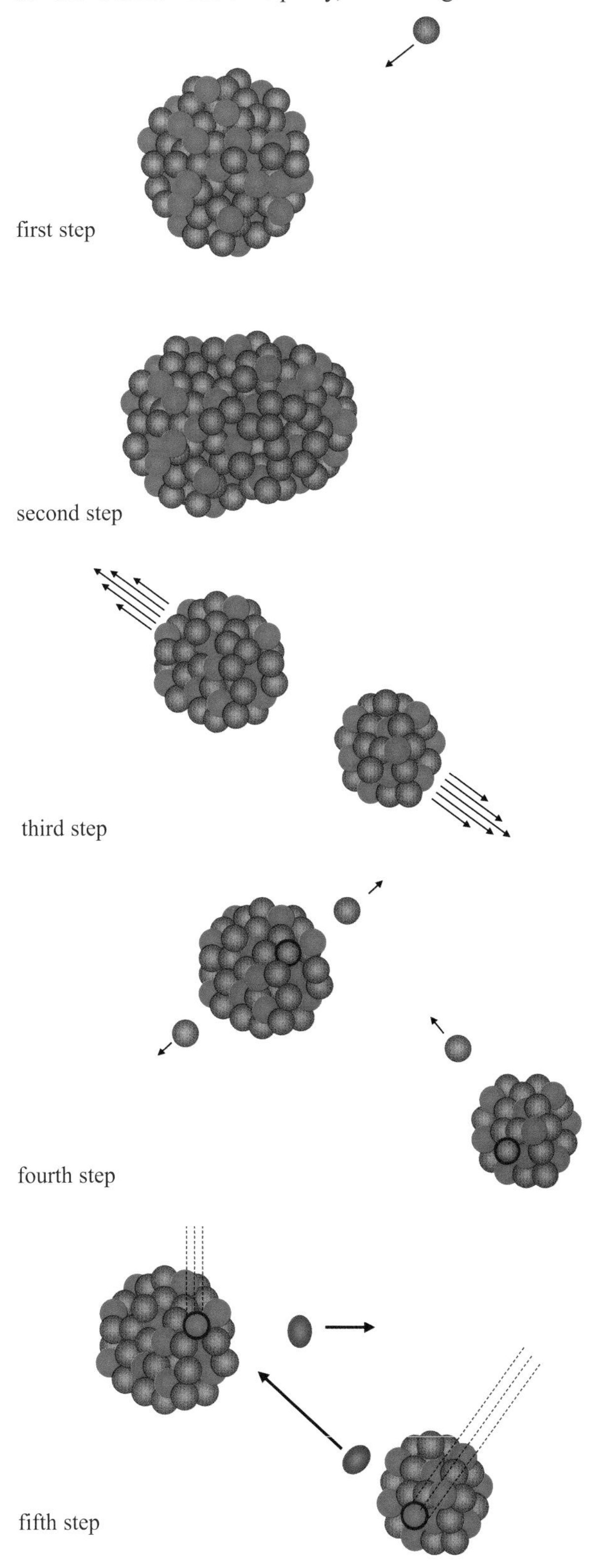

Within a few months he succeeded in producing about 40 new radioactive elements. In addition, he showed that the neutrons, slowed down by paraffin, are more efficient in inducing a transmutation than the fast neutrons directly emitted in nuclear reactions. Intuitively and without a clear reason, he interposed a block of paraffin to slow down these neutrons but he was able to interpret the result immediately. This discovery, which earned him the Nobel Prize, thus opened a significant pathway toward the use of energy produced by nuclear reactions in the civil and military fields. During his life Fermi touched on a host of phenomena concerning microscopic physics: he not only innovated experimental physics but also interpreted the phenomena very widely. In 1922, he obtained his Ph.D. at the University of Pise for work on X - rays (discovered by **Wilhelm Röntgen**). After a short stay in Göttingen under the direction of **Max Born**, he returned to Florence and taught physics from 1924 until 1926. With **Paul Dirac** he created a new concept of the statistical distribution of particles evolving under the form of anti-symmetrical wave functions which were later called fermions. Then he was appointed Professor in Rome, where he carried out the main part of the work which earned him the Nobel Prize. At the end of the prize-giving ceremonies in Stockholm he did not return to Italy, whose fascist regime he did not approve of, and joined Columbia University in New York. A short time after his arrival, he learned that **Otto Hahn** (1944 Nobel Prize for Chemistry) and Fritz Strassmann had produced a new phenomenon. It concerned the fission of uranium under the impact of slow neutrons, a reaction which released considerable energy while simultaneously producing new neutrons. These secondary neutrons could then be retrieved to produce new fissions and to initiate a chain reaction. Fermi had himself perceived this phenomenon in 1933 but had not understood it. Reproducing the experiments at Columbia, Fermi and Szilard confirmed Hahn's results. At once sensing the military applications and fearing that the German regime would use the process, Fermi used all his persuasive powers to convince **Albert Einstein** to use his fame to convince President Roosevelt to urgently launch an American program in order to gain a lead before any competing project. The first controlled fission, using graphite to slow down the neutrons between each step of the chain reaction, was achieved on December 2, 1942 in the research group that Fermi managed in Chicago. It was maintained for 28 minutes. The Manhattan Project, set up in 1943 in Los

The phases of the fission of uranium 235. The peaceful use of this fission principle highlighted for the first time by O. Hahn (Nobel Prize for Chemistry) has allowed the creation of electrical power from a nuclear reaction. The fuel consumption is of the order of a millionth of that of chemical fuel (coal, petroleum, gas). **First step**:a slow neutron enters a U^{235} nucleus. **Second step**:^{236}U is formed in a high excited state and starts to oscillate. **Third step**:due to its high excitation, the ^{236}U explodes in two parts which carry a high kinetic energy. **Fourth step**:these two parts being too neutron rich may "evaporate" fast neutrons. **Fifth step**:the residual nuclei finally transform one of their neutron into proton by giving rise to a β^- and an antineutrino.

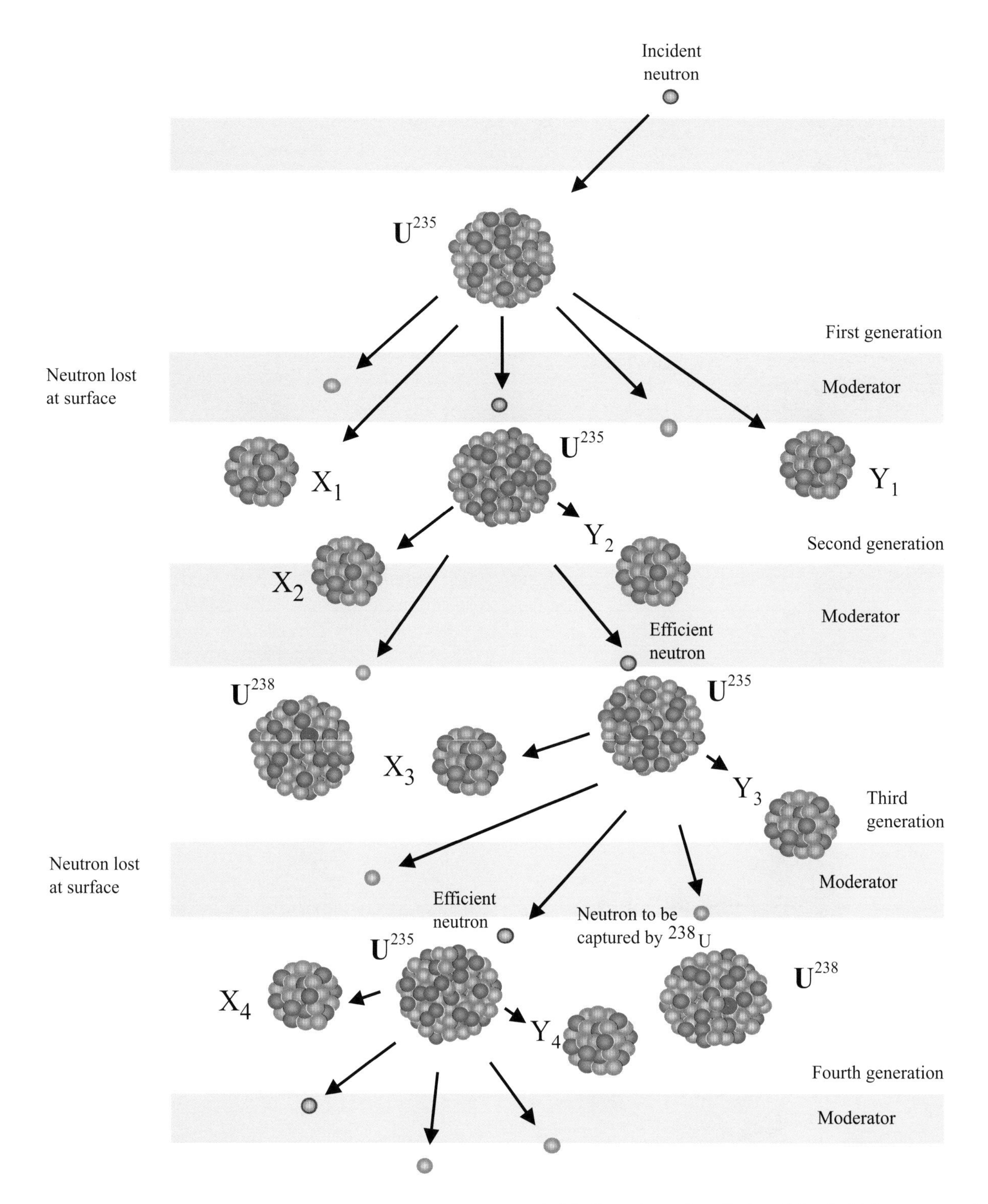

Chain fission of U^{235} requires the deceleration of the fast neutrons emitted in the fourth fission step. The moderator is water in the majority of the modern power stations but was carbon in Fermi's original version

Alamos, was vigorously supported, in particular by the theoreticians Fermi and **Hans Bethe** (both naturalized Americans), until the experimental explosion in the New Mexico desert on July 16, 1945, an explosion which they both witnessed. In 1946 Fermi returned to Chicago to become interested in the nature of emitted particles in high energy nuclear reactions, in particular the mesons. He was at the heart of the project for building a synchrocyclotron (whose energy would exceed that produced with **Ernest Lawrence's** cyclotron). Then he committed himself to a political campaign for the peaceful use of nuclear energy and was a passionate defender of Oppenheimer, the Director of the Manhattan Project who had been discredited in the United States for having refused to support the project to construct

The Tihange power plant (Belgium) .

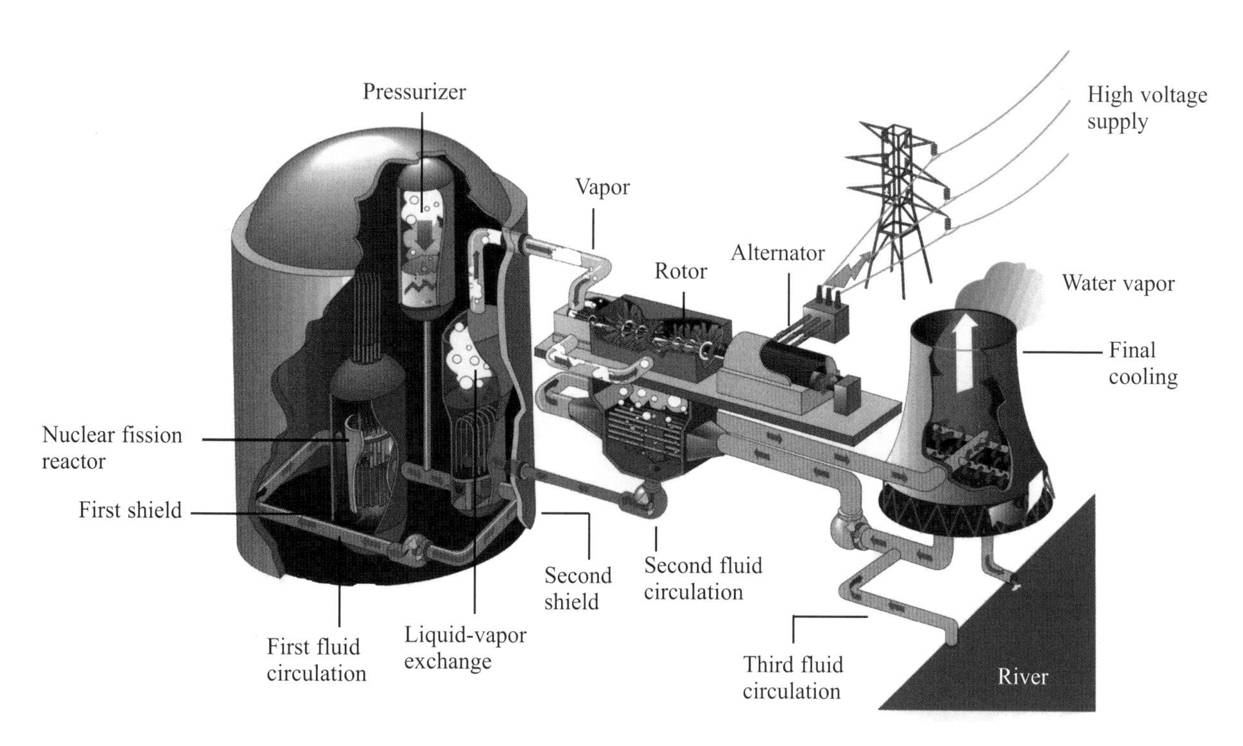

Scheme of a nuclear power plant .

the hydrogen fusion bomb. This uses the fusion phenomenon of light nuclei instead of heavy nuclei fission and releases significantly more energy. In honor of Fermi, the atomic element number 100 (containing therefore 8 protons more than uranium) was called fermium. Today, physicists still use the Fermi for characteristic length unit of the atomic nucleus radius, being equivalent to 10^{-15} m (one millionth of a billionth of meter). The Fermilab, the US center for research in particle physics opened in 1972 at Batavia (Chicago), was named after him.

Fermi took part in the Solvay Councils of Physics in Brussels in 1930 and 1933.

1939

Lawrence, Ernest Orlando (Canton, South Dakota, August 8, 1901 - Palo Alto, California, August 27, 1958). Ernest Lawrence, an experimental physicist, is one of the rare winners rewarded by the Nobel Committee for the construction of a device: the cyclotron. Moreover, the wording of the reason for the attribution states "and especially for the creation of radioisotopes by means of this instrument". This accelerator, the creation of which began in Berkeley in 1929, is based on the principle of the increase of energy acquired by a charged particle with each passage between two hollow electrodes during successive semicircular routes of increasing radius when tra- velling under the action of an electromagnet. At each half turn (in natural synchronism), the accelerating voltage is reversed to give this energy boost to the ions. The spiral - shaped trajectory allows the development of very compact devices in comparison with electrostatic accelerators like that of **John Cockcroft** and **Ernest Walton**. The synchronism in the passage between the two electrodes remains as long as the mass of the accelerated ions does not undergo an increase as postulated by the relativistic mechanics of **Albert Einstein**. To obtain higher energies Lawrence's successors would create, under **Enrico Fermi's** impulse, synchrocyclotrons and synchrotrons. The energy of the projectiles accelerated in cyclotrons enables them to penetrate the atomic nuclei despite their electric charge. This property was used

Ernest O. Lawrence

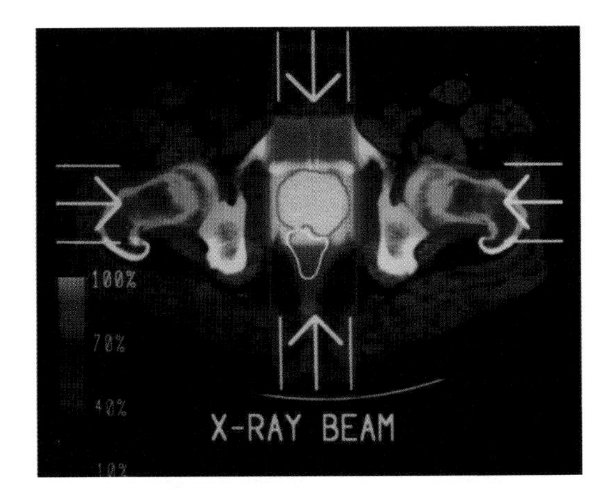

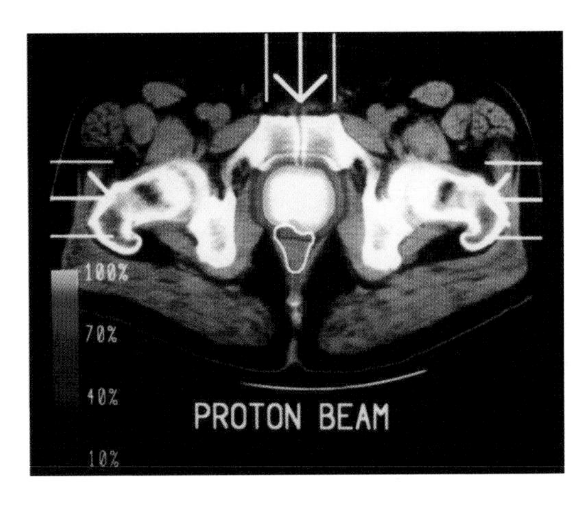

Comparison of the tomographies by 4 beams of X-rays and 3 beams of protons produced by an IBA cyclotron. In the tomography by protons, the deposited energy in the tumors to be treated is more important but, consequently, a lower amount is deposited in the nearby healthy tissues

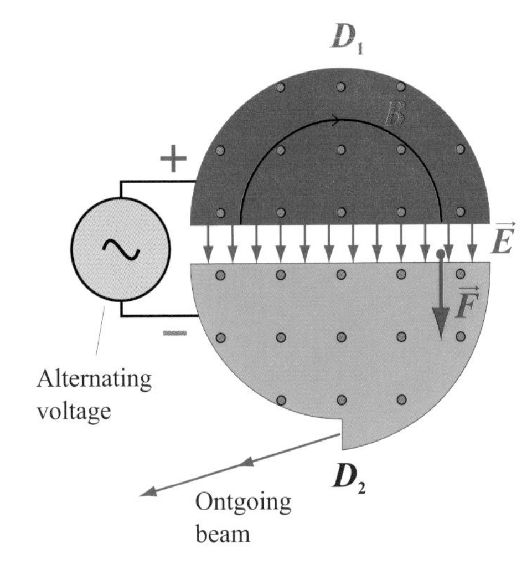

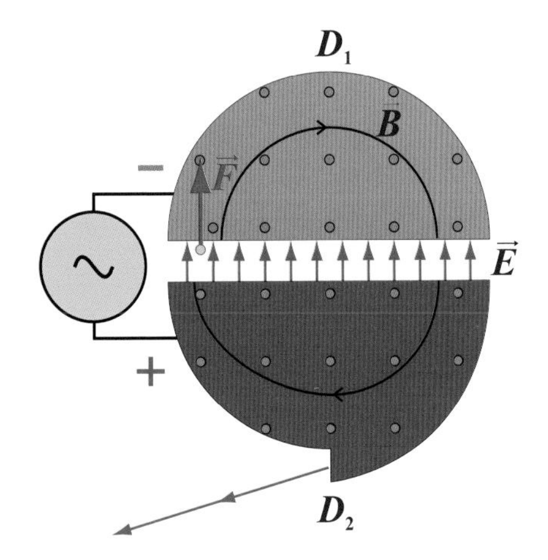

Alternate acceleration principle of particles in a cyclotron. Along a semicircular trajectory the particles have a constant speed. The acceleration only occurs between the two semicircular hollow electrodes. After each acceleration the radius of the trajectory increases. The acceleration voltage is reversed with each half-turn to ensure additional accelerations during the successive passes through the intermediate space between the electrodes. The magnetic field is normal to the sheet.

Modern version of the compact cyclotron conceived and built by IBA. (Louvain-La-Neuve, Belgium). The device is designed for medical tomography and therapy.

by Lawrence to produce radioisotopes of a different nature from those produced by Fermi with neutrons. The radioisotopes formed by the ions accelerated in cyclotrons are, in general, positron emitters; the radioisotopes formed by the neutrons are, in general, electron emitters. Ion beams extracted from a cyclotron are today commonly used in the treatment of cancer. In comparison with the use of radioisotopes or γ - ray beams, the irradiation of tumours with protons allows a better selection of the area to be treated in the patient, especially if it is loca- ted very deep within the body. Indeed, the protons produce minimal damage in the surrounding healthy tissues. IBA, a Belgian firm located in Louvain - la - Neuve, specializes in this. After his fundamental studies at the universities of South Dakota, Minnesota and Chicago, Lawrence joined the group at Yale University where, in 1925, he presented a Ph.D. on the photoelectric effect produced on potassium vapours. Then he taught in Chicago for 3 years. In 1928 he arrived at the University of California, Berkeley, and that is where he developed the cyclotron. He continued to design the same type of machine, increasingly bigger and more expensive. He was also at the root of the construction of increasingly more powerful magnets, able to separate major quantities of heavy elements according to their mass. One of the applications was the separation of isotopes 235 and 238 of uranium for the Manhattan Project, where **Arthur Compton** and **Harold Urey** (1934 Nobel Prize for Chemistry) were direct collaborators. The laboratory at Berkeley, which currently bears the name Lawrence Radiation Laboratory, is known to have produced a major quantity of new radioisotopes. Moreover, the elements of atomic number 103 (containing 103 protons and having a total mass between 255 and 260) bears the name of lawrencium. By means of the cyclotron irradiations, scientists continue to create nuclei of increasingly bigger atomic numbers. In 1996 all the elements of atomic number up to 110 had been created. With his brother John, a doctor, Lawrence applied neutron techniques to cancer treatment; he developed the radioactive iodine production for the diagnosis and the treatment of cancer of the thyroid, and the production of other radioactive isotopes for multiple selective diagnoses. He took part in the Solvay Council in Brussels in 1933. In 1958 he was part of the American delegation at the conference of Geneva responsible for East-West contacts on nuclear power. It was during his intervention at this conference that he became ill. He was quickly taken home but he did not recover and died one month later.

1940 to 1942 Not awarded

1943

Stern, Otto (Sohrau, Poland, February 17, 1888 - Berkeley, California, August 17, 1969). Describing himself as an "experimental theorist", Stern proposed, in 1920, to verify the reality of the quantum aspect of the interaction of a molecular beam with a magnetic field. Adopting Sommerfeld's idea (1868 - 1951) that several atoms had to possess a magnetic moment (a property associated, macroscopically, to electric current loops which interact with magnets), Stern conceived of sending, under a high vacuum, a narrow beam of silver atoms (and then small loops of electrons running around their nucleus) in a non-homogeneous magnetic field and to measure the orientation of these "loops". If the narrow beam simply expan- ded, the classical theory would be correct; if, on the other hand, the beam divided to follow two separate paths, the quantum theory would be the right one. The experiment that he

Otto Stern

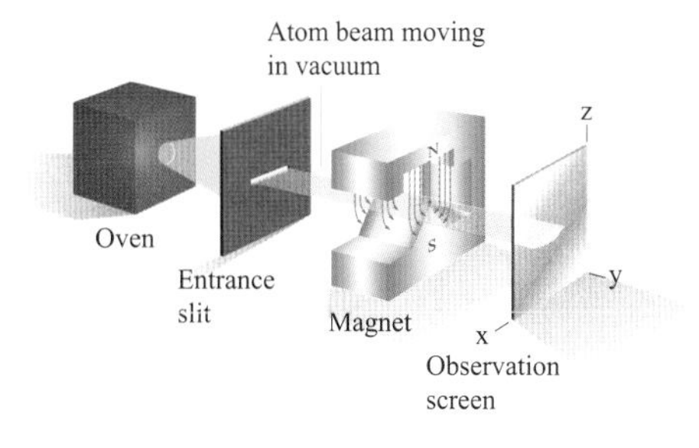

The experimental device of the STERN and GERLACH experiment. An atomic beam is forwarded along the y-axis. After transmission through a slit parallel to x, it enters into a non-homogeneous magnetic field. The electromagnetic force on the atoms with different dipole moments deflects them in different directions according to the particular dipole orientation.

carried out, with Gerlach's collaboration, demonstrated that the quantum theory was correct. Stern also had the opportunity to measure the magnetic moment of the proton, stating that its value was almost 3 times higher than that predicted by **Paul Dirac's** theory. At first a theorist, Stern was captivated by the writings of Boltzmann, Clausius and **Nernst** on thermodynamics, by Sommerfeld's talks, by those of **Einstein** and of **von Laue** both of whom he saw regularly in Zurich, and by those of **Max Born** with whom he worked in Frankfort. At the time of the discovery that earned him the Nobel Prize, he was "privatdozent" in Frankfort. His first position was at Rostock, from 1921 until 1923, the year when he was appointed Professor of Physical Chemistry in Hamburg. Regularly meeting **Niels Bohr** and **Wolfgang Pauli**, he carried out several key experiments to confirm the existence of particle waves (which **Louis de Broglie** had attached to electrons) and extended their range to atoms and molecules. He took part in the Solvay Council of Physics in Brussels in 1930. A German citizen, born in Sohrau (now Zory, Poland) his career was interrupted in 1933 when the Nazis came to power. He then moved to Pittsburgh, United States, where he equipped a new laboratory with a molecular beam. Finally he joined Berkeley Laboratory in 1946. **Isidor Rabi**, who was the 1944 Nobel Prize winner in Physics for having extended Stern's experiments (whom he met in Hamburg), claimed that it was from the latter and from Pauli that he learned what physics should be. Oskar Stern, Otto's father and a rich grain merchant, allowed his son to travel a lot before choosing his studies. Eccentric, he always appreciated not only great hotels and copious meals, but also the cinema and especially the theatre. Moreover, it was at the theatre that he died of heart failure aged 81 years old.

1944

Rabi, Isidor Isaac (Rymanow, Poland, July 29, 1898 - New York, New York, January 11, 1988). Experimental physicist. Isidor Rabi continued **Otto Stern**'s work, by compelling a fine atom beam to successively cross two high and non-homogeneous magnetic fields; the second field being used for re-focusing the beam. By adding an oscillator, which he could minutely adjust the frequency, in the intermediate space (where a weak magnetic field reigns), he could select and modify the quantum state of the atoms of the beam. The minute adjustment of the oscillator frequency induced a selective transition of one nuclear magnetic state toward another. This property is used today both for the fundamental study of the atomic structure as well as for medical and other applications, with the aim of identifying the variations in the chemical bonds of atoms with their environment. Refinements in this ultra fine identification (being able to reach an accuracy of 10^{-10}) would earn **Richard Ernst** the 1991 Nobel Prize in Chemistry. Rabi did his Chemistry studies at the universities of Cornell and Columbia (New York) where he defended his Ph.D. thesis in 1927. Holding various grants, he had the opportunity of working for two years in Europe: with Arnold Sommerfeld in Munich, with **Max Born** in Copenhagen, with Otto Stern in Hamburg, with **Wolfgang Pauli** in Hamburg and Zurich and with **Werner Heisenberg** in Leipzig. It was the latter who recommen- ded Rabi to the Columbia administration who entrusted him with the position of part-time lecturer. He remained loyal to Columbia where he received rapid promotions. In 1974 a bust of Rabi was installed there to commemorate the spirit of the person who inspired physics "twice" at Columbia: the first time, after 1930, when he completed his thesis, the second time, after the Second World War, when physics had "to be re-orientated" in American universities. During World War II, Rabi greatly contributed to the development of American radar devices. He was regularly consulted by Oppenheimer and later by several Presidents of the United States. He was very honoured to receive the Nobel Prize and considered that this honour "places any laureate on a pedestal, less to serve his own research than to allow him to play an influential role in the presentation of scientists to the public and in the media". He declared in particular: "the scientist does not defy the universe. He accepts it. This is the dish that he savours, the field that he explores. It is an adventure and an unending pleasure. It is pleasant, elusive and never boring. It is marvellous both on the small and the large scales. In short, the exploration of the universe is the primordial occupation of the gentleman". In Europe today, this important role of the ambassador of scientists to the media is carried out, in a masterful way, by **Georges Charpak**. Rabi was born in Rymanow (then in Austria), but he moved with his family to the United States when he was one year old.

Isidor Rabi

1945

Pauli, Wolfgang (Vienna, Austria, April 25, 1900 - Zurich, Switzerland, December 14, 1958). Follwing the theoretical work of **Heisenberg** on the uncertainty principle, Pauli is the author of the famous "exclusion principle", stated in 1924 to explain the electronic structure of atoms, Wolfgang Pauli simply proposed that "only two electrons of an atomic shell can be in the same state".

After the discovery of the spin (quantified entity connected with the rotational motion of the particles on themselves), this principle took a more general character. Thus, for atomic physics, "Pauli's principle" is expressed in the following, probably final, form: "a single atomic electron cannot have the same four quantum numbers". This principle rules not only the chemistry, but also the physics of the nucleus and of the more fundamental particles, of which it is composed. It finally allowed scientists to devise a full explanation of the **Zeeman** effect. Pauli was probably one of the youngest scientists to easily understand the basic written documents of **Albert Einstein's** relativity. In 1921, just after the presentation of his Ph.D. thesis, he was given the responsibility by Arnold Sommerfeld of writing a review on this difficult subject. Einstein himself was lost in admiration for this clear, synthetic, deep, critical presentation, which wonderfully integrated all the available knowledge of the time. To explain the energy taken away in the disintegration of radioactive nuclei by electrons or positrons, he postulated, in 1931, the existence of another particle, a neutral one, emitted simultaneously. In 1932 **Enrico Fermi** named this particle as the neutrino. The neutrino was experimentaly identified by **Frederick Reines** in 1953. Pauli extended his theoretical notions to the study of other particles (mesons), to the paramagnetic properties of gases, to the behavior of electrons in solids, but also to the way protons and neutrons cooperate within the nucleus of the atom. Wolfgang Pauli obtained his Ph.D. from the University of Munich in 1921 and pursued a postdoctoral work with **Niels Bohr**'s team in Copenhagen before carrying on his work as a "privatdozent" in Hamburg. He also collaborated with **Max Born** in Göttingen. After Heidelberg he taught in Zurich, as a professor of theoretical physics for 25 years. However, it was at Princeton that he spent the major part of his career (1935-1958). The influence of Pauli on physics is, indisputably, one of most important of any of the theoretical physicists of this century. He always stressed that a scientist needed the greatest possible freedom in order to stimulate his in-depth work and the need for exchanges and discussions to promote science. He took part in the Solvay Councils of Physics in Brussels in 1927, 1930, 1933, 1948, 1954 and 1958. A small sidelight tells the story that, in an experimental laboratory, Pauli's effect consists in deregulating all the equipment.

Wolfgang Pauli

The participants in the Solvay council of physics of 1958. Principal subject: The structure and evolution of the universe. From left to right and seated : W. McCrea, J. Oort, G. Lemaître, C. Gorter, **W. Pauli, W.L. Bragg,** J. Oppenheimer, C. Moller, H. Shapley, O. Heckmann, O. Klein, W. Morgan, F. Hoyle, B. Kukaskin, V.Ambarzumian, H. van de Hulst, M.Fierz, A.Sandage, W.Baade, J.Wheeler, H.Bondi, T.Gold, H.Zanska, L.Rosenfeld, L.Ledoux, A.Covell, J. Géhéniau (in bold: the Nobel laureates).

1946

Bridgman, Percy Williams (Cambridge, Massachusetts, April 21, 1882 - Randolph, New Hampshire, August 20, 1961). A researcher on high pressures (up to 3000 times atmospheric pressure from 1931), Percy Bridgman offered to both physicists and chemists not only compressors for multiple applications but also a selection of measuring devices for pressure, temperature and viscosity. He showed, in particular, that the viscosity of liquids increased with the pressure that was imposed on them, with the exception of water. The invention of powerful pressurizers made the creation of artificial diamonds possible around 1955. These apparatus allow geologists to recreate in the laboratory conditions similar to those existing several hundred kilometers under the earth's surface. During the Second World War, **Enrico Fermi** consulted Bridgman about the compression of components containing uranium and plutonium and the conception of framework necessary for the Manhattan Project. He spent all his

Atomic bomb containers similar to those experimented by Bridgman.

career as student, researcher and professor at Harvard until 1954. He welcomed **John Bardeen** in his research team for two years, the latter being the only physicist to have received two Nobel Prizes in physics. He published many books in the field of the philosophy of science (Reflections of a Physicist, 1950) and in the field of logic. He won the fiftieth Nobel Prize for Physics. Condemned by an incurable illness, he preferred to commit suicide, aged nearly 80 years old.

1947

Appleton, Edward Victor (Bradford, England, September 6, 1892-Edinburgh, Scotland, April 21, 1965). In 1924, while carrying out experiments on the transmission of radio waves between Bournemouth and Cambridge, Edward Appleton discovered, by varying the frequency of the emitter, the interference of that direct wave with another wave, reflected from high altitude. This reflective layer (which was thought to contain a considerable quantity of ions) had been claimed by Heaviside and Kennelly, since 1902, to explain the success of **Guglielmo Marconi's** experiments. Appleton then established the link between the daily variations of the interactions with this layer and the sunspots. He determined the altitude of Kennelly and Heaviside's layer (90 km) and identified a second, even higher, layer (240 km). These experiments led Watson-Watt to build equipment capable of detecting the approach of an aircraft as a result of the ionization caused when it passes through the lower atmosphere. Appleton is also one of the principal architects of our knowledge of the ionosphere. He was a student in Cambridge where his professors were **Joseph J. Thomson** and **Ernest Rutherford** (1908 Nobel Prize for Chemistry). He taught experimental physics at King's College, London (from 1924 until 1936) and at Cambridge (UK) (from 1936 until 1939). During the Second World War he was British Secretary of the Department of Industrial and Scientific Research, in charge of several research institutes, on which depended, in particular, the department dealing with atomic energy; after the war he continued this interest, in particular being involved in the establishment of the Center at Harwell. On this subject he opposed the ideas supported by **James Chadwick**. From 1949, he dealt with questions related to Education and Research administration. Passionate about sport, he almost did not become a physicist: his youthful ambition nearly led him exclusively to play cricket.

Edward V. Appleton

1948

Blackett, Patrick Maynard Stuart (Baron) (London, England, November 18, 1897 - July 13, 1974). Patrick Blackett, an English researcher, improved the Wilson cloud chamber with the principal aim of continuing the study of cosmic rays. The improvement consisted of a device making it possible to photograph the track left in the chamber by the addition of two auxiliary counters laid out on both sides of this principal chamber: the passage of a particle through the telescopic device triggered the photographic device. The process restored the complete efficiency of the device: without these two auxiliary counters, plates could be taken only in a purely random way. Blackett resigned as a Naval officer in 1919 to study in Cambridge, under **Ernest Rutherford**'s supervision. Graduating in 1921, he presented a thesis on the interaction of the a particles emitted by a radioactive source which transmuted nitrogen 14 nuclei into oxygen 17. He recorded nearly 400,000 tracks of a particles in 23,000 photographs. In only eight of them, he clearly highlighted the track of the incidental a particle, that of the emitted proton and that of the recoiling oxygen 17 atom produced in the reaction. This was a preliminary

Patrick M. Blackett

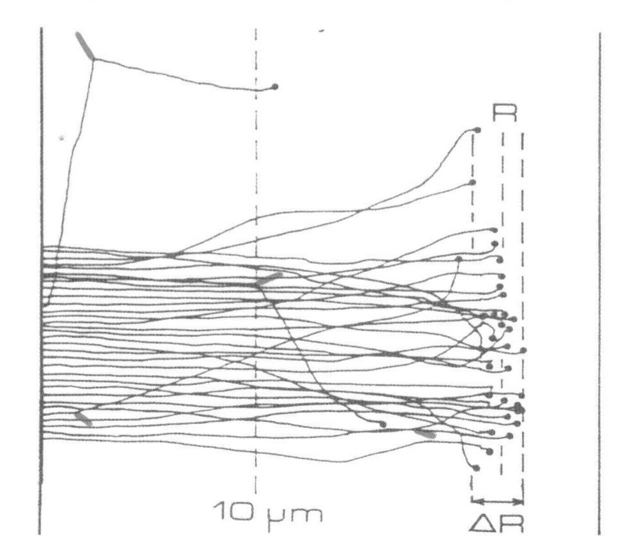

Calculated trajectories of deuterons in a photographic plate. The red lines correspond to traces of recoiling nuclei. The end of the deuteron trajectory is indicated by a point.

experiment to the research which would earn him the Nobel Prize. Then he completed the device by the addition of a magnetic field to curve the trajectory of the particles to detect and thus identify their charge and their momentum (which is the product of their mass by their speed). In 1925 he worked for one year in **James Franck's** laboratory in Göttingen. In 1933, while he was teaching in London, he was able to confirm the discovery of the positron announced a few months earlier by **Carl Anderson**. Observing the "shower" of trajectories induced in one point of the chamber by a particle resulting from cosmic radiation, he identified the production of pairs of electrons and positrons whose existence had been predicted by **Paul Dirac**. He was also the first to carry out experiments in a laboratory buried more than 30 meters under ground, in order to check the expected penetration of several particles of cosmic radiation. At the same time he was interested in the properties of the earth's magnetic field (and of its evolution in time in order to understand the continental drift), and of the magnetic fields around the sun and other stars. During World War II Blackett played a key role in the improvement of defense technologies against enemy aircraft and submarines. From 1937-1953 he taught at Manchester University before returning to London, where he worked until his retirement in 1963. After the war, he made it clear that he was firmly against the development of nuclear weapons. Blackett was a person of great charm. He took part in the Solvay Councils of Physics in Brussels in 1933 and 1948.

1949

Yukawa, Hideki (Tokyo, Japan, January 23, 1907 - Kyoto, Japan, Sept. 8, 1981). Basing his studies on the value of the **Planck** constant, on the mass - energy equation of **Albert Einstein**, Hideki Yukawa was the first Japanese physicist to receive the Nobel Prize in Physics, for his prediction, made in 1935, that the short range force (10^{-15} m) binding protons and neutrons in an atomic nucleus was due to an exchange of particles (which he first called meson), whose mass was close to 200 times that of the electron or 1/6 that of the proton or the neutron. The relevance of this statement was comparable with the one which **Paul Dirac** had expressed concerning the existence of the positron. The same year, **Carl Anderson** discovered (in cosmic rays) a particle (the μ - meson) having the expected mass, but it was quickly acknowledged that this was not the particle exchanged in the strong interaction binding the nucleons of the nucleus. It was only in 1947 that the π - meson (a particle of intermediate mass between that of the electron and that of the proton, hence its name) was identified. At the present time, the distinction between the meson of cosmic radiation (muon: a lepton and then a fermion) and Yukawa's particle (pion: a hadron with a property of boson) is completely understood. This pion, in its disintegration, gives rise to a muon and to a neutrino. The pion (a boson) is the vehicle of the strong interaction; the muon (a fermion) has properties close to those of the electron and undergoes the weak interaction and the electromagnetic interaction. Born into a family which produced several university professors, Yukawa studied in Kyoto and Osaka where he obtained his Ph.D. in 1938. A turning point in his life as a young scientist occurred in the autumn of 1929 when he attended the seminars of **Werner Heisenberg** and Paul Dirac, then on tour in Japan. From 1932 to 1938 he

Hideki Yukawa

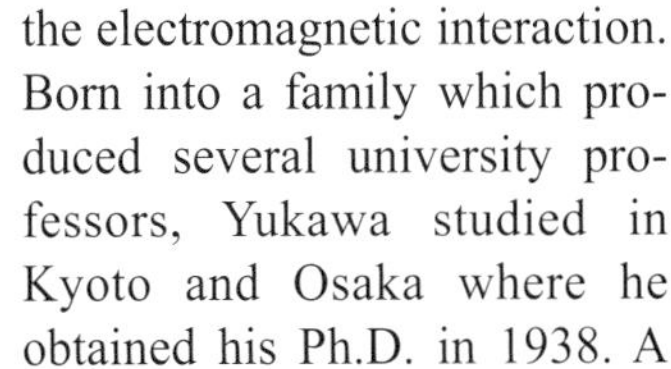

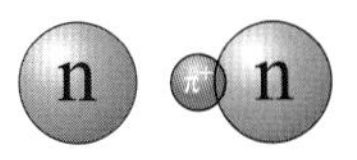

Exchange of pions in the neutron-proton interaction.

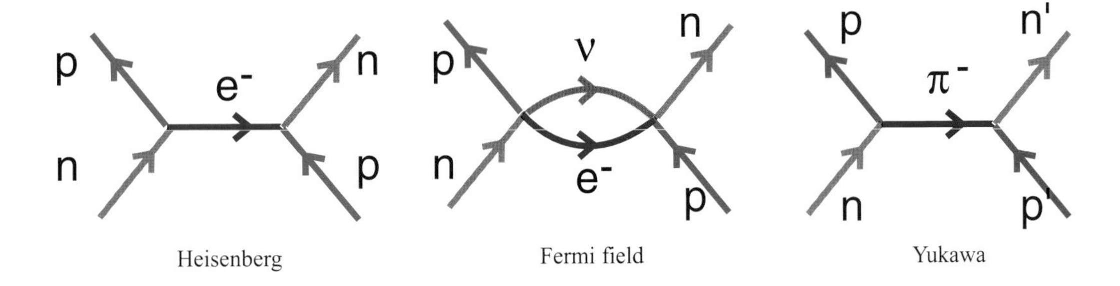

Heisenberg

Fermi field

Yukawa

Evolution of the concepts of nuclear interaction proposed by Heisenberg, Fermi and Yukawa

worked as a lecturer at the University of Kyoto and was appointed Professor of theoretical physics from 1939 to 1970. His choice of theory came from his incapacity to operate any laboratory apparatus. In 1949, Oppenhei-mer, who called Yukawa's proposition one of most significant of the 1935 - 1945 decade, invited him to join the group at Princeton for one year. He was invited to the Solvay Council for Physics in Brussels in 1961.

1950

Powell, Cecil Franck (Tonbridge, England, December 5, 1903 - Bellano, Italy, August 9, 1969). Experimenter physicist. Involved in the study of nuclear reactions, Cecil Powell developed, between 1935 and 1945, a simple photographic device for the identification of the trace left, in an emulsion, by particles resulting from these reactions. In comparison with the **Charles Wilson** chamber, this method allows the reduction of the range of the particle being detected (as its stop occurs in a solid and not in a gas) and also the detection of high energy particles, for example cosmic rays. During the decade 1940 - 1950, he sent this type of instrument in balloons inflated with hydrogen up to an altitude of 30,000 m. The traces left in the negatives were closely studied using a microscope and the numerous plates were also sent to many laboratories, where large teams reviewed them in minute detail. This method of detection is widely used today, but the interpretations of the traces is often made using automatic treatment. In 1947, this technique allowed the identification of the pion, the particle exchanged in strong interactions, whose existence had been postulated in 1935 by **Hideki Yukawa**. When accelerators replaced cosmic radiation for the study of interactions of fundamental particles and when the bubble chambers (invented by **Donald Glaser** in 1952) were preferred to the photographic plates, Powell resolutely turned to teaching. In addition, he involved himself in the definition of the social responsibility of scientists. Born in Kent, he received his basic schooling there before studying in Cambridge. He improved his knowledge at the Cavendish Laboratory, under the supervision of Charles Wilson and **Ernest Rutherford** (1908 Nobel Prize for Chemistry). He completed his Ph.D. in 1927 and settled down at the University of Bristol. He took part in the Solvay Council of Physics in Brussels in 1948.

1951

Cockcroft, John Douglas. (Todmorden, England, May 27, 1897-Cambridge, England, September 18, 1967). A researcher and joint holder of the Nobel Prize with **Ernest Walton**, John Cockcroft conceived a device able of producing a very high electrostatic voltage to accelerate ions up to a sufficient speed to allow them to induce nuclear transmutations in light nuclei. The first experiment, carried out in 1932, made it possible to transfer to protons an energy of 600,000 eV (energy acquired by an elementary charge of 1.6 x 10^{-19} Coulomb under a voltage of 600,000 volts). The speed reached by these protons is then higher than 10,000 km/sec and makes them able to break the Li^7 nucleus and to transform it into two a particles carrying together a total energy of 18 x 10^6 eV. In this interaction, the kinetic energy of the two a particles represents 30 times the energy imparted to the proton sent to the

John D. Cockcroft.at High Voltage Engeneering Facility in Amersfoort (The Netherlands).

fixed Li target. A part of the mass of the particles brought together (proton and Li^7) is thus transformed into energy according to Einstein's famous equation: $E = mc^2$. The energy of the emitted a particles was measured by the traces that they produced in a **Wilson** chamber. Cockcroft and Walton continued their experiments by propelling protons and deuterons to induce other nuclear reactions in light nuclei like boron and carbon. The deuterons (nuclei

The LARN accelerator (Notre-Dame de la Paix University Namur, Belgium) used by the author of the physics part of this dictionary and based on a modern version of Cockcroft's device.

of deuterium, that is of heavy hydrogen) were isolated by **Harold Urey** (1934 Nobel Prize for Chemistry) at Berkeley. The era of nuclear reaction was launched. **Ernest Lawrence** had already obtained the 1939 Nobel Prize for the invention, in 1928, of another type of accelerator: the cyclotron. In the cyclotron, the particles describe a spiral trajectory to reach, at the end of the acceleration, more important energies than with Cockcroft's apparatus. In their original structures, these cyclotrons provided only ion beams confined inside a magnetic field. They were then initially used to produce radioactive materials and not to induce nuclear reactions on targets placed outside the accelerator. At the end of the 1950s, devices were invented to extract these ions from the cyclotrons and to send them to external targets. These two generations of accelerators produced the major part of our current knowledge of atomic nuclei. The later generations of much larger accelerators (synchrocyclotrons and synchrotrons) of energy are being used for the study of more fundamental components of nuclei: that is, the internal structure of protons. The particle accelerators of the type invented by Cockcroft are used today in applied sciences, in particular for the analysis of materials. The Laboratory of Analysis by Nuclear Reactions (LARN), created in 1969 within the University of Notre-Dame de la Paix in Namur, has, as its principal purpose, the application of fundamental knowledge of atomic and nuclear reactions to the elemental characterizaiton of materials. The first accelerator built by Cockcroft and Walton is currently exhibited at the Science Museum in South Kensington. Cockcroft, a native of Lancashire, did his engineering studies in Manchester and graduated after the First World War. He devoted himself then to studies of fundamental sciences in Cambridge and joined **Ernest Rutherford** (1908 Nobel Prize for Chemistry) at the Cavendish Laboratory. There he created the electrostatic accelerator. After 1939 he devoted the major part of his time to the management of research. In 1944 he supervised a team of researchers involved in the study of uranium fission in Canada. In 1946 he became Director of the new Harwell Research Center, created by **Edward Appleton**, a center of excellence in reactor and accelerator technologies and, in 1949, became Director of Rutherford HEL (called the "British Brookhaven") in Chilton, close to Harwell. He was an ardent supporter of the establishment of a large European research centre equipped with powerful accelerators: CERN. From 1961 until 1965 he was Chancellor of the University of Canberra and he ended his career as President of several English universities. He took part in the Solvay Councils of Physics in Brussels in 1933 and 1948.

Walton, Ernest Thomas Sinton (Dungarvan, Ireland, October 9, 1903 - Belfast, Ireland, June 25, 1995). Associated with **John Cockcroft**, Ernest Walton created, in Cambridge, the plans for the first particle accelerator intended to transmute light atomic nuclei in order to study the mechanisms of nuclear reactions. Details of this invention will be found in the description of Cockcroft's activities, whom he met in **Ernest Rutherford's** laborato-

The zig-zag arrangement of the electrical power supply of the Cockcroft-Walton Accelerator in the garden of CERN Geneva.

ry. Walton studied in Dublin and Cambridge, where he defended his thesis before returning to Dublin. At the Cavendish Laboratory, he was also interested in the acceleration of electrons: a device which was developed by Wideroe and Kerst, and which would become the current Betatron. His other scientific concerns relate to hydrodynamics and microwaves. Walton took part in many scientific, governmental, religious and medical committees. He took part in the Solvay Council of Physics in Brussels in 1933. He was known as a quiet, undemonstrative and not very loquacious man.

1952

Bloch, Félix (Zurich, Switzerland, October 23, 1905 - Zurich, Switzerland, September 10, 1983). A theoretical as much as an experimental physicist, Felix Bloch shared the Nobel Prize with **Edward Purcell** for their methods of precision measurement (but each with a different detection process) of nuclear magnetism. Bloch was inspired by the proposals made by **Wolfgang Pauli** in 1924 on the explanation of the fine structure of the spectrum - lines of atoms, utilizing the intrinsic kinetic moment of the nucleus (spin) and thus its mag-

Félix Bloch

netic moment. In 1946 Bloch formulated the equations (which bear his name) allowing the precise determination of these magnetic moments. He proposed to make the nuclei interact with a combination of two magnetic fields: one constant and the other variable. He specified, in particular, how the neutron (a globally neutral particle), could also undergo interaction with an external magnetic field. He proposed, in particular, the use of the magnetic properties of iron to check this. This experiment was carried out in 1937 at the University of Columbia (the famous Nobel Prize seedbed) where the team of **Isidor Rabi** and Edward Purcell worked at that time. Bloch measured the magnetic moment of the neutron in 1940 together with **Luis Alvarez**, who would receive the 1968 Nobel Prize for the invention of the hydrogen bubble chamber. The work of Bloch is legendary in several fields of physics. We note in particular the description of the differences between me- tals and insulators, which led to the understanding of electric conduction processes in materials with intermediate resistivity (semiconductors). Five laws and original concepts in atomic physics bear his name. Bloch enrolled at the Federal Institute of Technology of his birthplace to follow the engineering curriculum there, but branched off towards physics. He had **Peter Debye** (1936 Nobel laureate for chemistry) and **Erwin Schrödinger** as professors. After receiving basic education in Zurich, Bloch wrote his doctorate, related to metal conductivity, in Leipzig, where he worked in **Werner Heisenberg's** team. Then he was granted scholarships, which enabled him to work in the **Pauli**, **Niels Bohr** and **Enrico Fermi** groups. In 1934 Bloch immigrated to the United States and got a position as a teacher at Stanford in 1936. He also carried out expe- rimental and theoretical research in cooperation with the Los Alamos and Harvard groups. In 1954 he was appointed Head of CERN, Geneva. He also taught the undegraduate students and derived a lot of pleasure from this activity. He considered teaching a great honor and at the same time a very difficult task to accomplish, as it always called his knowledge in question, and improved his teaching skills to keep up-to-date.

Purcell, Edward Mills (Taylorville, Illinois, August 30, 1912 - Cambridge, Massachusetts, March 7, 1997). Having spent the period of the Second World War improving the operation of airborne radar at the Massachusetts Institute of Technology (MIT), Edward Purcell returned to Harvard to use his skill in a group, which also included **Isidor Rabi** and some experts from Columbia University, to test the magnetic behavior of the proton and the electron of the hydrogen atom. The tiny magnetic moment of these particles is oriented along the magnetic field. Purcell had the idea of changing this orientation, by submitting the atoms to radio waves of well-defined frequency. As the frequency causing the reorientation is a characteristic of each atom capable of feeling it, we have at our disposal a nondestructive method of ultra fine pro- bing of atomic behavior of substances, both in their solid and their liquid or gaseous form. It was for a new development of this technology that **Richard Ernst** was awarded the Nobel Prize for Chemistry in 1991. Purcell applied his discovery to a wide range of applications in physics. In 1951, with H. Ewen, he built a telescope able to detect the signal emitted in the hydrogen clouds of space when these atoms are reoriented after having randomly undergone an inversion of their magnetic moment. He could thus follow the movement of hydrogen in the Milky Way, by detecting this signal with the frequency of 1420 megahertz. At the same period two Dutch and Austrian groups made the same observation. The work was then continued mainly by the Dutch group which reviewed a major part of our galaxy. Purcell obtained a diploma in electrical engineering from Purdue. Turning then toward the study of fundamental physics, he was interested, during his "new" first cycle, in the diffraction of electrons, corpuscles to which **Louis de Broglie** had associa- ted a wavelength. Having spent a year in Karlsruhe, he returned to Harvard where he obtained his Ph.D. in 1938. He was appointed professor at this university in 1949. In addition to his research activities, he was involved in the revision of the physics program at all teaching levels. He became famous through various televised program courses. One of his students, **Nicolaas Bloembergen**, was awarded the Nobel Prize for Physics in 1981. Purcell felt little motivation to transform the results of his discoveries on fundamental research toward applications, but he encouraged his students, and Bloembergen in particular, into this route.

Edward M. Purcell

1953

Zernike, Frits (Amsterdam, The Netherlands, July 16, 1888 - Naarden, The Netherlands, March 3, 1966). Zernike, a great specialist in optics and attached, at the beginning, to the department of astronomy of the University of Gröningen, modified the traditional use of the microscope. Rather than being interested in the intensities of the light coming from the various areas of the microscopic object being observed, he turned his attention to the interference of light waves which, passing through mediums of different refractive indexes, lead to a series of contrasted zones. His invention (phase-contrast microscopy) is still widely used in medicine because it allows the observation of cells without killing them.

These cells, entirely transparent, are made up of different density and refractive index areas and are identifiable by the contrast given, in the focal plane of a lens, by the interference of the light waves which cross the cell with those which passed through its periphery. The same technique can be used in metallurgy for the observation of tiny details. This invention, initially ignored by the Zeiss company, expanded very rapidly after 1942 in Germany as well as in England and the United States. Born into a family of mathematicians, Zernike was very fond of ma-thematics from an early age. He graduated in Gröningen where he defended his docto- ral thesis in thermodynamics in 1913. He spent his entire career at this university: first in the Astronomy Laboratory and then as a professor of theoretical physics and mathematics until his retirement in 1958. He always remained aloof from the "modern physics" which won so many of the Nobel Prizes after 1922, preferring to devote himself to the construction of demonstration equipment in optics, thermodynamics and chemistry.

Fritz Zernike

1954

Born, Max (Breslau (Wroclaw) Poland, December 11, 1882 - Göttingen, Germany, January 5, 1970). It may seem surprising that Max Born, who had **Wolfgang Pauli** and **Werner Heisenberg** as assistants in Göttingen in the 1920's, received the Nobel Prize well after his two "pupils". Heisenberg even sent him a touching message in reply to the Born's letter of congratulations, which Born had sent him for his 1932 Nobel win. Heisenberg informed his former professor how depressed he was that he was honored alone for a work which had been jointly realized. He assured Born that the scientific community had an accurate understanding of his own contribution. Twenty - two years later Born shared the 1954 Nobel Prize with **Walter Bothe**. although their work was achieved in completely different fields. It was during his long stay in Göttingen (from 1909 to 1933) that Born became interested in the model of the atom resulting from **Ernest Rutherford** (1908 Nobel Prize for Chemistry) and **Niels Bohr's** work, and also in the particle-waves formulated by **Louis de Broglie** combined with the complementary quantities of the simultaneous measurement of two so - called complementary entities (position and linear momentum, energy and time, angular momentum and phase) formulated by Heisenberg. By integrating all these data of modern physics (then in its infancy), he concluded that the wave functions which were, in particular, solutions of **Erwin Schrödinger's** equation, provided, by calculating the square, a measurement of the probability of the presence of the particle with which the wave was associated. It is this part of Born's work that was awarded by the Nobel Committee. Born had previously undertaken research in solid - state physics (crystal dynamics). He took part at the Solvay Council of Physics in Brussels in 1927. Native from Poland, he was brought up in his grandmother's house after the death of his mother when he was no more than four years old. He was educated in a great number of universities: Breslau, Berlin, Heidelberg, Zurich, Cambridge and finally Göttingen where in 1907 he defended his doctoral thesis. Then he went to Berlin and Frankfort, where he taught for a short period of time, and on returning to Göttingen he headed the Laboratory of Theoretical Physics, an institute with no rival except that of Bohr Institute in Copenhagen. In 1933 Hitler's regime made Born abandon Germany and he went to teach at Cambridge University. After his retirement in 1953, Born continued writing and was vigorously opposed to the development of nuclear weapons. Born's interest in new physics was a great deal stronger than that of **Albert Einstein's**. In 1971 Irene Born published their vast correspondence on this theme. Born particularly stressed the following: " To my mind, such notions as absolute certainty, complete exactness and final truth are pure imagination and are totally inadmissible in science." Being a musician, he played the piano in duet with Heisenberg.

Max Born

Bothe, Walter Wilhelm Georg (Orianienburg, Germany, June 8, 1891 - Heidelberg, Germany, February 8, 1957). A theoretical physicist at the beginning, it is for experimental work that Walter Bothe shared the Nobel Prize with **Max Born**, although for entirely unrelated research. The innovation, developed in 1924 by Bothe, consists of the triggered detection of a photon in a detector: namely only if a second detector records the arrival of a particle. The detection of the two events is thus carried out in coincidence. The development of this device had been conceived in order to decide if the conservation laws of momentum and energy in the photon - electron interaction were absolutely true (as was assumed by **Albert Einstein** and **Arthur Compton**) or if these laws were valid only on average, thus in a statistical way as **Niels Bohr's** supporters thought. In the Physikalische-Technische

Reichsanstalt of Berlin and with the collaboration of Geiger (who had developed a very powerful yet simple gas detector), Bothe concluded that the conservation of the quantities was maintained in these interactions. The coincidence technique was very widely used in the study of cosmic rays: to check if (in a **Charles Wilson** cloud chamber, or in **Cecil Powell's** photographic method, or again in a cascade of detectors aligned with the many lenses of a telescope) the primary particle really arrived from the upper atmosphere. This method is also used for the study of the nuclear reactions induced on various targets by projectiles produced in accelerators, for the identification of the properties of atomic nuclei emitting radiations in cascades and for the certification of the synthesis of new (or no longer existing) radioactive nuclei. In his lecture, which Bothe had prepared for the ceremony of the Nobel Prize presentation, he paid homage to Geiger, who, probably could have shared the prize with him. At that time Bothe had been already suffering from pains and could not participate in the ceremony in Stockholm. So he sent his lecture to Sweden, where it was read out on his behalf. Bothe studied in Berlin, where in 1914 he defended his doctoral thesis. At that time he studied under **Planck**'s direction. A prisoner in Siberia during World War I, Bothe returned to Berlin in 1920 and started teaching. Then he undertook the construction of the coincidence method. With his reputation established in this field, he was appointed Professor in Giessen in 1930 and there he stu died the particles emitted during the bombardment of a beryllium target with a particles, which were emitted by a radioactive source of polonium. He believed that the radiation, which was emitted, consisted of photons. It was **James Chadwick** who corrected this interpretation by identifying the neutron in 1932. As Head of the Max Planck Institute in Heidelberg, Bothe supervised the construction of the first German cyclotron, which was brought into operation in 1943. He was then commissioned with a task to direct research in the field of atomic energy in Germany. He taught in Heidelberg until his death. He attended the Solvay Concil of Physics in Brussels in 1933. He was known for his remarkable capacity for working quickly and efficiently. Bothe was a perfect musician, and his favorite composers were Bach and Beethoven.

Willis E. Lamb

1955

Lamb, Willis Eugene (Los Angeles, California, July 12, 1913 -). Willis Lamb, a great specialist in electrodynamics, showed at the University of Columbia (in 1947), that the fundemental state of the hydrogen atom, as described since 1928 by **Paul Dirac's** theory, was subdivi- ded into two very close sub-states: the splitting is known as the "Lamb shift". By means of a microwave emitter Lamb observed this very slight difference: 0.3 x 10^{-6} of the energy necessary to induce the ionisation of the hydrogen atom. As a result of this mea- surement Lamb opened the way to refining the description of the spectral lines of hydrogen as predicted by **Niels Bohr**'s theory and by **Erwin Schrödinger's** equation. At that time he followed **Gerhard Herzberg's** work (1971 Nobel Prize for

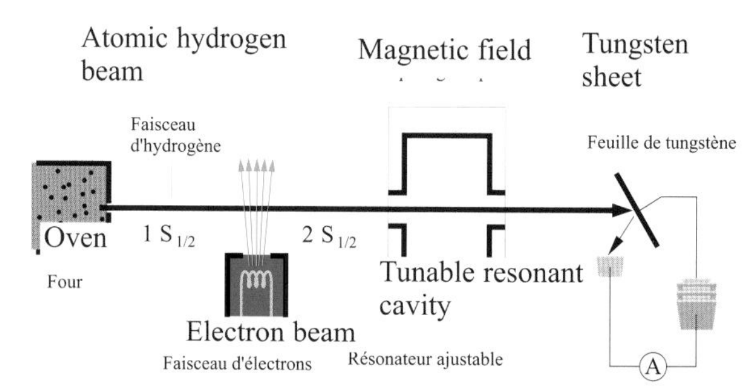

Lamb's displacement measuring device (Lamb shift). A beam of hydrogen atoms emitted by a furnace passes into an excited state thanks to electron bombardment. Then, atoms passing in a resonant cavity can undergo an electromagnetic transition, which leads to decrease of the number of the excited atoms. In this case the tungsten sheet emits fewer electrons.

Chemistry) as the basis of his physics course. Another measurement of prime importance in physics is the constant of fine structure: a quantity represented by a pure number $\alpha = e^2 / 2 \varepsilon_0 hc$, where e is the electric charge carried by the electron (thus also that of the proton), ε_0 is the constant of electrostatics connected with the force between the charges, h is **Planck**'s constant (the one that took physics into the quantum era) and c is the speed of light that **Albert Einstein** imposed as a constant in all inertial frames of reference. The reason for the value of this constant of fine structure (given with an extreme precision by Lamb and being equivalent to 1/137.0365) still remains today a major riddle for physicists. When we will understand the reason why this constant has this value, we will perhaps have at our disposal one of the keys necessary for the general unification of the forces that govern the behavior of the whole universe. Lamb was a student in Berkeley where he graduated in chemistry, in 1934, and doctor of physics, in 1938, under the supervision of J. Robert Oppenheimer. In 1938 he joined the physics group at Columbia in New York and was appointed Professor there in 1948. From 1943 until 1948 he was also part of the Columbia Radiation Laboratory. He successively occupied a position at Stanford (1951 - 1956), at Oxford (1951 - 1956), at Yale (1962 - 1964) and finally at the University of Arizona in 1974. He liked teaching at all levels of university courses. A well - balanced man, he also practiced sport, photography and chess.

Kusch, Polykarp (Blankenburg, Germany, January 26, 1911 - Dallas, Texas, March 20, 1993). Like **Willis Lamb** who received the Nobel Prize the same year, Polykarp Kusch was a member of the University of Columbia (New York). Both worked independently, carrying out measurements of microscopic effects in the field of quantum electrodynamics. In 1947, while he was part of the research group led by **Isidor Rabi**, Kusch and his colleague Foley showed that the magnetic moment of the electron (a property similar to a small magnet allowing it to undergo the action of a magnetic field) had a value 1% higher than that predicted by **Niels Bohr**. This "anomaly" was explained by the theorists **Sin-itiro Tomonaga**, **Julian Schwinger** and **Richard Feynman** and earned them the 1965 Nobel Prize in Physics. Born in Germany, Kusch emigrated with his family to Cleveland (United States) at the age of one. At 15 years old, he assiduously frequented libraries. He received his Ph.D. from the University of Illinois. He taught there from 1931 until 1936, then at the University of Minnesota before joining Columbia from 1937 until 1941. During the Second World War he did research for the Westinghouse and Bell Telephone companies. In addition to the research which earned him the Nobel Prize, he devoted a considerable time to the teaching of physics to nonscientists, both at Columbia and at the University of Texas from 1972.

1956

Bardeen, John (Madison, Wisconsin, United States, May 23, 1908-Boston, Massachusetts, January 30, 1991). A solid state physicist, mainly involved in the study of the electric properties of materials, John Bardeen is the only one to have received the Nobel Prize in physics twice (with a gap of 16 years between). It will be noted that **Marie Curie** also received two Nobel Prizes: one in physics, the other in chemistry. Bardeen shared the prize in 1956 with **William Shockley** and **Walter Brattain**, for the discovery of the transistor effect (which brought a spectacular revolution in the performances of electronic devices and data processing "for all") and, in 1972, with **Léon Cooper** and **John Schrieffer**, for the theoretical explanation of superconductivity (which will perhaps bring another revolution in the transport of electric current without thermal loss). The BSC theory (initials of the 3 authors) has boosted the technological interest in superconductors, somewhat neglected since the work of Heike **Kamerlingh-Onnes** in 1911. It was in 1947 in Bell Telephone's research laboratories in Murray Hill (New Jersey) that Bardeen, in collaboration with Brattain and under the principal supervision of Shockley, made the discovery which earned him the Nobel Prize for the first time. The first transistor effect (for transfer-resistor) consisted in particular of amplifying electric signals in an assembly made up of two small metallic contacts that "sandwich" a thin layer of germanium in which electrons did nothing but "transit." The first demonstration of this effect was carried out on December 27, 1947, with a power amplification of almost 20. The applications of these circuits are innumerable: they invaded all fields of technology thanks to a continuing decrease in cost, a low energy consumption for maximum efficiency, a minimum overcrowding and a need for only a very low voltage power supply. Afterwards, at the University of Illinois, together with Schrieffer and with the theoretical support of Cooper, he contributed a decisive explanation of the behavior of several materials which suddenly do not resist the passage of an electrical current as soon as the temperature is reduced below a critical point, as Kamerlingh-Onnes had noticed. This second discovery happened very quickly after he had received his first Nobel Prize. At the beginning of the 21st century, a very significant part of physics research is represented by solid state physics, to which these two major contributions of Bardeen's career are connected. It is necessary to go back to the great period (1914 - 1924) of discoveries on the interactions of X - rays to find other Nobel Prize winners in this field. We had to wait from then until 1962 (**Lev Landau**), 1970 (**Louis Néel**), 1973 (**Leo Esaki**, **Ivar Giaever** and **Brian Josephson**), 1977 (**Nevill Mott** and **Philip Anderson**), 1978 (**Pyotr Kapitza**), 1985 (**Klaus von Klitzing**), 1987 (**Karl Müller** and **J. Georg Bednorz**) to find laureates in this significant field of fundamental physics, which is inextricably linked to its applications. Son of an anatomy professor, Bardeen graduated from the University of Wisconsin in 1929 and defended his doctorate at Princeton in 1936 after having shown an interest in geophysics. Then he spent two years in **Percy Bridgman's** group at Harvard, and was an assistant at the University of Minnesota until 1945. Later he joined Bell Telephone until 1971, when he was appointed Professor at the University of Illinois. He took part in numerous developments at Xerox and General Electric. He was elected to the Academy of Sciences of the United States in 1954 and was President of the American Society of Physics (1968-1969). He received many scientific prizes and was mentioned by Life magazine as one of the 100 outstanding scientists of the 20th century.

John Bardeen

Brattain, Walter Houser (Amoy, China, December 10, 1902 - Seattle, Washington, October 13, 1987). Walter Brattain was the initiator, as far back as 1931 at Bell

Telephone, of research on electric conduction in semiconductors (materials whose resistance to the passage of a current is intermediate between that of insulators and that of conductors). This discovery made with Bardeen and Shockley relates to the transistor effect. Current applications extend far beyond the miniaturized radio sets with which general public associates the term "transistor." This electronic device has, during the past few decades, largely replaced the vacuum tubes (more cumbersome, slower and shorter lifespan) used before the 1960s. The transistor circuit is certainly the device that has experienced the most spectacular development of this century. Brattain's principal contribution was the determination of the surface properties of semiconductor materials. Of the three 1956 prizewinners, Brattain is apparently the least known: he is however the pioneer of this research. When he learned that the Nobel Prize had been attributed to him, he announced modestly: "My luck was to be in the right place, at the right time, and in the best team." Since the 1903 Nobel Prize in physics, attri-buted to the three French scientists, **Henri Becquerel** and **Pierre** and **Marie Curie**, who did not in any case all work as a single team, it is the first time the prize was awarded to a team for a truly collective work. Although Brattain was born in China, he spent his youth on a family ranch, in the state of Washington. A new step towards the miniaturisation of electronic devices took place in the 1980's with the development of heterostructures and the invention of integrated circuits. **Alferov**, **Kroemer** and **Kilby** were awarded the Nobel Prize for Physics for their research in this area. In 1926 Brattain received a first diploma from Oregon University and in 1928 he was conferred a doctoral degree at Minnesota University. In 1929 he got his job at the Bell Telephone Company, where he worked until his retirement in 1967. Then he taught at Whitman College, the university where both he and his parents had begun their schooling. During World War II he worked in the American Navy and helped detect submarines.

Walter H. Brattain

Shockley, William Bradford (London, England, February 13, 1910 - Palo Alto, California, August 12, 1989). Member of a brilliant team which also included **John Bardeen** and **Walter Brattain**, William Shockley showed, in 1939, that an assembly of two or three semiconductors could rectify alternating electric currents and amplify signals in a more efficient and economical way than the traditional electronic vacuum tubes. Their team work, interrupted during the Second World War, started again at Bell Telephone in 1945. After the first success gained by the pinpoint electric contact transistor over a germanium crystal, they invented, one month later, the junction transistor. They injected impurities inducing electron holes (by substituting gallium to germanium for example) in a semiconductor initially "doped" with atoms providing, on the contrary, a surplus of electrons (atoms of arsenic, for example). This device, which involves the motion of the electrons of the whole semiconductor vo- lume and not only surface electrons, considerably improved the performance obtained with the first assembly, which led to a lower effect due to the transport of charge at the surface only (a device created by Brattain and developed by the latter and Bardeen). Shockley was born in London but was granted American citizenship at age three. He lived in California and obtained his B.S. degree in 1932. He received his Ph.D. from the Massachusetts Institute of Technology, Cambridge. He then joined, in 1936, the research group of Bell Telephone where he became director of the semiconductors division in 1953, before starting up and managing several other companies for the production of transistors. He taught at the Stanford Faculty of Applied Sciences while keeping a position as a consultant at Bell Telephone. He left all of these positions in 1975. He took part in the Solvay Council of Physics in Brussels in 1961. After 1965 he was at the heart of many conflicts for his ideas advocating a personal conception of

William B. Shockley

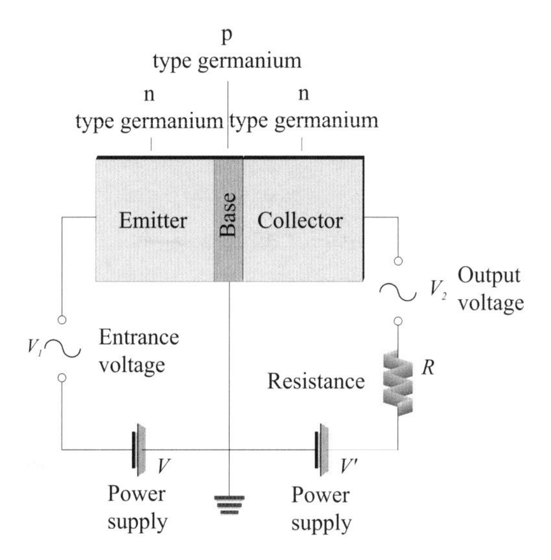

The transistor. Sketch of a connection in common base. The thickness of the base (B) is so small that the electric charges coming from the emitter do nothing but "transit" there on their way towards the collector.

"the improvement of the human race". He declared, among other things, during televised shows, that people of the Black race were genetically inferior to Whites. An animated reaction was not long in coming and, in 1984, Shockley was "crowned" "Nazi of the Year" in the United States. In 1980, he had proposed a plan aiming to "financially reward" "genetically handicapped" persons who would subject themselves to sterilization. The United States Academy of Sciences took an official position against the genetic theories of the brilliant physicist. In 1979, at the time of the creation of a sperm bank, proposing to only "young and brilliant women" the possibility to have themselves inseminated by a "Nobel Prize seed", Shockley was the only one to reveal that he was among the twenty eminent scientists having proposed a "donation".

1957

Lee, Tsung Dao (Shanghai, China, November 24, 1926 -). A theoretical physicist, Tsung Lee shared the honour of the 1957 Nobel Prize with **Chen (Franck) Yang**, a colleague also of Chinese origin, whom he met in **Enrico Fermi's** group in Chicago, where both of them defended their Ph.D. theses. They are also the only Chinese physicists to have received the Nobel Prize in Physics to date. Their proposal stipulated that, in weak interactions, the left is not equal to the right. This proposal was rapidly confirmed experimentally. Their forecast made **Isidor Rabi** say in 1957: "A theoretical structure which appeared fairly complete has just undergone a serious attack at its base. Nobody can say if it will be possible to put back the parts back into place". Lee began his university studies in 1945 (date of the Japanese invasion), in several universities of the south of his native country, and, in 1946, obtained a grant to continue his studies in Chicago, where he met Yang in a university residence. Their only concern at the time was: "to think". In 1950 Lee presented a Ph.D. thesis on an astrophysics problem: the hydrogen content of white dwarves, a research subject tackled by **Subrahmanyan Chandrasekhar** in Chicago. After a short time at the University of California in Berkeley, he became a member of the Advanced Study Institute of Princeton from 1951 until 1953. At the end of his stay the then director, Robert Oppenheimer, regretted the departure "of the most brilliant theoretician that he had had the occasion to meet, for a collection of work in statistical mechanics, in nuclear and sub-nuclear physics". Lee joined Columbia in 1953 but continued his profitable collaboration with Yang. He is among the 5 youngest physicists of the 20th century to have received the Nobel Prize for Physics.

Tsung Dao Lee

Yang, Chen Ning (Hefei, China, September 22, 1922 -). As a 35 year old theoretician, Chen Yang shared the 1957 Nobel Prize with **Tsung Lee**, four years his junior, for highlighting the "nonconservation of the parity in weak interactions". Since the invention of quantum physics by **Max Planck** to explain the study of the interaction of light with matter, the attribution by **de Broglie** of a wave property to particles and the formulation of new physical ideas by **Werner Heisenberg**, **Paul Dirac**, **Erwin Schrödinger** and **Wolfgang Pauli**, the proposition of Yang and Lee contradicts the affirmation (widely admitted) that, in all interactions, the left is equal to the right. They proposed experiments permitting the verification of their views to two teams at Columbia University. First, it was the physicist Chien-Shiung Wu who confirmed their proposal by studying ß radiation; emitted by a source of Co^{60} cooled to a very low temperature (a necessary precaution to lower as much as possible the disordered movement of the emitting nuclei) and placed in a magnetic field able to orient the quasi totality of these cobalt nuclei. Less than 10 years later, Leon Lederman and Richard Garwin (also from Columbia) opened a new path of investigation in the research of the classification of the interactions of physics by also verifying that the left was not equal

Chen N. Yang

to the right in the behavior of muons artificially produced in large particle accelerators. Yang was, from an early age, introduced to fundamental science by his father, an eminent mathematician. He obtained his first higher diplomas in China, but it was at Chicago, in Enrico Fermi's group, that he defended his Ph.D. thesis in 1948. He was appointed Professor at Princeton in 1949 and at Stony Brook (New York) in 1966. He is better known to his close coworkers under the first name of Franck.

1958

Cherenkov, Pavel Alekseyevitch (Voronehz, Russia, July 28, 1904 - Moscow, Russia, January 6, 1990).It's the spirit of curiosity, characteristic of the good researcher, that allowed Cherenkov to discover a particular effect, which today bears his name. Studying at the Lebedev Physics Institute in Moscow, he was invited by the academician Sergei Vavilov to study phenomena related to the absorption of various types of radiations in fluids. Many observers before him had noticed that transparent liquids became a source of bluish light when they were excited by various radiations. The majority of them attributed the phenomenon to a fluorescence in the impurities contained in the liquid. Cherenkov was not convinced and used, as a liquid, water which had been distilled several times so that all traces of impurity were eliminated. Repeating the experiment (after a long adaptation time to darkness in order to increase his vision), he noticed the persistence of the phenomenon and showed that the emitted light had a polarization direction related to the direction of the incident radiation. Using a radium source whose emission was filtered so as to tolerate only the penetration of the electrons in the tank, he concluded (in 1934) that very fast electrons were the cause of this blue gleam. His mathematician colleagues, **Ilya Frank** and **Igor Tamm**, explained the phenomenon in detail in 1937. The special detector developed by Cherenkov is an instrument that measures the kinetic energy of particles whose speed, in that material, is greater than the speed of the light in this same medium. It develops a (light) bow wave rather si- milar to the wave front which is produced by the passage of a projectile moving through the air more quickly than sound. The visible blue color in any pool of a nuclear reactor is due to this phenomenon. This detector was used, in particular for the identification of the antiproton, by **Owen Chamberlain** and **Emilio Segrè**. It was also installed in several artificial satellites to determine the speed of the particles resulting from cosmic radiation. Cherenkov initially studied in Voronezh, his native region. He had to work to pay for his higher education, which he started only at the age of 20 years old. In 1930, he joined the Institute of Physics of the Academy of Sciences in Moscow and received a doctorate there in 1940. In 1946 the three 1958 Nobel Prize winners, together with Vavilov (who died in 1951), had already received the Stalin Prize for the same discovery. After the Second World War, Cherenkov continued to improve the detector and conti- nued his research career in high - energy physics, nearby large accelerators. He aroused the interest of many of his young compatriots in this discipline which makes it possible to gain, gradually, knowledge of the infinitely small and, consequently, of the infinitely large.

Alekseyevitch P. Cherenkov

Blue radiation signatures of the Cherenkov effect induced by very fast electrons in a nuclear reactor pool.

Frank, Ilya Mikhailovich (St. Petersburg, Russia, October 23, 1908 - St. Peterburg, Soviet Union, June 22, 1990). A soviet physicist, Frank Ilya Mikhailovich stu died, as did **Pavel Cherenkov** under Vavilov. He defended his doctoral thesis on molecule dissociation and photochemistry. He became Director of the Laboratory of Nuclear Physics at the Lebedev Institute (Soviet Academy of Sciences). It was there that he was involved with the mathematical explanation of the discovery of the Cherenkov effect together with **Tamm**.

Tamm, Igor Evgenievich (Vladivostock, Russia, July 8, 1875 - Moscow, Russia, April 12, 1971). Soviet theoretician physicist. Igor Tamm spent a good part of his research career trying link **Einstein**'s theory of relativity to quantum mechanics. This difficult subject already studied by Einstein was regarded as anti-Marxist, but was considered of great interest for Tamm who did not hesitate to criticize the bureaucracy that paralyzed all sectors of activity in his country. With **Ilya Frank**, he managed to explain correctly the way in which a wave was propagated in a medium crossed by particles whose speed was higher than the speed of light in this same medium. Tamm is also well known for the explanation of the ß emission (an emission controlled by weak nuclear interaction), an explanation which he proposed in 1934 and whose model was used by **Ideki Yukawa** in his study of strong nuclear force. He also carried out theoretical work concerning the physics of surfaces and solids, the physics of semiconductors and the discharge in gases immersed in strong magnetic fields. With **Andrei Sakharov**, he pioneered research on the control of thermonuclear reactions (fusion). He stayed for two years in Scotland before starting his university studies. Graduating from the University of Moscow in 1918, he simultaneously taught in Simpheropol, Odessa and Moscow. From 1934 until his death, he was a Professor at the University of Moscow and Director of the Department of Physics of the Lebedev Institute.

1959

Chamberlain, Owen (San Francisco, United States, July 10, 1920 -). The discovery of the antiproton, which earned the Nobel Prize in Physics for **Emilio Segrè** and his pupil Owen Chamberlain, is the result of thoroughly designed, long term research. The construction, in Berkeley, of the Bevatron (an accelerator delivering an energy of almost 6 billion eV to protons) had been programmed for this precise objective. The antiproton is a particle having the mass of the proton, and a charge and a magnetic moment of signs opposite to that of the proton. **Dirac** had built a theoretical model predicting, for each particle, the existence of a "mirror" particle, namely its antiparticle. A first confirmation of Paul Dirac's proposition came with the discovery by **Carl Anderson**, in 1932, of the positron, antiparticle of the electron. Nearly 20 years after its operational start-up, the Bevatron produced, in 1954, a multitude of particles (pions, kaons, etc.) and also some antiprotons: only a few tens for one million pions. These antiprotons were produced by the interaction of protons of more than 5.6 GeV (5.6 billion eV) on a hydrogen target following the formula $p+p \rightarrow p+p+p+\bar{p}$: an incident proton sent to a hydrogen target produces three protons and one antiproton. The creation of a pair p and $\bar{p}$ lies in the transformation into masses of two of the six GeV of kinetic energy according to the famous formula of **Einstein**: $E = mc^2$. Segrè and Chamberlain used a clever assembly of shields and electromagnets to select the antiprotons among the multitude of negatively charged particles and in particular the π^--pions. One of the detectors was of the Cherenkov type, developed by a Soviet team, which received the 1958 Nobel Prize for Physics. Chamberlain graduated from Dartmouth College in 1941. He then joined Segrè's group involved in the Manhattan Project for the production of chain fission of uranium by slow neutrons. Between 1946 and 1949 he tackled some fundamental research, which led to his docto- rate. The subject was related to the diffraction of neutrons by crystals, a subject for which **Bertram Brockhouse** and **Clifford Shull** received the Nobel Prize in 1994. In 1948 Chamberlain was a teacher at the University of California at Berkeley and a researcher in the team which exploited the proton-proton interaction. In 1957, he obtained a grant to undertake the study in Rome of antiparticles (and in particular the antineutron, another antiparticle also discovered with the Bevatron of Berkeley) and of their interactions with matter.

Owen Chamberlain

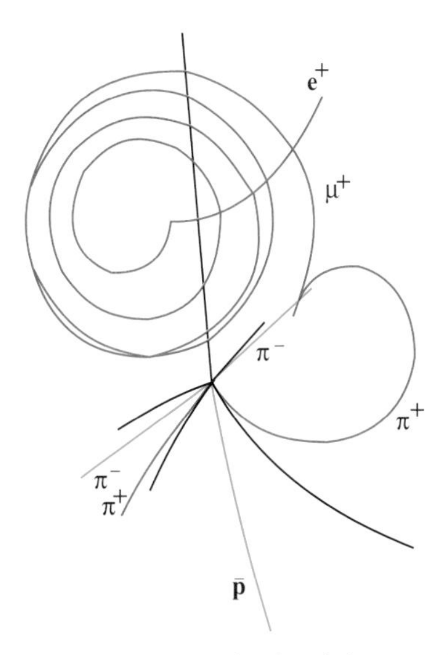

Identification of an antiproton. Coming from below, an antiproton produces, in a bubble chamber, a series of pions following trajectories curved by a magnetic field. Deviations in opposed directions depend on the sign of the charge.(Drawing made from an original document of the Education Development Center, Newton, MA)

Segrè, Emilio Gino (Tivoli, Italy, February 1, 1905 - Lafayette, California, April 22, 1989). Coming from the famous **Enrico Fermi** school, in Rome, Emilio Segrè was forced to leave Italy for political reasons. He pursued his career in the United States as early as 1938. In 1955 he discovered the antiproton, in collaboration with **Owen Chamberlain**, and two of his students, C. Wiégand and

Thierry Ypsilantis. It was Segrè who proposed the building of the Bevatron, the high energy accelerator (the highest of the time) which was in operation at huge cost in Berkeley, 20 years after the start of its construction. Born into a family of industrialists, Emilio Segré first entered the university as a student of engineering before devoting himself to physics. He received his Ph.D. in 1928. An assistant at the University of Rome, he had two stays abroad after 1930; one in Hamburg in **Otto Stern's** group and one in Amsterdam in the group of **Peter Zeeman**. He was appointed Professor at the University of Rome in 1932 and collaborated with Fermi in neutron physics. From 1936 to 1938, he headed the Ins-titute of Physics of Palerme. In 1937 he "created" a element non-existent in nature (technetium) be-aring the atomic number 43, which he synthesized by irradiation of molybdenum with deuterons (hea-vy hydrogen nuclei, made up of a proton and a neutron). In 1940 he created the element bearing the atomic number 85 (astate) and the element bearing the atomic number 93 (plutonium). **Glenn Seaborg** continued this type of research and was awarded the Nobel Prize in Chemistry in 1951. Segrè took part in the determination of numerous half-life times of unstable nuclei. From 1943 until 1946 he played a major role in the Manhattan Projet for developing the first atomic bomb. He was granted American citizenship in 1944. In 1974 he returned to the University of Rome and taught physics.

Emilio G. Segrè

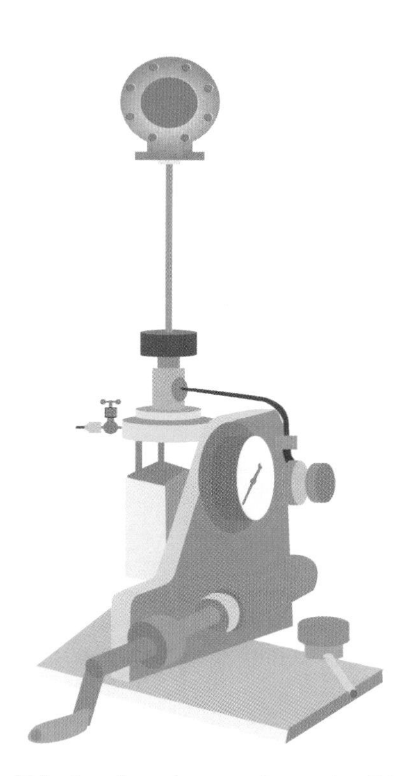

The first bubble chamber whose maximum size did not exceed 5 cm. See Alvarez for improvements.

1960

Glaser, Donald Arthur (Cleveland, Ohio, September 21, 1926 -). Not very satisfied with the performances either of the cloud chambers (invented by **Charles Wilson** in 1910) or of the photographic emulsions (invented by **Cecil Powell** between 1930-1940) for the detection of "strange" particles produced by cosmic radiation and around high energy accelerators, Donald Glaser created a new technique that he used after 1952: the bubble chamber. The Wilson chamber is based on the generation of droplets around an ion produced by the passage of a charged particle in a gas kept under conditions (of temperature and pressure) close to liquefaction. The density of gas is so low that the paths of the high-energy particles would have required detectors of enormous volume. Photo-graphic emulsions (of solids which are clearly more dense than gases), on the other hand, being too thin, can be used for the detection of fast particles only if they are piled up to reach a sufficient thickness. Glaser consequently invented a device based on a phenomenon similar to that which occurs in the Wilson chamber: the generation of bubbles around an ion produced by the passage of a charged particle in a liquid kept under conditions (of temperature and pressure) close to its vaporisation point. The tracks left along the trajectories of the fast particles were photographed (in stereo) in order to precisely

Donald A. Glaser

The Gargamelle bubble chamber of CERN which procured (in 1972 - 1973) more than 250,000 photographs.

reconstitute these trajectories thanks to a peripheral device invented for the cloud chamber by **Patrick Blackett**. The most useful liquid for the bubble chamber is hydrogen cooled to 27K. It was used to realise the most powerful bubble chamber: the one for which **Luis Alvarez** would receive the 1968 Nobel Prize. The volume of the detector passed quickly from 3 cm^3 (first test) to several m^3. At the end of the 20th century these detectors, of an even larger volume, are controlled by sophisticated electronics and computer technology and are frequently used in research laboratories involved in the study of elementary particles. Their response times are sometimes still too slow for particular needs. Only with the invention of the spark chamber, by **Georges Charpak**, would improvements be seen in these time constraints. Glaser was born in Cleveland, Ohio, where he received his early education in mathematics and physics. In 1946 he obtained his Ph.D. for work carried out at Cal Tech, whose first director (in 1920) was **Robert Millikan**. While remaining in contact with Cal Tech, it was at the University of Michigan that (in 1952) Glaser began the work that earned him the Nobel Prize. Thereafter, he applied his knowledge as a physicist to microbiology, cellular biology and molecular biology. He was a musician and played as part of a symphony orchestra from the age of 16. Also a sportsman, he loved climbing, skiing, sailing and underwater photography.

1961

Hofstadter, Robert (New York City, New York, February 5, 1915 - Stanford, California, November 17, 1990). Robert Hofstadter, an American experimenter, improved the description of the atomic nucleus principally given by **Ernest Rutherford** (1908 Nobel Prize for Physics) via a new route: the scattering of electrons by atomic nuclei. Hofstadter shared the 1961 Nobel Prize in Physics with **Rudolf Mössbauer**, also a nuclear physicist, but whose discovery is connected with the modification of a signal emitted, or absorbed, by a nucleus depending on its molecular or crystalline environment. Using high-energy electrons (10^9 eV) and therefore of very low associated wavelength, as anticipated by **Louis de Broglie's** model, Hofstadter was able to highlight the distribution of the electric charge and magnetic fields inside atomic nuclei. The spatial resolution of the device that he created allowed him to confirm that the proton and the neutron have similar dimensions and forms and are not point particles. In particular there exists, inside the proton, a non-uniform distribution of the electric charge. Hofstadter then opened the way for sub-nuclear research, which would lead to the description of hadrons in quarks by **Murray Gell-Mann** and to their experimental verification by **Jerome Friedman**, **Henry Kendall** and **Richard Taylor**. With a Bachelor in Sciences from the College of the City of New York, with the honor "magna cum laude", he continued his studies in Princeton, which he left as a Ph. D. in 1938. Then he moved to the University of Pennsylvania where he built a large accelerator based on Van de Graaff's model. The first device was installed in 1931 in Princeton. As the inventor of this accelerator, Hofstadter could have been also associated with **John Cockcroft** and **Ernest Walton** for the 1951 Nobel Prize in Physics. After undertaking activities outside theoretical research environments from 1940 until 1946, Hofstadter returned to Princeton. There he developed the most used detector of g rays in the world: namely, the crystal of sodium iodide activated with thallium, which allowed the realization of almost the complete g spectrometry of the nuclei before the arrival of

Robert Hofstadter

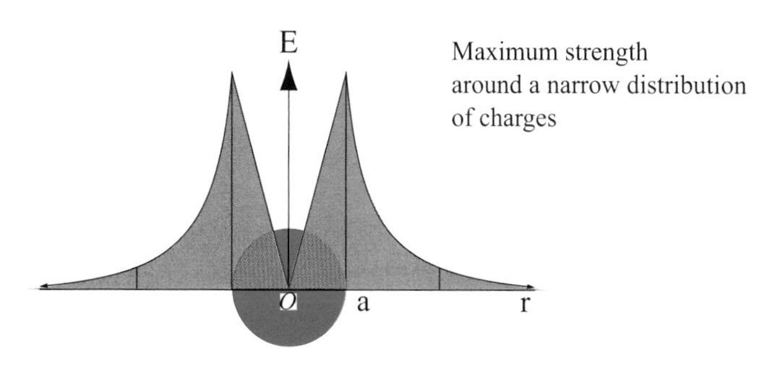

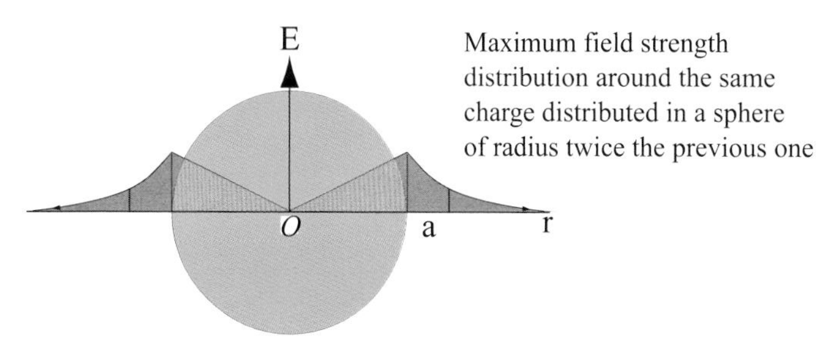

The maximum electric field in the vicinity of a spherical distribution of charge is the most important as the charge collapses in the smallest volume. The same electric charge uniformly distributed in a volume of double radius produces a maximum field 4 times weaker.

modern semiconductor detectors developed in the 1960s. In 1950 he joined the research group at Stanford involved in the construction of the large electron accelerator, an essential device that served as a "canon" for his study of the atomic nucleus structure.

Mössbauer, Rudolf Ludwig (Munich, Germany, January 31, 1929 -). The emission of a signal always leads to the recoil of the emitter. Thus an atomic nucleus (in an exci-

Rudolf Mossbauer

ted state) emitting g radiation borrows part of the energy of excitation to carry out this recoil. The emitted radiation is left with an energy lower than the energy of excitation. This radiation cannot, therefore, be reabsorbed by an identical nucleus (in its fundamental state and therefore of maximum stability). Rudolf Mössbauer tried to reduce this recoil effect by cooling a crystal containing the emitting nucleus. The recoil of the crystal (of a noticeably higher mass than the mass of the single emitting nucleus) borrowed, this time, an extremely weak energy and left a maximum energy to the γ radiation which was therefore sufficient to undergo the absorption. The Mössbauer spectrometry is based on variants of this recoilless emission or compensated recoil. This compensation can be obtained by moving either the emitter or the absorber at generally very low speeds; so weak that a few millimetres an hour are sometimes sufficient. The very weak gravity force is sometimes used to highlight this nuclear effect. The Mössbauer effect (the finest of the nuclear effects because it is able to differentiate energies of 1 part out of 10^{12}), is used for the measurement of the magnetic fields surrounding magnetic materials (a spectacular application in materials science) but also in more fundamental fields like time variation when "clocks" move at a speed close to that of light, as was predicted by **Albert Einstein**. Mössbauer was born in Munich and studied there in several institutions before obtaining his Ph.D. in 1958. Dissatisfied with budgetary restrictions and administrative problems, he accepted a position at Cal Tech (1961-1964). Despite everything he returned to the Technical University of Munich from 1965 until 1971, became Director of the Laue Langevin Institute, founded by Néel, in Grenoble (a reactor with a high flow of neutrons) from 1971 until 1977, before finally coming back to the University of Munich.

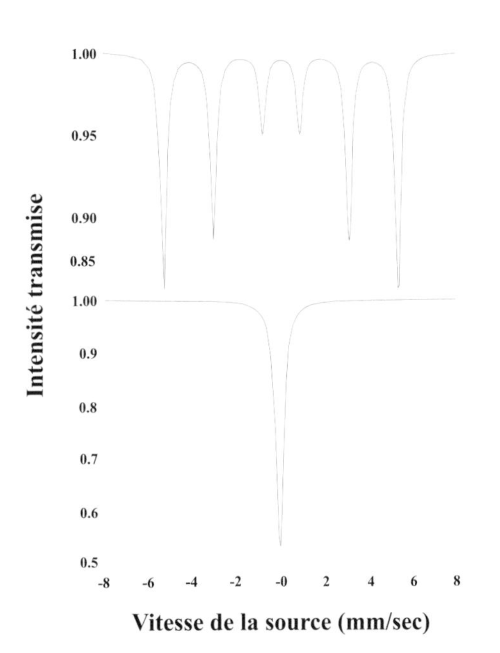

Mössbauer spectra of two iron compounds: metallic ion (α) and steel (γ).

1962

Landau, Lev Davidovich (Baku, Azerbaïdjan, January 22, 1908 - Moscow, Soviet Union, April 3, 1968). A Soviet physicist, universally recognised as one of the greatest theoretical physicists, Lev Landau was part of a research group on low temperatures (headed by **Pyotr Kapitza**) when, at the beginning of the 1940s, he provided a mathematical explanation of the reasons why liquid helium 4 became super- fluid (no resistance to flow) and showed a thermal conductivity almost 1000 times better than copper at a normal tempe- rature. He also predicted that superfluid helium would propagate sound at two different speeds. This prediction was verified by V. Peshkov in 1944. Landau won the Nobel Prize twenty

Lev D. Landau

years after the publication of this work for his contributions to the study of solids as a whole. A child prodigy, he began his university studies when he was 14 years old and graduated from the University of Leningrad aged 19, on presentation of a significant work on quantum mechanics which described quantum states by means of matrix densities. After 1929 he continued his studies (perhaps we should add his influence) in Denmark, in Switzerland, in Germany, in The Netherlands and in England. His contacts and exchanges with **Niels Bohr** brought a new dimension to quantum mechanics. From 1932 until 1937 Landau taught theoretical physics at the Physico-Technical Institute of Kharkov. He performed wonders there in the creation of mathematical solutions to complicated physics problems. He also supervised another young prodigy there: E. Lifshitz with whom he produced a monumental work: "the Landau and Lifshitz monographs", to which they provided numerous personal contributions (in quantum mechanics of course, but also in atomic and nuclear physics, astrophysics, thermodynamics, low - temperature physics and quantum electrodynamics). In the last field his successors were **Richard Feynman**, **Julian Schwinger** and **Sin-itiro Tomonaga**, who would also receive the Nobel Prize. The Landau and Lifshitz monographs were translated into many languages and earned them the 1962 Lenin Prize. Landau made visits to Bohr in Copenhagen and also kept up frequent debates with **James Franck**. They took pleasure in discussing not only the sciences but also art and religion. Landau was liberal and progressive in his political reflections. His professor, Pyotr Kapitza (who had summoned Landau back to Moscow in 1937 in order to create, together, the Institute for Physical Problems), also received the Nobel Prize in 1978. Kapitza was then 84 years old. Seriously injured in a car accident on January 7, 1962 in the Dubna area, Landau was plunged into a deep coma lasting 57 days. Neurosurgeons from France, Canada and Czechoslovakia joined their Soviet colleagues to save him, even though several times he was thought to be clinically dead. His wife (born K. Drobanezna, whom he married in 1937) and his son Igor were invited to receive the Nobel Prize on his behalf in November 1962. His health never really improved. He survived only 6 years after this terrible accident and died following a new attempt at surgery.

1963

Wigner, Eugène Paul (Budapest, Hungary, November 17, 1902 - Princeton, New Jersey, January 1, 1995). A theoretical physicist of genius, Eugène Wigner was recognised for his multiple contributions to quantum mechanics which permitted the elucidation of many problems in molecular, atomic and nuclear physics. He was also a philosopher and a visionary in his interpretation of the laws of nature: in particular, he had a great admiration for the symmetries that it presented. Wigner lived his youth during the influential years of German physics. He was delighted to have close and on his immediate horizon all those who had created quantum physics. With von Neumann he developed the theory of the atomic energy levels and with Seitz he proposed a description of the behavior of electrons in solids. In 1927, he introduced the concept of parity in quantum physics, a concept which he proposed as invariant in all interactions. This proposal remains valid today except for weak interactions as shown by the theorists **Chen Yang** and **Tsung Lee** and later confirmed by **Leon Lederman** and his team. Together with Breit, he explained the reason why, slow neutrons are absorbed in nuclei only if they have well - defined and quantified energies, that is to say in a resonant way. Wigner received his doctorate in applied sciences from the reknowned Technische Hochschule of Berlin (1925). He chose physics on his arrival at Göttingen. He taught physics at Princeton as early as 1938. With his Hungarian compatriots Szilard and Teller, and like **Enrico Fermi**, he persuaded **Albert Einstein** to use his influence with President Roosevelt in order to set in motion the development of the atomic bomb (The Manhattan Project) at the end of World War II. His training as an engineer, linked to his theoretical knowledge, enabled him to grasp all the complexities of the functioning of the first nuclear reactor which Fermi created. He used his prestige in the creation of several scientific societies and took an active role in 1975, with 32 fellow Nobel Prize winners, in the request for Andrei Sakharov to be authorized to collect the Nobel Peace Prize in person in Stockholm. The request was not successful. Wigner was born the same year as Paul Dirac. Moreover, he willingly called the latter his "famous brother-in-law": rightly so, since Wigner's sister married Dirac. He took part in the Solvay Council of Physics in Brussels in 1961. He died from pneumonia in the university hospital of Princeton.

Eugène P. Wigner

Goeppert-Mayer, Maria (Kattowitz, Poland, June 18, 1906 - San Diego, California, February 20, 1972). Maria Goeppert-Mayer, the only woman besides **Marie Sklodowska-Curie** to obtain the Nobel Prize in Physics, shared half of the prize with **Johannes Jensen**, the other half having been attributed to **Eugene Wigner**. After an already long scientific career (which was not easy for a woman), Maria Goeppert arrived in Chicago in 1946 and joined the group of Edward Teller (recently seconded to the Los Alamos Laboratory where he contributed to the Manhattan Project for the manufacture of the atomic bomb and still better known as the "father of the hydrogen bomb"). At that time Teller was studying the way in which the abundance of chemical elements and their isotopes were distributed in nature. Among them, those that contain a number of protons or neutrons being equivalent to 2, 8, 20, 28 were definitely more abundant. Numbers 50, 82 and 126, determined afterwards, provided the panoply of "magic numbers", a somewhat trivial designation given at this period and which was a good reflection of the ignorance of the reason why nuclei, containing these numbers of protons or neutrons, had a particular stability. In Chicago, benefiting from a very broad database on atomic nuclei and having access to another databank in Argonne (a research centre that she had frequented before), she built a model of the atomic nucleus based on a transposition of the model that **Niels Bohr** and **Wolfgang Pauli** had developed to explain the electron structure of the atom. These magic numbers, mentioned above, were thus the nuclear transpositions of particular atomic numbers of rare gases, themselves atoms with a particularly stable configuration in chemistry. Independently, and during the same period, Jensen and his colleagues Haxel and Suess from Hamburg made the same observation. Goeppert and Jensen met for the first time in 1950 and decided to continue their investigation together to produce their famous treatise on the shell model of the atomic nucleus in 1955. Their model was not easily accepted by their peers but Goeppert convinced **Enrico Fermi** while Jensen convinced **Werner Heisenberg** and victory was won. This model is currently part of the program of all nuclear physics courses taught at the university level. It was improved by **Aage Bohr**, **Ben Mottelson** and **Leo Rainwater**, who would receive the Nobel Prize in Physics twelve years later. Maria Goeppert was born in High Silesia. Still very young, she followed her parents to Göttingen where she studied, under the supervision of private tutors, to present her "Abitur" in Hanover in 1924. Enrolling at the University of Göttingen to study mathematics, she was quickly attracted towards physics and mainly toward quantum mechanics, then a very new subject, but held in particular honour at Göttingen. There, she was guided by Born and presented her Ph.D. in front of a prestigious jury including, in addition to Born, two other Nobel Prize laureates: **James Franck** in physics and Adolf Windaus in chemistry (1928). A brilliant physicist, also called the "Beauty of Göttingen", she married J. E. Mayer, an American student working under Franck's supervision. Going to the United States in 1930, she worked in the physics department of Johns Hopkins University (in a voluntary job: nothing more could be offered to a "lady" physicist!) in the group of her husband, a physical chemist. In 1939 she first joined Columbia and then **Urey**'s (1939 Nobel Prize for Chemistry for disco- very of deuterium) group. In 1960 she (in physics) and her husband (in chemistry) were appointed Professors at the University of California in La Jolla. A good skier in her youth, passionate about gardening, she was also inte- rested in archaeology.

Maria Goeppert-Mayer

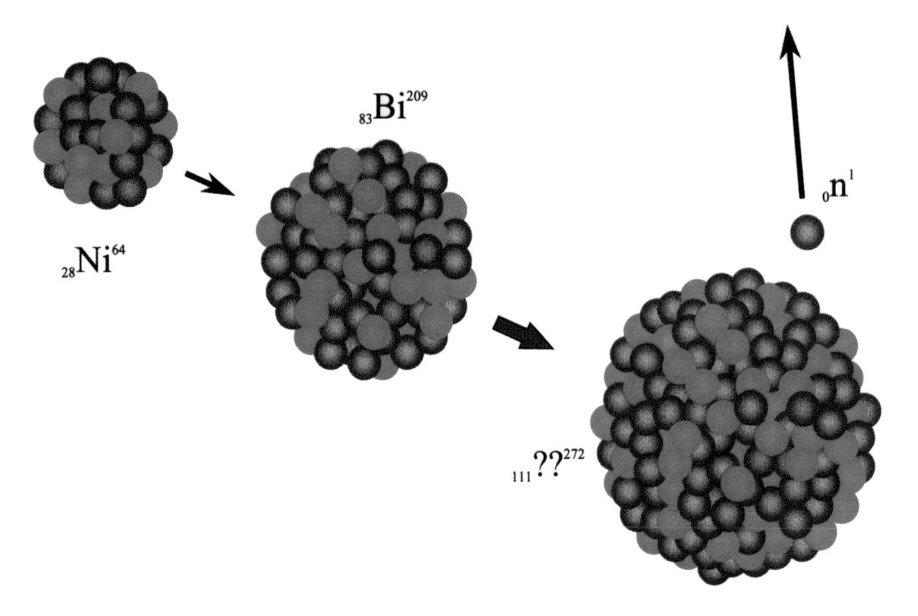

The creation of super heavy atoms. The bombardment of an atom of ^{209}Bi with ^{64}Ni ions can give rise, by capture (followed by an emission of neutron), to a nucleus of atomic weight 272. This nucleus contains 111 protons. It is only one step towards the search for still heavier nuclei (but of exceptionally high stability) which should contain 126 protons.

Jensen, Johannes Hans Daniel (Hamburg, Germany, June 25, 1907 - Heidelberg, Germany, February 11, 1973). Joint winner of the Nobel Prize in Physics with **Maria Goeppert-Mayer**, Johannes Jensen with his colleagues Suess and Haxel from Hamburg developed a shell model of the atomic nucleus, the description of which will be found under Goeppert's name. Before this period the only description of the atomic nucleus was the liquid drop model. Several ideas are common to these two models: the (spherical) form of the nucleus, the tendency to associate an equivalent number of protons and neutrons to ensure symmetry (at least as long as the Coulomb repulsion of too great a number of protons does not lead to rupture) and the association of the protons and neutrons by pairs. The characteristic of the new shell model provides an explanation of the remarkable stability associated with various typical numbers of protons or neutrons: 2, 8, 20, 50, 82 and 126.

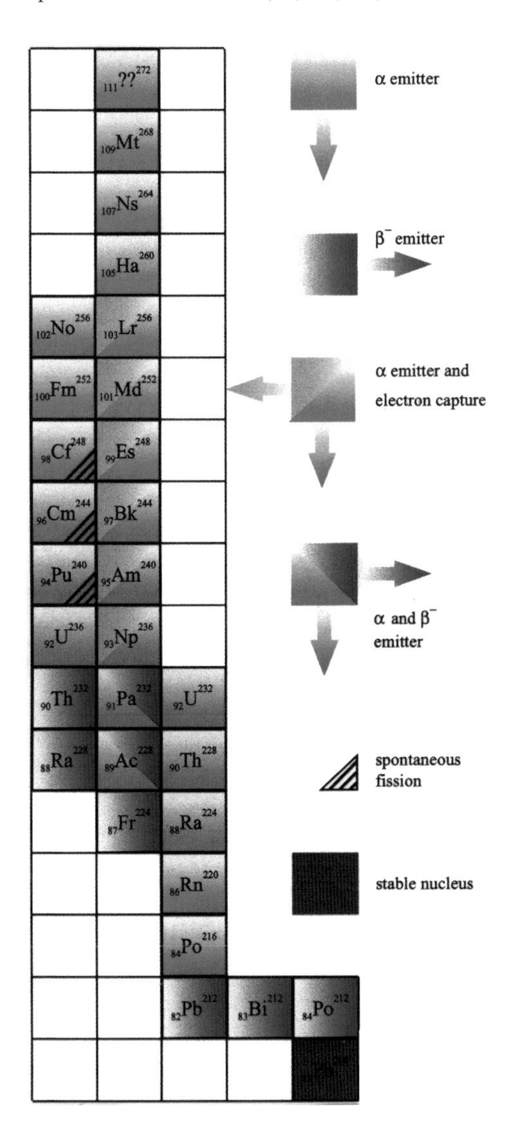

The identification of the disintegration chain of the nucleus of atomic weight 272 towards other lighter nuclei allows, by studying the spectroscopy of the emitted radiations, the guarantee of the reality of the production of this nucleus.

The success of this model today governs the research paths on the fusion of new nuclei in order to associate (in a metastable form) 126 protons, 43 more than in the last stable nucleus existing on earth: bismuth. These nuclei would certainly be unstable but definitely more stable than their neighbours. Jensen studied at the University of Freiburg, Germany, but he received his Ph.D from the university of his native town, Hamburg, in 1932. He stayed there till 1941 and then joined the Technische Hochschule of Hannover. He was appointed Professor in Heidelberg in 1949. He was invited to teach for short periods at Princeton, at the University of California, at the University of Indiana, at the California Institute of Technology and at the La Jolla University where Maria Goeppert was also working. **Eugène Paul Wigner** was also awarded the Nobel Prize for Physics in the same year, not for having participated in the discoveries of Goeppert and Jensen, but for his general work in quantum mechanics.

Johannes H. Jensen

1964

Townes, Charles Hard (Greenville, South Carolina, July 28, 1915 -). The electronic tubes, which equipped oscillators until the 1950s, imposed limits in the frequency at which they could work. Charles Townes therefore turned toward the exploitation of other devices. Combining the laws of distribution of energy in a disordered system established by Ludwig Boltzmann, **Albert Einstein's** concept of the emission of radiation first elaborated by **Max Planck** and the rules of selection of quantum mechanics (indicating in which conditions radioactive transitions are permitted), Townes, an American experimentalist, developed a system called microwave amplification by stimulated emission of radiations ("MASER") in 1953. Its extension into the range of visible radiation would later give the "laser". **Nicolaas Bloembergen** and **Arthur Schawlow** would receive the Nobel Prize, in 1981, for improvements of the laser developed the first time by Theodore Maiman in 1960, thus several years before the award of the Nobel Prize to Townes. Einstein had proposed, in 1917, that if atoms were in an excited state and bathed in a flow of radiation having the same frequency as the one they could emit while being unexcited, they would be stimulated to pass again from the excited state toward the fundamental state by emitting, in phase, a radiation of equal frequency. Townes used the NH_3 molecule to realize his experiment. The maser had immediately, and under the impulse of Townes himself, important applications in communications by radio and television and in the construction of highly accurate clocks, known today as atomic clocks. These were initially conceived by Townes to check proposals of Einstein's special relativity theory. Townes carried out his construction, based on the use of rubis, at Columbia University, where he was brought by **Isidor Rabi**. Born in Greenville, South Carolina, United States, Townes simultaneously followed courses in science and in foreign languages. He received his master's degree from the University of Duke in 1937 and his doctorate from the California Institute of Technology in 1939. He then became a member at Bell Telephone Research Laboratory from 1939 to 1947. He worked during World War II in designing radar and navigation control systems. He entered the University of Columbia in 1948 and was named professor in 1950. He then joined the Massachusetts Institute of Technology in Boston in 1967 and the University of California in 1967. In 1968, along with other colleagues at Berkeley, he participated in the detection of nitrogen and water molecules in space. This work was continued by his pupil

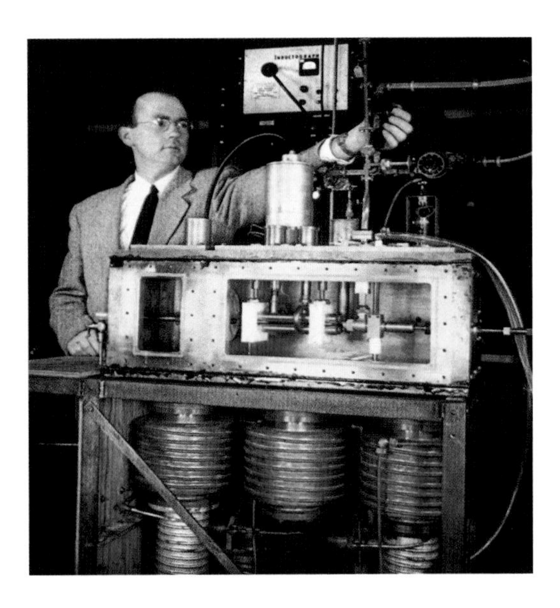

Charles H. Townes

Arno Penzias. This fundamental study is of primary importance for knowledge of interstellar clouds. Just like **Basov**, Townes was involved in politics but as an independent thinker. He also had excellent knowledge of Latin and Greek. He was passionate about walking and skiing. Although a well-rounded person, he was, above all, a scientist because he liked science: "In scientific discovery there is an enormous emotional experience that can be compared with a religious experience, a revelation to some extent".

Basov, Nikolai Gennediyevich (Voronezh, Russia, December 14, 1922- Russia, July 2, 2001). The use of atomic oscillators was a preferred research subject in the

Nikolai G. Basov

1950s. Nikolai Basov settled down to this subject at the Lebedev Institute of Moscow, in collaboration with **Alexander Prokhorov**. **Charles Townes** did the same, independently, in the United States. All three tried to use the property that several atoms are able to remain in an unstable state for a rather long lifetime (called a metastable state) before emitting radiation in an organized

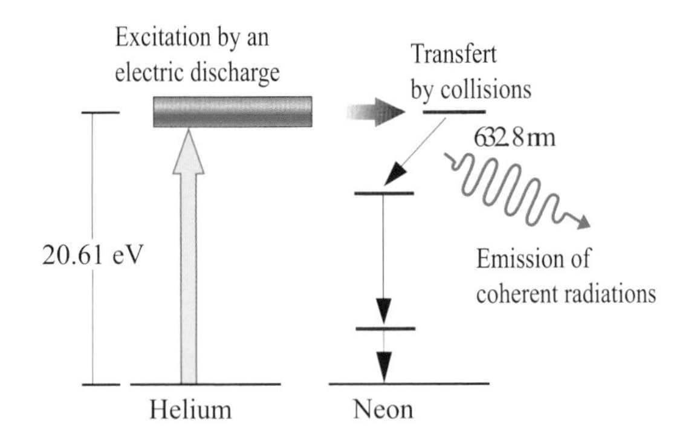

Optical pumping. An electron beam carries He atoms in their excited state to 20.61 eV. A transfer of energy through collisions of an excited He atom towards a neon atom allows the latter to emit radiations of a wavelength of 632.8 nm (red color) in phase with those of the same wavelength, already present in the environment.

way to ensure their stimulated return towards a stable state. It is a radiation of identical frequency to that which allows the inverse transition that stimulates the emission, hence the name MASER or LASER (Microwave (or Light) Amplification by Stimulated Emission of Radiation). In 1958 Basov proposed that a laser could work in the same way by using the two states of a semiconductor, by raising the electrons to the appropriate excited state by an electronic or photonic excitation. Later developments, leading to extremely significant intermittent powers of emission, allowed Basov to induce thermonuclear reactions (of fusion), from 1968. Basov joined the Army after finishing his secondary education. He obtained his graduate diploma at the Physical Engineering Institute of Moscow and joined the Lebedev Institute in 1948. He held various positions there after defending his doctorate in 1956. He became Director of the Institute in 1973. Entering politics in 1951, he was Member of Parliament (Deputy) of the Supreme Soviet in 1974 and took part in various international councils preaching world peace. The author of numerous publications, Basov was also the editor of foremost scientific journals and of popular books.

Richard P. Feynman

Prokhorov, Alexander Mikhailovich (Atherton, Australia, July 11, 1916 - January 2, 2002). In collaboration with his colleague **Nikolai Basov** of the Lebedev Institute in Moscow, Alexander Prokhorov, also a Soviet physicist, studied the spectroscopy of gases in the area of microwaves with the aim of creating precise standards of time intended for navigation systems. Around 1965 they developed an amplification system of electromagnetic waves, known under the names of maser and then laser, at the same time as **Charles Townes** in the United States. The research orientations of both Russians remained centred on quantum electronics. They jointly received the Lenin Prize in 1959 for this work. Prokhorov graduated from the University of Leningrad in 1939 before serving in the Red Army during the Second World War. He then joined the Lebedev Institute, climbing through all the grades until becoming Director in 1968. He taught in Delhi for one year and in Bucharest for another year. He was made an honorary member of the American Academy of Sciences and Arts. In 1970 he received the title of Hero of Socialist Labour.

1965

Feynman, Richard Phillips (New York City, New York, May 11, 1918 - Los Angeles, California, February 15, 1988). What image will the renowned Richard Feynman leave in the history of physics? That of a brilliant electrodynamics specialist or of a talented and original professor?

(Addison Westley Publishers)

The Nobel committee honored him for his contribution to the theory of quantum electrodynamics (the discipline which describes fundamental interactions between charged particles and photons) at the same time as **Julian Schwinger** and **Sin-itiro Tomonaga**, who worked independently in the same field. At Cornell and at the Institute of Technology of the University of California, Feynman tackled, in a personal and completely original way, the reconstruction of the way in which quantum mechanics can explain electromagnetic phenomena. The starting point for his thinking is the completely original description of what the electrostatic force (which varies, with the distance r, according to a law in $1/r^2$ and tends therefore toward an infinite value in $r = 0$) becomes at the "point" where the elementary charge of an electron is found. Freeman Dyson, a colleague of Feynman at Cornell demonstrated the concordance of the views of Feynman,

Schwinger and Tomonaga. He reports that Feynman could not, at that time, "understand quantum mechanics and that he had to reinvent it." In Feynman's theoretical approach, the probability that an event will occur (as was described by **Niels Bohr**) can be calculated by making a sum of all the paths, which can lead to this event. In 1949, in a publication entitled "Space-time approach to quantum electrodynamics," he proposed a method of calculation based on a set of evocative graphs known since then as the

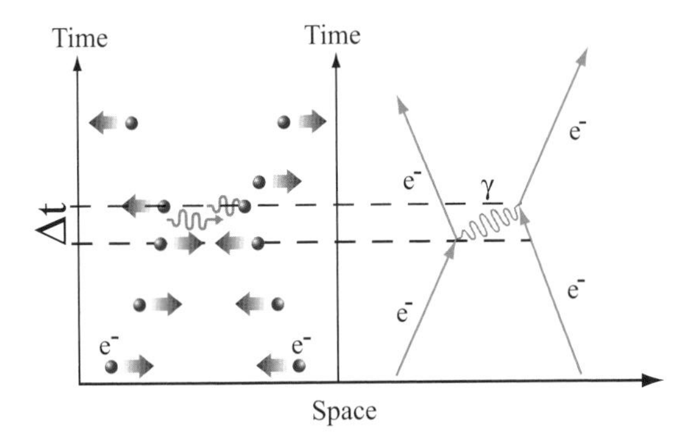

Feynman's diagram. During the approach of two electrons, the repulsive electrostatic interaction gradually increases and leads to a mutual repulsion. The modern explanation brings into play the exchange of electromagnetic radiation.

Feynman diagrams. His role was again decisive in the outcome of **Murray Gell-Mann's** work on weak nuclear interaction but also those of **t'Hooft** and **Veltman**. Besides this original work, he wrote a fundamental physics treatise: the "Feynman Lectures on Physics", intended for university students during their first two years at Cal Tech, a masterpiece with the dimensions of its author: grandiose and original. Translated into ten lan-

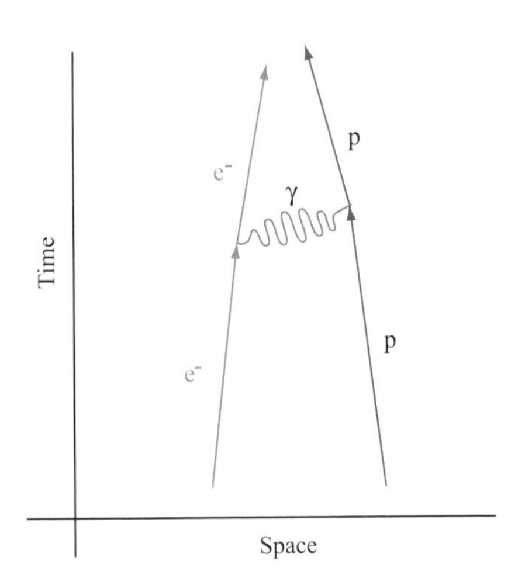

Feynman's diagram of the electron - proton attractive force.

guages, this course has a special place in the library of every physicist. During his life he was also an admirable professor of specialized courses. The appreciation of his students could reach adoration. In 1943, Feynman was

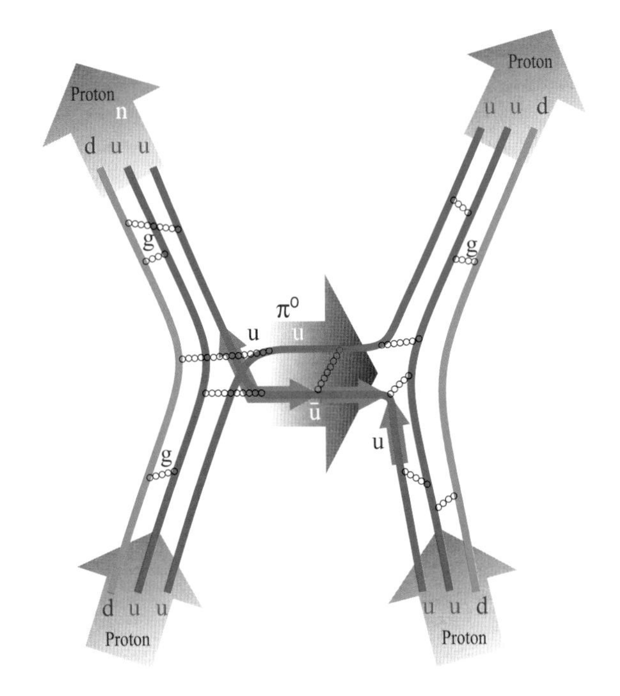

A modern view of the Feynman's diagram for proton - proton interaction taking into account the quarks composition of protons (Gell-Mann). The exchange of a pion π^0 (explained by Yukawa) is equivalent to an association of a quark u with its antiparticle $\bar{u}$.

called to Los Alamos to be part of the theoretical team, directed by **Bethe**, associated with the Manhattan Project. In 1955 he proposed a new model for the structure of liquid helium: a problem already tackled by **Lev Landau**. In 1968 he opened new ways of using "his" diagrams for the study of strong nuclear interactions. We also owe him

The participants in the Solvay Council of Physics of 1961. Principal subject: Quantum Mechanics. From left to right and seated: **S. Tomonaga**, W. Heitler, Y. Nambu, **N. Bohr,** F. Perrin, J. Oppenheimer, **W.L. Bragg**, C. Moller, C. Gorter, **H. Yukawa**, R. Peierls, **H. Bethe; I. Prigogine**, A. Pais, **A. Salam**, **W. Heisenberg**, F. Dyson, **R. Feynman**, L. Rosenfeld, **P. Dirac**, L. Van Hove, O. Klein; A. Whightman, S. Mandelstam, G. Chew, M.Goldberger, G.Wick, **M. Gell-Mann**, G.Källen, **E. Wigner**, G. Wentzel, **J. Schwinger**, M. Cini (in bold: the Nobel laureates).

the concept of the "parton" as an attempt to explain the internal structure of protons. These "partons" would be called quarks a few years later, by Gell-Mann. When he learned that the Nobel Prize in Physics had been awarded to him, Feynman showed himself initially hostile, calling it "a new error of Alfred Nobel" but finally let himself be convinced to accept this honor. He took part in the Solvay Council of Physics in Brussels in 1961. In the 1980s, having become very popular, he agreed to make a series of television programs and to publish a very popular book under the title: "Surely you are joking, Mr. Feynman". The French edition was published in 1985. In 1986, he again became well known as a member of the presidential commission set up to inquire into the accident which caused the in-flight explosion of the Challenger space shuttle. He identified the cause of the accident: the rupture of a polymer seal in a booster rocket. The reason for the loss of elasticity in the joint was the frost that occurred during the night, which preceded the launch. During the commission meeting he made a spectacular demonstration of this by plunging a sample of this type of material into a glass filled with ice. He also was very much against the idea of the nuclear winter, predicted by the astronomer Sagan in the event of an explosion of a modest-sized hydrogen bomb. Several chapters of Feynman's writings formed the basis of a film scenario (Infinity) which came out in October 1996 in the U.S.A. Infinity is an intimate portrait, a true love story, which unites the brilliant scientist Feynman and the lady of his dreams Arline Greenbaum, a pretty and talented artist, who gave a new dimension to the destiny of the physicist by opening his eyes to art, music and the beauty of nature. In 1942, in spite of family reticence, Richard married Arline, who was ill with tuberculosis. She died in a sanatorium in Albuquerque in 1945 during the time her husband was part of Bethe's team, in charge of the theory of the A bomb, in Los Alamos.

The explosion of the space shuttle was explained by R. Feynman: the rupture in flight of a joint which lacked its elasticity because of frost during the night before the launch.

Schwinger, Julian Seymour (New York, New York, February 2, 1918 - August 16, 1994). Theoretician physicist. Talented in mathematics from a very young age, Schwinger is the father of the fusion of the electromagnetism theory with quantum mechanics, the solution of a problem already dreamed about by **Paul Dirac**, **Wolfgang Pauli** and **Erwin Schrödinger**, and which also contained **Albert Einstein's** relativistic ideas. Simultaneously, but independently, **Richard Feynman** and **Sin-itiro Tomonaga** were involved in solving the same problem. **Willis Lamb** had already shown experimentally, in 1947, that the fundamental state of the hydrogen atom was divided into two sub-states, of very slightly different energies: which Dirac's fundamental theory did not foresee. Schwinger explained the reason for this by restoring quantum electrodynamics to its fundamental place, after the doubts caused by Lamb's experiments. In addition, Schwinger (by squeezing it out of a very complex mathematical development) justified the exact value of the magnetic moment of the electron, also determined by Lamb but slightly higher than the expected theoretical value. On **Isidor Rabi's** initiative Schwinger, as a young man, joined Columbia University (bachelor's degree at 17 years old, and doctor at 21). He was appointed Professor at Harvard when he was only 29 years old. From 1940 to 1942 he worked under Oppenheimer at Berkeley and conducted research at the Massachusetts Institute of Technology during World War II. He was present at the Solvay Council of Physics in Brussels in 1961. A loner and a relentless worker, he sometimes, during the night, completed a demonstration left on the board by his young associates. A theoretical genius, Schwinger also involved himself in the experimental physics of electron accelerators. He is also known for his pioneering work in distinguishing between the neutrinos connected with the electron and the muon (a problem which would receive an experimental confirmation from **Leon Lederman**, **Melvin Schwartz** and **Jack Steinberger**) and for the study of the magnetic moments of charged particles. Before receiving the Nobel Prize, Julian Schwinger had already been the winner of the first Einstein Prize (1951) and of the American Medal for Science (1964).

Tomonaga, Sin-itiro (Kyoto, Japan, March 31, 1906 - Tokyo, Japan, July 7, 1979). Deprived, during the 1940s, of contacts with American science, Sin-itiro Tomonaga had undertaken, on his own in Japan, to solve the difficult problem of the behavior of electromagnetic forces at very short distances from a charge: classical electrodynamics could only postulate an infinite value of the field at the place of a point charge. With a "renormalization" of the concept of charge and mass at short distances, he was able to reconcile classical electromagnetism with quantum mechanics and, at the same time, explain the tiny differences between theory and experimental values. Independently in the U.S.A., **Richard Feynman** and **Julian Schwinger** tackled the same problem. Western physicists discovered Tomonaga's work in 1948 in an article published in English in Kyoto entitled "An invariant and relativistic formulation of the quantum fields theory". Tomonaga studied in Kyoto where his father was teaching philosophy. In 1929, he worked as **Hideki Yukawa's** assistant before rejoining Yoshio Nishina's group in Tokyo for undertaking the study of quantum electrodynamics. In 1934 he was the author of works on the production of electron - positron pairs by photons. From 1937 until 1939 he was influenced by the work of **Werner Heisenberg** whom he had met during a stay in Leipzig. In 1941 he was appointed Professor in Tokyo where he conducted the major part of his work in quantum electrodynamics. In 1949, he studied American work in the same research field as a result of an invitation, sent by J. Robert Oppenheimer, to stay for one year in Princeton. He directed the Nuclear Study Center of the University of Kyoto and was named Chairman of this establishment from 1956 to 1962. He held the office of President of the Japan Science Council from 1956 to 1962. He took part in the Solvay Council of Physics in Brussels in 1961.

1966

Kastler, Alfred (Guebwiller, France, May 3, 1902 - Bandol, France, January 7, 1984). A French experimental research physicist, Alfred Kastler received the Nobel Prize in Physics for the meticulous care that he brought to optical experiments. France had waited 37 years for a successor to **Louis de Broglie**. The phenomena updated by Kastler are understandable only to initiates. This was the opinion even of a representative of the Stockholm Academy of Sciences. The New York Times headline which announced, on November 6, 1956, the attribution of the Nobel Prize said meaningfully on the subject: "Incomprehensible Science!" Combining light waves and polarised microwaves (the double resonance method), Kastler studied the population distributions of electrons in various quantum sub-states of atoms. The results obtained led to the construction of very precise references of frequencies, of clocks (known as atomic) which malfunction by only one second in 10,000 years and of measuring devices of very weak magnetic fields. This work should be compared with that of Charles Townes. The concept of optical pumping, so useful in the physics of lasers, is a designation invented by Kastler. Born in Guebwiller when Alsace was still part of Germany, Kastler was at first a student in Colmar before benefitting from a special procedure to enter the Ecole Normale Superieure in 1921. He left his studies for five years, between the first and the second cycles, to teach successively in Mulhouse, Colmar and Bordeaux. It was in Bordeaux that he obtained his degree and his Ph. D. in 1936. Then he taught at the Universities of Clermont-Ferrand and Bordeaux. George Bruhat summoned him to the Ecole Normale Superieure in Paris to lead a team of researchers in **Raman** spectroscopy. Reserved and modest, he was deeply interested in the work of his numerous colleagues. Gentle, but obstinate and firm on matters of principle, he publicly came out against the atomic bomb, against the Algerian and Vietnam Wars and against all forms of fascism. A convinced European, in 1971 he published a collection of poems in German; including, among others, "Europa, my homeland", "German Songs of a French European."

Alfred Kastler

Hans Bethe... oldest of the world's physicists.

1967

Bethe, Hans Albrecht (Strasbourg, Germany, now France, July 2, 1906 -). It took almost 30 years for the theory on the energy production in the sun and stars, via nuclear reactions, to be recognized by the Nobel Committee. Hans Bethe received this prize for two of his major research fields: nuclear reactions and their contribution to energy generation in stars. On one hand Bethe calculated, in collaboration with C.L. Critchfield, how the collisions between protons can give rise to the creation of helium, and how, on the other hand, the carbon nucleus takes part as a "nuclear catalyst" in another mode of helium production, while following a CNO (carbon, nitrogen, oxygen) cycle. The first reaction predominates in stars, like the sun, where the temperature does not exceed 107 K; the second takes over in hotter stars. Bethe did not seek an explanation for the way in which the first carbon nuclei were created to initiate this reaction and it was Hoyle, in 1950, who was able to demonstrate this. Bethe's concern resided in the knowledge of the processes releasing energy by fusion. This discovery was the result of the insistence of Edward Teller, an Hungarian emigrant and professor in Washington at that time. He managed to convince Bethe, considered the great specialist in nuclear reactions during the 1930s, to become interested in the process of nuclei production in stars so that he could make a synthesis of it for a scientific meeting on astrophysics organized in 1938. This nuclear astrophysics work would be continued by **William Fowler** who would show the

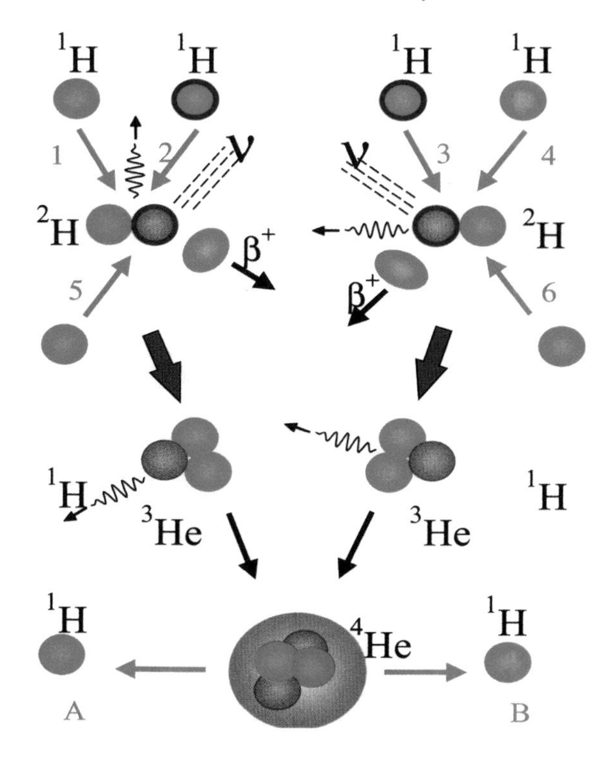

Cosmic synthesis of helium (fundamental mode). Two pairs of protons (**1** and **2**) and (**3** and **4**) collapse to give each of them a deuterium (^{2}H) nucleus, protons **5** and **6** interact with these two deuterium nuclei to form two ^{3}He nuclei which themselves, when colliding, give rise to a ^{4}He nucleus. Two positrons are also produced as well as two neutrinos. At the end of this process, two of the six protons A and B are restored.

Carbon cycle. Clockwise: a proton (**1**) interacts with ^{12}C, to give a ^{13}N nucleus, which, unstable, is transformed into ^{13}C. A second proton (**2**) transmutes the ^{13}C into ^{14}N. A third proton (**3**) transforms it in turn into unstable ^{15}O. The latter is then transformed into ^{15}N, which, absorbing a fourth proton (**4**), becomes ^{16}O, so unstable that it explodes into ^{4}He and ^{12}C. Four protons thus gave rise to a ^{4}He nucleus. The original ^{12}C served as "catalyst" and is found again at the end of the cycle. As in the principal creation mode of the ^{4}He, two positrons are produced (or two atomic electrons are captured) at the same time as two neutrinos. Several γ- rays are also emitted during intermediate stages.

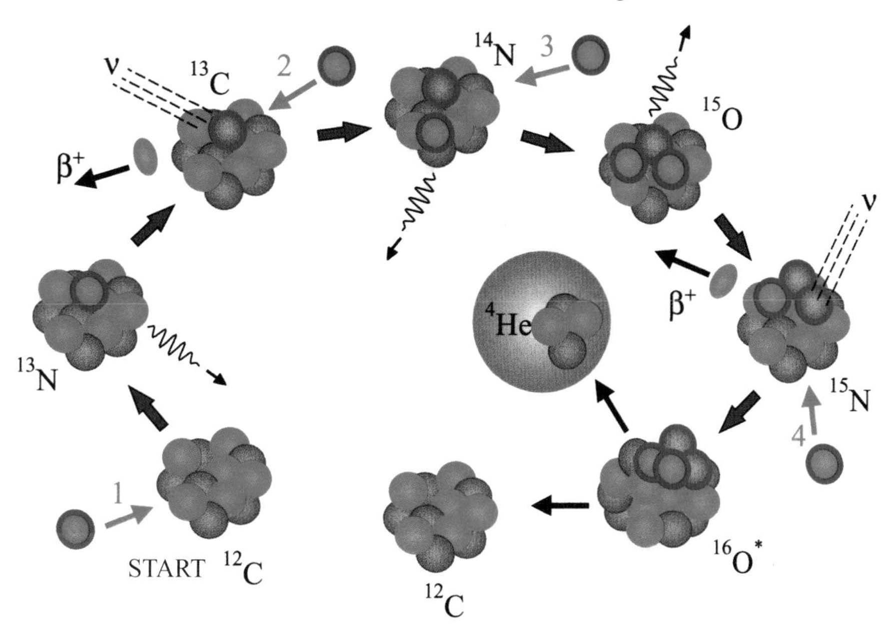

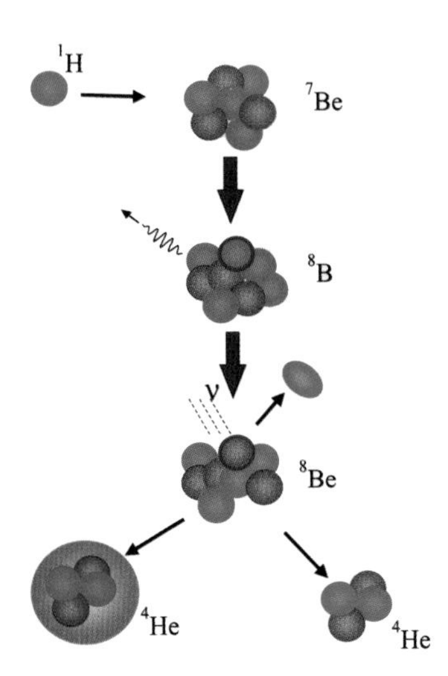

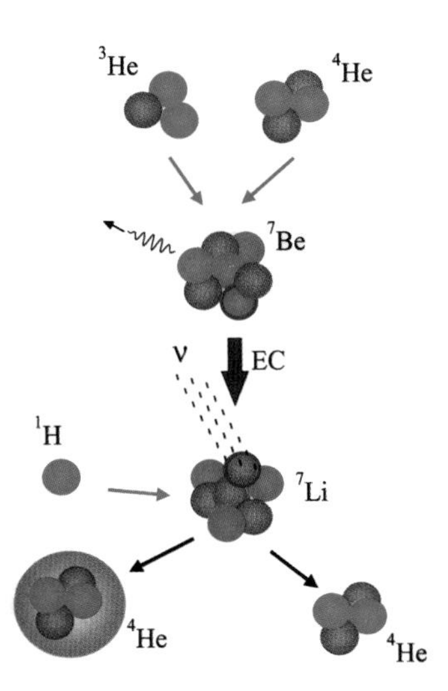

Two secondary modes of formation of He^4

manner by which the heavier elements were synthesized. The scientific output of Bethe is considerable. His fundamental contributions relate to nuclear reactions and their use in the description of nuclei of course, but also to the theory of radiation, the production of electron-positron pairs, the calculation of the displacement of the atomic lines highlighted by **Willis Lamb**, the theory of the order and disorder in alloys. He is also known by chemists as the father of the crystal field theory. One of his great qualities was to find simple and general explanations to very complex problems. Born in Strasbourg (a German city at that time), Bethe studied physics under the direction of Arnold Sommerfeld in Munich, where he gained his Ph.D. He first taught in Frankfort, Stuttgart and Munich. He joined the **Ernest Rutherford** (1908 Nobel Prize for Chemistry) group in Cambridge in 1930 and 1931 and then, that of **Enrico Fermi** in Rome. Later on, he returned to Tübingen and taught physics for two years. Bethe immigrated to the United States in 1935 and became Head of the Department of Physics at Cornell in 1937. He held this post for 38 years until his retirement in 1975. He attended the Solvay Council of Physics in Brussels in 1961. Indeed, he celebrated his 80th birthday by beginning to study the gauge theory, currently one of most promising in physics. During the war, he provided a contribution to the development of radar and was also in charge of the theoretical physics group engaged in the military application program of nuclear fission: the Manhattan Project launched in Los Alamos. His collaborators there included **Richard Feynman** and **Robert Wilson**, both of whom he invited to follow him to Cornell after the war. He clashed openly with Teller (father of the hydrogen bomb) concerning the need to build this bomb. Their contradictory positions were the object of divergent declarations lasting more than a decade. Bethe founded the Federation of American Scientists, with the help of which he fought against the spread of knowledge related to weapon manufacturing and advocated the peaceful use of atomic energy. Bethe was not completely satisfied with his status of outstanding scientist. He was also politically and socially active, and this enabled him to take his own stand in the major problems of our time and to make other scientists conscious of their responsibility. In the New York Book Review of 1992, Bethe made an official appeal to the Soviet Union and the United States, in which he demanded that they should do away with a good deal of their nuclear missiles. Bethe was a highly educated person, he made a great contribution to the development of science and possessed encyclopedic knowledge. He was fond of walking and enjoyed mountain scenery. He may be considered as the dean of American physicists, as well as for physicists all over the world. According to Silvan Schweler, one of Bethe's biographers, he was an incomparable physicist, one of the most brilliant and admired researchers of the 20th century. Silvan Schweler also asserted Bethe's honesty and integrity: "I have never met anyone, except, perhaps, for Edward Teller, who spoke badly of Hans Bethe."

1968

Alvarez, Luis Walter (San Francisco, California, June 13, 1911 - Berkeley, California, September 1, 1988). It was during lunch in the company of **Donald Glaser**, at a scientific congress organized in 1953 in Washington, that Luis Alvarez had the idea of associating a bubble chamber with a semi-automatic mechanism (controlled by computer means) for the recognition of the traces left by particles resulting from high - energy nuclear interactions. The

Luis W. Alvarez

6.2 billion electron volt accelerator at Berkeley, put into operation in 1953, produced a harvest of "strange particles." This name was given to them on account of their very long lifetime (a tenth of nanosecond!) in relation to the value that we could, a priori, allocate to them (namely millions of millions times shorter). The explanation of this lifetime would form the basis for the Nobel Prize that **Murray Gell-Mann** received one year after Alvarez. The electronic counters of the time were not helpful in the study of these particles. **Georges Charpak's** wire chambers didn't exist yet. **Charles Wilson's** cloud chambers could not be used to record the totality of the very long pathway of these particles due to the too low density of gases. The principal disadvantage of the **Charles Powell** photographic plates was the impossibility of allowing the visualization of the products of the reactions initiated by neutral particles. The arrival of the device developed by Alvarez allowed the identification of numerous resonance states, states connected with the production of a whole new particle zoo (a good hundred are currently known). Under Alvarez's leadership the conception of the Glaser bubble chambers grew rapidly from a length of 10 cm in 1952 to more than 2 meters in 1959. They are currently completely automated and coupled with powerful magnetic fields that curve the pathways of the charged particles according to the product of their mass by their speed (namely their momentum). They occupy an essential place in the experimental study of subatomic particles. Trained at the University of Chicago under the supervision of **Arthur Compton**, Alvarez received his graduate diploma in 1934 and his PhD in 1936. An expert in the construction of Geiger's counter (which he installed on the roof of a hotel in Mexico City), Alvarez was interested initially in cosmic radiation. He showed, in particular, that cosmic rays contained a high proportion of protons. He also explained why the isotope of mass 3 of helium was stable while the isotope of mass 3 of hydrogen (tritium) was unstable. Then he joined the University of Berkeley and took part in the conception of a linear accelerator of protons using resonant cavities. He collaborated in the construction of a machine of the same type for the Massachusetts Institute of Technology (Boston) in 1941. Currently, this type of accelerator serves particularly as an injector for proton synchrotrons. In 1943, for six months, Alvarez was a collaborator of **Enrico Fermi** in the construction of the first nuclear reactor. He also was the first to observe that besides the emission of positrons by nuclei, it was possible that the same transmutation would take place by the capture of an atomic electron. In 1939, with **Felix Bloch**, he made the measurement of the magnetic moment of the neutron. In 1952, during a visit to the site of Giza (Egypt), Alvarez had the idea to use muons, particles that are particularly penetrating in materials, to probe the Chephren pyramid in order to discover possible cavities. His project took six years before receiving the authorization to carry out the experiment.The first results seemed to indicate the pres-

Chephren's pyramid that Alvarez tried to probe, without success, by using the muons of cosmic radiation, to discover possible cavities.

The main vessel of the BEBC

The piston of the BEBC

Bubble chamber. This large dimension model bears the name of BEBC (Big European Bubble Chamber). It has a diameter of 3.7m and, between 1973 and 1984, recorded more than 6 million photographs of particle pathway. It is now exhibited in the gardens of CERN in Geneva. Below : the piston which activated the BEBE for 13 millions times in 11 years.

ence of such cavities, but a more meticulous examination of the data would invalidate this conclusion. Not content with this digression from physics towards archaeology, he was again involved, in 1977, in company with his son Walter, a geologist, in an interpretation of the reason why dinosaurs disappeared abruptly from our planet 65 million years ago. A concentration of iridium, three hundred times larger than normal, in a layer of clay, formed over approximately a thousand years and locked between two rock layers, one from the Cretaceous period and the other from the Tertiary period, was first considered as the cause of their disappearance, owing to supernova activity which could have been the source of this unusual iridium content. The absence of uranium 244 (which should have been simultaneously present) in this clay forced him to give up this interpretation. Another hypothesis adopted by him in 1980 involved the fall of an iridium rich meteorite, about 10 km in diameter, which would have raised such a quantity of dust that the sky would have been darkened for several years, preventing the reproduction of vegetation and therefore causing the brutal extinction of the herbivorous giants. Today, this interpretation still has both its followers and its opponents! Music, flying and golf were among his favorite recreations.

The distribution of quarks in mesons with respect to the property of strangeness.

1969

Gell-Mann, Murray (New York City, New York, September 15, 1929 -). Murray Gell-Mann, a theorist, shares with Richard Feynman (in the opinion of a majority of physicists worldwide) a position similar to the one occupied by **Albert Einstein**. Gell-Mann's principal contribution concerns a classification of elementary particles based on a group theory created by the Norwegian mathematician Sophus Lie. This way of providing order in the jungle of particles gives Gell-Mann a position in physics similar to that which Mendeleev holds for the periodic classification of chemical elements. It was at Cal Tech, during the 1950s, that Gell-Mann felt the need for this classification. Collisions in high energy particle accelerators (in particular, the Bevatron at Berkeley near Cal Tech) of species as simple as nuclei of hydrogen produce entities (like kaons, particles of intermediate mass between that of the proton and the electron and therefore part of the meson group) having a lifetime (10^{-10} second) distinctly longer than that anticipated by the theories of this period (10^{-23} second). Gell-Mann suspected that a new property, that he called "strangeness", must be attached to the particles

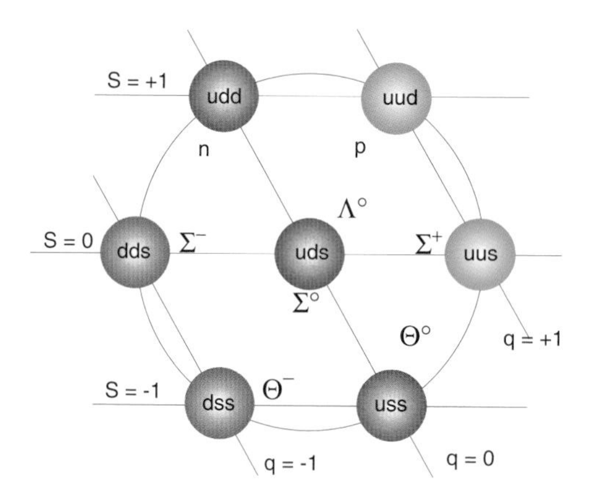

The distribution of quarks in hadrons with respect to the property of strangeness.

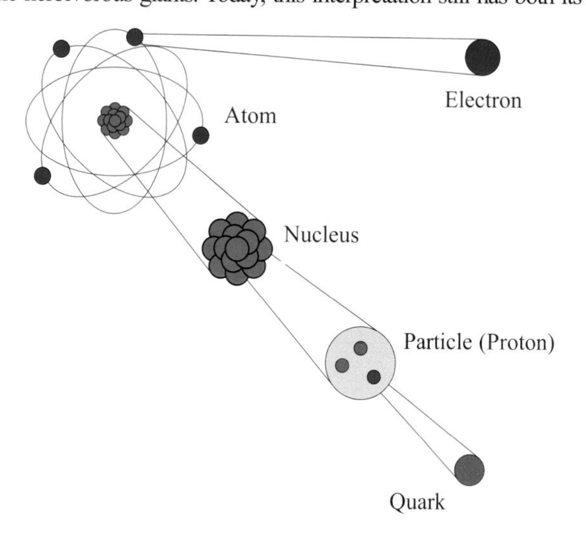

Sub-nuclear model. The atom is composed of a nucleus and peripheral electrons (as explained by Rutherford). The nucleus is itself composed of protons and neutrons: baryons. Each of these baryons is itself composed of 3 quarks, namely uud for the proton and udd for the neutron. The quark u has a positive charge being equivalent to 2/3 of the value of the elementary electron charge, the quark d carries a negative charge being equivalent to -1/3 of this elementary charge. Quarks are bound together by several types of gluons.

(hadrons) undergoing the strong nuclear interaction (this force had already been interpreted by the Japanese **Hideki Yukawa**). This peculiarity, S, was subsequently defined as being equivalent to 2Q - B: twice the charge minus once the baryonic number. In 1954, the Japanese K. Nishijima had a very similar idea. In strong interactions, this quantity S should keep a constant global value (law of conservation). In the 1960s, the predictions of Gell - Mann, and independently also of the Israeli Y. Neeman, were confirmed to give rise to an algebra with eightfold SU(3) symmetry. Bearing in mind the particles still missing from this proposed structure, experimental research was undertaken with the aim of bringing these particles to light and, as expected, new hadrons (baryons and mesons) were identified. Gell - Mann also postulated the existence of quarks: particles of fractional charge (1/3 and 2/3) of that of the electron. With the concepts of mass, of electric charge, of spin (quantum number connected with the rotational motion of the particles on themselves), Gell - Mann proposed a new way of organizing the classification of particles from "simple" numbers: assemblies of quarks in triplets give baryons, doublets of quark and antiquark form the mesons. These quarks are an alternative to the "partons" proposed by Feynman. On their side, leptons, muons and neutrinos in particular undergo only weak nuclear interaction and form part of another classification. With the later discoveries of **Sheldon Glashow**, **Leon Lederman** and **Martin Perl**, and the addition of two new characteristics defined as "charm" and "colour", we are probably today very close to a generalised understanding, but probably only until the next revolution. From 1956 Gell - Mann also proposed a model to explain weak nuclear interaction. He was nicknamed by his colleagues "the architect of the standard model". Born of a family of Austrian Jewish emigrants Gell - Mann entered Yale University at age 15 and received his Ph.D. at age 22. After joining the University of Chicago and working there for four years, he moved to the California Institute of Technology in 1955. As early as 1967 he took over the chair of physics created in honour of **Robert Millikan**. In 1966 he provided, on his own and in a way universally recognized as remarkable, a review of the state of knowledge on the new classification of elementary particles during a congress on High Energy Physics in Berkeley. He took part in the Solvay Council of Physics in Brussels in 1961. In the 1960s, he was named an advisor to the Pentagon, to draft an agreement between the United States and the Soviet Union for the reduction of the number of nuclear missiles. A man of culture, competent in 15 languages, Gell - Mann was also interested in "complex systems" and founded an institute in Santa Fe (New Mexico) to study them. He was also interested in the history of linguistics, in natural history, in environmental protection, in archaeology, psychology, in creative thought and in the stability of political systems.

1970

Alfven, Hannes Olof Gösta (Norrkoeping, Sweden, May 30, 1908 - Stockholm, Sweden, 1995). Hannes Alfven was a magnetohydrodynamician, convinced that the major fundamental principles of theoretical physics had to be verified through experiment. He devoted himself to plasma physics (a very strongly ionized gas having very high conductivity which forms probably 90% of the matter in the universe) and to experiments with large particle accelerators, because those instruments can reproduce, on earth, the conditions of the interactions of particles in the cosmos. In 1942, while studying sunspots, he postulated the existence of waves in plasmas (today known as Alfven's waves). This proposal took more than six years to be agreed by the physicists' community: it was finally accepted during a series of lectures that Alfven gave at the University of Chicago. When **Enrico Fermi** listened to those lectures, he could only conclude "it is obvious" and, in his wake, the scientific community concluded: "of course, it cannot be otherwise." In the 1950s, Alfven published two books: "Cosmical Electrodynamics" and "On the Origin of the Solar System." He applied his results to the description of the pathways of charged particles in the earth's magnetic field and provided the nuclear fusion specialists with methods of calculation of the confinement of plasma. More recently, he called into question the manner of interpreting the "Big Bang", but his proposals (in particular that the universe contains as much matter as anti-matter) met opposition, this time from the cosmologist camp. Born of doctor parents, Alfven studied in Uppsala where he held a teaching position after finishing his doctorate in 1934. He was then a research worker at the Nobel Institute of Physics (1937), and Professor at the Royal Institute of Technology (1940) and at the University of California in San Diego (1967). He was a vigorous supporter of the development of nuclear energy until 1970 when he changed his opinion completely to become a committed anti-nuclear supporter. His position of adviser to the Swedish government is probably not unconnected with the decision taken by Sweden to develop no more

Hannes O. Alfven

Louis E. Néel

nuclear power plants and to provide for the closure of existing ones at the beginning of the 21st century.

Néel, Louis Eugène Félix (Lyon, France, November 22, 1904 - Brive, France, November 17, 2000). At the beginning of the 1930s three behavior-types of materials introduced in a magnetic field had been identified: on one hand those that interact in a mutually independent manner with the external field, namely the paramagnetic, which slightly tighten the lines of the induction field, and the diamagnetic, which slightly spreads them out, and, on the other hand, the ferromagnetic, which show a collective alignment behavior and consequently interact more strongly with the external field. In 1932 Louis Néel showed that, in several crystals, parallel and anti-parallel alignments of elementary magnetic moments completely cancelled their effects. He foresaw a model of this new phenomenon and predicted that the particular order of orientation would disappear at low temperature. He also explained the reasons for the strong magnetic field found in ferrites (insulating oxides) and proposed the name of ferrimagnetism for this phenomenon. The everyday applications of these ferrites are currently numerous in telephony, magnetic tapes, computer memories, high frequency techniques and in the manufacture of a whole range of magnets such as those in microphones and loudspeakers. By his fundamental work Néel also provided valuable information on the history of the earth's magnetic field. He shared the Nobel Prize in physics with Hannes Alfven, himself rewarded for his work in magneto-hydrodynamics. Néel graduated from the École

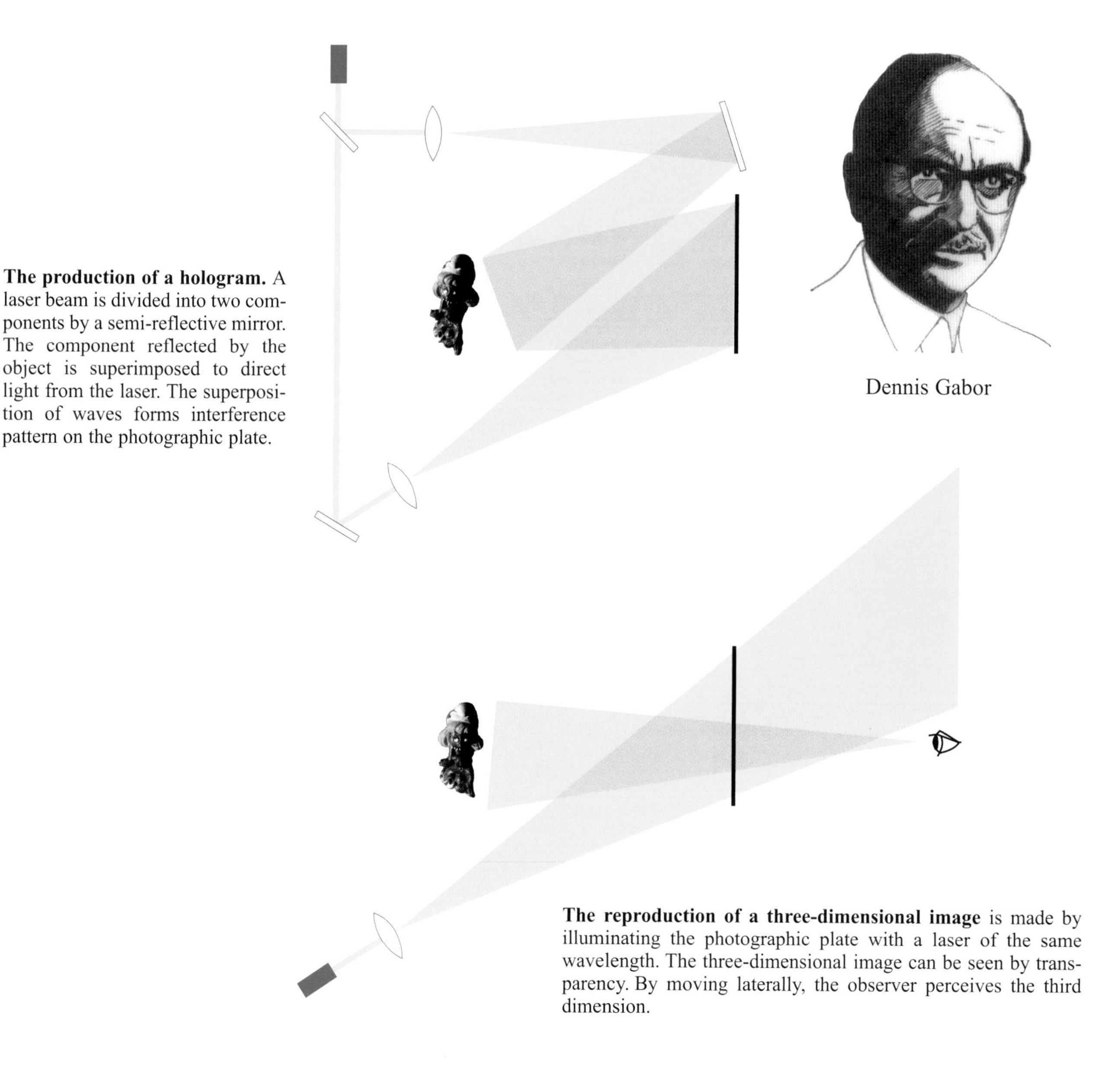

The production of a hologram. A laser beam is divided into two components by a semi-reflective mirror. The component reflected by the object is superimposed to direct light from the laser. The superposition of waves forms interference pattern on the photographic plate.

Dennis Gabor

The reproduction of a three-dimensional image is made by illuminating the photographic plate with a laser of the same wavelength. The three-dimensional image can be seen by transparency. By moving laterally, the observer perceives the third dimension.

Normale Supérieure in 1928 and registered for his doctorate at Strasbourg. He defended his thesis in 1932. In 1937 he became Professor at the Faculty of Sciences of the Univesity of Strasbourg and he was named to the same position at Grenoble in 1945. There he directed the Laboratory of Electrostatics and Metal Physics which was his creation in 1946, the Polytechnic Institute in 1994, and the Center of Nuclear Energy which began as his creation in 1986. Today more than 3000 people work around accelerators and reactors at this center. Néel also contributed to the establishment of a high-flux nuclear reactor at Grenoble (1967) in collaboration with Germany. The choice of the site of the European Synchotron (ERSF), set up in the same area, adds to the importance of this scientific park. Néel directed the CNRS, the French Physics Society, and the International Union for Pure and Applied Physics. He took part in the Solvay Council of Physics in Brussels in 1954.

1971

Gabor, Dennis (Budapest, Hungaria, June 5, 1900 - London, England, February 8,1979). Dennis Gabor, the inventor around 1948 of "three-dimensional photography", would only see the spectacular development of this technique with the arrival of lasers in 1961 and particularly, ten years after that, when these laser sources were more widely available. Originally, Gabor intended to develop a method to improve the resolution of electron microscopes, not by improving the optics of the instrument, but by trying to extract more information from the images taken by the existing instruments. With a view to preparing this device he began his investigations by using the light emitted by a vapour lamp (developed subsequently for street lighting) and passing through a very narrow diaphragm. Contrary to the direct registration of an image giving, on a flat document, a two-dimensional representation of a three-dimensional scene, the holographic technique involves the reconstruction of the third dimension from a photoplate which records not only the intensity of the reflected light by the object, but also the simultaneous information on the amplitude and the phase of the incident wave. The holographic plate contains, on the whole of its surface, information coming from each point of the object or the lighted scene, but also information coming from the light source, without having interacted with the object itself. E. Leith and J. Upatnieks were the first to use laser light for this application. The holographic method is still usable from multiple coherent waveforms, from acoustic waves in particular. Gabor was influenced, in his childhood, by **Gabriel Lippman's** photographic method. He discovered it in the museum of inventions of his birthplace. At that time, as physics was not a discipline which gave access to a career in Hungary, Gabor did engineering studies in Berlin, where he graduated in 1927 with a work on an oscilloscope with cathode ray tubes. During this basic training he attended lectures given by **Albert Einstein**, **Max Planck**, **Walther Nernst** (1920 Nobel Prize for Chemistry) and **Max von Laue** in the nearby Faculty of Sciences. In 1933 he left Hitler's Germany and, after a training period in a company producing gas-discharge lamps, he joined the Imperial College of Science and Technology in London. He continued to express enthusiasm and show experimental skill to crowds of assistants and young researchers until his retirement in 1967. Having become a British citizen, he behaved as an elegant gentleman. During the last ten years of his life he was involved in deep thought on the future of our heavily industrialised society. It is noted that this 1971 Nobel Prize was attributed to Gabor alone. Not until 1982 was there a similar situation: from 1972 to 1981 the Nobel committee shared the prize between two or three winners.

1972

Bardeen, John received his second Nobel Prize for Physics in 1972. Please refer to the description of his complete work in the bibliography of Nobel Prize winners of 1956.

Léon Cooper.

Cooper, Léon N. (New York City, New York, February 28, 1930-). Rewarded, with the Nobel Prize in Physics for his theoretical work on the explanation of the superconductivity phenomenon (a property which, below a given temperature, confers on several materials an infinite electric conductivity), Léon Cooper worked from 1955 to 1957 at the University of Illinois (Urbana-Champaign). Superconductivity had already been observed in 1911 by

Heike Kamerlingh-Onnes. Bardeen had already distinguished himself in 1956 with a first Nobel Prize for physics for a study with **William Shockley** and **Walter Brattain** on semiconductors. The close collaboration of the three 1972 Nobel Prize winners led to a reliable explanation of superconductivity, gathering together a significant quantity of experimental information: (a) the repulsion of a magnet by superconductors known as the Meissner effect, (b) the Ohm law relating the voltage to the electrical current, replaced by another law related to diamagnetism, and (c) the dependence of the superconductive transition temperature with respect to the molecular mass of the species. Their explanation used the following theoretical ingredients: (a) the Wolfgang Pauli principle of exclusion requiring that, even at a temperature close to absolute zero, all the electrons cannot have zero kinetic energy, and (b) the existence of a forbidden energy gap for the electrons in regularly structured materials. Cooper, a specialist in quantum field theory, imagined that two electrons (which normally repell each other because of their electric charges of the same sign), would have, in a crystalline solid, the possibility of forming pairs (today called Cooper pairs) able to move, together, in a well ordered motion under the effect of an electric field, even a very weak one. The BSC theory (initials of the three authors) is today universally accepted. It led other teams to become interested in superconductivity and to research the operating conditions at higher and higher temperatures. The **J. Georg Bednorz** and **Karl Müller** discovery in 1986 is therefore an illustration of superconductivity for the perovskytes. Cooper received his Ph.D. from Columbia in 1954 after having presented work on nuclear physics. After his work at the University of Illinois on BSC theory, he went to Brown University where he has been since 1958. He was also a consultant in industries and educational establishments, including several universities in the United States, Norway, Italy and France.

Schrieffer, John Robert (Oak Park, Illinois, May 31, 1931 -). Associated with John Bardeen and Leon Cooper for his Nobel Prize, John Schrieffer had already chosen, as his Ph.D. subject, superconductivity and especially the mathematical description of the ordered motion of "**Cooper** pairs". The problem was so difficult that he wished resign to carry on this field. He was then convinced by his teacher, **J. Bardeen** to continue for the short time it took to go to Stockholm to receive the 1956 Nobel Prize. Schrieffer accepted for this short time and invented a method of calculating the global wave function, which provided an explanation of all the known experiments connected with superconductivity. A special session of the meeting of the American Physics Society in March 1957 was devoted to the presentation of this almost complete theory. Schrieffer began his university studies at the Massachusetts Institute of Technology (MIT) in Boston

John R. Schrieffer

where he enrolled in the Faculty of Applied Science to become an electronics engineer, but he continued his studies as a physicist at the University of Illinois. After his thesis he undertook work in the same field at the University of Birmingham and at the Nobel Institute, Copenhagen, Denmark. He was appointed Professor at the University of Pennsylvania in 1964 and also taught physics for six months at Cornell in 1969. From 1980 until 1991 he held post as Professor at the University of California, Santa Barbara. Since 1992 he has taught physics at the Florida State University. Schrieffer is also known for his theoretical works on alloys, ferromagnetism (a domain in which he had thought of reorienting his doctoral thesis when he had problems with the calculations mentioned above) and solid-state physics in general.

1973

Esaki, Leo (Osaka, Japan, March 12, 1925 -). In 1957 Leo Esaki, a graduate researcher at the Sony Company in Tokyo, was studying the currents passing through germanium semiconducting junctions. He verified that, in extremely thin p - n junctions, electrons could "jump" a potential barrier: a situation prohibited by the laws of classical physics but allowed by quantum mechanics. (The specialists speak of the tunnelling effect). Pursuing this study on junctions very strongly "doped" with impurities, he showed that a current could cross the barrier, with an intensity which decreased when the voltage was increased instead of following the usual Ohm's law : $V = Ri$ (V: applied voltage = resistance (R) x current intensity (i)). For those junctions the differential resistance (dV/di) of the device became negative. The principal applications concern very fast electronic circuits and, therefore, computer technology. Esaki presented a thesis on this subject at the University of Tokyo in 1959 and was then hired by the IBM Research Center in Yorktown Heights (United States) in 1962. He shared half of the Nobel Prize in physics with **Ivar Giaever**, the author of parallel research in superconductors, with the other half going to **Brian Josephson** for a (theoretical) work in the same field. All three had completed the foundations of their work while they were still students. Although he was living in the States since 1962, Esaki retained his Japanese citizenship. He works now in close collaboration with the Russian **Zhores Alferov**, 2000 Nobel Prize winner for Physics.

Giaever, Ivar (Bergen, Norway, April 5, 1929 -). Ivar Giaever received the Nobel Prize for his work on tunnelling through a thin alumina contact, at the interface of an aluminium - lead structure, at a temperature where lead becomes superconductive. He shared half of the prize with **Leo Esaki** for a parallel discovery on semiconductors and with **Brian Josephson** for his theoretical contribution. He had a very short career as a solid state physicist, "just what is needed", he declared, "to benefit at the right place, at the right time, from good advice from a group of excellent experts in various scientific fields". The applications of his discovery were especially directed toward the measurement of the forbidden band (not easily measurable by another way) of superconductor materials and, therefore, toward great progress in the understanding of the superconductivity phenomenon previously explained by the BCS theory of **John Bardeen**, **Leon Cooper** and **John Schrieffer**. Giaever was very surprised by the attribution of the Nobel Prize. He had only started in the subject a few years earlier ... by chance! He finished his mechanical engineering studies at the Institute of Technology in Trondheim (Norway) in 1952. After having finished his military service, he was employed in the Patents Department of the Norwegian government. In 1954, Giaever joined a laboratory of General Electric's Advanced Engineering Program in Canada. In 1958, he was transferred to an American laborato-

Ivar Giaever

ry of the same company. Finally, he obtained his P.D. in 1964, while studying at Rensselaer Polytechnical Institute. Rather than continue in the field of solid-state physics, he directed his research toward biophysics (like Glaser) and, in particular, immunology and the behavior of proteins at the surface of solids.

Josephson, Brian David (Cardiff, Wales, January 4,1940 -). A solid state theoretical physicist, Brian Josephson is among the youngest winners of the Nobel Prize in Physics. He undertook his research under the supervision of **Philip Anderson**, seconded to England by Bell Telephone laboratories. By calculation, he predicted that, in the tunnel effect disco- vered by **Ivar Giaever**, it was the **Cooper** pairs that were responsible for the current and that a current can even circulate without any applied voltage (DC Josephson effect). When applying a power supply he calculated that the "super current" could oscillate at a frequency connected with the given voltage (AC Josephson effect). These two effects were effectively measured a short time after the publication of Josephson's calculations in Physics Letters. Thanks to this new data, fundamental physics has at its disposal the possibility of a precise measure of the ratio e/h. It concerns the ratio of two fundamental constants respectively determined by **Robert Millikan** in 1911 (e: charge of the electron) and by **Max Planck** in 1900 (h: Planck constant). In applied physics, the Josephson effect is used to construct, with a high degree of accuracy, reference voltages. These effects (DC and AC) were quickly tested by IBM for the design of computer circuits. Born in Wales, Josephson obtained his degrees from Cambridge and a Ph.D. from there in 1964. Still in Cambridge, he climbed the ladder to become a professor in 1974. He spent one year at the University of Illinois in 1965-1966. His research interests are currently directed toward the study of intelligence.

Brian D. Josephson

1974

Hewish, Antony (Fowey, England, May 11, 1924 -). In 1967 Antony Hewish and his Ph.D. student Jocelyn Bell (only 24 years old at that time), having a radio telescope made up of 2048 antennae covering a surface with 18.000 m^2 at their disposal, studied the twinkling of the stars (in the radio frequencies range). The observations made by Hewish in 1964, with a noticeably inferior instrument, had highlighted scintillations from interplanetary medium, of irregular frequency, in the same radio frequency range. In 1967, Hewish identified, from the observations of a regular variation of such signals, with a very precise period of

1.337 301 13 seconds. Hewish explained the phenomenon by postulating the existence of pulsating neutron stars. Today, the majority of astronomers think that these pulsars are stars that have a mass comparable with that of our sun but which are very small (~ 20 km in diameter) and therefore very dense (= 10^{15} gr/cm^3), so dense that the structure of the atoms would be completely changed by the enormous gravitational force. These pulsars would be essentially made up of neutrons and could be only the final remains of a supernova explosion. Rotating quickly, a pulsar would have the possibility of emitting very directional radio waves in the universe, with a frequency equal to its rotational frequency. Thus, an observer would receive an impulse (hence the English name of pulsar) whenever the beam moves toward it. These waves are sometimes associated with the gravitational waves predicted by Albert **Einstein's** general theory on relativity. Following the discovery of the first pulsar in 1967, the regularity of the emissions even provided populist pundits with ideas. Some people claimed that the existence of "small green men" was finally confirmed. At the beginning of the 1980s, over one hundred pulsars had been discovered and generations of small green men shelved indefinitely. Hewish shared the Nobel Prize with **Martin Ryle**. It was the first time that two pure astronomers had been honoured by this prize. In 1974 **Russel Hulse** and **Joseph Taylor** highlighted the existence of binary pulsars. Born in Cornwall, Hewish was educated at Trinity College, Cambridge. During the war, he had to stop to serve in a telecommunications corps. There he met his colleague Ryle. Returning to Cambridge, he obtained his Ph.D. in 1952 and occupied several temporary positions there before being appointed part-time lecturer (1961) then as Professor of Astronomy (1971). Jocelyn Bell-Burnell is currently professor of Astronomy and Astrophysics at the University of Oxford and is involved in the management of science in England.

Aage Bohr

Deformed atomic nucleus. The protons (red) and the neutrons (grey) oscillate around the centre of the mass of the atomic nucleus.

Ryle, Martin (Brighton, England, September 27, 1918 - Cambridge, England, October 14, 1984). The hundredth physicist to receive the Nobel Prize, Martin Ryle shared it with Antony Hewish, whom he met during the Second World War while they were both serving in a British telecommunications corps. Ryle's talent in his later work, in the Cavendish Laboratory, then managed by J. A. Ratcliffe, lies in the construction of enormous instruments, requiring tooling to a precision of one cm on overall dimensions of the order of kms. Indeed, the construction of a radio telescope requires diameters of around several km, in view of the large wavelength of the signals to be studied. Ryle got around the difficulty by using separate elements simulating a wide angle telescope, the positions of all the elements being monitored by a sophisticated computing technology. He was vigorously encouraged in his research by **William Lawrence Bragg**, the pioneer in radio crystallography and who designed equipment which was used as a model for the radio telescope. Ryle received his education at the Universities of Bradford and Oxford. He joined the Cavendish Laboratory of the University of Cambridge in 1948 and was appointed Professor of Astronomy there. As early as 1972 he was interested (with Fred Hoyle) in the birth of the universe. While Hoyle argued for a continuous creation of new matter driving the expansion of the universe, Ryle was a firm supporter of an explosively expanding universe, the so-called Big Bang Theory.

1975

Bohr, Aage Niels (Copenhagen, Denmark, June 17, 1922 -). Two models of the atomic nucleus were known in 1950. The liquid droplet model established around 1936 explained the global behavior of the protons and neutrons in the essentially spherical nucleus. It was likely to fluctuate around this average shape. This model knew a period of success in the explanation of the fission of the uranium

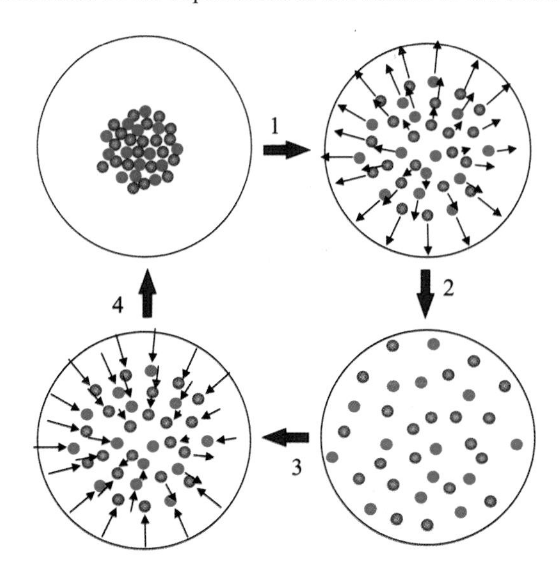

atom under the impact of a slow neutron. The shell model, established by **Maria Goeppert - Mayer** and **Johannes Jensen** in 1949, makes reference to the individual movement of each proton or neutron in a presumed spherical structure formed by all the other nucleons. In 1953, Aage Bohr and **Ben Mottelson** proposed a variant combining the successes of these two models. Moreover, they explained the continuous rotation including a deformation of the nuclei in ellipsoids of revolution. Independently, **Leo Rainwater**, at Columbia, studied the same properties. **Niels Bohr**'s fourth son Aage had been born several months before his father received the Nobel Prize. Aage Bohr spent his first years of scholarship in Copenhagen, where his father headed the Institute of Theoretical Physics. Since his early childhood he was in contact with the greatest physicists of the beginning of the 20th century. In 1943, Bohr fled from the Nazi regime at first to Sweden, then to the United States. In 1945, on his return to Denmark, he completed his basic education and was afterwards enlisted with the students of the Institute for Advanced Studies at Princeton. Together with his father, Aage Bohr participated in the construction of the atomic bomb in Los Alamos (The Manhattan Project). In 1962 he replaced his father and became the Head of the Niels Bohr Institute (until 1970). The Bohr family was not the only one, where both the father and the son were awarded the Nobel Prize. There are also such families as the **Braggs**, the **Curies**, the **Siegbahns** and the **Thomsons**.

Mottelson, Ben Roy (Chicago, Illinois, July 9, 1926 -). Ben Mottelson is four years younger than **Aage Bohr** with whom he built the most complete model, known to date, of the structure of the atomic nucleus. Reconciling the (collective) drop model and the (individual) shell model

Ben R. Mottelson

The collective oscillation, but in opposite directions, of the protons and the neutrons, provides another model of the behavior of nucleons in the nucleus.

proposed by **Maria Goeppert-Mayer** and **Johannes Jensen**, Bohr and Mottelson conceived that nucleons (protons and neutrons) are organised to carry out collective movements (oscillating and whirling), inside a volume often of ellipsoidal shape. Mottelson was, at that time, very influenced in his thinking on the nucleus by the model which **Leon Cooper** had developed to describe the orderly motion of pairs of electrons which had led to the superconductivity model. The treatise "Nuclear Structure" that he wrote with Aage Bohr, the result of the research that earned them the Nobel Prize, is a classic for all nuclear physicists. Mottelson graduated at the University of Purdue in 1947 and earned his Ph.D. in nuclear physics at Harvard in 1953 under the supervision of **Julian Schwinger**. He then rejoined, for a period of three years, the Niels Bohr Institute at Copenhagen, Denmark, where he collaborated with Aage Bohr. From 1953 until 1957 he conducted research at CERN (European Center for Nuclear Research, Geneva). After that he went back to Copenhagen as a professor of theoretical physics at Nordita, an institute adjacent to the **Niels Bohr** Institute. In 1946, he became a naturalized Danish citizen. Mottelson is recognised as an excellent teacher, a man of great culture, who was able to look intelligently at a wide variety of subjects.

Rainwater, Léo James (Council, Idaho, December 9, 1917 - Yonkers, New York, May 31, 1986). Around 1950, Leo Rainwater, offered an overall formulation of the behavior of protons and neutrons in the atomic nucleus. After two years of basic studies at Cal Tech, he was primarily attached to the University of Columbia where he gained his principal diplomas. Furthermore, at the end of his thesis, he became assistant there from 1939 until 1942, instructor in 1946, assistant professor in 1947, associate professor in 1949 and professor since 1952. During World War II he took part in the development of the atomic bomb (Manhattan Project). He had an impressive line-up of

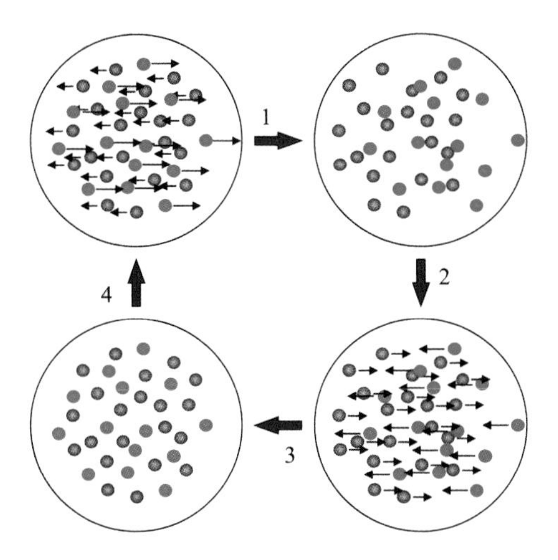

Nobel Prize winners as professors and colleagues. In 1936, **Carl Anderson** (the physicist who identified the positron) was his professor of physics at Cal Tech (then

Léo J. Rainwater

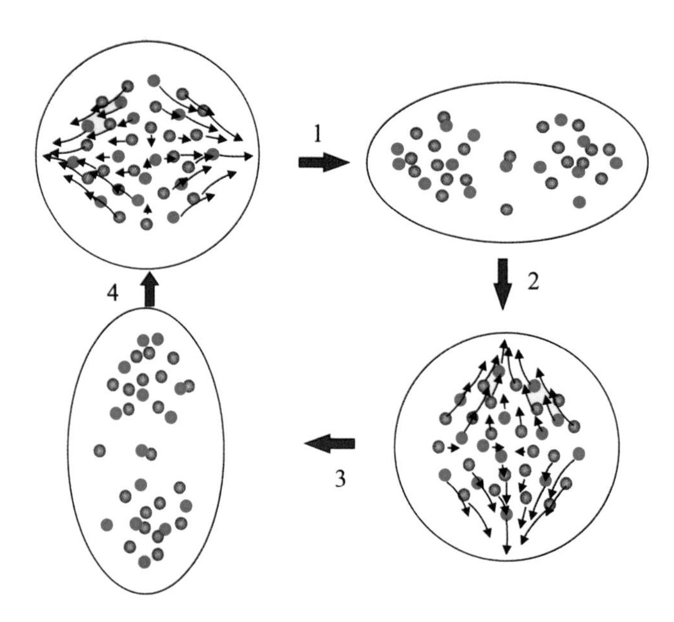

A third possibility of collective behavior in an atomic nucleus associating central and lateral movements.

directed by **Robert Millikan**) and **Thomas Morgan** (1933 Nobel Prize for Medicine) was his professor of biology. At Columbia he studied under such leading figures as **Enrico Fermi**, **Isidor Rabi** and Edward Teller (the father of the H-bomb). In the same institution he also met **Hans Bethe**, **Willis Lamb**, **Charles Townes**, **Tsung Lee** and **Polykarp Kusch**. In 1954, he published a theoretical article on the muon-nucleus interaction with **Leon Cooper** (who defended his thesis, not on superconductivity which earned him the Nobel Prize in 1972, but on a nuclear physics subject). In 1949, during a seminar given by **Charles Townes** (Nobel Prize for his study of the Maser effect) on the quadrupolar moments of the nuclei of rare earths, Rainwater discovered an additional argument to confirm the validity of the nuclear model that he prepared. It was also about this time that he interested **Aage Bohr** and **Ben Mottelson** in this model. From 1946 until 1978, he was associated with the experimental project to construct the synchrocyclotron at Columbia, an instrument designed, in particular, for the study of the interactions of pions and muons with nuclei.

1976

Richter, Burton (New York City, New York, March 22, 1931 -). In 1974 Burton Richter in Stanford, and jointly **Samuel Ting** in Brookhaven, confirmed, experimentally, a new property, to distinguish the fundamental building blocks of matter, namely the quarks. **Muray Gell-Mann** had already created the concept of strangeness (S) to increase the family of the (u) and (d) quarks, constituent parts of the two best known baryons: the proton and the neutron. The identification of a new particle (having a mass slightly larger than three times the mass of the proton) led the Stanford team to give it the name of y, while the Brookhaven team designated it by J. The publication of the results, in November 1974, in the same number of the famous journal Physical Review Letters, gave, for posterity, the double name J/ψ. This particle is a meson, made up of a quark and an anti - quark, namely $c\bar{c}$ (c owing to the new concept of charm). The lifetime of J/ψ, 5000 times longer than that expected for such a particle,

Burton Richter

required the invention of a new property known today as charm: the terminology was proposed by **Sheldon Glashow**. The two papers of November 1974 were signed by 35 and 15 authors, but only the names of the team-leaders, Richter and Ting, were retained for the Nobel Prize. Two other quarks t and b (top and bottom or truth and beauty) would be necessary to complete the description of the component parts of the hadrons known at the end of the 1970s. Remembering that the attribution of a Nobel Prize in October was decided after the consultation of a jury, which gives its opinions in February, thus there was a very short lapse of time between the announcement of the discovery and the decision to award the Nobel Prize. Born in Brooklyn, Burton Richter was educated at the Massachusetts Institute of Technology, Boston, where he

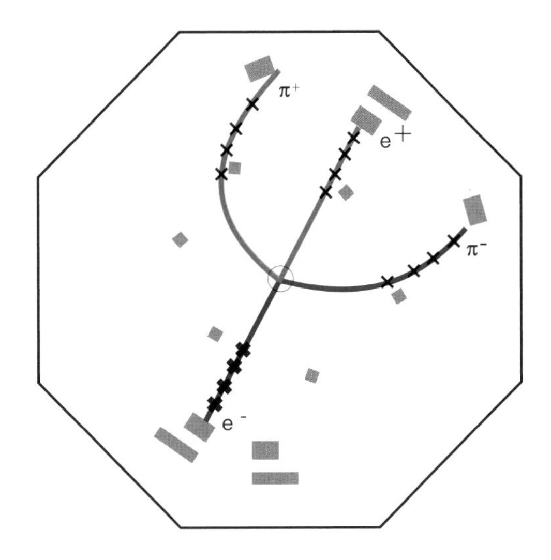

The J/ψ particle simultaneously discovered (but in very different ways) by the teams of B. Richter and S. Ting leaves here the proof of its very short existence (10^{-20} second) through the trace of two pions (π^+ and π^-) and an electron-positron pair. This very short lifetime is, however, regarded as sufficiently long to confirm the reality of the concept of charm introduced by Sheldon Glashow.

obtained his Ph.D. in 1956. Attracted by the study of elementary particles, Richter was employed at Stanford and, in the 1960s, took part in the construction there of the electron-positron collider (SPEAR - Stanford Positron Electron Accelerator Ring), a large instrument of 1 GeV (10^9 eV) with which he and his team made the discovery of the ψ. He has been professor at Stanford since 1967 and contributed to the research which also earned the Nobel Prize for **Martin Perl**.

Ting, Samuel Chao Chung (Ann Arbor, Michigan, January 27, 1936 -). While **Burton Richter** discovered the ψ meson in the frontal collision of electrons and positrons in the rings of the Stanford accelerator, Samuel Ting identified, at the same time in Brookhaven, the same particle (that he called J) by sending protons at a beryllium target. Several experiments were crowned with success between August and November 1974. The two groups simultaneously announced their discovery in November 1974. The equipment at Brookhaven included a large spectrometer (protected by a shielding of 10,000 tons of concrete and 5 tons of borax) adjacent to the synchrotron of 33 GeV (1 GeV = 10^9 eV). Sophisticated electronics and informatics were linked to it in order to select one event out of hundreds of millions which did not fit with all the parameters required for the identification of the sought-after particle. Ting was born in Ann Arbor while his father was studying at the University of Michigan. He spent his chilhood and his youth in China and in Taiwan, but returned to Michigan where he obtained his B.S. (1959) and master's degree (1960) and then his doctorate degrees (1962). He then spent one year at CERN and five more at Columbia. In 1969 he joined the Massachusetts Institute of Technology, while at the same time directing research groups at CERN, at Brookhaven and at the synchotron DESY in Hamburg (Germany).

1977

van Vleck, John Hasbrouck (Middletown, Connecticut, March 13, 1899 - Cambridge, Massachusetts, October 27, 1980). Together with **Nevil Mott** and **Philip Anderson**, one of his first students, John van Vleck received the Nobel Prize in Physics for their contribution to the fundamental knowledge of the electronic structure of disordered magnetic systems: a physical theory that van Vleck refused to explain to nonspecialists! He is regarded as the founder of the quantum theory of magnetism, in keeping with the line of **Paul Dirac's** work. He is, in particular, known for a concept of susceptibility (independent of the temperature) in paramagnetic materials, a property to which he gave his name. Other works are related to the behavior of atoms and ions in crystals and to the correlated motion of electrons for the creation of local magnetic moments in metals. During World War II he joined the Army and contributed to the development of radar devices. There he carried out several fundamental research studies,

John H. van Vleck

which later had impacts in astronomy and astrophysics. He first graduated in sciences at the University of Wisconsin but obtained his master's degree and his doctorate at Harvard in 1921 and 1922, respectively. He then taught at the Universities of Minnesota and Wisconsin before definitively settling down at Harvard in 1934. In quantum physics, he is recognized as the first American to reach the level of his European colleagues. His work, completed in the years 1920-1930, led him to study quantum systems of great interest in chemistry and molecular biology. The author of several major treatises on electric and magnetic susceptibilities, his works still sell very widely, nearly half a century after their first publication. He took part in the Solvay Councils of

Physics in Brussels in 1930 and 1954. Known for his prodigious memory, he also possesed a rare elegance.

Mott, Nevill Francis (Leeds, England, September 30, 1905 - Aspley, Guise, England, August 8, 1996). Prolific in advanced work in the field of theoretical solid-state physics, Nevill Mott was awarded the Nobel Prize, jointly with **John van Vleck** and **Philip Anderson**, for his studies on the electronic structure of disordered magnetic systems. He had previously won fame through books on the theory of atomic collisions (Mott's scattering is a phenomenon dominated by the importance of symmetries in physics), on the theory of metals and alloys (with extensions into the study of defects), on semiconductors and on photography methods. He was also the first to demonstrate the dual role of electrons in conductors: those that take part in the current and those that ensure the magnetic properties and are responsible for the scattering. As President of the Association of British Atomic Scientists, he expressed himself on the dangers but also the benefits of the emerging atomic era on several occasions during the 1940s. He did his university studies in Cambridge, but never presented a Ph.D. After having taught physics in secondary school he was appointed Professor of Theoretical Physics at the University of Bristol from 1933 until 1954. The same year, he went back to Cambridge and taught there until his retirement in 1971. He took part in the Solvay Council of Physics in Brussels in 1933 and 1954. He was knighted in 1962. Francis Nevill Mott was very fond of photography and archaeology, mainly stained glass windows and Byzantine coins, and was very interested in history and religion.

Anderson, Philip Warren (Indianapolis, Indiana, December 13, 1923 -). A student of **John van Vleck** at Harvard, it was with the latter and **Nevill Mott** that Philip Anderson shared the Nobel Prize for Physics for their studies on the electronic structure of magnetic and disordered systems. Anderson made two major contributions to the group: a) the concept of "Anderson localization" describing an impurity in a crystal and b) the theory of the "super exchange" explaining the coupling of the spins of two atoms having magnetic properties, separated by a nonmagnetic atom. In addition, he was the instigator of the work that would lead **Brian Josephson** to the 1973 Nobel Prize. He was also interested in superfluidity and in superconductivity. His work would thus provide valuable ideas to **Robert Schrieffer**. Born in Indianapolis, he had to interrupt his college studies to serve in the army during World War II. He achieved his Ph.D. at Harvard (1949) under the supervision of **van Vleck**. He worked as a researcher and Assistant Director at Bell Telephone (Murray Hill, New Jersey), while also teaching as a Visiting Professor at Cambridge, from 1967 to 1975. Since 1975 he has been a professor at Princeton.

Philip W. Anderson

1978

Kapitza, Pyotr (Peter) Leonidovich (Kronshtadt, Russia, July 9, 1894 - Moscow, Soviet Union, April 8, 1984). Pyotr Kapitza was awarded half of the 1978 Nobel Prize for his large contribution to low - temperature experimental physics, thus 16 years after his disciple **Lev Landau** received the same honour for the theoretical justification of his teacher's discoveries. **Arno Penzias** and **Robert Wilson** shared the second half of the prize for their work on the continuous background of radiation resulting from the Big Bang. In 1938, working on liquid helium, Kapitza highlighted that at a temperature lower than 2.17 K (thus 271 K below the melting point of ice), a phase (known under the name of helium II) presented surprising properties: complete absence of viscosity and extraordinarily large thermal conductivity. He gave the name of superfluidity to this phenomenon. Thus, he was the inventor of the physics of quantum liquids. Liquid helium had already been produced at the beginning of the century by **Heike Kamerlingh - Onnes**. During his very long scientific career Kapitza also became famous in various fields like: high - power microwave generators (175 kw), nuclear physics, hydrodynamics, plasma physics, light and solid-

Pyotr L. Kapitza

state physics in general. The research on thermonuclear fusion underwent rapid technological progress thanks to the devices that he developed. He was also much appreciated not only for his organisational and leadership capabilities but also for his acute perceptions of the possibilities of the industrial applications of his fundamental results. His new method of liquefaction of helium made the field of ultra - low temperatures more easily accessible. He also developed a very profitable cryogenic system of liquid nitrogen and oxygen production. He received

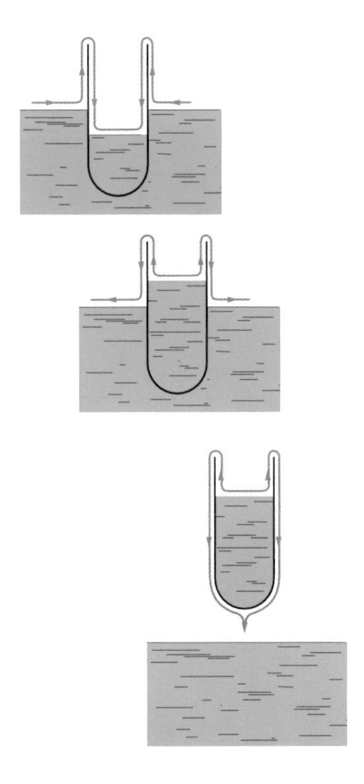

The superfluidity phenomenon explains the easy flow of "superfluid" in apparent disagreement with the laws of gravitation.

his basic schooling at the Polytechnic Institute and in the Industrial and Physics Institute of Petrograd. During this period he collaborated with **Nikolay Semionov** (who earned the Nobel Prize for Chemistry in 1956) in the study of the mechanisms of chemical reactions. Then, he spent three years at the Cavendish Laboratory of the University of Cambridge where, under the supervision of **Ernest Rutherford** (1908 Nobel Prize for Chemistry), he obtained his Ph.D. in 1923. He continued his activities in England and became Director of the Royal Society Mond Laboratory in Cambridge in 1930. In 1934, returning to his native country, he was allowed to remain there by the Soviet authorities. However, he obtained the necessary credits for transferring the equipment (in particular, a powerful magnet) that he had developed in Cambridge to the Institute for Physical Problems in Moscow. He was director of this Moscow Institute until 1946. The record for the intensity of magnetic fields that he had established at 500 kilogauss in 1924 was beaten only in 1956. In 1938 he had to use all his persuasive power to clear his pupil Landau of the accusation of spying for Germany. In 1946 he had to spread comparable energy to resist the pressure that Stalin tried to impose on him to devote himself to research on the atomic bomb. In 1970, with Andrei Sakharov, he again led a campaign to liberate the biologist Medvedev. After the death of Stalin he became Director of the Vavilov Institute and contributed to the fruitful Soviet effort for the launching of the first artificial satellite. He took part in the Solvay Council in Brussels in 1930. The recipient of many honours and distinctions in the Soviet Union, his international stature was also very widely recognised by eleven titles of doctor honoris causa.

Penzias, Arno A. (Munich, Germany, April 26, 1933 -). In 1964, in the astrophysics laboratory financed by Bell Telephone in Holmdel (New Jersey) and together with his colleague **Robert Wilson**, Arno Penzias discovered a continuous distribution of radio waves, which he identified as those that a black body would emit at a temperature of 3.5 K, according to the law established by **Wilhem Wien** 70 years earlier. The following year the theoreticians James Peebles and Robert Dicke interpreted the phenomenon while postulating that it concerned fossil signals of the Big Bang going back 15 to 18 billion years. In 1929 Edwin Hubble noted the expansion of the universe by the 'red-shift' of light wavelength emitted by remote galaxies. Two years before, Georges Lemaître, a Belgian astronomer, expressed the hypothesis of the primitive atom, a precursory model of the Big Bang. The paternity of the Big Bang was, however, given to Georges Gamow, who evoked this name in 1946. The low temperature (3.5 K) associated with this signal is the only link with the 2.17 K of the helium superfluid state, which earned **Pyotr Kapitza** the Nobel Prize in physics the same year. The researchers at Bell Telephone used the antenna developed to follow the progress of the Echo and Telstar satellites for this discovery. In fact, they were interested in radiowaves emitted by the gas halo surrounding our galaxy, and it was by trying to get rid of the background noise (which partially masked the signal they were looking for) that they discovered its isotopy: whatever the direction of the observation, the radiation spectrum maintained a constant shape. Later, the anisotropies discovered in the range of high frequencies would provide precious information for cosmology. Penzias and Wilson highlighted the presence of 8 molecular shapes in intergalactic

Arno A. Penzias

gases, including CO. Arno Penzias left his native country, Germany, and moved to the United States when he was only four years old. He graduated from Brooklyn Technical High School and received his B.S. from City

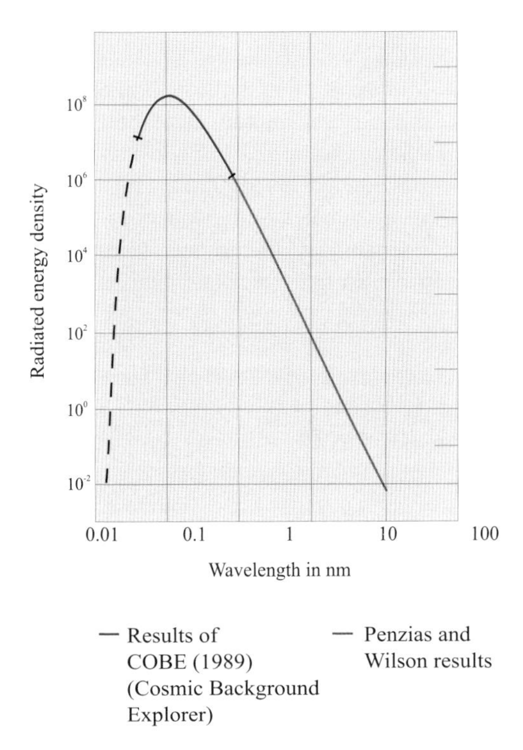

Continuous background at approximately 3 K. The radiation spectrum of the continuous background of the sky is identical in whichever direction we look. It corresponds to the radiation of the black body to approximately 3 K. This observation led many astrophysicists to see this as proof of the expansion of the universe starting from a Big Bang. In 1986 the COBE satellite provided more precise results establishing this temperature at 2.735 K.

College in New York in 1954. His aim was to do studies which would lead him to a remunerative occupation. To the question of knowing if "physics would pay its way", his professor answered him: "you could do the same things as an engineer but you will do them even better as a physicist". He continued his thesis studies in Columbia under the supervision of **Charles Townes** who had already discovered several molecular species in intergalactic gases.

Wilson, Robert Woodrow (Houston, Texas, January 10, 1936 - Ithaca, New York, January 16, 2000). The reasons for the Nobel Committee's choice will be found in the bibliography of Arno Penzias. Penzias and Wilson were working at the Bell Telephone Astronomy Laboratories in Holmdel (New Jersey). Their discovery permitted a giant step forward in the understanding of the process which occurred 15 to 18 billion years ago: the appearance followed by the expansion of the universe. Passionate about electronics from a very young age, this is how Wilson earned his spurs in science:first at Rice University, then at Cal Tech, where he defended his doctoral thesis in radioastronomy in 1962. Wilson continued along this path throughout his career, the research of relative abundances of the isotopes of the chemical elements in interstellar space. Betsy, his wife, affirms that she was able to share in the joys of her husband's discoveries because of a childhood lived with a scientist father.

1979

Glashow, Sheldon Lee (New York City, New York, December 5, 1932 -). Together with **Steven Weinberg** and **Abdus Salam**, Sheldon Glashow received the Nobel Prize in Physics for his contribution to the understanding of the link between electromagnetic force and weak nuclear force. Glashow's original and additional contribution to the work of his co-winners lies in the extension of the model which he proposed for all elementary particles and, in particular, the new property that he conferred on some of them and which he called "charm". This property was used by **Samuel Ting**

Sheldon L. Glashow

and **Burton Richter** as a basis for the experimental research leading to the discovery of J/ψ particle. Glashow and Weinberg both originated from New York City. He and Weinberg began their scientific studies at the Bronx High School of Sciences and both obtained their first degrees from Cornell University. However, Weinberg received his Ph.D. from Princeton (in 1957) while Glashow received his from Harvard (in 1959). They both occupied the position of professor at Harvard. However, they carried out their research in an independent way. Glashow was, at the end of his thesis work, attached

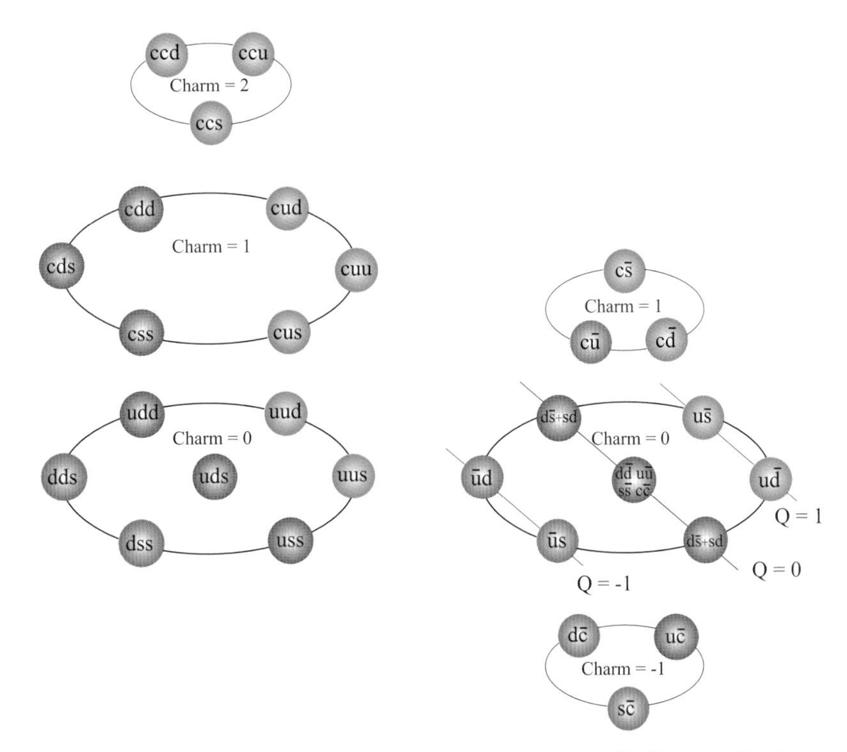

The classification of quarks with respect to their property of charm: on the left the distribution of hadrons, on the right the distribution of mesons. For the common names of those associations of quarks, refer to the bibliography of Gell-Mann (Nobel Prize winner in 1969).

to the Theoretical Research Institute of Copenhagen (1958 - 1960). He was a member of the Faculty of Sciences at Berkeley from 1961 until 1967. At that time he rejoined Harvard and he is still there.

Salam, Abdus (Jhang, Pakistan, January 26, 1926 - Oxford, England, November 21, 1996). In parallel with the work of **Weinberg** and **Glashow**, Salam was the author of the model which associates electromagnetic force and weak nuclear force in the same description. This unification is known as the Weinberg-Salam theory. Nevertheless the two researchers worked independently: Weinberg in the United States and Salam in Europe.Their theoretical model received international interest at the time of the discovery of particles W and Z by the team at CERN managed by **Carlo Rubbia** and **Simon van der Meer**. Salam first studied at Penjab University, then at St John College in Cambridge, England, where he defended his doctoral thesis. After teaching for three years in Pakistan, he was appointed Lecturer at Cambridge in 1954, then Professor at the Imperial College, London, in 1957. Finally he headed the International Center for Theoretical Physics at Trieste, Italy, from 1954. He took part in the Solvay Council of Physics in Brussels in 1961. He still holds a large number of scientific responsibilities in the learned societies of America, Europe and in his native country. A convinced and practicing Muslim, he defended his scientific and religious ideals in numerous articles. He became the spokesman of an ideology which advocated that the less economically favored countries should be able to benefit from resources necessary to eradicate the idea of the Third World. An essential step would be to bring them out of their scientific isolation by creating, in their own country, research centers freely open to the international community. Suffering from Parkinson disease, he spent the end of his life in London (UK).

Weinberg, Steven (New York City, New York, May 3, 1933 -). Steven Weinberg received the Nobel Prize, together with his two theorists colleagues, for a first decisive step in the achievement of **Albert Einstein's** dream: to see unified, in a single model, the four fundamental forces that control the physical phenomena. This concerns the gravitational force, the weak nuclear force, the electromagnetic force and the strong nuclear force so classified from the weakest to the strongest. This first hurdle was actually crossed with the unification of weak nuclear forces and electromagnetic forces. Weinberg contributed in particular to overcoming the difficulty in approaching this unification on account of the enormous difference in the range of these two interactions: the electromagnetic force appears throughout the space surrounding the electric charges whereas the weak nuclear force has a range only in a very restricted area: 10^{-17} m. In 1967 he applied himself to solving the incompatibility by proposing a spontaneous rupture of symmetry, a property which he had tried, since 1960, to attribute to strong interactions. In order to do this, Weinberg and **Salam** introduced the concept of the neutral current. Weinberg subsequently extended his thoughts to create the theory today called quantum chromodynamics. His work can be compared to that which Maxwell realized in 1865 by unifying electricity and magnetism. The electromagnetic interaction appears through the exchange of photons (quanta of energy without mass), the weak interaction through the exchange of W and Z bosons. The identification of the W and Z couple brought the Nobel Prize in Physics to **Carlo Rubbia** and **Simon van der Meer** in 1984. According to Weinberg, the difference between massless photons and massive bosons was probably very small immediately after the Big Bang when the temperature was enormous. During the cooling, there could have occurred a phenomenon parallel to a known phenomenon in solid-state physics: namely that iron, gradually cooled below 770°C, becomes ferromagnetic. At this stage, a magnetic field settles in the matter, pointing in a particular direction and thus breaking the original directional symmetry. Weinberg graduated from Cornell in 1954 and received his doctorate from Princeton in 1957. He successively held positions at Columbia (1957 - 1959), Berkeley (1959 - 1969) and at the Massachusetts Institute of Technology in Boston (1969 - 1973). He was appointed Professor at Harvard in 1973. From 1961 until 1976 he produced a considerable number of publications and was the physicist most cited in the work of his colleagues. He is the author of several highly specialized books but also of a best seller, translated into more than 15 languages: "The first three minutes", a presentation, for the general public, of the first moments which followed the Big Bang. He is a well - rounded person, passionate about history, and probably the only physicist member of the American Medieval Academy.

1980

Cronin, James Watson (Chicago, Illinois, September 19, 1931-). James Cronin and **Val Fitch** received the Nobel Prize in 1980 for the explanation of the decay mode of electrically neutral K° mesons, thus continuing along the path opened by **Chen Yang** and **Tsung Lee**. Yang and Lee had shown that, in a weak interaction (one which governs in particular the decay of the free neutron into a proton, an electron and an antineutrino, as well as the decay of a proton bound in the nuclei into a neutron, a positron and a neutrino), the parity was not preserved. The parity is a fundamental concept of quantum mechanics related to the exchange of the left and the right sides in mirror reflections. It was firmly believed in the 1960s that the CP pro duct (charge, parity) was preserved even if the parity P was not universally preserved. The conclusions of Cronin and Fitch invalidated this prediction. They showed this in 1964, by interpreting the two modes of decay of the K° meson with half-life times of 10^{-10} and 10^{-7} second. After his undergraduate scientific studies at Southern Methodist University (1951), Cronin continued his master's degree and his doctorate at the University of Chicago (1955). There he met **Enrico Fermi**, **Maria Goeppert-Mayer**, as his professors, and also **Murray Gell - Mann** who had just proposed his idea on strangeness, a decisive concept in the work that led to the Nobel Prize. He stayed for three years at Brookhaven National Laboratory before returning to Chicago, where he shared his time between the Institute of Theoretical Physics of High Energy and the Fermilab experimental laboratory in Batavia (Illinois). A lover of nature, he claimed to have devoted a part of the Nobel Prize award to fitting out a cabin in a Wisconsin forest.

Fitch, Val Logsdon (Merriman, Nebraska, March 10, 1923 -). Together with **James Cronin**, nine years his junior, Val Fitch demonstrated that the CP product (charge, parity) did not have the expected symmetry. The properties of K° particle are remarkable: it can be made half of matter and half of antimatter. The identification of this K° has an important implication in cosmology in respect of what occurred after Big Bang. The newborn universe was probably symmetrical with regard to the relative abundance of matter-antimatter. At the current stage of our knowledge, antimatter seems definitely less present than matter. The sole process of matter-antimatter annihilation by electromagnetic radiation, which ought to finally lead to a complete conversion of the pair, has undoubtedly to be supplemented by the model of Fitch and Cronin. If, at the start, a minute part of matter exceeded that of antimatter, then today matter can largely dominate antimatter after a partial disappearance by annihilation equivalent to both species (matter-antimatter). A U.S. Army soldier during the Second World War, Val Fitch was sent to Los Alamos to collaborate on the Manhattan Project. It was a chance for him to meet **Niels Bohr**, **James Chadwick**, **Enrico Fermi** and **Isidor Rabi**. It was then suggested to him that he continue his basic studies at McGill University (1948) and to complete his studies with a doctorate, which he presented in 1954 at Columbia. His thesis supervisor was **James Rainwater**. At that time Fitch was a pioneer in the study of the atoms in which a μ meson was substituted for an electron. While at Princeton, where he obtained a position in 1954, he branched out, for a duration of more than twenty years, towards the study of K mesons. Modest and realistic, during the Nobel Prize-giving ceremony, he expressed two opinions, which received considerable attention: "It is highly improbable, a priori, to begin one's life on a ranch in Nebraska and to find oneself, here in Stockholm to receive the Nobel Prize in Physics" and "Each time that a new frontier is crossed, we inevitably discover new phenomena which force us to modify substantially our former concepts".

1981

Siegbahn, Kai Manne Börje (Lund, Sweden, April 20, 1918 -). Son of **Karl Siegbahn**. Kai Siegbahn used the fine spectroscopy of the electrons emitted by matter, under the impact of monochromatic radiation, to provide a very precise means of identification of the chemical bonds, by using the equation which earned Albert Einstein the Nobel Prize, namely: $h\nu = E_k + B$ (the energy of the incident photon ($h\nu$) is converted into kinetic energy of the electron ejected (E_k) after rupture of its connection (B) with the medium). In 1954, Kai Siegbahn built a device allowing the measurement of this binding energy with a precision of one part in 104. At the beginning of the century, **William Henry** and **William Lawrence Bragg**, **Louis de Broglie**, and also Robinson and Williams, had already been interested, but with less precision, in this physical-chemical methodology. **Linus Pauling** (1954 Nobel Prize for Chemistry and 1962 Nobel Prize for Peace) was one of the first beneficiaries. His successors could appreciate the new data provided in the 1960s by Siegbahn's device. At the present time, tables index the binding energies for almost all the molecules and the solids. A technique of primary importance in the science of surfaces and therefore

Kai M. Siegbahn

of new materials, the production of devices for routine analyses constitutes a market which reached billions of dollars annually since the 1980s. The name given by its inventor to the technique, namely ESCA (electron spectroscopy for chemical analysis) is currently known under new initials XPS (X - ray photoelectron spectroscopy). Kai Siegbahn received his doctorate from the University of Stockholm in 1944 and was appointed Professor at the Royal Institute of Technology in the Swedish capital from 1951 until 1954. From 1954 until 1984 he taught at the University of Uppsala. He joins the families of **Bohr**, Bragg, **Curie** and **Thomson** whose parents and children have received the award of the Nobel Prize.

Bloembergen, Nicolaas (Dordrecht, The Netherlands, March 11, 1920 -). Nicolaas Bloembergen shared half of the 1981 Nobel Prize with **Arthur Schawlow**, also a specialist in laser spectroscopy, The other half was awarded to **Kai Siegbahn**. In 1950 Bloembergen started research to progress from the stimulated emission of microwaves (maser) to the stimulated emission of visible light (laser). This type of emission had already been predicted by **Albert Einstein**. The maser effect (discovered in 1953) earned the 1964 Nobel Prize for **Charles Townes**, **Nikolai Basov** and **Alexander Prokhorov** and the use of the laser for holography brought **Denis Gabor** the 1971 prize.

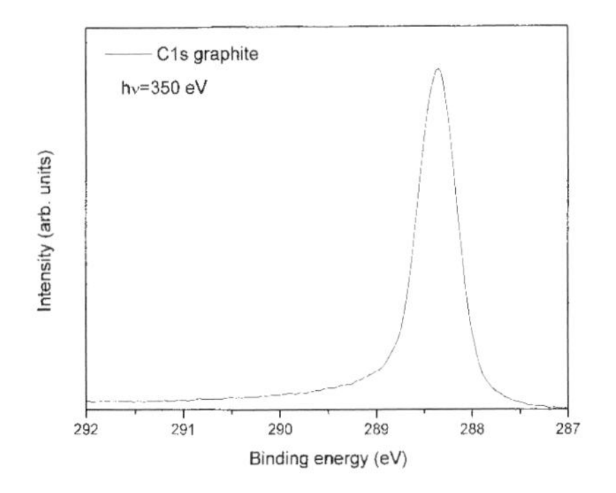

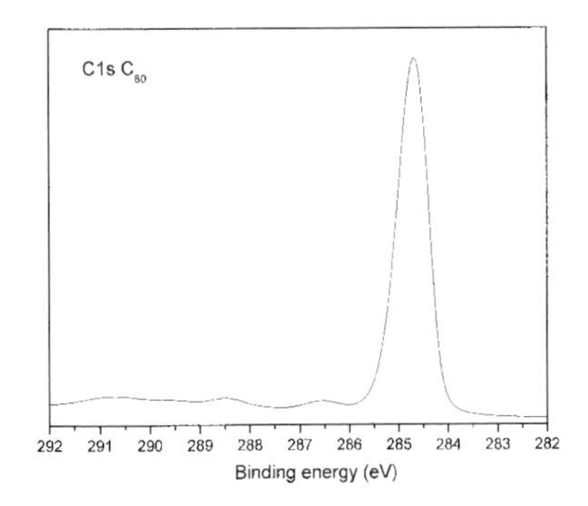

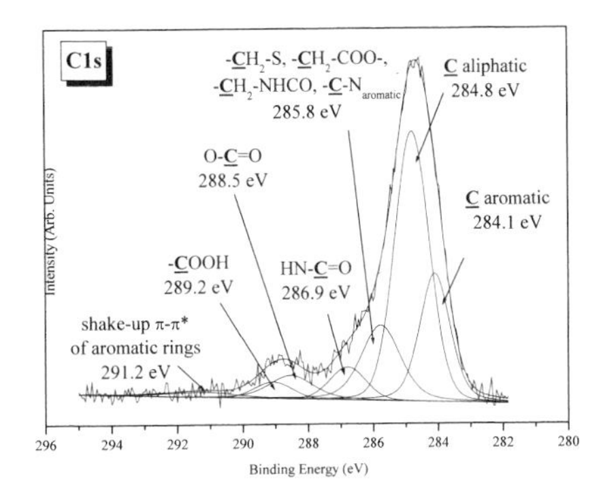

XPS spectra showing bonding energy of various carbon compounds. (a) : graphite with its 1s state at 288.3 eV; (b): fullerene C_{60} with the same state at 284,6 eV; (c) an organic molecule with its various bounds of the carbon atoms.

Nicolaas Bloembergen

Bloembergen endeavoured to develop sources of coherent polarized light of very high intensity, in order to produce non-linear effects or to break chemical bonds for a high resolution spectroscopy. This last application provides the reason for the simultaneous attribution of the 1981 Nobel Prize to Siegbahn. The latter developed, in fact, a high resolution electron spectroscopy (XPS) for chemical analysis (ESCA). Bloembergen was **Edward Purcell**'s assistant during part of his thesis which he prepared in Harvard. Their common work on the non - linear effects in nuclear magnetic resonance, published in 1952, led to the application of masers and lasers in molecular spectroscopy. Bloembergen obtained his first degrees in Utrecht and defended his thesis in 1948 in Leiden. He held post as a professor at Harvard University in 1951. He was granted American citizenship in 1958. It is said that Mrs. Bloembergen, who had met Mrs. Townes, who had been given a ruby in recognition of her husband's work, asked Nicolaas what she could have received in similar circumstances. Bloembergen could only answer: "Nothing good for you ."My laser functions on cyanide!" Very active in scientific societies, Bloembergen was also very widely consulted by organizations like NASA, the laboratory of Los Alamos and various companies involved in the construction of sophisticated scientific material.

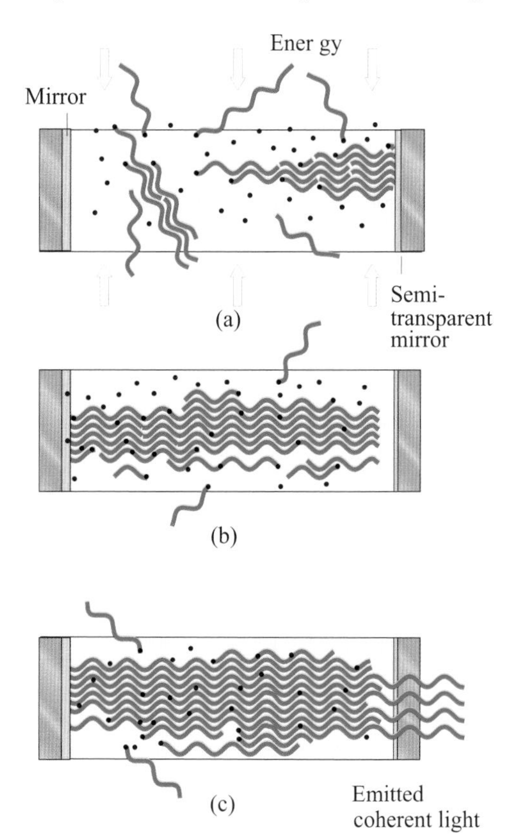

Stimulated emission. The radiation emitted by the excited atoms contained in an active environment are reflected by a pair of parallel mirrors (**a**). They amplify (**b**) the emission of radiation of the same wavelength. A beam of coherent light comes out through the partially transparent mirror (**c**).

Schawlow, Arthur Leonard (Mount Vernon, New York, United States, May 5, 1921- Stanford, California, April 4, 1999) An experimental physicist in the field of lasers, Arthur Schawlow was the associate of **Charles Townes** who shared the 1964 Nobel Prize with two Soviets for the discovery of the maser effect. Recognized as the inventor of the laser (first proposed in 1958), he shared half of the 1981 Nobel Prize with **Nicolaas Bloembergen** for the developments in molecular spectroscopy, made possible by the use of this coherent light. This proposal of Arthur Schawlow and Charles Townes was industrially exploited in June 1960 by Maiman, from the Hughes Laboratories, who built the first instrument to operate in the range of visible light wavelengths. In December of the same year, A. Javan, at Bell Telephone, made operational the first gas laser: an inexpensive device which is widely used today. Numerous forms of lasers have been proposed since then (quartz, semiconductors, CO_2, noble and chemical gases, dyes and eximeres). Powerful means are currently being used to try to produce lasers using X - rays. Schallow obtained all his university degrees at Toronto, where he defended his doctoral thesis in 1949. Having spent two years at Columbia, he then worked at Bell Telephone Laboratory until 1961. There he conducted the major experiments for which he was awarded the Nobel Prize. He was appointed Professor at Stanford in 1961. A remarkable teacher, he made himself world famous through a demonstration which he carried out in front

Arthur L. Schawlow

of his students. A small balloon of a dark color is inflated and placed inside a large transparent balloon. The whole is then irradiated by a laser beam. Whereas the transparent balloon allows the beam to pass through without damage, the colored balloon bursts. Schawlow's interventions in televised programs were very largely appreciated for their educational aspect in the United States, Canada and the United Kingdom. A holder of numerous honors, Schawlow was also a very active and very influential member in learned societies. In 1951, he married Townes's sister.

1982

Wilson, Kenneth Geddes (Waltham, Massachusetts, June 8, 1936 -). Theoretical physicist. Kenneth Wilson is the worthy successor of Johannes van der Waals. During the 75 years that separates the award of these two Nobel Prizes, the scientific community added to the three familiar fundamental states (solid, liquid, gas) other names like superfluid, ferromagnetic, anti-ferromagnetic, ferroelectrics, superconductive, nematic, and smectic. Under the effect of a rise (or a fall) in temperature, of pressure, of composition, of an electric or magnetic field, a substance can undergo various phase transitions. In 1971, Wilson described, in the journal Physical Review, a very complete theoretical model of the behavior of numerous materials in the vicinity of these transition states known as critical phenomena. The mathematical methods which he developed were also applied to the study of the flow and turbulence of fluids, in the propagation of fractures in materials. He was the first to win the Nobel Prize as a single winner since 1971. Contrary to the way of thinking of many physicists, who appreciated simple models, Wilson showed a particular interest for complex problems requiring very powerful computers to treat them. He was also active in the field of the description of sub-nuclear structures. The eldest son of a family of six children, whose father was an eminent chemist at Harvard, Kenneth entered this university aged 16 years old and graduated from there at the age of 20. He defended his doctoral thesis at the California Institute of Technology in 1961, under **Murray Gell-Mann**. In addition to research in the physics of elementary particles, the latter proposed solutions for the theoretical resolution of the difficulties appearing each time that a physical property presents an abrupt discontinuity. Wilson entered the Department of Physics at Cornell and then became a Professor at Ohio State University. He willingly describes himself as a workaholic, while appreciating frequent breaks during which he devotes himself to folk dancing, walking and the oboe.

1983

Chandrasekhar, Subramhanyan (Lahore, Pakistan, October 19, 1910 - Chicago, Illinois, August 21, 1995). Subramhanyan Chandrasekhar, an Indian astrophysicist, was honored by the Nobel Committee for his work on the evolution and structure of stars in connection with their internal transfer of energy. The death of a star can follow two diametrically different processes: an explosion in supernova or an implosion into a white dwarf. Chandrasekhar fixed the limits (1.44 solar mass and average molecular mass equal to 2) beyond which a star cannot end in a white dwarf. He began his studies in Madras, went to Cambridge (UK) in 1930 and received his doctorate there in 1933. Then he emigrated to the United States to hold various positions at the University of Chicago. From 1952 until 1986 he was Professor of Theoretical Astrophysics there. An extremely orderly man, he conceived his scientific career in steps during which he methodically devoted himself to the resolution of a particular problem: the theory of white dwarfs (1929-1939), stellar dynamics (1938 - 1943), radiative transfer including the theory of the stellar atmosphere (1943-1950), the stability of ellipsoidal structures (1961 - 1968), relativistic astrophysics (1962 - 1971), and the mathematical theory of black holes (1974 - 1983). He became an American citizen in 1953. Hanna Gray, President of the University of Chicago, described him as "a matchless scientist, the professor of professors, an honest collaborator, altruistic and full of humanity in what we undertook together. He took pleasure in teaching those who intended to take up history and music. He invested in creativity, that divine human spark, which he used in his writings but which, even more, he personified."

S. Chandrasekhar

Fowler, William Alfred (Pittsburgh, Pennsylvania, August 9, 1911 - Pasadena, California,April 27, 1995). In 1957 William Fowler, an American astrophysicist, published with Fred Hoyle, Margaret and Geoffrey Burbidge (British astrophysicists) a masterly demonstration of the way in which chemical elements could be produced in the stars. However, he shared the Nobel Prize with **Subramhanyan Chandrasekhar**, his three British colleagues were not entitled to this honor, probably because of the clearly demonstrated differences with some of their astrophysicist colleagues, in particular, their adversity towards the Big Bang theory. In 1938 **Hans Bethe**, in his research on energy production in the stars, had already shown how the synthesis of hydrogen in helium could occur in the CNO (carbon, nitrogen, oxygen) cycle. William Fowler was a member of the team that produced

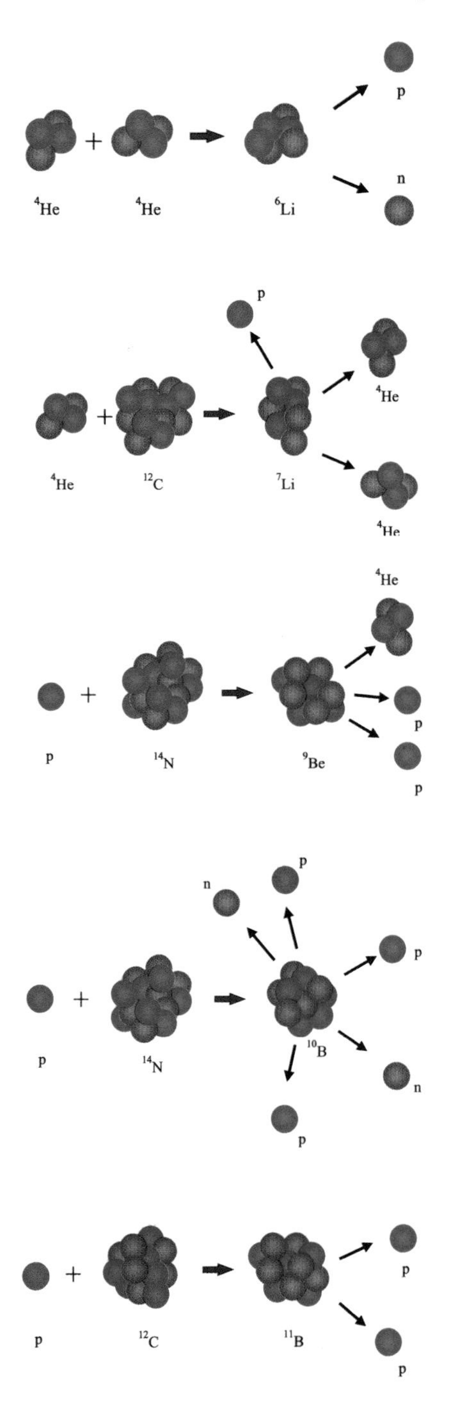

Cosmic creation of some elements heavier than helium. The production of Li^6, Li^7, Be^9, B^{10} and B^{11} in the stars.

the equipment (an electrostatic generator performing under high pressure) able to simulate this process. After 1958 he tackled the study of the generalization of the CNO cycle to conclude that it was operational only for stars whose mass is 1.2 times lower than that of the sun. Chandrasekhar fixed this limit at 1.44 solar mass. These small stars, after having passed through the stage of red giants, lose their outer shells and finish in planetary nebula, after which the remains of the star become a white dwarf. The synthesis of elements heavier than helium would then be realized in larger stars (novae and supernovae), with the final explosion of a supernova releasing the heavy elements into the universe. Fowler received his first degrees at the Ohio State University in 1933 and defended his doctoral thesis at the University of California in 1936. He had an extraordinary scientific output: excellent publications in 37 different journals and author of almost 50 books. During the Second World War, he was attached to the Defense Department of the Navy. He also took part in the manufacture of non - nuclear components for the construction of the first atomic bomb. After the war he worked at the Kellog Radiation Laboratory at Cal Tech (California) where he managed a very important team of nuclear physicists involved in cosmology and "nucleocosmochronology": the science which deals with researching the age of the structure of the universe from the nuclear mechanisms of production and disappearance of the isotopes of all the chemical elements. At the beginning of the 1950's, during a short stay in Cambridge (Great Britain), he met Burbidge and Hoyle who joined him at Cal Tech.

1984

Rubbia, Carlo (Gorizia, Italy, March 31, 1934 -). Carlo Rubbia and **Simon van der Meer** were chosen from among the 126 authors (from 11 laboratories) of the publication related to the identification of particles W^+ and W^-, followed, a few months later, by Z^0. The existence of these particles had been proposed by **Steven Weinberg**, **Abdus Salam** and **Sheldon Glashow** who formulated the electroweak theory. Their mass, expected to be around 80 to 90 times the mass of a proton, could only be produced by the interaction of an antiproton of 150,000 billion eV on a static target containing hydrogen. Antiprotons had been produced for the first time by **Emilio Segrè** and **Owen Chamberlain** in the bevatron at Berkeley. To reach the projected collision energy level, Carlo Rubbia proposed to modify an existing accelerator at CERN to create the device suggested by two Soviet physicists (Budker and Skrinsky from the Novosibirsk Laboratory) in order to send the proton and the antiproton toward each

Carlo Rubbia

other. The production of the W and Z bosons would then be more easily achievable with existing accelerators. The American researchers from the Fermilab at Batavia (Illinois) refused this type of modification, but the CERN management (Geneva) decided in 1978 to follow the idea expressed by Rubbia in 1976. The only things missing

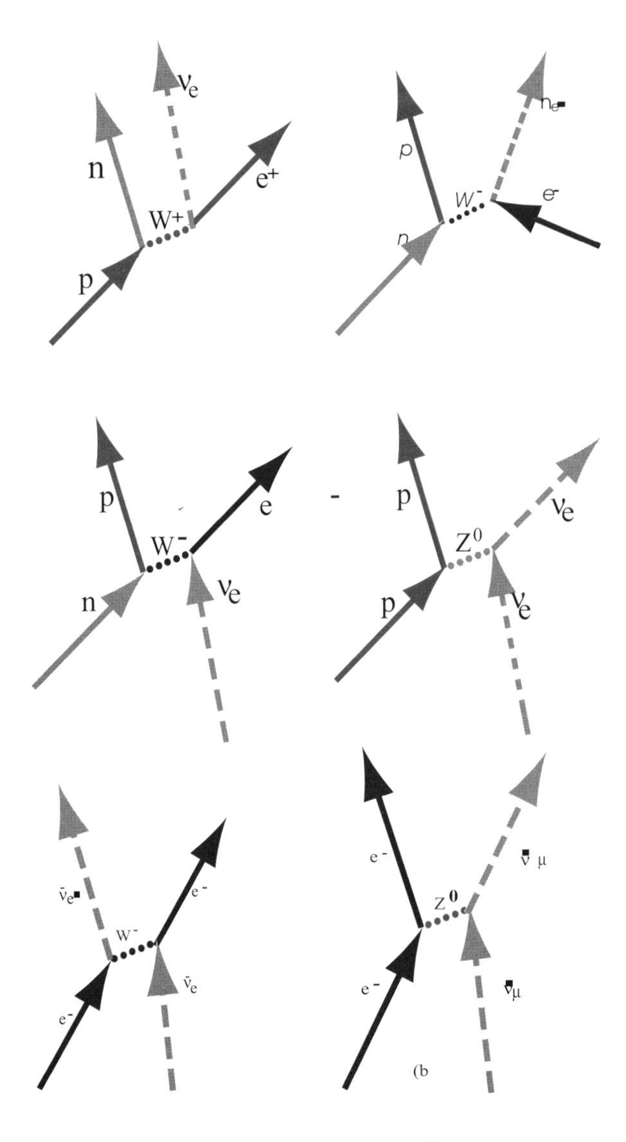

Feynman diagrams showing Z^+, Z^- and W^0 bosons exchanged during the interaction.

were a suitable confinement of the antiprotons (this problem was solved by **van der Meer**) and an appropriate detector (**Georges Charpak's** multiwire proportional chamber) to sort out information on the disintegration of these very short-lived particles: 10^{-20} second. Originally from Northern Italy, Rubbia left his native region in 1949, when the Yugoslav troops invaded, for Venice and later Udine. Rubbia received his university education at Pisa (1958). He then held the post of professor for three years at Columbia and joined the Research Group of the European Center for Nuclear Research in Geneva in 1961. He was also enlisted in high energy physics at the Fermilab and at Brookhaven. From 1970 to 1988 he taught at Havard while pursuing his research in Geneva. This situation imposed on him one transatlantic flight per week every other semester. The French students called him "turbo prof", a name which did not displease him. From 1989 until 1993, he was Director of CERN with, as a flagship, the development of the LEP, the inauguration of which took place in November 1989. The coincidence of this inauguration with the fall of the Berlin wall was exploited by Rubbia to open the doors of CERN to a still growing scientific community and to encourage collaboration with the scientists of the ex-Soviet Union. In 1993, having understood the continuing lack of industrial opportunities for nuclear fusion, he had the idea of applying the techniques of the accelerators to a more reliable control of nuclear fission power plants. The accelerator - reactor coupling would ensure a remarkable stability and safety for the system and would generate less radioactive waste.

van der Meer, Simon (The Hague, The Netherlands, November 24, 1925 -). Simon Van der Meer, an engineer of accelerators, is the brilliant inventor of the confinement processes of the antiprotons in the high-energy collider of

Simon van der Meer

CERN, to check the existence of 3 ultra-heavy particles, exchanged during weak nuclear interaction. The existence of these particles, W^+, W^- and Z^0 bosons, had been predicted by **Steven Weinberg**, **Abdus Salam** and **Sheldon Glashow**, in their theory unifying the weak interaction and the electromagnetic interaction. The essential condition for producing these bosons is to transform, into mass, a total energy of 540 GeV: in particular in a head - on col-

CERN and its infrastructure of large accelerators.

lision of a proton with an antiproton, each one having an energy of 270 GeV. Van der Meer had already installed, in 1981, a device able to store up 1.5 10^{11} antiprotons per day on a circular orbit. It remained to make them collide with protons, circulating in the opposite direction, in the same magnetic field. These antiparticles thus had to be confined, in sufficiently tight packages, so that the local density of the colliding protons and antiprotons be sufficient to give the possibility of creating one of these bosons in an area of the big accelerator where the detector could highlight their production. The process is known as stochastic cooling, Such was the experimental achievement of van der Meer. Graduating as a physics engineer from the University of Delft in 1952, he first worked for three years on electron microscopy, at Philips in Eindhoven before rejoining CERN until his retirement in 1990. The Rubbia-van der Meer tandem worked wonders, probably because of their extremely different characters, an impetuous Italian associated with a calm Dutchman.

1985

von Klitzing, Klaus (Schroda, Poland, June 28, 1943 -). Klaus von Klitzing, highlighted the quantum aspect of an effect discovered for the first time by the American H. Hall (1855-1938) in 1879. The latter could have been honoured with the Nobel Prize if he had made his discovery less than a quarter of a century later. Hall had observed that a current, passing through a conductor subjected to a magnetic field perpendicular to this current, induced a voltage perpendicular at the same time to the magnetic field and to the current. The direction of this voltage allows the identification of the carriers (positive or negative) of the current. By dividing the voltage by the induced current that it produces, Hall's classical value of resistance is obtained. One century later, von Klitzing started the experiment again, but this time making the current pass through a very thin (1 nm) layer: At very low temperature and in a powerful magnetic field, he noted that the ratio of the current on the voltage (that is to say, the conductance or the inverse of the resistance) did not vary in a continuous manner but in steps: each of the steps corresponding to a value of the current/tension ratio decreased in whole units of e^2/h. To the minimal value of this ratio, the resistance corresponds to 25,812.8 ohms. The constant of the fine structure $\alpha=e^2/2\varepsilon_0 hc$, established

Klaus von Klitzing

with good precision by **Willis Lamb's** experiments in 1947 (value 1/137.0365), can, thanks to von Klitzing's discovery, be determined with even better precision. The optimal value accepted today is 1/137.03604. The maximum error is less than 10^{-6}. von Klitzing studied at the University of Brunswick but he earned his doctorate at the University of Würzburg in 1972. He stayed there until 1980 and was named Professor at the Technical University of Munich the same year. He then worked at the Max Planck Institute for Solid State Physics in Stuttgart.

1986

Ruska, Ernst August Friedrich (Heidelberg, Germany, December 25, 1906 - Berlin, Germany, May 30, 1988). German experimental physicist. Ernst Ruska was awarded half of the Nobel Prize, more than 50 years after the development of his research: the design of the electron microscope. **George Thomson** (son of the famous Joseph J. Thomson who showed in the beginning of the 20th century that the electron was a particle), had already expressed this idea in 1927. Since **Louis de Broglie** had shown that a wavelength (inversely proportional to their energy)

Ernst A. Ruska

could be associated to electrons inmotion, the idea of building a microscope where the light beam would be replaced by an electron beam excited the curiosity of several physicists. Ruska demonstrated the first prototype of it in 1928 and, thanks to improvements in the electromagnetic lenses, he managed, in 1939, to distinguish objects whose dimension was below 0.05 micrometer. This resolution was already of an order of magnitude better than that of optical microscopes. The performance of today's instruments are a further two orders of magnitude better.

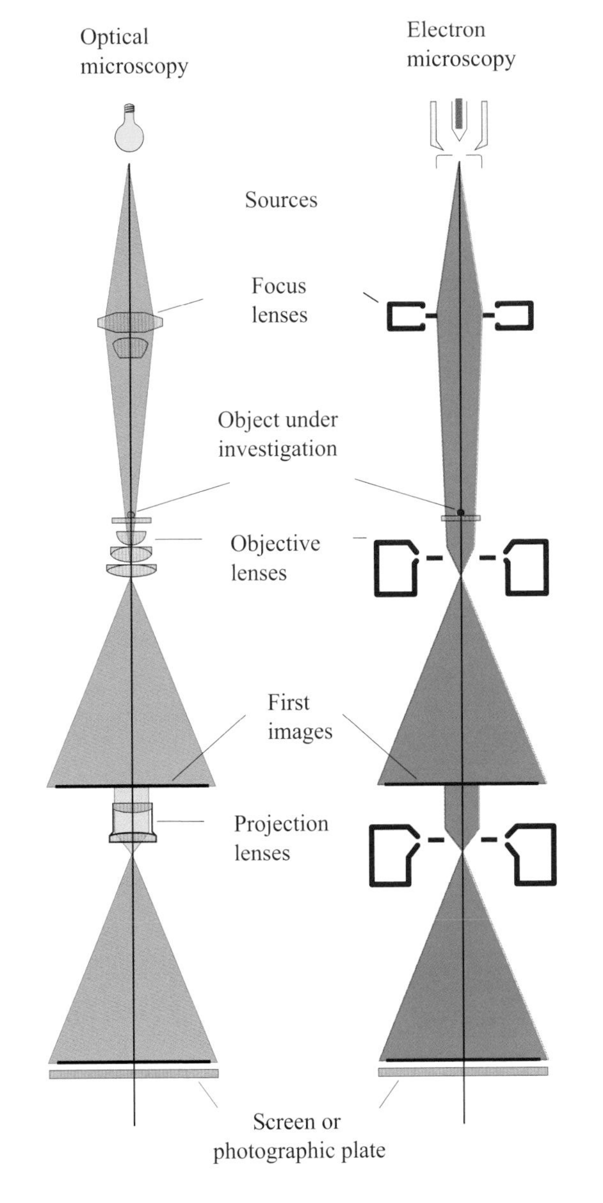

Compared principles of optical and electron microscopes

Scientific applications of electron microscopy cover all the fields. By 1954, 1200 research devices had been built by Siemens, the company which had engaged the services of Ruska. **Christian de Duve** and **Albert Claude** (1974 Nobel Prize for Medecine) used electron microscopy to demonstrate the existence of cellular organelles containing digestive enzymes. Ruska would probably have been forgotten by the scientific community (which suggests the award of the Nobel Prize) without the work of **Heinrich Röhrer** and **Gerd Binnig** (on the scanning tunneling microscopy), to whom the second half of the 1986 prize was awarded. Ruska was educated at the University of Berlin and earned his doctorate at the University of Munich in 1934. His thesis already concerned the use of magnetic fields to develop electronic lenses. He first took up his career in industry (most notably with Siemens) and headed the Fritz Haber Institute of the Max-Planck Gesellschaft in Berlin-Dalem until his retirement. He lived just over one year after receiving his Nobel Prize.

Binnig, Gerd (Frankfurt, Germany, July 20, 1947 -). Physicist and German experimenter. In 1978 Gerd Binnig, together with Heinrich Röhrer, created the first scanning tunneling microscope (STM), the only device today which makes it possible to give an image of the structure of surfaces atom by atom. The description of the device will be found in the bibliography of Röhrer. They received half of the 1986 Nobel Prize, the other half being given to Ernst Ruska for the invention (in 1928!) of the electron microscope. STM microscopy is applicable only to conducting surfaces; its AFM (atomic force microscope) version is used for insulating samples, biological samples in particular. Since his early childhood Binnig has been keen on physics and at the same time he has been fond of classical music and that of the Beatles and the Rolling Stones. He studied at the University of Frankfort, where he obtained his doctorate in superconductivity in 1978. He immediate-

Gerd Binnig and Heinrich Röhrer.

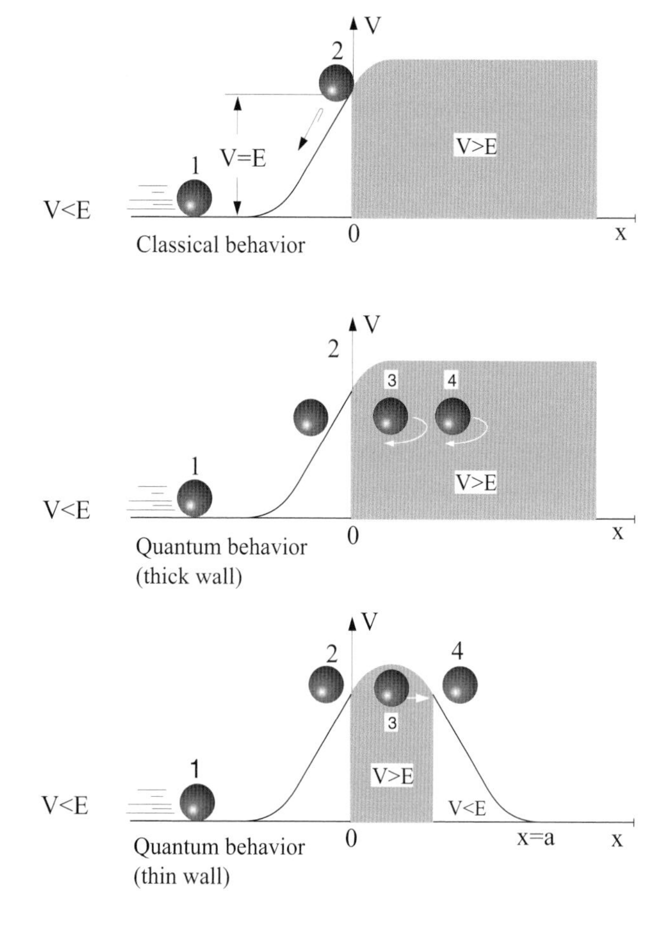

Tunnel effect. In the classical situation, a ball launched with an energy E along an inclined plane towards the higher platform located on a V>E floor cannot reach this platform. It falls down again at its arrival in 2 when it has exhausted its incident energy E (through climbing). In the quantum situation, the ball could penetrate the thick wall and reach levels 3 or 4, but would be compelled to fall back. For a thin wall, the preceding situation 4 enables it to be found behind the obstacle while "penetrating" it, hence the name "tunnel effect".

ly joined the IBM research center, where he carried out investigation in the field of STM. It was on the insistence of his wife (a doctor), whom he readily used to call "his private psychotherapist", that he accepted the position that was offered to him. He recognized that he made a good choice: "stimulating scientific environment, above all average technical resources, effective collaboration with Röhrer, good team atmosphere."

Röhrer, Heinrich (Buchs, Switzerland, June 6, 1933 -). Swiss physicist. Working at the famous IBM research laboratory near Zurich (Rüschlikon), it was in association with **Gerd Binnig** that he developed a new tool for investigating the conduction of semiconductor solids having a space resolution of about 0.1 nm: the scanning tunneling microscope (STM). Within the framework of their research on the surface of semiconductors, the essential components of micro - electronics, they wished to obtain spatial pictures with a surface resolution of 10 nm, a performance unattainable with the electron microscope developed by **Ruska**. In 1978 they developed a system based on a quantum process: the tunnelling effect. It was a matter of inducing, through a very thin potential barrier, a current between the material under investigation and a metal tip. Quantum mechanics stipulates that, if the tip is very close to the surface, a current can occur, even if the classical energy of the electrons is not sufficient to enable them to cross the barrier, which keeps them in the material. The quantum wave functions associated with the presence of the electrons of the tip and of the close by surface interpenetrate sufficiently to allow the passage of electrons from one to the other. The intensity of the current is a function which is very quickly variable with the gap between tip and surface. By forcing this current to remain constant (by acting on the tip to surface distance), Röhrer and Binnig invented a very sensitive method of measuring this interdistance. A monitored scanning of the tip in the vicinity of the surface therefore permits the measurement of the surface irregularities with an accuracy of 0.1 nm. Even better, as the binding of the electrons to the studied surface varies according to their own affinity with regard to neighbouring atoms, we have at our disposal a technology able to determine the depth of the chemical bound normally to the surface with a precision of about 10^{-3} nm. The hardest technological problem is to avoid the weakest vibrations of the device. Röhrer, educated in his young childhood in a rural environment, became interested in technology after his family settled in Zurich in 1949. He

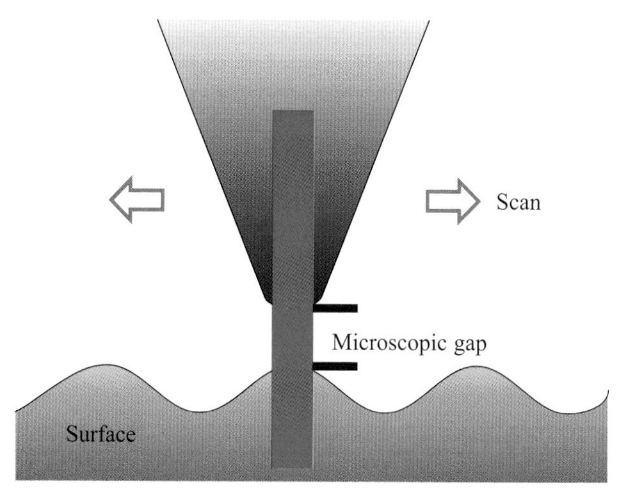

The principle of a STM microscope (scanning tunneling microscopy). Using the electronic version of the above mechanical model, a tunnel current can cross the small gap between a very sharp tip and the surface of a conductor when a voltage is applied. The intensity of the current strongly depends on the tip-plane distance. By controlling the system to work at constant current, the scanning allows the reproduction of the topography of the surface with a resolution of atomic level.

was educated at the Swiss Federal Institute of Technology. **Pauli**, the theoretical physicist, was among his professors. But Röhrer was most attracted by experiments and first worked on superconductivity. In 1963, after a working

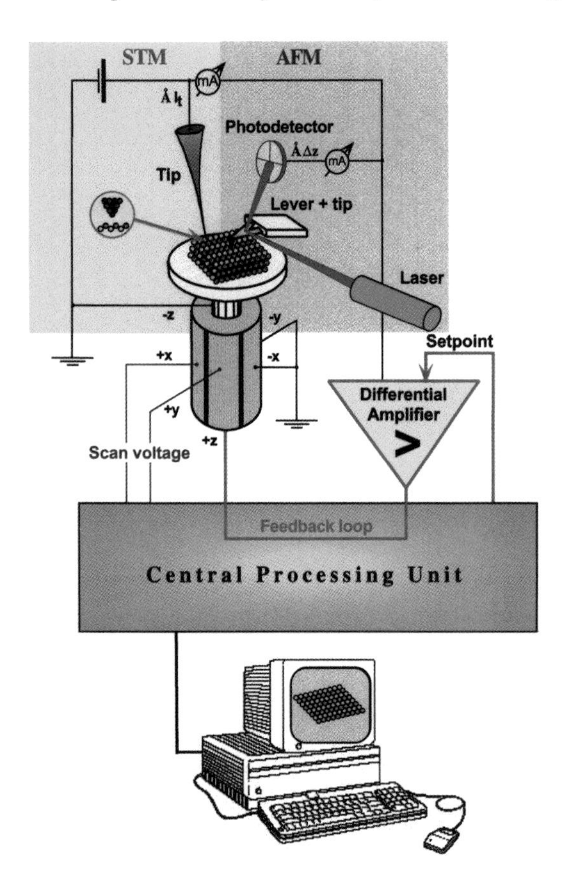

The principle of tunneling microscopy (STM) for the study of conducting surfaces at atomic scale and the corresponding AFM (atomic force microscope) for insulators.

stay in the United States, he was invited to join the IBM Research Group in Zurich. Still associated with Binnig, he had frequent contacts with Karl Müller, who was a user of the STM in his research on superconductivity. This research later earned the 1987 Nobel Prize for its author and **J. Georg Bednorz**.

1987

Bednorz, J. Georg (Neuenkirchen, Germany, May 16, 1950 -). A German physicist and currently the youngest among the Physics Nobel Prize winners, Georg Bednorz was rewarded together with **Karl Müller** for their discovery, made one year only before the decision of the Stockholm jury, of a mixed oxide which presents no resistance to the passage of electric current at a temperature of 35 K. This was noticeably higher and then closer to the normal temperature than that at which **Heike**

J.Georg Bednorz

Kamerlingh-Onnes had to work in the case of lead and mercury (4 K) in 1911 and even at the time when the **John Bardeen**, **Leon Cooper** and **John Schrieffer** team provided the theoretical explanation at the end of the 1950s. The discovery was achieved in December 1986 at the IBM Zurich Research Laboratory on an oxide $(La\text{-}Sr)_2CuO_4$. The potential repercussions of this discovery encouraged many mixed teams of physicists and chemists to set to work to try to develop chemical

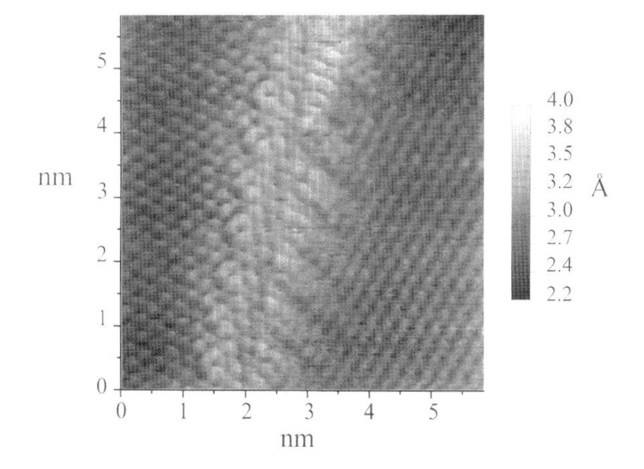

A STM view of a defect on a carbon surface

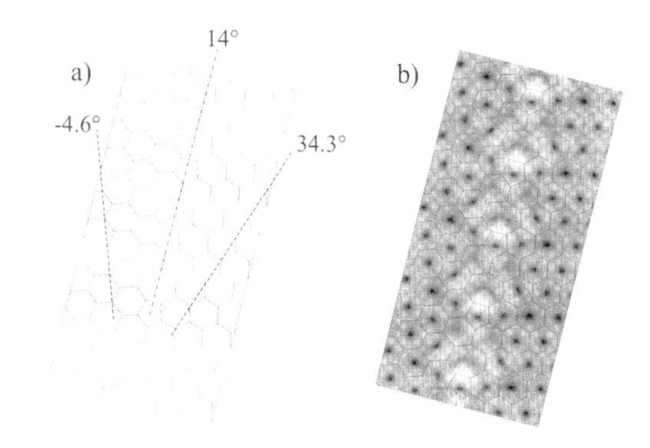

Reconstruction of the defects at the surface of a carbon surface

structures, which could become super conductive at a higher temperature. At the University of Houston, Ching-Wu Chu showed that the compound $Y_1Ba_2Cu_3O_{6.7}$ became super conductive at 84 K; when cooling in liquid nitrogen (and not in liquid helium, much more expensive) became possible. Bednorz began his university studies in

Munster. In 1982 he defended his doctorate at the Federal Technological University of Zurich and was immediately offered a position with IBM. Just like Karl Müller, he devoted part of his time to teaching at his former institute, as well as at Zurich University.

Müller, Karl Alex (Basel, Switzerland, April 20, 1927 -). Swiss physicist. A researcher at the IBM research centre in Zurich (Rüshlikon), Karl Müller with his young colleague **J. Georg Bednorz**, made superconductive a mixed oxide earned his Ph.D. at the Federal Institute of Technology in Zurich, Switzerland, in 1958. He first spent five years at the Bagatelle Institute of Geneva before joining the Rüshlikon Center. In 1962 Müller began to teach at the University of Zurich where he was appointed Professor in 1970, while still continuing his activities at the IBM research centre. Since 1985 he has devoted himself essentially to superconductivity.

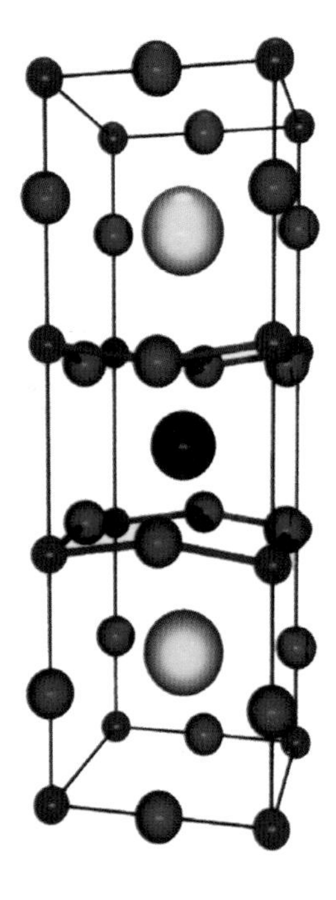

The YBaCuO superconductor. The $Y_1Ba_2Cu_3O_{6.7}$ oxide does not have any resistance to an electric current at temperatures lower than 90 K. In blue, the atoms of yttrium; in yellow, those of barium; in green, those of copper; in red, those of oxygen.

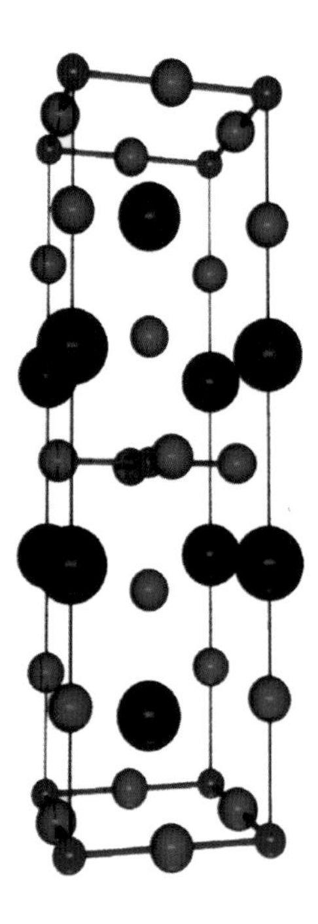

The $La_{(2-X)}Sr_X O_4$ superconductor with a Tc temperature of 35K. In blue, the La or Sr atoms, in green, copper atoms; in red, oxygen atoms.

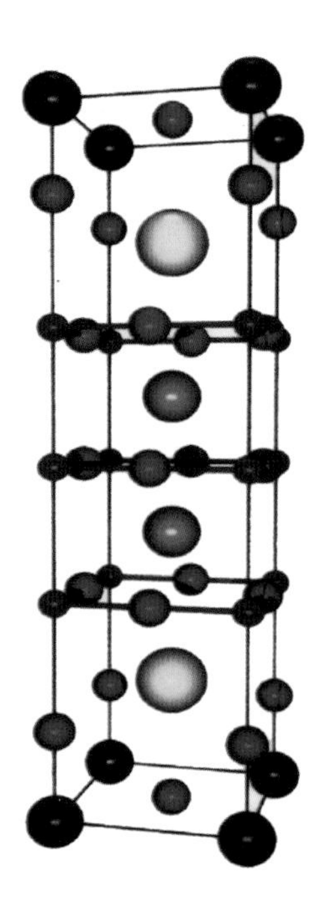

The $HgBa_2Ca_2Cu_3O_9$ superconductor discovered in 1993 whose critical temperature is currently the highest (135K). In blue, mercury atoms; in yellow, barium atoms; in orange,calcium atoms; in green, copper atoms; in red, oxygen atoms.

compound of barium, lanthanum and copper at a temperature of 35 K. In 1973, niobium-germanium compounds also presented an almost zero resistivity at a temperature of 23.3 K The Zurich la- boratory, which had already been rewarded the Nobel Prize for the works of **Heinrich Röhrer** and **Gerd Binnig** in 1986, was honoured once again the following year for the work of both of its researchers. Karl Müller

Karl A. Müller

1988

Lederman, Léon Max (Buffalo, New York, July 15, 1922 -). A high energy physicist and member of that "famous nursery" for Nobel Prize winners, the University of Columbia, Lederman formed a team with Melvin Schwartz and Jack Steinberger. In 1959 the team was approached by **Tsung Lee** who asked them to extend his own studies on weak interaction towards the high energies, where they risked being masked by phenomena connected with other, more intense interactions. A part of these experiments was to highlight whether the neutrino (actually its antiparticle) emitted simultaneously to the emission ß- (in the disintegration of the neutron, for example) was the same one as that produced at the time of the disintegration of the pion into muon and neutrino. The experiment carried out by the three Columbia researchers would allow them to make the distinction between the two neutrinos ν_e and ν_μ. It was carried out in 1962 with the

Léon M. Lederman

synchrotron of Brookhaven, which provided an intense beam of 30 GeV protons. Sent to a beryllium target, these protons generated a large quantity of pions. Their disintegration leads to the emission of the required neutrino but, owing to the weak interaction of the neutrino with matter, a detector of more than 10 tons had to be built to try to make some of them interact. The 10^{14} neutrinos produced in 10 days of experimenting showed that, in 40 events, 34 induced the creation of muons and only 6 the creation of electrons (to the last ones resulting from background noise in the detector environment, probably of cosmic origin). It was proved that the neutrinos, accompanying the muons, resulting from the disintegration of the pion are not of a comparable nature with those resulting from the disintegration of the neutron. The authors of this superb experiment, later confirmed in other laboratories, had to wait 26 years before receiving the recognition of the Nobel Prize. Lederman did not stand on his laurels after this success. In 1977, in the Fermilab of Batavia, he identified a new particle almost ten times heavier than the proton (this particle was called upsilon by him) and concluded the obviousness of the existence of a fifth quark (a component of the hadrons postulated by **Murray Gell-Mann**). It was given the name of quark b (b for bottom or beauty). Its doublet partner would be called top or truth and identified in 1995. The theoretical prediction of its mass has been determined by Veltman and 't-hooft. Lederman obtained his Ph.D. at the University of Columbia in 1951, occupied the position of Professor from 1958 until 1979, before becoming Director of the Fermilab from 1979 until 1989. He undertook his experiments in large high - energy physics laboratories, namely (in addition to Fermilab): Nevis, Brookhaven, CERN, Berkeley, Rutherford and Cornell. He holds many prizes and medals. He was also heavily involved in the rewriting of science courses in the State of Illinois. In his book "The God Particle", translated into French under the title "Une sacrée particule", he relates his life as a physicist... and that of some of his colleagues who "discovered" what he had already found!

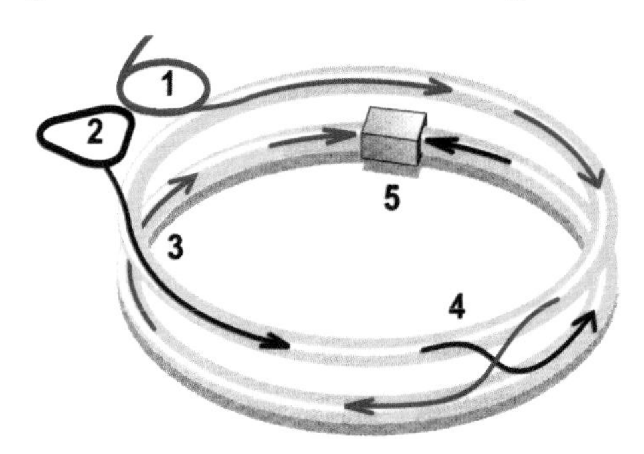

The principle of the Fermilab accelerator in Batavia which allowed the identification of a fifth quark b.(bottom or beauty).**1**): proton synchroton; **2**):storage ring; **3**) and **4**):tevatron accelerator; **5**):detector. Storage of protons and antiprotons respectively in (**1**) and (**2**) are re-accelerated in two different (**3**) and (**4**)) circular tubes (radius of the whole device :1km) before colliding (**5**) into a particle detector.

Schwartz, Melvin (New York City, New York, November 2, 1932 -). Experimenter physicist. In 1962 Schwartz showed, together with **Leon Lederman** and **Jack Steinberger**, that the neutrino, simultaneously emitted with the μ meson, was not the same as the one which accompanied the β^- disintegration of the neutron. Schwartz was, at this time, a member of Columbia University. He received his graduate diploma in 1953 and then worked as a researcher at the Brookhaven National Laboratory for two years. During this period he prepared his thesis. He obtained his Ph.D. in 1958. After conducting research at Columbia, where he was appointed Professor in 1963, he taught at the University of Stanford. In 1983 he left fundamental research, tired by the bureaucracy connected to the effort which had to be deployed in order to obtain the necessary funding. He then settled down in "Silicon Valley" and founded his own company, called "Digital Pathways" which specializes in computer security systems. Despite everything, he returned to Columbia in 1991 where he is an excellent professor, paying particular attention to awakening the interest of young people in science. He has pro-

Melvin Schwartz

duced excellent works on electromagnetism and electrodynamics.

Steinberger, Jack (Bad Kissingen, Germany, May 25, 1921 -). Physicist of German origin. Steinberger shared the honor, with **Leon Lederman** and **Melvin Schwartz**, of having started the classification of elementary particles in the form of triplet pairs: electron and electron-neutrino, muon and muon-neutrino, which would be followed by the tau (τ) (identified by **Martin Perl**) and by the tau-neutrino.These particles undergo weak nuclear interaction. As a Jew, Steinberger left Germany with his family, when he was 13 years old, and moved to the United States. He studied chemistry at the Illinois Institute of Technology, where **Enrico Fermi** was one of his professors. During the war he worked at the Massachusetts Institute of Technology,

Jack Steinberger

defended his doctoral thesis in 1948 at Columbia, and was appointed Professor at this university from 1950 to 1968. Then he returned to Europe and worked at CERN until 1986. The same year he was appointed Professor at the University of Pisa, Italy. It was there that he learned that he had been awarded the Nobel Prize.

1989

Dehmelt, Hans (Görlitz, Germany, 1922 -). Hans Dehmelt, a physicist of German origin, shared half of the Nobel Prize with **Wolfgang Paul**. (The other half of the prize was awarded to **Norman Ramsey**). Dehmelt and Paul's pioneering work concerned the attempt to "trap" a single electron, in order to isolate it completely from the influence of its environment and to be able to measure its magnetic moment with great precision. They invented the apparatus and then adapted it to the isolation of a single positron for months. He could thus fix, with an accuracy of 4 parts per billion, the value of the electron's magnetic

Hans Dehmelt

moment. After this success in the 1970's, Dehmelt did the same thing to isolate a single ion. In 1940, finishing his secondary studies, Dehmelt enlisted in the German Army and was taken prisoner by the American Army at Bastogne (Belgium). After the war he began studying physics at Göttingen University, receiving his doctorate in 1950. Holder of a specialization grant in 1952, he joined the group of physics at the University of Duke (North Carolina) from 1952 until 1955. He then accepted a position at the University of Washington, Seattle, and became a full professor in 1961 (the year in which he became an American citizen). He continues his scientific activity today in New Zealand.

Paul, Wolfgang (Lorenzkirch, Germany, August 10, 1913 - Bonn, Germany, December 7, 1989). German physicist. Working in Bonn on the same subject as **Hans Dehmelt** in Seattle, Wolfgang Paul designed a device, known as the Paul trap, which was able "to trap", and therefore to isolate, a single electron. This device is formed from 3 electrodes: two electrodes in the form of rounded pins located on the axis of a third electrode of toric shape. The circular ring is excited by an oscillating voltage, which allows the movement of an electron back and forth between the tip and the ring in order to finally stabilise it. Paul was also involved in mass spectrometry and in particle physics. In Bonn, he was the first director of the electron synchrotron. He was also the director of German laboratories of nuclear physics and of particle physics which included mainly teams working in Julich, in CERN (Geneva) and in DESY (Hamburg). For 10 years he was President of the Alexander von Humboldt Foundation (1979 - 1989). He retired scientifically in 1981. A few months before his death he was elected Honorary Senator of Bonn University. Paul graduated from the University of Munich but received his Ph.D. from the University of Berlin in

1939. After the war he taught physics at Göttingen until 1952 before teaching at the University of Bonn.

Ramsey, Norman Foster (Washington D.C., August 27, 1915 -). Norman Ramsey, an American physicist, devo - ted all his scientific career to the measurement of time with a great accuracy. He continued experiments already developed by his professor **Isidor Rabi** from Columbia University. The Nobel Committee awarded him half of the 1989 prize for his invention of the method of the separate oscillating fields and its use in the development of the hydrogen maser and other atomic clocks. We refer to **Charles Townes** for the invention of the maser, the basic

Norman F. Ramsey

instrument used in Ramsey's research. The latter is part of a long line of Nobel Prize winners who have worked in the field of ultra precise measurements for atomic and molecular physics, in the footsteps of Rabi. Let us mention, in particular, **Edward Purcell**, **Felix Bloch** and **Nicolaas Bloembergen** for physics and **Richard Ernst** for chemistry. Ramsey's method led to the construction of the cesium atomic clock: a device still used for the precise measurement of time. A second of time represents exactly 9 192 631.770 cycles associated with a transition between two energy states of the cesium atom. The same method, invented by Ramsey, this time using a molecular hydrogen beam, improves the performance, further leading to an accuracy in the measurement of the frequencies of one part in 10^{12}. These precise clocks are used to correlate the observations from telescopes located at long distances from each other and for navigation. Ramsey's work is not limited to molecular physics, but includes numerous phenomena involving the interaction of variable magnetic fields with the magnetic moment of elementary particles, the search for a possible electric dipole moment of the neutron. He was also involved in thermodynamics and statistical mechanics of very low temperature systems, and even the possibility of conceiving a negative thermodynamic temperature (below the "absolute zero") for several nuclear spins. After his official retirement he continued his research activities, announcing that the amount of the Nobel Prize would be valuable in providing for his travelling expenses to Grenoble, to use the high neutron flux reactor, in order to continue his fundamental experiments on the internal structure of the neutron. Ramsey graduated from Columbia and Cambridge (United Kingdom) and obtained his Ph.D. at Harvard in 1940. Then he held a post at MIT and was sent to Los Alamos where he contributed to the Manhattan Project. Finally he joined the University of Harvard as a professor from 1947 until his retirement. An excellent teacher, he was "adored" by his pupils who moreover considered him as a father. He was one of the founders of the B.N. Laboratory and, in 1946, the first President of the Physics Department of this research center. He was a visiting professor at the Universities of Oxford, Virginia, Colorado and Michigan. He was consulted, as an expert in ballistics, in the case of President Kennedy's assassination and in the debates on cold fusion.

1990

Friedman, Jerome Isaac (Chicago, Illinois, March 28, 1930 -). In the 1970's, together with **Henry Kendall**, a colleague from the Massachusetts Institute of Technology (Boston) and **Richard Taylor**, a compatriot from Stanford, Jerome Friedman resumed work similar to that which **Ernest Rutherford** (1908 Nobel Prize for Chemistry) had carried out almost 70 years earlier. Whereas Rutherford tried to come closer to measuring the size of the atom nucleus, this trio tackled the more difficult task of probing the internal structure of protons and neutrons. They wanted to clarify the structure of the nucleons (protons and neutrons). **Robert Hofstadter** had already confirmed, by the distribution of scattered electrons, that protons did not have a point structure. Using electrons of very high energy (8 to 15 GeV) produced by the Stanford linear accelerator (SLAC) as projectiles, Friedman, Kendall and Taylor highlighted the presence of substructures (quarks) in the proton (the nucleus of the hydrogen atom being used as a target) by measuring the inelastic scattering of the electrons. The concept of quark had already earned the Nobel Prize for Murray Gell-Mann. The information on the internal structure was

Jerome I. Friedman

more precise than that given by Hofstadter. It is currently agreed that the energy necessary to extract the quarks from their confinement in a nucleon must increase with the distance between these quarks. But, if the energy is increased, new quark - antiquark pairs are created and the situation becomes complicated: only groups of quarks or signs of the association quark-antiquark can be detected, never a single

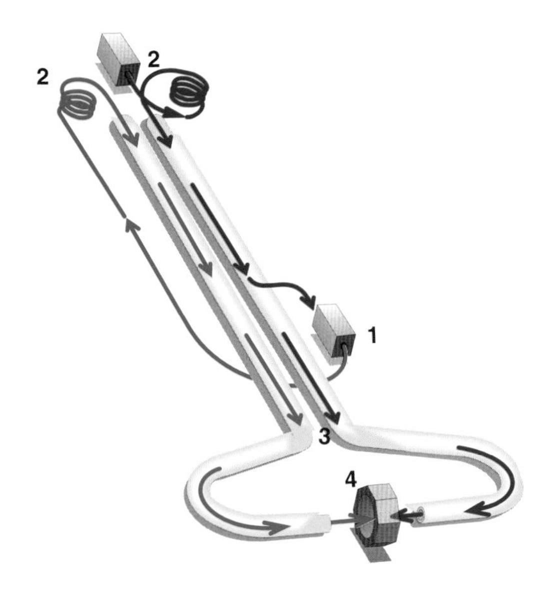

The principle of the linear accelerator of Stanford: **1)** positron source; **2)** storage rings of positrons and electrons; **3)** acceleration of both types of charges along 3 km; **4)** head on collision region.

quark. The experiment showing the presence of quarks in nucleons resembles one, which we could do by shaking a hollow and opaque ball, which has marbles inside. By shaking the whole, we could deduce that three hard internal clusters exist, without needing to open the ball. The "shake" is caused here by the electrons that are sent to the protons and whose characteristics are measured after inelastic scattering. Friedman was educated at the University of Chicago and earned his Ph.D. in 1956. He spent 3 years at Stanford University before moving to the Massachusettes Institute of Technology where he was later appointed to the chair of physics in 1967. He was associated with many teams in Cambridge (Massachusetts) and Princeton. Learning, via a phone call from his wife, that he was a Nobel Prize winner Friedman, then at a scientific meeting in Texas, was so surprised that he declared to his entourage: "it was so incredible that I believed I was still asleep and that it was part of my dream". **Steven Weinberg** who was also at the same meeting in Texas, declared that, for his part, the experiment carried out by Friedman, Kendall and Taylor marked a turning point in the acceptance of the quarks model.

Kendall, Henry Way (Boston, Massachusetts, December 8, 1926 - Wakula Springs, Florida, February 15, 1999). Together with **Jerome Friedman** and **Richard Taylor**, Henry Kendall showed, experimentally, that the existence of quarks inside nucleons (protons and neutrons) was more than a theoretical model that had been proposed by **Murray Gell-Man** and, before him, **Richard Feynman**. The latter was also very closely connected with the experiment which was carried out in 1967 at the powerful Stanford accelerator (SLAC). A member of an influential Boston family, Kendall studied at Massachusetts Institute of Technology (MIT) (Ph.D. in 1955). He then had a career parallel with that of Friedman: first in Stanford for three years, then director of MIT. They had the same promotions at the same time. Kendall was also involved in ecology, criticising the administration of President Bush for not having taken the necessary measures to avoid global warming of the atmosphere. In particular, he reproached the lack of initiative in the research on energy-saving technologies. He also militated to make nuclear power plants safer and to abolish the arms race in the program known as "Star Wars". On the second matter he was seconded in Europe by **Georges Charpak**. He drew the attention of the political authorities to the danger of an unintentional shooting of a missile at a civil target. is also recognised as a generous man: part of his considerable family inheritance was distributed to charitable works. He tragically died during a diving to make submarine photography, one of his main hobbies.

Taylor, Richard (Medicine Hat, Canada, November 2, 1929 -). Canadian physicist. Taylor collaborated with **Jerome Friedman** and **Henry Kendall** for revealing the existence of quarks inside nucleons (proton and neutron). He graduated from the University of Alberta, Canada, and in 1962 he defended his doctoral thesis at Stanford, California. He then entered a research group at the University of Berkeley for two years before definitively returning to Stanford University where he was appointed Professor of Physics in 1970.

1991

de Gennes, Pierre-Gilles (Paris, France, October 24, 1932-). A theorist, but at the same time very close to industry, de Gennes can be called a complete physicist. His approach to complex phenomena, governing the behavior of partially ordered molecular systems, like liquid crystals, superconductors and polymers, earned him the recognition of the Nobel Prize. de Gennes has never hesitated to tackle complex problems but has always endeavored to give simple explanations both in his writings and in the numerous conferences that he held for varied audiences. He came to the study of the "soft matter" while moving from solid-state physics to the physical chemistry of liquid crystals, polymers and any material apparently in disorder. He was also interested in many

Gilles de Gennes

materials that we all handle daily like the suspensions that form inks, the dyes in painting oils, mayonnaise, but also in more sophisticated materials like the ferro fluids with magnetic properties, polymers generating a grafting process, and copolymers made of two chains welded in a point with each link having different properties. To account for the movement of a long molecule chain, inter-

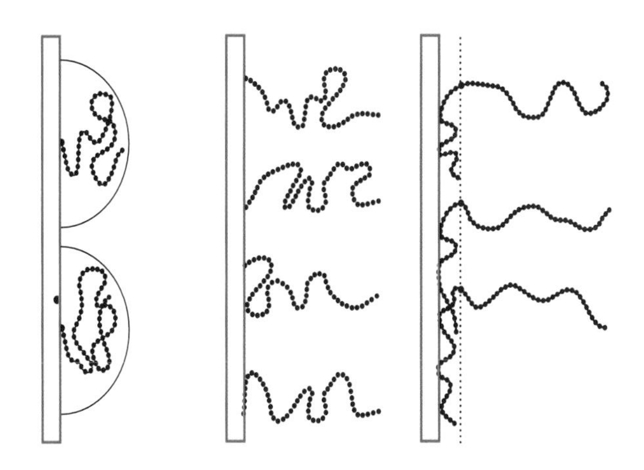

The snaking phenomenon (a term invented by de Gennes) to explain the behavior of a polymer chain which clings to a surface.

twined among a crowd of similar molecules, de Gennes proposed the concept of "snaking" similar to the way in which a worm manages to move in the chinks left between twigs and blades of grass, a description that allows, in particular, a good understanding of the mechanism of electrophoresis. Not satisfied to propose models for complex structures, and in order to quantify the concepts of gluing, adhesion and rupture, de Gennes also participated in experimental teams to determine the properties of these materials by optical methods (polarized light, ellipsometry) by neutron diffraction or measurement of the contact forces. His approach to physics classes him among visionary physicists like **Lev Landau** and **Richard Feynman.** De Gennes studied at the Ecole Normale and began research with the French Atomic Energy Commission on the magnetism of materials. He obtained his doctorate in 1955. He then went to Berkeley to work with Kittel, a world-renowned solid-state physicist. In 1961 he was welcomed at the Orsay Faculty of Science in the solid-state physics group where he was initially interested in superconductivity, which had recently been explained by **John Bardeen**, **Léon Cooper** and **John Schrieffer**. Subsequently, he set up a research group on liquid crystals. Since 1971 he has been a Professor and Director at the College of France where he formed a new research group on polymers.

1992

Charpak, Georges (Dabrovica, Poland, August 1, 1924 -). Persistent research worker, permanent member of CERN in the search for the furtive information that subnuclear components can leave in materials, Charpak is the inventor of a new type of particle detector: the multiwire proportional chamber which he created in 1968. These ultra-

Georges Charpak (at left : Guy Demortier)

fast detectors make it possible to visualize, in three dimensions, the trajectories of high energy particles when interacting with the gas of the device . They gave many teams of research workers, spread throughout the world, the means to understand the behavior of matter on a scale of 10^{-17} m and consequently the "shape" of this matter such as it existed a short time after the Big Bang. Charpak's detector is made up of thousands of tight wires in a very big volume filled with gas. These wires, of 20 μm diameter and placed approximately 1 mm apart from one another, form a plane. They are connected to a positive voltage (anode). On both sides of this plane of wires, and par-

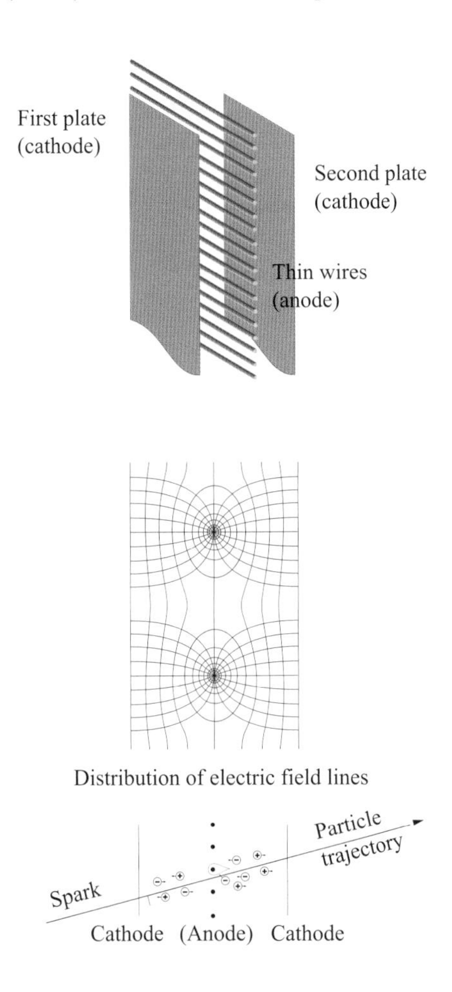

The multi-wires detector. a) The principle: an assembly of tight wires forming a plane and plates in parallel. b) The distribution of the electric field: very strong close to the wires, weaker further away from the wires. c) The migration of the sparks of electric charges produced during the passage of a fast charged particle.

allel to it, there are two plane plates (cathodes), which face the plane of wires. In the space between the two cathodes, the areas close to the wires are regions of an intense electric field. On the other hand, the areas far away from the wires are regions of a weak electric field. A particle entering this detector induces electron - ion pairs. The (light) electrons attracted by the anode produce, in its vicinity, a multiplication of the pairs (electron - ion) resulting in an avalanche. This in turn leads to a local amplification of the signal, limited to the single area close to the concerned anode wire. Each anode can thus be considered as an individual detection site. A main property of this assembly is that wires act as individual detectors even their strong coupling property. The global detector, made up of a multitude of these individual sites, makes it possible to simultaneously follow the trajectories of all the induced debris in a collision with one of the atoms of the detector and thus to identify each one of these debris. Multiple alternatives of this basic device are at present installed near big accelerators. The description of other detectors can be found in the biographies of **Charles Wilson**, **Donald Glaser**, **Walther Bothe**, **Luis Alvarez** and **Pavel Cherenkov**. Assemblies of several types make up the technological park for particle physicists and astrophysicists. **Burton Richter** and **Samuel Ting**, for their discovery of the J/ψ, as well as **Carlo Rubbia** and **Simon van der Meer** for their discovery of the W^+, W^- and Z^0 bosons, used the technology developed by Charpak. Charpak also developed other detectors for biology, medicine and astrophysics. In medicine, the performance of the detectors built by Charpak makes it possible to decrease by a factor of 100 the exposure time for X - ray examinations. The St. Vincent de Paul hospital in Paris was the first purchaser of the system in the Western countries. A simplified model had already been installed at Novosibirsk Hospital in 1987 by Schektmann. Of Polish origin, Charpak arrived in Paris in 1931. After having studied in Montpellier, he obtained an engineering degree at the School of Mines in Paris in 1946, the year he became a naturalized French citizen. Involved in the resistance, he was arrested, before the end of his studies, by the Vichy authorities in 1943 and was sent to Dachau, where he stayed until 1945. He defended his doctoral thesis in 1955 in the laboratory of **Frédéric Joliot** at the College of France; seconded by the CNRS to the synchrocyclotron laboratory of CERN (Geneva) in 1959. He became a staff physicist of CERN in 1963. In 1984 he was also holder of the J.-Curie Chair at the Higher School of Industrial Physics and Chemistry of Paris, directed by **Pierre-Gilles de Gennes**. Author of three recent books written for the general public , he is very popular in the media, much appreciated for his lucid views on the use of nuclear energy and radioactivity in general and for the fight which he started, with Garwin, against excessive arms build-up by the Great Powers. Fascinated by archaeology, he dreams of being able to grasp a "fossil song" left by a potter in a pot he had thrown.

1993

Hulse, Russel (New York City, November 28, 1950 -). At the time of his discovery in 1974, Russel Hulse, an astrophysicist and young graduate of the University of Massachusetts (Amherst), was a doctorate student under

Taylor's supervision. The Nobel Committee awarded them the 1993 prize for their identification of a new type of pulsar (by means of a receiver of radio waves built in their laboratory with re - employed equipment). Contrary to the case of J. Bell, a student (in Cambridge, U.K.) of **Antony Hewish** who discovered the first pulsar in 1967 but who was not included in the 1974 Nobel Prize, this time there was recognition of the common work undertaken by the professor and his assistant. With their receiver installed on the largest radio telescope in the world (Arecibo, Puerto Rico), Hulse and Taylor observed that the signals did not reach them in a regular way (like those detected by Jocelyn Bell and explained by Hewish), but that the intervals between the flashes increased and decreased in turn. The double periodicity was interpreted by the presence of binary neutron stars (PSR B 1913+16) orbiting one around the other: the difference in the observed periodicity coming from a Doppler effect. When the binary star is furthest away from the earth the pulses seem more spaced out and when it is nearer the earth the pulses then seem more frequent. For a period of this particular

Russel Hulse

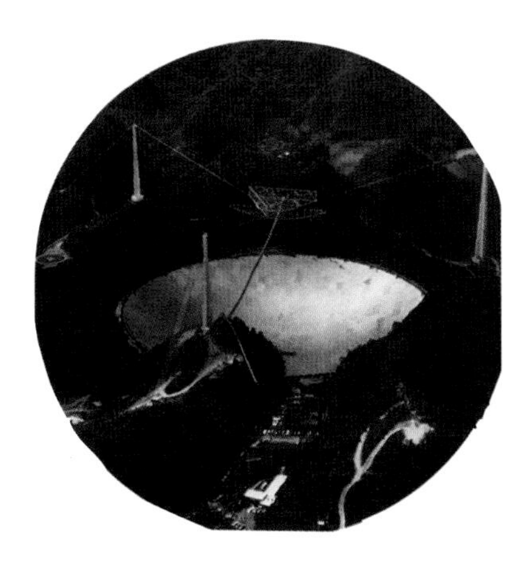

The big telescope facility of Arecibo.

pulsar (0.059 second), Hulse observed some variations of 5 microseconds from one day to another, a complete variation cycle recurring every 7 hours 45 minutes or, more exactly, every 27906.9807804 seconds. In the opinion of specialists, this observation provides another proof of the exactitude of **Albert Einstein's** theory of general relativity and of complementary arguments establishing the existence of gravitational waves. Graduated from the University of Massachusetts, Hulse defended a doctoral thesis to the study of binary pulsars in 1975. Since then he has held a post at Princeton in the group of plasma physics.

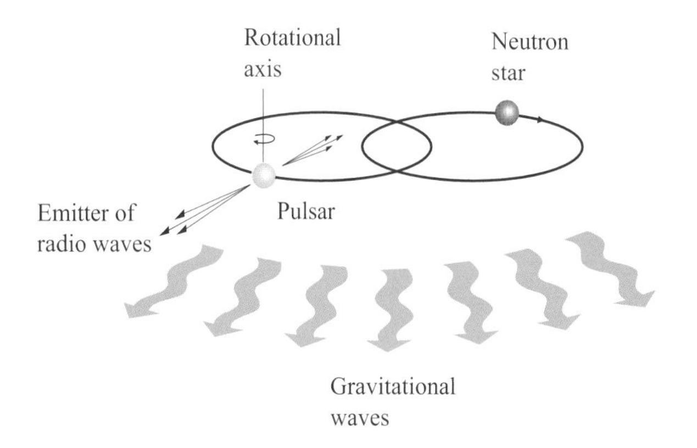

The principle of the emission of radio waves by a pulsar revolving on itself but associated to an (invisible?) companion:a neutron star.

Taylor, Joseph Hooton Jr. (Philadelphia, Pennsylvania, March 29, 1941 -). American astrophysicist. Joseph Taylor received the Nobel Prize, together with his former student **Russel Hulse**, for the identification of pulsar-emitting radio waves whose period was modified by its rotation around a companion of almost identical mass which did not emit radio waves. They had built a special detection device for the observation of (single!) pulsars with the large radio telescope of Arecibo (Porto Rico). The pulsar being 16,000 light years away, it is understandable that the radio transmission is heard very faintly. However, the device, developed cheaply by Hulse and Taylor, permitted the measurement of a daily variation of the pulse periodicity of only 5 microseconds. Repeated experiments, during several years, have shown a very slight reduction in the rotation period of the pulsar, attesting to a probable emission of gravitational waves, as predicted by Taylor in 1978. The Nobel committee, with its usual reserve, formulated the reason for the award of the prize as follows: for the discovery of a new type of pulsar, a discovery that has opened up new possibilities for the study of gravitation. Contrary to Hulse who joined the group of plasma physics in Princeton, Taylor continued his work on pulsars with significant data processing resources. In the radio noise emitted from space, Taylor's group discovered pulsars of very short period and conse-

Joseph H. Jr Taylor

quently heavenly objects whirling very quickly on themselves. In 1990, about twenty of these pulsars (five of which were discovered by Taylor) have a period of about 1.5 milliseconds. Joseph Taylor was educated at Haverford College, Pennsylvania, and received his Ph.D. at the University of Harvard in 1968. He then entered the University of Massachusetts, Amherst, and was appointed Professor in 1977. Since 1980 he has been teaching physics at the University of Princeton.

1994

Brockhouse, Bertram Neville (Lethbridge, Canada, July 15, 1918 -). Bertram Brockhouse, a Canadian researcher, devoted most of his scientific career to neutron spectroscopy and particularly to the inelastic scattering of neutrons. The neutrons, produced in significant numbers by a nuclear reactor (or with less intensity in a collision of a charged particle from an accelerator with a nucleus), can interact only with the atomic nuclei of the matter to which they are sent. As these neutrons are not charged, they do not undergo an interaction with the atomic electrons. Thus, after their interaction with a nucleus, the energy that they carry away is only a function of the isotope encountered. If part of the neutron's energy was given up to the nucleus, we talk about inelastic collision. The care taken by Brockhouse to design various apparatuses for the selection of the incident energy of the neutrons and the precise measurement of their energy after scattering allowed analytical developments in the knowledge of the structure of crystal at the submicroscopic level. Contrary to X - rays, neutrons make it possible to characterize the presence of light nuclei in molecular structures, and in particular in molecules of biological interest. Neutrons (having a magnetic moment: a property which seems astonishing for an electrically neutral particle but which was explained by **Felix Bloch**) are also able to probe magnetic structures. In "labelling" molecules with cadmium or samarium (nuclei having a particularly large capability to interact with neutrons), neutron spectroscopy can lead to the determination of the structure of large molecules like insulin. It is also extremely useful for determining the structure of superconductors and materials used in electronics and metallurgy. Before Brockhouse's work, neutron scattering had already played an essential part in the design of nuclear reactors. Brockhouse began his university studies only after the Second World War, during which he served as a reserve volunteer in the Canadian Navy. Previously he had been a laboratory assistant and radio repairer. He obtained his doctorate in 1950 at the University of Toronto on a solid - state physics subject: properties of ferromagnetic materials. Immediately after completing his thesis he joined the Canadian Atomic Energy Agency laboratory in Chalk River, which, at that time, possessed the nuclear reactor delivering the largest fluxes of neutrons. The Brookhaven Laboratory and the Institute Laue Langevin at Grenoble, founded by **Louis Néel**, are currently the most impressive in this field. Bertram Brockhouse was appointed Professor at McMaster University, Ontario, from 1952 until his retirement in 1984.

Bertram N. Brockhouse

Clifford G. Shull

Shull, Clifford G. (Pittsburgh, Pennsylvania, September 23, 1915 - Lexington, Massachusetts, March 31, 2001). American experimenter. Clifford Shull received the Nobel Prize 45 years after having carried out the decisive experiments. **Bertram Brockhouse**, the joint winner, had to wait almost as long. Both of them were pioneers in the characterization of materials by neutron diffraction but they worked independently. Their fundamental research has been exploited by numerous teams and led other solid-state physicists to the Nobel Prize. The objectives of their work are to be compared to that of **Max von Laue**, **William Henry** and **William Lawrence Bragg**, for the use, with a similar objective, of X - rays, to that of **Clinton Davisson** and **George Thomson** for the use of electrons and, to a lesser extent, to that of **Fritz Zernike** for his invention of a specific optical microscopy. These objectives concern the identification of atomic structures. Neutron crystallography is so promising that conferences exclusively devoted to neutron diffraction are regularly organized. The laboratory of Oak Ridge had, in project for that purpose, the installation of a reactor with a high flux of neutrons. The Rutherford Appleton Laboratory (Great Britain) built, for its part, an accelerator also capable of producing a very high flux of neutrons. Japan has projects also in this direction. The recognition of the work of Brockhouse and Shull should help the scientific community to convince politicians to release significant funding to this end. Clifford Shull graduated from the Carnegie Institute of Technology in 1937 and received his Ph.D. from the University of New York in 1941. He then worked as a research physicist at Texas Company (now Texaco), Beacon (New York) before joining Oak Ridge National Laboratory. Finally he joined the Massachusetts Institute of Technology where he taught from 1955 until his retirement in 1986. He held the office of President of several scientific societies involved in solid state physics. He took part in the Solvay Council of Physics in Brussels in 1954.

1995

Perl, Martin Lewis (New York City, New York, June 24, 1927-). American experimenter. Since the beginning of the 1970s Martin Perl, with a large team of high energy physicists, has pursued research to discover potential (super heavy) "cousins" of electrons (particles identified

Martin L. Perl

by Joseph J. Thomson one century ago) and of muons (identified by Carl Anderson in 1938). The muon has a mass equivalent to almost 207 times that of the electron; the tau a third member of that family would have a mass 3500 times greater than that of the electron. According to **Murray Gell-Mann's** standard model of classifying the leptons, the tau would be part of this third family, or more exactly, going very far back in the past, of the second generation having preceded the arrival of the electrons. Three neutrinos, of which **Leon Lederman** showed the nonidentity, are associated with these three particles. In parallel with these leptons, which only interact through weak nuclear interaction, another family, the quarks, interact via strong nuclear interaction. At the time of the iden-

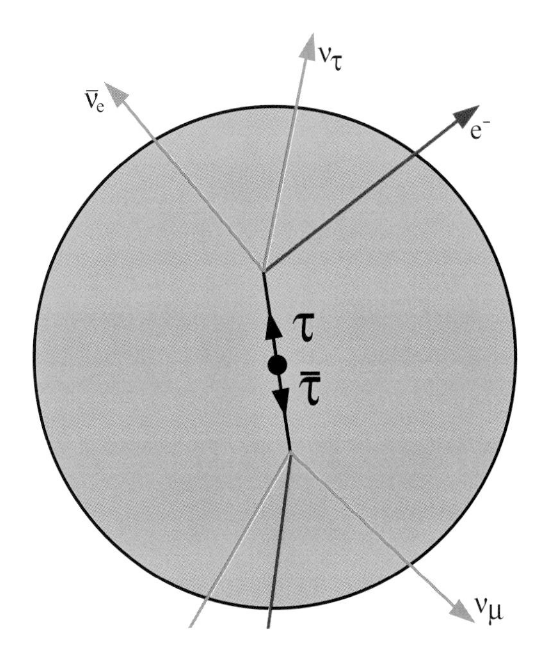

The detection of the disappearance of a pair of leptons $\tau,\bar{\tau}$ starting from the traces left in the detectors identifying a $\mu+$ and a e^- coming respectively from τ and $\bar{\tau}$. The initial leptons τ and $\bar{\tau}$ also produce neutrinos which cannot be detected.

tification of tau by Perl in 1979, we knew only two families of quarks, u (up) and d (down), which are the component parts of protons and neutrons, and s (strangeness) and c (charm), names given by Gell-Mann and **Sheldon Glashow**, respectively. But the third family has been recently identified: the quarks b (bottom or beauty) and t (top or truth). Of course, to each of the particles are associated the corresponding antiparticle. The discovery of the tau lepton (whose mass is close to twice that of the proton), has a half-lifetime of 5 x 10^{-12} seconds and can therefore only be identified through the detection of the particles created during the disintegration of this lepton. It was with the Stanford linear accelerator that Perl and **Burton Richter's** team identified the products of the disintegration of the tau and its antiparticle (the antitau) produced simultaneously. They were produced in the head - on collision of electrons with positrons. These particles had to have a kinetic energy sufficient to produce, in their collision, the mass of the desired particle and its antiparticle . The emitted particles then had to respect a similar rule, to produce the neighbouring "cousin", the muon. It is known that this gives rise to an electron, a neutrino associated with the ν_μ muon and an antineutrino associated with the electron ν_e.Thanks to impressive detection methods: namely, four concentric spark chambers inside an arrangement of scintillators responsible for beginning the operation of spark chambers, another detector arrangement able to detect the electromagnetic showers, an iron shield to stop the electrons and finally spark chambers for the identification of the muons, Perl and his team concluded that a tau and an anti - tau (produced simultaneously at the time of the annihilation of the electron and the positron) were emitted in opposite directions and themselves disintegrated (after having travelled less than one mm) in two ways: the one (the anti - tau) would be transformed in $\mu^+ + \nu_\mu + \bar{\nu}_\tau$, the other (the tau) another "triplet" $e^- + \nu_\tau + \bar{\nu}_e$. However only the e^- and the μ^+ were detected (because charged) and shared the appropriate momentum and mass: namely, those which can be deduced from the shape of curved trajectories they travel in the magnetic field applied throughout the region of detection. All the neutrinos ($\bar{\nu}_\tau$, ν_τ,ν_μ,$\bar{\nu}_e$)are invisible. Contrary to the electron, the positron and the muon, which were identified by chance, the search for the tau (a heavy lepton) was meticulously prepared by excluding any possibility of confusing it with any other hadron. The name tau was given by Perl himself as the third (τριτου: in Greek) of the family. Perl graduated in chemistry from the Brooklyn Polytechnic Institute in 1948. He then worked in the chemical industry but returned to college for further studies in physics and earned his Ph.D. at Columbia in 1955. From 1955 to 1963 he held a post at the University of Michigan and then was appointed Professor at Stanford.

Reines, Frederick (Paterson, New Jersey, March 16, 1918 - Irvine, Kentucky, August 28, 1998). Experimenter phycisist. The first to have highlighted the extremely weak interaction of neutrinos with matter. Reines seems to have been forgotten by the Nobel committees for many years! Indeed, it was in the middle of the 1950s that Reines, in partnership with Clyde Cowan (1919-1974), tried to track the antineutrino. This particle is simultaneously emitted with an electron when a neutron disintegrates by the way $n \longrightarrow p^+ + e + \bar{\nu}$. The existence of this antineutrino had been postulated by **Wolfgang Pauli**, in 1930, to guarantee the conservation of the linear momentum and the energy in β radioactivity. In the model of neutron disintegration, **Enrico Fermi** anticipated that this neutrino would have a zero mass. Having no charge and no mass, this particle, which undergoes neither the electromagnetic nor the strong interaction, was expected to be extremely elusive. Reines and Cowan's proposal was to use high flux of antineutrinos produced in a fission nuclear reactor (that of Hanford in the State of Washington) to make them interact with protons in order to give rise to a neutron and a positron. It is known that neutrons are particularly well absorbed in cadmium and that they induce γ-rays after their capture. It is also known that when the positrons are stopped in matter, they give rise to two γ rays of 511 keV each in opposite directions. By the detection of these γ rays of 511 keV followed by the detection (15 microseconds later) of the delayed γ rays emitted after capture of the neutron in cadmium, Reines and Cowan intended to identify the antineutrino. A complicated assembly of 90 detectors, containing cadmium, was installed in a shield designed to avoid any contamination by any ionising particle which might come from the outside. This first experiment, carried out in 1952, was a semi-failure. As the probability of this reaction of the antineutrino with a proton, calculated by **Hans Bethe** and Rudolf Pierls, could be about 10^{20} times weaker than the strong interactions that protons can undergo, they moved to a device containing 1500 kg of liquid scintillators distributed in three containers containing cadmium chloride and 110 solid state detectors of γ rays, all installed near the Savannah River reactor (South Carolina). With this gigantic device they succeeded (in 1956), in identifying that the antineutrinos

Frederick Reines

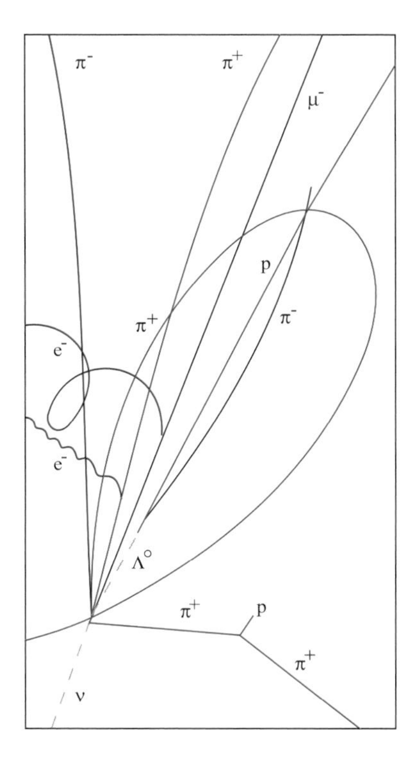

Interaction of a high energy neutrino with a proton in a bubble chamber gives rise to 5 particles: a muon (μ^-), four pions (3 π+ and 1 π^-) and also a Λ (neutral). In addition, one observes spiral-shaped trajectories of electrons having undergone a collision.

provided 2.88 ± 0.22 events per hour. The three types of neutrinos ν_e, ν_μ, ν_τ and their antiparticles known today could be only three partners of more complex phenomena. Numerous experiments to track them down are in progress, using the sun as a source. In each fusion in the sun, four protons give rise to one helium and two neutrinos, but many of them seem to become lost on the way to the Earth. In the disintegration of the pion into muon, **Leon Lederman**, **Melvin Schwartz** and **Jack Steinberger**, have also shown that the ν_μ was different from the ν_e. Speculations, in all directions, on the properties of the ν_τ are happening among the theoreticians. The story of the production and the detection of neutrinos is far from complete. Reines earned his M.A. in 1941 at the Steven Institute of Technology and he earned his Ph.D. at New York University in 1944. From 1944 until 1959 Reines was a member of the theoretical physics group of the Los Alamos Laboratory. In 1959, he headed the Department of Physics at the California Institute of Technology before moving to the University of California at Irvine. He worked there from 1966 until 1988, the year of his official retirement. After this date, he still remained scientifically very active. He was honoured by many prizes, was made a member of numerous scientific societies and became a member of the Academy of Sciences of Russia (1994).

1996

Lee, David M. (Rye, New York, January 20, 1931 -). David Lee, American experimenter, shared the Nobel Prize with his former student **Douglas Osheroff** and his colleague **Robert Richardson** for an experiment carried out, in 1971, at Cornell within the framework of Osherhoff's Ph.D. thesis. They showed then that the isotope of mass 3 of helium became superfluid at a temperature of 0.0026 K, no longer had any viscosity and flowed through microscopic chinks. In 1938 **Piotr Kapitza** and his pupil **Lev Landau** had explained the reason for the superfluidity of isotope 4 of helium. It seemed that helium 3 (a fermion), a particle of half integer spin could not become superfluid, a property reserved for bosons. Therefore, the passage to the superfluid state that helium 3 had to go through required the association of fermions in pairs to give bosons (particles with integer spin values), as proposed by **Leon Cooper** to explain an electrical phenomenon (superconductivity), similar to this mechanical phenomenon of superfluidity. The superfluid model of helium 3 is very special because of the association of the particular magnetic moments of both atoms of the pair, an association anti-parallel to that of the electrons in a Cooper pair. In the presence of a magnetic field, three different superfluid phases are observed. Lee graduated in physics from Harvard in 1952. He then joined the University of Connecticut in 1955 and gained his Ph.D. at Yale in 1959. Then he moved to Cornell where he is currently professor, with several positions as visiting professor at the Brookhaven National Laboratory and in the universities of Beijing, California and Grenoble.

David M. Lee

Osheroff, Douglas (Aberdeen, Washington, August 1, 1945 -). American experimenter. Douglas Osheroff was carrying out his thesis work in the laboratory directed by **David Lee** and in which **Robert Richardson** was assistant, when, in 1972, he discovered the superfluidity of helium 3 at very low temperature. Osheroff was beginning the study of the nuclear antiferromagnetism of helium 3 and in fact was not looking for the superfluidity pheno- menon. For his experiments he needed to use a large magnet, which was also used by several groups. It was when this equipment was not available that Osheroff, then working day and night (as is common when working on a thesis), to improve an experimental device intended

Douglas Osheroff

to produce solid helium, identified a phase transition in an experiment using nuclear magnetic resonance. In his "laboratory notebook", he wrote on the night of April 20, 1972: "we have discovered a BCS type transition in liquid helium tonight". The BCS theory (**John Bardeen**, **Leon Cooper**, **John Schrieffer**) describes the transition of a material to the superconductive state by the coupling of fermions to give bosons. When the result was submitted for publication it was rejected at first, but finally accepted after insistence. Osheroff graduated from Cal Tech in 1967 and obtained his Ph.D. at Cornell in 1973. From 1972 until 1987 he was part of the research team on solid state physics at the AT&T Bell Laboratory, of which he was Director from 1981 until 1987, when he was appointed Professor at Stanford. He received numerous prizes before his Nobel Prize honour.

Richardson, Robert C. (Washington D.C., June 26, 1937 -). American experimenter. Robert Richardson is the 150th laureate of the Nobel Prize in Physics.The discovery on the superfluidity of helium-3 dates back to 1973. Richardson began his scientific training in the V. P. Institute where he obtained his master's degree in 1960. Richardson gained his Ph.D. from Duke University, North Carolina, in 1966. Then he joined Cornell as a researcher and climbed through the ranks to be appointed Professor in 1975.

Robert C. Richardson

Currently, he is Director of the Laboratory of Atomic and Solid State Physics at the same university. In 1981, he was attached as visiting scientist to Bell Laboratories in Murray Hill (New Jersey). His research field remains the physics of very low temperatures (sub-millikelvin) in collaboration with Lee.

1997

Chu, Steven (St. Louis, Missouri, 1948-). Steven Chu was awarded the 1997 Nobel Prize in Physics, together with **Claude Cohen -Tannoudji** and **William Phillips**, for their work on ultracold atoms. We refer to the description of their joint scientific activity (which gives a new dimension to the study of light - matter interaction) in the C. Cohen-Tannoudji biography. Only one year after awarding the Nobel Prize to **Douglas Osheroff**, **David Lee** and **Robert Richardson**, 1996 Nobel Prize winners, the Nobel Committee once again honored low temperature physics. A graduate of Rochester University in 1970, Doctorate from the University of California at Berkeley in 1976, Chu worked on the AT&T Bell Telephone research team from 1978 until 1983, and became Director of the Quantum Electronics department of this company before accepting a position as Professor at Stanford Univer-sity in 1987, a position which he still holds today. He was visiting Professor at the College of France in Paris in 1990, where he collaborated directly with C. Cohen-Tannoudji. He has won many prizes and is a member of several Academies of Sciences. He continues high precision research in nuclear atomic physics and in polymers and is also interested in the description of the subtle movement of isolated proteins with a resolution of the scale of the molecule.

Steven Chu

Cohen-Tannoudji, Claude (Constantine, Algeria, 1933-). Claude Cohen-Tannoudji and his American colleagues **Steven Chu** and **William Phillips** were chosen by the Nobel Committee for the Nobel Prize in Physics for the development of a method permitting individual atoms to be brought to a state of quasi rest, by making them interact with a laser beam. It is known that the temperature of a disordered distribution of atoms is directly related to their average kinetic energy. By reducing the speed of these atoms, their temperature is thus automatically reduced. By subjecting a gas in an enclosure to the action of a powerful laser, it is possible to control subtly their disordered path and to bring these atoms nearly to rest in such a way that the temperature related to their residual disordered movement is as low

Claude Cohen-Tannoudji

as 1 micro Kelvin, thus only one millionth of degree above the temperature of absolute zero. Sodium atoms, in particular, would then have a speed of a few centimeters a second, whereas this speed is about 600 m/sec at the ambient temperature. It is not only a question of a technical achievement by itself, but above all of a method that allows the deepening of our knowledge of the interaction of light with matter and that leads in particular to the creation of markedly most efficient clocks. The basic principle of the cooling of atoms lies in a repeated exchange of (resonant) absorption of radiation emitted by a laser and the re-emission by the atom of a radiation of a slightly different frequency due to the recoil of the re - emitting nuclei. These repeated recoils (up to a hundred million times a second) lead to a global deceleration of the concerned atom, about a hundred thousand times higher than that which produces the action of earth gravity. Everything happens as if the atoms moved in a viscous medium. Atomic physicists, moreover, gave to this medium, bathed in an intense radiation, the name of optical molasses. In front of his students, Cohen - Tannoudji associated with this technique the idea of the dressing of the atom by radiation. The physics of cold atoms could still lead to the study of new states of matter. The 1989 Nobel Prize for Physics, awarded to **Hans Dehmelt** and **Wolfgang Paul**, had already crowned work in the field of atom trapping. C. Cohen - Tannoudji was the pupil of **Alfred Kastler**. He passed his university teacher examination in physics in 1957. He entered the CNRS in 1960. Since 1962 he has taught at the University of Paris VI and has held the Atomic and Molecular Physics Chair at the College of France since 1973, directed by **Pierre-Gilles de Gennes**. At the present time he manages a very high level international team, within the framework of the Centre of Atomic Physics and Optics of the College of the Ecole Normale Supérieure, called the Kastler-Brossel Laboratory. In 1996, he received the gold medal from CNRS. He is regularly asked to give advanced courses at Harvard and Yale, and universities in Leyden, New York, Pisa, and Toronto. He is the author of a remarkable treatise on quantum mechanics. He is an associate member of several academies, and in particular the Royal Academy of Sciences, Letters and Fine Arts of Belgium.

Phillips, William D. (Wilkes-Barre, Pennsylvania, November 5, 1948 -). William Phillips carried out experiments on "cooling atoms " reaching temperatures close to a few billionth of a degree above absolute zero, that is to say a level noticeably lower than that which could be encountered in any other part of the universe. To reach these low temperatures, the team of W. Phillips at first made the atoms interact with several crossed, high - intensity laser beams (several tens of millions of photons must interact with the same atom). He reduced the movement of these atoms even further by making them undergo a **Zeeman** effect in a magnetic field, whose intensity varies (in an appropriate way) along the atom's pathway. The use of the laser for cooling atoms had already been suggested, in 1975, by T. W. Hanseh and **Arthur L. Schawlow**. With these very low tempe - ratures, a new state of matter, known as the Bose-**Einstein** condensation, could be experimentally reached. This goal was achieved in 1995 by **Cornell**, **Keterlee** and **Wieman** who were Nobel Prize winners in 2001. William Phillips performed his work at the National Institute of Standards and Technology (Gaithersburg, Maryland). He received his Ph.D. from the Massachusetts Institute of Technology in 1976. He was the holder of the 1996 **Albert Michelson** Medal for the demonstration of the performance of the laser-atom interaction in order to cool and trap these atoms.

1998

Laughlin, Robert, B. (Visalia, California, November 1, 1950 -). A theoretical physicist, Laughlin was working in the Department of Physics of Stanford University (California) at the time of the Nobel Prize award. The prize was awarded jointly to him and two colleagues, **Horst Störmer** (University of Columbia) and **Daniel Tsui** (University of Princeton). His research, the results of which have been disputed (even controversial) since 1983, shows that, in a powerful magnetic field and at a very low temperature, electrons (fermions) can undergo condensation to bring them to a new quantum state completely unprecedented in nature. Two of Laughlin's Stanford colleagues, **Douglas Osheroff** and **Steven Chu**, had also very recently received the Nobel Prize in Physics, the first in 1996 for his work on the superfluidity of helium - 3 (another fermion), the second in 1997 for his study of the physics of light-matter interaction at very low temperature. Thus, for three consecutive years, Stanford physicists have seen themselves "nobelized" in a research field where condensed matter reveals its indisputable quantum reality, as experimentally verified by the two co - winners of 1998. Paradoxically, Laughlin showed that this new quantum, described by a general motion of electrons in an extremely dense state, also revealed the possibility of the existence of quasiparticle excitations carryng an extract fraction of the elemental electric charge (multiples of 1/3, 1/5, 1/7 of the electron charge). This behaviour can also be associated with the ideas of **Klaus Von Klitzing** on the quantum Hall effect, but it was Laughlin who gave the correct explanation of its exactness and who identified the implications. Laughlin is known for his tenacity in defen- ding his innovative views, an attitude which probably caused him not to participate in research carried out by large civil American companies (to the great satisfaction of many young people who wish to work under his supervision at Stanford). Presently Laughlin is working on high - temperature superconductivity, quantum phase transitions, the question of emergence in nature and the interface between physics and biology. Although a committed worker day and night when a new idea and its explanation were taking shape, he is greatly attached to family life. Laughlin gained his Ph.D. in 1979 at the Massachusetts Institute of Technology (MIT). He had already distinguished himself as a brilliant student, intuitive and confident in his ability, a theoretician first of all, but strongly committed to experimental verification of results.

Robert Laughlin

Tsui, Daniel, C. (Henan, China, 1939 -). Experimental physicist. With **Horst Störmer** and **Robert Laughlin**, Daniel Tsui showed and explained the behavior of quasi particles carrying a fractional elementary charge (in particular: 1/3, 2/5, 3/7, 4/7, 3/5, 2/3, 5/7, 4/5, 4/3, 7/5, 5/3) in the study of quantum jumps, explained previously in the Hall effect for integer charges by **von Klitzing.** Their experiments were carried out in slightly conducting (multi) layers, which are so thin that they can be considered, as "only" two - dimensional. The theoretical explanation was provided by Laughlin, only one year after the announcement of Tsui's and Störmer's results. These charge states (fractional of that of the electron) have values whose denominator is always an odd number. Of Chinese origin, Tsui did his preparatory studies in Hong Kong and emigrated to the United States in 1958 to continue his studies there (until 1961) in the Augustana College of Rock Island (Illinois). He received his doctorate at the University of Chicago in 1967, then joined Bell Laboratories in 1968 to work there together with Störmer and then occupied the position of professor at Princeton from 1982, the year of the discovery of this fractional quantum Hall effect. The three laureates of the 1998 Nobel Prize in Physics received, the same year, the famous Physics Medal of the Benjamin Franklin Institute. Tsui holds numerous prizes and is the inventor of several devices including semiconductors as detectors functioning in the field of infrared radiation.

Daniel Tsui

Störmer, Horst, L. (Frankfort, Germany, 1949 -). Experimental physicist of German origin. With **Robert Laughlin** and **Daniel Tsui**, Horst Störmer observed quantum jumps in the variation of the resistance in the presence of a magnetic field. These jumps are associated to fractional states of the elementary electric charge. Applications of this new phenomenon (which only appears at temperatures close to zero Kelvin and in intense magnetic fields) are expected in opto-electronics and, in particular, in the manufacture of solid lasers, and in the development of compact computers with large calculation power. The press of October 1998 announced the reasons for the award of the Nobel Prize with a series of headlines showing the extraordinary aspect of the particular beha-

Horst L. Störmer

viour of these "split" fermions that, when they meet, form bosons. "In the country of particles which do not exist" was the headline in Le Figaro; "The Nobel jury celebrates the success of quantum mechanics" announced Le Monde on its front page. Trained in the team of Professor Manfred Pilkun of the University of Stuttgart, Horst Störmer gained his Ph.D. in 1977 and quickly emigrated to the United States. He currently shares his activities between Columbia University where he has taught physics and applied physics since January 1998 and Bell Laboratories in Murray Hill (his first collaboration in the United States), where he is now director of the physics department. He had been spotted through an account of his Ph.D. work at the American Physics Society and it was his dynamism and his creative spirit that earned him immediate employment at Bell. With his two co-laureates, Horst Störmer benefited much from the support of MIT in starting the research that led him to the 1998 Nobel Prize.

Gerardus 't Hooft

1999

't Hooft, Gerardus (Den Helder, The Netherlands, July 5, 1946 -). A Dutch expert in mathematical physics. Nobel Prize for Physics along with **Martinus Veltman** for elucidating the quantum structure of electroweak interaction in physics. This subject had already been recognized by the Nobel jury, in 1965 (**Richard Feynman**, **Julian Schwinger**, **Sin-itiro Tomonaga**), in 1969 (**Murray Gell-Mann**), and in 1979 (**Steven Weinberg**, **Abdus Salam**, **Sheldon Glashow**). The experimental results obtained at CERN, and which also earned **Carlo Rubbia** and **Simon van der Meer** the 1984 Nobel Prize, were considered as one of the first decisive steps in the confirmation of the "standard model" with the discovery of the W and Z bosons. Gerardus 't Hooft and Martinus Veltman's work provided new perspectives in the confirmation of the standard model, including all the forces of physics and justifying, in particular, the need to finance the continuation of expe- rimental work, especially with the new accelerator being built at CERN in the hope to produce the Higgs boson, which is currently proposed as a key to achieve the unification of all the forces of physics. The Nobel Prize awarded to 't Hooft underlined once again the quality of fundamental research in The Netherlands, in the tradition of **Hendrik Lorentz**, **Peter Zeeman**, **Johannes van der Waals**, **Heike Kamerlingh - Onnes**, **Fritz Zernike**, **Nicolaas Bloembergen** and Simon van der Meer. 't Hooft received his doctorate from the University of Utrecht in 1972 and is at present a professor in the same institution. He was also a member of the scientific staff of CERN and taught physics at Cal Tech as well as at the universities of Stanford, Boston, Durham and Antwerp. Married to an anaesthetist and father of two girls, the younger one is studying veterinary sciences in Belgium, owing to the "lottery system" which gives access to this type of study in Holland. The 9491 asteroid, which was discovered in 1971, bears the name of 't Hooft and revolves between the orbits of Mars and Jupiter: a recognition of 't Hooft's work to introduce gravitation into the global model of physics. A great traveler, 't Hooft has a passion for jungle photography, for solar eclipses and still militates in favor of research to be conducted on the moon (in particular for its large reserve of isotope 3 of helium) and on Mars. He is a co-signatory of the Noordwijk Statement of July 14, 2000 for the intensification of the use of the moon as an experimental laboratory on the laws of the universe. He has received numerous prestigious prizes and distinctions.

Martinus J.G. Veltman

Veltman, Martinus, J.G. (Waalwijk, The Netherlands, June 27, 1931 -). Dutch theoretical physicist, involved since 1978 on elucidating the quantum structure of electroweak interactions in physics, Martinus Veltman published, in 1972, together with his young colleague **'t Hooft**, the first decisive work for the understanding of all the interactions of physics. The content of their results is from the field of mathematics, and avoids several pitfalls (like the "zeros" and "infinites" of older theories). Up to now, mathematical physics has been a field very poorly rewarded by Nobel Prizes in the past. This recognition of 't Hooft's and Veltman's work put an end to the criticisms expressed since the 1970s about the validity of the standard model. The present achieved theory was implemented since 1973 when more powerful computers at the University of Amsterdam were available. This Nobel Prize, boosts hope for the continuation of large scale experiments both through powerful accelerators and in space exploration. Martinus Veltman received his doctorate from the University of Utrecht in 1963. He taught there from 1966 until 1981 and, while there, developed the data-processing programs necessary for the algebraic developments essential for solving the extremely complex equations of the quantum mechanics of fields. The new impulse was provided after 1969, in collaboration with his young pupil, G. 't Hooft, only 23 years old, who was beginning a Ph.D. thesis in Veltman's team. One of their results permitted the calculation of the mass of the top quark, which was confirmed by the experimental researchers of the Fermilab in Batavia in 1995. In the opinion of the experts, a complete (and final?) confirmation of the standard model is expected around 2005. In 1981 Veltman rejoined the group of theoretical physicists at the University of Michigan until his retirement. He then

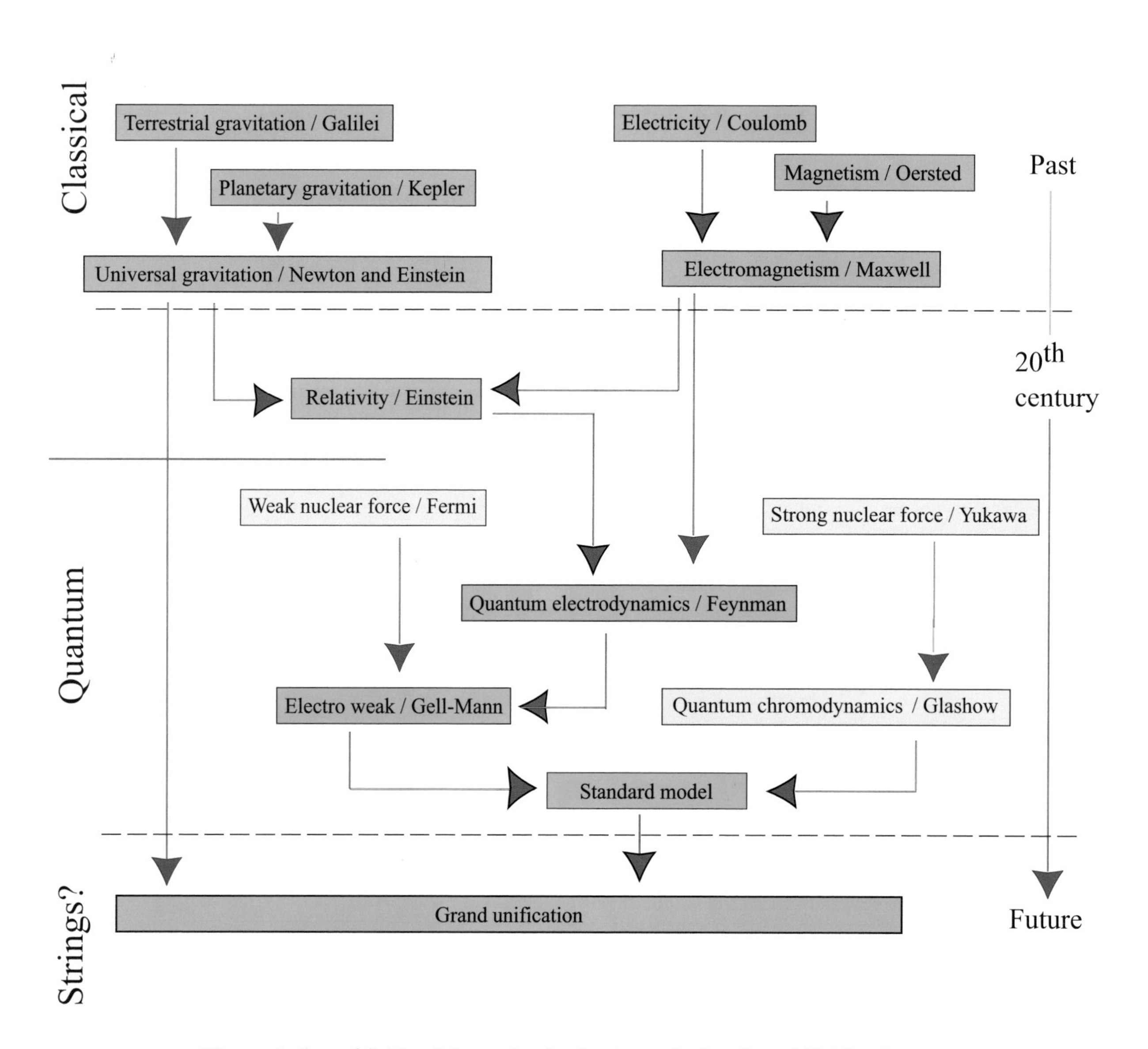

The evolution of fields of forces in physics towards the Grand Unification.

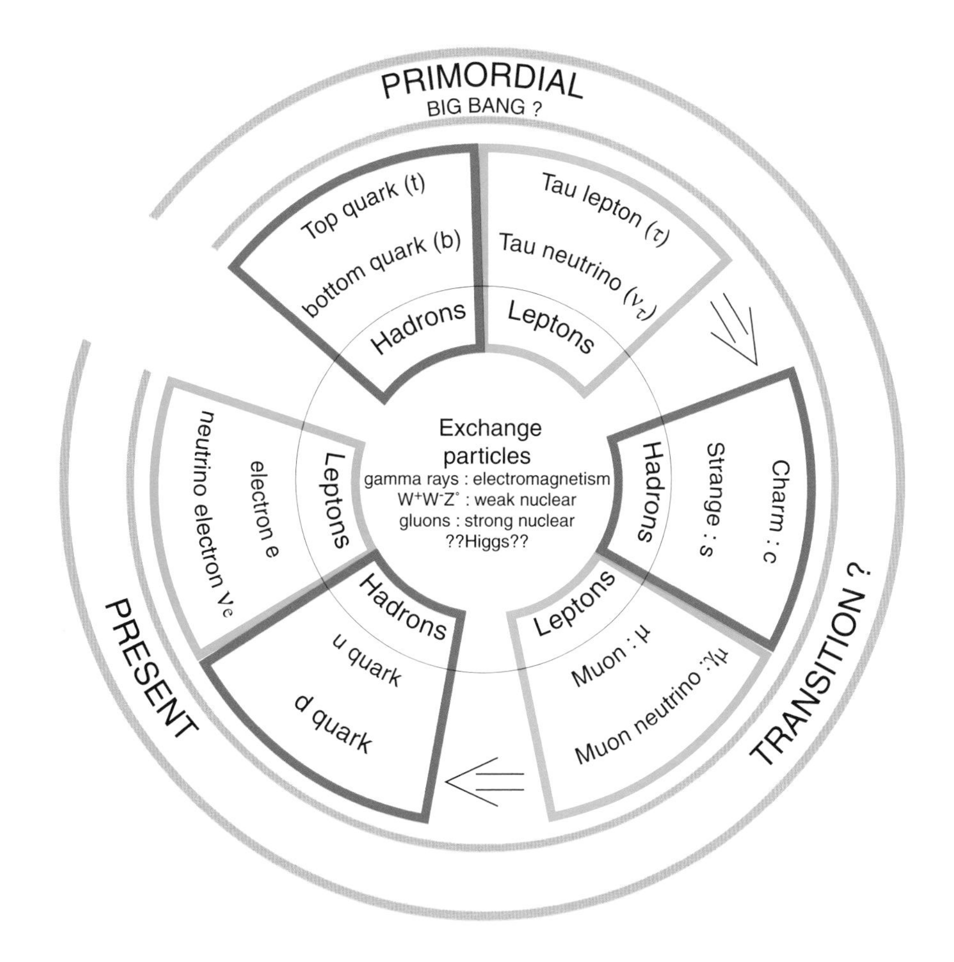

A tentative representation of known fundamental particles.

decided to return to Holland (Bilthoven), this extraordinary country, a nation rich in Nobel Prizes in Physics, as recalled in 't Hooft's biography.

Zhores Alferov

2000

Alferov, Zhores (Vitebsk, Belorussia, March 15, 1930 -). The prize was awarded with one half jointly to him and **Herbert Kroemer** for developing semiconductor heterostructures used in high-speed electronics and optoelectronics and the other half to **Jack St. Clair Kilby**. The merit of Zhores Alferov lies in the creation of semiconductor heterostructures intended for high-speed electronics and optoelectronics. It had been 22 years, that is to say since the award of the Nobel Prize in Physics to **Pyotr Kapitza**, since a Soviet physicist had been honoured. Kroemer and Alferov's work follows upon research (rewarded, moreover, by the 1973 Nobel Prize in Physics) on the concept of the superb - network expressed by **Leo Esaki**. Alferov's researches led to the manufacture of solid-state lasers using a gallium arsenide heterostructure manufactured by epitaxy. Without these devices based on heterostructures we would not have experienced, at the present time, the spectacular development of CDs, the fulgurating promotion of the mobile (cellular) telephone nor the improvement in the performances of the atomic force

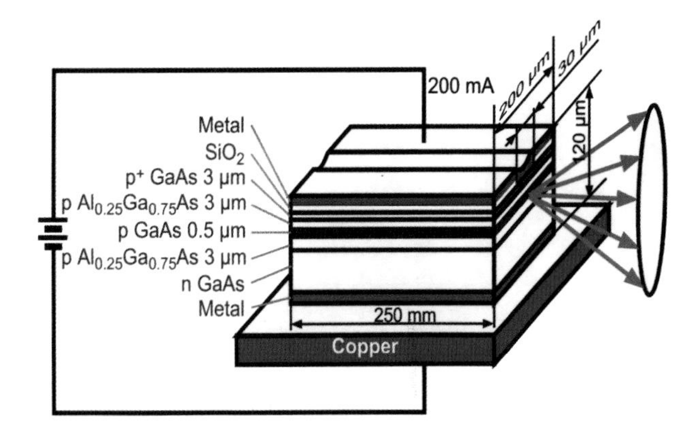

Schematic representation of the double heterostructure (DHS) injection >laser in the first CW-operation at room temperature

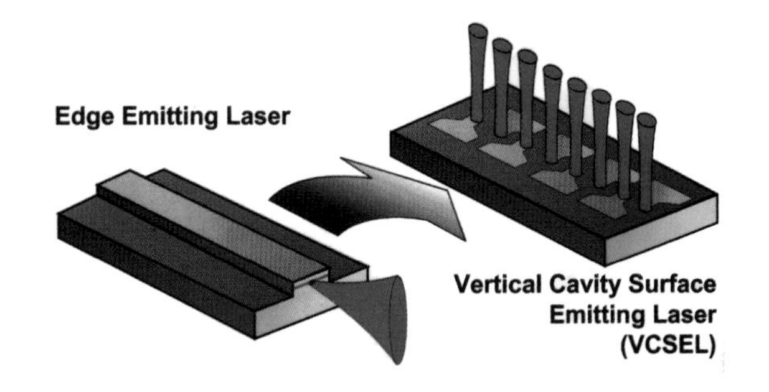

Vertical-cavity Surface-Emitting Lasers (VCSEL)

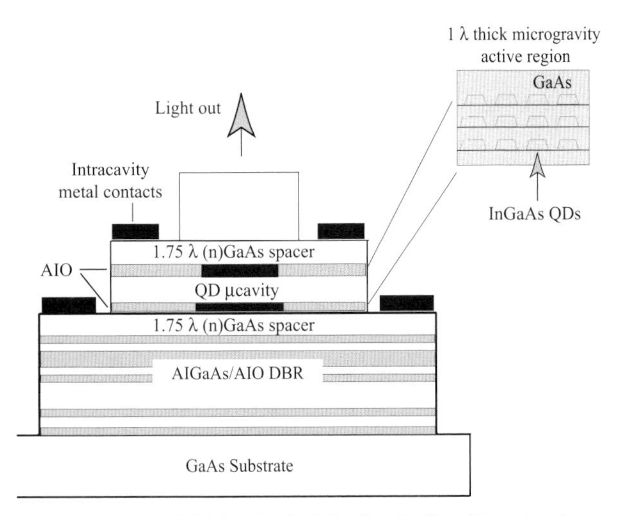

Quantum dots (QD) in Vertical Cavity Surface Emitting Laser (VCSEL).

microscopy (AFM) and scanning tunneling microscopy (STM) microscopes, already recognized by the 1986 Nobel Prize awarded to **G. Bining** and **H. Röhrer**. Since 1992 Alferov has organized an annual symposium on the physics and technology of nanostructures. This meeting, which attracts the elite researchers in this field (from Russia and other countries) is held under the sponsorship of the "Ioffe Physico-Technical Institute" of Saint Petersburg, which Alferov has directed since 1987. Since 1993 the co - president of the congress is none other than Esaki. The quality of this congress is the result of the limited number of invited participants (approximately 150) even though nanostructures currently represent the most studied field of solid-state physics. This technology, already present in Japan and in the United States in the 1980s, is today seeing its most spectacular development in Europe, starting with Russia and Germany. Zhores Alferov was educated at the Ioffe Institute, from which he received his Ph.D. in 1970. In 1973 he took over the Chair of Optoelectronics at the St. Petersburg State Electrotechnical University and was appointed Dean of the Faculty of Sciences at the St. Petersburg Technical University in 1998. Alferov was awarded a number of prizes in Russia, in the United States, and in Europe. He is Editor-in-Chief of a Russian journal dealing with advanced research called "Technical Physics Letters" and is the inventor of nearly 50 devices using semiconductor technology.

Kroemer, Herbert (Weimar, Germany, August 25, 1928 -). A theoretical physicist, but also much involved in applications, Kroemer's doctoral essay, submitted to the

Herbert Kroemer

University of Göttingen in 1952 already concerned the behavior of hot electrons in semiconductors. After several affiliations at German laboratories, Herbert Kroemer joined the Electronic and Engineering laboratories of the University of California Santa Barbara (California) in 1976. He then persuaded those responsible for this centre to develop materials other than silicon for the manufacture of high performance semiconductor devices, which he had already demonstrated in 1957. The epitaxic growth of GaP, GaAs, InAs, GaSb, AlSb structures on silicon supports would lead to spectacular progress in information technology and other means of fast communication. The principal contributions of Kroemer concern the theoretical and practical aspects connected with the association of superconductors: semiconductors structures and the transport of current in semiconductor super-networks in high electric fields. The miniaturisation of devices permits, in particular, frequencies in the terahertz regime to be reached. In addition to receiving the Nobel Prize, Kroemer has been honoured by German, American and Swedish academies and universities.

Jack St. Clair Kilby

Kilby, Jack St. Clair (Jefferson City, Missouri, 1923 -). American electronics specialist and engineer, Jack Kilby shared half the prize for his part of the invention of the integrated circuit. The other half was awarded jointly to **Zhores I. Alferov** and **Herbert Kroemer**. After the award of the 1999 Nobel Prizes for a very mathematical subject (**Gerardus t' Hooft** and **Martinus Veltman**), the Swedish jury's award was once again for a subject with high technological value. Like the prize awarded to **Ernst Ruska** in 1986 for the invention (in 1928) of the electron microscope, J. Kilby was "forgotten" for a long time since the first use of the integrated circuit, better known today as the "chip", which dates back to 1958. The miniaturisation of circuits led to greater speed in the processing and distribution of information: the mobile phone (GSM) being an outstanding example at the end of the twentieth century. The history of transistor electronics had begun with the work of **John Bardeen**, **Walter Brattain** and **William Shockley**, winners of the 1956 Nobel Prize in Physics, and of **Leo Esaki**, **Ivar Giaever** and **Brian Josephson** (Nobel laureates in 1973). The number of applications generated by the miniaturisation and resulting high speed of electronic circuits is in the thousands. As examples, let us mention household and automobile equipment, space probes and medical diagnosis equipment (tomography and nuclear magnetic resonance imaging). J. Kilby did his university studies at the Universities of Illinois and of Wisconsin. He began his professional career in 1947 at Globe Union (Milwaukee) and then moved, in 1958, to Texas Instruments in order to realize, in particular, his first integrated circuit. Integrated circuits would be the first of about sixty patents with which he would be credited. He demonstrated it on September 12, 1958, an historical date at the beginning of the microelectronic era. He partially left Texas Instruments in 1978 to teach at Texas A. and M. University until 1984, but he continued as a consultant to American industry and government. Well - known in industry milieu and the recipient of prestigious prizes, he retains great modesty and an astonishing enthusiasm.

2001

Cornell, Eric A. (Palo Alto, California, USA, December 19, 1961 -). Experimental physicist. Professor at the University of Colorado (Boulder) as well as a member of JILA (Joint Institute for Laboratory Astrophysics) and of NIST (National Institute of Standards and Technology), Eric Cornell is the youngest member of the Nobel Prize winners of 2001. They discovered a new state of matter: the condensate of Bose-Einstein. This new state is made of a well-organized association of several thousands of Ru^{57} atoms. This physical object, completely synthetic, was obtained for the first time on June 5, 1995. No other such object could exist or have existed in the natural universe .The temperature at which this condensate was obtained is far below the mean temperature in the universe. It was obtained at a temperature of a few hundredth of nano-

Eric A. Cornell

kelvins. The original results were published in the July 1995 issue of Science. The properties of this new state are not fully understood today but new developments are expected as a large number of scientific teams are now involved in this kind of research. The condensate of Bose-**Einstein** is a new behavior of quantum effect in large systems like superconductivity and superfluidity. The first condensate was realised with rubidium atoms, but other ones have already been produced with sodium atoms (by **Ketterle** at MIT) only a few months after the work at JILA, and later with potassium (in Florence), with lithium (at Rice University and ENS

- Paris), with helium (at Orsay - France) but probably also with hydrogen in 2001. Aggregates of millions of atoms were announced in 1998. From May 1997 to December 2001, more than 20 research teams from over the world were involved in such activities. American teams from several American universities are now working on strategic applications. Eric Cornell received his bachelor degree at the University of Stanford in 1985 (where he was given the Firestone Award thanks to the quality of his research) and his Ph.D degree at MIT in the team of Ketterle. He has accumulated high level scientific awards for more than 15 years.

Ketterle, Wolfgang (Heidelberg, Germany, October 21, 1957 -). German experimentalist, W. Ketterle migrated to MIT in 1990, just at the moment Eric Cornell left Cambridge to join the team of **Carl Wieman** at JILA. Independently, by using different techniques and working on different alkaline atoms, both groups began research on a common subject to be awarded the Nobel Prize of Physics together in 2001. The MIT team which Ketterle joined was managed by three main senior scientists: David Pritchard (the supervisor of Cornell), Daniel Kleppner and Thomas Greytack who supported the difficult project of Ketterle. The experimental set-up created by those physicists enabled Ketterle to achieve the goal of creating a Bose-Einstein condensate with sodium atoms on September 30, 1995: about 100 days after the successful experiment of the JILA group with rubidium atoms. Educated in Germany in low temperature physics and newly integrated in the MIT team, Ketterle recognized, when he was awarded by the Nobel Committee, that the honor reflected on the whole MIT team. He is indeed the 51st physicist of MIT to be honored in 58 years. The condensate of Bose - Einstein is a corpuscular version of the maser and the laser recognized in the past by the Nobel Prizes winners in 1964 (**Townes**, **Basov**, **Prokhorov**), and in 1981 (**Bloembergen**, **Schawlow**). Furthermore, those lasers were used by **Cohen-Tannoudji**, **Chu** and **Philips** (Nobel Prize winners in Physics in 1997) to cool atomic beams. The temperature achieved to produce this Bose-Einstein condensate is nevertheless 100 times lower than that achieved by the 1997 Nobel Prize winners. For standard conditions in the laboratory (20°C), the **de Broglie** wavelength of moving particles are so small in comparison with the atom dimension that their wave properties are not observable. The possible applications of low temperature physics involving the use of the Bose-Einstein condensate are being studied at MIT for precise investigations in nanotechnology, lithography and holography. One of the main performances of the MIT group was to have created observation means which do not disturb the condensate and which enable them to observe the condensate many times. Ketterle graduated from the Technical University of Munich in 1982 and Ph.D of the Maximilian Universität, a part of the Max Planck Institüt fur Quantumoptik of Garching in 1986.

Wolfgang Ketterle

Wieman, Carl E. (Corvallis, Oregon, March 26, 1951 -). Experimental physicist, C. Wieman is presently a member of the University of Colorado (Boulder) but simultaneously a fellow of the JILA (Joint Institute for Laboratory Astrophysics). This first Nobel Prize in physics of the new century will probably open up a new world of quantum physics. The first Nobel Prize of the 20th century attributed to **W. Röntgen** for his discovery of X - rays was fortuitous, as admitted by Röntgen himself who did not recognize the origin of those X - rays. The first Nobel Prize of this new century awards the achievement of a long and patient improvement of experimental conditions to demonstrate the proposal of Satyendra Bose dating from 1924 and immediately supported by **A. Einstein**. This theoretical proposal had to wait for 71 years to get an experimental verification. The first Bose-Einstein condensate was observed by Wieman and **Cornell** on June 5, 1995 at 10:54. The researchers of the JILA group concentrated on the subject for more than 15 years before reaching their goal, often under the skeptical appreciation of their colleagues. The Bose - Einstein condensate is for the physics of particles the "brother" of the laser in the physics

Carl E. Wieman

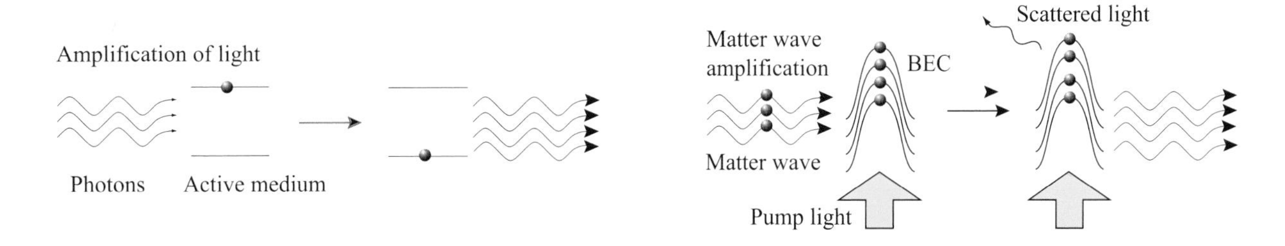

A comparison of laser emission (coherent light) and its particle dual process (Bose-Einstein condensate).

of waves. A laser is a large association of quanta of light propagating in phase and the Bose - Einstein condensate (BEC) explains in the same way the associative behavior of a large number of atoms, propagating together with the same phase: a terminology of **Louis de Broglie** (Nobel Prize winner in Physics in 1929) who associated a wave property to a moving particle. To produce this particular state of matter, the JILA group achieved a very, very low temperature in a vapor of rubidium (Rb^{57}): 170 billionths of a kelvin. Such a low temperature had never been reached in the past. Starting from room temperature, they first lowered it down to 10 millionths of a Kelvin by slo - wing down the atoms with convergent laser beams. A further decrease of the mean velocity of those atoms by a factor of 100 was then realised by allowing the faster atoms to escape. For this last decrease of temperature, they used the properties of spin interaction with finely oriented magnetic fields. From a velocity of 450 m/s at room temperature the velocity of the BEC is only several hundreds of a micron per second...nearly at rest. Fundamental applications of those BEC are expected to be used to study the weak gravitational field. After spending his youth in the forests of Oregon, Carl Wieman first gra- duated from MIT in 1973 and received his Ph.D at the University of Stanford in 1977. He was appointed assistant and then professor at the University of Michigan and since 1984 he has gradually been promoted to increasing responsibilities at the University of Colorado. He has been director of the JILA from 1993 to 1995. Since 1973 he has received lots of scientific honors and awards, at least one every year. He is much appreciated as a teacher and research leader by generations of students.

The Nobel Laureates in Medicine (1901 - 2001)

Francis Leroy

The Nobel Laureates in Medicine

Artistic rendering of the interior of an animal cell. The nucleus is surrounding by a landscape containing elements of the endoplasmic reticulum as well as a Golgi body and various membrane vesicles. The discoveries of cellular ultrastructures, the function of metabolism, and the regulation of cellular activities have been the object of numerous presentations of the Nobel Prize in medicine over the course of the second half of the twentieth century.

Introduction

The progress in biology and medicine due to the laureates in science and to many other scientists spans essentially three periods. The first was marked by the flourishing development of bacteriology, which began towards the end of the 19th century thanks to Louis Pasteur, Robert Koch, Paul Ehrlich, and many other microbiologists. After a few decades, this discipline was to yield major innovations such as vaccines and antibiotics, remarkable for their efficacy against bacterial infections.

The 1940s were a time of growing interest in molecules such as hormones and vitamins, as scientists were discovering the (often crucial) role of enzymes in directing and regulating cell metabolism. During and right after World War II, there were some major breakthroughs in general physiology.

Finally, as chemistry made its appearance in biology with the elucidation of the structure of DNA (carrier of genetic information and of elements that regulate its expression) and with the discovery of the intimate architecture of proteins and cells, came the dawn of molecular and cell biology. These are the disciplines on which most current knowledge in general biology, physiology, and medicine is based. Remarkably, they have also introduced countless applications into the field of health.

A colony of staphylococci. Pathogens and the illnesses they cause have been a major subject of medical research since the work of Louis Pasteur.

1901

A. von Behring and one of his students

Behring, Emil Adolf von (Hansdorf, West Prussia, March 15, 1854 - Marburg, Germany, March 31, 1917). German physician and microbiologist. Nobel Prize for Medicine for his work on serum therapy, especially its application against diphtheria. Emil von Behring came from the family of a schoolmaster. He was the eldest of 13 children. He began his studies in the field of microbiology at a time when scientists were enthusiastically pursuing research in this new area because of the realization that many diseases were of bacterial origin. Behring started out as an assistant of Robert Koch and he is associated with the demonstration that the blood of an infected individual contained antitoxins capable of neutralizing bacterial toxins. These studies were conducted along with the Japanese researcher Shibasaburo Kitasato, who was a member of Koch's team. One of their first important successes was the development of a serum for the treatment of diphtheria, a paralysing respiratory disease responsible for many infant deaths at the time. The first anti - diphtheria vaccine was successfully tested out on a sick child on Christmas night 1891. The work of Behring and his co - workers led the way for serotherapy as well as the development of more efficient diptheria vaccines later by Paul Erlich, another colleague of Koch. In 1895 Adolf von Behring became Professor at the University of Marburg. He was interested in combatting tuberculosis and, along with his co - workers, contributed to the successful immunisation of cattle against this disease.

1902

Ross, Sir Ronald (Almora, India, May 13, 1857 - London, England, September 16, 1932). British physician. Nobel Prize for Medicine for his work on malaria. Had he followed his inclination, Ronald Ross could just as well have been a painter, poet, or mathematician, but the irrevocable decision of his father, an English officer in the Indian Army, led him to undertake medical studies. Once they were over, however, he wrote no less than 15 literary works and an algebra treatise. Yet Ross's name remains associated with the confirmation that the vector of the haematozoan responsible for malaria in man is a mosquito. Sir Patrick Manson was the one to suggest this, but Ross practically proved it by discovering the plasmodium in the bodies of mosquitoes that had fed on the blood of people suffering from malaria. He then identified the parasite that causes malaria in birds, in the saliva glands of *C. fatigans*, the mosquito vector for this group of animals. Later it was shown that the plasmodium's reproductive cycle includes a stage in the salivary glands of the dipteran, and that the latter transmits the parasite to a healthy bird through a bite. The intervention of mosquitoes as vectors of human malaria was also confirmed, by an Italian team headed by Giovanni Battista Grassi. The discovery that a mosquito could carry an infectious germ was new at the start of the 19th century, even though this had already been mentioned by the Cuban biologist Carlos J. Finlay as early as 1881. Finlay's observation that a dipte-

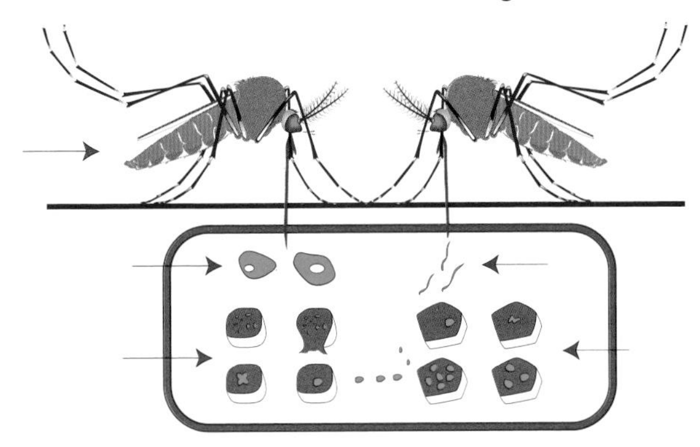

The life cycle of the malaria parasite Plasmodium. Plasmodium's gametocysts produced during the sexual phase of the parasite are picked up when a mosquito bites an infected person. Sporozoites are produced and migrate to the salivary gland of the insect. Finally, the infected mosquito injects sporozoites into a healthy person via the skin.

Ronald Ross

ran was the vector of yellow fever failed to attract the attention of his fellow scientists, and his results were not confirmed until years later by an American commission meeting in Cuba. He became Vice - President from 1911 to 1913. Ronald Ross was knighted in 1911.

1903

Finsen, Niels Rydberg (Thorshavn, Denmark, December 15, 1860 - Copenhagen, Denmark, September 24, 1904). Danish physician. Son of a civil servant. Nobel Prize for Medicine for the application of light in the treatment of skin diseases. Thanks to Finsen's discovery, it was shown that ultraviolet radiations have a bactericidal effect. The use of ultraviolet light proved to be effective in treating lupus vulgaris, a disfiguring skin disease caused by bacteria. Although this light therapy was not recommended ulteriorly, the Nobel Committee decided to reward this researcher in 1903. Some applications of ultraviolet radiation have proved very useful: it is used to induce mutations experimentally and to kill bacteria in hospitals and research laboratories. Niels Rydberg Finsen received his Ph.D. in medicine from the University of Copenhagen in 1890.

1904

Pavlov, Ivan Petrovich (Ryazan, Russia, September 26, 1849 - Leningrad, Soviet Union, February 27, 1936). Russian physiologist. He was awarded the Nobel Prize for Medicine for research on digestion and on the conditioned reflex. In 1891, Ivan Pavlov became Director of Physiology in the St. Petersburg Institute for Experimental Medicine. Having first been noticed by the scientific community for major work on food digestion, he then focused on the neural aspects of this function. He notably showed that the mere sight of food, with the subsequent emission of neuronal signals from the brain to the organs concerned, is sufficient to trigger secretion of gastric juices. Ivan Pavlov then showed how a signal not related to food could make a dog salivate, once it had been associated with the initial signal. Every time he offered food to a group of experimental dogs, he also rang a bell. The dogs soon learned to associate the sound of the bell with food. After a sufficient number of repetitions, Pavlov showed that the dogs would secrete saliva in response to the bell alone; the taste, sight, smell of food was no longer necessary. The Russian scientist thus demonstrated scientifically the existence of so - called "conditioned reflexes", actually already known to animal trainers. This type of associative learning became the basis of major works in experimental psychology and gave rise to the idea that several human behaviours could be conditioned. From this type of experiment Pavlov concluded that "reflex activity of the brain adapts a living organism to its environment through the creation of new nerve connections, and species evolution transforms certain acquired reflexes into innate reflexes". For this scientist, the link between psychic activity and the brain was paramount. He even wondered whether it might be possible to detect an elementary manifestation of the psyche that could be considered purely physiological. According to him, such a discovery would have made possible the development of scientifically rigorous and objective studies on the conditions of appearance of the psyche. Furthermore, the fact that training programs based on the principles of conditioning animals to learn quite complicated behavioral routines quickly suggested that the phenomenon might also account for the complex patterns of behavior found in man. Indeed Pavlovian conditioning became an extremely influential school of thought in the study of animal behavior. In addition, they led to the elaboration of hypotheses suggesting that several forms of human behaviour could be conditioned. His open criticism of communism did not prevent him from keeping his job, and Pavlovian psychology remained very popular in the Soviet Union.

Ivan P. Pavlov

1905

Koch, Robert (Klausthal, Germany, December 11, 1843 - Baden - Baden, Germany, May 27,1910). German microbiologist and physician. Son of a mining engineer. Nobel Prize for Medicine " for his discoveries in the field of tuberculosis". Koch is considered as the greatest of all pure bacteriologists. Robert Koch received his medical education from the University of Göttingen in 1862. After earning his medical degree in 1866, he moved to the University of Berlin where he studied chemistry. From 1872 to 1880, he worked in Wollstein, a small town in Polish Prussia, as a humble practitioner without any contacts with the academic world. Yet his work led to the development of techniques enabling him to isolate the anthrax bacillus that could be seen in the blood of infected cats. Later, after being appointed to a position in Berlin, he developed a rigorous method for creating gelatine - based - and later agar - based culture media, combined with the use of aniline dyes and favourable to the growth of bacterial colonies. These techniques revolu-

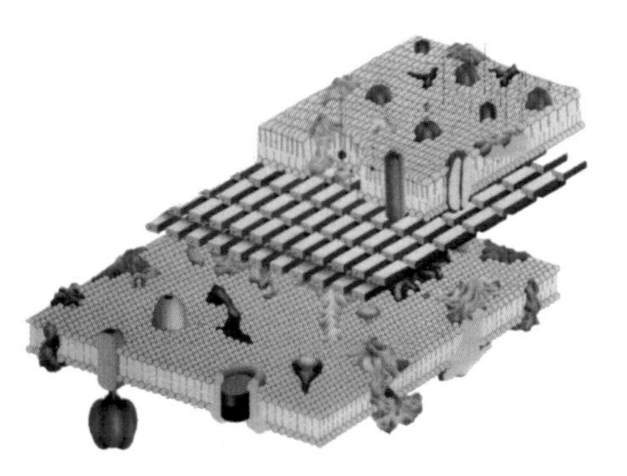

Gram - negative bacteria. Their wall are thin but complex, having a lipopolysaccharide layer outside it. These bacteria are resistant to the lytic action of lysozyme.

A dividing bacteria. Note the punching in of the cell wall between the two newly formed daughter cells.

Robert Koch

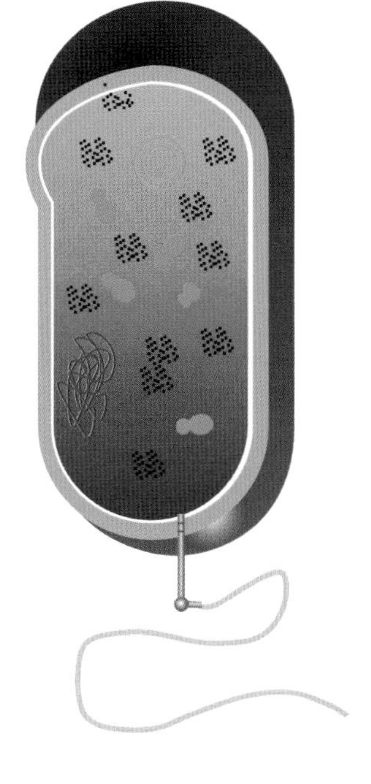

Bacterium. A three - dimensional representation.

" **Gone with the wind** ". Vivien Leigh, who played Scarlett, the celebrated heroine of the film, died of tuberculosis at the age of 53.

tionized bacteriology. Yet Koch is famous above all for having isolated the bacterium that causes tuberculosis in humans. In the second half of the 19th century, this disease was the cause of nearly one-seventh of all deaths. His failure to treat the disease or to vaccinate against it with tuberculin - a sterile liquid containing a concentrate of dead tuberculosis bacilli - was offset by the great utility of tuberculin in detecting patients who had been previously exposed to the disease. This is the famous tuberculin test. Koch's growing reputation brought him into competition with another giant in bacteriology: the French scientist Pasteur. They carried this rivalry to Egypt, where Koch's team and a French one led by Roux both attempted to discover the cause of cholera. Koch was the one to demonstrate the link between the disease and a bacillus. He correctly concluded that cholera is caused by a bacterium, which he succeeded in isolating and growing. This remarkable scientist was also a great traveller. During his trips to South and West Africa, as well as to India and

Indonesia, he carried out research and published many papers on illnesses such as bubonic plague, malaria and sleeping sickness, which at one time were widespread.

1906

Cajal, Santiago Ramon Y. (Navarre, Spain, May 1, 1852 - Madrid, Spain, October 17, 1934). Spanish physician and histologist. Son of a physician. Nobel Prize for Medicine, along with **Camillo Golgi**, for establishing the neuron, or nerve cell, as the basic unit of nervous structure. In his youth Santiago Ramon Y Cajal showed little interest in his studies, except for fine art. He was first a barber's apprentice then a cobbler. However, he had a change of heart and took up his studies. Alongside **Ivan Petrovitch Pavlov** and **Charles Sherrington**, he became considered as one of the great specialists in the anatomy and physiology of nerve tissue. Using an improved silver nitrate staining technique established previously by Golgi, he demonstrated the extraordinarily intricate interactions of nerve cells in grey matter. Cajal, who had dreamed of becoming a painter, was excellent at drawing and made extremely precise sketches describing the anatomical relationships between nervous fibers and neuron bodies and showed that nerve cells were rarely in direct contact with each other. He also established that neurons were anatomically independent and destroyed the belief that existed since Galien of a continuity among them. Another major observation was the cellpede and cellfuge direction of transmitting nervous impulses in dendrites and axons. He also contributed towards the understanding of the structural organization of neurosensory retinal cells. Even today, the technique of nerve tisue identification established by Cajal enables tumours in nerve tissue to be detected. Santiago Ramon y Cajal taught at the Universities of Valence and Barcelona and up until his death worked at the University of Madrid, in an institute constructed during the reign of Alphonse XIII and which is named after him.

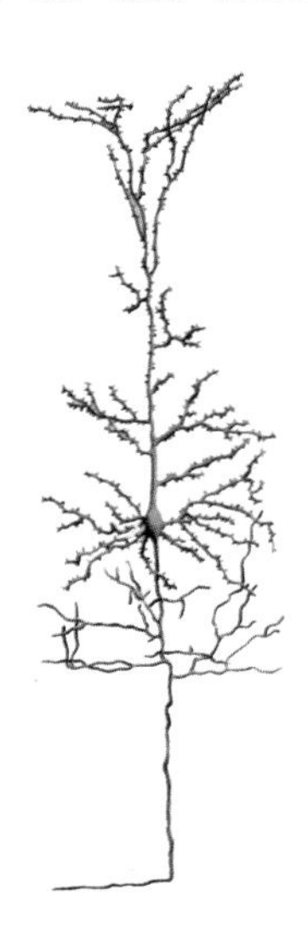

Pyramidal cell of the cortex.

Santiago Ramon Y Cajal

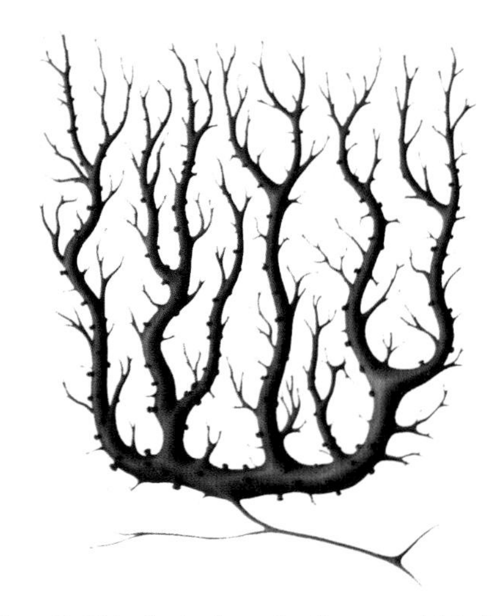

Purkinje cell. This flask-shaped cell type, which forms a single layer in the cerebellar cortex, was described by Camillo Golgi. It was discovered by Johannes Purkinje, a Czech naturalist.

Glial cells. An artistic drawing showing the relations of this cell type with blood vessel and neurons. Glial cells were also observed by S. Cajal as well as by C.Golgi.

Golgi, Camillo (Corteno, Italy, July 7, 1843 - Pavia, Italy, January 21,1926). Italian physician and histologist. Nobel Prize for Medicine with **Santiago Ramón y Cajal** for discovery of the minute structure of the nervous system. Like Ramón y Cajal, he was the son of a modest physician. Golgi made his first steps in the field of science at Pavie University. There he studied under Oehl, who was especially interested in the study of cellular structures by microscopy. Later he became an assistant in a psychiatric clinic, which was undoubtedly a defining factor in his later research on the delicate makeup of the nervous system. Golgi's research on the delicate structure of this system had positive results, largely due to the specific coloration method refined in the early 1870s. Thanks to this method, he was able to identify cellular elements of nerve tissue by dyeing them black with silver nitrate after dipping the tissue in potassium dichromate or ammonium dichromate. It was this discovery that started him on the path toward sharing the Nobel Prize. His method of impregnating nerve tissues with silver nitrate later knew great success because it made possible, for the first time, the discovery of a certain order in the apparently chaotic universe of nerve tissue. Several cell types have been identified using the techniques developed by Camillo Golgi, for example cells of the olfactory bulb, Golgi Type I cells found in the cerebellar cortex ans Golgi Type II cells, nerve cells that

Golgi apparatus. A three-dimensional representation. An essential component of the internal membranes of the cell. Its name recalls the Italian histologist, Camillo Golgi, who discovered it. This apparatus is essentially involved in glycosylation as well as in membrane flow, secretion or production of intracellular components

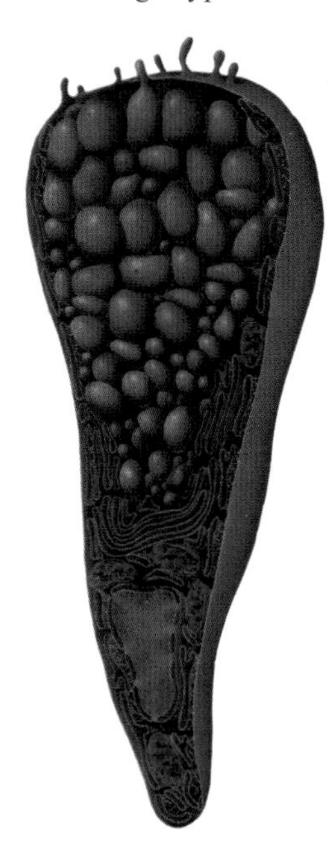

Goblet cell. A cell of the gut epithelium. These mucous cells contain a well - developed Golgi apparatus, located near the nucleus. The figure shows numerous round membrane - bound mucous droplets which determine the goblet - shaped appearance of the cell

Camillo Golgi

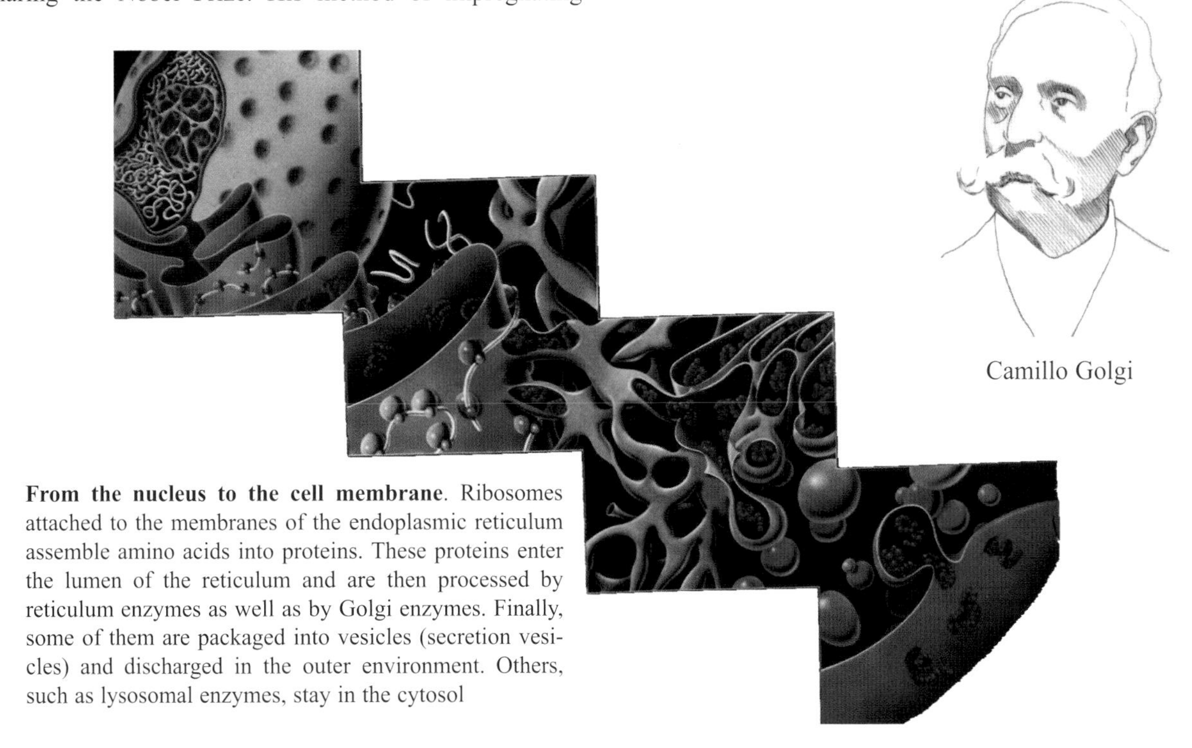

From the nucleus to the cell membrane. Ribosomes attached to the membranes of the endoplasmic reticulum assemble amino acids into proteins. These proteins enter the lumen of the reticulum and are then processed by reticulum enzymes as well as by Golgi enzymes. Finally, some of them are packaged into vesicles (secretion vesicles) and discharged in the outer environment. Others, such as lysosomal enzymes, stay in the cytosol have short axons that do not extend beyond the vicinity of the cell body. Another very important discovery made by Golgi in 1898 was the cell structure now called the Golgi apparatus or Golgi body (but first named by Golgi as a "new reticular apparatus"), which is a characteristic of most eukaryotic cells. This membrane-bound apparatus is necessary for the processing and sorting of proteins and lipids. Golgi's discovery was a real breakthrough in cytology and cell biology but was only confirmed in the mid - 1950s by the use of the electron microscope. Golgi's name is also mentioned in connection with the study of malaria. Malaria is a widespread human disease caused by *Plasmodium falciparum*, a protoan parasite. This parasite had been discovered by Alphonse Laveran, a French military physician. His studies led to the confirmation that this protozoan completed several reproduction cycles, each of which periodically caused a high fever due to the synchronous reproduction of the parasite.

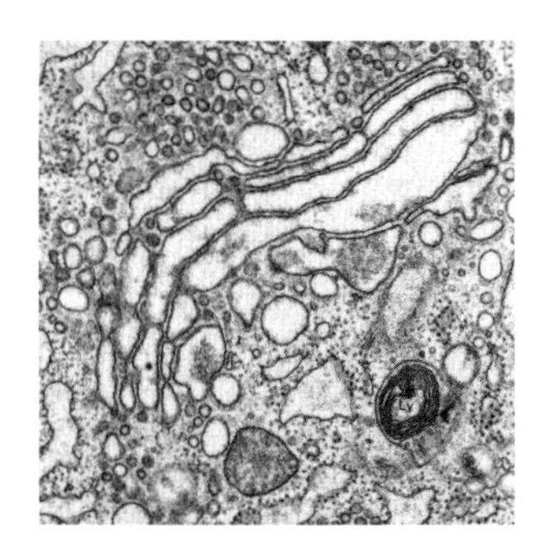

Golgi body Electron microscopy

1907

Laveran, Charles-Louis-Alphonse (Paris, France, June 18, 1845 - May 18, 1922). French military physician and pathologist. Nobel Prize for Medicine for his work on protozoal diseases. A teacher of hygiene at the "Val de Grâce" hospital. It was during the 1880s that Laveran, himself son of a military surgeon and posted in the colonies, identified a ciliate, flagella-bearing protozoan as the causative agent of paludism, also known as malaria. This was one of the most spectacular discoveries of 19th - century medicine. The protozoa multiplied in red blood cells according to a cycle that triggered periodic bouts of fever. Laveran showed that the haematozoan could be transmitted from a person sick with malaria to a healthy individual. As the story goes, one morning in 1884 he came to fetch Pasteur and Roux at their laboratory on rue d'Ulm, and accompanied them to the Val de Grâce where he showed them the parasite in a smear of fresh blood from a hospitalised malaria patient. The British doctor Ronald Ross was to confirm the parasite's presence in the blood of the vector. First called *Oscillaria malaria* by Laveran, the parasite finally received the Italian name *Plasmodium*. The French scientist suspec- ted, but could not formally prove, that the vector of the haematozoan was a mosquito. The Nobel Prize money which had been granted to him for this work, was used for the creation of the Tropical Medicine Laboratory at the Pasteur Institute. Charles Laveran received his medical education from the University of Strasbourg. Later, after teaching military medicine until 1897, he joined the Pasteur Institute in Paris.

Charles L.A. Laveran

1908

Ehrlich, Paul (Strehlen, Prussia, March 14, 1854 - Homburg, Germany, August 20, 1915). German physician and microbiologist. Nobel Prize for Medicine for his work on immunity. Paul Ehrlich first became famous for formulating the hypothesis that injection of bacterial toxins into an animal induces in that animal the formation of toxin - neutralizing antitoxins. Ehrlich's first sketches illustrating the neutralisation mechanism prefigure the way the reaction between antigen and antibody was later represented, showing the union of two chemically complementary surfaces. This was thus the first attempt to explain immunisation in molecular terms. A second success resulted from the German

Paul Ehrlich

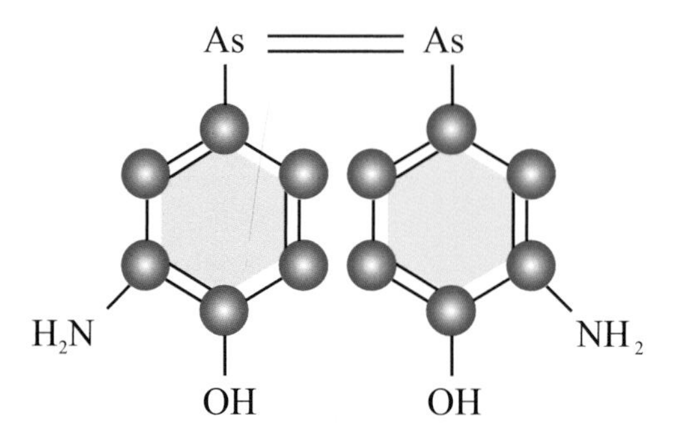

Salvarsan. One of the first chemotherapeutic drugs. Also named by Ehrlich "magic bullet", it was useful in the treatment of syphilis, as well as in the treatment of a form of sleeping sickness caused by trypanosomes. Nevertheless, it proved to be rather toxic and has ben replaced by other drugs.

scientist's research aimed at using dyes that specifically stain bacteria for therapeutic purposes. His demonstration that fuchsine specifically stains the tuberculosis bacterium was much appreciated by Koch, who used it with success. Aware that many of these dyes kill the bacteria to which they bind, Ehrlich ingeniously altered some of them chemically, turning them into 'magic bullets' capable of targeting bacteria without damaging the cells of the host organism. In 1909 Ehrlich discovered the so - called "salvarsan" - a compound relayed to arsenic, which was found a usefull drug to treat syphilis. In 1910, this discovery made a major advance in medicine, and, as such, Ehrlich became a renowned scientist. This molecule was the first substance to be used for the treatment of syphilis until it was replaced by penicillin. This led the way for the treatment of diseases using chemical therapy.

Metchnikoff, Ylia Ilytch (Ivanovka, Ukraine, May 15,1845 - Paris, France, July 15, 1916). Russian - French bacteriologist. Nobel Prize, along with the German bacteriologist **Paul Ehrlich**, for his works on immunity. It was 1882 when this Russian researcher became interested in certain blood cells which, without seeming to be directly involved in digestion, had the capacity to absorb bacteria. He called this process phagocytosis. He notably showed that the pus appearing in infected wounds resulted from the accumulation of dead phagocytic white blood cells that had been engaged in the fight against infection. These results led him to emphasize the role of phagocytosis in

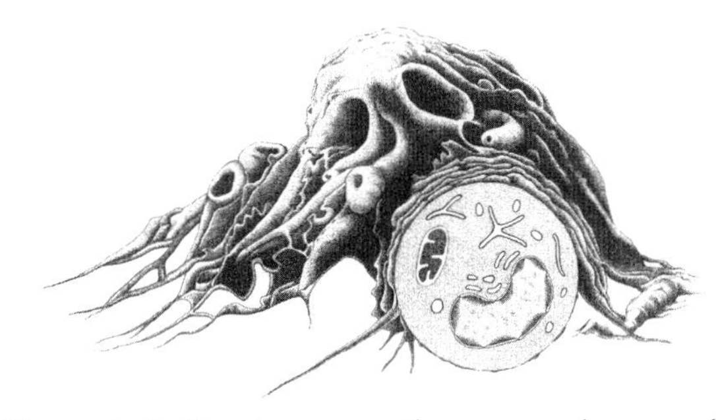

Phagocytosis. Drawing representing a macrophage engulfing a cell. (seen here in cross - section).

immunity. The idea that such a mechanism could support the fight against infection was so new that the great German microbiologist Rudolf Virchow, even though he was given a chance to see phagocytosis occurring before his very eyes, shook his head and refused to accept that such a phenomenon could be part of our antimicrobial defence system. Louis Pasteur, however, showed interest in Metchnikoff's discovery. He invited him to Paris, where the latter remained until the end of his life. After Louis Pasteur's death he took over the direction of the Pasteur Institute and carried on his work on antibacterial immunity. Nevertheless, during the last decade of his life Metchnikoff mainly turned his attention to the role of lactic acid - producing bacteria in increasing human longevity. Son of an officer of the Imperial Guard, Ylia Metchnikoff was educated at the University of Karkhov, Ukraine. After having studied bacteriology in Germany

Macrophage engulfing bacteria. Note the spikes and the filamentous expansions of the phagocyte cell membrane.

under the direction of Carl Theodor Ernst von Siebold he returned to Russia where he held post as a lecturer at the University of Odessa.

1909

Kocher, Emil Theodor (Bern, Switzerland, August 25, 1841 - Bern, Switzerland, July 27, 1917). Swiss surgeon and son of an engineer. Kocher was awarded the Nobel Prize for medicine for his work on the thyroid gland. Using antiseptic techniques developed by Lister in London, Kocher became the first surgeon to excise this organ to treat goiter. He showed that, contrary to total excision, partial resection of the thyroid could avoid the appearance of cretinism or at least attenuate its symptoms. Over five thousand operations on the thyroid gland made this doctor a confirmed expert, renowned in the medical

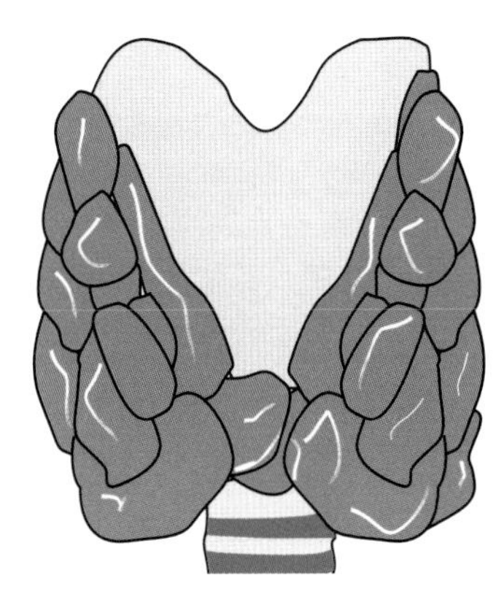

The thyroid gland. The gland secretes thyroglobulin, which is then taken up by follicular cells and broken down to active thyroid hormones.

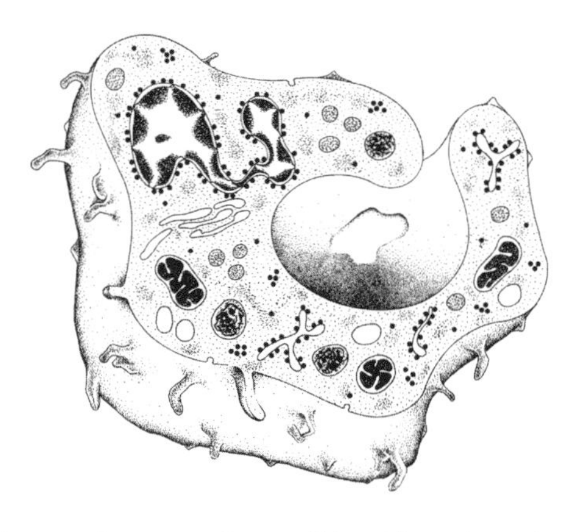
Cut trough a macrophage cell ingulfing a bacterium

Macrophage cell. A three - dimensional drawing.

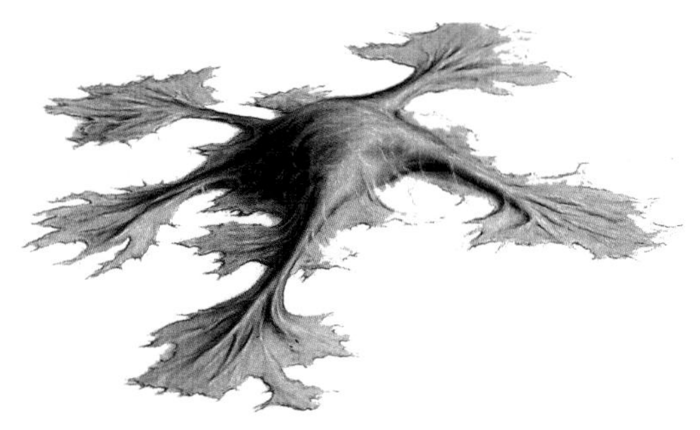
Macrophage in cell culture

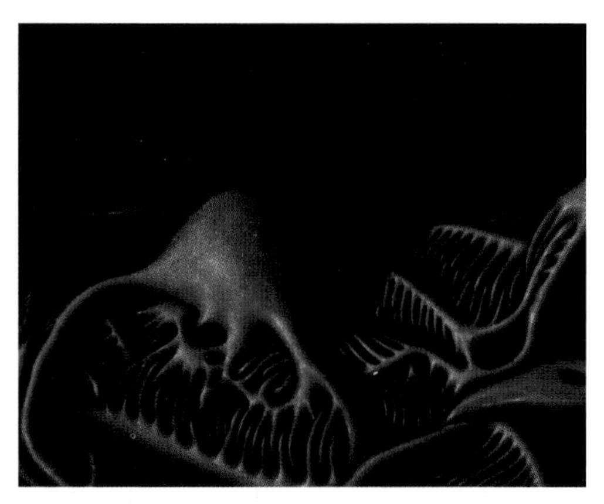
Kidney macrophages (podocytes)

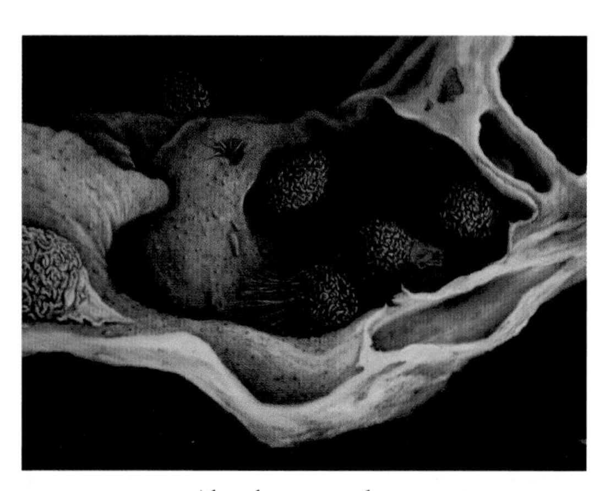
Alveolar macrophages

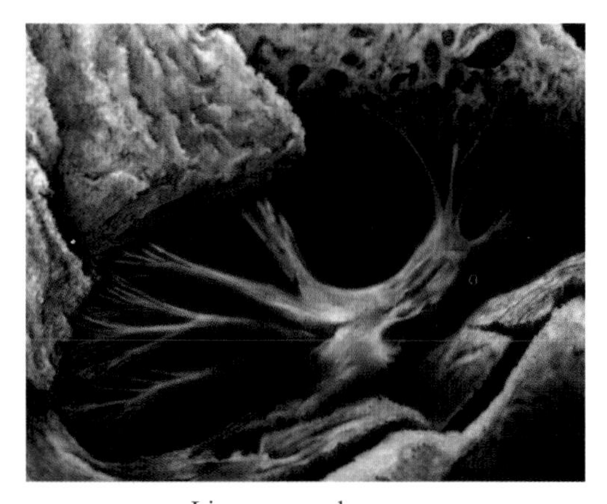
Liver macrophage

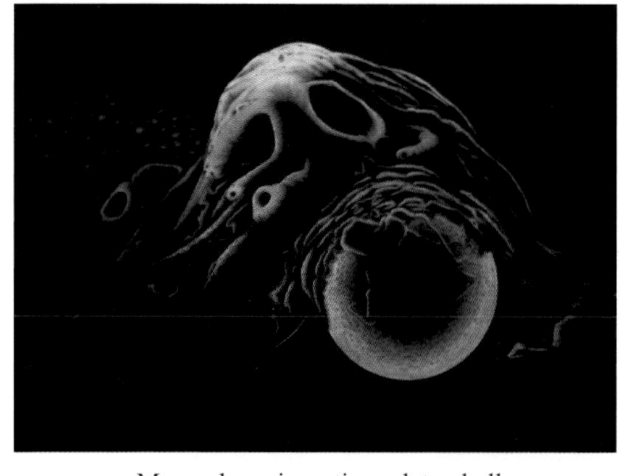
Macrophage ingesting a latex ball

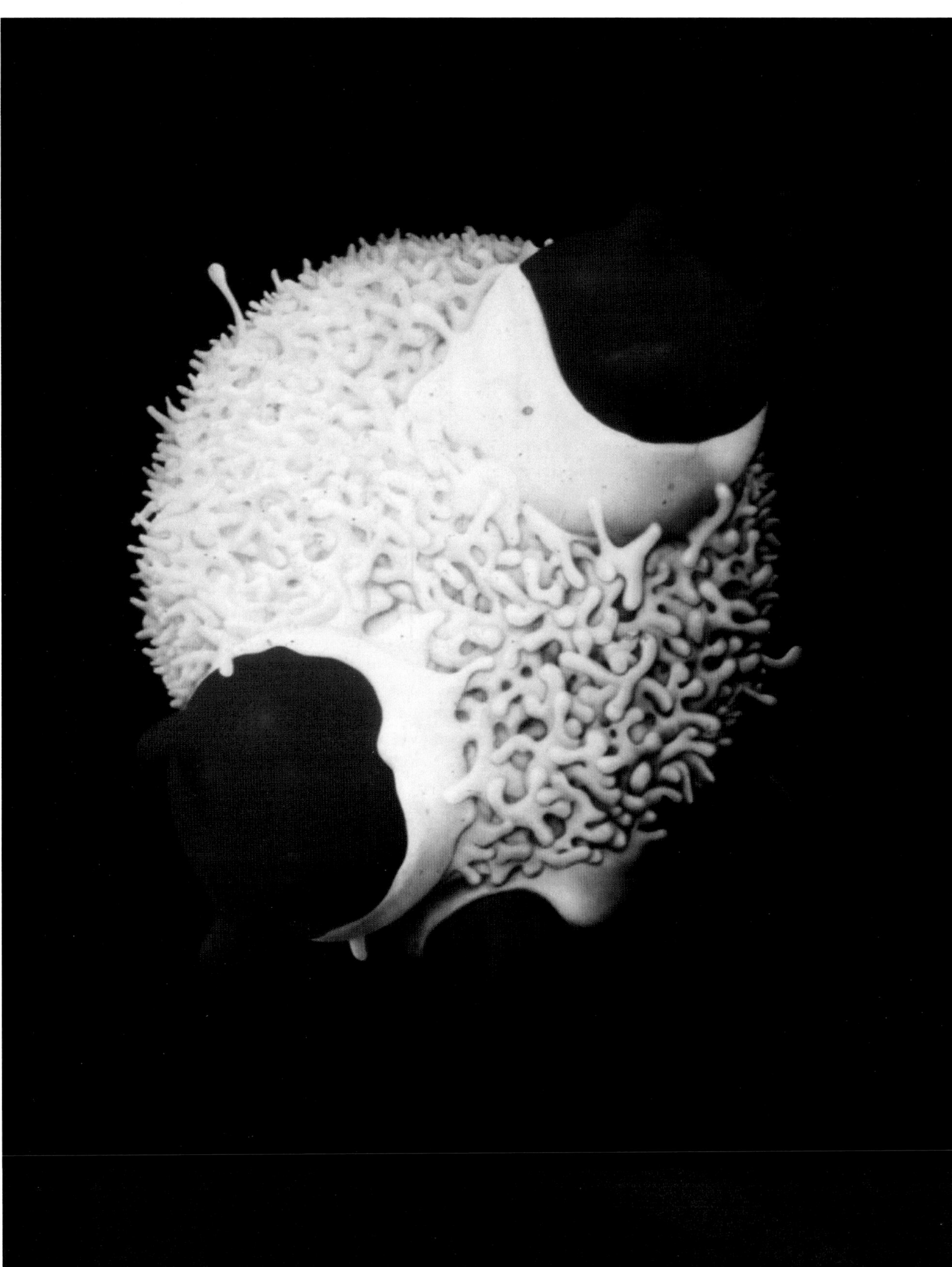

Spleen macrophage engulfing red blood cells. Owing to its abundance of phagocytic cells and the close contact between the circulating blood and these cells, the spleen represents an important defense against microorganisms which penetrate the circulation, and is also the site of the destruction of many red cells.

community. The similarity between the symptoms of cretinism and those displayed by patients having undergone total ablation of the thyroid gland is what made him sus-

Emile T. Kocher

pect the existence of a secreted thyroid hormone. Emil Theodor Kocher also distinguished himself by developing many surgical instruments, among which are the forceps, still used today. We are also indebted to this physician for

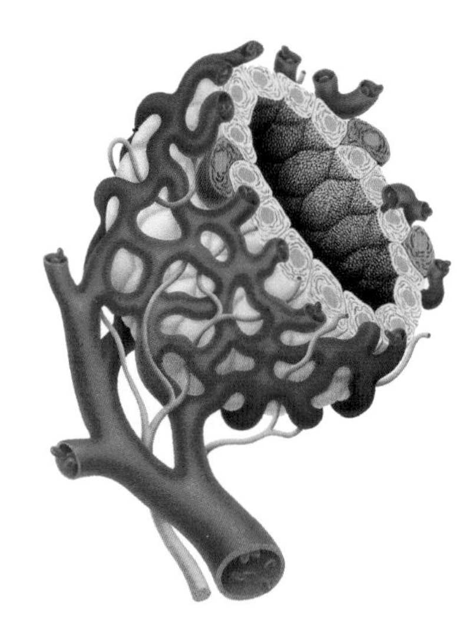

Thyroid follicles and their vascularization.

the development of surgical techniques in such varied fields as stomatology, pneumology and neurology. Emil Theodor Kocher received his medical education from the University of Bern in 1865. He studied further at the University of Berlin under the famous surgeon Theodor Billroth before being appointed Professor of Surgery in Bern at the age of 31 years.

1910

Kossel, Albrecht (Rostock, Germany, September 16, 1853 - Heidelberg, Germany, July 5,1927). German biochemist. Nobel Prize for Medicine, for his contributions to understanding the chemistry of nucleic acids and proteins. When he received the prize, both Albrecht Kossel and the Nobel Jury members were unaware of the extraordinary impact that nucleic acids would have on

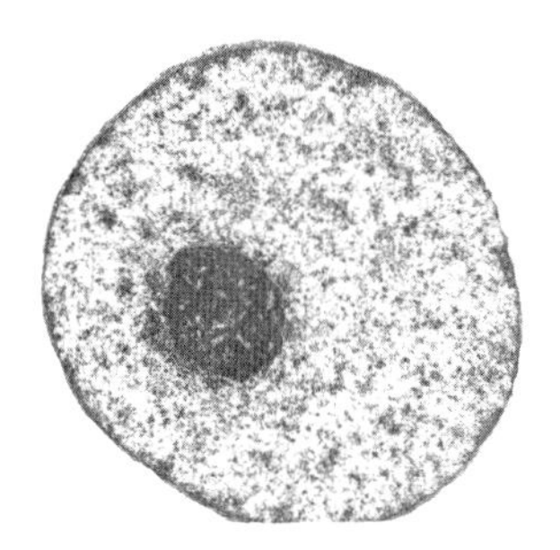

Cell nucleus. The nuclein is arranged in more or less dense masses, with an important part situated up against the nuclear membrane.

scientific and medical research. We are indebted to Kossel for having shown that nuclein is composed of nucleic acids and proteins. With the unsophisticated analytical methods of the time, he also managed to establish that nucleic acids are themselves composed of nitrogen - containing substances (purines and pyrimidines) and carbohydrates. Kossel also succeeded in discovering the amino acid histidine. Moreover he conducted advanced studies on the basic amino acids in general. Albrecht Kossel received his medical education from the University of Strasburg (now Strasbourg, France) in 1878 before conducting research there and at the University of Berlin. He then taught cell physiology at Marburg before joining the University of Heidelberg in 1901. His son, Walther Kossel, was a well-known physicist.

1911

Gullstrand, Allvar (Landskrona, Sweden, June 5, 1862 - Uppsala, Sweeden, July 28, 1930). Swedish ophthalmologist. Son of a physician. Nobel Prize for Medicine for his research on the eye as a light-refracting apparatus. Towards the end of the 19th century, Helmholz had shown that eye accommodation, which enables us to distinguish objects both near and far, depends on the curvature of the eye's crystalline lens, imposed by certain muscles. The convexity of the lens increases when we look at an object that is close to the eye, and it decreases when we look at one that is distant. Gullstrand revealed the existence of yet another mechanism, called the "intracapsular mechanism", by which fibres located behind the lens contribute to accommodating the eye to the perceived object. He collaborated with the German optician Zeiss and invented the ophthalmoscope in 1911. With this instrument it is possible to observe the retina at the back of the eye. Gullstrand has also allowed the understanding of optical images formation. A Chair of

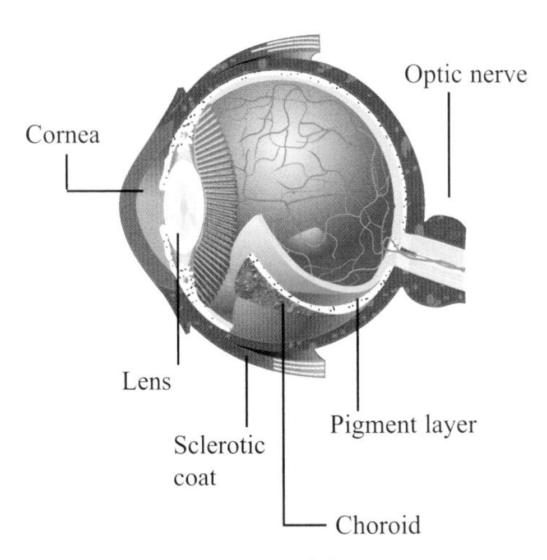

Anatomy of the eye

Physiological Optics was created on his behalf at the University of Uppsala in 1913. Gullstrand studied ophthalmology at the universities of Uppsala, Vienna, and Stockholm, and earned his Ph.D. in medicine in 1890. Some years later he began teaching ophthalmology at Uppsala University.

1912

Carrel, Alexis (Ste. Foy-les-Lyon, France, June 28, 1878 - Paris, France, November 5, 1944). French physician. Son of a textile manufacturer. Nobel Prize for Medicine for his work on ligating blood vessels and transplanting organs. In spite of his promising results obtained in the area of blood vessel surgery in France, Alexis Carrel could not accept the lack of interest shown by the French scientific community in his work. Moreover, his colleagues considered him as a mystic. In fact, Carrel reported that on one of his trips to Lourdes, a French town renowned for pilgrimages to the Virgin Mary, he had witnessed a case of a miracle cure and stated that some hea-ling was beyond scientific explanation. The report caused the rupture of his contact with his scientific colleagues. His scientific work was largely undertaken in Canada and the United States, where he emigrated in 1906, and can essentially be divided into two areas. The first is his Nobel Prize work in which he successfully established a technique for suturing the two ends of sectioned blood vessels. This technique led the way for transplanting damaged organs, such as a kidney, by correctly ligating the bood vessels following surgery. During the First World War, along with the British biochemist Henri Dakin, Carrel developed a method for treating wounds with sodium hypochlorite solution. At the end of the war, he went back to the Rockefeller Institute in New York, where he developed an in vitro system for culturing animal tissues. In spite of the difficulties involved, he was successful in establishing a perfusion technique which relied on the provision of a blood supply to the cultured organ. This approach proved to be invaluable later for the culturing animal tissues and for transplanting organs. In this respect it is noteworthy that he succeeded in setting up the in vitro culture of an embryonic chick heart which continued to beat for 36 years, long after his death, and was deliberately stopped subsequently. The last scientific contribution of this eminent surgeon was towards developing a sort of artificial heart capable of sustaining an animal's life for several days, if not weeks. He set up this technique with the help of his great friend and famous aviator Charles Lindbergh. In 1935 he published a book entitled 'Man the Unknown' which provoked controversy because some of his philosophical ideas were in favour of racial selection which were considered in the United States to be anti-democratic. As a result, he was forced to resign from his position at the Rockefeller Institute. Moreover, the job he held during the Second World War at the foundation for the study of human problems, was managed under German occupation and supported by the Vichy authorities. As a result, he was accused of collaborating with the Germans, however, his premature death following a heart attack in 1944 meant that he was never incriminated.

Alexis Carrel

1913

Richet, Charles Robert (Paris, France, Augustus 26, 1850 - Paris, France, December 4, 1935). Nobel Prize for Medicine for his work on anaphylaxis. As Professor of Physiology at the Faculty of Medicine in Paris, Charles Richet first distinguished himself with studies on the physiology of heat production. His next major work was in the dawning field of immunology. In collaboration with the French scientist Portier, he showed that dogs could survive injection of a moderate dose of poison, but that they died rapidly if they later received an injection of the same poison, even in lesser amount. To describe their observations, the two scientists coined the term anaphylaxis, meaning "non - protection", and wrongly concluded that the first dose suppressed certain immunological reactions, thus rendering the organism sensitive to a subsequent attack. Later advances in immunology showed clearly that the reverse is true: the absence of a reaction after the first injection reflects the latency period required to develop an adequate immune response. In reality, the initial contact with the foreign substance

Pollen grain. A coloured scanning electron micrograph. Such plant cells cause allergic reactions in numerous individuals.

"teaches" the organism to identify it and to respond to its presence by producing antibodies against it. The second time the foreign substance (also called "antigen") is injected, the immune response takes place and normally, the antigen is neutralised. In many cases this reaction is "normal" and cures the organism, but in the case of certain so - called "anaphylactogenic" poisons, this reaction is very violent and inappropriate, causing mastocytes to release diverse highly active molecules (serotonin, histamine, etc.). These induce harmful physiological disturbances that can even lead to death. Shortly after Charles Richet's work, these anaphylactic poisons were called allergens and it was shown that their ability to trigger an allergic reaction was not linked to toxicity.

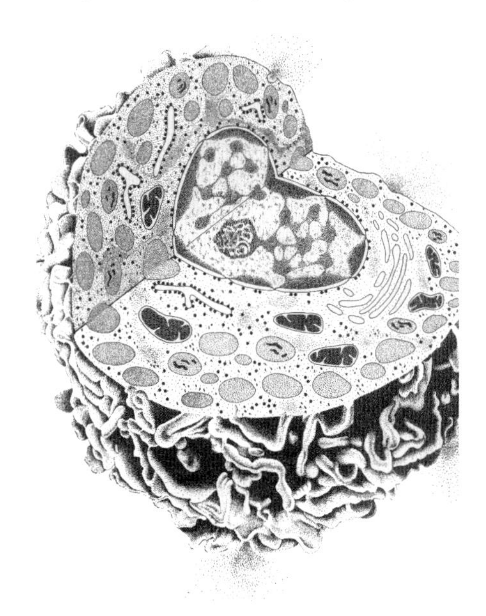

Mastocyte. A three - dimensional representation. This cell type stores pharmacologically potent mediators (mainly histamine or serotonin) that are released in the outer environment after stimulation.

1914

Barany, Robert (Rohonc, Austria, April 22, 1876 - Uppsala, Sweden, April 8,1936). Austrian physician and otologist. Son of a banker. Nobel Prize for Medicine for his work on the physiology and pathology of the vestibular apparatus of the inner ear. After 1917, professor at the University of Uppsala. Robert Barany owes his scientific reputation to the development of tests to determine the physiological state of the functions requiered for equilibrium. In one of these, the ears of a patient were irrigated separately, one with warm water, the other with cold water. In normal patients, this treatment provoked nystagmus, that is, rapid and involuntary movement of the eyes. In patients whose fuction of equilibriun was abnormal, a delay in the onset of these movements was observed. Similarly, in the test of "the rotating armchair" in which the head of the patient was bent slightly forward, a deviation from the normal nystamus reaction revealed problems in the the function of equilibrium. Barany, a soldier in the Polish Army in World War I, was taken prisoner in 1915 and deported to Central Asia. As a result, it was only in 1916 that he learned that he had been granted the Nobel Prize. Upon his return to Austria, Barany's colleagues gave him a very cool welcome, being upset that the Nobel Commitee failed to mention what they felt was their contribution to Barany's work. Although the Nobel Committee found no serious grounds for his colleagues' claims, Robert Barany, suffering greatly from his colleagues' jealousy, decided to settle in Sweden where he continued his research in the field of the physiology of equilibrium.

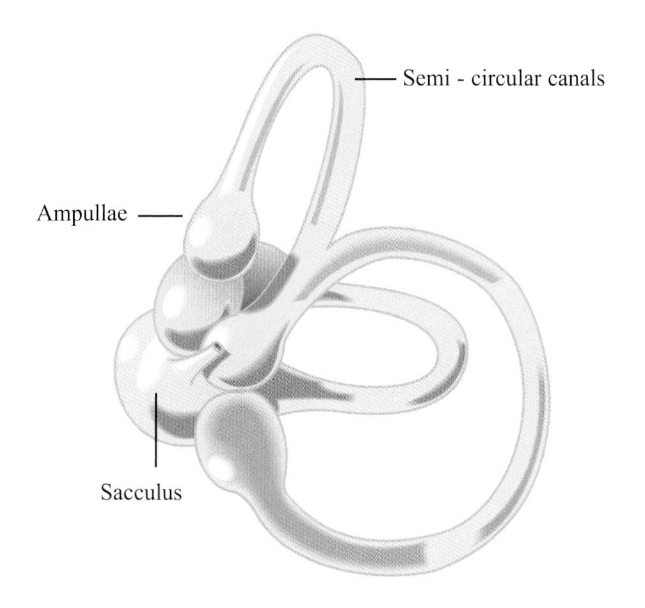

Vestibular system. This sense organ contains mechanoreceptors which detect changes in both motion and position of the head. Robert Barany devoted his scientific carrier to the study of its physiological properties.

1915 to 1918 Not awarded

1919

Bordet, Jules Jean Baptiste Vincent (Soignies, Belgium, June 13, 1870 - Brussels, Belgium, April 6, 1961). Belgian physician and immunologist. Son of a schoolmaster. Nobel Prize for Medicine for his discovery of immunity factors in blood serum. Jules Bordet was a researcher at the Institut Pasteur in Paris and a teacher of bacteriology at the Free University of Brussels. After his graduation from high school, he became interested in science and spent a lot of time carrying out experiments in a small laboratory he had set up at home. Bordet began his medical studies when he was 16 and was awarded his degree of Doctor in Medicine in 1892. He obtained a grant from the Belgian government allowing him to go on a seven year training course at the Pasteur Institute in Paris. There he started working in the laboratory of **Ilya Ilitch Metchnikoff**, a Ukranian biologist. It was here in the same institute that he made major discoveries which led to the Nobel Prize. Along with Octave Gengou, Bordet discovered that when animals were immunized with bacteria, the antibodies produced could only destroy the bacteria in conjunction with another substance which he called 'alexin', today known

Jules Bordet

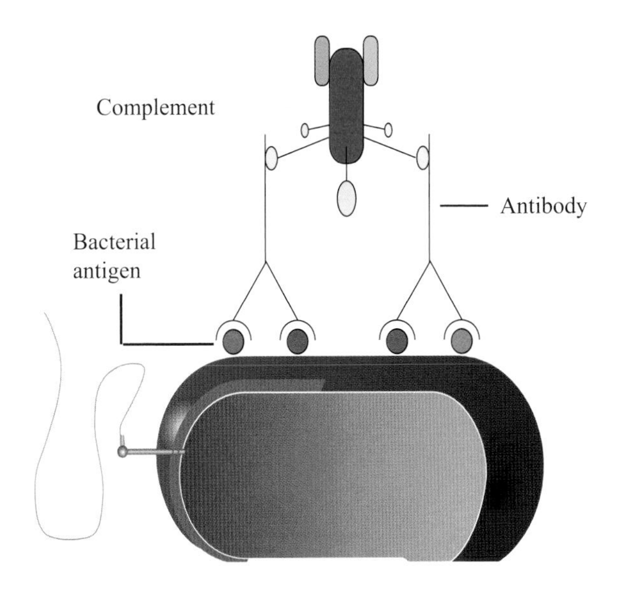

Scheme showing the relationship between the complement, the antibodies and a bacterium.

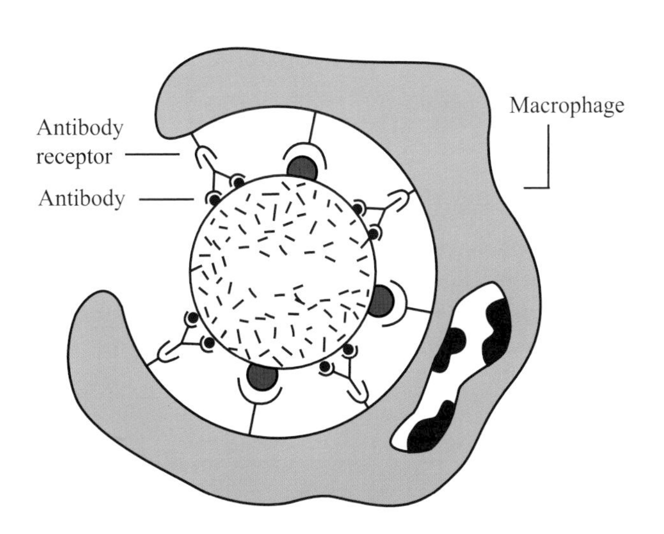

Phagocytosis of a bacteria is helped by complement. Macrophages and neutrophiles bear complement receptors that act in the same way. Here, a macrophage binds to the bacterium by both antibody receptors and complement receptors (in red). These receptors synergize in inducing bacterium uptake and macrophage stimulation.

as 'complement'. Bordet showed that this substance (which is in fact a mixture of molecules) was present not only in immunized animals but also in non - immunized animals. The discovery of this immune defence system against the invasion of foreign cells led to the development of the complement fixation test, a very sensitive method of detecting the presence of pathogenic microorganisms. This test was later modified by the German bacteriologist August von Wasserman for the detection of syphillis. Bordet and his co - workers used the test to detect infection by bacteria that caused typhoid fever and swine fever. On his return to Brussels, he ran a new institute for bacteriology which two years later became the Pasteur Institute of Brussels. In 1906, while still collaborating with Gengou, he identified the bacterium that caused whooping cough and made a vaccine to treat the disease. A year after the First World War, he departed for the United States to try and raise funds to rebuild the University of Brussels, which had been damaged during the conflict. It was while he was abroad that he learned that he had been awarded the Nobel Prize in Medicine in 1919. Bordet was also interested in bacteriophages, viruses that infect and lyse bacteria during their replication cycle. He also worked on blood clotting.

1920

Krogh, Schack August Steenberg (Grenaa, Denmark, November 15, 1874 - Copenhagen, Denmark, September 13, 1949). Danish physiologist. Krogh's father was engaged in shipbuilding.Nobel Prize for Medicine for his discovery of a motor-regulating mechanism of capillaries. It was August Krogh who, for the first time and with the help of his wife Marie, proved that the oxygen pressure is higher inside the alveoli than in the blood. He concluded that in the lungs, the absorption of oxygen and elimination of carbon dioxide occur through diffusion. The work for which he received the Nobel Prize, however, was his work on the capillary system. Krogh showed that the number of functional capillaries in tissues (such as muscles) and their degree of dilation or contraction increase or decrease according to whether they are active or not. Having com-

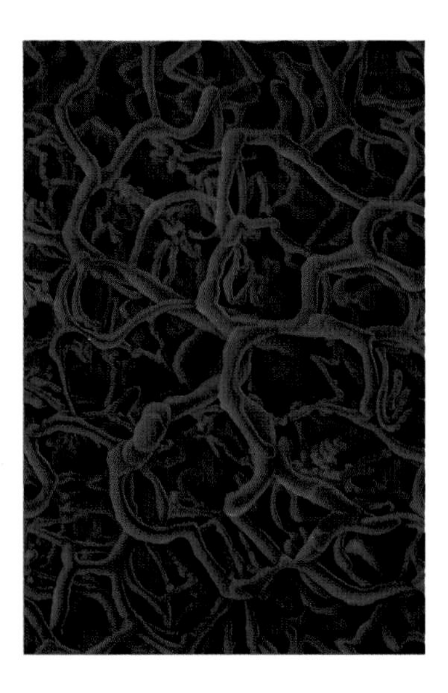

A microcapillar lattice

pleted his secondary education at the Aarhus Lyceum, Krogh entered Copenhagen University. At first he tackled physics and then took to studying zoology, to which he devoted his career until his retirement in 1945.

1921 Not awarded

1922

Hill, Archibald Vivian (Bristol, England, September 26, 1886 - Cambridge, England, June 3, 1977). British physiologist and biophysicist. Nobel Prize for Medicine, shared with **Otto Meyerhof**, for discoveries concerning the production of heat in muscles. Robert Hill was a professor of physiology at the University of Manchester and later at University College, London. He perfected a thermocouple that permitted the measurement of tiny variations in the heat released by a muscle as a consequence of a specific amount of work. His research showed that the oxygen consumed by a muscle occurred after and not during contraction, thus, indicating that oxygen was not

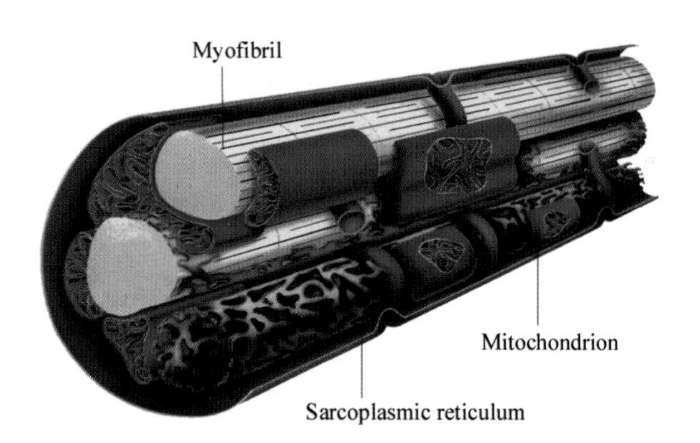

Part of a muscle cell, a three-dimensional representation. The activity of this organ generates much energy in the form of heat, of which precise measurements were made by A. Hill.

required for the contraction itself. Hill was also a pioneer in the study of oxygen-binding by hemoglobin and he established the so - called "Hill plot", that expresses the myoglobin saturation by oxygen. Sir Archibald Hill taught physiology at the University of Manchester from 1920 to 1923 and at University College in London from 1923 to

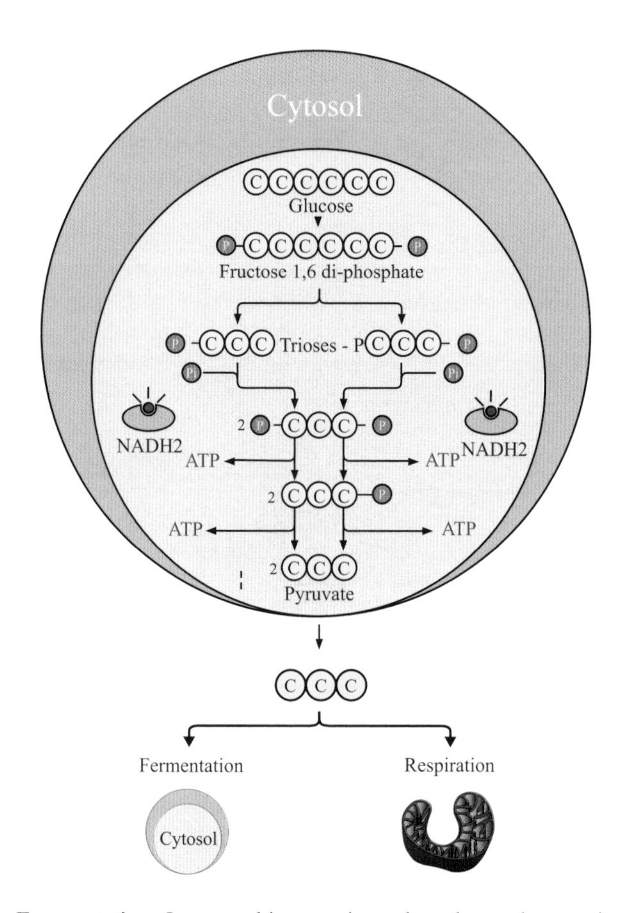

Fermentation. In anaerobic organisms, but also under certain conditions in muscles, glycolysis is a major source of the cell's energy. In such organisms, pyruvate is not metabolized in mitochondria but is converted to ethanol plus carbon dioxide (in yeast), or into lactate (as in muscle). (See Meyerhof).

1925. Dr. Hill is the author of renowned scientific books, including Muscular Activity, Muscular Movement in Man and Living Machinery, all three published towards the end of the 1930s.

Meyerhof, Otto Fritz (Hannover, Germany, April 12, 1884 - Philadelphia, Pennsylvania, October 6, 1951). German-American biochemist. Nobel Prize for Medicine "for his discovery of the relationship between oxygen consumption and the rate of lactic acid metabolism in muscle". Most of Meyerhof's scientific career was devoted to the metabolism of muscles. His most significant work was the demonstration that in anaerobic conditions lactic acid is formed in muscle tissue as a result of glycogen breakdown. This process, which supplies energy to the muscle, is called anaerobic glycolysis. Meyerhof thus showed that the glycolytic metabolic pathway is not specific to a few microorganisms such as yeast, but that it occurs also in other eukaryotic cells. It became more and more clear that all the living cells could be partially supplied in energy and carbon atoms required for their synthesis by this metabolic pathway, the so-called Embden - Meyerhof pathway. People were beginning to accept the idea that certain basic biochemical processes are common to all living organisms. Meyerhof's research led him to become aware - along with **Fritz Lipmann** and several other biochemists - of the importance of phosphoesters in capturing the energy contained in food and delivering it to the cell. Otto Meyerhof taught physiology at Kiel and Heidelberg Universities before directing the Department of Physiology at the Kaiser Wilhelm Institute in Berlin. In 1938 he fled the Nazis and moved to Paris. Finally he settled down in Pennsylvania, where he was made Chairman of the Department of Physiology. This biochemist, a lover of art and philosophy, wrote a doctoral thesis on the theory of knowledge.

1923

Banting, Frederick Grant (Alliston, Canada, November 14, 1891 - Newfoundland, February 21, 1941). Canadian physician. Son of a farmer. Nobel Prize for Medicine, along with **John McLeod**, for the discovery of insulin. This academic's research began in 1920 after recovering from serious wounds sustained on the Allied front in France. His work focused on the cause of sugar diabetes, a disease in which blood sugar levels become abnormally elevated, resulting in a slow but certain death. It had been known for some time that removal of the pancreas led to similar symptoms and it was suspected that this organ produced a hormone capable of regulating blood sugar levels. In 1901, work conducted by Edward Albert Sharpey-Schaefer and other researchers had indicated that diabetes could be caused by lack of a substance produced by specialized cells discovered in the pancreas by Paul

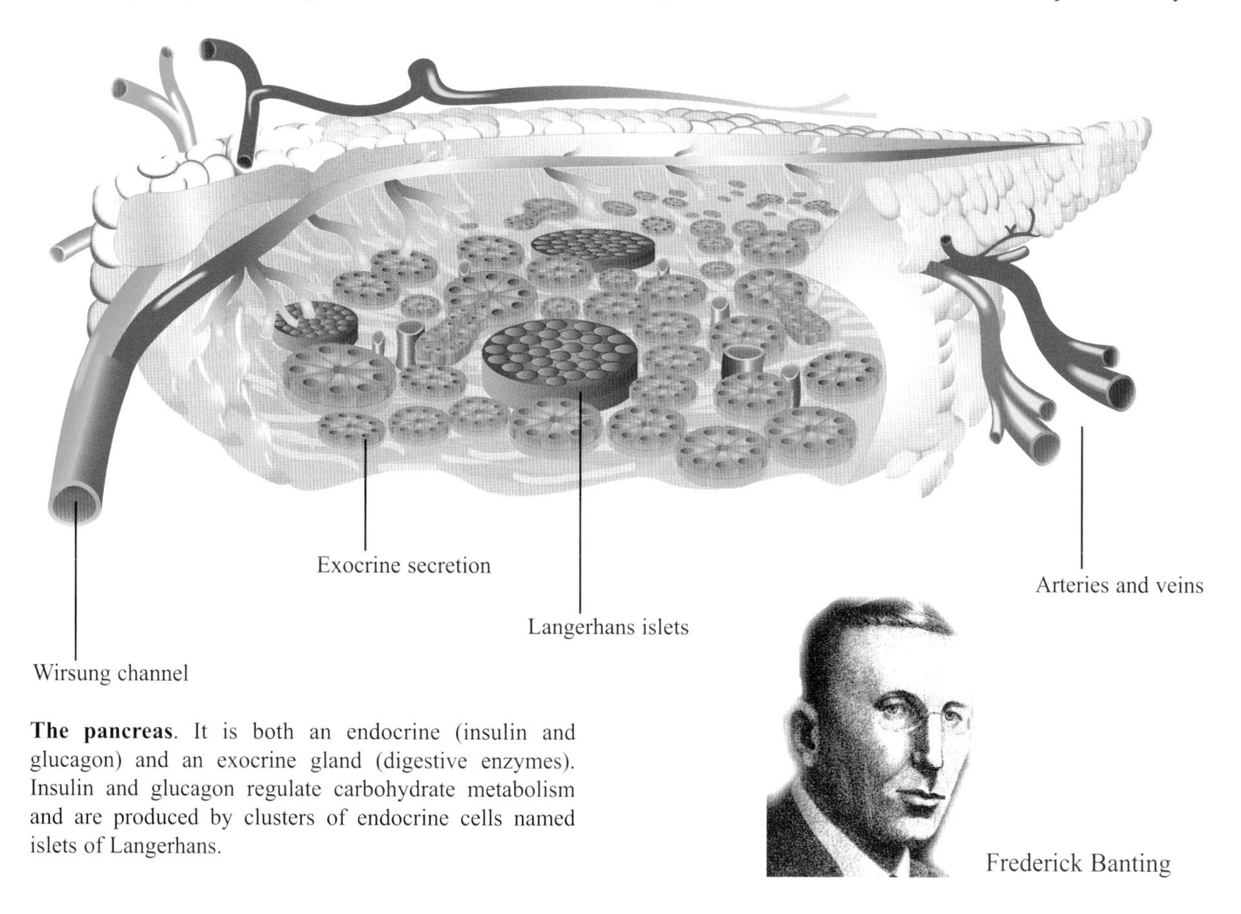

The pancreas. It is both an endocrine (insulin and glucagon) and an exocrine gland (digestive enzymes). Insulin and glucagon regulate carbohydrate metabolism and are produced by clusters of endocrine cells named islets of Langerhans.

Frederick Banting

Langerhans, a German histologist (Islets of Langerhans). While reading a medical paper on diabetes, Banting realized that it should be possible to extract insulin from the pancreas, provided that a ligature was first applied to the excretor canal. With this in mind, he contacted McLeod, who ran a research laboratory in Canada, and persuaded him to carry out the experiments in his institution. The difficult and arduous task of obtaining a sufficient quantity of insulin was carried out in collaboration with Charles Best, a young Canadian physiologist belonging to the same laboratory. To test the insulin - containing pancreatic extracts and to ensure that the procedure would not be harmful to diabetic patients, Banting and Best first injected themselves with the extracts. During January 1922, the first therepeutic test was carried out on a 14 year-old diabetic subject and, a few days later, on the same adolescent with more purified insulin. James B. Collip, McLeod's assistant, established a method for obtaining insulin in a highly purified form and in much larger amounts. A short time later, the young man's blood and urine sugar concentrations decreased, a testimony to the success of this treatment. Banting recognised the invaluable contribution of Best and was very unhappy when he learned that his colleague had not been jointly awarded the Nobel Prize. In fact, it was attributed to McLeod, who Banting felt had not been a great help in the discovery of insulin. In the end, Banting was persuaded to accept the prize he had initially refused, and shared the prize money with Best. In 1941 Banting died in an air accident on the way to the Ministry of Health of the Canadian Army. The plane crashed over Scotland. After Banting's death, Best took charge of the laboratory in Toronto. Thanks to the pioneering work of Banting and Best, millions of lives have been saved and people suffering from diabetes are able to enjoy a better quality of life.

McLeod, John James Richard (Cluny, Scotland, September 6, 1876 - Aberdeen, Scotland, March 16, 1936). Scottish physiologist. Nobel Prize for Medicine, along with **Frederick Banting**, for his work leading to discovery of insulin. McLeod specialized in the carbohydrate metabolism and headed a research laboratory in Toronto, Canada. Although he supervised Banting's research and gave him a precious collaborator (Charles Best), he did not really like Banting and was sparing in his assistance to him. When Banting and Best, after years of effort, succeeded in extracting in-sulin from the pancreas, he at last decided to add a member to their team (James B. Collip) in order to isolate the hormone chemically. Ill at ease about being awarded the Nobel Prize together with Banting for work to which he had contributed but little, he shared his portion of the prize money with Collip. We are also indebted to McLeod for experimental work on hyperglyceamia. In particular, he showed that from experiments done on rabbits, that the parasympathetic nervous system was involved in the enhancement of gluconeogenesis in the liver. One of his main writings was Physiology and Biochemistry of Modern Medicine.

John McLeod

1924

Einthoven Willem (Java, Indonesia, May 21,1860 - Leiden, The Netherlands, September 28, 1927). Dutch physician. Nobel Prize for Medicine for his discovery of the electrical properties of the heart through the electrocardiograph. Willem Einthoven was born in Java, one of the main islands of Indonesia. His father worked as a doctor and Einthoven was only 6 years old when he died. His mother emigrated along with himself, his five sisters and brothers to Holland. They settled down in Utrecht, where he studied. Einthoven was one of the first scientists who used physics for designing and constructing medical instruments. Einthoven is best known for his research detecting the relationships that exists between the electrical activity detected in the heart and the beating of cardiac muscle. At the time the instruments used to measure the heart's electrical activity were very imprecise and it was to this that Einthoven brought a much appreciated contribution. Einthoven invented and perfected the string galvanometer, an apparatus made up of a galvanometer and a thin wire of platinium able to measure the electrical signals produced by the heart. The wire was oriented in a magnetic field in such a way that it moved in response to

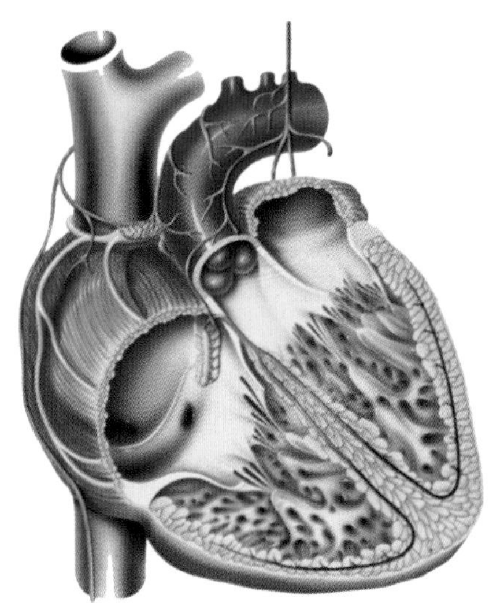

The human heart.

weak electrical currents generated by heart contraction. He succeeded in magnifiyng the movement thanks to an optical system and tracing this movement on paper termed an electrocardiogram. The galvanometer conceived by Einthoven allowed one, among other things, to measure different electrical waves associated with the beating of the heart, known as P, Q, R, S, and T waves. The precision of these measurements was such that he could even distinguish the electrical modifications that came from the contraction of the ventricles as opposed to those coming from the contrac-tions of the atria. Einthoven established the electrocardiogram of healthy subjects, which allowed him to compare it with the EKG of sick subjects and make a diagnosis. With this work, the man that Nature review described as a particularly good and likeable man founded the basis of electrocardiography. Einthoven's studies on electrocardio- graphy, along with Roentgen's X - rays discovery, played a vital role in the development of medicine.

1925 Not awarded

1926

Fibiger, Johannes Andreas (Silkeborg, Denmark, April 23, 1867- Copenhagen, Denmark, January 30, 1928). Danish pathologist. Nobel Prize for Medicine for achie- ving the first controlled induction of cancer in laboratory animals. Fibiger's father was a physician and his mother was a writer. Johannes Andreas Fibiger is known for his work on the experimental induction of cancer in healthy animals. Having finished his medical studies just two years earlier, he distinguished himself by developing a serum that proved effective against diphtheria, a common disease in the early 20th century. At that time, the idea was proposed that cancer could be caused by environmental agents such as chemical compounds. Fibiger was able to show that rats developed cancer when dyes were applied to their skin. He noted that cancer frequencies varied according to the species and even within species. The latter fact suggested that certain individuals could be predisposed to cancer. Earlier, he had formulated the idea that parasites such as certain worms could induce cancer in their hosts, but this hypothesis is no longer retained. Although Fibiger's experimental protocols showed some weaknesses and his results were criticised by his peers, he received the Nobel Prize for the impetus he had given to anti-cancer research. Johannes Fibiger studied bacteriology at the University of Berlin under the direction of Robert Koch and **Emil von Behring**. In 1900, he became professor of pathological anatomy at the University of Copenhagen.

1927

Wagner-Jauregg, Julius (Wells, Austria, March 7, 1857 - Vienna, Austria, September 27, 1940). Austrian physician. Nobel Prize for Medicine, for treatment of syphilitic meningoencephalitis, or general paresis, by the artificial induction of malaria. The work of this Austrian psychiatrist centred essentially around an illness called GCI, for General Complication of the Insane. This is a fairly common complication of syphilis. Around 1887, this researcher had noticed that many remissions from GCI were preceded by a bout of fever. He concluded that the fever must somehow cause the remission. Wagner-Juaregg proceeded to inoculate about ten of his patients with the malaria agent, and observed a long-term remission in six of them. Another discovery due to this scientist

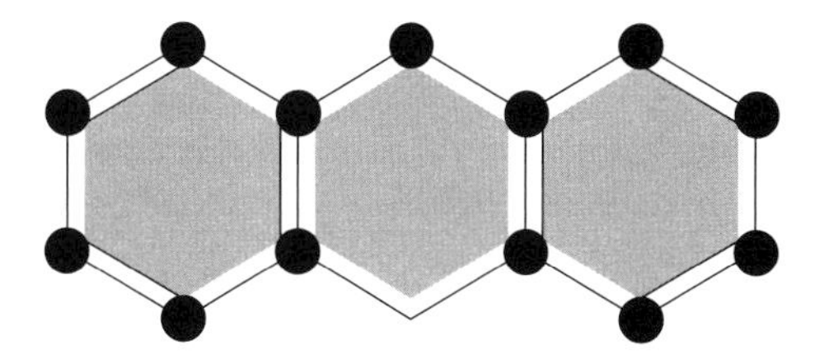

Vitamin B_2. A schematic structure of its isoalloxazine nucleus.

was his discovery, with Robert Kuhn and Albert Szent-Györgyi, of vitamin B_2, a deficiency of which causes eye and skin problems. Julius Wagner - Jauregg graduated in medicine from the University of Vienna in 1874.

1928

Nicolle, Charles Jules Henri (Rouen, France, September 21, 1936 - Tunis, Tunisia, February 28, 1936). French physician and bacteriologist. Son of a physician. Nobel Prize for Medicine for his discovery that typhus is transmitted by body louse. This director of the Pasteur Institute in Tunis, ex-collaborator of Roux, discovered the typhus vector. Although seemingly simple, this discovery was not at all obvious at the time. Like his fellow - doctors, Nicolle had noticed that people taking care of typhus patients outside the hospital often caught the disease themselves, whereas people caring for patients inside the hospital, where the conditions were more hygienic, did not. Nicolle concluded that the agent responsible for this must be the louse. Proof that typhus was indeed transmitted by a louse came from an ingenious experiment. In this experiment, blood from a sick monkey was injected into a healthy monkey. Lice were then deposited on the latter and transferred to a third monkey, which caught the disease. This showed that the lice were indeed the vectors of the disease and that they transmitted it through their bite. Nicolle obtained his medical degree in Paris in 1893. He then taught medicine at the faculty of the medical school in Rouen before moving to Tunis, where he served as director of the Pasteur Institute until his death.

1929

Eijkman, Christiaan (Nijkerk, The Netherlands, Augustus 11, 1858 - Utrecht, The Netherlands, November 5, 1930). Dutch physician. Son of a schoolmaster. Nobel Prize for Medicine. He shared the prize with **Frederick Gowland Hopkins** for his work on the recognition and study of vitamins. Eijkman thought along with colleagues in microbiology that microbial infection was the cause of many diseases. Nevertheless, a fortunate event would modify this idea. While he was directing a medical laboratory in Batavia, a town in the Dutch Indies, he noticed that the chickens serving as test subjects had suddenly contracted a disease whose symptoms closely resembled those of beriberi. Strangely, the illness disappeared some time later. Thus by chance he learned that these same chickens had been fed hulled rice for several weeks. The disappearance of the symptoms coincided with the return of whole grain rice. Eijkman concluded that the hull of a grain of rice must contain a substance necessary to maintaining a healthy organism. Christiaan Eijkman did not initially understand his discovery because he first believed that the beriberi was caused by a microbe (Eijkman was a student of Robert Koch, a bacteriologist, and he was convinced that all illnesses were due to microorganisms). It was not until later that the nutrient component that seemed indispensable was discovered in the husk of rice by the Pollack C. Funck. It was later found to be thiamine or vitamin B_1. (But it was John Hopkins who isolated it.) Funck coined the term "vitamin" for the unidentified substances present in food which could prevent the diseases scurvy, beriberi or pellagra. The discovery of vitamins was a major medical advance because severe nutritional deficiencies were common and led to deadly afflictions.

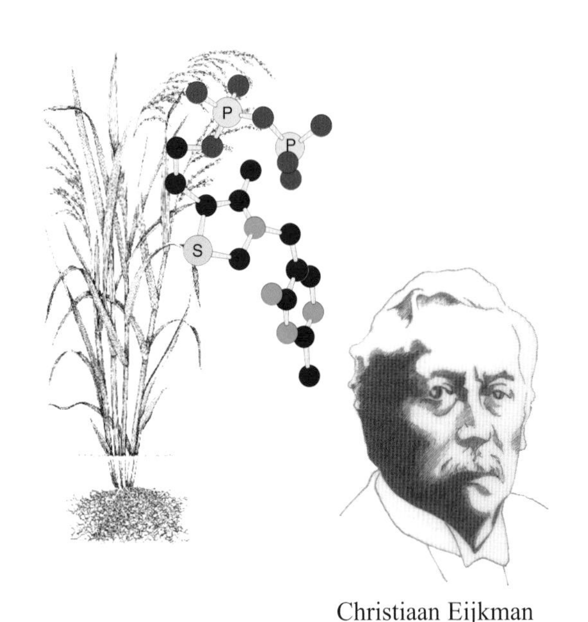

Rice and vitamin B_1. Vitamin B_1, or thiamine, mainly occurs in rice husks. Deficiency in this vitamin is common in areas where the staple diet is polished rice.

Indonesia, where C. Eijkman was made medical officer of health and where he conducted studies on beri-beri.

Hopkins, Frederick Gowland (Eastbourne, England, June 20, 1861 - Cambridge, England, May 16, 1947). British biochemist. Nobel Prize for Medicine, along with **Christiaan Eijkman** for the discovery of vitamins and nutritional factors. Fredrick Hopkins was the scientist who isolated the first known vitamin: B_1. Eijkman had shown that the absence of this vitamin from the diet is responsible for the symptoms of beriberi. Hopkins's work on the nutritional requirements of animals led him to establish a table of essential nutrients indispensable to the organism, such as substances that fulfil the requirement for nitrogen. Hopkins was also a renowned biochemist who discovered tryptophan, an amino acid. Moreover, he discovered glutathione, a tripeptide which acts as a hydrogen carrier in many biochemical reactions. Gluthathione is now consid-

Tryptophan, an amino acid discovered by F.G. Hopkins. In mammals it is an essential dietary amino acid.

ered as a very efficient antioxidant substance, the levels of which in cells are predictive of how long they will live. Another field of research was enzymic catalysis, which was then an intensive area of investigation. He considered that cells were a factory of numerous molecules undergoing complex interactions. This concept was a major contribution to science. Hopkins may be considered as the father of biochemistry. He showed that, contrary to a belief of the time, there is nothing "magic" about the chemistry of living organisms, but that "life" is merely more complex than "nonlife" chemistry. Frederick Gowland Hopkins really made biochemistry a recognized science. In 1912 he was Head of the Department of Biochemistry at Cambridge University. He was knighted in 1925.

1930

Landsteiner, Karl (Baden, Austria, June 14, 1868 - New York, New York, June 26, 1943). American immunologist and pathologist. Son of a journalist. Nobel Prize for Medicine for his discovery of the major blood groups and development of the ABO system of blood typing. Karl Landsteiner entered the University of Vienna in 1885 and studied medicine. Later he became interested in chemistry and one of his teachers was the famous chemist Emil Fischer. While he was teaching pathology at this same institution, he discovered the existence of blood groups A, B, and O. He showed that the red blood cells of individuals of groups A and B carry certain characteristic substances, which he named iso-agglutinogens. Group O

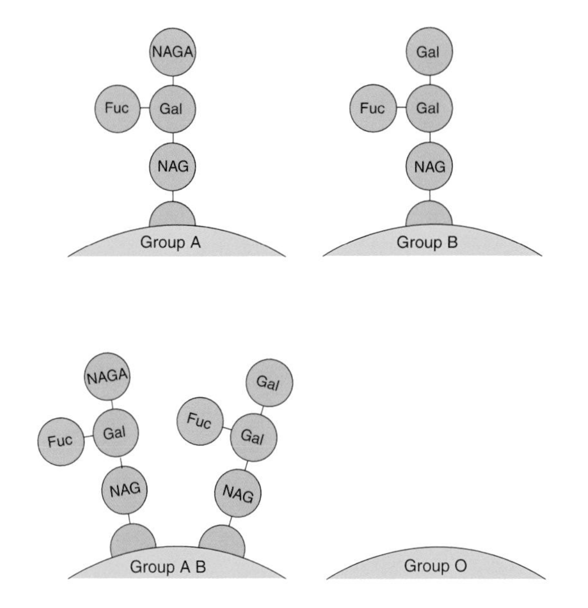

Blood groups A, B, AB and O.

individuals, he found, do not. After World War I Landsteiner was expatriated to Holland. Finally he moved to New York, where he settled down and resided until the end of his life. At the Rockefeller Institute he established, with Philip Levine and Alexander Wiener, the existence of

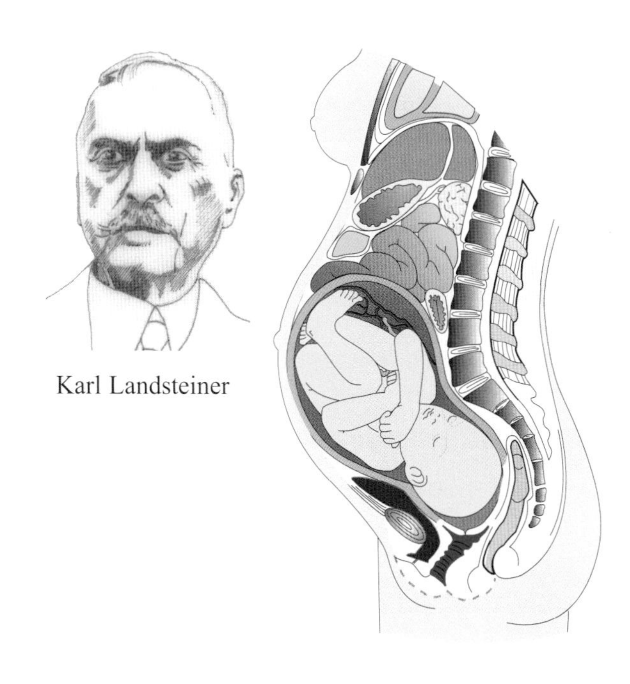

Mother and fœtus. Karl Landsteiner demonstrated the relationship between Rh factor and the erythroblastosis fetalis, from which it results that babies die after birth (for example, if an Rh - negative pregnant woman bears a child by an Rh - positive man).

agglutinogenic factors M, N, and Rh. It was thanks to this work that Levine demonstrated the relationship between the Rh factor and the haemolytic disease of the newborn: the latter is caused by Rh incompatibility between mother and fetus. Other advances due to Landsteiner notably include detection of the causative agents of syphilis (*Treponema pallidum*) and poliomyelitis (he was the first to isolate the poliovirus). In addition to their more obvious medical implications, the discoveries of this great specialist in blood immunology had deep repercussions in various areas of medicine such as the study of human migrations. It emerged, for example, that blood groups are distributed differently in different populations. Among Africans, the predominant groups are B (42%) and O (40%), and 85% of individuals are Rh+; among whites, the predominant groups are A (45%) and O (45%), and the percentage of Rh+ individuals is 99%.

1931

Warburg, Otto Heinrich (Freiburg-im-Breisgau, Germany, October 8, 1883 - Berlin, Augustus 1, 1970). German physician and biochemist. Nobel Prize for Medicine for his work on tissue respiration. The question of how the organism uses oxygen obsessed this great biochemist, student of one of the most eminent scientists of his time, the chemist **Emil Fischer**. Otto Warburg conducted the first in - depth studies on cytochromes, whose involvement in oxygen consumption he demonstrated and which he suspected of containing iron. The Swedish biochemist **Axel Theorell** would later also do major research

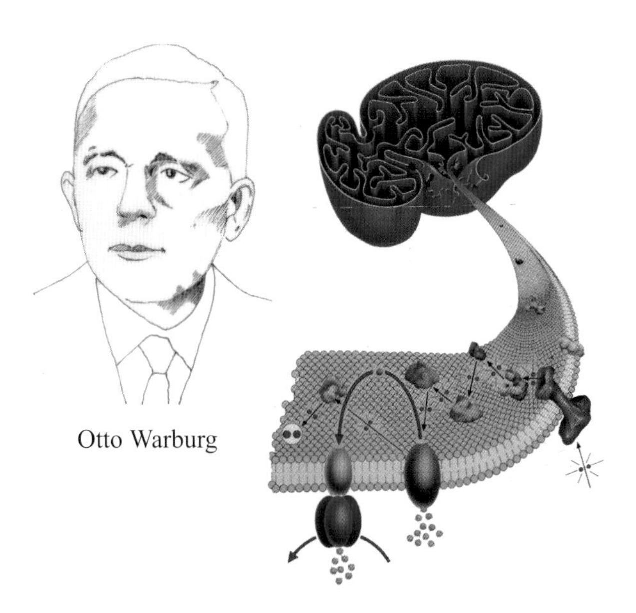

Otto Warburg

The electron transport chain. This drawing clearly shows that the process of electron transport involves the participation of several substances included in the inner membrane of the mitochondrium. Among these substances, cytochromes were a main subject of investigation by Warburg.

in the field of cellular oxidations. Warburg succeeded in isolating a substance which was formely called "yellow ferment", which takes up electrons from nutrients such as glucose and fatty acids. It could be shown that this substance was a flavin coenzyme and that it was bound to a protein. We are also indebted to Warburg for the discovery of an enzyme called glucose 6 - phosphate dehydrogenase and a nicotinic coenzyme, nicotinamid dinucleotide phosphate (NADPH), involved in hydrogen transport to biosynthetic pathways. People became aware of the importance of vitamins in the 1930s, when research on coenzymes began to flourish and people realised that most

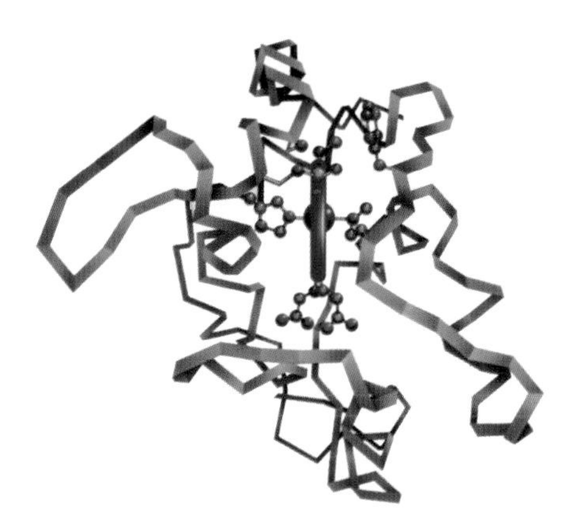

Cytochrome C. A ribbon model. This molecule plays the role of electron transporter in the electron transport chain of aerobic organisms.

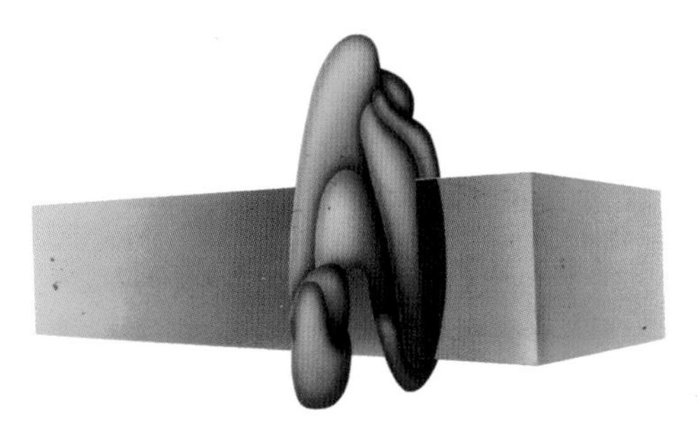

Cytochrome oxidase. An enzyme complex of major importance in aerobic respiration. This complex accepts electrons from cytochrome C and turns them to oxygen.

coenzymes derive from vitamins. Being half-Jewish, Warburg probably owed his research on cancer in avoiding being sent to a Nazi extermination camp. It seems that Hitler himself, fearing he had developed throat cancer, allowed the scientist to continue his work during World War Two. The Nazi police, however, had forbidden any German citizen to receive the Nobel Prize, so Warburg had no hope of being awarded a second prize, even though some people envisaged this possibility. This great researcher endowed with exceptional technical skill also became known for his discoveries in the field of photosynthesis. Otto Heinrich Warburg headed the Kaiser Wilhelm Institute for Cell Physiology in Berlin from 1931 to 1953.

1932

Adrian, Edgard Douglas (Lord) (London, England, November 30, 1889 - London, England, August 4, 1977). British neurophysiologist. Son of a lawyer and descendant of the scottish philosopher and historian David Hume. Nobel Prize for Medicine, shared with **Charles Sherrington**, for discoveries regarding the nerve cell. After military service as a member of the medical staff of British Army du- ring World War I, Edgard D. Adrian spent the major part of his scientific career at Cambridge University where he became a Fellow of Trinity College at the age of 28 and later, rector of the university. Adrian's main research concerned the transmission of nerve impulses. In 1905 he was able to record the electrical discharges in single nerve fibers and he demonstrated that a nerve impulse obeys the "all-or-none" law. For this work he was elected to a Fellowship of Trinity College, Cambridge. Adrian conducted intensive research on the electrical activity of the brain. He is also known for having shown the importance of areas of the brain that are specialized for particular functions such as touch or smell. He also studied abnormal brain electrical activity known under the name of "Berger's rythme." This work opened up new fields in the study of neural disorders such as epilepsy as well as in the location of cerebral

Douglas Adrian

damages. Edgard Douglas Adrian was President of the Royal Society from 1950 to 1955. He was knighted in 1965.

Sherrington, Charles Scott (London, England, November 27, 1857 - Eastbourne, England, March 4, 1952). Nobel Prize for Medicine, along with **Douglas Adrian** for works that establish the foundation for an understanding of integrated nervous function in higher animals. The first works of Sherrington related to bacteriology. We are indebted to him for precious studies on cholera and for the first use of the diphtheria antitoxin on a patient who was none other than his own nephew. During World War I, he tested a tetanus antitoxin on wounded soldiers. Yet, the most important part of his career concerned an entirely new field. Before Ivan **Petrovich Pavlov**, the Russian physiologist famous for his work on conditioned reflexes, he demonstrated the

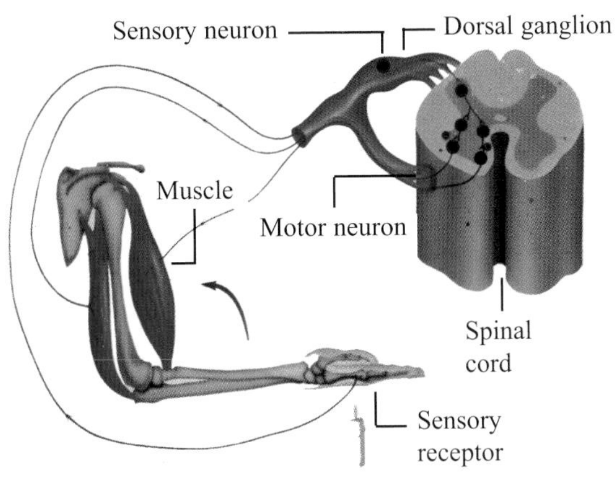

Reflex arc

existence of a movement - triggering reflex arc functioning as a unit. He also did important work on spinal cord degeneration and on the topographic anatomy of nerve fibres issued from the anterior and posterior roots of the spinal nerves. Several celebrated books, particularly "Man on his Nature" (1941) and "Goethe on Nature and Science", gave evidence of his interest in the human and philosophical aspects of science. Charles Sherrington graduated from the University of Cambridge and St. Thomas' Medical School where he obtained his medical degree. He also studied in Europe, in particular in Berlin under the direction of the German pathologist Rudolf Virchow and **Robert Koch**. He was knighted in 1922. He is considered as one of the fathers of neurophysiology.

The four chromosomes of *Drosophila melanogaster*. An electron micrograph.(See Thomas Morgan).

1933

Morgan, Thomas Hunt (Lexington, Kentucky, September 25, 1866 - Pasadena, California, December 4, 1945). American zoologist. Son of a diplomat. Nobel Prize in Medicine "for his discoveries concerning the role played by chromosomes in heredity". The name of Morgan is indelibly linked to the remarkable research he carried out on the genetics of the fruit fly *Drosophila melanogaster*. In the years around 1910, a lively controversy developed over sex dermination. Those who thought that a male or female sex was determined by environmental factors such as temperature or nutrition were opposed by others who maintained that sex was inherited like any other genetic character. Proponents of the latter view (Mendelism) were influenced by the results of Mendel according to which transmisible characters of an individual were carried on discrete entities. The Czech monk did not know the nature of these entities but at the beginning of the 19th century a number of biologists suspected that these were chromosomes. Although he thought that chromosomes must have some relation to heredity, Morgan was at first reluctant to accept Mendelism and his studies with *Drosophila* caused him to revise his point of view. Morgan had discovered the first Drosophila mutant, a fly that unlike the normal red-eyed fly, had white eyes. In crosses of white-eyed flies with wild-type flies, the flies in

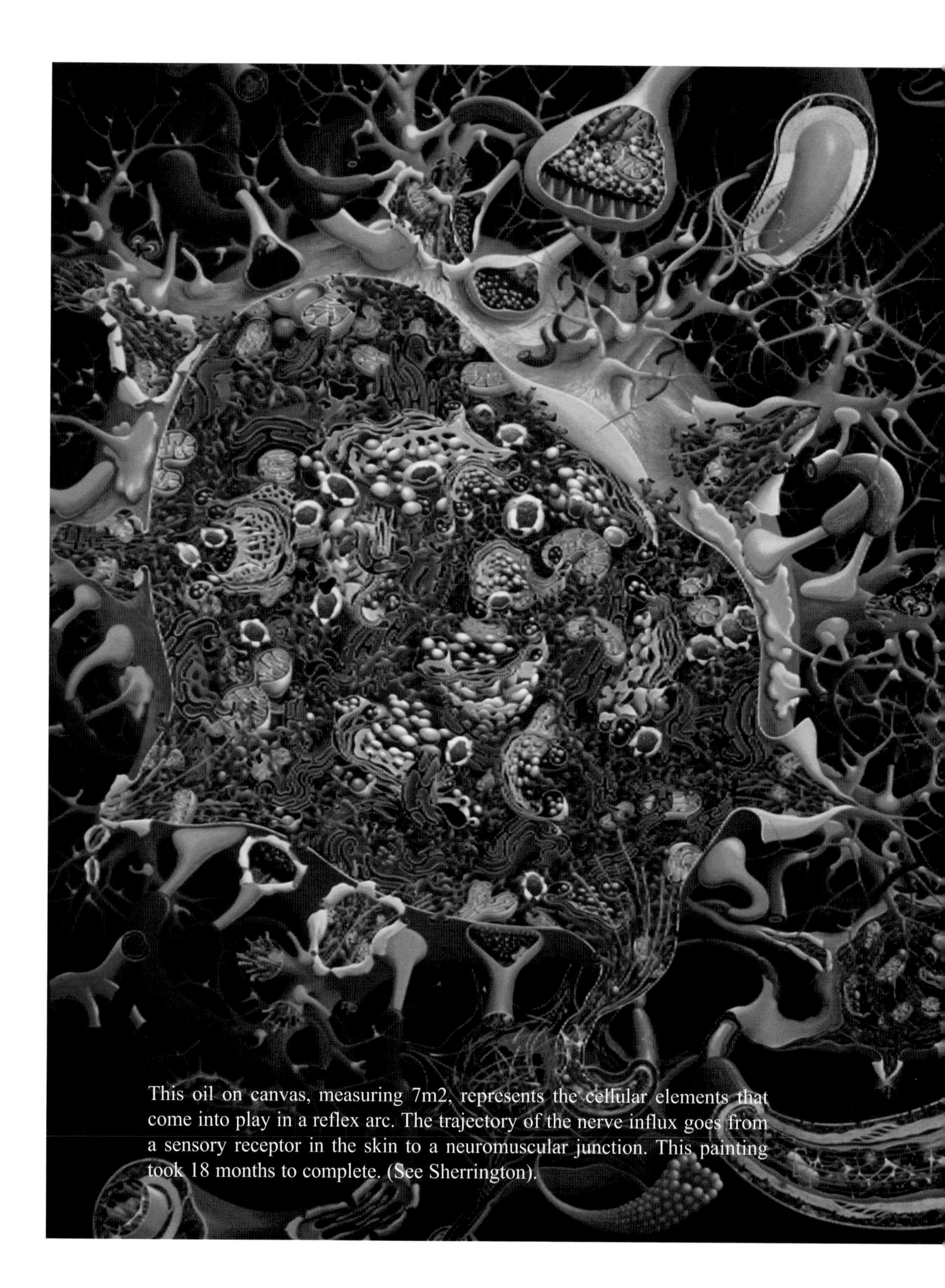

This oil on canvas, measuring 7m2, represents the cellular elements that come into play in a reflex arc. The trajectory of the nerve influx goes from a sensory receptor in the skin to a neuromuscular junction. This painting took 18 months to complete. (See Sherrington).

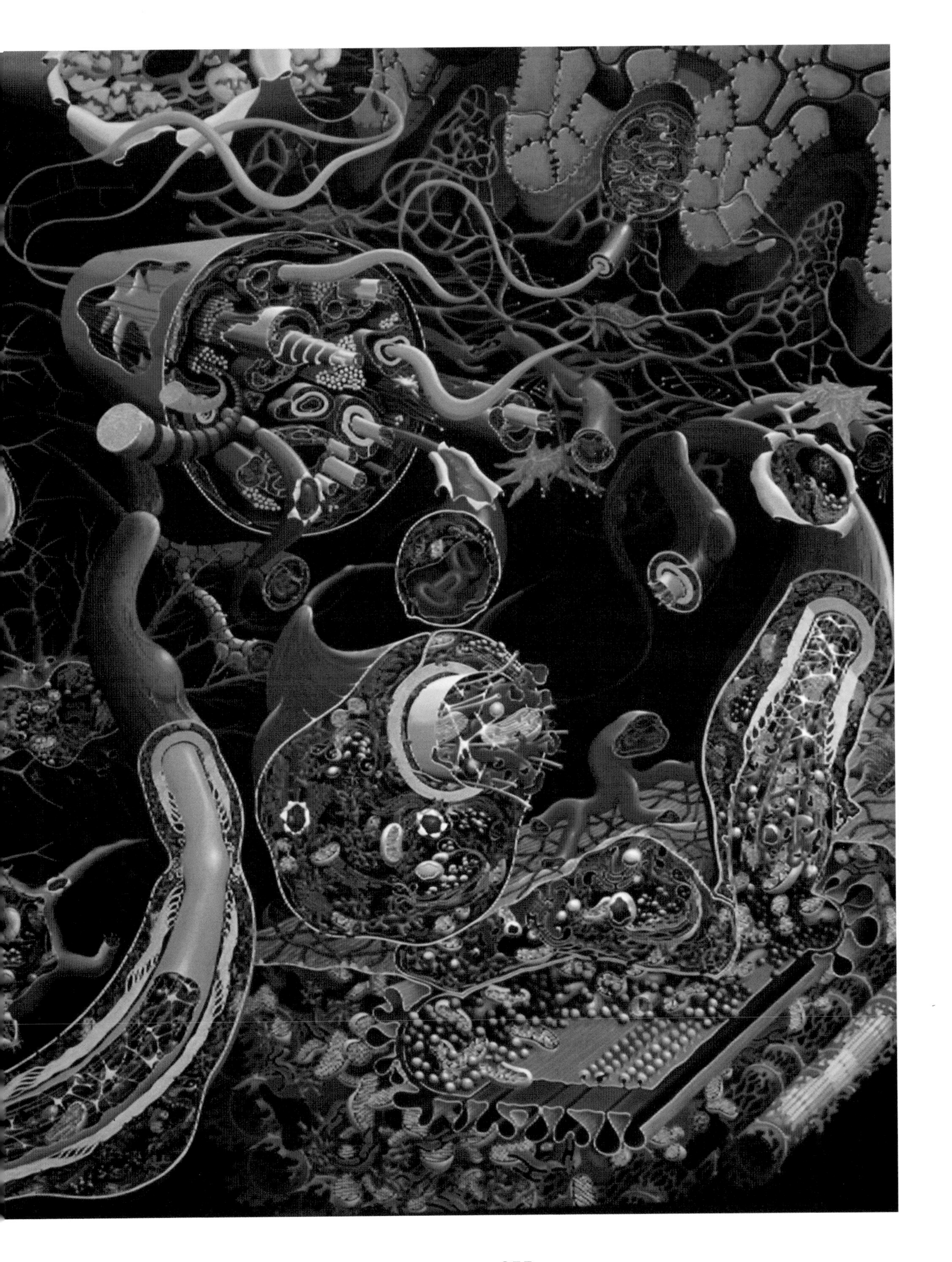

Gregor Mendel. He was the father of genetics (Moravian Museum, Brno, Czech Republic).

Thomas Morgan chose the fruit fly, *Drosophila melanogaster*, for his studies because it matures through an entire development cycle from fertilized egg to adult in about two week.

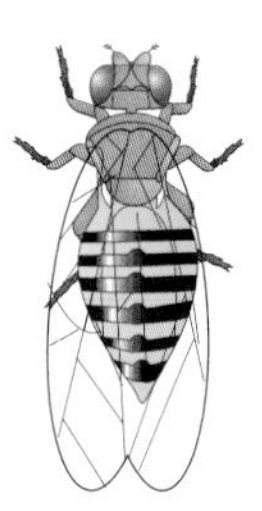

Parental generation

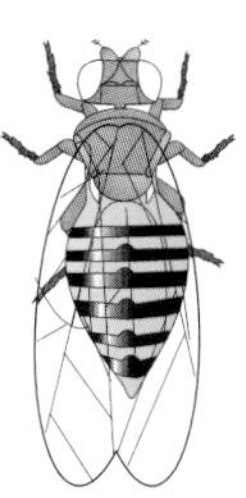

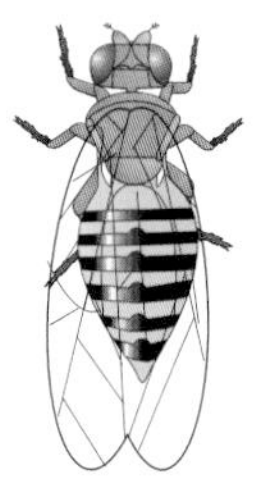

F_1 generation

X

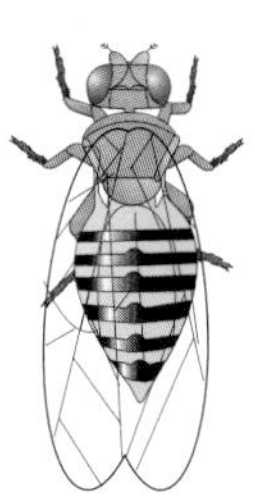

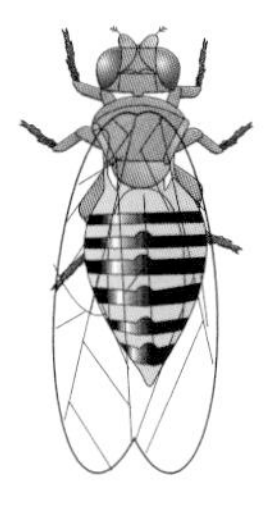

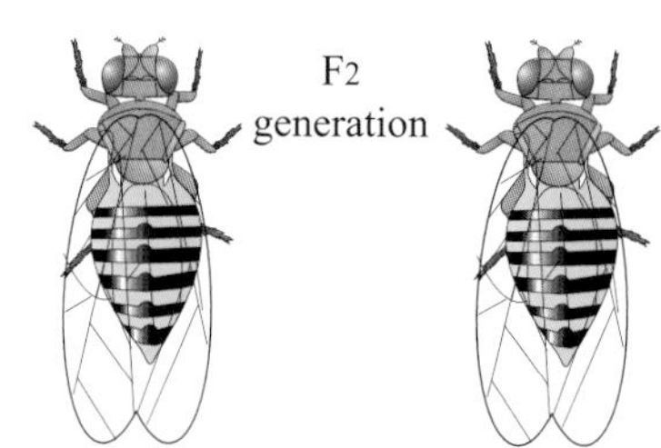

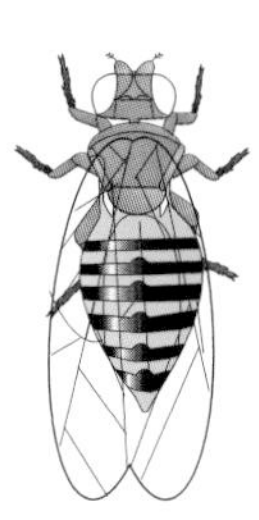

Sex linkage in the fruit fly. Genes controlling eye colour are located on the X chromosomes and are therefore sex - linked. The result of crossing red - and white - eyed flies depends on which parent is red - eyed and which white.

the first generation had red eyes, indicating that red eye was dominant. Progeny of matings of these F_1 flies with each other revealed that red eyes were associated with females and white eyes with males. At the time of these experiments, it was known that the male flies had only one

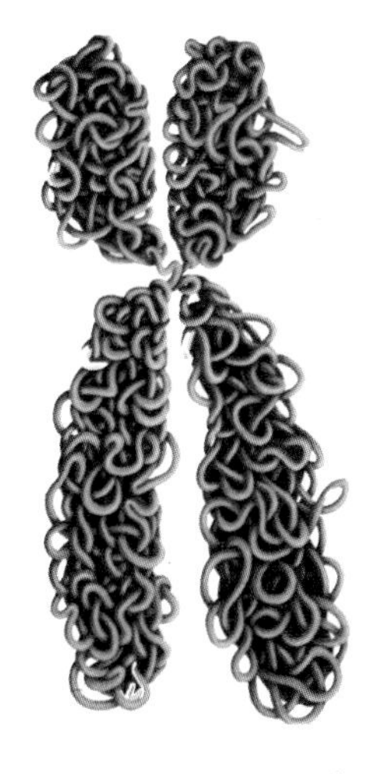

Thomas Morgan

The chromosome. A schematic representation with the accent on the high degree of winding of DNA fibres and their associated proteins.Thomas Morgan demonstrated that the chromosome was the main site of heredity in animals and plants.

X chromosome while females had two. Morgan hypothesized that the recessive character "white eyed" resided on the X chromosome, thus explaining why the white eyed progeny were males. This constituted the first proof that a hereditary character was linked to a chromosome. Characters on the X chromosme were subsequently called sex linked characters. Other work on *Drosophila* genetics was carried out by Morgan and the remarkably talented researchers (Alfred Sturtevant, Calvin Bridges and Hermann Joseph Muller) who were attracted to Morgan's laboratory. The Belgian cytologist F. Jansens had presented a theory of recombination involving material exchange between homologus chromosomes in meiosis. Subsequently, the Morgan group discovered genetic linkage among mutants on the same chromosome and showed that crossover frequencies could be used to construct a linear genetic map of a chromosome. These results were interpreted as being consistent with the Janssens' in a general theory of linkage elaborated by Morgan. The work of Morgan and his group on chromosomes and their genetic structure provided the basis for modern genetic analysis and, in addition, provided a chromosomal equi- valent for the gemmules in Darwin's theory of pangenesis. After the years at Colombia University, the last part of Morgan's career was spent at the California Institute of Technology in Pasadena. It was there that he created the research group that Georges Beadle joined in 1931 as a post doctoral fellow. Morgan and his group established an ideal environment in which **Max Delbrück** and **George Beadle** (who had returned to Cal Tech as Chairman of the Division of Molecular Biology) were to make significant contributions to the beginnings of molecular biology.

1934

Minot, George Richards (Boston, Massachusetts, December 2, 1885 - Brookline, Massachusetts, February 25, 1950). American physician. He was awarded the Nobel Prize for Medicine along with **George Whipple** and **William Murphy** for discovery of liver therapy in anemias. Basing his research on results obtained 10 years earlier by Whipple demonstrating the role of the liver in the regeneration of hemoglobin, George Minot observed that liver extracts cured patients suffering from pernicious anaemia. This disease is characterized, among other symptoms, by a deficiency in red blood cells, by lesions in the gastric lining, and by disturbances of the nervous system. It was only in 1948 that vitamin B_{12}, the liver substance responsible for the disappearance of the disease symptoms, was isolated and purified. Later it became clear that pernicious anaemia was due not only to a lack of vitamin B_{12} but also could result from a genetic incapacity to synthesize a factor normally secreted by gastric cells and reabsorbed in the form of a complex with vitamin B_{12}. George Richards Minot graduated in medicine from Harvard in 1912. After working at Massachusetts General Hospital, Boston, and at Johns Hopkins Hospital and Medical School, he moved to Harvard where he taught medicine. He finally became director of the Thorndike Memorial Laboratory in Boston from 1928 until his death.

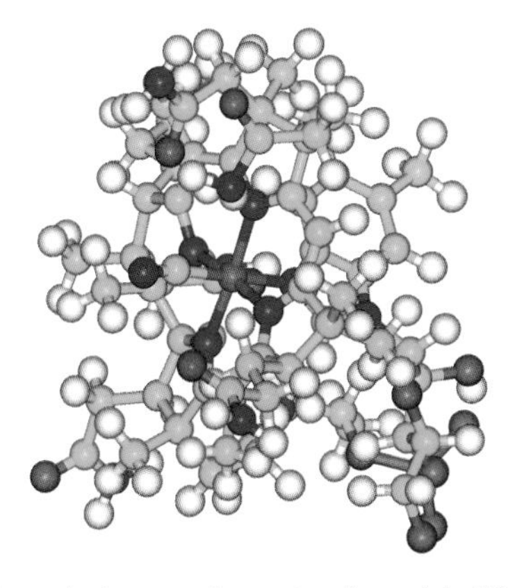

Vitamin B_{12}. A three - dimensional model. This vitamin is required by some enzymes to function. Its lack, or more often the lack of a substance required for its efficient absorption by gastric mucosa, leads to pernicious anemia.

Murphy, William Parry (Stoughton, Wisconsin, February 6, 1892 - Brookline, Massachusetts, October 9,

1987). American physician. Son of a Congregational minister. Nobel Prize for Medicine shared with **Georges Minot** and **George Whipple**, for success in the treatment of pernicious anemia with a liver diet. William Murphy received his education at the University of Oregon and later at Harvard Medical School where he earned his Ph.D. in medicine in 1922. After having practised medicine for a time he engaged in research on diabetes mellitus. He, Minot and Whipple succeeded in treating pernicious anemia by using intramuscular injections of extracts of uncooked liver (see Minot and Murphy). In 1939, he published a book entitled:Anemia in Practice: Pernicious Anemia.

Whipple, George Hoyt (Ashland, New Hampshire, Augustus 1878 - Rochester, New York, February 1, 1976). American physician and pathologist. Son of a physician. Nobel Prize for Medicine along with **George Minot** and **William Murphy**, for his discovery concerning the treatment of pernicious anemia. George Whipple first focused his attention to the bile pigments such as bilin, biliverdin and bilirubin which are the main products of hemoglobin degradation and he conducted experiments in order to show how hemoglobin was synthesized. In particular, he succeeded in showing the key role of iron for the synthesis of this substance and other hemoproteins. These studies led him to research which nutrients would most effectively regenerate new blood cells, particularly erythrocytes. In collaboration with Minot, he demonstrated that the addition of liver to the diets had a therapeutic value in the treatment of pernicious anemia, a form of anemia characterized by defective production of erythrocytes in the bone marrow. George Hoyt Whipple graduated from Phillips Academy, Andover, Massachusetts, in 1896, Yale University, in 1900, and Johns Hopkins University, where he took his Medical degree in 1905.

1935

Spemann, Hans (Stuttgart, Germany, June 27, 1869 - Freiburg-im-Breisgau, September 12, 1941). German zoologist. Son of a bookseller. Nobel Prize for Medicine for his discovery of the effect now known as embryonic induction. Exceptionally, the Nobel Prize in Medicine was attributed that year to a scientist who had devoted his career to the embryology of animals. In a famous experiment conducted with his co-worker Hilde Mangold on frog embryos, Hans Spemann noticed that when he transplanted the dorsal lip of the blastopore of an early gastrula into a different region of the gastrula, this induced a second gastrulation zone, resulting in the formation of a double larva possessing two nervous systems. As transplants from other areas of the gastrula did not

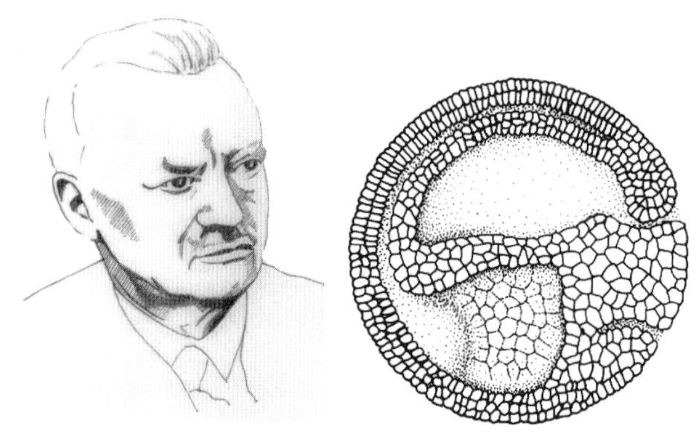

Hans Spemann

Gastrula, a major step in the development of animal eggs. Spemann's research showed that the dorsal lip of the blastopore was a region of strong influence in determining subsequent embryonic growth.

lead to similar results, Spemann concluded that the dorsal lip of the blastopore must play an inducing role, particularly in the formation of the nervous system and in establishing the antero-posterior axis of the embryo. The dorsal lip of the blastopore was thus called the organiser. Spemann proposed that the com-

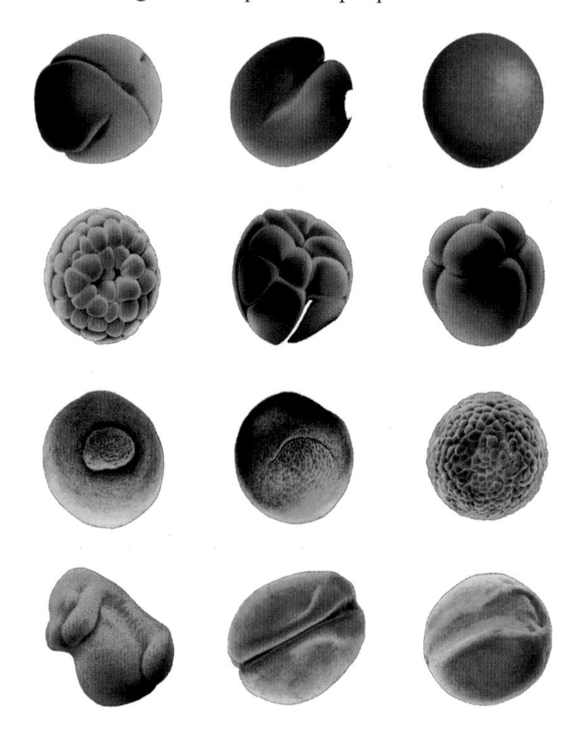

Overall development of the frog. Scanning electron micrographs provide a complete and dramatic picture of the changes in structure as the frog embryo develops to the stage where the central nervous system forms.

plete development of amphibians is determined by a succession of organisers, each induced by the previous one and capable of inducing morphogenetic movements. This work laid the foundation of causal embryology, a science in which the Brussels school was later to dis-

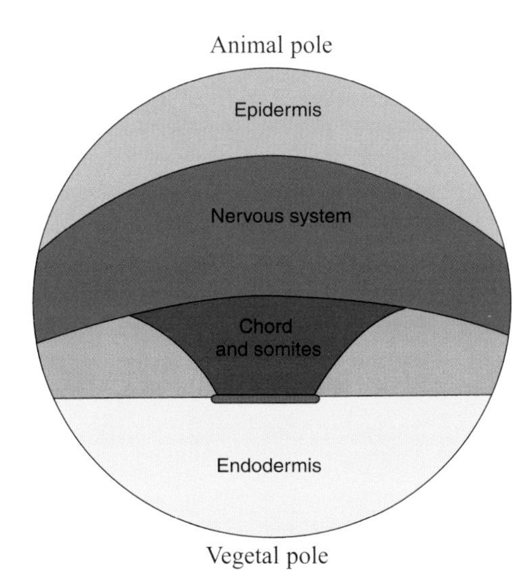

Determination of the fate of tissues in embryonic development. It was ascribed to "organizers" by Hans Spemann. As shown in this figure, epidermis and neural plate will derive from ectoderm while somites and notochord will derive from mesoderm.

tinguish itself. Spemann also showed great interest in literature and philosophy. His students and close

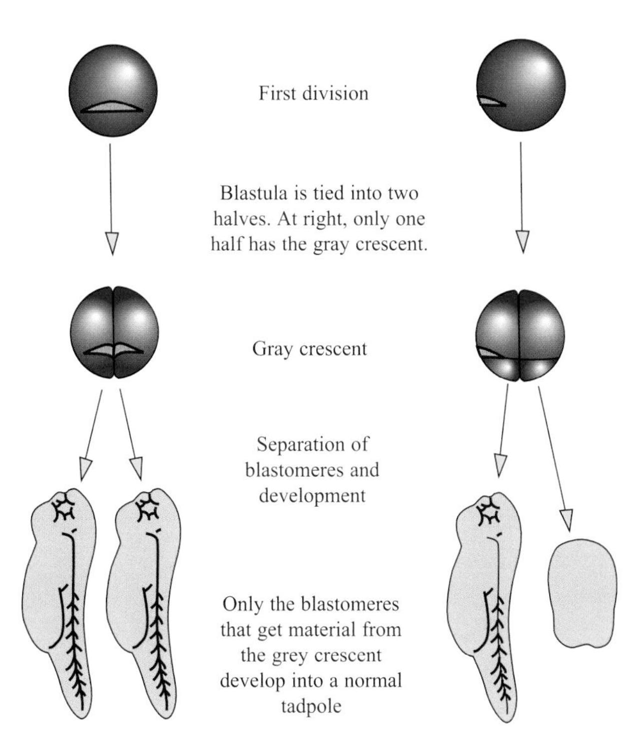

A Spemann experience. By tying a fertilized salamander egg into two halves so that a portion of the gray crescent extended into each half, Spemann obtained two normal embryos. But, by tying the egg so that only one half had the gray crescent, only this half developed normally. He deduced that the gray crescent contained groups of cells (primary organizers) that were able to organize other tissues into more differentiated regions of the embryo.

collaborators venerated him for his modesty, simplicity, and profound intellectual integrity. Hans Spemann received his education in zoology and botany at the Universities of Heidelberg, Munich, and Würzburg. After having worked at the Zoological Institute of Würzburg he taught zoology at Rostock before heading the Kaiser Wilhelm Institute for Biology in Berlin during World War I. Finally, between 1919 and 1935, he was made Chairman of the Department of Zoology at Freiburg.

1936

Dale, Henry Hallet (London, England, June 9, 1875 - July 23, 1968). British physiologist and pharmacologist. Son of a businessman. Nobel Prize for Medicine with Otto Loewi for isolating acetylcholine and for his discoveries on the chemical trans-mission of nerve impulses. Henry Dale was one of the great figures in British medi-cine of the 20th century. His eldest daughter Alison married Alexander Todd, who was awarded the Nobel Prize in Chemistry. Dale's work began in 1902 under the supervision of the British physiologists Ernest H.Starling and William Maddock Bayliss at University College London on the potential action of secretin, a hormone that had recently been isolated from the small intestine, on the pancreas. Although this was not the main theme of his research, the work exemplifies the focus of his investigations, namely understanding how a chemical substance could elicit a physiological response in an individual. At the begining of the 20th century, this question was largely unanswered, the more so because it also applied to the action of artificial substances such as drugs. Many renowned scientists helped in his research to resolve this problem, such as the German microbiologist Paul Ehrlich, in whose laboratory Dale worked. There was also T.R. Elliot, one of his colleagues in Cambridge who was trying to isolate noradrenaline, a compound which stimulates the parasym-pathetic nervous system. This hormone is also responsible for the fright-induced acceleration of heart rate and for the increase in vasodilation following exercise. The pharmaceutical laboratories of the Wellcome group also contributed by asking him to undertake research on active compounds from the fungus ergot, used by obstetricians to help during childbirth. He also studied certain effects of histamine. For example, he investigated its effect on the entry of plas-

Henri Dale

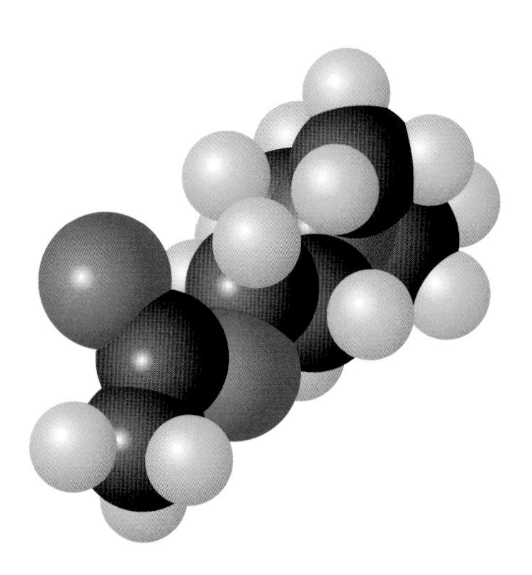

Acetylcholine. This substance was the first neurotransmitter discovered. It plays a role in many synapses in the peripheral nervous system and in the central nervous system, including in neuromuscular junctions.

matic fluid in tissues, which can cause a sudden drop in blood pressure leading to an anaphyllactic shock and even death. Yet another of his research areas, in collaboration with Henri Dudley, was the isolation of a substance called acetylcholine involved in transmitting nerve impulses from one neurone to another. Loewi had also come to the conclusion from his work that such a molecule must exist. The discovery of acetyl-

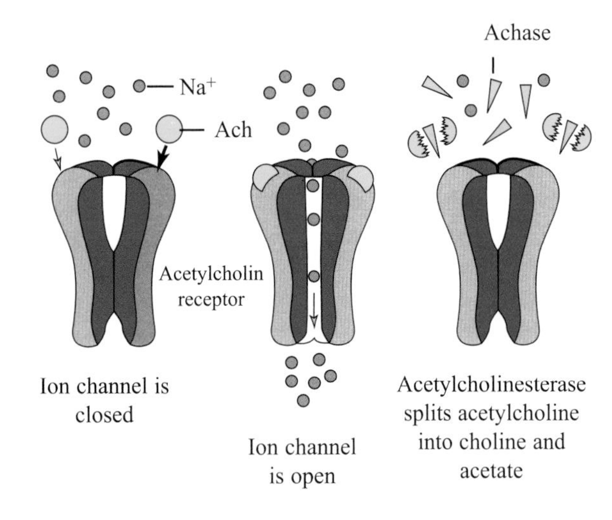

Chemical transmission at the synapse. The binding of acetylcholine to the postsynaptic membrane is followed by its degradation to acetate and choline by the enzyme acetylcholinesterase.

choline, the first neurotransmitter identified, was made when Dale joined the National Institute of Health in 1914. The understanding of the role of acetylcholine meant that hypotheses initially proposed by **John Eccles** and other neuropsychologists, stating that nerve impulses were transmitted by an electrical mechanism, could be discounted.

Loewi, Otto (Frankfort, Germany, June 3, 1873 - New York, New York, December 25, 1961). German-born American pharmacologist. Nobel Prize, with **Henri Dale**, for his work on the chemical nature of nerve impulse transmission. We are indebted to this great physiologist for having demonstrated that the transmission of nerve influx through synapses required chemical compounds. In 1921 he discovered that a chemical substance (subsequently shown by Henri Dale to be acetylcholine) is released after stimulation of the vagus nerve, a nerve of the frog's heart. Proof of the intervention of a chemical neurotransmitter was provided by experiments in which he first showed that when a frog's parasympathetic nerve ends were stimulated, it's heart rate decreased. Then he transferred blood from this amphibian to another, and noted a similar slowing of the heartbeat in the recipient. Otto Loewi concluded that parasympathetic stimulation in the first frog had caused secretion of a substance responsible for the effect. It was the proof that a nerve influx could be transmitted by a chemical compound. Otto Loewi studied medicine first in Stras-burg, then in Munich. Afterward he taught at the universities in Vienna and Graz , where he became the Head of the Pharmacology Department. Some time later he moved to London, where he worked in close collaboration with the physiologist Starling. He fled from the Nazi regime to Austria, and then he immigrated to the United States, where he became an American citizen.

Otto Loewi

1937

Szent-Györghyi von Nagyrapolt, Albert (Budapest, Hungary, September 16, 1893 - Woods Hole, Massachusetts, September 16, 1983). Hungarian-American biochemist. 1937 Nobel Prize for Medicine for

Albert Szent-Györghyi

Three-dimensional representation of vitamin C.

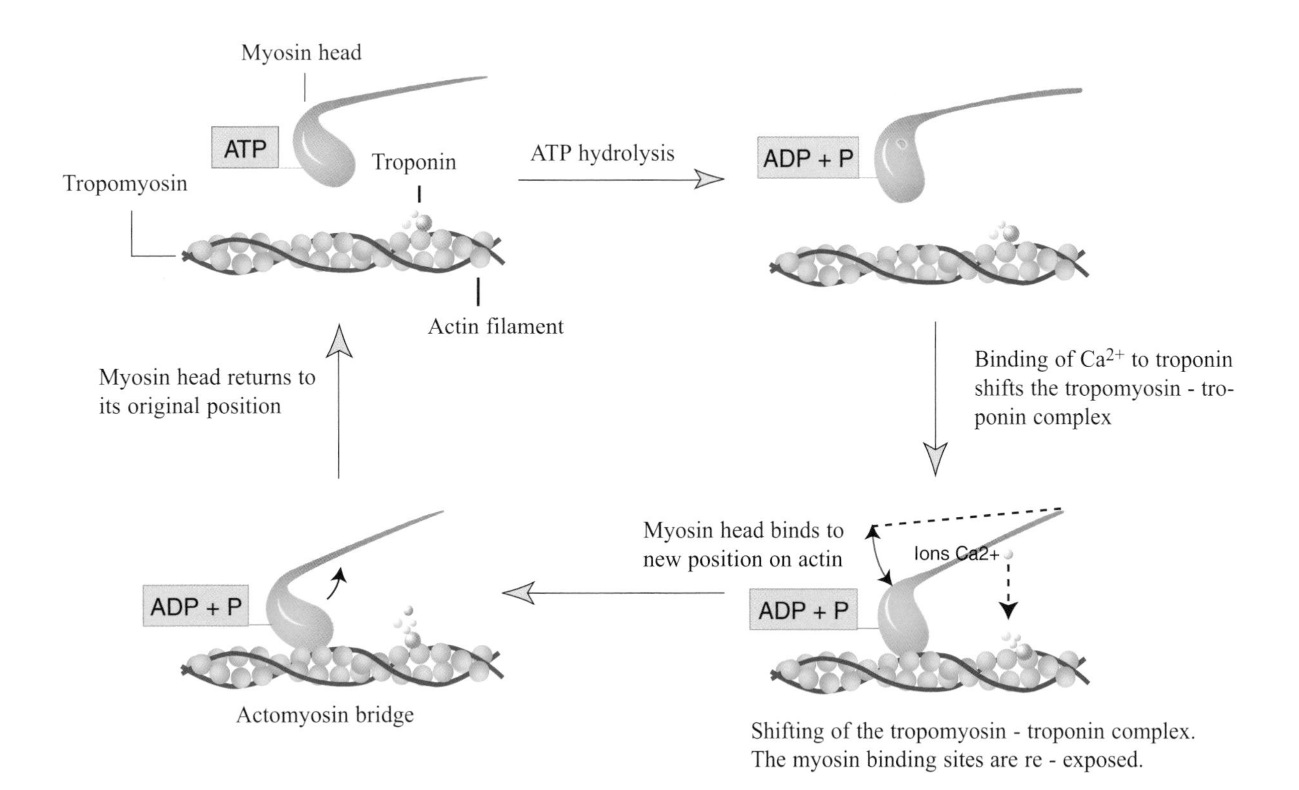

Interactions between actin and myosin during muscle contraction (four steps).

discoveries concerning the roles played by certain organic compounds, especially vitamin C, in the oxidation of nutrients by the cell. The first part of this scientist's research was marked by the isolation from the adrenal gland of a substance that could easily accept and donate electrons. At the time, Albert Szent-Györghyi was working in the laboratory of the famous Frederick Gowland Hopkins in Cambridge, and he suspected that the substance could be an electron transporter. He called it hexuronic acid because it had six carbon atoms. As it could also be obtained from oranges or cabbages, foods known for their action against scurvy, he assumed it was vitamin C. Before he mentioned his discovery, his assumption was confirmed by two other researchers, Waugh and King. In a-greement with Szent-Györghyi, they called it ascorbic acid (ascorbic means 'against scurvy', scorbutus being a synonym of scurvy). Another important discovery, performed on pigeon muscles, was that at least four organic compounds, respectively succinic acid, malic acid, fumaric acid and oxalacetic acid, activated the uptake of oxygen. From this work Szent-Györghyi concluded that these substances stimulated the aerobic degradation of sugars.

Pyruvate
Glycolysis
Pyruvate deshydrogenase
Acetyl-coenzyme A
Acetyl group
Krebs cycle
Intermediary metabolites
NADH2
NADH2
NADH2
FADH2
CO_2
CO_2
ATP
Reduced coenzyme
Electron transport chain

The contribution of A. Szent Györghyi to the comprehension of the Krebs cycle was its recognition of the catalytic function of the C4-dicarboxylic acids in this process (**** in the drawing).

Muscle fiber and neuromuscular junction. A three - dimensional representation. Note the distribution of the sarcoplasmic reticulum around the myofibrills. Golgi apparatus has not been represented in this illustration. Mitochondria are most abundant near the neuromuscular junction and they lie immediately beneath the fiber membrane.

Since the effect disappeared when either of them was added separately, he concluded that they must interconvert in cells. This discovery was the first step toward the elucidation by **Hans Krebs** of the complete cycle that bears his name (the Krebs cycle, a cycle of reactions in which pyruvate ends up being oxidised to carbon dioxide via the formation of some ten metabolic intermediates). Szent - Györghyi also made a major breakthrough in understanding the physiology of muscle contraction:he discovered actin and demonstrated the ultrastructural relationship between myosin and actin in actomyosin. Engaged in the fight against Nazism, he was persecuted and managed to escape with the help of Sweden, a neutral country. In 1974, with his country being occupied by the Soviet Army, he chose to emigrate to the United States, where he continues to conduct research on thymic secretions while remaining an active defender of humanitarian causes.Albert Szent-Györgyi graduated in medicine from the University of Budapest in 1917. Later he worked at the University of Cambridge and at the Mayo Foundation in Rochester, Minnesota.

1938

Heymans, Corneille Jean François (Ghent, Belgium, March 28, 1892 - Knokke, Belgim, July 18, 1968). Belgian physiologist. Nobel Prize for Medicine for his discovery of the regulatory effect on respiration of sensory organs associated with the carotid artery in the neck and with the aortic arch leading from the heart. Heymans was the son of a teacher at the University of Gent, where he studied and followed in his father's steps at the pharmacological department. Heymans gave the first scientific explanation of the relationship between respiration and blood pressure. He showed that blood pressure is recor- ded by specialised receptors such as those present in the carotid sinus, a nasal branch of the carotid artery, and that these receptors then transmit the information to the nerve centres. The latter then adjust respiration, accelerating

Corneille Heymans

or slowing down its rhythm according to whether the pressure is low or high. Corneille Heymans also discovered the glomus caroticum, a structure resembling the receptors just mentioned and which act as a chemoreceptor responding to changes in the ratio of oxygen to carbon dioxide in the bloodstream.

1939

Domagk, Gerhard (Lagow, Germany, October 30, 1895 - Burberg, Germany, April 24, 1964). German bacteriologist and pathologist. Nobel Prize for Medicine for his discovery of the antibacterial effects of Prontosil, the first of the sulfonamide drugs. Gerhard Domagk discovered Prontosil rubrum, a reddish - orange compound with potent biological effects. The substance was used to combat streptococcal infections in strains of mice. This was the first demonstration that a chemical compound could be used to fight bacterial infections in intact animals. In the wake of his discovery, laboratories throughout the world started working on the discovery of other potential therapeutic agents. **Alexander Fleming** himself threw aside his own work on penicillin to study the effects of Prontosil. It was later shown by the French scientist, **Daniel Bovet**, working at the Pasteur Institute in Paris, that the biological effects of Prontosil were due to its sulfonamide group. By chemically modifying this group, a whole range of derivatives were synthesized, known collectively as the sulfamides. These compounds were used to target a wide spectrum of bacterial species. It is likely that the scientific research of Domagk was influenced by the work of **Paul Ehrlich**, who had used arsenical compounds for treating sleeping sick-ness and syphillis. As well as Ehrlich, Domagk invented thousands of potential antimicrobial

Gerhard Domagk

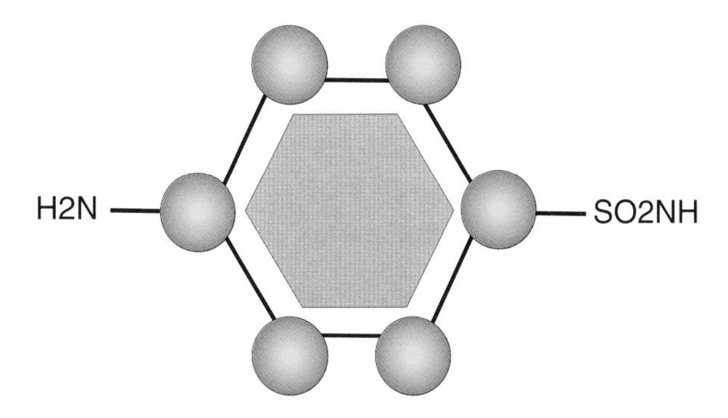

Protonsil, the first sulfa drug. It was discovered by **G. Domagk** and used by him to treat infections caused by *streptococcus.*

compounds. He gave Prontosil to his four-year old daughter and was able to save her life from a serious bacterial infection. Franklin D. Roosevelt and Winston Churchill (1953 Nobel Prize for Litterature) also benefited from treatment with the drug. Since the Nazis refused his candidature for the Nobel Prize, the Nobel Prize Committee awarded him the Nobel Prize Medal. The financial revenue was transferred back to the foundation, in agreement with internal policy. Domagk also discovered Conteben, an active drug to fight tuberculosis. He also worked on chemotherepeutic agents for treating cancer. Although he was unsuccessful in his last venture in the field of cancer treatment, his work on sulfamides led the way for the development of new drugs. Domagk remained active until his death. He passed away suddenly on the 4th of April, 1964. Gerhard Domagk received his Ph.D. in medicine from the University of Kiel in 1921. He then successively taught medicine at the universities of Greifswald and Munster and was appointed Director of the new Research Institute for Pathological Anatomy and Bacteriology (I.G Farben-industrie Laboratory for Experimental Pathology and Bacteriology) in Wuppertal in 1932.

1939 to 1942 Not awarded

1943

Dam, Henrik Carl Peter (Copenhagen, Denmark, February 21, 1895 - April 1976). Danish biochemist. Nobel Prize for Medicine, along with **Edward Adelbert Doisy**, for research into antihemorrhagic substances and the discovery of vitamin K. Henrik Dam graduated from the Polytechnic Institute, Copenhagen in 1920. In 1925, he studied chemistry in Graz (in Austria, under **Fritz Pregl** (1923 Nobel Prize for Chemistry) and received his Ph.D. in biochemistry from the University of Copenhagen in 1934. During his research work between 1929 and 1934, he demonstrated that a disease found in chickens was characterised by abundant bleeding and problems in blood coagulation due to a deficiency in a hemorrhagic factor. He identified the latter as a liposoluble substance found in fats and also present in vegetables. Along with Doisy, but working independently, Henrik Dam isolated the compound from the plant alfalfa and named it vitamin K.

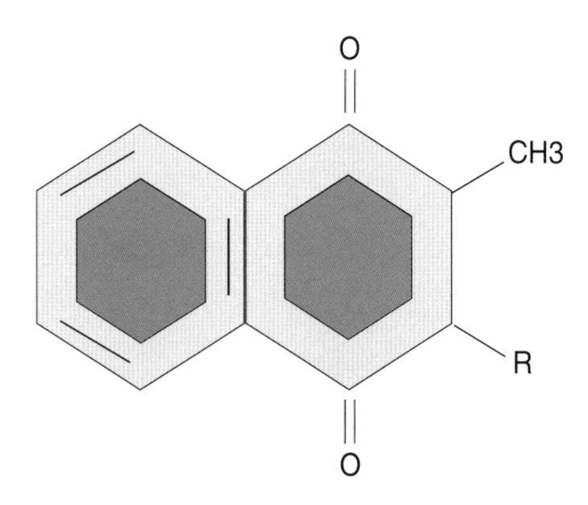

Vitamin K. This substance is required for the synthesis of prothrombin and certain other blood coagulation.

Doisy, Edward Adelberg (Hume, Illinois, November 13, 1893 - St.Louis, Missouri, October 23, 1986). American biochemist. Son of a commercial traveller. Nobel Prize for Medicine with Henrik Dam for the isolation and synthesis of the antihemorrhagic factor, vitamin K. Edward Doisy graduated from Harvard University and earned his Ph.D. in biochemistry in 1920. He started out his scientific career by participating in the crystallization of a steroid hormone, the female sex hormone oestrone, discovered at that time by the German scientist, **Adolf Butenandt**. (1939 Nobel Prize for Chemistry) In 1923 he was appointed Professor of Biochemistry at the University of St. Louis and headed the Department of Biochemistry in 1924. Later, in 1939, he embarked on the purification of an antihemorrhagic factor identified by the Danish biochemist, Dam. Working in Doisy's research team, Dam discovered the existance of two types of the vitamin, vitamin K_1 extracted from the plant *alfalfa*, and which Doisy succeeded in chemically synthesizing, and vitamin K_2 which he purified from rotting fish. Todays, large quantities of vitamin K are produced commercially, thanks to the

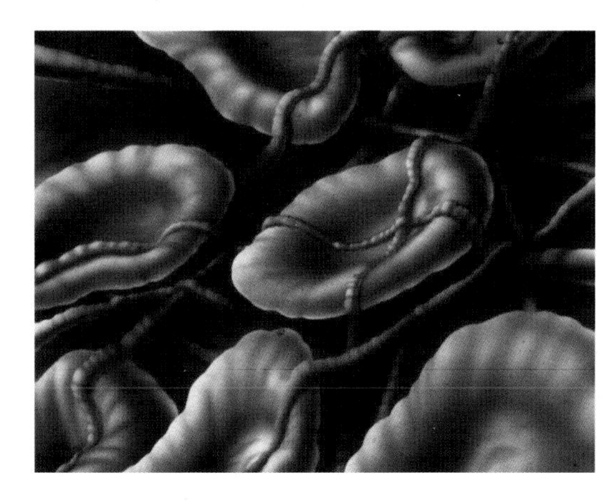

Scanning electron micrograph of red blood cells entangled by fibrin. Vitamin K is a key factor in such a process.

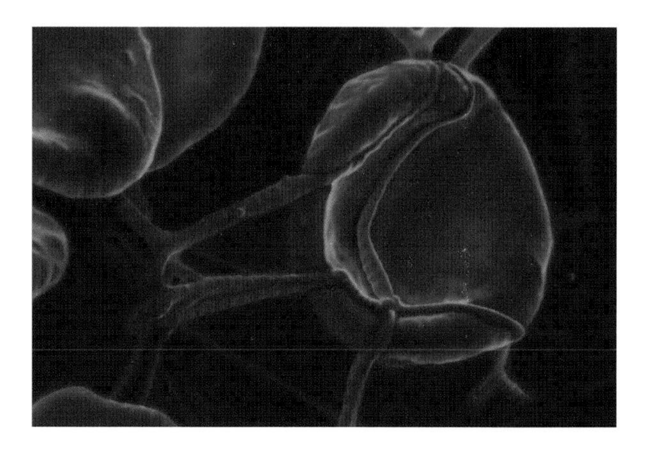

Scanning electron micrograph of red blood cells entangled by fibrin. Another image.

procedure set up by Doisy and his associates. The purified vitamin is used to cure hemorrhages and as a consequence thousands of lives are saved every year.

1944

Erlanger, Joseph (San Francisco, California, January 5, 1874 - St Louis, Missouri, December 5, 1965). American physiologist. Nobel Prize for Medicine, along with **Herbert Spencer Gasser**, for their discoveries relating to the differentiated functions of single nerve fibres. Joseph Erlanger was a teacher at Wisconsin and Washington Universities. His parents were German immigrants who came to America during the gold rush. As a child, he sho-wed a great interest in sciences and was fascinated by cardiac physiology. While working at Johns Hopkins University, he created a sphygmomanometer, an instrument for measuring blood pressure. It was the arrival of **Gasser**, one of his former students, that ignited his interest in the study of the conduction of nervous impulses by inviting him to try to understand with him how nerves transmitted electrical impulses. They performed their research on a humidified frog nerve kept at a constant temperature. One of the main problems was that the electrical impulses generated by nerves are very weak and only last a very short time, so much so that it was practically impossible to measure them. Nevertheless, their collaboration finally led to the discovery, with the aid of a cathode oscillograph perfected by Western Electric, and thanks to an amplifier invented by S. Newcomer, of the means of amplifying impulses up to 100,000 times and recording them. Thanks to the use of this apparatus, they noticed that following nerve stimulation, nervous impulses were sent out at different speeds as a function of their diameter. As a rule, the conduction velocity increased with increasing axonal diameter. Moreover, the proposed the hypothesis that thin nervous fibers conducted nervous impulses related to pain, whereas the thickest were used for motor impulses. These discoveries constituted considerable progress for that time period.

Joseph Erlanger

Gasser, Herbert Spencer (Platteville, Wisconsin, July 5, 1888 - New York City, New York, March 11, 1963). American physiologist. Son of a physician. Nobel Prize for Medicine, with **Joseph Erlanger**, for fundamental discoveries about the functions of nerve fibres. Herbert Spencer Gasser studied at Washington University in St. Louis, Missouri, under Erlanger. Under his supervision he invented the cathode-ray oscilloscope, an instrument for measuring nerve impulses. By using the oscilloscope, he and Erlanger were able to show that the velocity of a nerve impulse increased with the diameter of the nerve fibre. This was the discovery of a correlation between the velocity of a nerve impulse and nerve fibre diameter. Gasser was appointed Director of the Rockfeller Institute for Medical Research in New York from 1935 to 1953. He held this position until he retired in 1953. Later, Gasser became interested in observing nerve fibres using electron microscopy.

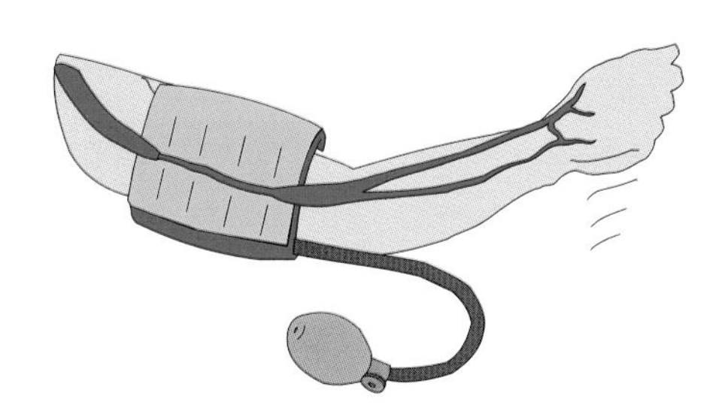

Sphygmomanometer. A device designed by Erlanger. It consists of an inflatable cuff and a pressure gauge for measuring blood pressure.

1945

Chain, Ernst Boris (Berlin, Germany, June 19, 1906 - London, England, August 12, 1979). German-born British biochemist. Nobel Prize for Medicine, along with **Alexander Fleming** and **Howard Florey**, for his pioneering work on the isolation and production of penicillin. Ernst Chain's father was a Russian Jew, who emigrated to Germany and settled down in Berlin, working as an engineer. Although Chain was a gifted pianist and initially his cherished dream was to become conductor of an orchestra, he chose to study science at the Friedich-Wilhem University in Berlin. In 1933, shortly after Adolf Hitler had come to power, Chain fled Germany to settle down in England. It was there that he embarked on the study of enzymes under the direction of the great British biochemist **Frederick Gowland Hopkins**. Hopkins encouraged Chain to continue his work in collaboration with Florey, who was a pathologist from Australia. **Florey** became interested in Chain's work on the properties of lysozyme. Chain was familiar

Ernst B. Chain

with Florey's work on the therepeutic properties of the mould *Penicillium notatum* and worked alongside him in isolating extracts from this fungus that were enriched in a substance that he named 'penicillin'. He demonstrated its

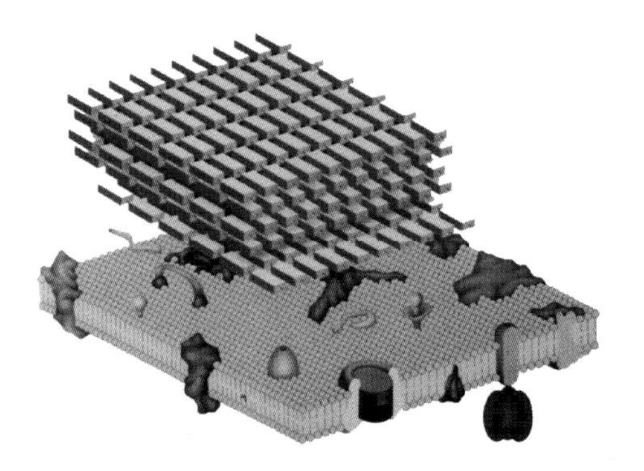

Gram positive bacteria. Their wall is relatively thick and consists of a network of peptidoglycans. This wall is readily digested by lysozyme.

powerful antibiotic effect on strains of *Streptococcus*. Florey discovered a system for culturing Penicillium in maize extracts and they produced large quantities of the antibiotic during the Second World War in the United

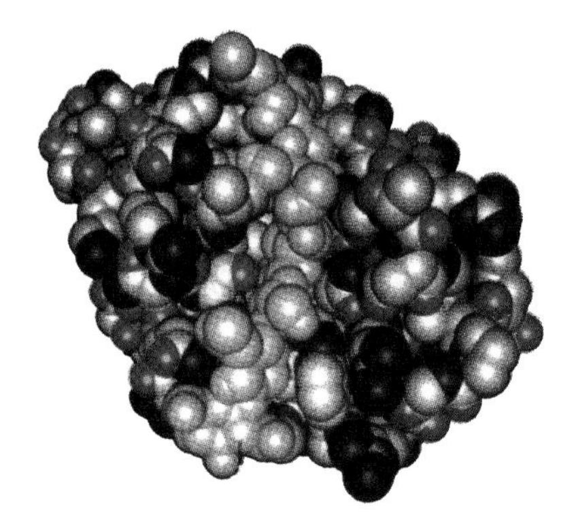

Three - dimensional structure of lysozyme. Also called muramidase. This enzyme was discovered by Alexander Fleming. Its properties result from its capacity to cleave glycosidic linkages between sugar components that are required for the building of the bacterial cell wall.

States. It was administered to wounded soldiers and saved hundreds of thousands of lives. We now know that the antibiotic works like lysozyme by inhibiting cell wall synthesis, explaining why it is only effective on growing bacteria. Nowadays, Penicillin is routinely used for treating bacterial diseases such as pneumonia, gonorrhea and impetigo. Chain left Oxford and it was only while he was in Rome in 1958 that Penicillin was isolated in a purified form. He returned to England and became Head of the Wolfson laboratories at Imperial College London. Even during times of intense activity, Chain managed to find time to develop his remarkable piano skills. He was knighted in 1969.

Fleming, Alexander (Lochfield, Scotland, April 6, 1881-London, England, March 11, 1955). British microbiologist. Son of a Scottish farmer. Nobel Prize for Medicine, along with **Ernst Boris Chain** and **Howard Walter Florey**, for his discovery of penicillin. This substance was an excellent substitute for antiseptics. These were widely used at the time, particularly during World War I, but people doubted their efficacy. Chance was an important factor in this discovery, which was due to the accidental contamination of a poorly closed culture dish by the mould *P. notatum*. It so happens that this species cannot develop effectively and produce its antibiotic unless the ambient temperature is low, and Fleming had left his laboratory unheated during a two-week holiday. He thus unknowingly favoured the growth of the fungus. Back in London he resumed heating, so that the temperature became favourable to bacterial growth. Again he was lucky: penicillin can only affect growing bacteria. This is when Fleming judiciously observed that the bacteria had died in the area surrounding the mould. He concluded that the mould must secrete a bactericidal substance. Having failed to isolate the substance, he tried to demonstrate its potential therapeutic effects by applying the mould directly to wounds. This approach also failed, so he briefly

Penicillium notatum.

Alexander Fleming

mentioned his discovery and it was forgotten. Later, Chain and Florey became interested in the finding and they were the ones to isolate, purify, and produce the antibiotic industrially. Penicillin turned out to be a very useful drug. The use of penicillin during World War II saved many soldier's lives. Interestingly, Fleming had previously shown that a substance called lysozyme was also an efficient bactericidal agent. It was later found that this substance was a lytic enzyme and that it was also present in body fluids, such as tears and other secretions. After taking his medical degree at the University of London in 1906, Fleming spent a major part of his career at St. Mary's Hospital. He worked there until 1948. At the same time he continued teaching bacteriology at London University. Fleming was knighted in 1944.

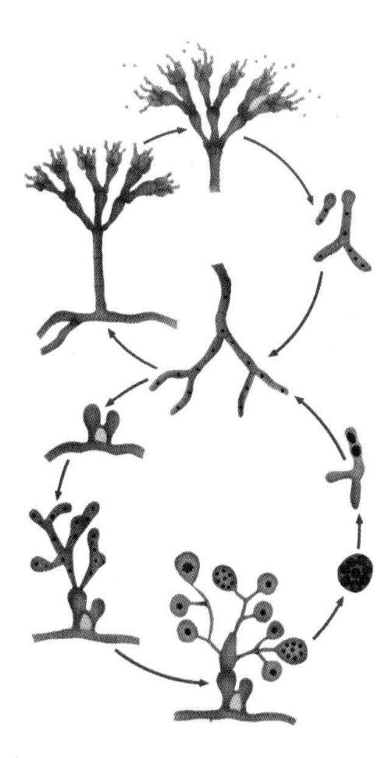

Life Cycle of Penicillium. *Ascomycota* reproduce sexually (through a reproductive structure called ascus) or asexually (by producing special spores called conidia, which appear as characteristic chains at the end of hyphae).

Florey, Howard Walter (Adelaïde, Australia, September 24, 1898 - Oxford, England, February 21, 1968). Australian pathologist. Florey's father was engaged in shipbuilding. Nobel Prize for Medicine, shared with **Alexander Fleming** and **Ernst Boris Chain**, for industrial production of penicillin. Florey studied medicine at Oxford, under the guidance of talented scientists. Among these were notably Professor **Sherrington**, a renowned neurophysiologist. He persued his studies at the University of Cambridge, England, and with the help of **John Hopkins**, he improved his education and his know ledge in biochemistry. We are indebted to Florey for his in depht study of lysozyme, the enzyme which had been discovered by Fleming. In the early 1930s, he worked with Chain and showed that this enzyme prevented bacterial growth by inhibiting cell wall synthesis. A crucial experiment carried out in 1940 was paramount in showing the effectiveness of penicillin. Convinced that this antibiotic had great therapeutic potential, Florey wanted to obtain it in large quantities by crystallising it in the cold, but he lacked the necessary financial means. A visit to the United States, however, enabled him to collect the necessary funds, and from 1943 onward, penicillin was used massively to treat soldiers wounded on the fronts. The year 1959 marked the beginning of industrial production of a variety of semi - synthetic penicillins. In 1944 he received his knighthood in recognition of his major scientific contribution. After World War II, Florey returned to his native country and continued developing medical research.

Howard W. Florey

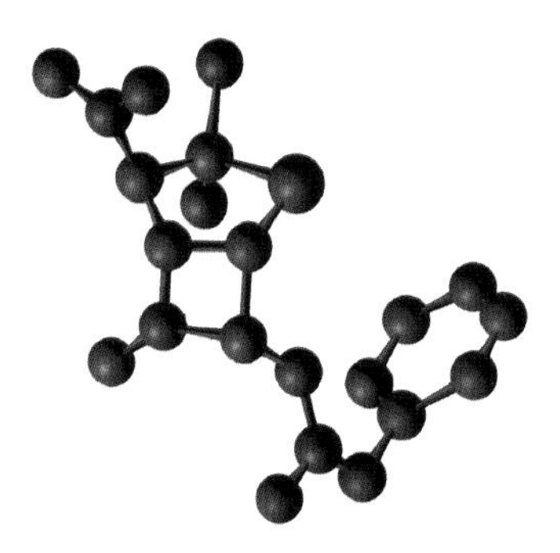

Penicillin. Its discovery by Alexander Fleming, its purification and the clinical proof of its extraordinary antibacterial efficiency by Howard Florey and Ernst Chain, were honoured by the Nobel committee in 1945.

1946

Muller, Hermann Joseph (New York City, New York, December 21, 1890 - Bloomington, New Jersey, April 5, 1967). Americn geneticist. Nobel Prize for the discovery that X - rays could induce mutations. His cleverly designed experiments showed that the rate of mutation in *Drosophila* could be increased more than 10,000 fold by expo-

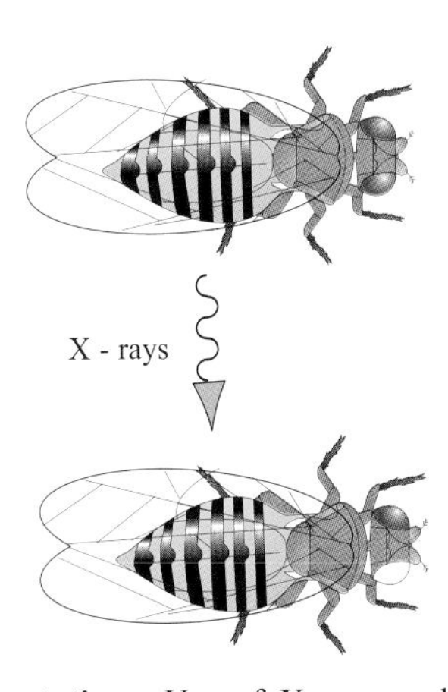

X - rays and mutations. Use of X - rays allows to induce changes in the genetic material of the fruit fly *Drosophila*. Such experiments enabled P.H. Muller to carry out genetic studies with the fruit fly.

sure of the flies to X - rays. This discovery of induced mutation was considered by many of his contemporaries as one of the most important contributions to genetics in the first part of the 20th century. He also made many theoretical contributions to the science of genetics. For example, his concept of the gene as an entity that could changed by a physical agent and could reproduce those changes, focused attention on the nature of gene structure. He also proposed that genes with their autoreplicative properties were the origin of life on earth. According to Muller, all selective mechanisms acted on genes, so that evolution was tightly linked to the expression of genes selected in sucessive generations of an organism. Although it did not have much impact at the time, it is interesting that Muller proposed that bacteriophage might be ideal material for the study of genes.

1947

Cori, Carl Ferdinand (Prague, Czechoslovakia, December 5, 1896 - Cambridge, Massachusetts, October 20, 1984). American physician and biochemist born in Czechoslovakia. Nobel Prize for Medicine, along with his wife, **Gerty Theresa Cori**, for their work on glycogen metabolism and with **Bernardo Houssay**, who was awarded the Nobel Prize for research on sugar metabolism. After studying medicine at the German University of Prague where his father taught zoology, Carl Cori was a research worker in Vienna, as well as in Graz. In 1922 he was invited to Buffalo University in the United States, where he conducted research in the field of sugar metabolism. He pursued his investigations in close collaboration with his wife, Theresa Gerty Cori, who joined him in the United States some time later. In 1929, the Coris became American citizens and accepted positions at the University of St Louis in Missouri. It was here that they established the biochemical mechanisms underlying the interconversion of glycogen and glucose.

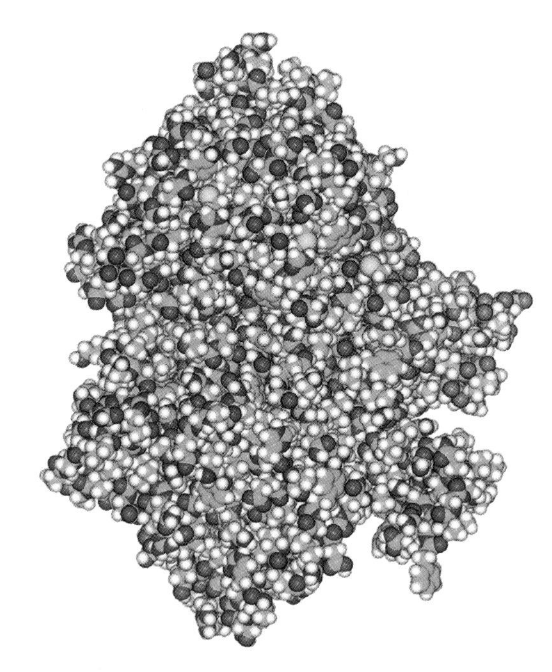

Glycogen phosphorylase. This enzyme catalyses the removal of the non reducing terminal glucose from glycogen.

Cori, Gerty Theresa (Prague, Czechoslovakia , August 15, 1896 - St Louis, Missouri, October 26, 1957). American physician and biochemist born in Czechoslovakia. Nobel Prize for Medicine, along with her husband **Carl Ferdinand Cori**, for their studies on glycogen metabolism and with **Bernardo Houssay**, who was awarded the Nobel Prize for research on sugar metabolism. Gerty Theresa Cori was the daughter of a Czech businessman. In 1914 she went to the German University of Prague to study medicine and it was here that she met

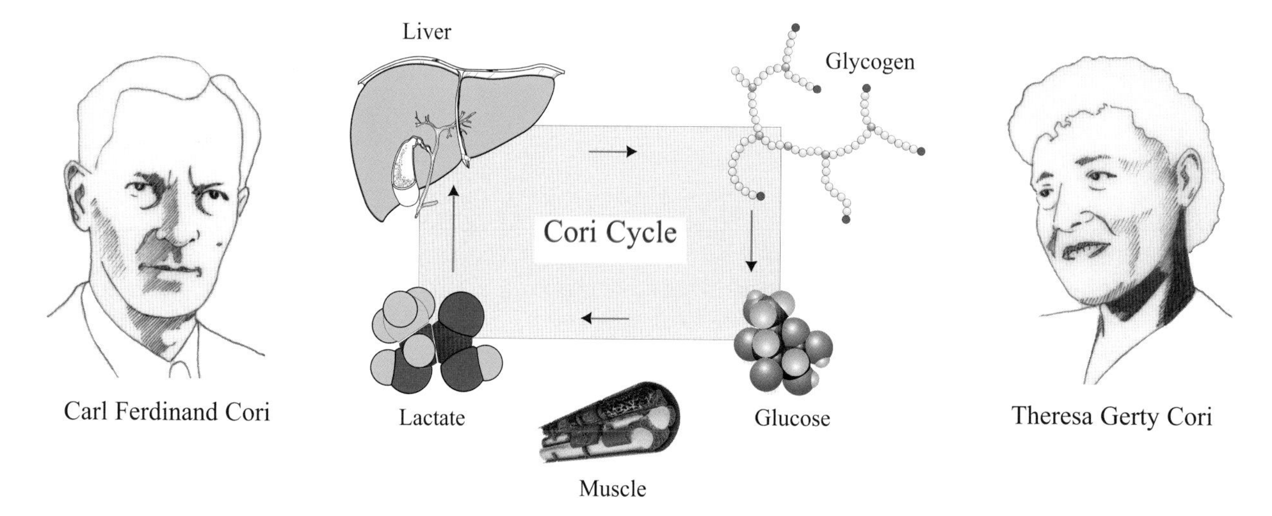

Cori cycle. A schematic representation. As shown, muscle glycogen becomes available as blood sugar through the intervention of the liver, and blood sugar in turn is converted into muscle glycogen.

her future husband Carl F. Cori. She joined him in the United States and studied the effects of X - rays on cells. In addition, she investigated the effects of nutrient deprivation on metabolism and thus became interested in sugar metabolism. From that moment on, her research interests became inevitably linked to those of her husband. The Coris first showed that liver glycogen was sequentially and reversibly degraded to glucose 1 - phosphate by an enzyme that they discovered called glycogen phosphorylase. They went on to show that this intermediate was converted into glucose 6 - phosphate by the enzyme phosphoglucomutase. They then studied the effects of insulin on liver and skeletal muscle glucose metabolism. They established a cycle of reactions whereby muscle glucose was transformed into lactate, which was then reconverted into glucose by the liver. This cycle subsequently became known as the Cori Cycle, which provided an improved understanding of sugar metabolism and in particular the role of sugar phosphates as an energy source. The work of the Coris also led to the understanding of hereditary diseases of glycogen metabolism, called glycogenoses, due to defects in the glycogen storage system. One such disease is known as Coris disease, which is the inability to produce functional amylo - 1,6 - glycosidase, one of the enzymes necessary for glycogen breakdown. The Coris also discovered the cause of Von Gierke's disease, a congenital defect in glucose -6 -phosphatase, leading to an abnormal increase in liver glycogen as well as hypoglycaemia. Gerty Theresa and Carl Ferdinand Cori were the third couple to be awarded the Nobel Prize after the **Curies** and the **Joliot-Curies** in France.

Houssay, Bernardo Alberto (Buenos Aires, Argentina, April 10, 1887 - September 21, 1971). Argentine physiologist. Nobel Prize for Medicine along with **Carl Ferdinand** and **Gerty Theresa Cori** for his discovery of the role of hormones produced by the anterior lobe of the hypophyse in regulating the blood sugar level. By injecting pituitary extracts into normal dogs Bernardo Houssay showed that this organ is responsible for secreting adrenocorticotropic hormone, which causes the blood sugar level to rise and thus antagonises the effect of insulin (this hormone acts by affecting the permeability of cell membranes to glucose and this leads to a reduction of blood sugar). Moreover, it also plays a role in increasing the rate of glucose-to-glycogen conversion in the liver, thereby further lowering blood sugar levels. Houssay got secondary education at Colegio Britanico in Buenos-Aires and studied medicine at the university of the same city. Then he taught physiology at the Buenos-Aires Institute of Physiology, which was at that time famous all over the world, until he was discharged from his post when Juan Perón came to power. Bernardo Houssay took great interest in physiology, ranging from cardiovascular to gastrointestinal diseases. But it was his investigation in the field of endocrinology that pushed him toward the Nobel Prize. This research was encouraged by the discoveries of **Frederick Banting** and **Charles Best**.

1948

Müller, Paul Hermann (Olten, Switzerland, January 12, 1899 - Basel, Switzerland, October 12, 1965). Swiss chemist. Nobel Prize for Medicine for discovering the potent toxic effects on insects of dichloro-diphenyl-trichloroethane (DDT). Paul Hermann Müller became interested in chemistry as a simple laboratory technician, and then worked for Geigy in a dye factory in Basel. In 1935 he became famous for synthesising DDT. This insecticide has proven particularly valuable against insect disease vectors, in particular mosquitoes such as Anopheles. The DDT molecule acts on the peripheral sensory organs of insects and causes convulsions leading to death. It permitted the control of several diseases over large parts of the world. In the early 1960s DDT was used by the World Health Organization as the major weapon in their war against malaria.It also saved agriculture from uncounted millions of dollars of damage from a variety of plant-eating insect pests. In reality, DDT was first synthesized the German chemist Zeidler in 1874 but its insecticidal properties were not discovered until 1938 by Müller. Yet Paul Hermann Müller recognised that the high stability of this insecticide could lead to its gradual accumulation in animals. This was later confirmed and raised fears of serious ecological consequences. Consequently, the use of DDT has now been banned.Paul Hermann Müller graduated from the University of Basel and earned his Ph.D. there in 1925. He conducted research in a Swiss firm (J.R. Geigy) before heading the Department of Scientific Research on Substances for Plant protection in 1946.

1949

Egas Moniz, Antonio Caetano de Abreu Freire (Avanca, Portugal, November 29, 1874 - Lisbon, Portugal, December 13, 1955). Portuguese neurologist. Nobel Prize for Medicine for his work on psychiatric therapy and mainly for the development of prefrontal leucotomy. Egas Moniz started out as a politician and was Head of the Neurological Department of the University of Lisbon. His first successes in scientific research were on the structure of the cardiovascular system. He mapped out the distribution of the major blood vessels in the brain. This was a daunting task, relying on the injection of radioactive pigments into the blood followed by imaging on photographic films. He proposed that modifications of this network could be caused by tumours and thus might be used in the diagnosis of cancer. His ideas were later proved right. However, Egas Moniz is rather known for his work on prefrontal leucotomy, a delicate operation involving the insertion of a thin blade between the prefrontal lobes to cut the connection between them. This surgery also cut the

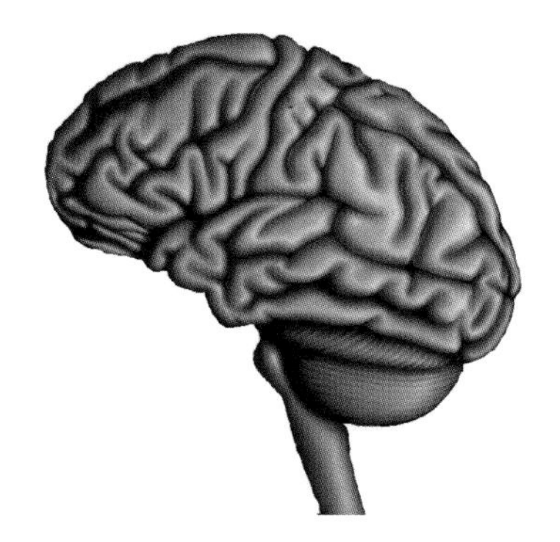

The brain. A side face.

connection between the prefrontal lobes and the hypothalamic system. The procedure had previously been carried out by other doctors on chimpanzees and the operation was used to check the progress of certain psychotic symptoms. Although the Nobel Committee had approved of his research and considered it to be one of the most important contributions in the development of psychiatric surgery, his therapy turned out later to be detrimental and was gradually abandoned.

Hess, Walter Rudolf (Frauenfeld, Switzerland, March 17, 1881 - Locarno, Switzerland, August 12, 1973). Swiss physiologist. Nobel Prize for Medicine

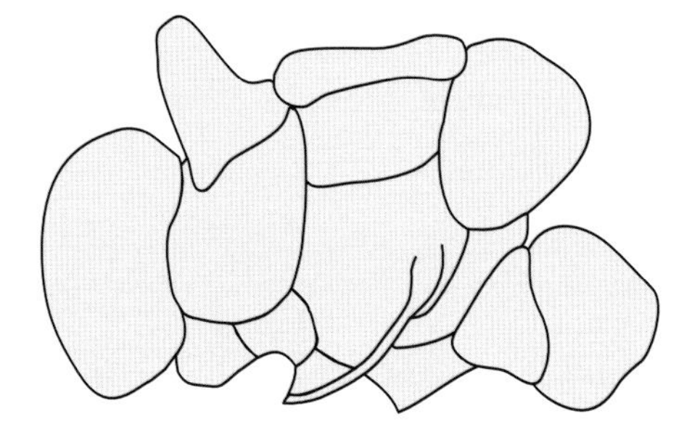

Hypothalamus. This organ is composed of several lobes and is involved in secreating various releasing factors into the hypophysis.

for discovering the role played by certain parts of the brain in determining and coordinating the functions of internal organs. Walter Rudolf Hess is known for his abundant work on the autonomous nervous system. He showed that the main centres of this system are located in the brainstem and diencephalon, particularly in the hypothalamic regions. He mapped this part of the brain in such detail that by applying electrodes to certain parts of the hypothalamus of a cat confronted with a dog, he was able to modify specifically the cat's behaviour. He was famous for his work: "Biology of the Mind." Walther Hess was Professor and head of the Department of Physiology at the University of Zurich in 1917.

1950

Hench, Philip Showalter (Pittsburgh, Pennsylvania, February 28, 1896 - Ocho Rios, Jamaica, March 31, 1965). American physician. Nobel Prize for Medicine, along with **Edward Kendall** and **Tadeus Reichstein**, for his clinical appplication of hormones of the adrenal cortex, especially cortisone. A major part of this scientist's work focused on the causes of rheumatoid arthritis (RA). Two facts made Hench suspect that this was not a bacterial disease but rather, possibly, a metabolic disturbance. Firstly, the

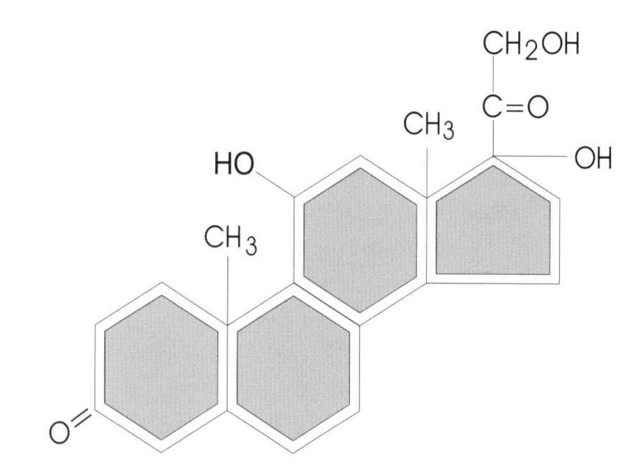

Cortisone. An hormone of the adrenal cortex. One of the 29 steroids isolated from this gland by Tadeus Reichstein. It is mainly known for its anti-inflammatory effects.

symptoms of the disease tended to diminish or even disappear temporarily in pregnant women and hepatitis patients. Hench thus proposed investigating the possibility that a substance common to both conditions might be the cause of this relief. Secondly, stimulation of the adrenal glands also brings relief to patients. These findings led Hench to develop a fairly effective treatment for the disease: injection of the adrenal hormone cortisone discovered by Kendall. Cortisone was the first synthesized hormone used to treat arthritis. Philip Hench was a teacher at Minnesota University.

Kendall, Edward Calvin (South Norwalk, Connecticut, March 8, 1886 - Princeton, New Jersey, May 4, 1972). American chemist. Nobel Prize for Medicine, along with **Tadeus Reichstein** and **Philip Showalther Hench** for discovery of adrenal cortex hormones and works on their phy- siological effects. The first essential contribution made by Edward Kendall was the resounding discovery (in 1915) of thyroxin, the active component of thyroid hormone. The existence of this substance secreted by the thyroid gland had been proposed by the Swiss doctor **Theodor Kocher**. Although quite similar to the amino acid tyrosine, thyroxin was later found to be peculiar in that it possesses four iodine atoms. Kendall's research also focused on identifying the many substances made by the cortical part of the adrenocortical gland. In the first studies aiming to understand what causes Addison's disease, characterised by bilateral destruction of the adrenal glands, patients had received adrenalin, synthesised by the adrenal medulla. As this treatment had no effect, investigators turned to the cortical part of the gland. Independently of Reichstein in Switzerland, Kendall discovered no less than 28 hormone-type compounds in the adrenal cortex. These compounds were called corticoids because of their site of synthesis. During World War II, research into these compounds received a major impetus and even became a priority when the rumour circulated that the Germans were about to isolate, from adrenals obtained in large quantities from Argentine slaughterhouses, active substances enabling aviators to fly at much higher altitude than the aviators of the Allies. Although this rumour proved unfounded, it nevertheless enabled research to progress. No less than 32 compounds were identified by the end of the war. One of those discovered by Kendall, cortisone, turned out to be active against rheumatoid arthritis. It was Kendall's co-worker Hench who first introduced cortisone into therapy. Over 70 hormones, all derived from the cholesterol nucleus, have since been isolated from the adrenal cortex. An interesting aspect of Kendall's work is that it led to the industrial synthesis of several natural hormones and to their gradual introduction into clinical medicine. No less remarkable was the realisation that certain hormones of the adrenal cortex resemble hormones produced by the gonads. Progesterone, for example, is produced both by the *corpus luteum* of the ovary and by the adrenocorticals. Likewise, androsterone is found in both the testicles and the adrenal cortex. Edward Calvin Kendall received his first medical degree from Columbia University, New York, and further specialized in chemistry. He successively headed the Graduate School of the Mayo Foundation in Rochester (1914) and the Department of Biochemistry of the University of Minnesota (1915) before conducting research at St. Luke's Hospital, New York. In 1951, he held the position of Visiting Professor at Princeton University.

Edward Kendall

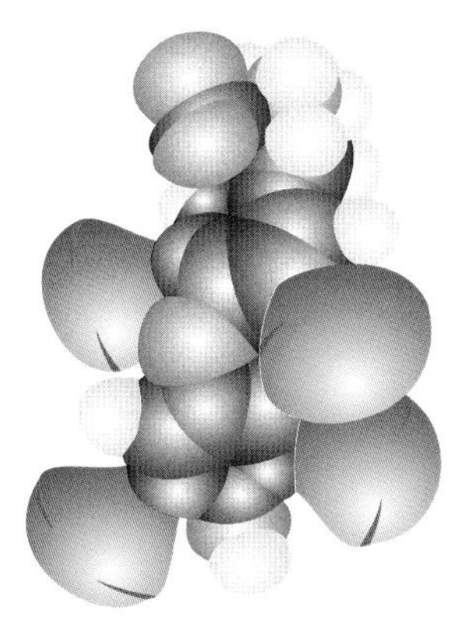

Thyroxine. A three-dimensional representation

Adrenal gland. The discovery of nearly 30 hormones in this tissue, in the cortex in particular, was a surprise for the endocrinologists in the first half of the twentieth century. While the inner medulla produces adrenaline and noradrenaline, the outer cortex is responsible for the biosynthesis and secretion of steroid hormones.

Reichstein, Tadeus (Wloclavek, Poland, July 20, 1897 - Basel, Switzerland, August 1, 1997). Polish biochemist. Nobel Prize for Medicine, along with **Edward Kendall** and **Philip Showalter Hench**, for his discoveries concerning hormones of the adrenal cortex. Tadeus Reichstein was educated at the Federal Institute of Technology in Zurich and taught there before moving to Basel where he headed the Institute of Organic Chemistry. Later he turned his attention to the adrenal glands. These endocrine glands are located just above the kidneys and each one consists of two regions:an outer layer, the adrenal cortex and an inner layer, the adrenal medulla. The adrenal cortex secretes many hormones, mainly cortisol, aldosterone and androgens. Reichstein succeeded in isolating most of them.

Tadeus Reichstein

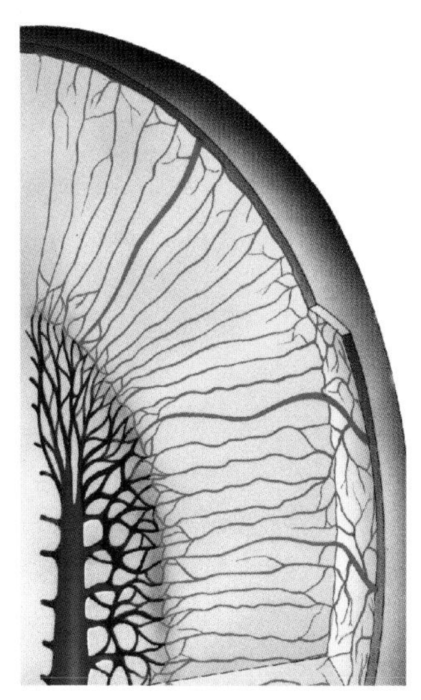

Among the most active were cortisone, aldosterone, corticosterone and hydroxycorticosterone. Reichstein also succeeded in synthesizing desoxycorticosterone, a mineralocorticoid hormone. Thanks to Reichstein's work in Basel and to Kendall's in Rochester, the chemistry of adrenocorticoid hormones developed considerably. A particularly interesting aspect of this research was that it led to industrial - scale synthesis of some of these hormones, although the glands produce them only in trace amounts. This production greatly facilitated their therapeutic use. Apart from hormone research, we are also indebted to Reichstein for the synthesis of vitamin C in the beginning of the 1930s, a process performed about the same time by Walter Haworth in England, but independently. Toward the end of his career, Reichstein's investigations turned to the biochemical properties and the therapeutic effects of compounds extracted from plants in tropical rain forests.

1951

Theiler, Max (Pretoria, South Africa, January 30, 1899 - New-Haven, Connecticut, August 11, 1972). 1951 Nobel Prize for Medicine for developing a vaccine for yellow fever. Before finishing his studies at the University of Capetown, Max Theiler persuaded his father to let him complete his education in England. Yet upon his arrival in London and upon discovering that he would have to start his studies all over from the first year, he preferred to take a job at Saint Thomas's Hospital in the same city and to begin training in practical medicine. He finally ended up in the United States where, after holding a position in the Harvard Department of Tropical Medicine, he went to the Rockefeller Foundation in New York where he continued his research on yellow fever. A major part of his scientific work was devoted to this disease. Yellow fever is a viral disease caused by the yellow fever virus. The vector is a biting mosquito. Inoculating the virus into a monkey and then transferring it to successive mice, Theiler obtained an attenuated form of the virus with which he produced a vaccine capable of conferring immunological protection against the harmful form of the virus. This vaccine was largely used in French Africa during the 1930s and the 1940s. Max Theiler made his career at the Rockefeller Institute for Medical Research in New York.

1952

Waksman, Selman Abraham (Priluka, Russia, July 22, 1888 - Hyannis, Massachusetts, August 16, 1973). Russian-American biochemist. Nobel Prize for Medicine, for discovery of the antibiotic streptomycin, the first specific agent effective in the treatment of tuberculosis. This illness is caused by *Mycobacterium tuberculosis*, a gram-negative bacteria. This 1943 discovery was an outstanding breakthrough because antibiotics had just been discovered and the known ones were active only against gram-positive bacteria. Streptomycin, a water-soluble antibiotic produced by soil bacteria *Streptomyces griseus*, inhibits initiation of translation and induces misreading of messenger RNA. It is active against gram-positive and gram-negative bacteria. Experiments have shown that it acts by binding to a pro-

Abraham Waksman

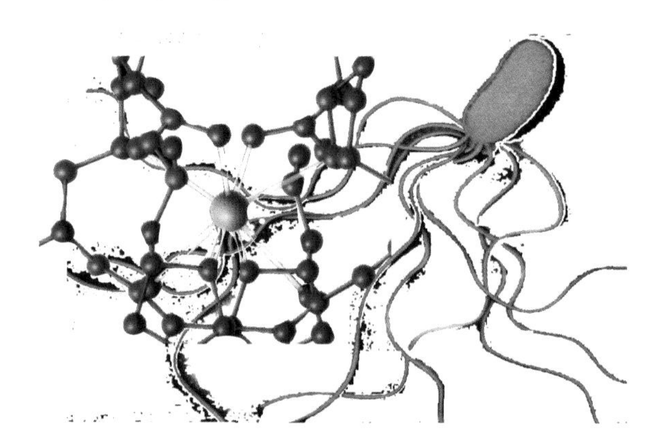

Bacteria and the antibiotic valinomycin. The cyclic antibiotic, which can act as an ionophore, is specially active against *mycobacterium tuberculosis*.

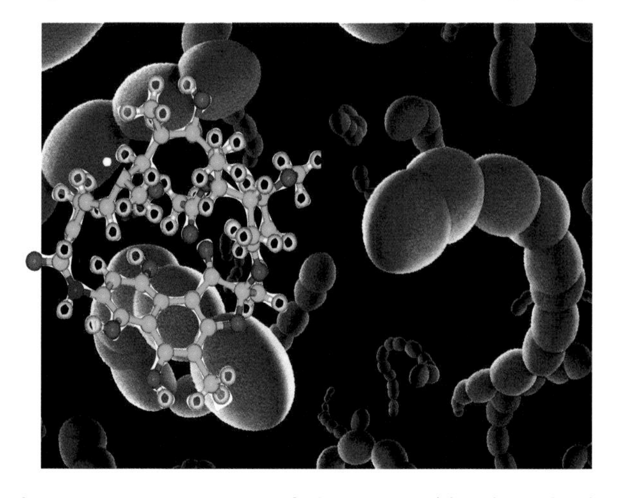

Streptococcus. A genus of Gram - positive bacteria that produces antibiotics such as streptomycine, which acts at the ribosome and causes miscoding of messenger RNA. In the box : rifamycin. An anti-transcription antibiotic produced by *Streptomyces* that is active against Gram - positive bacteria and *Mycobacterium tuberculosis*..

tein in the 30S subunit of the bacterial ribosome. Today, streptomycin is used in combination with other drugs such as isoniazid and aminosalicyclic acid. Still more antibiotics were discovered in the wake of this discovery, notably neomycin and actinomycin D, the latter having both anti - bacterial and anti - cancer properties. In 1940 while teaching microbiology at the Rutgers University, Selman Abraham Waksmann headed the Department of Microbiology. He was granted American citizenship in 1916.

1953

Krebs, Hans Adolf (Hildesheim, Germany, August 25, 1900 - Oxford, England, November 22, 1981). German-born British biochemist. Son of a surgeon. Nobel Prize for Medicine, shared with **Fritz Lipmann**, for his contribution to the discovery of the series of chemical reactions known as the tricarboxylic acid cycle. Hans Adolf Krebs received his medical education from the Universities of Göttingen, Freiburg - im - Breisgau, and Berlin. Later, he worked under the famous biochemist **Otto Warburg** at the Kaiser Wilhelm Institute for Biology at Berlin - Dahlem. In 1933, because of his Jewish origins he was forced to move to England, where he was invited by **Frederick Gowland Hopkins.** From 1935 to 1954, Krebs taught biochemistry at Sheffield University in England and later, from 1954 to 1967, at Oxford University, where he founded the famous Institute of Biochemistry, which bears his name. We are mainly indebted to this German scientist for our knowledge of the energetic metabolism. Krebs's research was probably influenced by the work of Warburg, who had already shown that oxygen consumed during aerobic life contributes importantly to creating an energy supply. A first and important part of his scientific work was the discovery of a cycle of reactions, now known as the urea cycle, by which ammonia is converted to urea, a less toxic waste product, in terrestrial vertebrates. Urea is synthesized in the liver and then sequestered by the kidneys for excretion in the urine. The other major breakthrough, the one for which he was awarded the Nobel Prize, was the discovery of a cycle of reactions which accounts for the major portion of carbohydrate, fatty acid and amino acid oxidation. This cycle is alternatively known as the Krebs cycle or the tricarboxilic acid cycle. As a matter of fact, the Hungarian biochemist **Albert S. - Györgyi** had already demonstrated that cellular respiration was stimulated by catalytic amounts of four - carbon compounds, in particular succinate, fumarate, malate and oxalacétate. On the other end, Martius and Knoop showed that citric acid, a six-carbon compound, was decarboxylated and conver-

Adolf Krebs

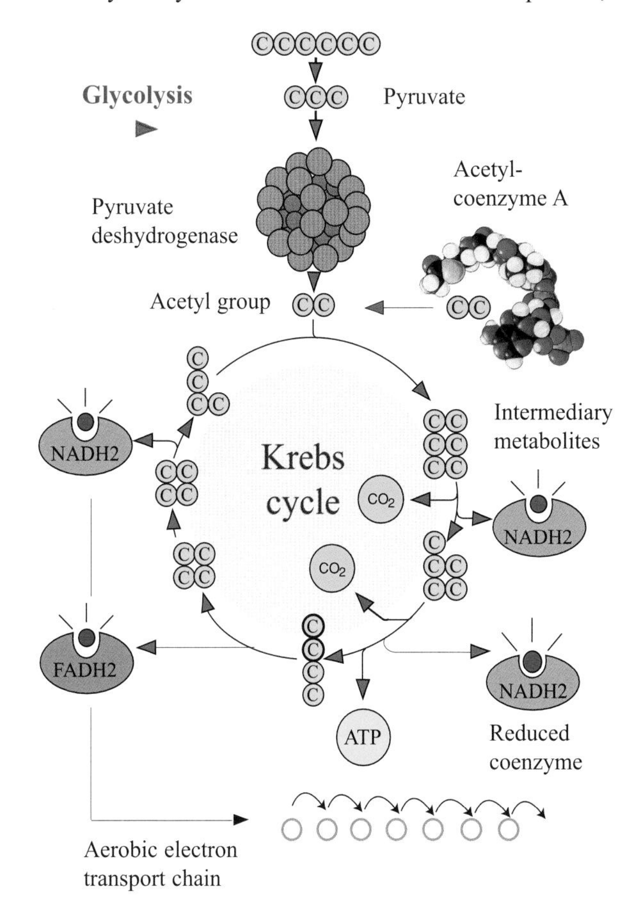

Citric acid cycle This metabolic cycle is carried out in mitochondria. It is a more recent discovery than glycolysis and is found only in organisms that respire and oxidize their foods to completion. The citric acid cycle is the final common pathway for the oxidation of pyruvate, fatty acids and the carbon chains of amino acids. Its other name, the Krebs cycle, was given in honour of Hans Adolf Krebs, who described important steps of this metabolic pathway and realized that it functioned as a cycle.

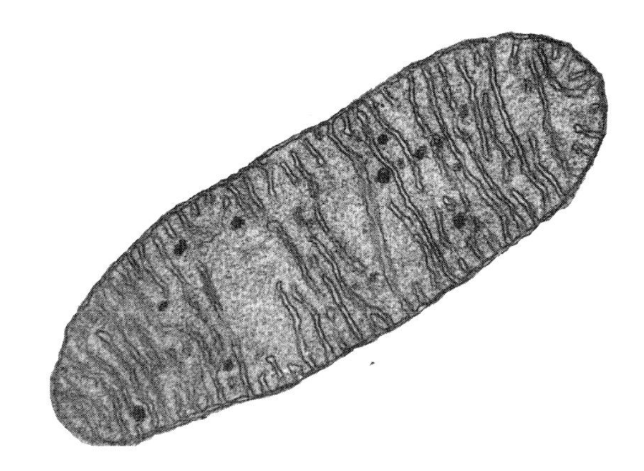

Mitochondria. An electron micrograph.

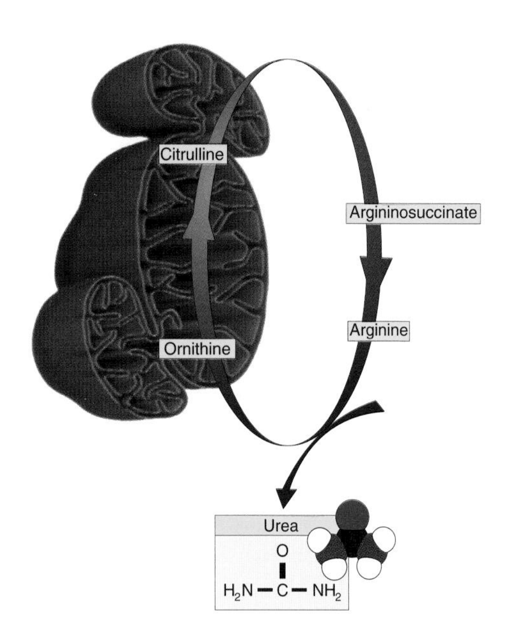

Urea cycle. Also called ornithine - citrulline cycle.Urea constitutes the main excretory product in urotelic animals. It is produced by the liver from deaminated excess amino acids. This cycle was recognized by H. A. Krebs. As shown, some steps occur in the mitochondria.

ted into alpha - cetoglutarate and that this process was accompanied by the coupled reduction of NAD to NADH2. Other research established the subsequent conversion of alpha-cetoglutarate into succinate. Krebs's merit was to show that citric acid is formed through the combination of oxalacetic acid with a two-carbon chain arising through pyruvate degradation. The details of this reaction are known thanks to Lippmann. All of this work revealed the energetic importance of what Krebs had accurately described as a cycle. The cycle involves two oxidative decarboxylations in which the four pairs of electrons taken from the metabolites undergoing oxidation are channelled into a transport chain coupled to ATP productions.

Lipmann, Fritz Albert (Königsberg, Prussia, now Kaliningrad, Russia, June 12, 1899 - Pough-Skeepsie, New York, July 24, 1986). German - born American biochemist. Nobel Prize for Medicine, shared with **Hans Krebs**, for his discovery of coenzyme A, an important ca talytic substance involved in the cellular conversion of food into energy. Two major contributions characterise this researcher's work. First, he resumed work that had been started by scientists such as **Otto Meyerhoff**, **Arthur Harden**, and the **Cori** couple on the in-tervention of phosphate in metabolic sugar degradation. His own contribution was to reveal the energetic importance of phosphoesters, intermediates of oxidative metabolism, and to show that transfer of phosphate groups from ATP can create 'energy-rich' phosphoester bonds. Later he demonstrated the determining role of ATP in ensuring the continuous energy

Fritz Lipmann

supply that cells need. In 1947 Lipmann was the first to show that a coenzyme was required to transfer the acetyl group (the acetyl group is derived from aerobic oxydation of pyruvate) from one compound to another. He named this substance as coenzyme A. Lipmann also discovered that this coenzyme contained vitamin B (panthotenic acid) and that it carried an activated acetyl group into the Krebs cycle. The complex formed between the acetyl group and coenzyme A was first isolated by **Feodor Lynen**, a German biochemist who collaborated with Lipmann in that field. It

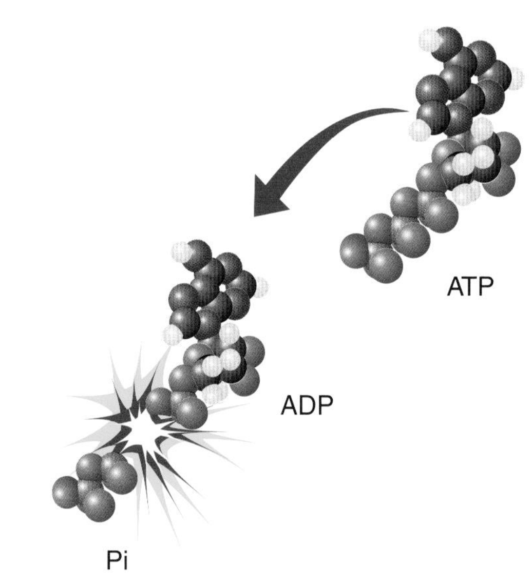

ATP hydrolysis. It provides most of the energy required by cell activities. This energy is used to synthesize small and large molecules, to contract muscles, to transport substances into or out of the cell.

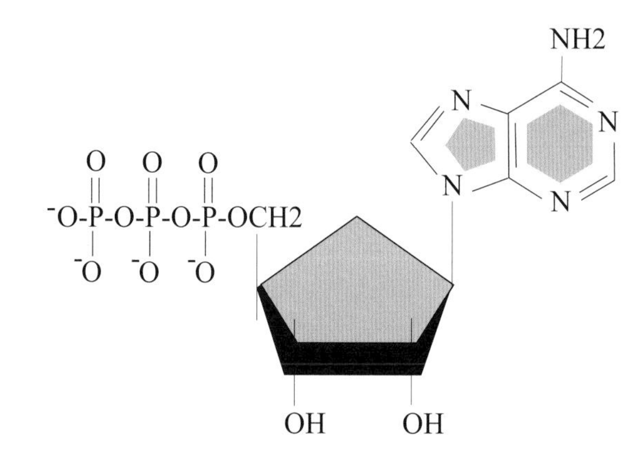

Adenosine triphosphate (ATP). A schematic structure. Note the bonds between the phosphate groups, the breaking of which release energy useful for cell activity.

was later found that coenzyme A played a crucial role in biochemical pathways which breakdown sugars, lipids and several amino acids. Moreover, coenzyme A also interferes in the anabolic pathway that converts sugars into lipids. Fritz Albert Lipmann was one of the greatest pioneers in biochemistry. During the rise of Nazism in

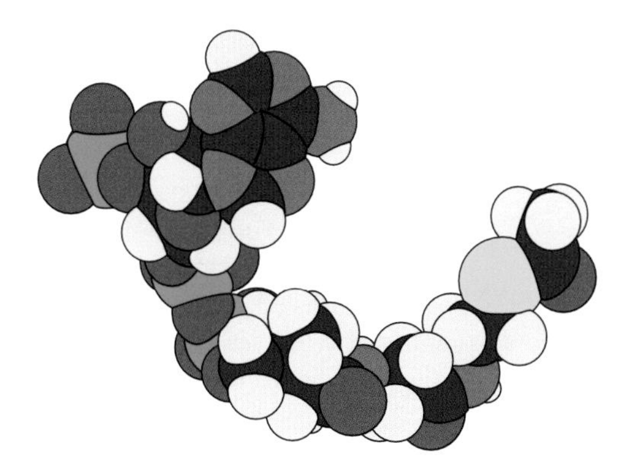

Acetyl-CoA. This substance, which is a derivative of coenzyme A, is a central metabolite. It is derived from catabolic pathways namely glycolysis, fatty-acid oxidation and the degradative metabolism of several amino acids. It is also a key intermediate in lipid biosynthesis and other anabolic pathways.

Germany, Lipmann, like many other Jewish scientists, was obliged to flee from his native country. First, in 1932 he immigrated to Copenhagen, and finally settled down in the United States, where in 1939 he found a job and five years later was granted American citizenship. After he had worked with **Vincent Du Vigneaud** at Cornell University, Lipmann was engaged in medical research at the Rockefeller Institute for Medical Research in New York.

1954

Enders, John Franklin (West Hartford, Connecticut, February 10, 1897 - Waterford, Connecticut, September 8, 1985). American virologist. Son of a banker. Awarded the Nobel Prize jointly with **Thomas Weller** and **Frederick Robbins** for discovering that the poliomyelitis virus could be grown in culture in a variety of tissues. John Enders's work formed the basis for a vaccine against poliomyelitis, a paralytic disease caused by a virus that infects the brain and nervous tissue. In the first part of the 20th century, the disease was prevalent, mainly affecting young children. Enders was aided in his research by Weller and Robbins, two graduates of Harvard Medical School who had joined his laboratory at Childrens Hospital in Boston. It was through a lucky accident that Enders became involved in research that proved essential for the creation of a vaccine against this terrible illness. Wellers had prepared more human embryonic tissue than was required for an experiment on chicken pox virus, one of the sujects of research in the laboratory. Following an impulse, Enders suggested that Weller infect the surplus tissue with polio virus and to his surprise, the virus grew beautifully on the cells. When in 1949, this discovery was made known to the medical world, the scientific community was convinced that the culture of polio virus was possible only in nervous tissue. Studies of the virus had been carried out in nervous tissue obtained from monkeys imported from the Indies and available only in limited quantities. Futhermore, culture of the virus in the monkey tissue presented a certain number of additional problems. Thus, the results obtained by Enders and his collaborators were remarkable in that they permitted the culture of the virus on a number of tissues different from nervous tissue. Polio virus could now be produced on a large scale and so, could be better studied. Although Robbins and Weller tried hard to convince Enders, he refused to attempt to develop a vaccine against poliomyelitis even though he appeared to be in an ideal position to succeed. The reasons for his refusal have remained obscure. The vaccine was developed by the microbiologists Albert Sabin and Jonas Salk. However, it is clear that this would probably not have been possible without the contributions of Enders and his collaborators. Enders, a brillant researcher, did develop vaccines against chicken pox, mumps, and measles. He was still an assistant professor when he received the Nobel Prize.

John Enders

Robbins, Frederick Chapman (Auburn, Alabama, August 25, 1916 -). American physician. Son of a plant physiologist. Nobel Prize for Medicine, along with **Thomas Weller** and **John Enders** for successfully cultivating poliomyelitis virus in tissue cultures. Frederick Robbins first studied at the University of Missouri and later graduated from Harvard Medical School. Finally he was appointed Resi-dent Physician in Bacteriology at the Children's Hospital Medical Center in Boston where he met Weller There, in collaboration with him, he succeeded in cultivating polio virus in various tissues, in particular, on human embyonic tissue. Previously, it had been possible to cultivate the virus only on nerve tissue from chimpanzees, tissue that could not be obtained in large quantities. The deve- lopment of techniques for efficient culturing of polio virus was essential to the production of a vaccine effective in preventing poliomyelitis

Weller, Thomas Huckle (Ann Arbor, Michigan, June 15, 1915 -). Son of a pathologist. Nobel Prize for Medicine, along with **John Enders** and **Frederik Robbins**, for the successful cultivation of poliomyelitis virus in tissue cultures. Thomas Weller graduated from the University of Michigan in 1937 and became Professor at the Harvard Institute of Tropical Medicine after working at the Boston Children's Hospital. In collaboration with Franklin Neva, he first succeeded in propagating the rubella virus in tissue cultures. He did the same with the chicken - pox virus and isolated it from human embryonic tissues (muscle and skin). Other work carried out in close collaboration with Enders (one of his professors) and Robbins led to cultivating the poliovirus in tissue cultures. A vaccine of living strains of poliomyelitis virus of greatly attenuated virulence was introduced by Albert Sabin, a Polish-American physician and microbiologist. Thomas Weller is also known for having isolated the viruses that cause varicella and herpes.

Thomas H. Weller

1955

Theorell, Axel Hugo Theodor (Linköping, Sweden, July 6 1903 - Stockholm, Sweden, August 18, 1982). Swedish biochemist. Nobel Prize for Medicine for his discoveries on oxidation enzymes. Axel Hugo Theorell showed that the Warburg enzyme (yellow ferment), isolated from yeast by **Otto Warburg**, was in reality composed of a coenzyme including vitamin B_2 and an apoenzyme. In addition, he demonstrated the enzyme oxidized glucose, taking up two atoms of hydrogen. We know today that this enzyme is a member of a large group of enzymes called dehydrogenases, involved in cellular oxidations, and that the coenzyme associated with the yellow ferment is a non-protein, flavin-type substance. Theorell and his team conducted meticulous studies on the mechanisms of action of certain dehydrogenases. They focused particularly on alcohol dehydrogenase from liver, describing its enzymatic mechanism with outstanding precision. This was the first time that an enzymytic mechanism was described in such detail. Theorell also made extensive studies of cytochrome C, showing that the enzyme center was occupied by a porphyrin ring which bound an iron atom. This structure is crucial to the participation of cytochromes in cellular oxidations. He is also known for achieving the first crystallization of myoglobin. Axel Hugo Theorell received his

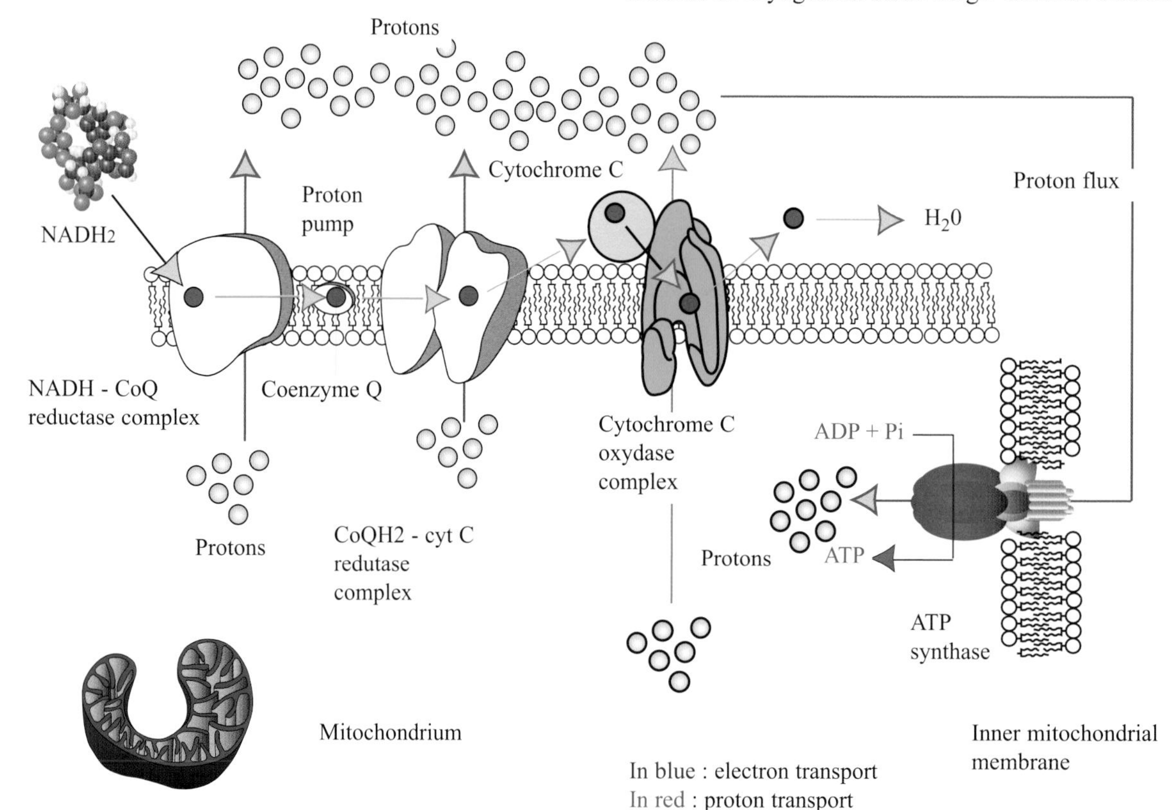

The electron transport chain. Several protein complexes involved in the transport of electron (from NADH2 to oxygen) are embedded into the membrane. Cytochrome C intervenes in this process by transferring electrons to cytochrome oxydase complex. As shown, electron transport is coupled to proton pumping.

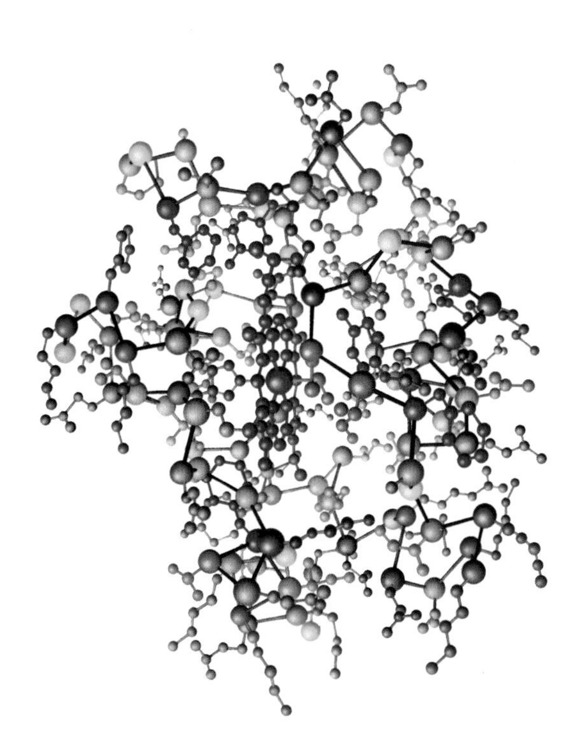

Cytochrome C.A three-dimensional representation of an electron transporter studied by Axel Theorell.

education from the State Secondary School in Linkköping and graduated later in medicine from the Karolinska Institute in Stockholm. He also studied bacteriology at the Pasteur Institute in Paris under the French microbiologist Albert Calmette.

1956

Cournand, André Frédéric (Paris, France, September 24, 1895 - Great Barrington, Massachusetts, February 19, 1988). French - American physician and physiologist. Nobel Prize for Medicine with **Dickinson Woodruf Richards** and **Werner Forssmann** for discoveries on heart catheterization and circulatory changes. André Cournand worked with Richards to perfect the cardiac catheter for studying the nature of heart diseases. This instrument had previously been developed by Forssmann, Cournand is also known for his remarkable work on intrapulmonary blood pressure. He also worked on the effects of drugs, such as digitalin, on cardiac function. In 1945, Cournand designed a two-branched catheter allowing the blood pressure across two adjacent chambers of the heart to be measured. His investigations contributed to the establishment of more precise techniques for the diagnosis of various heart conditions. André Frédéric Cournand joined the College of Physicians and Surgeons of the University of Columbia, New York in 1934 where he gave lectures on heart physiology until retiring as Emeritus Professor of Medicine in 1964. He became an American citizen in 1941.

Richards, Dickinson Woodruff (Orange, New Jersey, October 30, 1895 - Lakeville, Connecticut, February 23, 1973). American physiologist. 1956 Nobel Prize for Medicine, along with **Werner Forssmann** and **André Cournand** for his works and discoveries concerning heart catheterization and pathological changes in the circulatory system. Dickinson Woodruff Richards actively collaborated with Cournand in order to refine this technique; Werner Forssmann had inven - ted cardiac catheterization some years earlier in 1929. Richards also conducted research on the properties of digitalin, a glycoside extracted from *Digitalis purpurea* that increases the contractile power of the weakened heart. As early as in the 1940s, he showed that by analysing blood taken from the heart by means of this catheter it was possible to detect congenital

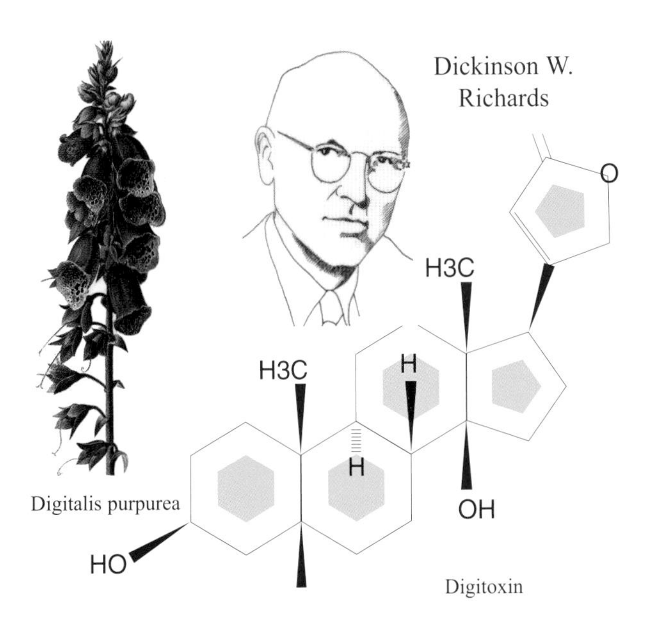

Digitalis purpurea. The plant produces a family of cardiac glycosides such as digitotoxin (at right). This substance is used as a chemotherapeutic drug to increase the force of the systolic contractions.

heart diseases and to measure the effects of various drugs on the heart muscle itself. His collaboration with Cournand also led to the study of blood circulation in the lungs and how it changes in response to trauma. Richards studied medicine at Yale University and received his medical degree from the College of Physicians of Columbia in 1923. He also studied the effects of digitalis, a drug given to patients to stimulate their weakened heart as well as to slow the rate of their heartbeat.

Forssmann, Werner (Berlin, Germany, August 29, 1904 - Schofheim, June 1, 1979). German surgeon. Son of an employee of a Berlin insurance company. Nobel Prize for Medicine, along with **André Cournand** and **Woodruff Richards Dickinson**, for his development of cardiac catheterization. He chose to study heart diagnosis techniques, and the considerable improvements he made in this area summarise his career. At the beginning of the 20th century, some techniques were available for measuring the heart's electric and muscle activity, but in addition to their relative imprecision, they were not safe enough. One such technique was catheterisation, used in the se cond half of the 19th century to measure cardiac pressure in the left and right chambers of the heart. Werner

Drawing of a fiber-optic endoscope for *in situ* visual examination of cardiovascular system.

Forssmann used this device experimentally on the horse to reach the heart through the jugular vein. From his studies on cadavers, he concluded that it was possible to introduce the catheter into a vein in the shoulder, at the site where intravenous injections were traditionally performed. He informed the hospital Director of his intention to demonstrate the effectiveness of his method experimen-tally on himself, but the Director refused to let him conduct the experiment at the hospital. Forssmann convinced one of the hospital nurses to help him perform the operation at home, by means of a device he had prepared. Although the nurse was willing to submit to the procedure, Forssmann refused and maintained his decision to try the experiment on himself. By inserting a catheter into a vein of his left elbow, Forssmann was able to enter the right ventricle of the heart and observe its function by means of a fluoroscope. Then, calmly, he had pictures taken to prove the success of this great and brave experiment. Although he was to make a variety of measuring instruments related to blood circulation, this is the invention that brought heart surgery in his time to its pinnacle. He was severely criticized by his colleagues for this rather unorthodox experiment. He then abandoned this scientific work and took up a position of a surgeon in a small city in south Germany. His pioneering work was further perfected by Richards Dickinson and André Cournand and used as a tool for heart diagnosis. While performing successful operations on the kidney, he was amazed to learn that he had been awarded the Nobel Prize. Forssmann received his medical education from the University of Berlin in 1928. He then directed the Department of Surgery at the city hospital in Dresden - Friedrichstadt and in 1958 he was appointed Director of the Surgical Division of the Evangelical Hospital in Düsseldorf.

1957

Bovet, Daniel (Neuchatel, Switzerland, March 23, 1907 - Rome, Italy, April 8, 1992). Italian pharmacologist born in Switzerland. Son of a lecturer of pedagogy at the University of Geneva. Nobel Prize for Medicine for his discoveries of certain chemotherapeutic agents. One contribution of this rese-archer, along with Gerhard Domagk, is the discovery of Protonsil, an efficient antibacterial chemical compound for combatting streptococcal infection, such as pulmonary meningitis. Daniel Bovet showed that the action of Protonsil was due to its sulfonamide - like group. Based on this knowledge, he and his co-workers synthesized a variety of sulfonamide derivatives, many of which were equally effective and became known as 'miracle drugs'. The compounds were produced on a large scale and saved millions of lives during the Second World War. Another field of research came from an idea he had in the 1930s, namely that the effects of some molecules could be opposed by those of related molecules. For example, if a molecule reacts with a substance by first binding to it, a similar molecule could recognize this same substance without reacting with it. After many years of hard work, Bovet synthesized a series of mimetic molecules which counteracted the inflammatory effects of histamine. Today these drugs are known as antihistamines, one of which is pyralamine. Later, he became interested in the physiological effects of curare, a highly toxic mixture of various alkaloids. Curare acts on the central nervous system. The purified form, tubocurarine chloride, isolated from the bark and stems of a South American vine, is used in medicine as a muscle relaxant. He spent some time with the Indians of South America who used this drug. When administered in small doses, it had a muscle relaxing

Daniel Bovet

effect allowing certain forms of surgical intervention. Bovet synthesized some 400 analogues of curare, one of which, succinylcholine, is still used today in surgical oper-

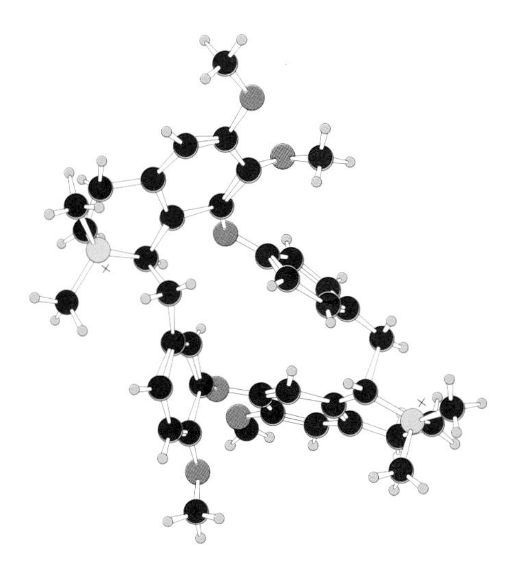

Curare. This deadly South American Indian arrowhead poison blocks neuromuscular transmission in voluntary muscle and is useful as a muscle relaxant.

ations. He also worked on mental illnesses, which he thought could be treated by chemicals, but was without notable success in this domain. His wife Filomena Nitti was the daughter of an Italian diplomat, who had fled from

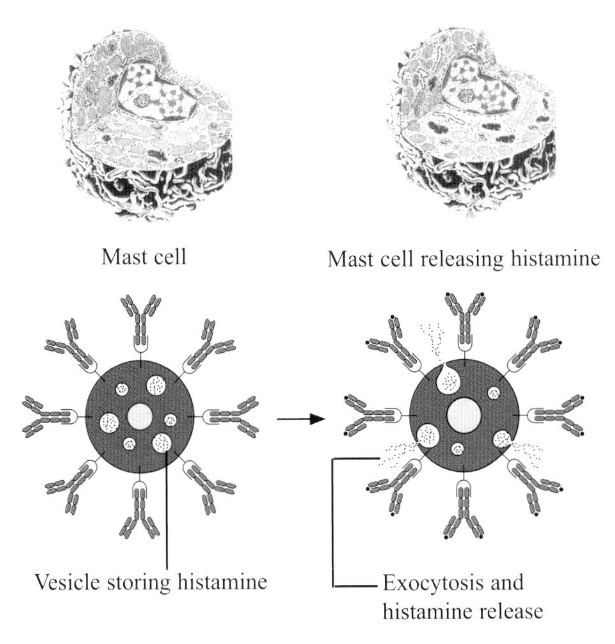

Histamine and allergy. Following an exposure to an appropriate allergen (for example, a pollen), mast cells degranulate and release histamine, a mediator in allergic reactions.

Mussolini's regime. She was actively involved with Bovet's research and was co-author on a number of important scientific publications.

1958

Beadle, George Wells (Wahoo, Nebraska, October 22, 1903 - Pomona, California, June 9, 1989). American geneticist. Son of a farmer. Nobel Prize for Physiology or Medicine with **Edward Tatum** and **Joshua Lederberg** for establishing the relationship between genes and proteins. Along with Tatum, Georges Beadle was the founder of modern molecular biology. He was Professor at the University of Stanford from 1937. After his graduation, Beadle thought of going back to farming, however he became interested in science and was persuaded to do a Ph.D. on the genetics of maize at Cornell University. He began his research in the laboratory of **Thomas Hunt Morgan** and soon realized that genes should determine the phenotype of an organism via chemical mediators. In 1935, he went to Paris to work with Bernard Ephrussi on the development of eye pigments in the fruit fly, *Drosophila melanogaster*. On his return to the United States, he met Tatum, with whom he developed a

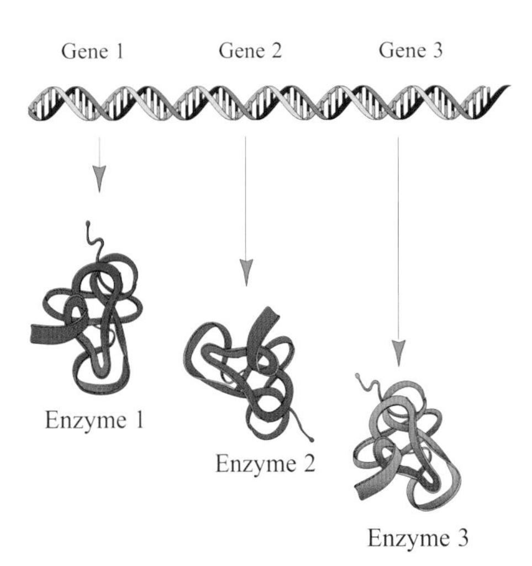

One gene - one enzyme hypothesis. It was formulated by G. Beadle and E. Tatum following their work on auxotrophic mutants of Neurospora crassa and stated that each naturally occuring enzyme is synthesized under the control of a different corresponding gene.

method for selecting yeast mutants unable to grow on various culture media because of deficiencies in certain metabolic enzymes. For example, one of these mutants was unable to synthesize arginine due to the absence of, or defect in one of the enzymes necessary for the biosynthesis of this amino acid in yeast. The outcome of these studies with Tatum was that the role of a given gene was to specify the synthesis of a particular enzyme within the context of a metabolic pathway. They proposed the concept of "one gene, one enzyme". Accordingly, the gene was defined as a functional unit or chromosome segment capable of specifying a particular protein structure. Although this concept has been developed considerably since its original proposal, it represents a landmark demonstrating how genetic manipulation can be used as a powerful tool in biochemistry. At the end of his career, George Beadle became involved in a scientific dispute known as "the corn war". Different hypotheses were put forward as to the origin of this cereal. According to him, its cultivation went back as far as the Maya and Aztec civilizations.

Tatum, Edward Lawrie (Boulder, Colorado, December 14, 1909 - New York City, New York, November 5, 1975). American biochemist and geneticist. Nobel Prize for Medicine shared with **George Beadle** and **Joshua Lederberg**, for research that helped demonstrate that genes determine the structure of particular enzymes. One of the pioneers of biochemical genetics, he collaborated with George Beadle in the studies of nutritional mutants of the fungus *Neurospora crassa* (bread mold) that demonstrated the functional relationship between genes and enzymes. Tatum extended the isolation of biochemical mutants to *Escherichia coli*, an important contribution to the elucidation of biochemical pathways. Later, using nutritional mutants of *E. coli* that permitted the selection of rare recombinants, he and Lederberg demonstrated that bacteria were capable of genetic recombination, a discovery that was essential for the realization of the enormous potential of bacteria for the study of basic genetic phenomena. Edward Lawrie Tatum earned his Ph.D. in medicine from the University of Wisconsin in 1934. He was a professor and a researcher at Stanford University and later, at the Rockefeller Institute of Medical Research in New York.

Lederberg, Joshua (Montclair, New Jersey, May 23, 1925 -). American microbiologist. Nobel Prize in Medicine shared with **George Wells Beadle** and **Edward Tatum**, "for his discoveries concerning genetic recombination and the organization of the genetic material of bacteria". Recipient of the Nobel prize at the age of 33, Joshua Lederberg presented the proof that, like organisms whose cells have a nu-cleus, bacteria have genes as well as a sex. The decisive experiments done in collaboration with Tatum at Yale University, consisted of growing together different gene-tically marked strains of *Escherichia coli*. The selective detection of bacteria recombinant for the genetic markers sho-wed that mating accompanied by the exchange of genetic material had occurred. This was the discovery of the phenomenon of conjugation, a kind of sexual reproduction particular to bacteria. A second discovery, made with Norton Zinder in 1952, revealed that temperate bacteriophages could transfer genetic mate-rial of one bacterial host to a se-

Joshua Lederberg

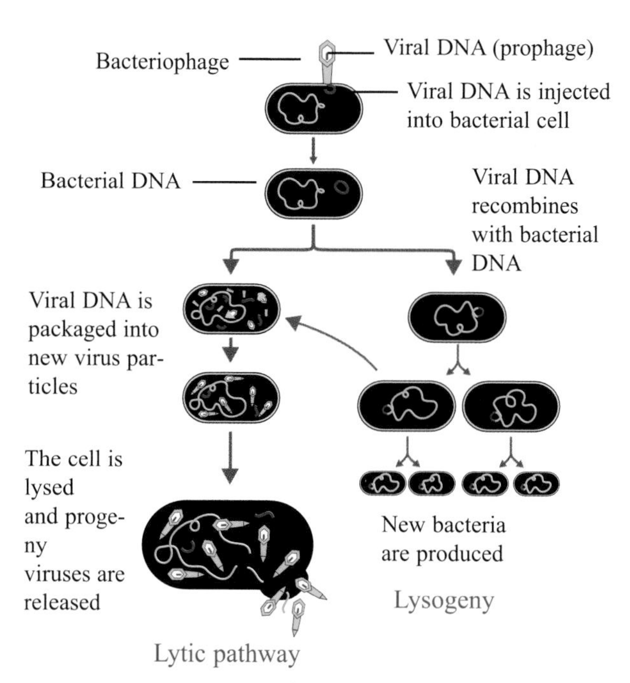

The lysogenic and lytic reproductive cycles of a virus. After binding the surface of a bacteria, a phage injects its DNA. Phage DNA can either integrate into the bacteria chromosome (lysogenic cycle) or initiate the production of numerous progeny phage particles (lytic ccle).

cond host strain and thus give rise to bacterial re-combinants in the absence of any direct contact between the do-nor and recipient bacterial strains. The discovery of this phenomenon, called transduction, raised the possibility of using a viral vector as a general tech-nique for gene transfer from one organism to another, a technique that has developed into an important feature of present day genetic engineering. Joshua Lederberg graduated from the College of Physicians and Surgeons of Columbia University Medical School, New York. He then moved to Yale University, New Haven, Connecticut, where he studied microbiology under Tatum and received his Ph.D. in 1948. He then became Professor of Genetics at the University of Wisconsin before joining Stanford Medical School from 1962 to 1978. The same year, he was appointed President of Rockefeller University and held this post until his retirement in 1990.

1959

Kornberg, Arthur (Brooklyn, New York, March 3, 1918 -). American biochemist and physician. Nobel Prize for Medicine, along with **Severo Ochoa**, for discovering the

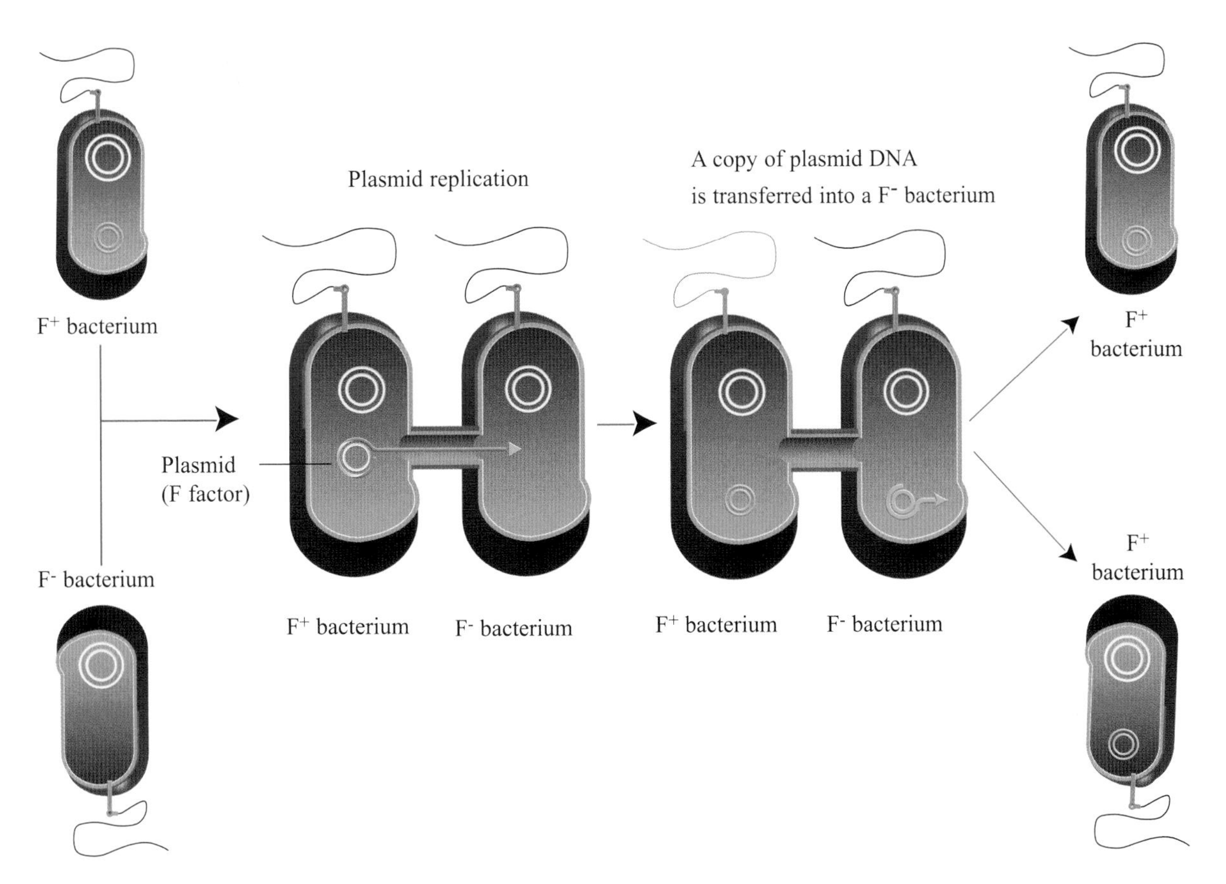

Bacterial conjugation. The discovery that bacteria had a sexuality and could recombine their chromosomes allowed new experimental processes in the field of molecular biology to be developed.

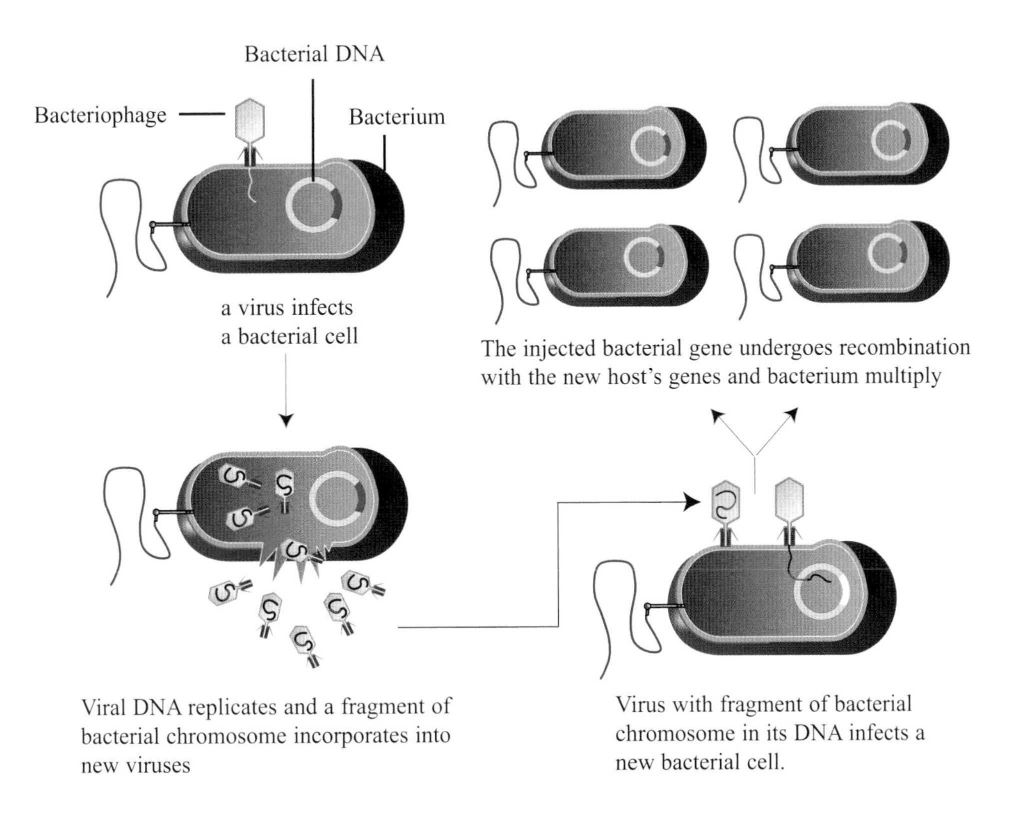

Transduction. As shown, a genetic information is translocated from a bacteriophage (which serves as a vector) to a bacterium. After recombination with the bacterial host genome, new virus particles are produced and can infect other bacteria.

means by which DNA molecules are duplicated in the bacterial cell. From 1959 to 1988 Arthur Kornberg was professor at the Department of Biochemistry at Stanford University School of Medicine. He accepted the title of Professor Emeritus in 1988 and has been on active status to the present. From his early studies of the mechanisms of the enzymatic synthesis of coenzymes and inorganic pyrophosphate, he extended his interest to the biosynthesis of the nucleic acids, particularly DNA. After elucidating key steps in the patways of pyrimidine and purine nucleotide synthesis, including the discovery of phosphoribosyl pyrophosphate (PRPP) as an intermediate, he found the enzyme that assembles the building blocks into DNA, named DNA polymerase. This ubiquitous class of enzymes make genetically precise DNA and are essential in the replication, repair and rearrangement of DNA. Many other enzymes of DNA metabolism were discovered responsible for the start and elongation of DNA chains and chromosomes. These enzymes were the basis of discovery of recombinant DNA which helped ignite the biotechnology revolution. Since 1991, he switched his research from DNA replication to inorganic polyphosphate (poly P), a polymer of phosphates that likely participated in prebiotic evolution and is now found in every bacterial, plant and animal cell. Neglected and long regarded a molecular fossil, he has found a variety of significant functions for poly - P that include responses to stress and stringencies and factors responsible for motility and virulence in some of the major pathogens. Although the pursuit of research has been his primary concern, other interests include the formal teaching of graduate, medical and postdoctoral students, and the authorship of major monographs:DNA Synthesis in 1974, DNA Replication in 1980, For the Love of Enzymes : The Odyssey of a Biochemist in 1989 and, in 1995, The Golden Helix Inside Biotech Ventures, which provided an insider's view of biotechnology. Among Kornberg's honors are memberships in the National Academy of Sciences, the Royal Society, American Philosophical Society, a number of honorary degrees, the National Medal of Science (1979), the Cosmos Club Award (1995) and other medals and awards.

Arthur Kornberg

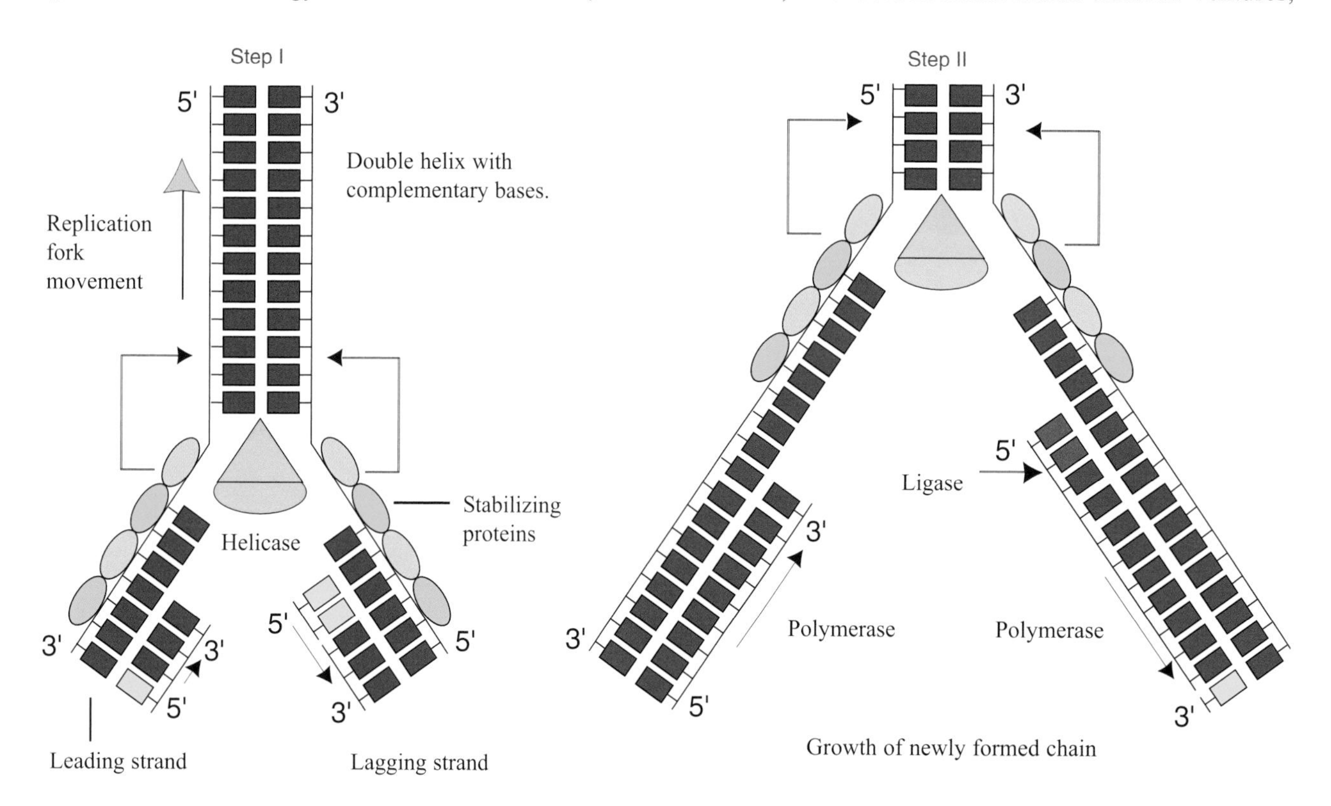

DNA replication.

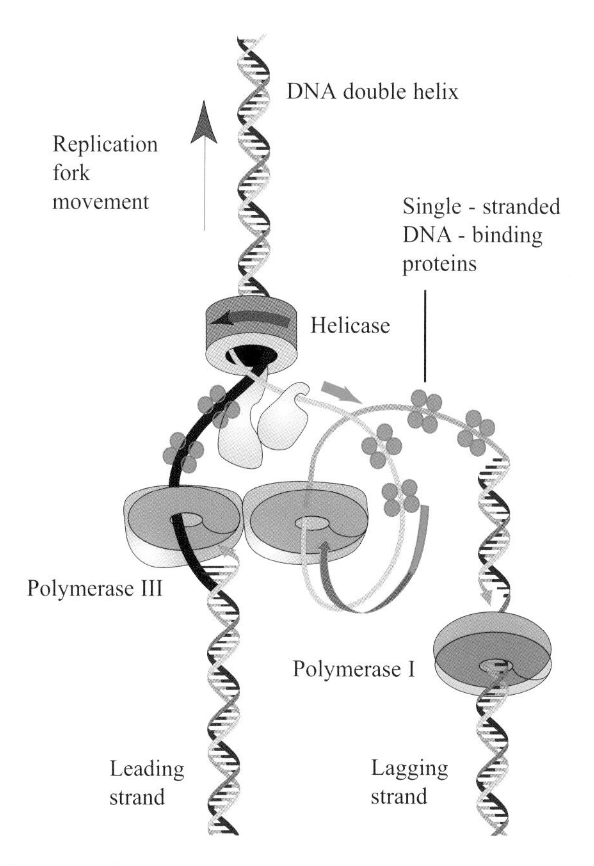

DNA replication, general features of a replication fork. DNA gyrase and helicase unwind the DNA duplex and the single strands are coated with DNA binding proteins. Short fragments of RNA serve as primers for DNA replication while topoisomerases are needed to unwind DNA. RNA primers are removed by DNA polymerase I and the gaps are then filled by the same enzyme. Ligase acts behind the fork and joins Okasaki fragments together.

Ochoa, Severo (Luarca, Spain, September 24, 1905 - Madrid, Spain, November 1, 1993). Spanish - American biochemist. Nobel Prize for Medicine shared with **Arthur Kornberg** for discovery of an enzyme in bacteria that enabled him to synthesize a polynucleotide of ribonucleic acid. The first part of his scientific work dealt with the metabolism of energy-rich compounds, in particular acetyl - Coenzyme A, which enters the Krebs cycle. In 1951, with **Feodor Lynen**, he showed that acetyl - CoA was a major intermediate in the catabolism of sugars, lipids and several aminoacids. Nevertheless the work for which he received the Nobel Prize was the isolation of the enzyme polynucleotide phosphorylase from bacteria. This work was done in collaboration with the French biochemist Mariane Grundberg - Manago. They showed that this enzyme was able to catalyse the synthesis of polyribonucleotides from nucleoside diphosphates in vitro, a reaction that required short RNA primers but did not utilize a template. Moreover their experiences proved to be a most valuable tool in deciphering the genetic code. For example, poly U was synthetized by adding high concentrations of uridine diphosphate in the presence of polynucleotide phosphorylase. They also stimulated new research in the molecular biology of bacteria, eucaryotic cells and viruses. Severo Ochoa was educated at the University of Madrid, Spain, where he received his Ph.D. in 1929. After his medical studies he moved to Germany and studied biochemistry under **Otto Meyerhoff**. Following a stay at Oxford University he was finally named professor of pharmacology at New York University. He became an American citizen in 1956.

1960

Burnet, Franck McFarlane (Traralgon, Australia, September 3, 1899 - Port Fairy, Australia, August 31, 1985). Australian immunologist and virologist. Son of a banker. Nobel Prize for Medicine, along with **Peter Medawar**, for the discovery of acquired immunological tolerance to tissue transplants. At the begining of his career he was interested in bacteria and bacteriophages and proposed that the phage genome could integrate into the host genome and thus become an almost whole replicative entity. These ideas were later confirmed by **André Lwoff** in Paris. Later, Burnet studied immune mechanisms for combatting viral infection. He is credited for the demonstration that fever viruses could be grown in chick embryos. Today this technique is widely used for culturing viruses. During the course of this work, he made a surprising discovery. When antigens were administered to newly born chicks hatched from these eggs, they did not produce the expected antibodies. By contrast, adult chickens from the same eggs produced the relevant antibodies. Therefore, he put forward the hypothesis that specific antibody production against antigens was not hereditary but was acquired during embryonic development. According to his hypothesis, chick embryos infected with a virus, but which did not yet possess a complete immune system against the virus, were immunologically 'illiterate'. At this stage, they were unable to detect the viral antigens and so were tolerent towards infection. Burnet's work was also the basis for the theory of clonal selection.

Franck McFarlane Burnet

According to this theory, a wide variety of lymphocytes are present in organisms producing different antibodies, a given lymphocyte making a single specific antibody. This variety is such that whatever the invading antigen, there will always be at least one lymphocyte able to recognize the antigen and develop into a clone of cells able to produce antibodies

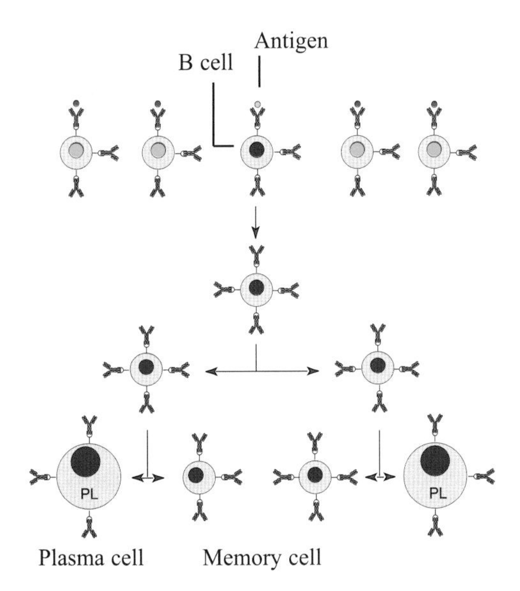

Clonal selection. By this process, an antigen selects the appropriate antibody from a preexisting pool of antibody molecules. This concept was developed by **Niels Jerne**

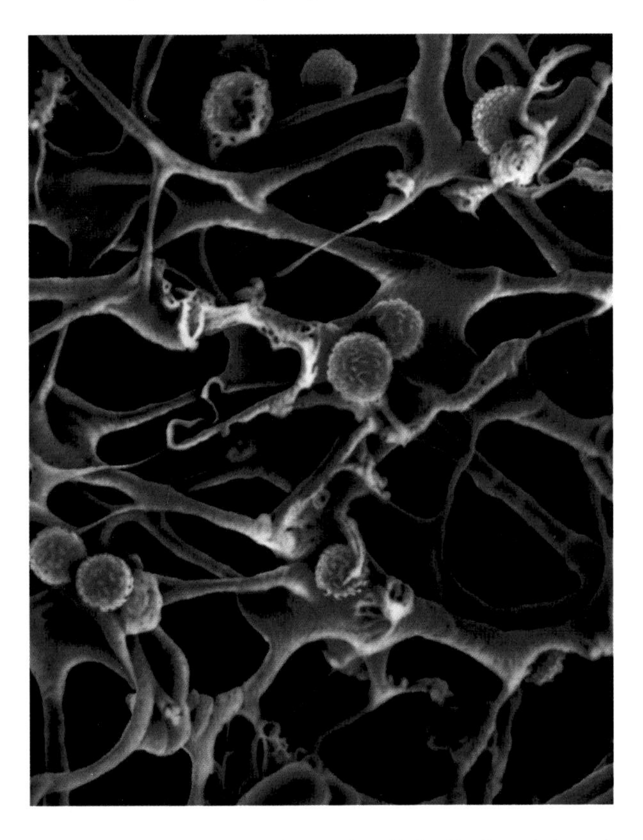

Lymph node. This type of organ is widely distributed throughout the lymphatic system of vertebrates. Several kind of cells take up residence in specialized regions of a lymph node's body, among them, B and T lymphocytes, macrophages and dendritic cells. Lymph nodes function as filters along the course of the lymph vessels. When a foreign material enters the body, specific immune cells can be stimulated.

against it. Franck McFarlane Burnet studied medicine in Melbourne and bacteriology in London. In 1944 he became Head of the Walter and Eliza Hall Institute in Sydney.

Medawar, Peter Bryan (Rio de Janeiro, Brazil, February 28, 1915 - London, England, October 2, 1987). British immunologist. Son of a Lebanese businessman. Nobel Prize for Medicine, along with **MacFarlane Burnet**, for the discovery of acquired immunological tolerance. After having worked under **Howard Florey** and conducting research on blood coagulation, Peter Medawar revealed a great interest for the study of transplant rejection. He noticed, among other things, that non - rejection of skin grafts depended on how closely related the donor and recipient were. The effect was most striking in the case of identical twins, but it was also observed when the donor and recipient of the graft were genetically similar individuals, such as mice belonging to a pure strain. Strangely, he observed that when a chick embryo received transplants from various adult chickens, it later developed into an individual capable of tolerating grafts from those same adults. Moreover he showed that the continuous presence of nonself antigens during the development of the immune system leads to an unresponsiveness to the specific nonself antigens. From this type of observation stemmed the

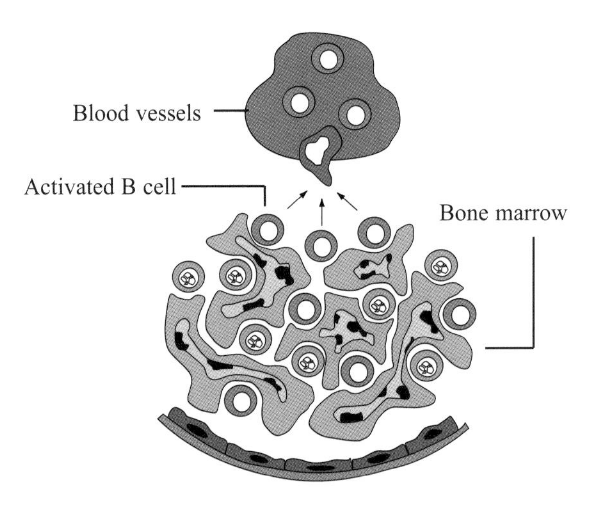

Selection of B lymphocytes in bone marrow. B cells are continually formed by the bone marrow throughout the life. Most of them survive only a few days. But if during their circulation B cells encounter an antigen to which their surface binds correctly, it is activated to growth, division, and the secretion of antibodies. Moreover, long - lived memory B cells are also produced.

notion of acquired specific immunotolerance. The idea was also supported by Burnet, who is seen, with the Danish immunologist **Niels Jerne**, as one of the "fathers" of the clonal selection theory. Burnet showed that the production of antigen-specific antibodies does not result from an inherited capacity of the organism but develops during embryogenesis. Peter Medawar taught zoology at

Birmingham University, then at Oxford University College. In 1962 he headed the Bethesda National Health Institute in the United States.

1961

Békésy, Georg Von (Budapest, Hungary, June 3, 1899 - Honolulu, Hawaii, June 13, 1972). American physicist and physiologist born in Hungary. Nobel Prize for Medicine for his discovery of the physical means by which sound is analyzed and communicated in the cochlea. Georg Von Békésy was the son of a Hungarian diplomat. He began his studies in Switzerland and then went to the University of Budapest where he studied physics. His particular interest was in the field of sound transmission. Von Békésy was appointed Director of the Hungarian State Telephone Company, a position he held for more than 20 years. Although he remained in Hungary during the Second World War, he emigrated to the USA during the Russian occupation. There he continued his research on hearing. He worked at Harvard University, where he was provided with all the necessary facilities to pursue his research work on the mechanism of hearing.

Georg Von Békésy

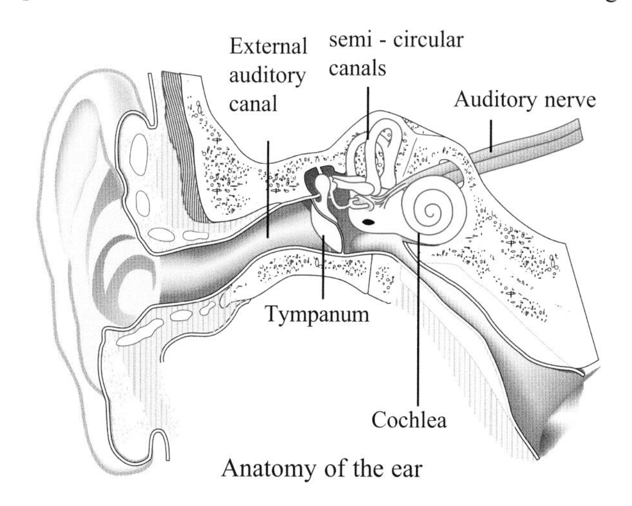

Anatomy of the ear

At that time, it was known that the hearing apparatus was located in a spiral tube called the cochlea, a structure in the inner ear divided into two parts by the basolateral membrane. The predominant theory on hearing at the time was that of Helmholtz who thought that some 24,000 hairs attached to this membrane were each tuned to a particular frequency, and that sound itself was a combination of pure sounds transmitted to the brain via these sensitive hairs. Von Békésy's discovery questioned this hypothesis when he showed using his models of the cochlea, along with miniaturized surgical instruments, that when sound crossed the fluid of the cochlear canal, a perturbation in the form of a wave was transmitted to cilia on the basolateral membrane. According to Georg Von Békésy, the shape of the wave varied according to the volume of sound as well as its quality. This information was then transmitted to the central nervous system.

1962

Crick, Francis Harry Compton (Northampton, England, June 8, 1916 -). British biophysicist and crystallographer. Nobel Prize for Physiology or Medicine, along with **James Watson** and **Maurice Wilkins**, for their determination of the molecular structure of deoxyribo-nucleic acid (DNA). Francis Crick's research originally concentrated on X - ray diffraction studies of proteins under the direction of **William Lawrence Bragg** (Nobel Prize for Physics in 1915). With three other scientists, Maurice Wilkins and the late Dr. Rosalyn Franklin at King's College, London, together with James Watson, he was responsible in 1953 for the discovery of the molecular structure of deoxyribonucleic acid (DNA), the biological structure which makes possible the transmission of inherited characteristics. In the course of two years of enthusiastic collaboration, he and Crick showed that it was possible to describe genes in molecular terms. These results were largely due to the research led by Avery before 1950, which showed that the genetic material of phages was DNA, as well as to the work of Todd, which established that DNA was made up of a chain of deoxyribose sugars held together by phosphate bonds and linked to organic bases. Moreover, using paper chromatography, the Austrian-American Erwin Chargaff had showed that the number of adenine units were equal to the number of thymine, and the number of units of guanine was equal to the number of cytosine units (Chargaff's rules). Furthermore, helix structures were beginning to gain recognition, especially in proteins, thanks to Pauling's work. Finally, the independent research of Wilkins and Franklin allowed them to analyze interesting negative photographic images obtained by DNA analysis by X-ray diffraction. From these data, Watson and Crick built their model of DNA. The central features of this model are that DNA is a double helix with the sugar-phosphate backbones on the outside of the molecule and the bases packaged on the inside and oriented in such a manner that hydrogen bonds are established between purines and pyrimidines. The article that they published in the review Nature noted, among other things, "It has not escaped our notice that the specific pairing we have postulated immediately suggests a possible copying mechanism for the genetic material." Aside from that, they showed how

DNA could assist in encoding proteins. Crick proposed the DNA-RNA-protein relationship, which would become the cornerstone of genetic expression.Subsequently he worked on problems connected with protein synthesis and the genetic code. In 1961, Francis Crick and **Sidney Brenner** discovered that groups of three nucleotide bases, which make up a codon, specify amino acids. **Marshall Niremberg**, Heinrich Mathaei, and **Severo Ochoa** determined what codon sequences specified each of the twenty

Cavendish laboratory. A major centre of scientific reasearch in England. A number of well-known scientists worked here, among them W.H. Bragg and his son W.L Bragg, both of them Nobel laureates for Physics for their work in the field of crystallography. It was also here that the discovery by Francis H.C. Crick and James D. Watson of the double helical structure of DNA was made.

amino acids. With his teammates he also originated the hypothesis of an adaptor, a nucleic acid molecule known today as tRNA that acts as an intermediary between messenger RNA and amino acids during protein synthesis.

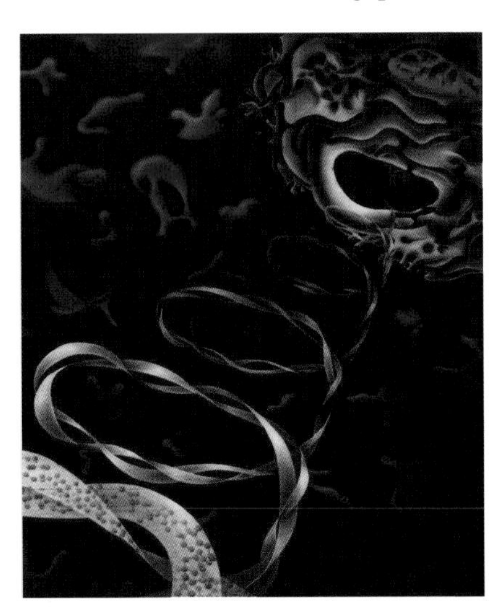

Artistic drawing showing that DNA is located in the cell nucleus.

After 1966 his interest turned to development biology and chromatin structure. Since joining The Salk Institute in 1976, Crick's work has been entirely theoretical. He hopes

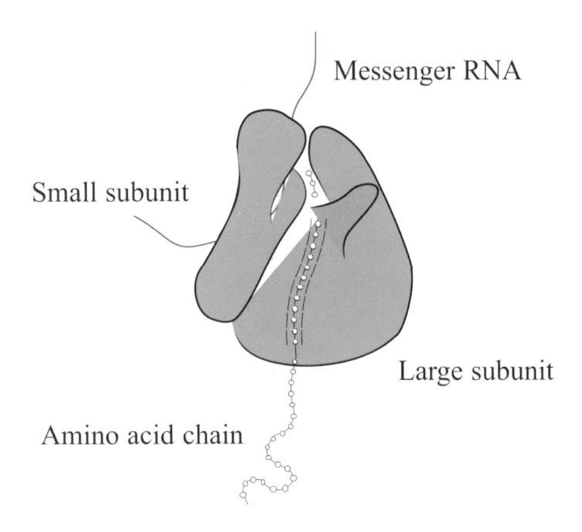

Ribosome, messenger RNA and polypeptide. As shown, mRNA (in red) serves as an intermediate between the information provided by DNA and the resulting amino acid chain. In 1961, Francis Crick and Sidney Brenner discovered that amino acids were in reality specified by one of a group of three nucleotide bases, which make up codons.

to bring together the molecular and cellular aspects of neurons, the observations of neuroanatomy and neurophysiology and the behaviour of organisms as studied by psychologists. His work has been mainly in neurobiology, and especially in the visual system of mammals. He has also written on the neural basis of attention, rapid eye movement sleep (REM sleep), and visual awareness. His current research is focused on discovering the neural correlate of consciousness. In 1992, Francis H.C. Crick

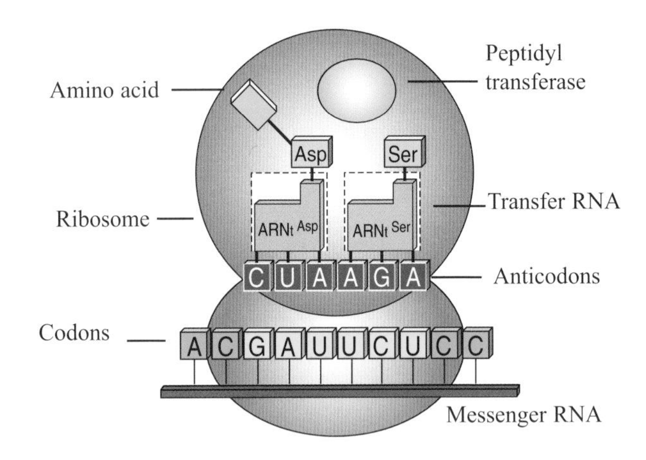

Codons and anticodons. A ribosome has binding sites that ensure mRNA and the tRNA-amino acid complexes will be brought together. The binding site for mRNA is located on the smaller subunit and is arranged in such a way that its nucleotide sequence (codons) is brought into direct contact with anticodons of tRNAs. This arrangement allows peptide bond formation between amino acids that are incorporated in the growing polypeptide chain.

received the Order of Merit from the Queen of England. This is a lifetime honor given to only 25 living English people at any one time. Francis Harry Compton Crick was educated at Mill Hill School and University College, London, majoring in physics. He worked with the British Admiralty for seven years during and after World War II where he was a scientific civil servant working mainly on the design of such devices as magnetic and acoustic mines. In 1947, he turned to biological research, going first to the Strangeways Laboratory, Cambridge, as a student. In 1949, he joined the Medical Research Coucil Unit in the Cavendish Laboratory at Cambridge University as a Laboratory Scientist. He received a Ph.D. in 1954.

Watson, James Dewey (Chicago, Germany, April 6, 1928 -). American molecular biologist. Son of a businessman. Nobel Prize for Medicine along with **Francis Crick** and **Maurice Wilkins** for discovering the structure of deoxyribonucleic acid. In 1953, James Dewey Watson, with Francis Crick, successfullly proposed the double helical structure of DNA, a feat described by **Peter Medawar** as "the greatest achievement of science in the twentieth century". When they proposed their model of the hereditory molecule, many researchers immediately recognized that the DNA molecule was admirably suited to a genetic role. Watson and Crick pointed out the central features of the DNA molecule that could explain processes as replication, a fundamental property of genetic material. In their joint paper, published in "Nature" in 1953, Watson and Crick stated that:"it has not escaped our notice that the specific pairing we have postulated immediately suggests a possible copying mechanism for the genetic material". According to their hypothesis, DNA molecules could be replicated precisely through the synthesis of new complementary strands for each of the parental strands in the double helix structure. It is now well established that DNA is the molecule of heredity in nearly all organisms, from bacteria and viruses to mammals and human beings. James Dewey Watson, who was first interested in bird watching, graduated in Zoology from the University of Indiana in Bloomington and earned his Ph.D. in zoology in 1950. It was **Salvador Luria** who advised Watson to go to Cambridge, England, where he began his first discussions with Crick on the structure of DNA. Without tarnishing the glory of Watson and Crick, credit is also due to other researchers, who obtained X - ray diffraction images of DNA, among them, Rosalyn Franklin and **Maurice Wilkins** of King's College, London, who succeeded in obtaining much sharper X - ray diffraction patterns than had previously been obtainable. On another part of DNA, the specificity of base pairing (adenine - thymine and guanine - cytosine) accounts for the earlier results of Erwin Chargaff, who established the base composition of various DNA molecules and discovered that the amount of guanine was always equal to that of cytosine, and the amount of adenine was equal to that of thymine. While a Professor at Harvard, Watson commenced a writing career that generated the seminal text, Molecular Biology of the Gene, and the best-selling autobiographical volume, The Double Helix. Later, while leading the Cold Spring Harbor Laboratory, he was a driving force behind setting up the Human Genome Project, a major factor in his receipt in 1993 of the Copley Medal from the Royal Society that elected him a member in 1981.

James D. Watson

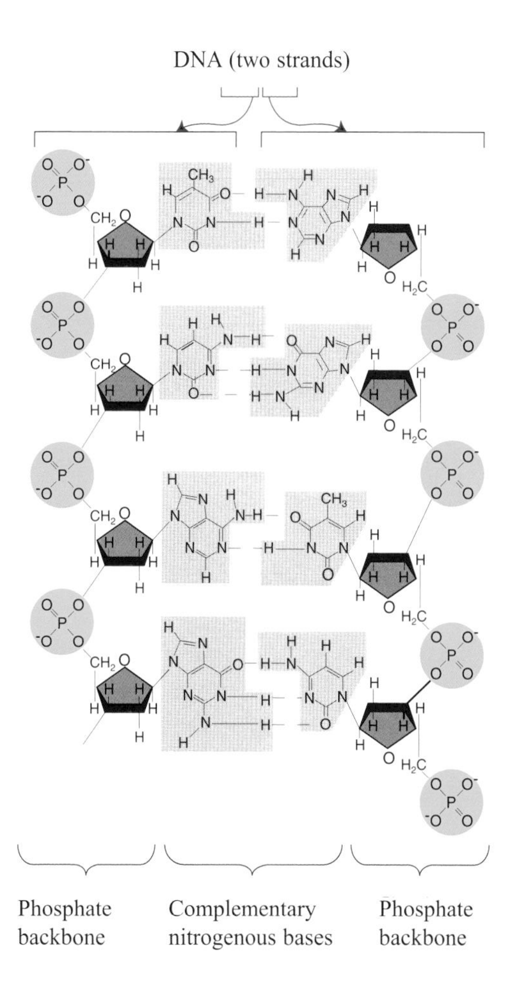

DNA double helix. The five nitrogenous bases found in nucleic acids and two of the five nucleotides they form when united with a phosphate group and a pentose sugar. The joining together of many different nucleotides makes a DNA double helix.

Wilkins, Maurice Hugh Frederick (Pongaroa, New Zealand, December 4, 1916 -). British physicist and crystallographer. Nobel Prize for Medicine along with **James**

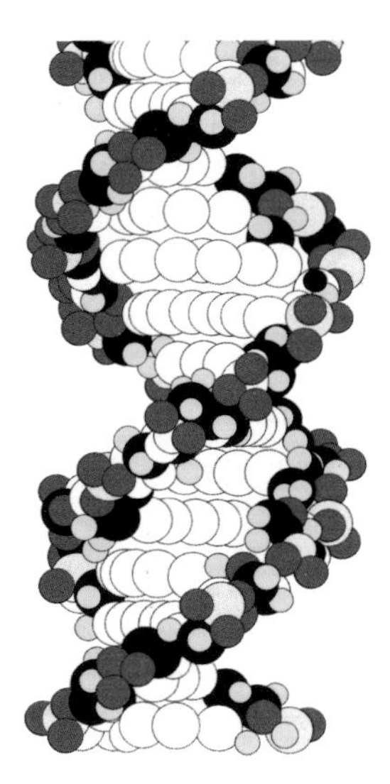

DNA double helix.

Watson and **Francis Crick** for X-ray diffraction studies of deoxyribonucleic acid. During World War II, Maurice Wilkins was assigned to the Manhattan Project and development of the atomic bomb. Yet frightened by the impli-

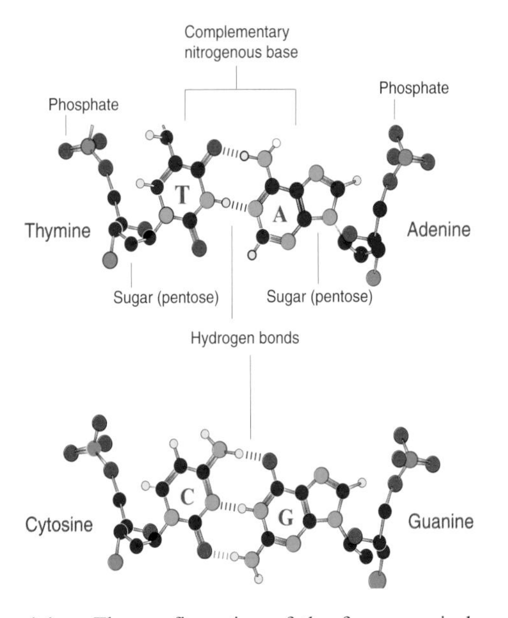

Base pairing. The configuration of the four organic bases in DNA is such that two hydrogen bonds are formed between adenine and thymine, whereas three are formed between guanine and cytosine.

cations and consequences of this project's success, he turned to Biochemistry and set out to apply his knowledge of physics to making a contribution in this field. This is when he learned that the nature of genes had just been discovered notably thanks to Avery's research and that genes were made of deoxyribonucleic acid. By chance, it appeared that DNA fibers could be easily studied by X - ray diffraction analysis. It so happened that Wilkins had become one of the great specialists in this technique. Through a trip to Naples, Italy, Wilkins met Watson and showed him X-ray images of DNA preparations. Watson decided then to join the Cambridge Laboratory, England, to learn about X - ray crystallography and study the structure of nucleic acid, especially DNA. It was at Cambridge that he and **Crick** proposed their model of the double helical structure of the molecule of heredity. It is a well-known fact that Wilkins and Rosalyn Franklin brought an important contribution in that field. By X - ray diffraction analysis, Franklin obtained excellent pictures of various crystalline forms of DNA. Furthemore, she showed that the phosphate groups of DNA are located at the outside of the double helix. Maurice H.F. Wilkins graduated in physics from King Edward's School in Birmingham and from St. John's College in Cambridge.

Maurice Wilkins

1963

Eccles, Sir John Carew (Melbourne, Australia, January 27, 1903 - Contra, Switzerland, May 2, 1997). Australian physiologist. Nobel Prize for Medicine along with **Alan Hodgkin** and **Andrew Huxley** for his discovery of the chemical means by which impulses are communicated by nerve cells. John Eccles came from a family of Australian teachers. He is known for his work on the understanding of the mechanism by which nervous impulses are transmitted from one neurone, or nerve cell, to another or to a different cell type. In particular, he studied how one neurone could display an inhibitory effect on the activity of an adjacent cell with which it is in contact. Eccles believed wrongly at that time, that this transmission was purely electric in nature. Sometime later, **Bernard Katz** showed that the transmission of nervous impulses and especially the transmission of inhibitory effects, required the secretion of a chemical compound. Based on the discovery of Katz, Eccles forwarded the hypothesis that the molecule involved in nerve stimulation at the neuro-muscular junction must induce membrane depolarization of the muscle cell due to the extrusion of ions. Towards the 1950s, his experiments with microelectroded glass proved the involvement of a chemical mediator in nerve transmission across synapses. Eccles also established the existence of post-synaptic channels in the muscle membranes of the motor plate, a micro-anatomical structure he had discovered in the 1930s in collaboration with the famous British physiologist, Charles Sherrington. Eccles is considered as the forefather of cell neurophysiology and was knighted in 1958. John Eccles was educated in biology and medicine at the University of Melbourne in 1925. After working in diverse areas of science, he taught physiology at the

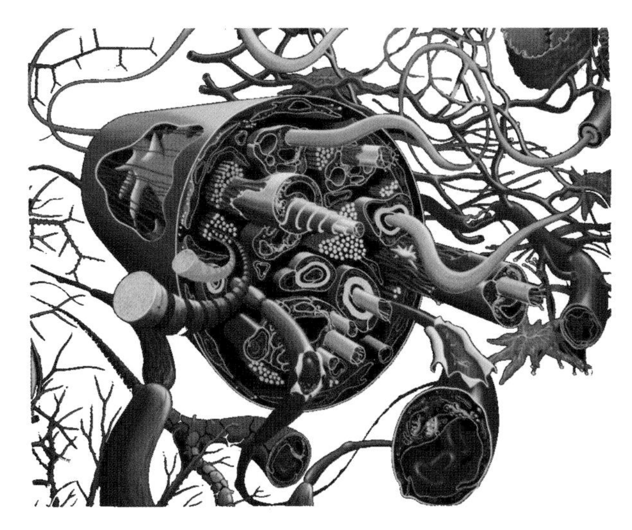

Cut through a nerve. A nerve contains many nerve fibers. Some of them are wrapped with myelin ; others are not. Nerves also contain blood vessels and connective tissue (collagen).

Australian National University from 1951 to 1966. He then moved to the State University of New York, Buffalo, and conducted research there from 1968 to 1975.

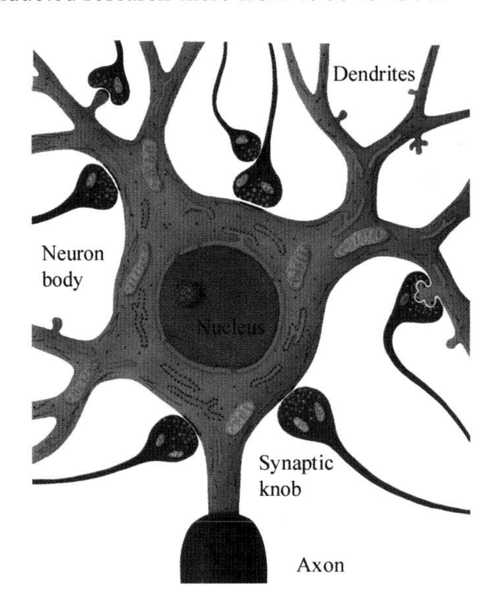

Neuron body and synapses.

Hodgkin, Alan Lloyd (Banbury, England, February 5, 1914 - Shipston-on-Tour, July 29, 1994). British physiologist and biophysicist. Nobel Prize for Medi-cine, shared with **Andrew Fielding Huxley** and **John Eccles** for the discovery of the chemical processes responsible for the passage of impulses along individual nerve fibres. Using the squid giant axon, Alan Hodgkin with Huxley and **Bernard Katz**, demonstrated that the transmission of the nerve impulse along the fiber required the unequal exchange of sodium ions for ions of potassium at the cytoplasmic membrane. These exchanges resulted in the propagation along the length of the nerve of a membrane potential larger than that measured in the system at rest. Consequently, the transmisson of a nerve impulse depends upon the propagation of an ionic potenetial along the nerve membrane. Hodgkin showed that the ion exchanges occur through the action of a "pump" (sodium - potassium pump) present in the nerve fiber membranes. Alan Hodgkin received his education at Trinity College, Cambridge. After World War II, he joined the University of Cambridge and conducted research on the transmission of the nerve impulse along individual nerve fibres. In 1970, he was appointed Professor of Biophysics at Cambridge and some time later Chancellor of the University of Leicester. Alan Hodgkin was knighted in 1972 and the following year was appointed to the Order of Merit, a British honour.

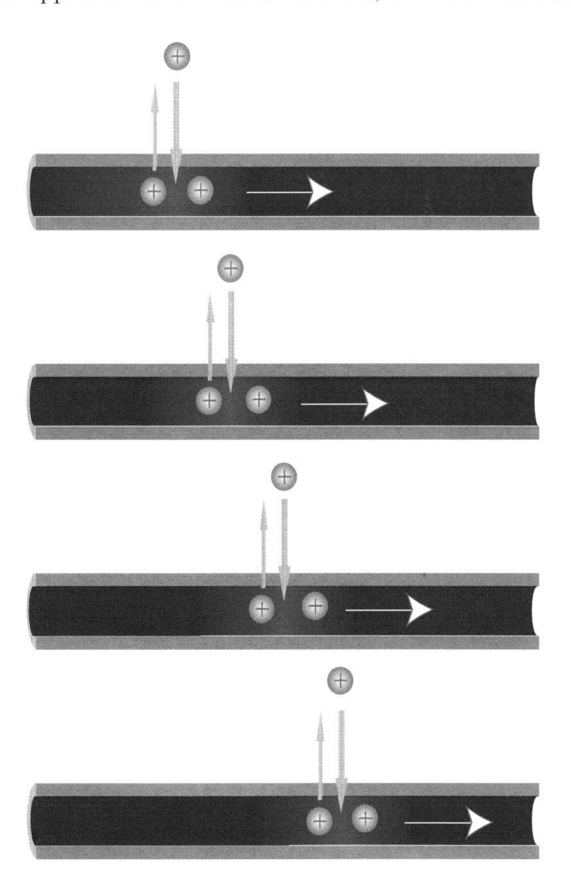

Nerve impulse. Nerve impulse that passes along a fiber, an axon or dendrite, involves both chemical and electrical changes. As the impulse passes along the plasma membrane, the electrical charges of this membrane are reversed. These changes affect adjacent parts of the fiber, so the impulse moves along.

He also served as President of the Royal Society from 1970 to 1975. Hodgkin married the daughter of the Nobel laureate Peyton Rous.

Huxley, Andrew Fielding (London, England, November 22, 1917-). Nobel Prize for Medicine, shared with **Alan Hodgkin** and **John Carew Eccles**. His research centred on nerve and muscle fibres and dealt particularly with the chemical phenomena involved in the transmission of nerve impulses. Andrew Fielding Huxley came from a well-known family of scientists. His grandfather was Thomas Henri Huxley, a British biologist who was a vigourous defender of the Darwinian theory of evolution. He started his career at the Plymouth Marine Laboratory, where he met the biophysicist Alan Hodgkin. They showed that the transmission of the nerve influx was accompanied by the exchange of potassium ions and sodium ions through a membrane. Huxley and Hodgkin established that the message carried by a nerve fibre consists of a sequence of electrical impulses, each of which lasts about one thousandth of a second. The work for which Huxley and Hodgkin received the Nobel Prize was precisely a combination of experiments in which they measured the currents carried into or out of a nerve fibre by ions of sodium and potassium, by which they showed that these currents would account well for the characteristics of the impulse, such as its shape and the speed it travels along the fibre. For their experiments, they used the very large nerve fibres found in squid. We are also indebted to Huxley for important works on muscle contraction as well as for the development of the ultramicrotome and the differential interference - contrast microscopy. Huxley taught experimental biophysics at the University of Cambridge and held a post as a professor at the University College,

Andrew F. Huxley

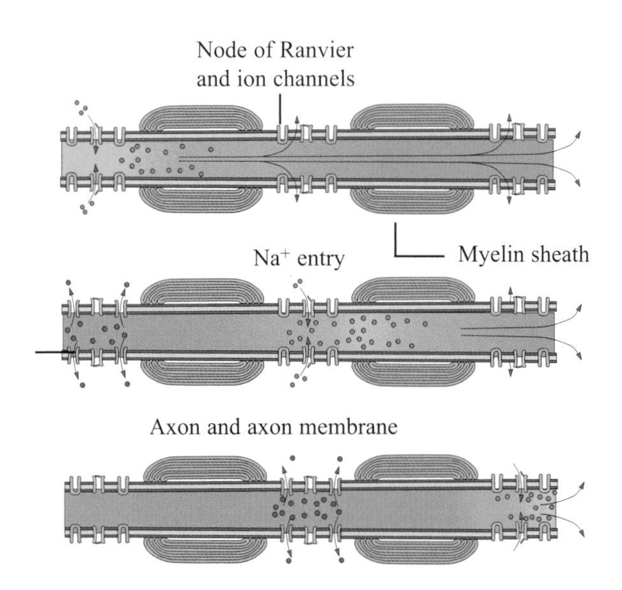

Saltatory conduction. The myelin sheath acts as an insulator so that the local current propagating the action potential " jumps " between nodes.

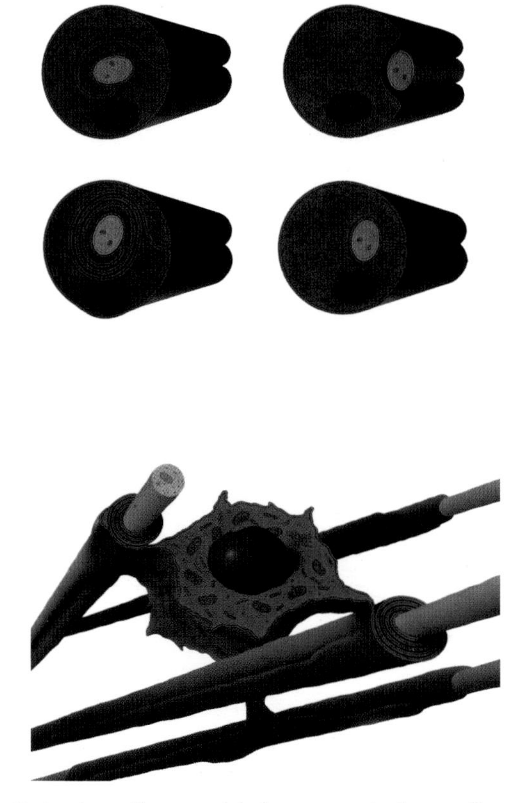

Myelinization. Above and below respectively, myelin - filled Schwann cells form a spiral around processes of many nerve cells in the peripheral and in the central nerve system. Vertebrates evolved myelinated fibers as an alternative adaptation for increasing conduction speed.

London, from 1960 until 1969. In 1980, he succeeded Todd as President of the Royal Society. He was knighted in 1974.

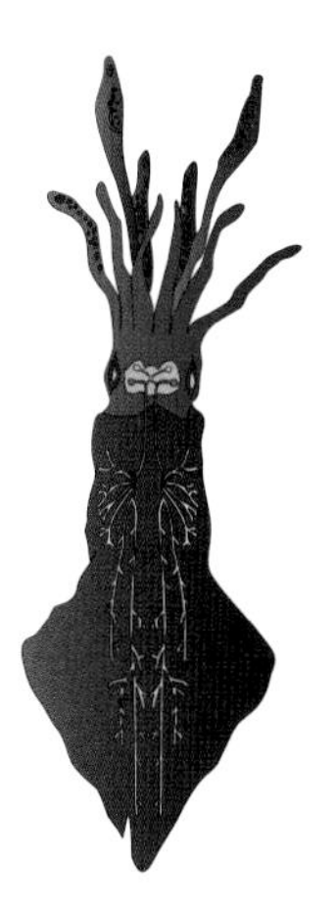

The squid. Their giant axons may be as much as a millimeter in diameter. A.L.Hodgkin and A.F. Huxley succeeded in inserting a very thin microelectrode in such fibers and in establishing a relation between membrane ionic change and the passage of a nerve impulse.

1964

Bloch, Konrad Emil (Neisse, Germany, January 21, 1912 -) German - born American biochemist. Nobel Prize for Medicine, shared with **Feodor Lynen**, for his discoveries concerning the natural synthesis of cholesterol. Because of his Jewish origin, Konrad Bloch had to flee from Germany in 1934. At first he immi grated to Switzerland; then, in 1936 he settled down in the USA and was granted the American citizenship in 1944. At Columbia University in New York, Bloch met David Rittenberg and Rudolf Schoenheimer, the inventors of a procedure using radioactive labelling that made it possible to trace the fate of different compounds in cells. In collaboration with Rittenberg, Bloch, using C14 labelled acetate, discovered that this compound, central to intermediary metabolism, constituted the basic element furnishing the 27 carbon atoms in a molecule of cholesterol. An impressive series of experiments allowed him to identify most of the 36 steps in the biosynthesis of the molecule. Bloch evidenced the hinge elements of this complex process. He showed that squalene is a key intermediate deriving from mevalonic acid (converted into an isoprenoid structure) and forming cholesterol *via* another intermediate, lanosterol. Other studies carried out by Bloch established the important role of cholesterol in normal cell physiology, in particular, its stabilizing effect in membranes and its role as a precursor in the biosynthesis of substances such as the sex hormones and other corticosteriods, and in the synthesis of bile acids. These studies also suggested a possible relation between excess levels of cholesterol and other fats in blood and the origin of the characteristic vascular fat deposit in arteriosclerosis. In 1936, Konrad Emil Bloch left Switzerland for the United States where he earned his Ph.D. in medicine at the College of Physicians and Surgeons, Columbia, in 1938. From 1946 to 1954 he taught biochemistry at the University of Chicago. He was later appointed Professor of Biochemistry in the Department of Chemistry, Harvard University.

Konrad E. Bloch

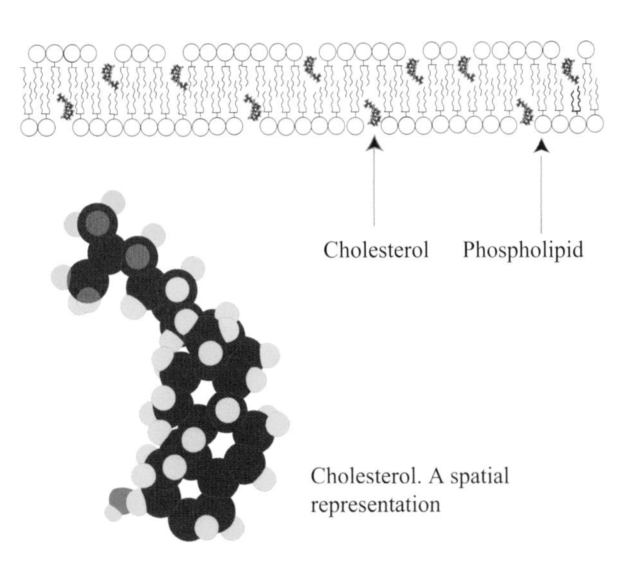

Cholesterol and membrane. Cholesterol inserts into the lipid bilayer of the membrane. Its polar group is oriented towards the polar heads of phospholipids.

Lynen, Feodor (Munich, Germany, April 6, 1911- August 8, 1979, Munich, Germany, 1979). German biochemist. Son of a professor. Nobel Prize for Medicine, along with **Konrad Bloch**, for his research on the metabolism of cholesterol and fatty acids. Lynen was educated at the University of Munich. Feodor Lynen prepared his doctorate under Professor **Heinrich Wieland**,(1927 Nobel Prize for Chemistry) who had been awarded the Nobel Prize for Chemistry and who became his father-in-law. It was Wieland who stimulated his interest for the metabolism of acetic acid. In 1942 Konrad Bloch had shown that this compound was a major building block in cholesterol biosynthesis. Lynen demonstrated that the usable form of acetate, a 2-carbon compound precursor of this molecule, was acetyl-coenzyme A, a complex resulting from its

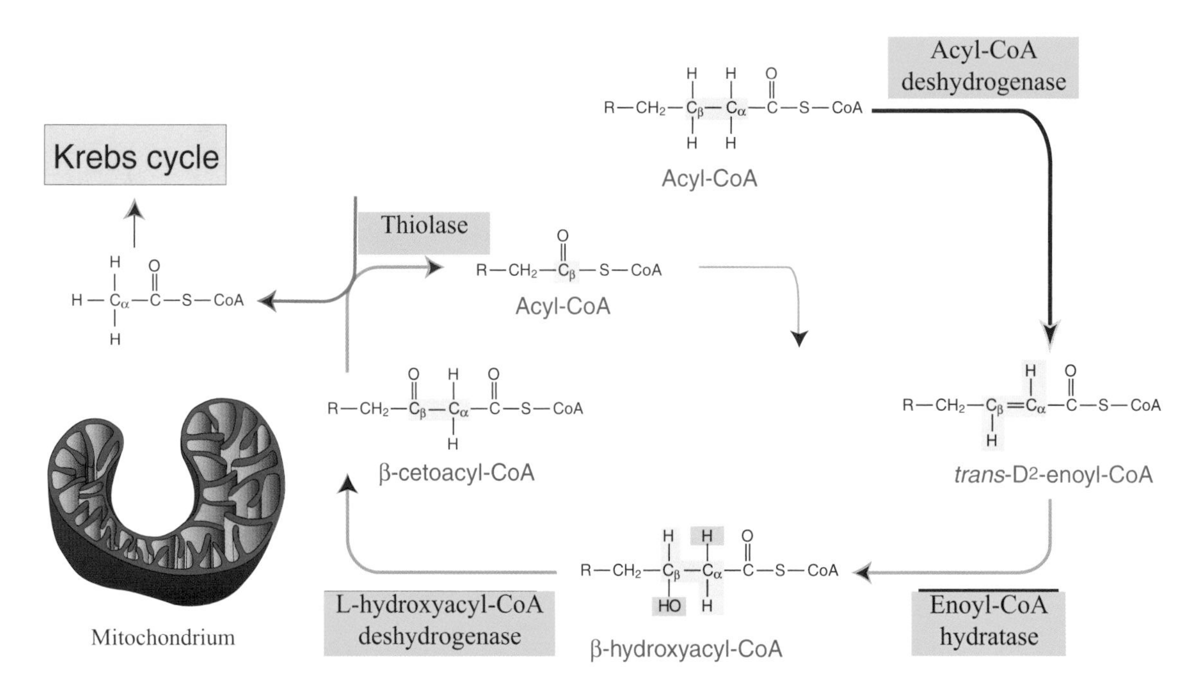

Beta - oxidation. A process that removes two carbon units at a time from linear fatty acid until breakdown is complete. It was elucidated by Feodor Lynen and Franz Knoop.

association with coenzyme A. Coenzyme A was discovered by **Fritz Lipmann**, another German biochemist. Independently of Konrad Bloch, he elucidated several steps leading to the formation of squalene, a major intermediary in cholesterol biosynthe-sis. Nevertheless, Lynen's best known work is the description of a biochemical pathway called the fatty acid cycle or Lynen cycle. During this process, fatty acid chains are sequentially broken down to yield acetate molecules which enter the Krebs cycle. Feodor Lynen showed that activated acetic acid (acetyl-coenzyme A) was the central crossroads of the metabolism. Indeed, not only does this small intermediary

This artwork represents a lipid globule inside a cell. At upper left , a molecule of acetyl coenzyme A. At upper right, a molecule of acetic acid .

Mitochondrion shown in longitudinal section. Mitochondrion is covered by an outer membrane and lined by an inner membrane that is infolded into numerous partitions, called cristae. Two main functions of mitochondria (cell respiration and lipid synthesis) remained unknown until the mid-1930s. Here the mitochondrion is associated with a lipid globule.

result from the breakdown of sugars, lipids and some amino acids, but it is also a key intermediary in fatty acid synthesis as well as in other anabolic reactions. Lynen taught biochemistry at the Cellular Chemistry Institute of Munich, which would later become the Max Planck Institute of Munich.

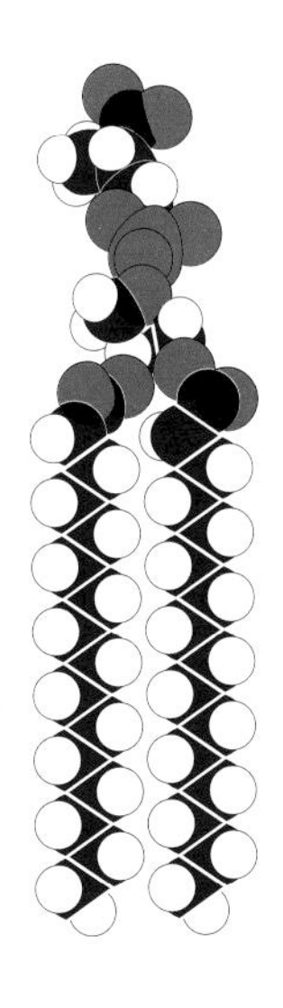

Phospholipid. A compact molecular model.

1965

Jacob, Francois (Paris, France, June 17, 1920 -). Physician and molecular biologist. Nobel Prize in Medicine, shared with **Jacques Monod** and **André Lwoff**, "for his discoveries concerning the mechanisms of genetic control of the synthesis of bacterial enzymes". Admitted to the Pasteur Institute in 1950, he carried out his first research under the direction of André Lwoff. In 1964, he was made a professor in the Collège de France where a chair of genetics was created for him. Francois Jacob made important contributions to the genetics of bacteria and their viruses, the bacteriophages, and to an understanding of the biochemical effects of mutations. His research concerned, among other subjects, the properties of lysogenic bacteria and, in collaboration with Elie Wollman, the analysis of genetic recombination in bacterial conjugation. In this phenomenon discovered by **Joshua Lederberg**, DNA of "male" bacteria is transferred to "female" bacteria and recombines with the recipient cell chromosome. The work for which Jacob received the Nobel Prize was carried out in collaboration with **Jacques Monod** and led to the discovery that the activity of the lactose operon of the bacterium *Escherichia coli* was regulated by a specific gene coding for a "repressor" protein. The lac repressor was shown to determine the level of transcription of the genes coding for enzymes involved in lactose metabolism. The concept of genes (regulatory genes) whose function is to control the expression of other genes (structural genes) was an entirely new concept and represents one of the most fundamental contributions to our knowledge of living systems. Jacob (in collaboration with **Sidney Brenner** and François Cuzin) is also known for the concept of a replicon, defined at the time as a unit of replication (a plasmid, for example) that possesses an origin of replication and a gene that controls replication.

François Jacob

Lwoff, André Michael (Ainy-le-Château, France, May 8, 1902 - Paris, France, September 30, 1994). French biologist. Professor at The College de France and member of the Pasteur Institute in Paris. Nobel Prize for Medicine, shared with **Jacques Monod** and **François Jacob**, for his works on lysogeny. André Lwoff, who was to become one of the great figures of the Pasteur Institute, became known in the scientific community, in large part, for his research on the phenomenon of lysogeny. In lysogeny, the DNA of

Plasmid. A coloured scanning electron micrograph.

a temperate bacteriophage is integrated in the bacterial chromosome and is replicated passively by the bacterial host replication system. In this state (prophage state) most of the genes of the virus are not expressed. Lwoff showed that induction of viral development ocurred spontaneously in a small number of cells in a population of lysogenic bacteria. He also demonstrated that upon exposure to

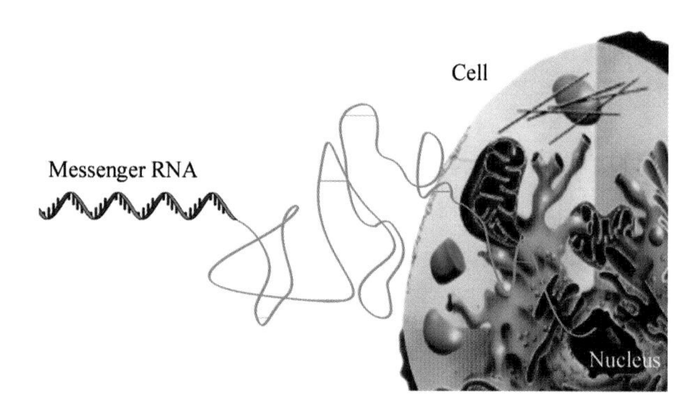

Messenger RNA. François Jacob and Jacques Monod postulated the existence of this important intermediary of protein synthesis. Messenger RNA carries the information embodied in the genes of DNA to ribsomes, where they direct protein synthesis. This information is carried in the form of a succession of codons.

ultraviolet light, viral development could be induced in essentially the entire population of lysogenic bacteria. Thus, in lysogeny, the quasi totality of the bacterial population derived from cells infected by a temperate bacteriophage harbor the virus (viral genome) in the prophage state. Another important contribution of Lwoff was the link that he established between vitamins and coenzymes. Although one of the pioneers in molecular biology, he was initially a protistologist whose first work concerned the group of ciliated infusoria. In the 1920s, he showed that certain characters of these organisms, notably those in connection with the ciliated envelope, appeared to be controlled by genes located ouside of the nucleus. Lwoff also made important contributions to the understanding of the relation between vitamins and coenzymes. Although a pioneer in molecular biology, he was initially a protistologist, his first work concernimg a group of ciliated infusoria. In the 1920s, he showed that certain characters, in particular, those related to the distribution of cilia, appeared to be controlled by extranuclear genes. André Lwoff, born of Russian-Polish parents, graduated from the University of Paris. He then taught microbiology and molecular biology at the Pasteur Institute and at the Sorbonne. From 1968 to 1972 he headed the Cancer Research Institute at Villejuif.

Didinium phagocyting paramecium. André Lwoff devoted the first part of his scientific research to the study of these one-celled animals.

Monod, Jacques Lucien (Paris, France, February 9, 1910 - Cannes, France, May 31, 1976). French biologist. Nobel Prize, jointly with **François Jacob** and **André Lwoff** for discoveries concerning the genetic control of enzyme synthesis. Monod is famous for work that showed that certain bacterial genes were under the control of a genetic system that allowed expression only when needed. The genes

Amoeba. This common protozoa living in fresh water was also studied by A. Lwoff.

studied were involved in lactose metabolism in E. coli. He showed, in collaboration with **Jacob**, that the genes coding for enzymes (structural genes) related to lactose metabolism were present as a functional unit that was named an operon. In the case of lactose metabolism, the structural genes occupy adjacent positions in the bacterial chromosome and are preceeded by a gene (regulatory gene) that controls the level of expression of the operon. The regulatory gene codes for a protein called the repressor which, in the absence of lactose, binds strongly to

Jacques Monod

sequences in the operon promoter region. The binding of the repressor to these sequences blocks expression (transcription) of the operon genes. In the presence of small molecules that can act as inducers, a repressor - inducer

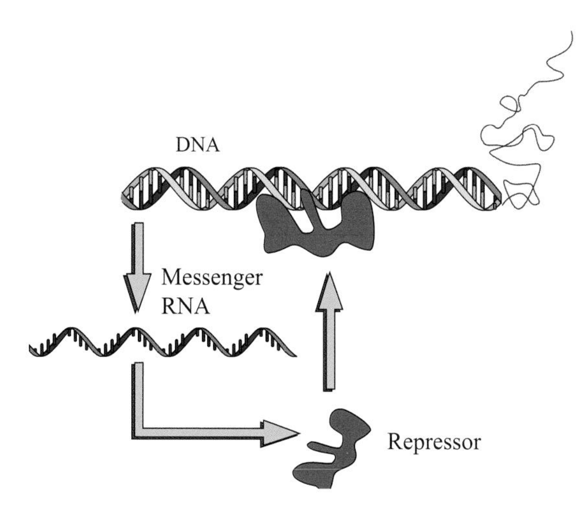

Lac operon. Discovered by **Jacques Monod** and **François Jacob**, it contains several genes involved in the induction of several enzymes that metabolize lactose (structural genes) as well as genes that control this metabolism. Among them are an operator gene which controls the rate of transcription of the structural genes, a promotor gene, which provides a signal for the beginning of transcription of the structural genes, and a regulator gene, which codes a specific repressor controlling the action of the operator gene.

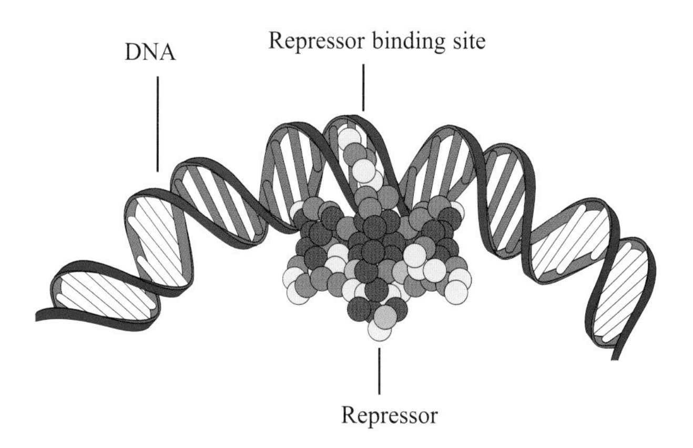

The repressor. This molecule specifically prevents the transcription of genes under control of an operator. When it binds to the operator, transcription cannot occur. The protein was initially isolated by **Walter Gilbert**, Nobel Prize for Chemistry in 1980.

complex forms that can no longer bind to sequences in the promoter, thus allowing RNA polymerase to bind to the promoter and initiate transcription of the operon. The discovery of regulatory genes represents one of the most important contributions to biology in the 20th century. We also owe to Monod and his group their contributions to the concept of messenger RNA. Still another contribution of Monod's was his work on allostery, a phenomenon in which the binding of a molecule to a specific site in a protein induces a change in the conformation of the protein. The formation of such complexes can lead to changes

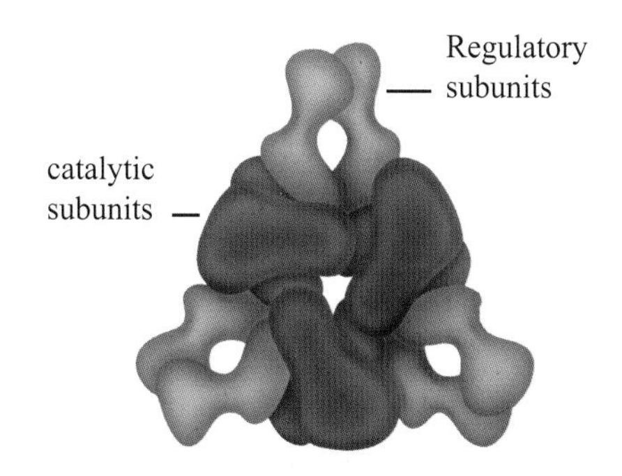

Aspartate transcarbamylase, an enzyme composed of several subunits. This enzyme is subject to allosteric regulation, a process by which it can modulate its configuration and its activity in response to its binding to a variety of small regulator molecules.The concept of allostery was developped by J. Monod, P.Changeux and F. Jacob.

in the activity of another site in the protein. Monod's book "Le Hasard et la Nécessité", his sole publication of scientific vulgarization, received wide public recognition and was the source of much discussion and controversy. After having been a member of the resistance during the German occupation, he joined the group of Lwoff at the Pasteur Institute. From 1967 on, he taught at the College de France. He was named director of the Pasteur Institute in 1967, a post that he occupied until his death.

1966

Huggins, Charles Brenton (Halifax, Nova Scotia, Canada, September 22, 1901- Chicago, Illinois, January 12, 1997). Canadian-born American surgeon and urologist. Nobel Prize for Medicine, along with **Francis Peyton Rous**, for his research demonstrating the relationship between hormones and certain types of cancer. Huggins gained certain significant results in the treatment of prostate (prostatic gland) cancer by means of testicle ablation. He had observed around 1939 that the prostate gland was controlled by androgens produced by the testicles. He concluded that it should be possible to treat cancer of the prostate by eliminating male hormones. This type of treatment was replaced with oestrogen administration, the oestrogen stilbestrol being able to neutralise the effect of testicular androgens. Although still rather low, the percentage of cures obtained demonstrated the relationship between hormones and cancer development. Charles Huggins studied medicine at Acadia University, Wolfville, Nova Scotia, Canada, and earned his Ph.D. from Harvard in 1924. He then worked at the University of Michigan where he was Instructor in Surgery and successively Assistant Professor and Professor of Surgery. In 1951 he founded the Ben May Institute for Cancer Research, Chicago, in order to understand the basis for hormonal regulation of cancer growth and differentiation. He headed this institute from 1951 to 1969.

Rous, Francis Peyton (Baltimore, Maryland, October 5, 1879 - New York City, New York, February 16, 1970). American pathologist and physiologist. Nobel Prize for Medicine shared with **Charles Huggins** for the discovery of cancer-inducing viruses. In studying a malignant tumor of chicken connective tissue, Francis Rous observed that a filtrate of disrupted tumor cells was capable of inducing formation of the same tumor if innoculated in healthy tissues. Since the filter used did not permit the passage of cells or bacteria, it appeared that the active tumorigenic agent could not be of a size larger than a virus. That observa-tion made in 1911, thus suggested that a virus could induce tumor formation. The hypothesis proposed by Rous was not immediately accepted by the scientific community for the symptoms observed by Rous in the chicken were seemingly quite different from those generally described in the cases of viral infections such as grippe or measles. It was only at the end of the 1920s that a certain number of animal tumors were accepted as being of viral origin. Curiously, it was not until 55 years after the publication of his first results that Rous was awarded the Nobel prize. Rous received his medical education from Johns Hopkins University in Baltimore, and from the University of Michigan. Most of his research was conducted at the Rockefeller Institute for Medical Research in New York City in 1909. He remained there until his retirement.

Francis Rous

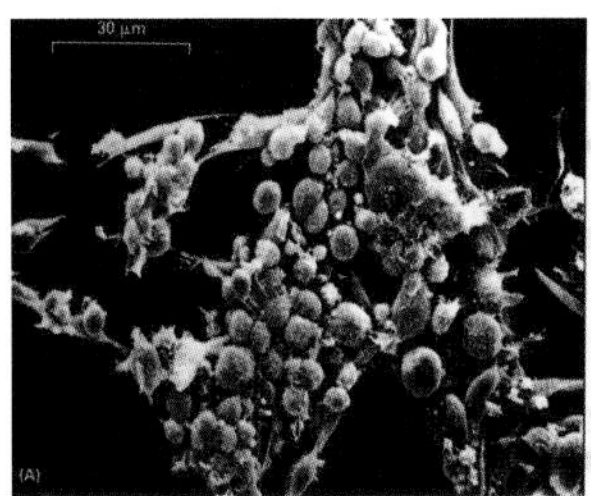

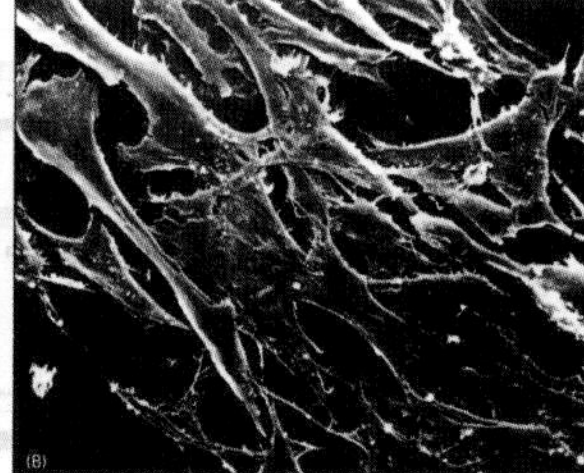

Rous sarcoma virus. This electron micrograph shows the effects of tumor - causing viruses on cultured cells. At right, cells infected by the viruses are said to be transformed and they continue to divide indefinitely.

1967

Granit, Ragnar (Helsinki, Finland, October 30, 1900 - March 12, 1991). Finnish neurophysiologist. Nobel Prize for Medicine, along with **George Wald** and **Haldan Hartline**, for his analysis of the internal electrical changes that take place when the eye is exposed to light. Ragnar Granit was a teacher of physiology at Helsenki University. He then taught neurophysiology at the Karolinska Institute in Stockholm. Soon after studying experimental physiology at the University of Helsinki, Granit turned to medicine. His particular interest was in ocular physiology. While at Oxford University he worked under the famous physiologist **Charles Sherrington** and there he learned techniques of neurophysiology. Later, at the University of Pennsylvania, he met Hartline and Wald, two specialists of eye neurophysiology. Granit revealed that colour vision depends on more than just the existence of three types of cone cells: certain ocular nerve fibres are sensitive to a broad spectrum of colours, whereas others are sensitive to a much narrower spectrum and are thus specific to a given colour. He also showed that light, although it stimulates some nerve fibres, can inhibit others. This discovery had considerable impact on our understanding of how the retina processes light and on the perception of contrast. Ragnar Granit graduated in medicine from the University of Helsinki in 1927, after which he conducted research at the University of Pennsylvania. He also conducted research under Sherrington at Oxford, in England. He was appointed Professor of Physiology at the University of Helsinki in 1937 and became Head of the Department of Physiology at the Karolinska Institute in Sweden until the end of his career.

Hartline, Haldan Keffer (Bloomsburg, Pennsylvania, December 22, 1903 - Fallston, Maryland, March 17, 1983). American physiologist. Nobel Prize for Medicine, shared with **George Wald** and **Ragnar Granit**, for his work in analyzing the neurophysiological mechanisms of vision. Hartline developed techniques that permitted the isolation of the sensory nerve fibers that propagate nerve impulses initiated in individual cells of the retina. These initial studies were carried out on the crab *polyphemus,* an animal with compound eyes. Each of the photoreceptor cells of this crab is connected to a nerve fiber, the ensemble of these fibers forming the optic nerve. Haldan Hartline developed a system that allowed the measurement of the electrical impulses arising in individual fibers following stimulation of the photoreceptors with light. Later work concerned the eyes of vertebrates and in particular, the measurement of electrical activity in individual nerve fibers attached to rods and cones. He demonstrated that that the differential processing of visual information begins at the moment of impact of light on the cells of the retina. Other important contributions of Hartline concerned studies of the phenomena of retinal inhibition, the perception of contrasts and forms, and the detection of movement. Haldan Keffer Hartline studied retinal elec-trophysiology at Johns Hopkins University, Baltimore, where he earned his medical degree in 1927. Later he became professor of biophysics and chairman of the department at Johns Hopkins in 1949. He then joined the Rockefeller University, New York City, and taught neurophysiology there.

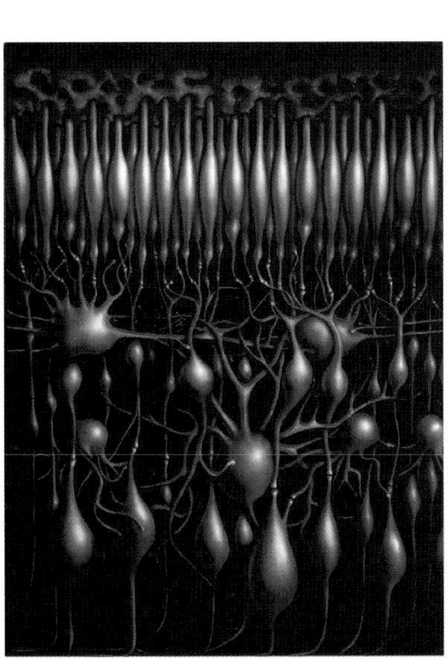

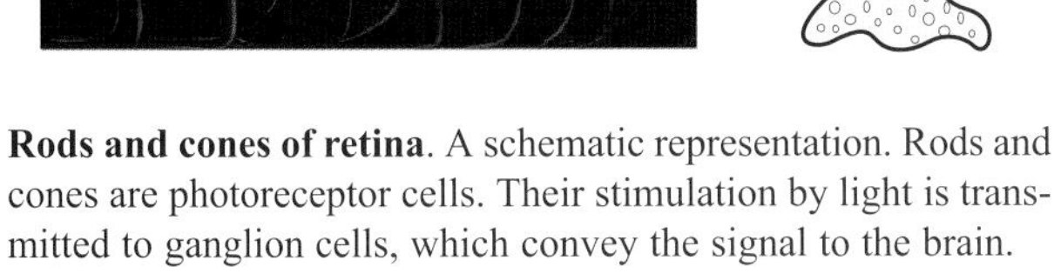

Rods and cones of retina. A schematic representation. Rods and cones are photoreceptor cells. Their stimulation by light is transmitted to ganglion cells, which convey the signal to the brain.

Wald, George (New York City, New York, November 18, 1906 - Cambridge, Massachusetts, April 12, 1997). American biochemist. Nobel Prize for Medicine along with **Haldan Hartline** and **Ragnar Granit** for his fundamental work on the chemistry of vision in rod cells. George Wald was first well-known for having shown that light isomerizes the 11-*cis* - retinal group of rhodopsin to all-*trans* - retinal and that this isomerization, which is the initial event in visual excitation, alters the structure of the molecule. A series of experiments, conducted in the begin of the 1950s, allowed him to identify the chemical reactions that take place in the vision process in rods, which are the type of light - sensitive cell containing rhodopsin and which are responsible for vision in dim light. Wald also succeeded in identifying a variety of pigments in the retina that are sensitive to particular electromagnetic radiations such as yellow-green light and red light. George Wald joined the University of

George Wald

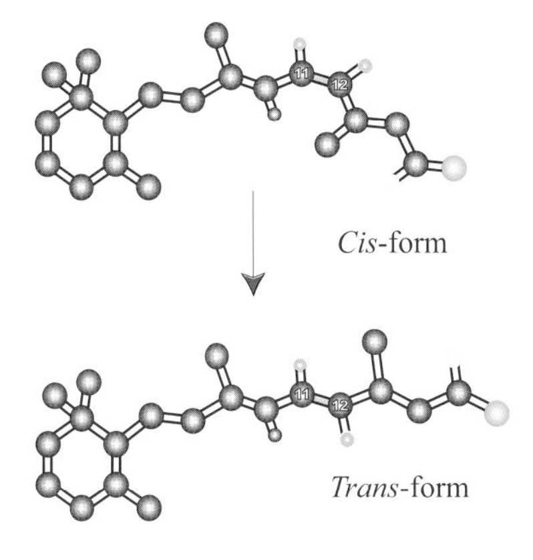

Isomerization of the *cis* isomer of retinal to the *trans* form

Harvard in 1934 and became professor of biology there in 1948. He stayed in this institution until his retirement in 1977 and was named professor emeritus the same year.

1968

Holley, Robert William (Urbana, Illinois, January 28, 1922 - Los Gatos, California, February 11, 1993). American biochemist. Nobel Prize for Medicine, shared with Marshall **Warren Nirenberg** and **Har Gobind Khorana**, for his elucidation of a transfer RNA sequence. Their research helped explain how the genetic code controls the synthesis of proteins. The research of these scientists helped to explain the relation between the genetic code and the sequences of amino acids in proteins. Robert Holley is known for the first determination of the nucleotide sequence of a transfer RNA, that of an alanine transfer RNA. The sequencing required the purification of a gram of this tRNA from some 90 kilograms of yeast, a process that took three years. In the course of establishing the sequence, Robert Holley revealed the presence of a large number of unusual nucleotides in the tRNA. The most common modification is the addition of a methyl group to specific nucleic acid bases, giving some nucleotides like 1-methylguanosine or pseudouridine. Optimum base pairing by hydrogen bonding subsequently suggested a clover - leaf structure made up of three unpaired loops and four base-pair stem regions. One of these regions contains a sequence of three bases called anticodon, complementary to the mRNA codon for the particular amino acid carried by the charged tRNA (amino acid is linked to the free 3'-OH group present at the 3'-terminus of the molecule). The tRNAs were later called "adaptors" by **Francis Crick** because they recognize both amino acids and a codon in the messenger RNA. Robert William Holley studied organic chemistry at the University of Illinois first and then at Cornell, Ithaca, where he received his Ph.D. Robert Holley was a teacher at Cornell Medical School and at the California Institute of Technology.

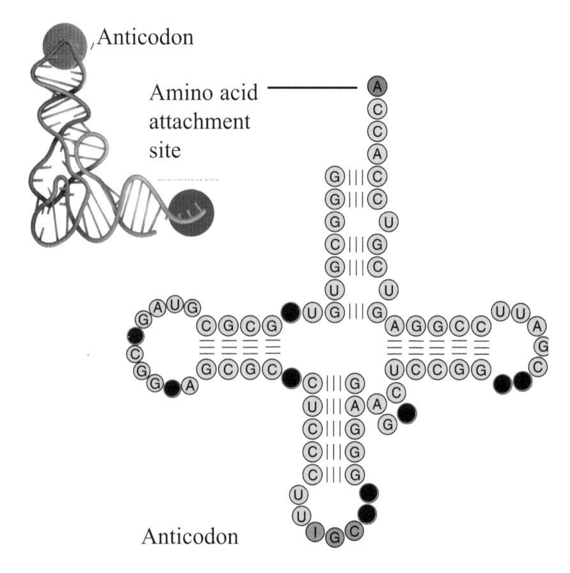

Alanine tRNA from yeast. Its nucleotide sequence was first established by R. Holley in 1965. At the upper left, a three-dimensional representation of a transfer RNA.

Khorana, Har Gobind (Raipur, India, January 9, 1922 -). Indian chemist. Awarded the Nobel Prize, shared with **Marshall Niremberg** and **Robert Holley** for research that established the correspondance between the nucleotide triplets of the genetic code and the animo acids that they specifiy in the synthesis of proteins. As in the experiments of Nirenberg, Khorana and his collaborators examined the amino acid composition of polypetides synthesized in an *in vitro* system where protein synthesis by ribosomes was directed by added RNA "messengers" they had synthesized. The polyribonucleotides made by the Khorana Laboratory had non-random sequences of nucleotides. For example, when a polyribonucleotide with repeating UC pairs was used, the polypeptides synthesized in the *in vitro* system was composed of repeating pairs of serine - leucine. This result provided proof that the genetic code was a triplet code and for the triplets UCU and CUC, that one must code for serine, the other for leucine. The results of similar experiments with synthetic messengers with other repeating sequences allowed Khorana to assign UCU as the codon for serine and UCU as the codon for leucine. In other experiments like those done in the Nirenberg laboratoy, Khorana and collaborators used specific tRNA binding to ribosomes provoked by trinucleotide "messengers" to establish correspondances between codons and the amino acids. As a result of the efforts of the Nirenberg and Khorana groups, the genetic code was almost entirely elucidated when presented at the Cold Spring Habor Symposium of 1966. Khorana is also responsible for the first laboratory synthesis of a gene, the DNA coding for an alanine tRNA molecule. Har Gobind Khorana worked with **Alexander Todd** at Cambridge University and he became a professor at Wisconsin University, and later, at the Massachusetts Institue of Technology.

Har G. Khorana

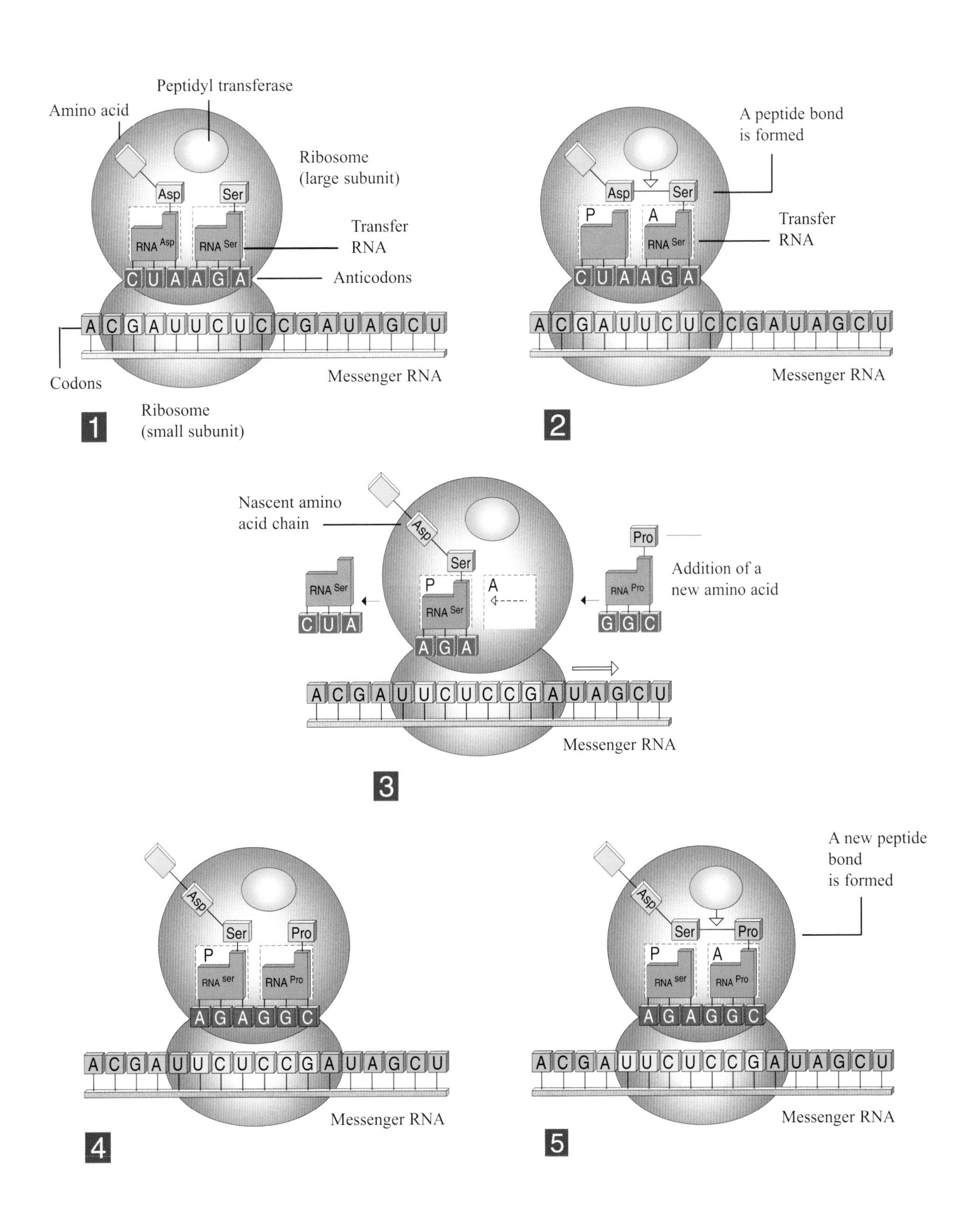

The synthesis of a amino acid chain. Five steps showing the relation between the ribosome, the messenger RNA, transfer RNAs and the nascent polypeptide chain du- ring the elongation phase of the process. The translocation from one codon to the next determines the entry of a new transfer RNA and the positioning of its anticodon opposite a complementary codon of the messenger. Insertion of a new amino acid in the polypeptide chain follows. (See Nirenberg).

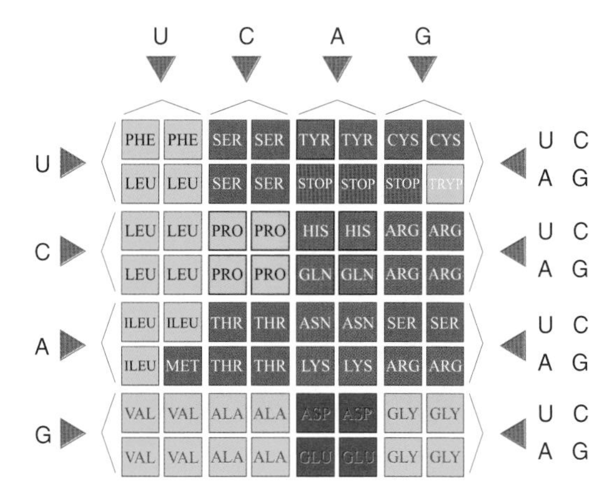

The genetic code The figure shows 64 triplets of nucleotides, each of them specifying one amino acid. Three triplets, UGA, UAA and UAG are called stop codons. They do not code for amino acids but instead specify the termination of a polypeptide chain. The genetic code is degenerate in the sense that most amino acids, except Met and Trp, are specified by more than one codon.

Nirenberg, Marshall Warren (New York City, New York, April 10, 1927-). American biochemist. Nobel Prize, jointly with **Har Gobind Khorana** and **Robert Holley** for his contributions to the elucidation of the genetic code and to a better understanding of the role of the code in protein synthesis. Making use of the technique developed by **Severo Ochoa** for the in vitro synthesis of RNA, Marshall Nirenberg, in collaboration with Johann Mathaei, synthesized an RNA containig only uracil. The only triplet possible in this polynucleotide is UUU, no matter where translation of the RNA is initiated. In a cell-free system for the synthesis of protein, translation of this RNA was shown to give rise to polypetide chains that contained only sequences of phenylalanine. Thus, for a triplet code (here, a code composed of all possible combinations of four bases taken three at a time) this result indicated that UUU coded for the amino acid phenylalanine. Further experiments, for example, with random sequences of two ribonucleotides in the synthetic RNA, provided additional insight into the base composition of other coding triplets (codons). In different experiments, Nirenberg devised a simpler method for the identification of the amino acid encoded by a triplet. Using synthetic ribotrinucleotides, he tested the ability of a given labelled aminoacyl tRNA molecule to bind to ribosomes carrying a given short trinucleotide "Messenger". Only combinations of tRNA and ribotrinucleotide that represented complementary codon - anticodon pairs produced binding. The results obtained in the experiments of Nirenberg and Khorana showed that excluding nonsense codons (triplets not co- ding for amino acids) each codon directed the incorporation of a specific amino acid in the synthesis of a polypeptide chain. Marshall Warren Nirenberg graduated in medicine from the University of Michigan where he earned his Ph.D. in 1957. He then joined the National Institutes of Health in Bethesda, Maryland.

Marshall Nirenberg

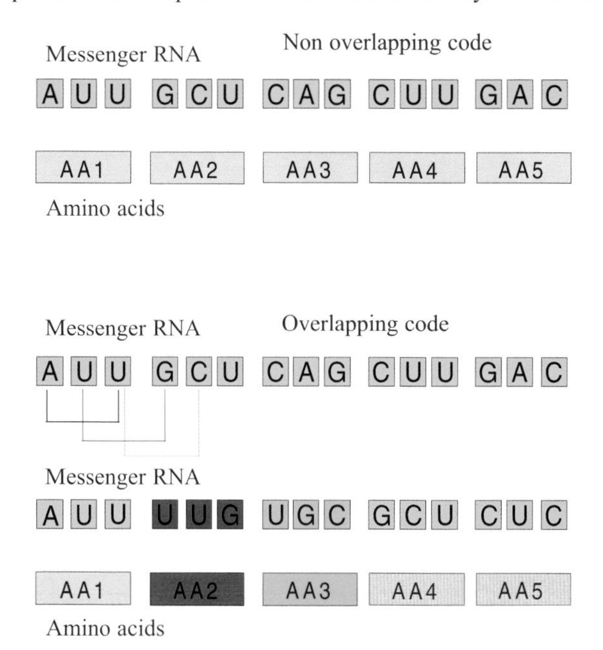

Overlapping and non-overlapping code. As the figure shows, no element of one codon forms part of an adjacent codon. On the contrary, overlapping code involves the participation of two adjacent codons in an encoded piece of information.

1969

Delbrück, Max (Berlin, Germany, September 4, 1906 - Pasadena, California, March 10, 1981). German biologist. Considered to be the founding father of molecular biology. 1969 Nobel prize for Medicine together with **Alfred Hershey** and **Salvador Luria** for discoveries concerning the genetic structure and mode of replication of bacteriophage. Son of a professor of history at the University of Berlin, his grand-father was Justus von Liebig, the founder of organic chemistry. First trained as a physicist, his evident talents brought him into contact with scientists at the forefront of research. He was able to work under the direction of the famous Danish atomic physicist **Niels Bohr** (1922 Nobel Prize for Physics) and later, with Lise Meitner, the celebrated German atomic physicist. The influence of **N. Bohr** in this conversion to biology arose perhaps indirectly from Bohr's own suspicions that the study of living organisms would be fecilitated if they were considered at the outset as associations of molecules. Max Delbrück rapidly adopted this point of view and chose bacteriophage as the most favorable object for study. His remarkable range of contributions include the development of basic methodology for the early studies of bacteriophage, that they undergo genetic recombination and that they mutate spontaneously to produce viral variants. In collaboration with Salvador Luria, he was able to show that mutation to phage resistance in bacteria was not the result of induction through contact with the virus, but occurred spontaneously in the absence of the selective agent. Eminent although merciless critic of scientific hypotheses offered by biologists of his time, Delbrück introduced a way of thinking among nascent molecular biologists that rejected any idea of vitalism.He considered that phages, and indeed, all living things, ultimately were governed by the laws of physics. Thus, h believed that it must be possible to understand their behavior uniquely in terms of molecules. Delbrück was the founder of the "phage group" whose first members were Luria, Hershey, and himself. He also was the originator of the "Phage Course" at Cold Spring Harbor and, for many of the students, their first contact with bacteriophage determined their choice of a career in molecular biology. The phage group then grew rapidly and Max Delbrück promoted meetings during which participants were encouraged to propose new hypotheses and methods that might provide a better a better understanding of viruses.

Max Delbrück

Cold Spring Harbor Laboratory (Long Island). It has become a world center for molecular biology. In 1953, phage group leader Max Delbruck organized a Symposium on Viruses.

Hershey, Alfred. (Owosso, Michigan, December 4, 1908 - May 22, 1997). American microbiologist. Received the Nobel Prize jointly with **Max Delbrück** and **Salvador Luria**, for his work on the genetic structure of bacteriophages and their mechanisms of replication. The second member of the phage group founded by Max Delbrück. Hershey devoted the major part of his research career to the biology of bacterial viruses. After isolating mutants of bacteriophage T2, he was able to demonstrate, for the first time, genetic recombination in a bacterial virus. These

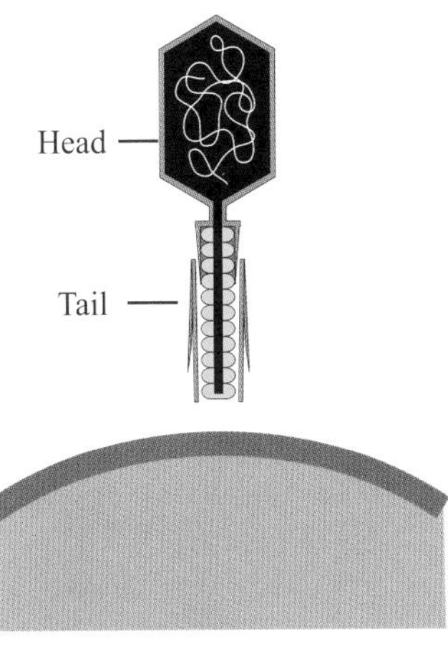

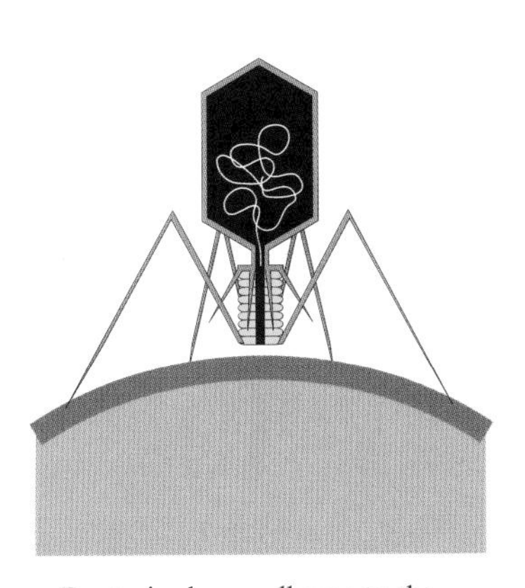

Bacteriophage adheres to the surface of the bacterium.

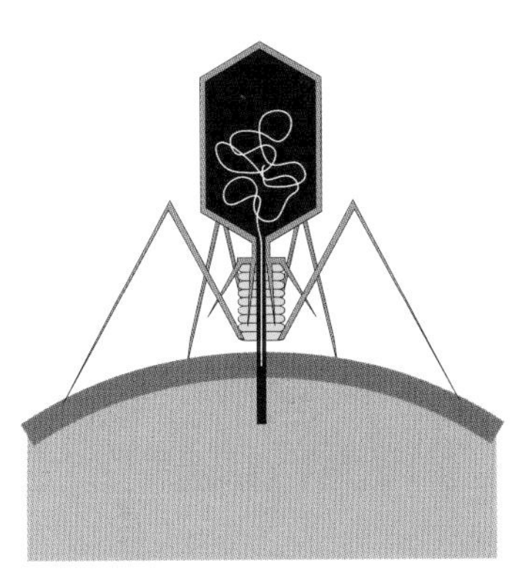

Inner core of the tail thrusts through the wall of the bacterium.

Infection of a bacterium by phage T4. This illustration shows the bacteriophage adsorbing onto the bacterium and puncturing the cell wall. In a second step, the phage injects its chromosome into the bacterium. This bacterium furnishes many enzymes required for phage reproduction. Research and studies by Max Delbrück and Salvador Luria gave us our first insights about viruses which infect bacteria.

studies showed that a bacterial virus possesed a formal genetic structure strikingly similar to that of eucaryotes. Hershey is also known for the celebrated experiment he carried out with the help of Martha Chase. The results of the "Hershey - Chase" experiment (1952) strongly sugges -ted that the genetic material of a bacteriophage was DNA, not protein as many thought at the time. That DNA was the genetic material was suggested by earlier experiments on bacterial transformation in *Pneumococcus* carried out by Oswald Avery, Colin MacLeod and Maclyn McCarthy. The Hershey - Chase experiment did not provide "irrefutable evidence" for DNA as the genetic material, but did convince many doubters who had previously accepted a genetic role for proteins. Alfred Hershey was Professor at Washington University, St. Louis, then member of the Carnegie Institute in Cold Spring Harbor, New York.

Luria, Salvador (Turin, Italie August 13 1912 - Lexington, Massachusetts, February 6, 1991). Italian-American biologist. Nobel Prize in Medicine, shared with **Alfred Hershey** and **Max Delbrück** for their discoveries concerning the the genetic structure of bacterial virus and the mechanism of their replication. In collaboration with

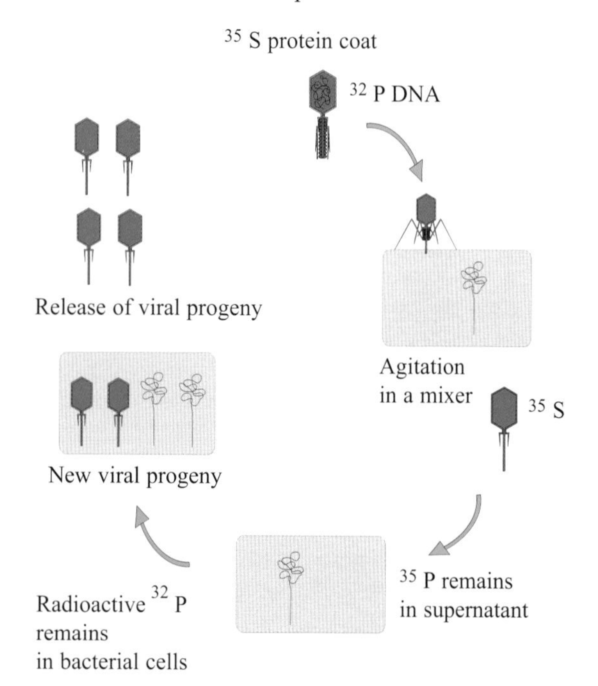

Martha-Chase and Hershey experience. This experience demonstrated that only the DNA component of bacteriophage was injected into bacteria. It was the proof that both the new protein coat and new viral DNA was provided by the parental viral deoxyribonucleic acid.

Delbrück, Luria attempted to understand how certain bacteria in a population developed resistance to infection by a specific bacterial virus. In 1943, they showed that the mutations to resistance were not induced by exposure to the virus, but occured spontaneously in individual bacteria in the absence of any contact with the virus and throughout the growth of the bacterial population. Thus, an event of mutation in individual bacteria gives rise, upon multiplication, to a clone of mutant bacteria. Luria also demonstrated the existence of mutations in bacteriophage and, in collaboration with Hershey, genetic recombination in bacteriophage. He also made contributions to studies of bacterial lysogeny, transduction, and host controlled modification of bacteriophage. **James Watson**, the future co-discoverer (with **Francis Crick**) of the structure of DNA, was a student of Luria and was encouraged by him to study the biochemistry of DNA. Salvador Luria was a professor at Indiana University, then at the Massachusetts Institute of Technology. After having worked at the Pasteur Institute in Paris for some time, he moved to the United States where he met Delbrück and Hershey with whom he founded the celebrated "phage group".

Salvador Luria

1970

Axelrod, Julius (New York City, New York, May 30, 1912 -). American pharmacologist. Nobel Prize for Medicine along with **Ulf Svante Von Euler** and **Bernard Katz** for discoveries concerning the humoral transmitters in the nerve terminals and the mechanism for their storage, release and inactivation. Julius Axelrod was the son of a Polish immigrant who supported the family as a basket-maker. Axelrod's mother wanted him to be a doctor, but his grades were not good enough and his Jewish heritage also made it difficult for him to get into medical school. After graduating from the City University of New York in 1933, Axelrod was a chemist for the New York City Department of Health and worked for a variety of university and government-sponsored laboratories through 1949. During this period, he helped to discover the pain-relieving medicine acetaminophen, better known by its brand name, Tylenol. Axelrod also lost the sight of his left eye in the mid - 1940s when a bottle of ammonia exploded in his face. Despite this accident, he continued to work in the laboratory. In 1949, Axelrod moved to the National Heart Institute, a part of the National Institutes of Health (NIH), Bethesda. Axelrod became increasingly interested in how drugs affect the nervous system, and was one of the first scientists to conduct full studies of caffeine, amphetamine, and mescaline. In 1954, Axelrod shifted offices to the NIH's National Institute of Mental Health. Over the next 30 years, until his retirement in 1984, he worked on research projects that sought to elucidate the relationship between drugs and behavior. In the late 1950s, for example, he studied how the body metabolizes the drug lysergic acid diethylamide-25, better known as LSD. Beginning in 1957, Axelrod focused his research on neurotransmitters, hormones that send chemical messages throughout the nervous system. In 1961 he announced that neurotransmitters such as norepinephrine, dopamine, and serotonin do not merely send information and terminate at the nerve ending. Rather, neurotransmitters are recaptured by the nerve that originally sent the message, and are used again for later transmissions. As Axelrod told the New York Times that year, "The decisive experiment took a couple of hours.... Working out all the details afterward took three years." Axelrod's discovery provided a new model for understanding the metabolism of neurotransmitters. His research also suggested that mental states were the result of complicated physiology and brain chemistry, rather than the sole result of psychological or environmental factors. This ushered in an era of pharmacological drugs that were designed to inhibit or stimulate neurotransmitters in the nervous system. Axelrod's work enabled researchers during the 1970s to develop a new class of antidepressant medications, especially SSRIs such as Prozac. In 1959, Axelrod's interest in neurotransmitters led him to study the workings of the tiny pineal gland located deep in the brain. Axelrod reported that melatonin, the chemical produced by the pineal gland, is actually the neurotransmitter serotonin that has undergone a chemical conversion. He also concluded that the pineal gland is a kind of "relay station" for serotonin in transit in the system. Cycles of serotonin secretion are responsible for what scientists call the body's "circadian rhythms," the natural rhythms that regulate the body's internal mechanisms for rest and sleep. After he received the Nobel Prize in 1970, Axelrod became a visible public figure. As he remarked in 1978, "I was always conscious of [political issues], but before no one asked me to sign petitions. A Nobel Laureate's signature is very visible." In 1973, for example, Axelrod joined other prominent U.S. scientists who decried the former USSR government's treatment of the dissident nuclear scientist Andrei Sakharov. The web site shows off a variety of documents and includes materials that span the various phases of Axelrod's life and career. These include examples from his extensive collection of laboratory notebooks showing his early experiments with caffeine and LSD, an unpublished manuscript from 1994, and a large sampling of his most important published articles. "Profiles in Science" was launched September 1998 by the National Library of Medicine, a part of the National Institutes of

Julius Axelrod

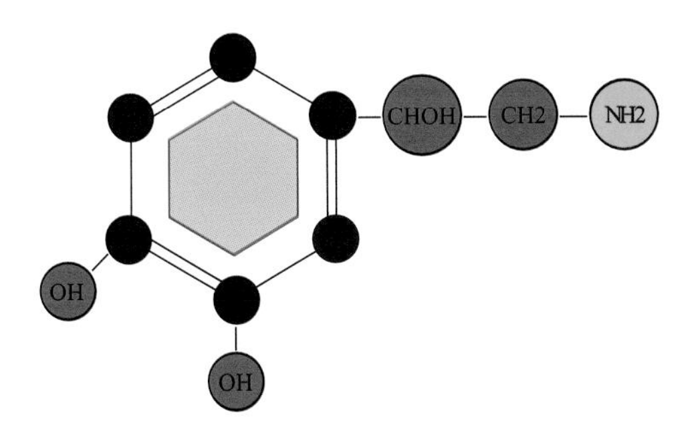

Noradrenaline. Also called norepinephrine. Apart from its role as a neurotransmitter, this substance is a hormone secreted by the adrenal medulla. It induces arteriolar constriction and raises arterial blood pressure.

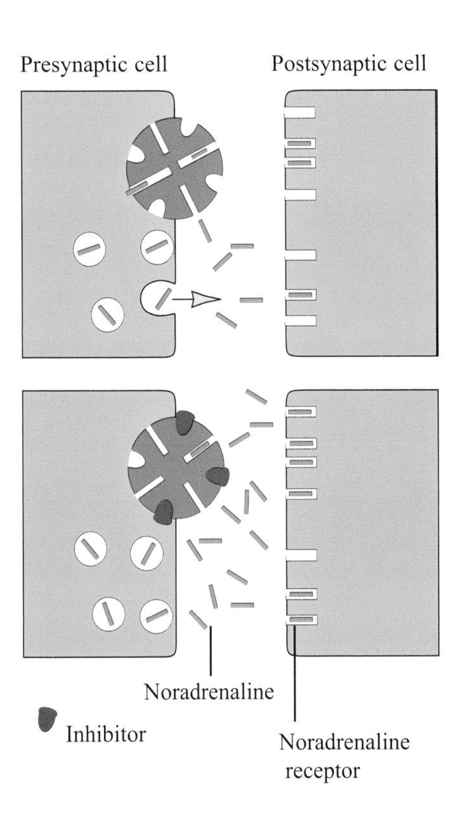

Reuptake inhibition of noradrenaline and its effects. Therapeutic antidepressants agents called tricyclics act by central inhibition of the reuptake of biogenic amines like noradrenaline. An antidepressive effect is produced due to accumulation of noradrenaline in the synaptic gap and to an augmentation of its effects on the post-synaptic neurone.

Health in Bethesda, MD. It is a continuing project and the Library plans to announce each new collection as it is added to the site.

Katz, Bernard (Leipzig, Germany, March 26, 1911-). German-born British physiologist. Nobel Prize for Medicine, along with **Julius Axelrod** and **Ulf Svante von Euler** for his studies on the release of the neurotransmitter acetylcholine at the level of synapses. Bernard Katz is famous among neurophysiologists for his quantal theory of nerve impulse transmission at synapses. His so-called vesicle hypothesis of the transmission of the nerve influx trough synapses postulated that at the arrival of the nerve impulse at the nerve ending, there is a synchronous relea-se of a number of minute vesicles of neurotransmitters (the quanta), which are libe-rated into the space separa-ting the postsynaptic membrane from the presynaptic membrane. He also showed that in a resting synapse some neurotransmitter vesicles may open spontaneously, producing a miniature end plate potential, each of which is attributed to the release of single quanta. Katz received his Ph.D. in Medicine from the University of London in 1938, under the supervision of **Archibald V. Hill**. He then moved to Sidney Hospital, Australia, and carried out research in collaboration with the neurophysiologist J. Eccles. He served in the Royal Australian Air Force as a radar operator during World War II. Afterwards, he taught biophysics at the Uni-versity of London.

Bernard Katz

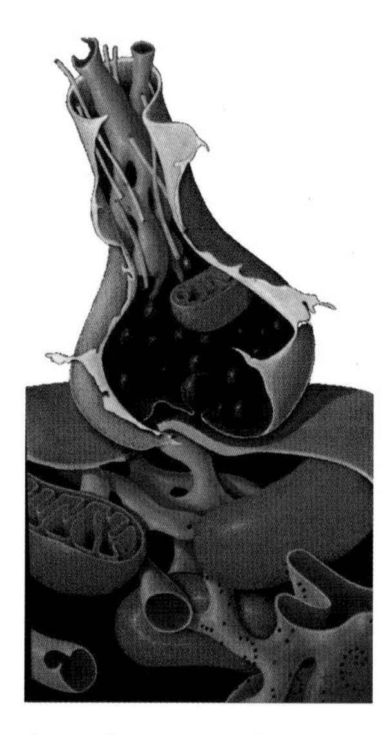

Synaptic knob. Artwork representing a presynaptic nerve ending. It contains synaptic vesicles which store neurotransmitters and release them by fusing with the presynaptic membrane.

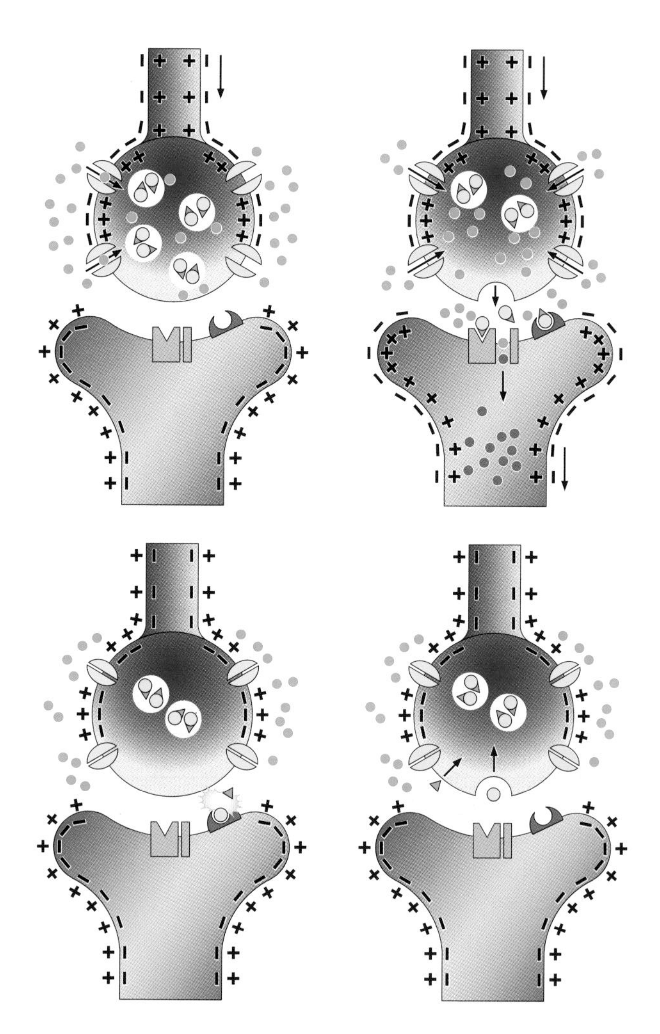

Release of acetylcholine at the synapse. Synaptic vesicles store neurotransmitters (acetylcholine) and release them by fusing with the presynaptic membrane. An enzyme, acetylocholinesterase, then breaks acetylcholine into acetate and choline. As shown, the release of acetylcholine is accompanied by the sequestration of Ca^{2+}. Acetate and choline reenter the axoplasm of the presynaptic fiber and are recycled.

Euler, Ulf Svante von (Stockholm, Sweden, February 7, 1905 - March 10, 1983). Swedish physician and physio-

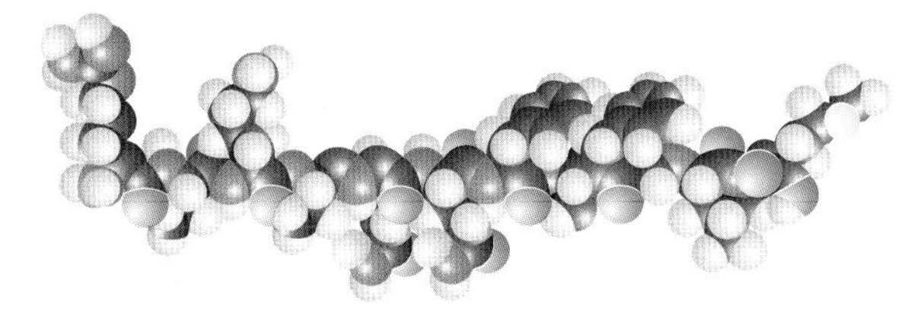

Substance P. Discovered by Ulf Svante von Euler, it is an 11-residue amide found in most tissues of the mammalian body, especially in nervous tissue.

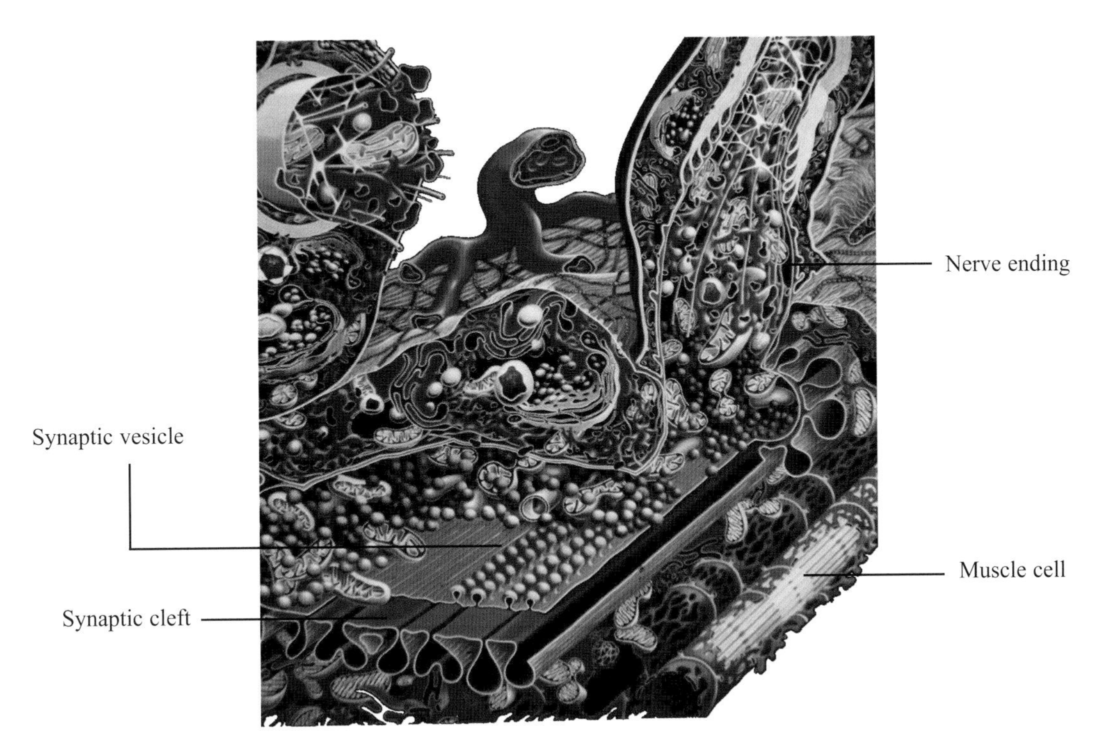

Neuromuscular junction. The place of contact between the motor end-plate of the fibre of a motor neuron and the membrane of a muscle fibre supplied by the neuron. Transmission of impulses requires release of a neurotransmitter by the presynaptic cell..(See B. Katz)

logist. Nobel Prize for Medicine, jointly with **Julius Axelrod** and **Bernard Katz**, two leaders in the field of neurotransmitters for his discoveries concerning the chemical transmission of the nerve impulse and for the mechanisms of storage, liberation and inactivation of the neurotransmitters. One of the many children of **Chelpin Von Euler**, himself a Nobel laureate, Ulf Svante Euler benefited in addition from a family environment that was comprised of other scientific personnalities. He first carried out research that led to the discovery of the substance P, a bioactive peptide secreted by the neurones of the central nervous system and by the digestive tract. This peptide stimulates intestinal contractions and lowers blood pressure. In 1914, he made another major discovery, that of prostaglandin, so named because he thought erroneously that it was secreted by the prostate gland. It was only 20 years later that the role of this substance was determined through the work of the Swedish scientist **Sune Bergström**. The award of the Nobel Prize to Euler was in recognition of his work on chemical signaling involved in the transmission of the nerve impulse. His interest in neurophysiology was a late development, begining after the Second World War.

Ulf S. von Euler

Possibly, he was influenced by **Henry Dale**, awarded the Nobel prize for his discovery of the neurotransmitter acetylcholine and with whom he worked in the 1930s. In particular, Euler is known for his discovery of noradrenaline (also called norepinephrine). A precursor of adrenaline, it is involved in the transfer of the nerve impulse between the sympathetic nervous system and muscle. Euler showed that noradrenaline was stored in the ends of nerve fibers and is liberated in response to stress that induces the "fight or flight" reaction. Ulf Svante von Euler graduated from the Karolinska Institute in Stockholm and taught there from 1930 to 1971. In 1966 he was apointed President of the Nobel Foundation.

1971

Sutherland, Earl Wilbur (Burlingame, Kansas, November 19, 1915 - Miami, Florida, March 9, 1974). American pharmacologist and physiologist. 1971 Nobel Prize for Medicine for his discovery of cyclic AMP. The main discovery of this researcher concerns cyclic adenosine monophosphate (cAMP), a smal molecule derived from adenosine triphosphate (ATP). This hormonal intermediate plays a crucial physiological role, not only in the conversion of glycogen into glucose but also in a number of chemical transformations occuring in the living cell, such as the regulation of gene expression and cell growth. In attempting to understand how the hormone epinephrine stimulated the production of glucose and its release in the

Earl W.. Sutherland

blood, Earl Sutherland and his collaborator T. Rall showed that the hormone promoted the activation of the enzyme adenylcyclase. This enzyme converts ATP into cAMP. Later, he showed that the pancreatic hormone insulin also used cyclic AMP as a second messenger.The term " second messenger " for cAMP was coined by Sutherland because it relays the initial message delivered by the hormone and triggers the complex process causing glycogen to be converted to glucose. Earl Wilbur Sutherland graduated in medicine from Washington University Medical School in

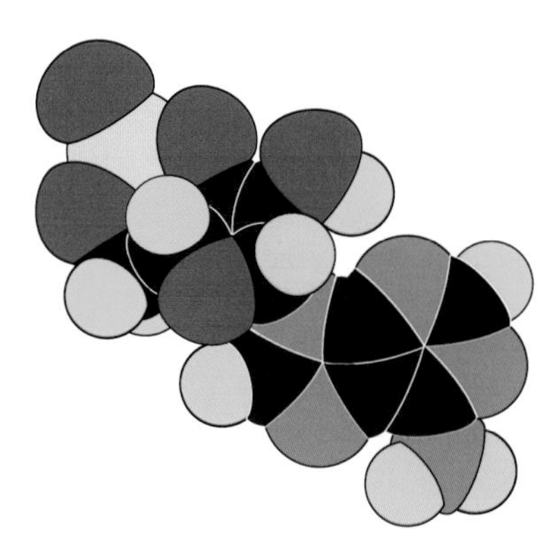

Cyclic AMP. A three-dimensional model.

St. Louis, Missouri in 1942. In 1953 he headed the department of medicine at the University of Cleveland, Ohio, where in 1956 he discovered cyclic AMP and where its

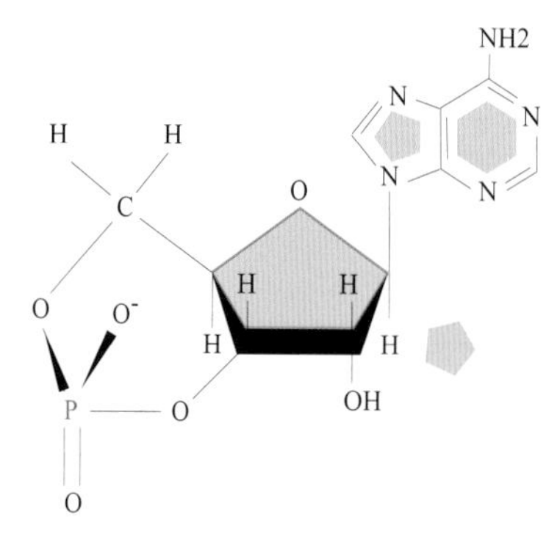

Cyclic AMP.

extraordinary role in regulating metabolic functions was established. In 1953, he taught physiology at Vanderbilt University, Tennessee before joining the University of Miami Medical School.

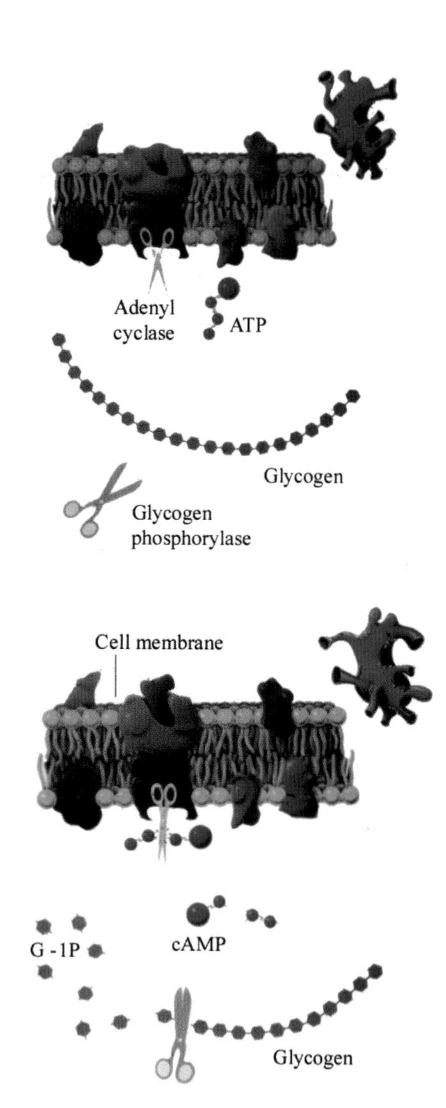

Information transfer by hormones. Glucagon has an opposite action to that of insulin. It increases the glucose concentration in the blood by stimulating adenylate cyclase, via a membrane receptor, and cyclic AMP synthesis. This process is followed by glycogenolysis.

1972

Edelman, Gerald Maurice (New York, New York, July 1, 1929 -). American biochemist. Son of a physician. 1972 Nobel Prize for Medicine, along with **Rodney Porter**, for his discoveries concerning the chemical structure of antibodies. Gerald Edelman has made significant research contributions in biophysics, protein chemistry, immunology, cell biology, and neurobiology. His early studies on the structure and diversity of antibodies led to the Nobel Prize for Physiology or Medicine in 1972. Antibodies are very large molecules and chemical analysis techniques at that time were not capable of mastering such substances. Edelman and Porter thought that it would be better to try to break them into fragments, which would make them easier to study.By using the enzyme papaïne, they cleaved the IgG molecule into three pieces, two of which were called Fab fragments (for Fragment antigen binding) because they are able to bind antigen. The other fragment contained no antigen - binding activity but crystallized readily and for this reason it was named Fc fragment (for fragment crystallizable). The two researchers also succeeded in localizing the antigen binding site at the amino terminal ends of the Fab fragments. Armed with this knowledge and the results provided by electron microscopy, in 1962 Edelman and Porter were able to define the base structure in Y of IgGs. Edelman then began research into the mechanisms involved in the regulation of primary cellular processes, particulary the control of cell growth and the development of multicellular organisms. He has focused on cell-cell interactions in early embryonic development and in the formation and function of the nervous system. These studies led to the discovery of cell adhesion molecules (CAMs), which have been found to guide the fundamental processes by which an animal achieves its shape and form, and by which nervous systems are built. One of the most significant insights provided by this work is that the precursor gene for the neural cell adhesion molecule gave rise in evolution to the entire molecular system of adaptive immunity. Most recently, he and his colleagues have been studying the fundamental cellular processes of transcription and translation in eukaryotic cells. They have developed a method to construct synthetic promoters and have also been able to enhance translation efficiency by constructing internal ribosomal entry sites of a modular composition.

Gerald Edelman

These findings have rich implications for the fields of genomics and proteomics. Edelman has formulated a detailed theory to explain the development and organization of higher brain functions in terms of a process known as neuronal group selection. This theory was presented in his 1987 volume Neural Darwinism, a widely known work. Edelman's continuing work in theoretical neuroscience includes designing new kinds of machines, called recognition automata, that are capable of carrying out tests of the self-consistency of the theory of neuronal group selection and promises to shed new light on the fundamental workings of the brain. A new, biologically based theory of consciousness extending the theory of neuronal group selection was presented in his 1989 volume The Remembered Present. A subsequent book, Bright Air,

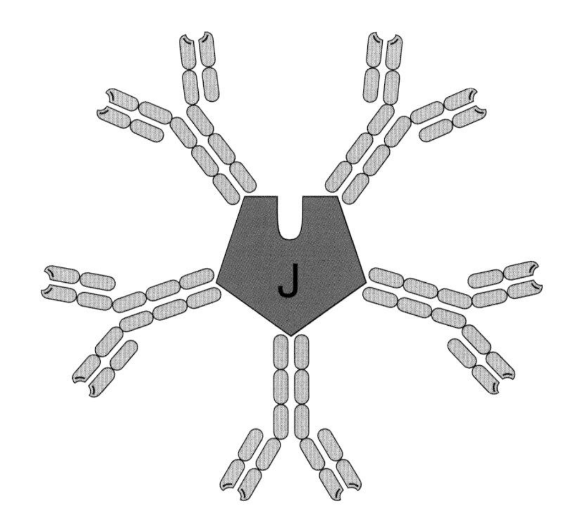

IgM molecule. The star-shaped polymer contains five basic four chain structures., each of them bearing two antigen recognition sites.

Brilliant Fire, published in 1992, continues to explore the implications of neuronal group selection and neural evolution for a modern understanding of the mind and the brain. His most recent book, published with Giulio Tononi, and entitled A Universe of Consciousness : How Matter Becomes Imagination, presents exciting new data on the neural correlates of conscious experience. Edelman earned his B.S. degree at Ursinus College and an M.D. at the University of Pennsylvania. He spent a year at the Johnson Foundation of Medical Physics, and after a medical house officership at the Massachusetts General Hospital, he served

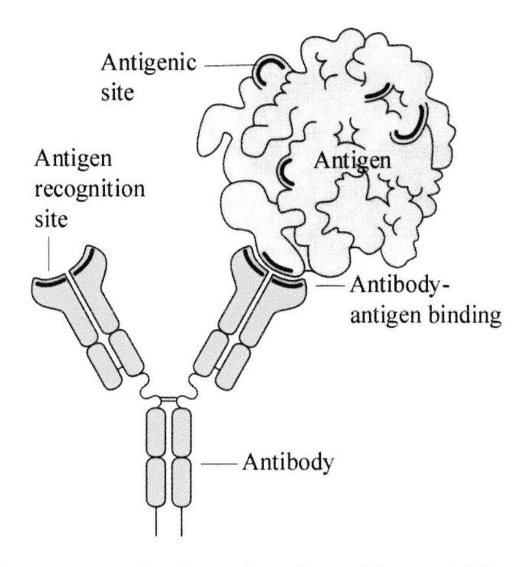

A complex between antibody and antigen. The variable region of the antibody chain is much like the active site of an enzyme molecule. It is specific for a particular antigen.

as a captain in the Army Medical Corps. In 1960 he earned his Ph.D. at the Rockefeller Institute (now University). Gerald Edelman is Director of The Neurosciences Institute and President of Neurosciences Research Foundation, the

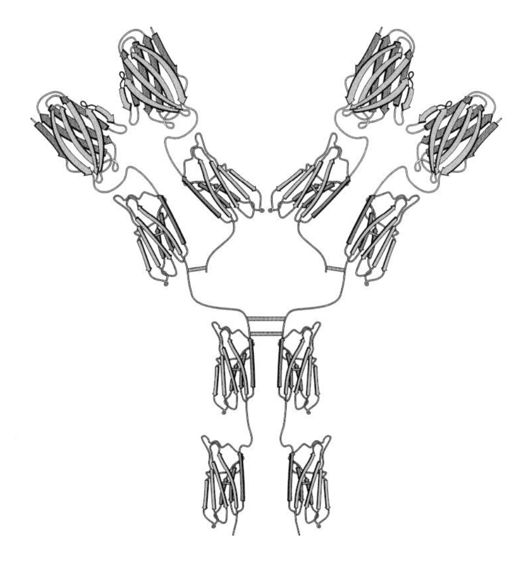

IgG molecule. A detailed structure showing the domains of an immunoglobin. The two terminal domains on each side of the polymer are variable and bear the antigen recognition sites of the antibody.

publicly supported not-for-profit organization that is the Institute's parent. Separately, he is Professor at The Scripps Research Institute and Chairman of the Department of Neurobiology at that institution.

Electron microscopy of an IgM antibody.

Porter, Rodney Robert (Liverpool, England, October 8, 1917 - Newbury, England, September 6, 1985). British biochemist. Nobel Prize for Medicine shared with **Gerald Edelman** for their work in determining the chemical structure of an antibody. By cleaving IgG antibody molecules with the hydrolytic enzyme papain, Rodney Porter and Edelman obtained two similar fragments still able to bind antigen. These fragments were called Fab (for fragment antigen binding). Another fragment was called Fc because it could be crystallized. Porter showed that the latter contained elements that are common to all IgG-type antibodies. By combining these results with those of Edelman and of electron microscope studies of antibodies, Rodney Porter was able to formulate the hypothesis that IgG antibodies have a Y structure composed of 4 chains

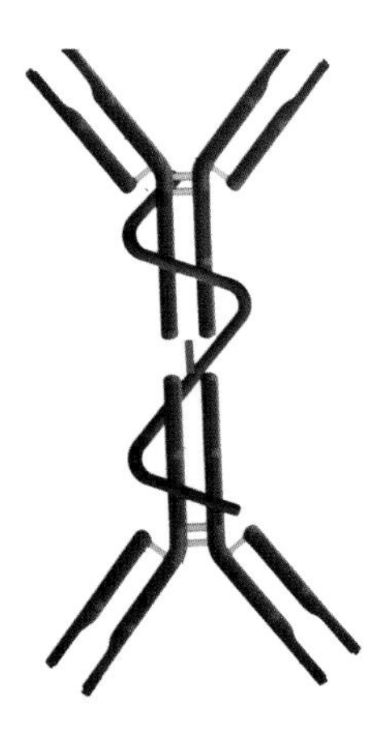

IgA molecule.

interlinked by disulfide bridges. After his studies at the Universities of Liverpool and Cambridge, where **Frederick Sanger** (Nobel Prize for Chemistry) was one of his teachers, Rodney Porter served as a professor of immunology at the National Institute for Medical Research (from 1948 to 1960), London, before teaching at St. Mary's Hospital Medical School and Oxford University.

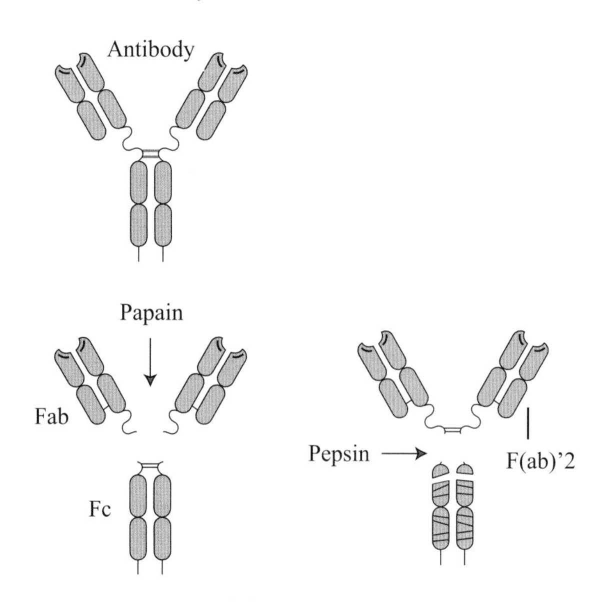

Papain and pepsin. Their lytic properties were used by George Edelman and Porter to establish the molecular structure of IgG antibodies. Papain is an endopeptidase extracted from the papaya tree. Pepsin is a proteinase found in the gastric juice of vertebrates.

1973

Frisch, Karl Von (Vienna, Austria, November 20, 1886 - Munich, Germany, December 6, 1982). Austrian zoologist and ethologist. Son of a physician. Nobel Prize for Medicine, shared with **Konrad Lorenz** and **Nikolaas Tinbergen**, for studies on chemical and visual sensors of insects, in particular bees. Although oriented by his father toward medical studies at the University of Vienna, Karl von Frisch was encouraged by his mother's brother, a physiologist in the same institution, to pursue his research in zoology and to study the location of various pigments in the eyes of insects and crustaceans. He then abandoned his medical studies and went to the University of Munich where he studied light and sound perception in fishes. There he distinguished himself with the discovery, in this species, of a small frontal sensory area, explaining why blind minnows nevertheless respond to light by changing colour. Afterwards, Von Frisch spent 40 years carefully studying the sense organs of bees together with their social behaviour. In ingenious experiments, he showed how bees find their way to their hive even in the absence of sunshine, thanks to their ability to perceive ultraviolet radiation. One of his most popular findings was the discovery of the various "dances" that "scouts" perform in order to signal the presence of a food source to foraging worker bees. He spent almost 20 years in defining the complexity of this dance behaviour. Studying the color vision of bees was also one of his favourites research endeavours. Here again, von Frisch surprised zoologists by proving that bees have incomplete colour vision: they cannot see red, for instance, but they can perceive ultraviolet radiation which has a wavelength greater than the maximum wavelength perceptible by humans. Toward the end of his career Karl von Frisch focused on the study of hearing in fish and he wrote many books on this subject for the general public. In spite of the fact that the Nobel Prize in Zoology did not exist at that time, the Nobel Committee recognized the importance of his studies on animal physiology and behavior, which could help in the understanding of human physiology and behavior.

Karl von Frisch

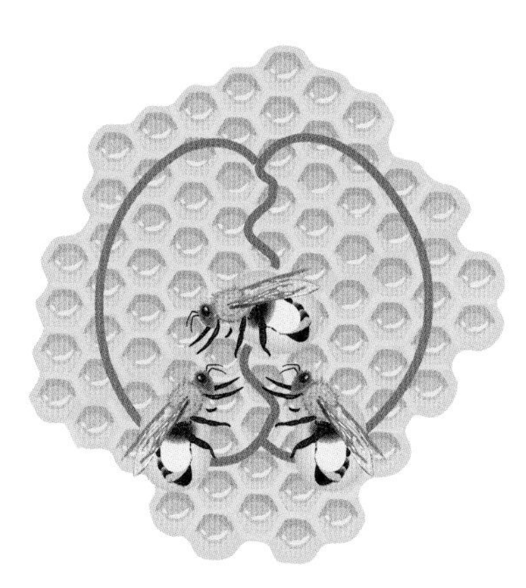

Honey dance. Most sophisticated mode of communication among bees is a stereotyped series of body movements known as a dance. Dance gives information such as the distance separating the hive and a food supply or the sugar content of the nectar.

Honeybees and colors. Experiments conducted by Karl von Frisch have clearly demonstrated that bees are able to distinguish color but they see a visual spectrum different from ours. For example, while they are less sensitive than man to red light, they can perceive ultraviolet.

Lorenz, Konrad (Vienna, Austria, November 7, 1903 - Altenburg, Austria, February 27,1989). Austrian zoologist. Shared the Nobel Prize for Medicine with **Nikolaas Tinbergen** and **Karl Von Frisch** for his discoveries concerning animal behaviour patterns. Son of an orthopedic surgeon. In his youth Konrad Lorenz was very keen on studying animal behavior. He investigated the origin of

Konrad Lorenz

the signals they produced and what they meant. Another important question was whether these behaviours could evolve. From his observations he devised a theory according to which complex behaviours appear without external stimuli and without prior learning. He hypothesised that instincts could be described in terms of a certain number of properties, i.e. their innate character and their species specificity. Lorenz tried to apply the analytical principles of comparative ethology in animals to human behaviour. Lorenz, who served in the German Army as a doctor, was taken prisoner by the Russians and was not sent home until 1948. Konrad Lorenz graduated from the University of Vienna in 1928 and earned his Ph.D. in zoology in 1933. In 1950 he founded the Department of Ethology of the Max Planck Buldern Institute in Westphalia. Konrad Lorenz proceeded with his research in Bavaria and finally settled down in Grunau, Austria, where he founded his own institute of research.

Tinbergen, Nikolaas (The Hague, Netherlands, April 15, 1907 - Oxford, England, December 21, 1988). Son of a grammar school master. Nobel Prize for Medicine, along with **Konrad Lorenz** and **Karl Von Frisch** for his work in the field of animal behaviour. Nikolaas Tinbergen, a Dutch ethologist, was the brother of Jan Tinbergen, laureate of the Nobel Prize in Economics. He specialised in the study of orientation mechanisms in animals. For instance,

Nikolaas Tinbergen

he studied how wasps belonging to certain species are guided to their nests by recognition of the spatial arrangement of their surroundings. He also did a major study on the regulation of sexual behaviour in sticklebacks. Tinbergen is even better known for his famous studies on the behavioural signals developed through evolution in various goose species and for an attempt to relate human behaviour to animal behaviour. Thanks to his studies on animal behaviour, characterised by rigorous objectivity, he is perceived as one of the founders of comparative ethology. Tinbergen's career was brutally interrupted in World War Two, when he was sent to a concentration camp for having refused the nazification of Leyden University. He wrote several books among which:"The Study of Instinct", published in 1951, and "The Animal in Its World", published in 1973. Toward the end of his career he conducted research on early chilhood autism. Nikolaas Tinbergen worked at the University of Leyden from 1936 until 1947 and received his Ph.D. in 1932. After teaching there until 1949, he moved to Oxford and conducted research on animal behaviour. He was granted British citizenship in 1955.

1974

Claude, Albert (Longlier, Belgium, August 24, 1898- Brussels, Belgium, May 22, 1983). Belgian-American cytologist. Nobel Prize for Medicine, along with **George Palade** and **Christian de Duve**, for developing methods of separating and analyzing components of the living cell. Albert Claude, born of a modest Ardennes family, carried out his primary studies at Athus. He then spent some years as an apprentice in the Athus - Grivegnée workshops. During the First World War, he enlisted in the British Intelligence Service. Twice put into concentration camps, he won various decorations and was mentioned in dispatches signed by Winston Churchill, then Minister of State for War. Although he had had no secondary education, the doors of the University Faculty of Liège were suddenly thrown open for him in 1922, when a circular from the Ministry of Defence allowed him, as a participant in World War I in the Allied armies, to enter without diploma or examination. Affected at a very early age by the loss of his mother who had died of breast cancer, he devoted the final years of his studies to fundamental research, directed especially to cancer. After obtaining his medical degrees in only six years, he turned directly to a career as

Albert Claude

a researcher which led him, after a year at Berlin where he was initiated to the technique of cell culture, to the Rockefeller Institute in New York where he entered the Murphy laboratory. His first work enabled him to isolate and characterize the agent of the Rous sarcoma, a virus of the ribonucleotic type discovered by the American microbiologist **Francis Peyton Rous**. It was on the conclusion of this work that the idea came to Claude to apply to normal cells the same technique of separation and isolation which had seemed to him so effective and promising in the course of previous experiments. Thus began a prodigious period of about 12 years which allowed him to establish an

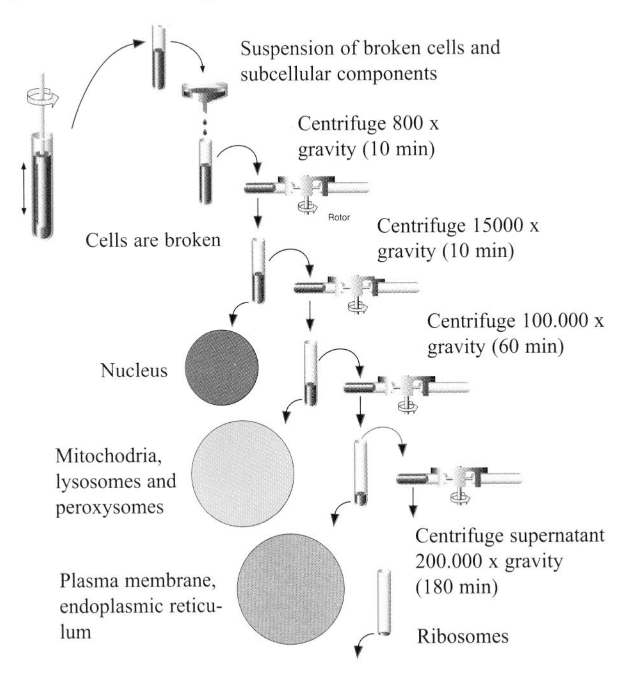

Subcellular fractionation. Cells are lysed and subcellular components are separated by a series of centrifugations at increasing speeds. Nuclei and other organelles can sometimes be purified completely by such a procedure.

analytical, morphological and biochemical balance of cellular components separated by differential centrifuging without destroying them, but after the "opening" of the cell. Thanks to this work, the cytoplasm of the normal cell revealed its nature for the first time, as well as the chemical composition and the enzymatic function of its fundamental components. Furthermore, Claude, genius of applied technique and demanding perfection, was to obtain the first slides worthy of the name of the entire normal cell in electronic microscopy. His work was to be summed up modestly in two fundamental communications in 1945 - 1946 which laid down the foundations of molecular biology and amply justified the Nobel Prize which was awarded to him. In 1949, he accepted the post of Director of the Institut Jules Bordet, Tumor Centre of the Free University of Brussels. In a few yeras, he had transformed this Centre into one of the pilot institutions of the European Continent for the integrated diagnosis and treatment of that scourge of humanity, cancer. He was Professor at the University of Brussels and left the Institute in 1970, after receiving in the meantime numerous national and international scientific distinctions. At

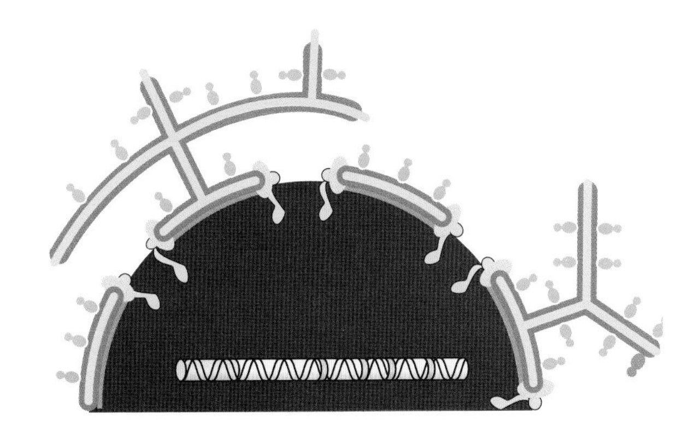

Nuclear pore-complex. This very large protein structure forms a membrane channel that allows small molecules as well as macromolecules to travel between the nucleus and the cytoplasm.

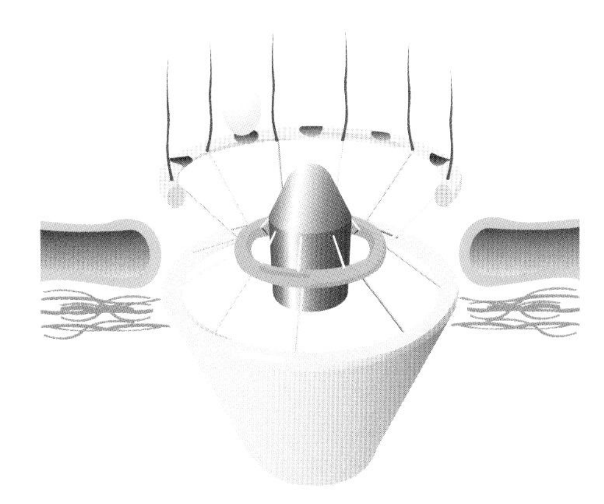

The membrane channel

the outset of his career, **Christian de Duve** had been initiated into the methodology and the spirit of Claudian research. After 1974, Albert Claude gradually withdrew

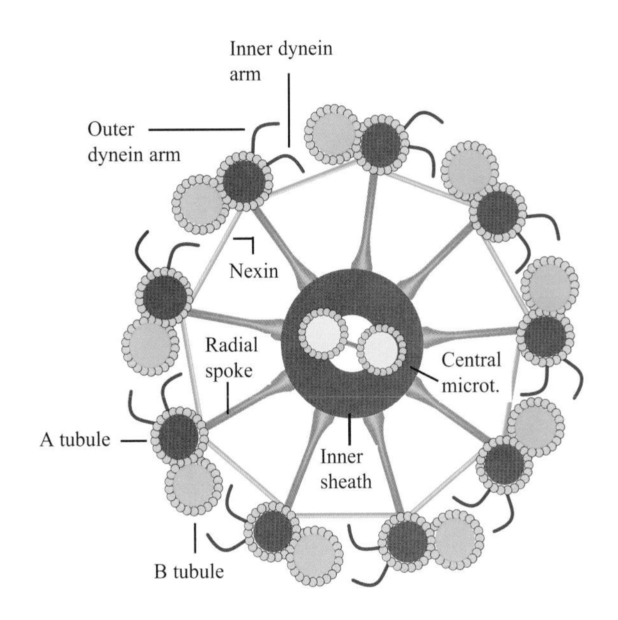

Cut through a cilium, a specialized eukaryotic organelle responsible for movement of many cells.

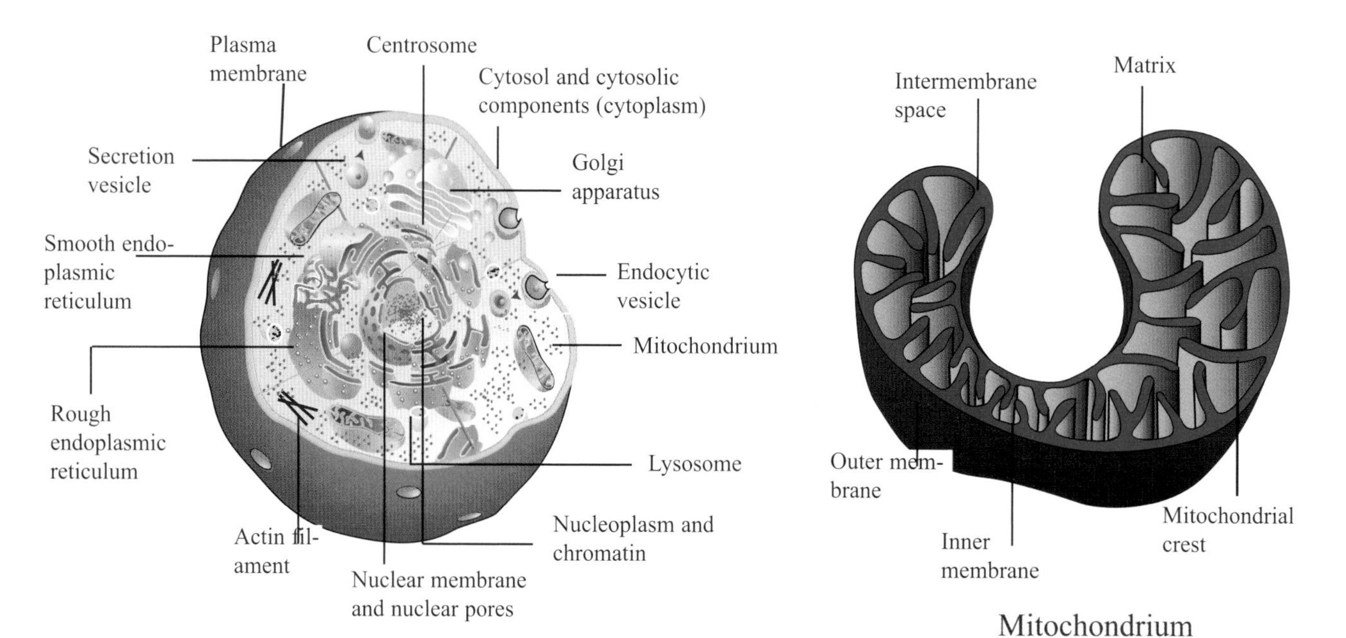

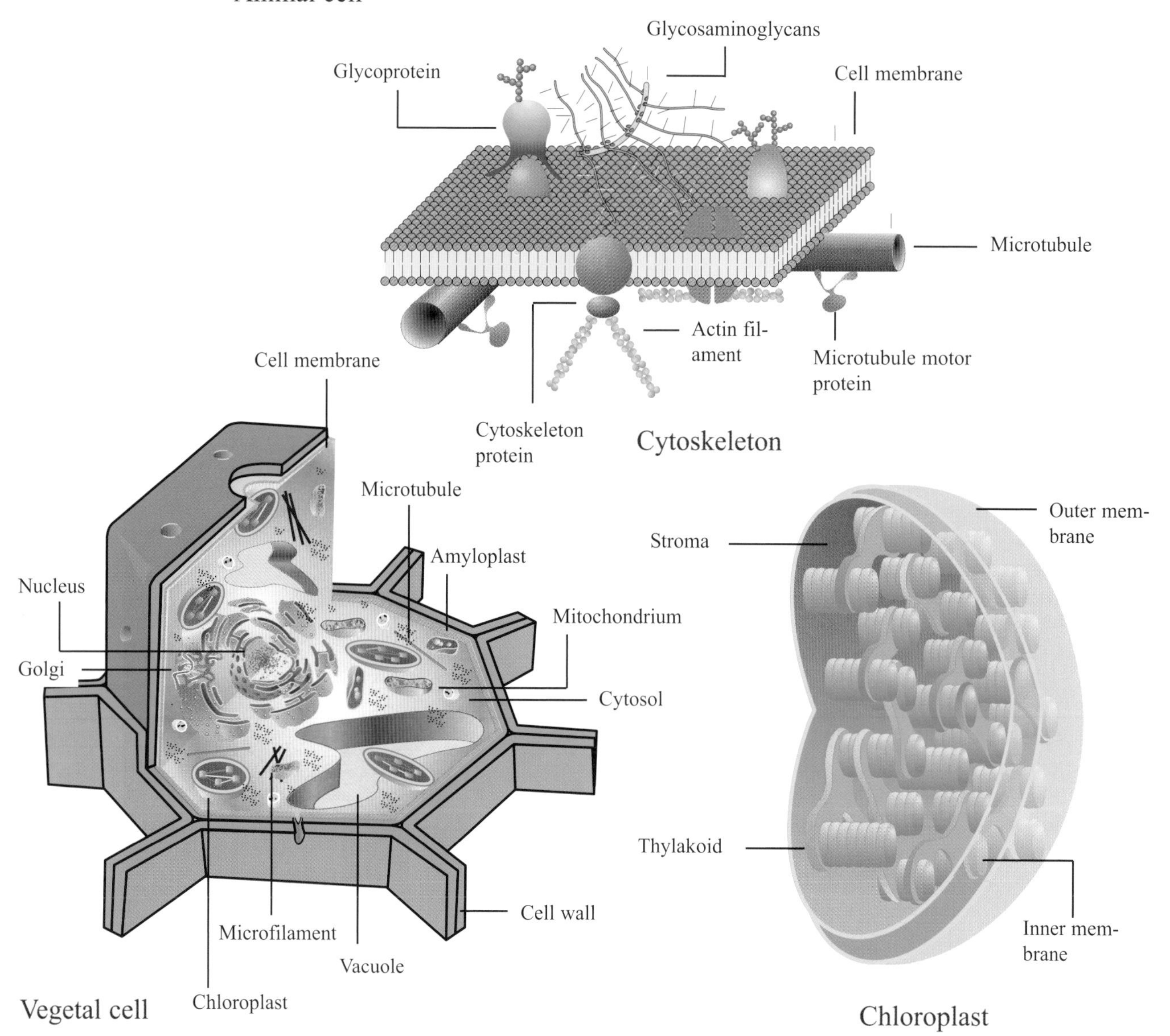

Ultrastructural elements of the living cell.

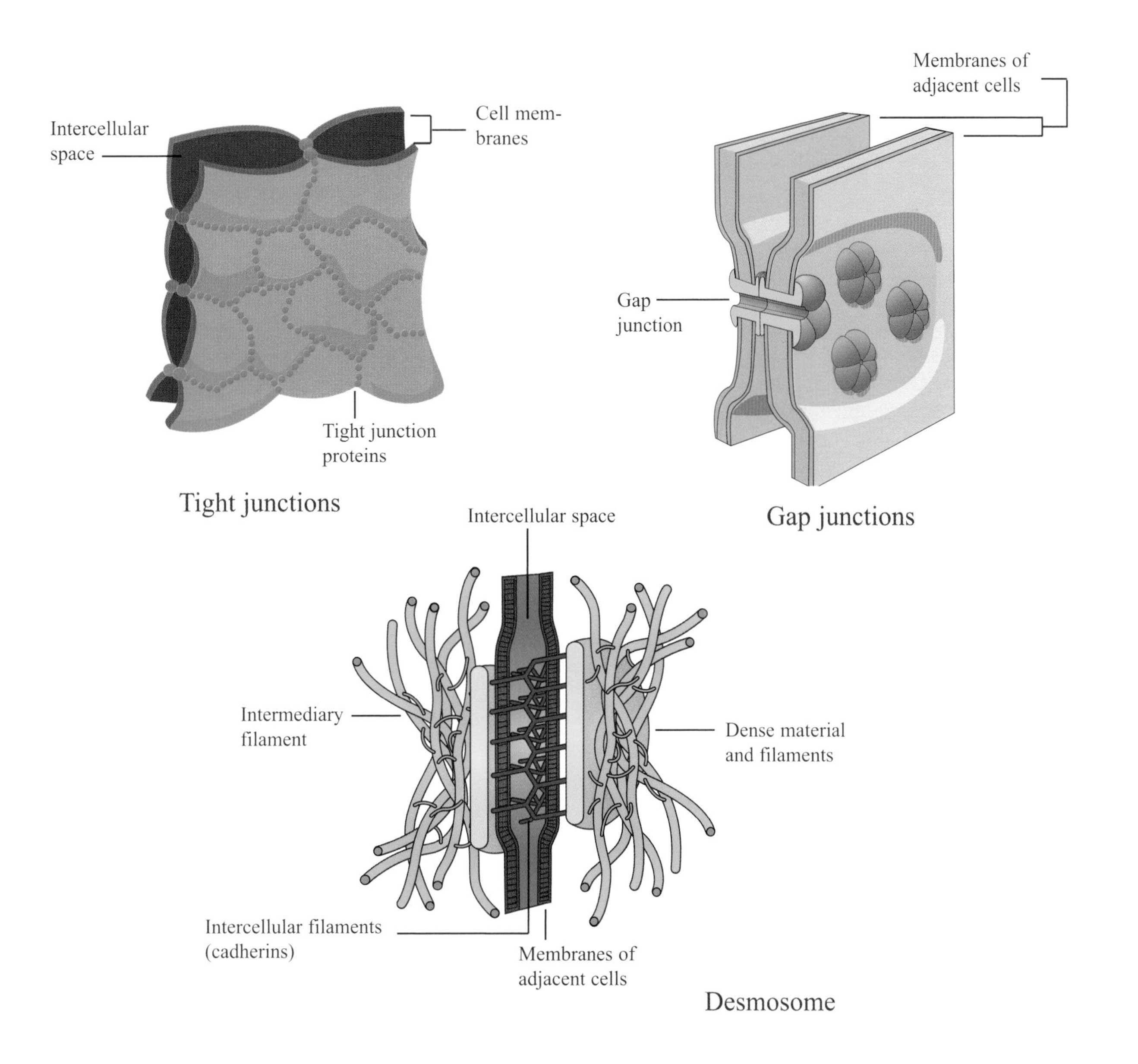

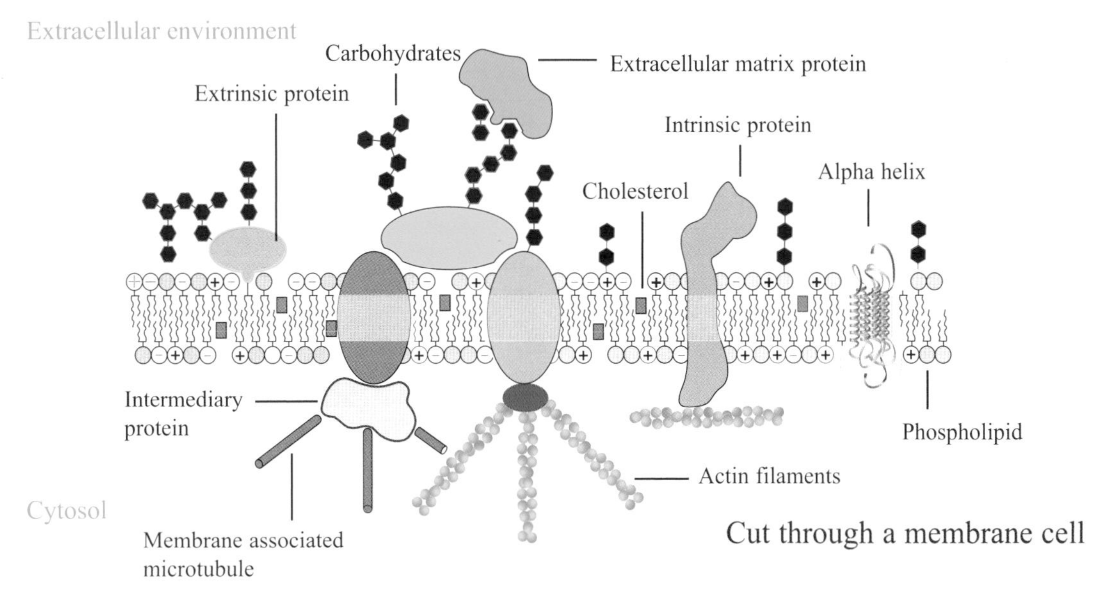

Ultrastructural elements of the living cell.

from active scientific and public life; during the Whitsun weekend of 1983, faithful to his personality and his concept of life, he passed gently on.

de Duve, Christian René (Thames-Ditton, England, October 2, 1917-). Belgian cytologist and biochemist. Nobel Prize for Medicine, along with **Albert Claude** and **George Palade**, for his discovery of lysosomes and peroxisomes. Christian de Duve embarked on his career as researcher when he began his medical studies (1934-1941) at the Catholic University of Louvain where he was appointed Professor in 1947. Initially, he became involved in the physiology of endocrinology and rediscovered glucagon, the insulin antagonist hormone. He then became interested in metabolic biochemistry and studied under **Hugo Theorell**, **Carl** and **Theresa Cori**, and **Earl Sutherland**, all Nobel Prize winners. From 1950, de Duve devoted himself to subcellular biochemistry, a new field which he was to pioneer with Albert Claude and George Palade. He particularly concentrated on refining fractionation techniques to isolate the various subcellular compartments and to establish their corresponding contribution to the cell's economy. His work has led to the discovery of lysosomes and peroxysomes. He divides his time between the Catholic University of Louvain and the Rockefeller Institute (now the Rockefeller University) where he was appointed Professor in 1962. de Duve is steadfast in his pursuit of a twin objective:to gain a greater in-depth knowledge of fundamental biology and to apply the most recent discoveries to the improvement of man's well-being and health. When, in 1974, he was joint winner of the Nobel Prize for Medicine with Claude and Palade, these concepts led him to found the International Institute

Christian R. de Duve

of Cellular and Molecular Pathology (ICP) with the backing of the authorities of the Catholic University of Louvain and its faculty of medicine. He headed this biochemical research institute until 1991. Holder of several awards, de Duve is also a member of a number of the most

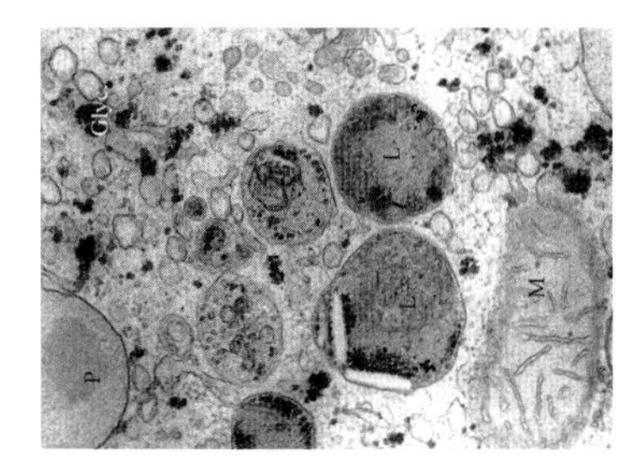

Lysosomes. As shown in the center of this electron micrograph, some lysosomes at different stages of activity. Left below, a peroxysome with its characteristic crystalline body.

prestigious academic societies in the world. In 1989 he was raised to the rank of Viscount by the King. This tireless reseacher revealed another aspect of his talents as a writer of "A guided tour of the living cell" in which he

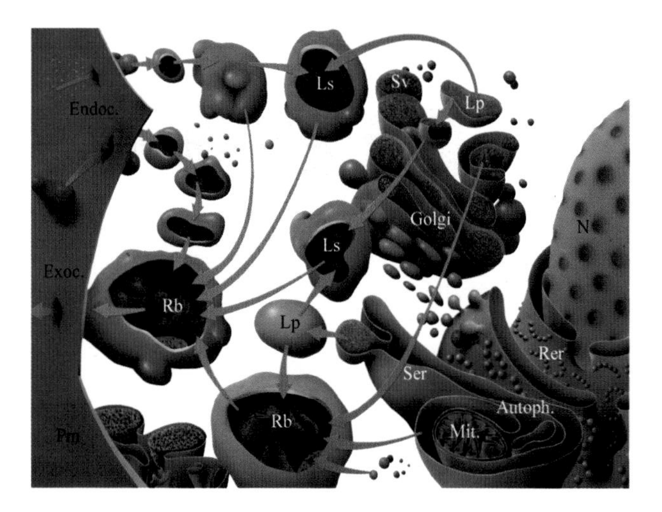

Primary and secondary lysosomes. Lysosomes are organelles containing enzymes that hydrolyse macromolecules taken up by cells during endocytosis. Primary lysosomes are produced by the Golgi apparatus. Secondary lysosomes (LS) result from the fusion between a primary lysosome with an endosome or an autophagic vesicle (Autoph.). G: Golgi apparatus; Mp: plasma membrane; N: cell nucleus; Endoc.: endocytosis; exoc: exocytosis; Cr: residual body).

summarises his vision of the cell, of "The Blueprint of a Cell" where he describes his bold theory on the origins of life, and of his recent "Vital Dust". Christian de Duve is currently pursuing his investigations into the origins of life and gives young ICP researchers the benefits of the scientific knowledge he has acquired during his long and brilliant career.

Palade, George Emil (Iasi, Romania, November 19, 1912 -). Romanian-born American cell biologist. Son of a professor. Nobel Prize for Medicine, along with **Christian René de Duve** and **Albert Claude** for developing technical tools that resulted in the discovery of several cellular structures. George Palade was a pioneer of modern cell biology. He helped lead the convergence of electron microscopy, cell fractionation and, with Philip Siekevitz, biochemistry in the study of cell structure and function. We are also indebted to Palade and his second wife, Marylin G. Farquhar (also a cell biologist), for the discovery of endocytosis, a phenomenon by which a cell takes in external materials in vesicles formed from the plasma membrane through the mechanisms of pinocytosis and phagocytosis. Later Palade investigated the intracellular

George Palade

membrane traffic along the exocytic and transcytotic pathways in eukaryotic cells as well as in intracellular aspects of the process of protein synthesis. During the 1950s, along with Gregory Roberts, he studied tumor microvasculature. At the time his research focused on determining the differences between tumor vessels and normal vessels in order to understand the mechanisms by which tumor-

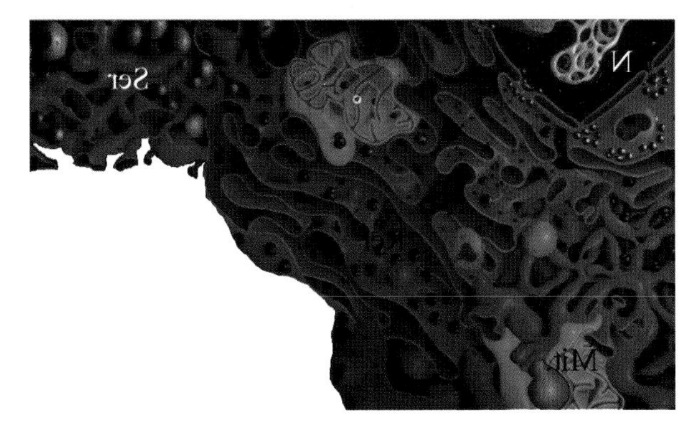

An artistic drawing of the rough and smooth endoplasmic reticula, a network of membranes in which glycoproteins and lipids are synthesized.

secreted growth factors cause these differences. His hope is that these studies will lead to therapies that can detect tumor growth earlier and selectively identify and destroy tumors by killing the blood vessels that maintain their existence. George Palade received his medical education

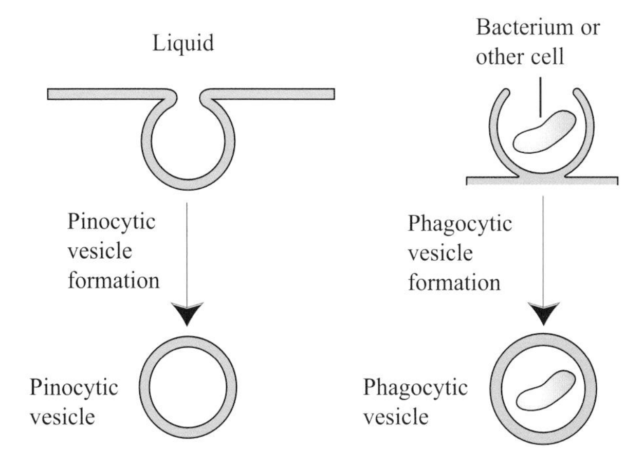

Endocytosis. A process by which a region of the plasma membrane envelops an external particle or a sample of the external medium to form an intracellular vesicle. Phagocytosis is a different process, which involves the expansion of the plasma membrane to engulf large particles such as bacteria.

at the School of Medicine of the University of Bucharest and earned his Ph.D. in medicine in 1940. In 1945 he immigrated to the United States for postdoctoral studies and was granted American citizenship in 1952. In 1946, he joined the Rockefeller Institute for Medical Research in New York and was appointed assistant professor there in 1948, and later full professor of cytology and head of the Department of Cell Biology until 1973. He then moved to Yale University Medical School as Professor and Chairman of the Section of Cell Biology. In 1990, he became Professor of Medicine at the University of San Diego, California. He is now a member of the National Academy of Science and of the American Academy of Arts and Science.

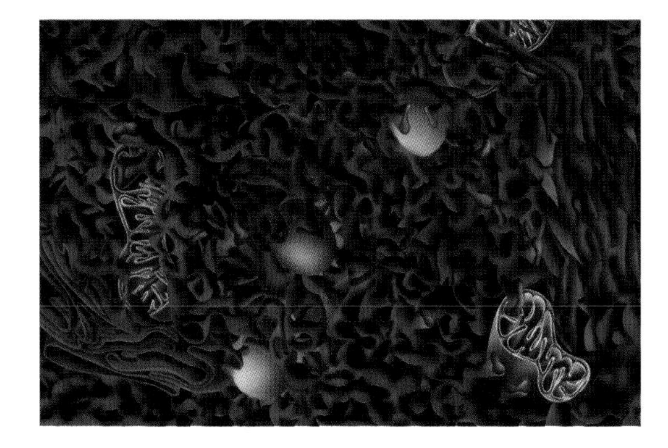

Smooth endoplasmic reticulum. That part of the internal membranes of the cell is free of ribosomes and have diverse functions like the production of lipids as well as in detoxification of drugs.

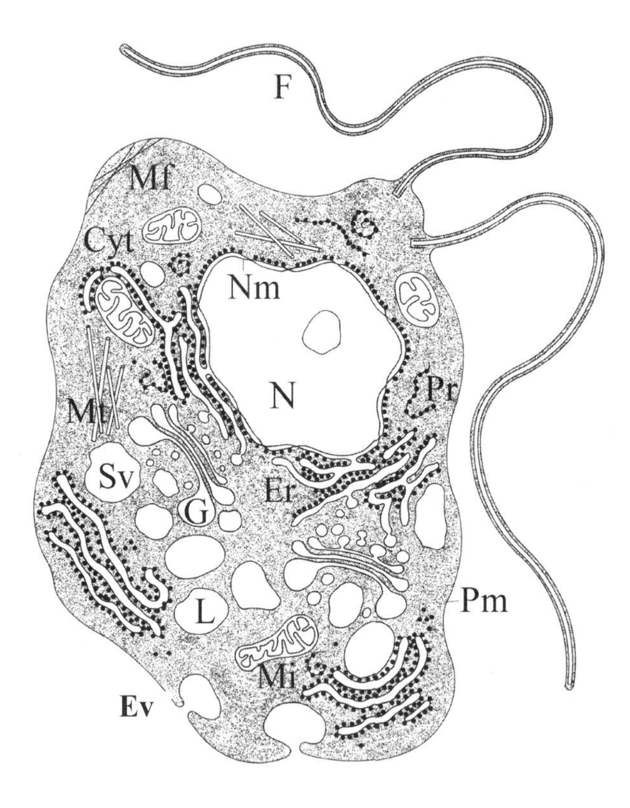

Cut through an animal cell (Cyt., Cytoplasm; Ev., Endocytic vesicle; Er., Endoplasmic reticulum; F., Flagellum; G., Golgi; L., Lysosome; Mf., Microfilament; Mi., Mitochondrion; Mt., Microtubule; N., Nucleus; Nm., Nuclear membrane; Pm., Plasma membrane; Pr., Polyribosome; Sv., Secretion vesicle.

Animal cell. A three - dimensional drawing. Among different ultrastructures are lysosomes, the organelle involved in intracellular digestion and the peroxisomes, a large variety of organelles which contain enzymes involved in the metabolism of oxygen and in the metabolism of lipids.

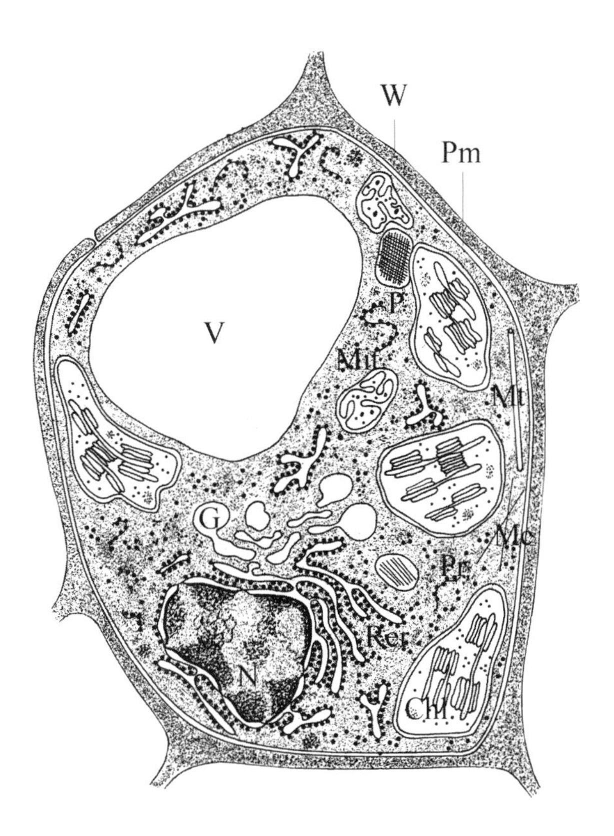

Cut through a vegetal cell (Chl., Chloroplast; Cyt., Cytoplasm; Er., Endoplasmic reticulum; F., Flagellum; G., Golgi; L., Lysosome; Mf., Microfilament; Mit., Mitochondrion; Mt., Microtubule; N., Nucleus; Nm., Nuclear membrane; P., Peroxysome Pm., Plasma membrane; Pr., Polyribosome; V., Vacuole; W, Wall.

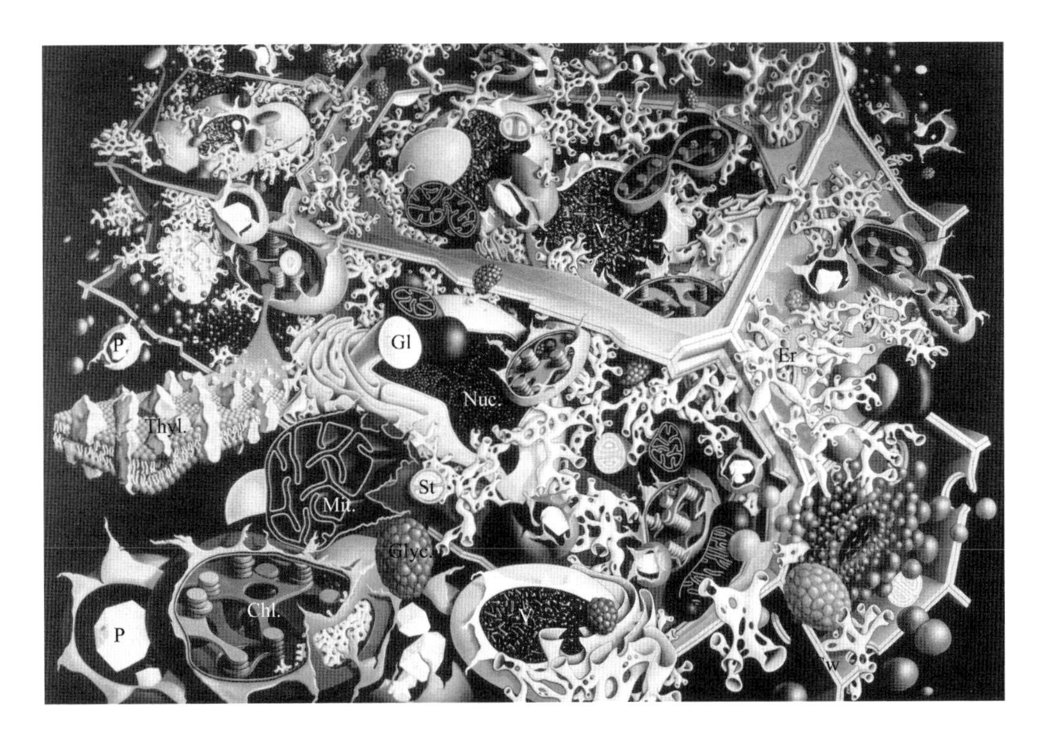

Vegetal cell. A tridimensional drawing. Apart from the organelles which are found in animal cells (such as nucleus, endoplasmic reticulum or mitochondria) a plant cell also contains numerous plastids, among which are numerous chloroplasts. Note the presence of a rigid and thick wall against which the plasma membrane presses internally.

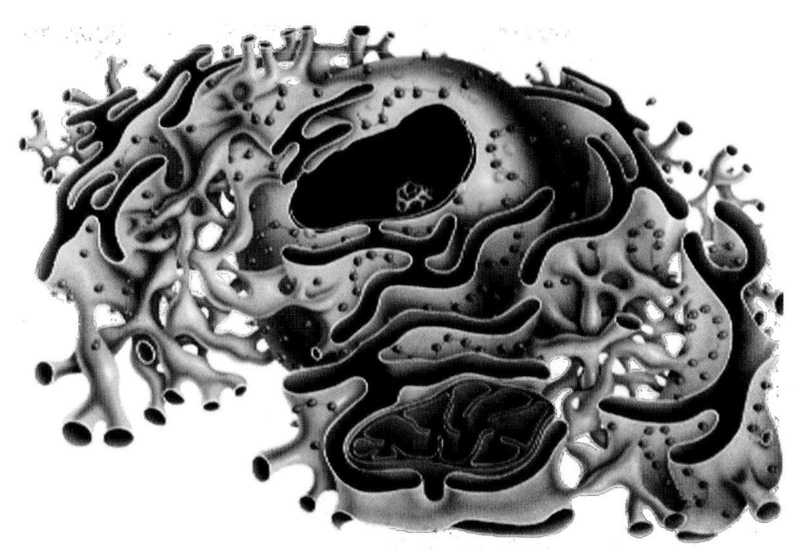

The cell nucleus and the endoplasmic reticulum. This artistic drawing shows some ultrastructural cell components in more details. Note the endoplasmic reticulum covered by ribosomes and polypeptide chains translocated into the endoplasmic cisternae. The body (in ochre) inside the nucleus is a nucleole.

1975

Baltimore, David (New York City, New York, March 7, 1938 -). American molecular biologist and virologist. Nobel Prize for Medicine, along with **Howard Temin** and **Renato Dulbecco** for contributing to an understanding of the role of viruses in cancer development. David Baltimore is associated with the discovery of certain RNA viruses that self-replicate in cells by first transforming their genome into DNA. In other words, he identified a viral enzyme that catalyses the transcription of RNA to DNA. The same enzyme was discovered independently by Temin. By chance, this finding was presented by each of these research workers within a few days of each other at two separate scientific meetings. Their discovery contradicted the central dogma in molecular biology which existed at that time, namely that the transfer of genetic information always involved the transcription of DNA into RNA. The enzyme responsible for the transcription of RNA into DNA was subsequently called 'reverse transcriptase' and the viruses concerned became known as the retroviruses. Baltimore also tried to propose a legal code of conduct concerning experimentation in the field of genetic engineering. During the 1970's, more and more elaborate techniques of manipulating recombinant DNA became available, which made scientific workers propose a set of rules to control experimentation in this area, which they felt could be detrimental to humanity. Unfortunately, Baltimore was later accused of falsifying results because one of his co-workers had fabricated experimental data. The results had previously been published in a joint scientific paper. Although he was never suspected of any malpractice himself, his name was inevitably linked to this fraud. Under pressure from members of the Massachusetts Institute of Technology, he was forced to resign from his position as Head of the Institute in 1991, a post he had held since 1983.

David Baltimore

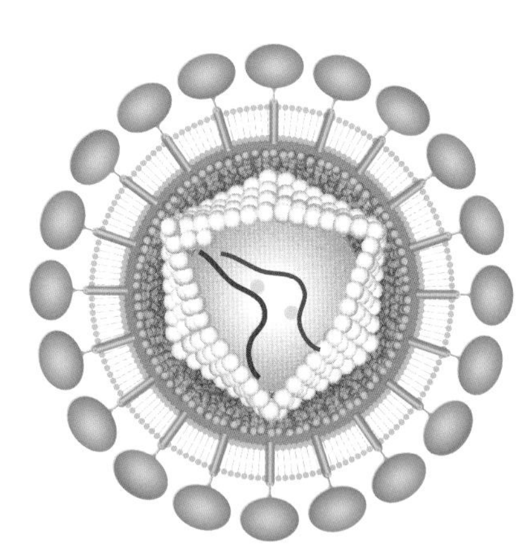

Retrovirus, a model. This virus belongs to the family of animal viruses. Its genome consists of a dimer of identical RNA chains. Howard Temin and David Baltimore independently discovered in 1970 that replication of this virus involved a reverse transcriptase which operates reverse transcription of the infecting RNA genome into complementary DNA strands.

Temin, Howard Martin (Philadelphia, Pennsylvania, December 10, 1934 - Madison, Wisconsin, February 9, 1994). American molecular biologist and oncologist. Nobel Prize for Medicine along with **Renato Dulbecco** and **David Baltimore** for his discovery of reverse transcriptase. Having noticed that the Rous sarcoma virus,

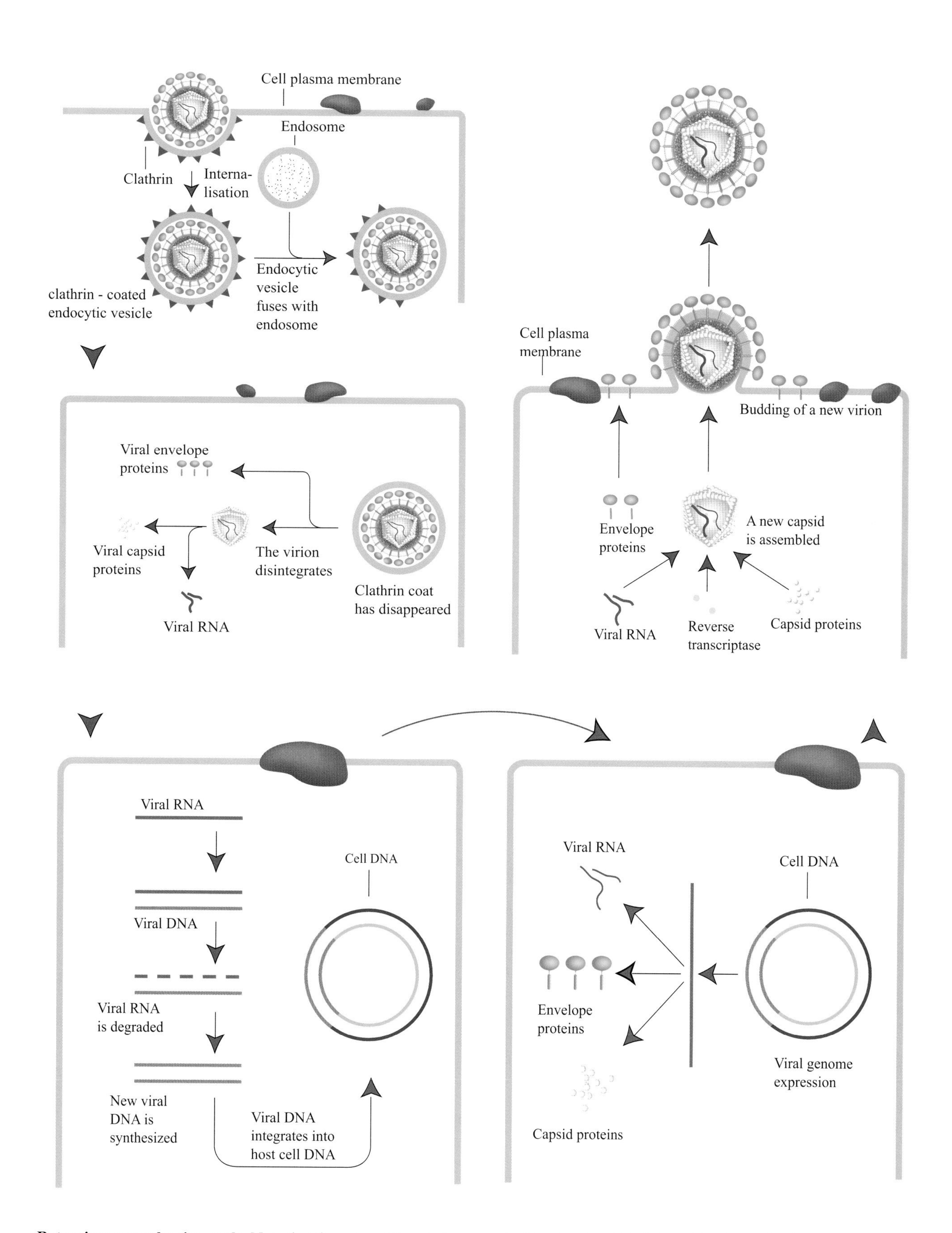

Retrovirus reproduction cycle. Note that the entry of the viral genome inside a host's cell is followed by the production of a DNA copy of the parental RNA. After being incorporated into the host chromosome, the host enzymes take over and accomplish replication, transcription, and translation of the viral genes. In this way, new retroviral particles are produced and can infect other cells.

an RNA virus, could not replicate in the presence of actinomycin D, an inhibitor of DNA synthesis, Howard Temin postulated the existence of a DNA intermediate in the cycle of this virus, and thus of an

Howard Temin

enzyme capable of synthesising DNA on an RNA template. The existence of such an enzyme, evidenced by Temin and independently by Baltimore, disproved a view that had become a dogma some twenty years earlier: that genetic information can flow only from DNA

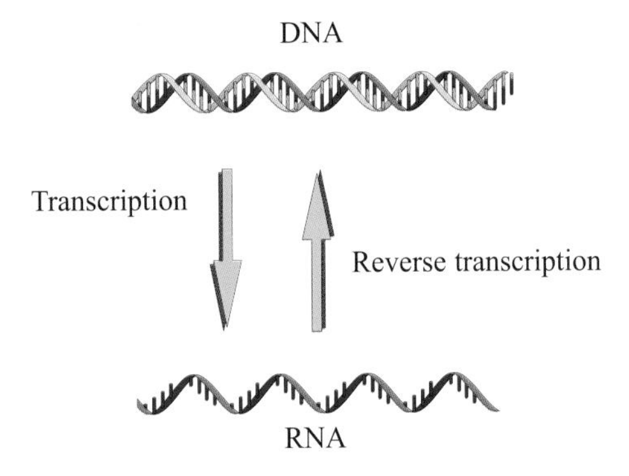

Reverse transcription. The process by which genetic information of RNA is passed to DNA. The enzyme responsible is a DNA polymerase called reverse transcriptase. This enzyme was discovered independently by H. Temin and D. Baltimore in 1970.

to RNA and not from RNA to DNA. Temin graduated from the University of California and earned his Ph.D. in virology in 1959. He taught molecular biology at the University of Wisconsin in Madison, where he conducted research until his death.

Dulbecco, Renato (Catanzaro, Italy, 1914). American virologist. Nobel Prize for Medicine shared with **David Baltimore** and **Howard Temin** for discoveries concerning the interaction of tumor viruses with the genetic material of host cells. The son of a civil engineer, Renato

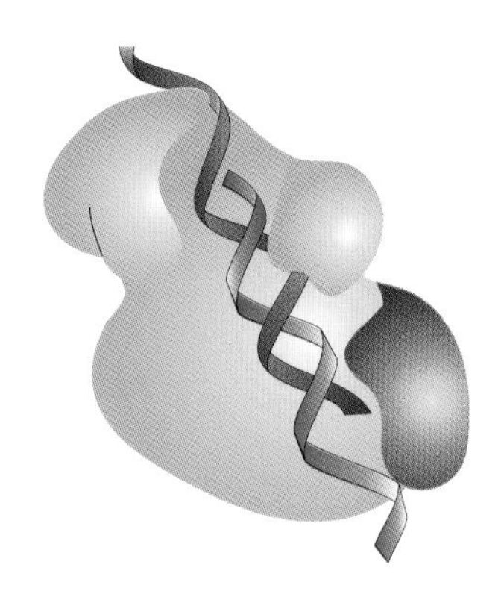

Reverse transcriptase. A DNA polymerase found in retroviruses, that uses either DNA or RNA as a template. Temin discovered that a tumor virus can use viral DNA as a template for DNA synthesis.

Dulbecco manifested a strong interest in physics in high school, constructing by himself one of the first electronic seismographs. Nevertheless, influenced by his mother, he studied for a medical degree. First recruited in the fascist army of Mussolini, he subsequently deserted and served as a doctor in the reistance. Thus, he began his research career only in 1946 when invited by **Salvador Luria** to work at Indianna University. At first, his interests were directed towards bacterial viruses. Later, probably influenced by **Max Delbrück**, he focused on the study of animal viruses, a subject that occupied the greater part of his scientific career. A first important contribution to animal virolgy was his adaptation of the plaque assay technique developed for the determination of titers of infectious bacterial viruses, to the quantitation of infectious animal viruses. (Assays of infectious bacterial virus particles by plaque assays were first developed by the Canadian-French physician Félix d'Herelle in 1917.) This highly significant technique became an essential element

Renato Dulbecco

in major discoveries in animal virology, for example, in the deve- lopment of the polio virus vaccine by Albert Sabin. Dulbecco then concentrated on the study of oncogenic viruses, in particular, polyoma, a virus that induces tumors in mice. One major contribution to an under standing of the mechanism of development of a cancer was the discovery that the viral genome integrated in the host cell chromosome in the course of infection. Thus, the viral genome established as a provirus is capable of transmission from one cell generation to another. It was subsequently shown that an oncogenic virus infecting cells in tissue culture, could transform those cells, inducing a state of anarchic growth. These studies of Dulbecco furnished oncologists with an experimental system that was to permit important advances in the understanding of the development of a cancer. Renato Dulbecco was a professor at the California Institute of Technology and subsequently, at the Salk Institute.

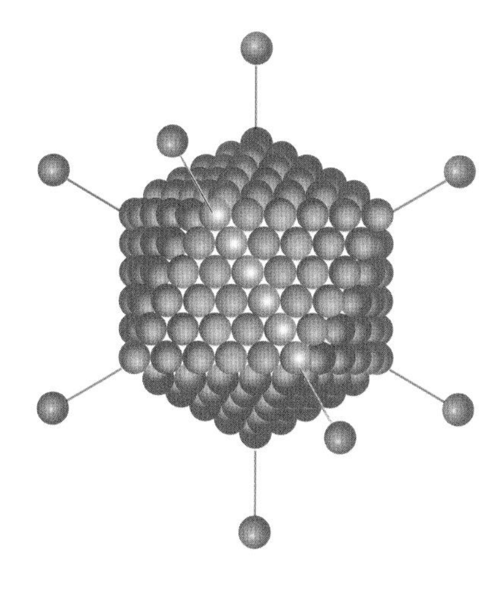

Polyoma virus. This virus has a wide distribution in nature. It has an icosahedral capsid with closed circular DNA. Polyoma virus causes cell transformation and tumors in animals. Its studies by Renato Dulbecco allowed considerable advances in our comprehension of cancer induction by viruses.

1976

Gajdusek, Daniel Carleton (Yonkers, New York, September 9, 1923 -). American physician and medical researcher. Son of a Slovak farmer. Nobel Prize for Medicine for his research on the causal agents of various degenerative neurological disorders. In 1963, this paediatrician, who had also studied viral immunology with **Franck McFarlane Burnet**, made a discovery that would later have deep repercussions in the medical world. While he was studying what caused the neurodegenerative disease "kuru " in an aboriginal population of New Guinea, Daniel Gajdusek noted that a healthy monkey could contract the disease upon receiving an injection of brain extract from a dead kuru patient. Some individuals in this population practised a form of cannibalism: they ate human brains in the hope of capturing the "spirit" of the deceased. This practice was actually a means of spreading the terrible disease. This marked the discovery of "slow virus" infections, thus named because the incubation period was very long. An analogous discovery was made in England by the American veterinary William Hadlow on sheep (the disease in sheep is scrapie, characterised notably by continuous trembling). Today it is known that both kuru and scrapie, and diseases such as Creutzfeldt-Jakob disease (leading to a form of presenile dementia), are not caused by viruses but by 'rogue proteins' called prions (see **Stanley Ben Prusiner**). Other pathological affections such as Parkinson's desease and multiple sclerosis are also invoked. Daniel Gajdusek graduated in physics from the University of Rochester, New York, in 1943. He then studied protein physical chemistry at Harvard and received his medical degree there in 1946.

Blumberg, Baruch Samuel (New York, New York, July 28, 1925-). American biochemist. Son of a lawyer. Nobel Prize for Medicine, along with **Daniel Carleton Gajdusek**, for his work on the origins and spread of infectious viral diseases. Baruj Blumberg became famous for the discovery of the so-called "Australian antigen", found in the blood of aboriginals as a result of hepatitis B infection. With a collaborator, Irving Millman, he developed a test that identified hepatitis B in blood samples. Thanks to Blumberg's discovery, scientists developed a vaccine to combat this virus. This vaccine has been administered to millions of people, in particular in Asia and in Africa. At the end of his career, Blumberg was Joint Head of the Research Cancer Institute at Pensylvania University in Philadelphia. He was also interested in medical anthropology and human genetics. Moreover, his intellectual curio sity took him to the field of evolutionary biology and early forms of life. Baruch Blumberg graduated from the College of Physicians and Surgeons of Columbia University in 1947.

1977

Guillemin, Roger Charles L. (Dijon, France, January 11, 1924 -). French born American physiologist. Son of a machine-tool manufacturer. Nobel Prize for Medicine, along with **Andrew Schally** and **Rosalyn Yalow**, for discovery of hormones peptides produced by the hypothalamus gland. The isolation of these then unknown substances controlling the secretion of all hormones of the pituitary gland has increased understanding of a variety of disorders, including thyroid diseases, problems of infertility, of statural growth, diabetes, diabetic reti-nopathy and several types of tumors. Pharmaceutical forms of these various brain hormones are now major medications in endocrinology and in the treatment of pituitary and other secretory tumors. That early work was followed by the isolation of the endorphins, characterization of cellular growth factors (FGFs), and inhibins and activins which led to recognition of the roles of all these novel entities in multiple physiological functions and developmental

Roger Guillemin

At right. **Thyrotropin - releasing factor (TRF)**. This substance is produced by the hypothalamus and regulates the secretion of the thyroid - stimulating hormone by hypophysis.

mechanisms. Roger Guillemin joined The Salk Institute in 1970, where he founded the Laboratories of Neuroendocrinology. He earned his M.D. from the University of Lyons, France (1948), and his Ph.D. in experimental medicine and surgery from the University of Montreal (1953). Since his retirement in 1989 from his career in the active pursuit of science, Guillemin has shifted his long-

Met - enkephalin. One of the pentapeptide endorphins found in organs like brain and spinal cord. This natural analgesic binds to the same receptors as morphine.

standing expertise with computers from science to art. He is using the Macintosh computer to create images / paintings that are eventually transferred to paper or canvas, either by classic lithography or a variety of ink jet processes. His computer paintings have been shown in a number of galleries and museums in the United States, Italy, France and England. Guillemin has served as president of the Endocrine Society (USA) and has been elected to the

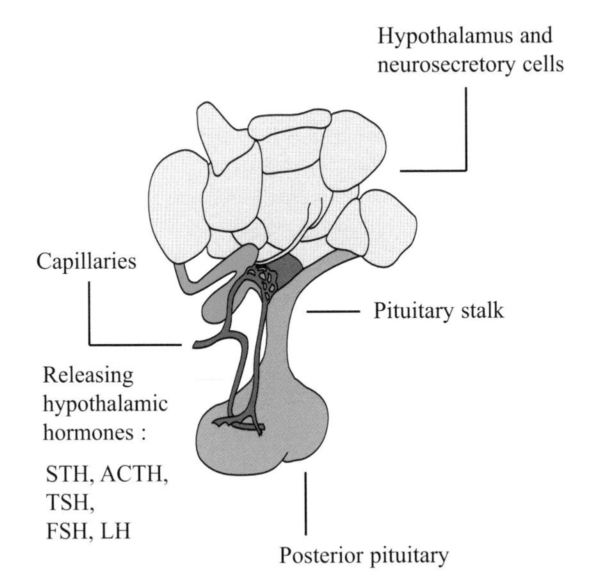

Hypothalamus. A schematic representation. It lies just above the hypophysis, to which it supplies a variety of factors.

National Academy of Sciences (USA) and the Académie des Sciences (Paris). He has been awarded many prestigious honors, including the Lasker Award in Basic Sciences and the National Medal of Science (USA). He holds honorary degrees from leading universities througouh the world.

Schally, Andrew Victor (Wilno, Poland, November 30, 1926 -). Nobel Prize for Medicine, along with **Roger Guillemin** and **Rosalynn Sussman Yalow** for work in isolating and synthesizing hormones that are produced by the hypothalamus. Andrew Schally carried out pioneer work in the field of hypothalamic hormones controlling the function of the pituitary gland. He first proved their existence by carrying out extensive physiological and biochemical studies which validated the concept that hypothalamic hormones regulate the secretion of the anterior pituitary gland. He then purified several hypothalamic hormones. After obtaining them in a pure state, he determined the structures of several of these hormones and, with the aid of his collaborators, carried out their synthesis. The existence of hypothalamic hormones had'been previously postulated by G.W. Harris but stayed unproved until 1955. At that time, Saffran and Andrew Schally showed that hypothalaic materials stimulate the release of ACTH in vitro. Later, Schally purified this corticotropin-releasing hormone and obtained its partial structure. He then started work on hypothalamic hormones controlling the release of thyrotropin (TSH), growth hormone (GH), lueteinizing hormone (LH) and follicle stimulating hormone (FSH). The purification of these hormones required stupendous efforts, since hypothalamic hormones are present only in minute quantities and hundreds of thousands of hypothalami had to be used to obtain enough material for determination of structure. In 1966 he isolated from porcine hypothalami the thyrotropin-releasing hormone (TRH), the first hypothalamic hormone to be obtained in a pure state, and published the its correct chemical compo-

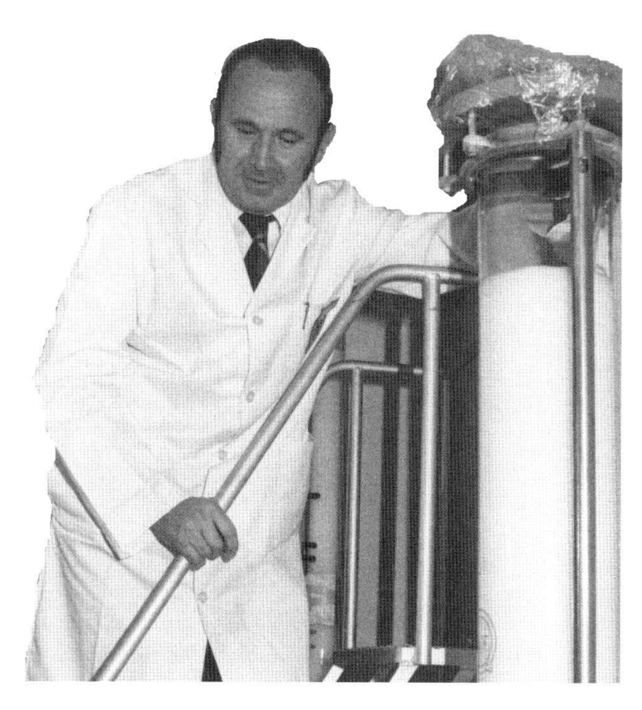

Andrew Schally

sition, that is, a peptide containing three amino acids, glutamic acid, histidine and proline. In 1969, he determined the structure of porcine TRH and carried out its synthesis. After that, he became more and more involved in the field of hypothalamic control of reproduction. Andrew Schally is a Senior Medical Investigator of the Department of Veterans Affairs (VA) and Chief of the Endocrine, Polypeptide and Cancer Institute at VA Hospital in New Orleans, as well as Head of the Section of Experimental Medicine at Tulane University School of Medicine. His research career began in December 1949 at the National Institute for Medical Research, Mill Hill, London, England. He then studied biochemistry and endocrinology at McGill University, Montreal, Canada.

Yalow, Rosalyn Sussman (New-York City, New York, July 19, 1921-). American medical physicist. Nobel Prize for Medicine, along with **Andrew V. Schally** and **Roger Guillemin**, for development of the technique of radioimmunoassay (RIA). This technique allows one to detect very small quantities of substances such as hormones in biological preparations. The method was first applied in determining insulin concentration in the blood of diabetics but also in diagnosis of thyroid diseases. Moreover, the amounts measured are extremely precise, since the method uses antibodies that specifically recognise their target. Endocrinology has benefited considerably from this discovery, which allows detection of quantities as low as one thousandth of a billionth of a gram. Rosalyn Yalow graduated from Hunter College in New York City in 1941 and obtained her Ph.D. in physics at the University of Illinois, Urbana - Champaign, in 1945. Her first research was conducted with the collaboration of Solomon Berson. She has headed the Solomon Berson Research Laboratory at the Bronx Veterans' Administration Hospital since 1973.

1978

Arber, Werner (Gränichen, Switzerland, June 3, 1929 -). Nobel Prize for Medicine, shared with **Daniel Nathans** and **Hamilton Othanel Smith** for discovery of enzymes that break DNA molecules and for their application to studies in molecular genetics. Arber studied natural sciences at the Federal Institute of Technology in Zürich. In 1953 he became assistant for biophysics at the University of Geneva, where he applied electron microscopy to morphological and to statistical studies of bacteria and of their viruses. This work directed him to microbial genetics, in particular the generation of hybrid recombinants between the genome of a bacterial virus and that of its host. This was the topic of the Ph.D. thesis accomplished in 1958. He then spent a postdoctoral research stage at the Department

Werner Arber

of Microbiology of the University of Southern California. In 1960 he returned to Geneva to become engaged in a research program on radiation effects on microorganisms. It was in the course of these studies that he discovered the molecular basis of a phenomenon called restriction. Many bacteria possess enzyme systems that enable them to identify, if DNA entering a cell originates from other types of organisms than the infected one. In the case of a foreign origin the DNA filament is cut into fragments by restriction enzymes. Then years later, when restriction enzymes

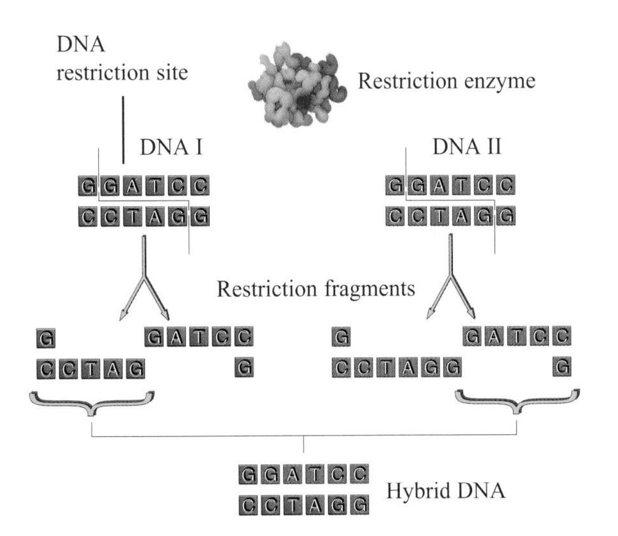

Action of a restriction enzyme. In this example, the restriction enzyme cuts DNA at specific recognition sites and yields restriction fragments with complementary single-stranded ends. Restriction fragments with such cohesive ends can associate by complementary base pairing with other restriction fragments generated by the same restriction enzyme.

had been isolated and purified, they revealed to be very useful tools in structural and eventually functional studies of genomes. In the meantime, from 1971 to 1996, Arber had a full professorship of molecular microbiology at the Biozentrum of the University of Basel. His research then mainly concerned studies of genomic rearrangments in bacteria and their viruses as mediated by genetic elements and by so-called site-specific recombination systems. This led him to develop an integral view on the spontaneous generation of genetic variations, which are the driving force of molecular evolution. It turns out that spontaneous mutagenesis is caused by a fine-tuned collaboration between several non-genetic factors and the products of so-called evolution genes acting either as generators of genetic variations or as modulators of the frequency of genetic variation. In later years, Arber was actively involved in national and international science politics, as rector of the University of Basel, as vice-president of the Swiss Science Council and as president of ICSU, the International Council for Science.

Nathans, Daniel (Wilmington, Delaware, 1928 - 1999) American biologist. Nobel Prize, shared with **Werner Arber** and **Hamilton Smith** for research that led to the discovery of restriction enzymes and the means to use them in the study of gene structure and function. Daniel Nathans demonstrated that the Hind II restriction enzyme discovered by Smith, constituted a valuable tool for mapping the genes of the simian virus SV40. Although the circular DNA molecule of this virus is one of the simplist known (only 5243 base pairs), it was difficult to map the location in the DNA of the SV40 genes using classical methods of genetic analysis. U-sing the Hind II enzyme, Nathans cleaved the SV40 DNA into 11 fragments. Then, through an analysis of partial digests with the enzyme, he was able to able to map the locations of the Hind II restriction sites and establish the order of the Hind II fragments. Similar studies with other restriction enzymes allowed him to determi-ne the location of other restriction sites and thus, produce a more detailed map of the SV40 DNA. The establishment of a restriction site map of the virus allows one to determine the location of other sites of biological interest. In other experi-ments, for example, Nathans was able to map the location of the SV40 replication origin to one of the Hind II restriction fragments.

Daniel Nathans

Daniel Nathans was educated at Delaware University in Newark, and at Washington University in St. Louis, Missouri where he earned his Ph.D. in medicine in 1954. He became a professor of microbiology at Johns Hopkins University in Baltimore, Maryland, in 1962 and was appointed Director of its Department of Microbiology in 1972. Nathans was a member of the National Academy of Sciences.

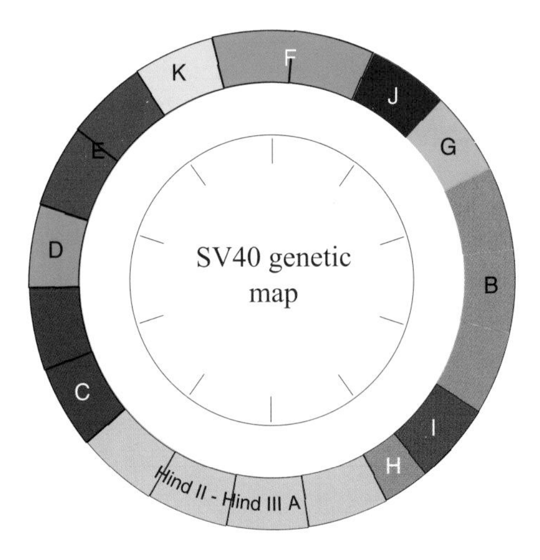

SV40 genetic map established by Nathans. The complete genetic map of an animal virus which was resolved with the use of restriction enzymes.

Smith, Hamilton (New York City, New York, August 23, 1931 -). American biologist. Nobel Prize for Medicine shared with **Daniel Nathans** and **Werner Arber** for work that led to the discovery of restriction enzymes and to the means to use them in the study of gene structure and function. Hamilton Smith was the first to purify and characterize a restriction enzyme, the Hind II of *Haemophilus influenzae*, a bacterium found in the respiratory tract of humans. As shown by him and collaborators, this restriction enzyme can cleave DNA molecules at specific sites in a specific sequence of bases (its recognition sequence). As shown by Arber, cleavage occurs provided the site has not been modified by methylation (specific modification enzyme). The Hind II enzyme differs from the Hind I enzyme studied by Arber in that the Hind I enzyme cleaves DNA at non-specific sites beyond its recognition site to produce DNA fragments with random terminal bases. Since the isolation of the Hind II enzyme, a great number of restriction enzymes have been discovered as a result of the pioneering work of Smith. Hamilton Smith graduated from the University of California at Berkeley in 1952 and received his medical degree from Johns Hopkins University Medical School in Baltimore, Maryland, in 1956. After internship and residency he joined the faculty of the University of Michigan in 1962. In 1967 he returned to Johns Hopkins where he was named professor of microbiology in 1973.

Hamilton Smith

1979

Cormak, Allan McLeod (Johannesburg, South Africa, February 23, 1924 -). South African-born American physicist. Nobel Prize for Medicine, along with **Godfrey Hounsfield**, for his work in developing the powerful new diagnostic technique of computerized axial tomography (CAT). Allan Cormack is a South African biophysicist who became an American citizen in 1966. This laureate is rather unusual in that he never studied to obtain a medical degree or Ph.D. in another science. Nevertheless, his qualities as a researcher enabled him to obtain a position at the Tufts University of Massachusetts. Cormack was interes -ted in the X - ray analysis of soft tissues or tissue layers of differing cell density. He designed a scanner (computer-assisted tomography) allowing images to be obtained for the diagnosis of cancer. The instrument could be used, for example, to localize tumours in the brain. The use of this scanner, equiped with its computer program, revolutionized the field of radio - diagnosis. Allan McLeod Cormak graduated in physics from the University of Cape Town, South Africa, in 1944. After working at Harvard University, he became Assistant Professor of Physics and later Professor at Tufts University until his retirement in 1980. He was granted American citizenship in 1966.

Hounsfield Godfrey N. (Newfold, England, August 28, 1919 -). British engineer and electronics specialist. Nobel Prize for Medicine along with **Allan Cormack** for the development of the computer tomography scanner. Godfrey Hounsfield invented the computer tomography scanner, a machine which scans the body for use of medical diagnosis. It produces a picture showing the soft tissue in a narrow slice through the body as one would see a "slice of a bacon". The value of each pixel of the picture varies as a density of the material it represents to a very high accuracy. An X - ray tube is collimated to irradiate a narrow strip across the body, the width of the slice to be taken. The emerging X - rays fall on a bank of detectors on the other side of the body. To take a picture of a slice through the patient both X - ray tube and detectors rotate around the body taking a multitude of intensity readings at all angles through the material within the slice. The readings from the detectors are fed to a computer which calculates a picture of the material within the slice only. The accuracy of the reading is high so that soft tissue can be clearly seen. Godfrey Newbold Hounsfield graduated in electronics from Faraday House Electrical Engineering College in London. He worked as a radar expert for the Royal Air Force during World War II.

X - Ray Computed Tomography. The development of computed tomography (CT) in the early 1970's revolutionized medical radiology. For the first time, physicians were able to obtain high-quality tomographic (cross sectional) images of internal structures of the body. The first practical CT instrument was developed in 1971 by Dr.G.N.Hounsfield in England and was used to image the brain.

1980

Dausset, Jean (Toulouse, France, October 19, 1916 -). French hematologist and immunologist. Son of a physician. Nobel Prize for Medicine, along with **George Snell** and **Baruj Benacerraf**, for studies of the genetic basis of the immunological reaction. Jean Dausset finished his studies of medicine and deepened his knowledge in haematology and in immunology at Harvard University. An important stage of his career was marked by his research on the so-called agranulocytosis disease. Dausset had previously noticed that the concerned patients had an abnormally low leukocyte count. A similar observation was also made on individuals who had undergone repeated blood transfusions. Dausset proposed the hypothesis that their immune system agglutinated the leukocytes and the platelets of the individual donors, while leaving their red blood corpuscles intact (1952). The defenses of these individuals consequently favored the development of antibody against the leukocytes and the foreign platelets. In widening these observations, he published, in 1958, an article in which he formulated the hypothesis that, each

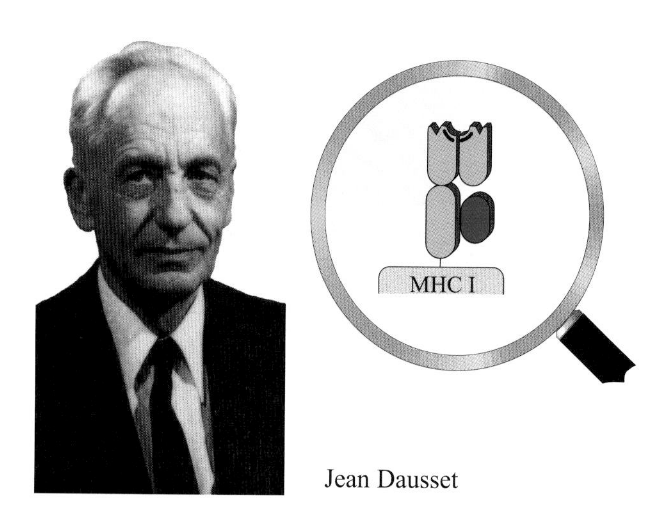

Jean Dausset

Structure of major histocompatibility antigens (MHC I). Class I MHC molecules (or transplantation antigens) are glycoproteins found on the surface of most nucleated somatic cells that enable the immune system to distinguish self from foreign cells.

individual's lymphocytes possessed a particular category of antigens on the surface membrane, which he called antigens Hu-1, later changed to HLA, to signify human leukocyte antigen, and which constitute a group of mar kers that identity the cells of an individual. These antigens, revealed afterwards in numerous hu-man groups, were called major histocompatibility antigens. They are specified by the genes of a system indicated by the abbreviation MHC (major histocompatibility complex). The HLA system discovered by Dausset is similar to the one that had been described in the mouse in the 1940s by Peter Gorer and which had been studied by Snell in the United States. The latter had called it the H-2 complex. The implications of this discovery for medicine, and in particular for transplantation, are numerous. One of them allows, for example, to determine by means of a simple blood test, if an organ intended to be transplanted is compatible or not with the recipient, or whether or not a transfusion presents risks for a recipient. Additionally, the works of Jean Dausset opened a new avenue of research which permitted the detection of a possible association between some diseases, such as juvenile onset diabetes and degenerative osteoarthritis.

Snell, George Davis (Bradford, Massachusetts, December 19, 1903 - Bar Harbor, Maine, 1996). American geneticist and immunologist. Nobel Prize for Medicine, along with **Jean Dausset** and **Baruj Benacerraf** for his studies of histocompatibility. Georges Davis Snell studied genetics at the University of Harvard. His work on organ transplants and tumours in mice, combined with the results of the British geneticist Peter Gorer, led to establishing the existence in mice of a group of genes called histocompatibility genes. We now know that these genes belong to a locus called the major histocompa-tibility complex (MHC) and that they are expressed on the cells of all higher organisms. They are responsible for the synthesis of proteins whose role precludes transplantation between a donor and a genetically different recipient. In such cases, the graft is rejected. Georges Snell showed that rejection is caused by mechanisms that recognise in the grafted tissue certain proteins characteristic of all of the donor's cells, not just those present in the graft. George Davis Snell earned his Ph.D. from Harvard University in 1930 and studied genetics at the University of Texas under Hermann J. Muller. He then entered the Jackson Laboratory in Bar Harbor, Maine, where he spent most of his career until his retirement in 1969.

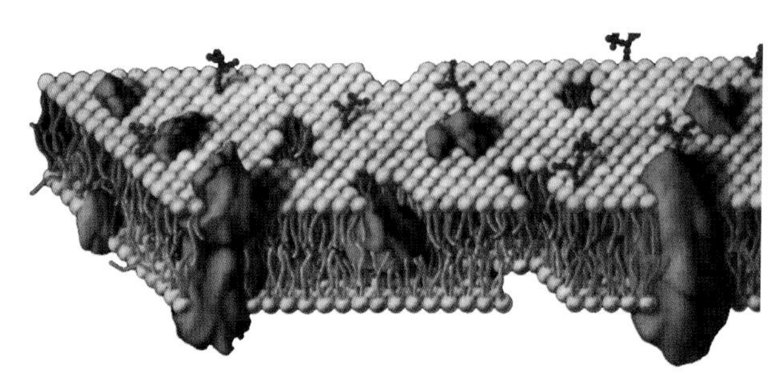

Membrane cell histocompatibility antigens. All the cells of an organism carry membrane - bound proteins which are encoded by the major histocompatibility complex, the so - called MHC. Antigens fixed and presented by these proteins are used by the immune system to distinguish body cells from invading substances or to distinguish immune system cells from other body cells.

Benacerraf, Baruj (Caracas, Venezuela, October 29, 1920 -). Venezuelan immunologist. Nobel Prize for Medicine, along with **George Snell** and **Jean Dausset**, for his discovery of genes that control the immune response. Baruj Benacerraf married Annette Dreyfuss, a descendant of the famous French captain, Henri Dreyfuss, and niece of **Jacques Monod**. Benacerraf was an old student of Kabat, an immunologist with whom he worked in the 1950s. Later he worked with **Gerald Edelman** on the structure of antibodies in relation to their specificity. Benacerraf discovered by chance that when he injected guinea pigs with an antigen, some of them developed an immune response while others did not. When he crossed animals that gave an immune response with those that did not produce antibodies, he isolated a gene that controlled the immune response. He called this gene Ir but in fact it turned out to be a member of a family of genes which make up the type II major histocompatibility complex. This complex is expressed in B lymphocytes, macrophages and several classes of T lymphocytes. Benacerraf also demonstrated the association between defective Ir genes and some hereditary immune disorders. Examples include psoriasis, Grave's disease and Reiter's disease in which the immune system makes antibodies against a person's own tissues (auto-antibodies). At the end of his career, Baruj Benacerraf was Head of the Dana Farber Cancer Institute in Boston.

Baruj Benacerraf

Structure of major histocompatibility antigens (MHC II). Classe II MHC molecules are found on the surface of only certain cell types. Among them, most B lymphocytes and macrophages.

Langerhans' cell, a dendritic cell of the immune system. After migrating from the skin to lymph nodes, its membrane becomes covered by molecules such as MHC I as well as MHC II.

1981

Hubel, David Hunter (Windsor, Ontario, Canada, February 27, 1926 -). Canadian physiologist. Nobel Prize for Medicine, along with **Roger Sperry** and **Torsten Wiesel** for his investigations of brain function, in particular for his studies of how visual information is transmitted through a complex network of nerve fibers from the eye's light-sensitive inner lining, the retina, to the brain. His association with **Wiesel** and Stephen W. Kuffler gave birth to the Neurology Department of Harvard University. Their work enabled them to analyse information flows from the eye to the visual centres in the occipital lobes of the brain. High-precision measurements with extremely fine electrodes enabled David Hubel to measure the electric discharges occurring in individual brain-cell fibres according to the response of the retinal cells to light. Hubel and **Wiesel** showed that many cortical cells have a specific, orientation preference, and respond to contours of either vertical, horizontal or oblique orientations (see **Wiesel**). David Hubel studied medicine and received his medical degree from McGill University, Montreal, Quebec. In 1959, he joined Harvard Medical School, Cambridge, Massachusetts, where he met Wiesel and where they did much to elucidate the functionning of the visual process.

Sperry, Roger Wolcott (Hartford, Connecticut, August 20, 1913 - Pasadena, California, April 17, 1994). American physiologist. Shared the Nobel Prize for Medicine, along with **David Hubel** and **Torsten Wiesel** for his his studies of the functional specialization of the cerebral hemispheres. Sperry's work began in the 1940s and focused on the hemispheres of the brain, first in fish, then in cats and apes. The aim of these experiments was to understand what might be the specific role of each hemisphere in learning. Sperry showed that the left hemiphere is more apt than the right at correctly assembling logical sequences appearing in language and writing, whe-

Roger Sperry

reas the right hemisphere is better at recognising faces, copying drawings, and distin-guishing objects through the sense of touch (see Wiesel, Torsten). He was recognised by part of the scientific community as one of the top specialists in brain physiology of the time. Roger Walcott Sperry graduated in psychology from Oberlin College, Ohio, and earned his Ph.D. in zoology from the University of Chicago in 1941. After working with the American psychologist and geneticist Karl Lashley at Harvard University, he conducted research at the University of Chicago in 1946. He then taught at the California Institute of Technology.

Wiesel, Torsten Nils (Uppsala, Sweden, June 3, 1924 -). Swedish neurobiologist. Nobel Prize for Medicine, along with **David Hunter Hubel** and **Roger Wolcott Sperry** for his investigation of brain function, in particular for his studies of how visual information is transmitted through a complex network of nerve fibers from the eye's light-sensitive inner lining, the retina, to the brain. For nearly two decades, beginning in the late 1950s, Torsten Wiesel and David Hubel explored the means by which visual information is collected and processed. Using cats and monkeys as test subjects, they identified many cell functions in the brain's visual cortex. Sensitive electrodes measured the electrical impulses emited by single neurons in the cortex while the animals were exposed to a range of visual stimuli. In the studies, it was found that many cortical cells have a specific, orientation preference, and respond to contours of either vertical, horizontal or oblique orientations. The observation that single nerve cells are specialized came as a surprize and led to futher work on the cellular basis of the perception of forms. Wiesel and Hubel discovered that information gathered by the two eyes comes together in the brain's visual cortex. Their research demonstrated that binocular cells in the cortex respond preferentially to signals from one eye or the other, and that cells with the same eye preference are organized in a series of columns, left- and right-eye columns, that extend across all cortical layers. Similarly, cells with the same orientation preference aggregate in columnar fashion. Scientists have shown since that the functional architecture of the visual cortex as described by Wiesel and Hubel apply to other parts of the cortex. In fact, columnar structure is important in other sensory mechanisms, the control of movement and certain higher brain functions. Wiesel's work has important clinical applications. In the 1960s and 1970s, he and Hubel demonstrated that the physiological mechanisms necessary for normal vision are present at birth, and that normal visual experience is necessary for full development of visual capabilities. When deprived of normal vision during a critical period in the first few monts of life, permanent visual impairment results because the deprivation causes dramatic changes in the physiological machinery. The experiments led to more aggressive treatment of children born with cataracts, many of whom now receive operations shortly after birth. Wiesel graduated in medicine from the Karolinska Institute in Stockholm in 1954. From 1992 to 1998, Wiesel was President of the Rockefeller University, New York.

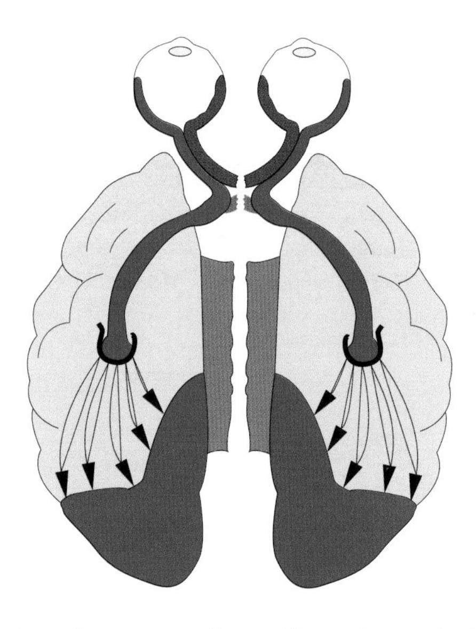

Research on the corpus callosum (in grey) caused scientists to postulate a view of the relative capabilities of the two halves of the human brain: that both hemispheres are involved in higher cognitive functioning, with each half of the brain specialised in complementary fashion for different modes of thinking

Torsten Wiesel

1982

Bergström, Sune (Stockholm, Sweden, January 10, 1916 -). Swedish biochemist. Nobel Prize for Medicine, along with

Bengt Ingemar Samuelsson and **John Robert Vane** for their discoveries of prostaglandins and related biological substances. In 1963 Sune Bergström was Dean of the Karolinska Institute Medical Department in Stockholm and where in 1969, he became Rector. In the 1930's, **Ulf Svante Von Euler** found a substance that was capable of both contracting and relaxing the smooth muscles of the uterus. Since this compound appeared to be a constituant of seminal fluid, he speculated that it must have come from the prostate gland and for this reason called it prostaglandin (it was later shown that the substance was produced by seminal vesicles). On Euler's advice, Bergström studied this phenomenon and in collaboration with **Samuelsson**, he isolated the first two prostaglandins, F and E, in 1957 and 1960. In 1964, he showed that prostaglandins belong to a family of lipids derived from arachidonic acid, a polyunsaturated fatty acid found in meats and vegetables. Prostaglandins are now known to affect a wide range of physiological processes, for example inflammation, the maintenance of body heat, muscle contraction and even allergic reactions. The very low concentration of prosta-glandins in organisms and their instability meant that they were difficult to identify. Bergström is also credited with the discovery of thromboxanes, derivatives of prostaglandins, that are responsible for platlet aggregation and vasostriction.

Sune Bergström

Bengt I. Samuelsson

Samuelsson, Bengt Ingemar (Halmstad, Sweden, May 21, 1934 -). Nobel Prize for Medicine along with Sune Bergström for his work on prostaglandins. Bengt Samuelsson discovered the prostaglandin PGA2, called thromboxane. This substance is synthetisized in platelets and strongly activates their aggregation. We are also indebted to this researcher for the discovery of PGE2, a substance which blocks the transmission of the nerve influx by inhibiting the release of epinephrine, a neurotransmitter of the sympatic nerve system. He also did important work showing how these substances are synthesised by the reaction of arachidonic acid with oxygen. He further elucidated their structure, classifying them in three groups, PG1, PG2 and PG3, according to the number of double bonds (one, two, or three) in their hydrocarbon chain. The major part of Samuelsson's scientific career took place at the Karolinska Institute in Stockholm.

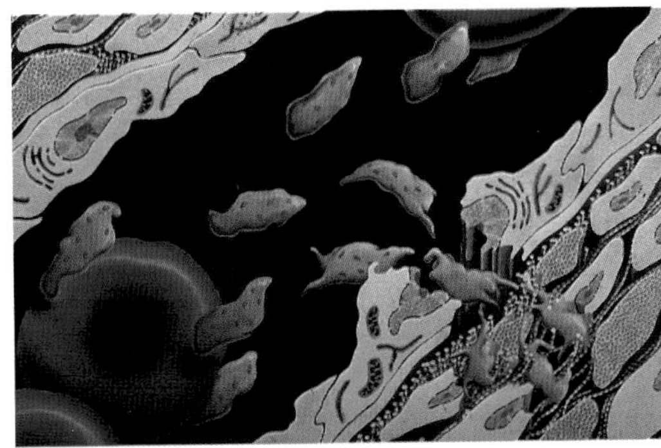

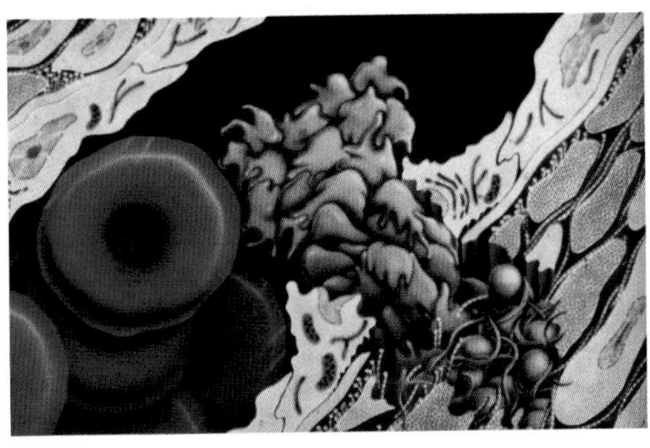

Blood clot formation, four steps. Platelets are required for clot formation. As the figure shows, their importance in blood coagulation is evident after injury of a vessel because they accumulate within seconds to form the platelet plug. During this process, they release several substances, among which are vasoactive prostaglandins.

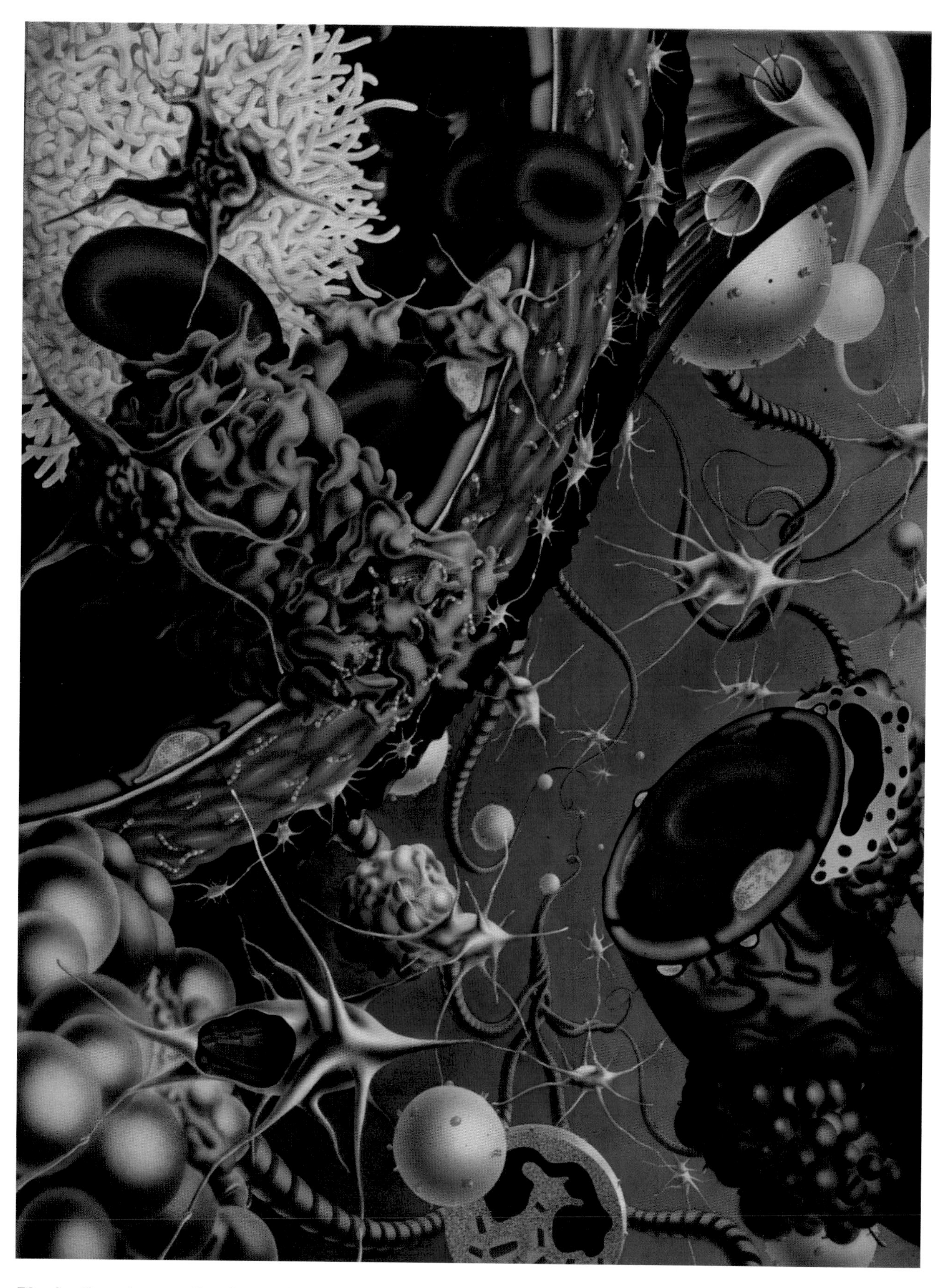

Blood cells and connective tissue cells. A platelet agregate can be seen in the aperture of the blood vessel. Thromboxane, a prostaglandine (discovered by Bergström) synthesized by platelets is one of the substances able to produce platelet aggregation.

Vane, John Robert (Tardebigg, Wortcestershire, England, March 29, 1927-). British pharmacologist. Nobel Prize for Medicine along with **Sune Bergström** and **Bengt Samuelsson** for his work on prostaglandins. Sir John Vane has had a distinguished career both in academia and industry. After receiving his bachelor's and doctorate in pharmacology from Oxford, he pursued academic research at Oxford and Yale University. He was a co-recipient of the Nobel Prize in Physiology or Medicine in 1982 for his work in prostaglandins and for discovering the mode of action of aspirin. His most highly regarded research was in the areas of cardiovas-

John Vane

cular disease and chronic inflammation. He spent eighteen years as a research scientist at the Royal College of Surgeons of England, the last seven of which were as Professor of Experimental Pharmacology. As a consultant to Squibb in the 1960s, he was the father of the company's angiotensin converting enzyme programme, which led to the block-busting drug, Captopril. During his twelve years as Group Research and Development Director at the Wellcome Foundation Ltd., he oversaw the development of a number of innovative drugs including Tracrium, Flolan, Zovirax and Lamictal. In 1986, Vane moved to St. Bartholomew's Hospital Medical School to found the Institute of Cardiovascular Research. He called it the William Harvey Research Institute because Harvey discovered the circulation of blood and had been a physician at Bart's in the 17th Century. In 1990, the Institute was awarded charity status. At the time of his retirement as Director General in 1997, the Institute had become one of the top 20 medical charities in the United Kingdom supporting a scientific staff of 130 people, including 17 different nationalities. It had trained many students, including more than 50 Ph.Ds. Over the years, Vane has jointly edited 20 books and published more than 900 papers. In 2000, the Institute was split into two entities, to enable the scientists to integrate with Queen Mary College, London University. He is now the Honorary President of the charitable arm, the William Harvey Research Foundation and holds a Professorship in the William Harvey Research Institute, Queen Mary College. He also acts as a consultant or Board member to several pharmaceutical and biopharmaceutical companies. John Vane became a Fellow of the Royal Society in 1974, was knighted in 1984 and has received numerous fellowships and honorary doctorates.

1983

McClintock, Barbara (Hartford, Connecticut, June 16, 1902 - Huntington, New York, September 2, 1992). American cytogeneticist. Daughter of a physician. Nobel Prize for Medicine for her works on "jumping genes". In several respects, this woman's remarkable work on the cytogenetics of maize is comparable to the famous work of **Thomas Morgan** on *Drosophila*. Like Morgan, she showed that chromosomal changes were responsible for phenotypic changes. During her long stay at Cold Spring Harbor, she first demonstrated the existence of linkage groups, as Morgan had done with the fruit fly. However, the work for which she was granted the Nobel prize opened an entirely new field in genetics. After several years of long, hard, solitary work, McClintock showed that certain genetic sequences can move on and between chromosomes, and thus influence the expression of genes determining certain characters (such as the colour of maize kernels). These sequences, called mobile genetic elements, transponders, or more commonly "jumping genes", can be occasionlly activated to move to another

Barbara McClintock

Corn

Barbara McClintock correlated visible rearrangements of chromosome segments with the redistribution of genetic traits in corn.

DNA site in the same cell by a process called transposition. McClintock worked on a variety of maize whose kernels were differently coloured because of unstable mutations. For instance, a mutation conferring the red colour appeared only in certain kernels, the others being colourless. The Cold Spring Harbor geneticist showed that the instability of this mutation resulted from the proximity of a transposable element which prevented its expression. If this transposable element is far from the mutant gene, the latter is expressed and a purple kernel appears on the cob. Barbara McClintock graduated in genetics from Cornell University in Ithaca, N.Y., and earned her doctorate in botany in 1927. After she taught at Cornell and

Missouri Universities, she started working at the Carnegy Institute in Long Island (Cold Spring Harbor), where she remained until her retirement.

1984

Jerne, Niels K. (London, Great Britian, December 23, 1911 -). Danish immunologist. Niels Jerne received the Nobel Prize in Medicine together with **Georges Köhler** and **César Milstein** for "theories concerning the developmental specificity and control of the immune system and for the discovery of the principle of clonal selection". Niels Jerne contributed one of the most fundamental ideas required for an understanding of the immune system. In 1956, he presented the hypothesis of clonal selection according to which an individual possesses at birth a multitude of lymphocytes, each one, when stimulated by an antigen, being capable of producing an antibody directed agaist the antigen. He was also one of the first to suspect

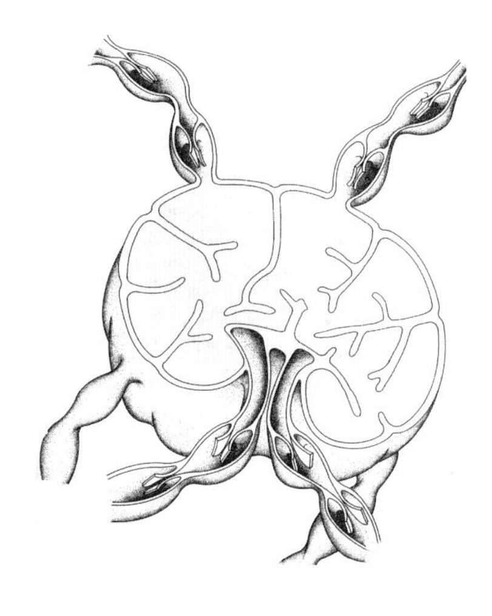

Lymph node. This type of organ is widely distributed throughout the lymphatic system of vertebrates. Several kind of cells take up residence in specialized regions of a lymph node's body, among them, B and T lymphocytes, macrophages and dendritic cells. Lymph nodes function as filters along the course of the lymph vessels. When a foreign material enters the body, specific immune cells can be stimulated.

that a mechanism of somatic mutation (of antibody genes) could contribute to the diversity of antibody response. The hypothesis of clonal selection was confirmed by the Japanese immunologist **Susumu Tonegawa** at the Basel Institute of Immunology. Tonegawa showed that the maturation of B lymphocytes was accompanied by sequence rearrangement in segments of genes coding for heavy and light chains of immunoglobulins. Jerne is also responsible for a theory of networks according to which the immune system autoregulates through antibody interactions. Niels Jerne was director of the Basel Institute for Immunology until his retirement in 1980.

Köhler, Georges J.F. (Munich, Germany, April 17, 1946 -). German immunologist. Nobel Prize for Medicine, along with **Niels Jerne** and **César Milstein**, for theories concerning the specificities in development and control of the immune system and the discovery of the principle for production of monoclonal antibodies. The research conducted in collaboration with Milstein at Cambridge University led to the creation of the first monoclonal antibodies. Monoclonal antibodies are antibodies with the same specificity, produced by a clone of genetically identical lymphocytes. To obtain mo-noclonal antibodies, individual B lymphocytes (able to produce antibody but having a limited life span) were fused with cells derived from B-lymphocyte tumors (unable to produce antibodies but "immortal"). This fusion results in the creation of hybrids cells called hybridomas, each of these able to produce a clone which in turn produces a single type of antibody. This discovery was hailed as a remarkable event by all immunologists and industrial biotechnologists. They were quick to see its potential as a means of producing specific antibodies in large quantities. Antibodies could be produced, for instance, that would recognise a certain cell type or even a cancer cell, or transport a therapeutic drug to its chosen target. George Köhler was awarded his doc-toral degree at the University of Freiburg-im-Breisgau in 1974. He went on to study at Cambridge and then at the Institute of Immunology in Basel. In 1985 he became the Head of the Max Planck Institute of Immunology in Freibourg.

Georges Köhler

Milstein, César (Bahia Blanca, Argentina, October 8, 1927 -). Argentinian-born British molecular biologist. Nobel Prize for Medicine, along with **Georges Köhler** and **Niels Jerne** for developing the technique in order to produce monoclonal antibodies. Milstein's research focused on how animals produce antibodies which specifically recognize and destroy invaders like viruses and bacteria. With collaborators, César Milstein discovered the way proteins are destined for secretion (leader peptide) but is best known as pioneer in the production of monoclonal antibodies. Few techniques have had such a sustained and profitable scientific and industrial impact as that developed by Milstein in 1975. Before then, it was practically impossible to obtain a pure preparation of antibody specifically recognising a given antigen. The reason was that an antigen usually elicits production of a great many different antibodies. This was troublesome for several reasons and notably hindered the study of antibodies themselves. His collaboration with **Köhler** made it possible to solve this problem. Cesar Milstein graduated

Cesar Milstein

from the Universities of Buenos Aires and earned his Ph.D. in medicine at Cambridge, England, in 1960. He was Deputy Director of the Molecular Research Laboratory (MRC), Cambridge, England from 1988 to 1995.

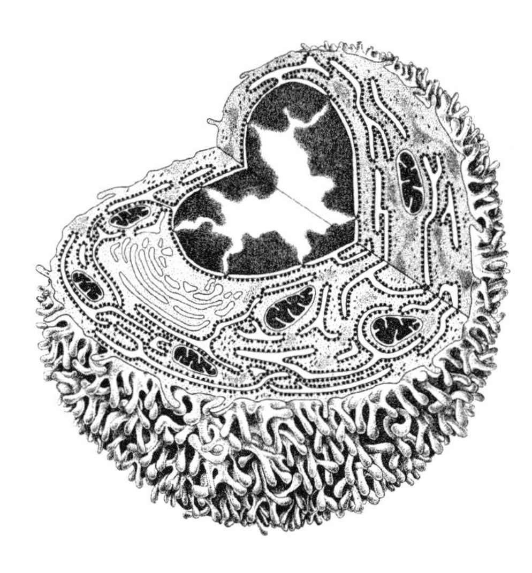

Lymphocyte B. A three - dimensional representation. The production of B cell clones, all producing the same type of antibody, was made possible by the work of Cesar Milstein and Georges Köhler.

1985

Brown, Michael Stuart (New York City, New York, April 13, 1941-). American molecular geneticist. Nobel Prize for Medicine, along with **Joseph Goldstein**, for their discovery of cholesterol cell receptors and the elucidation of their role in the body. Michael Brown and Joseph Goldstein, his long-time colleague, together discovered the low-density lipoprotein (LDL) receptor, which controls the level of cholesterol in blood and in cells. They showed that mutations in this receptor cause familial hypercholesterolemia, a disorder that leads to premature heart attacks in one out of every 500 people in most populations. The work of Brown and Goldstein established the first cause of heart attacks that could be traced to the molecular level. Their work also provided a strong scientific foundation for the theory that cholesterol-carrying LDL particles are a major cause of heart attacks. Brown and Goldstein also discovered receptor-mediated endocytosis, the fundamental process by which cells take up molecules from the blood. This dis-covery led to the realization that viruses, hormones and nutrient molecules enter cells by the same route. They have received many awards for this work, including the U.S. National Medal of Science. Michael S. Brown received a B.A. degree in Chemistry in 1962 and a M.D. degree in 1966 from the University of Pennsylvania. He was an intern and resident at the Massachusetts General Hospital, and a postdoctoral fellow with Dr. Earl Stadtman at the National Institute of Health. In 1971, he went to Dallas where he rose through the ranks to became professor in 1976. He is currently Paul J. Thomas Professor of Molecular Genetics and Director of the Jonsson Center for Molecular Genetics at the University of Texas South- western Medical School at Dallas.

Michael Brown

Goldstein, Joseph Léonard (Sumter, South Carolina, April 18, 1940 -). American molecular geneticist. Nobel Prize for Medicine, along with **Michael Brown**, for his elucidation of the process of cholesterol metabolism in the human body. Son of a clothier. After completion of his medical training, Joseph Goldstein spent two years at the National Institutes of Health, where he worked in the Laboratory of **Marshall W. Nirenberg** and also served as a clinical associate at the National Heart Institute. In Nirenberg's laboratory, Goldstein and his colleague C. Thomas Caskey isolated, purified, and worked out the mechanism of action of several proteins required for termination of protein synthesis. Here, he acquired scientific skills and taste, experienced the thrill of discovery and the excitement of science, and appreciated the power of a molecular biology approach to human disease. As a clinical associate, Goldstein served as physician to the patients of Donald S. Fredrickson, the Clinical Director of the National Heart Institute and an expert on disorders of lipid metabolism. His curiosity about hypercholesterolemia

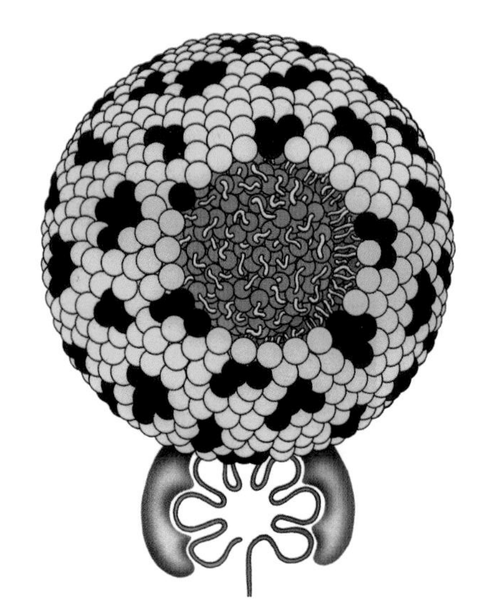

LDL vesicle. This kind of vesicle has a core of about 1500 cholesterol molecules esterified to long - chain fatty acids. This core is surrounded by a lipid monolayer containing a large protein molecule (not shown here).

was aroused when he cared for patients with the striking clinical syndrome of homozygous familial hypercholesterolemia. These patients were intensively discussed with Brown. In view of his and Brown's common interest in metabolic disease, Goldstein convinced his colleague to join him as a faculty member at the University of Texas Health Science Center at Dallas, where they would work collaboratively on the genetic regulation of cholesterol metabolism. While at the National Institutes of Health, Goldstein and Brown became avid duplicate bridge players. Their successful bridge partnership proved to be a valid testing ground for their future scientific partnership. Before returning to Dallas, Goldstein spent two years (1970 - 1972) as a Special NIH Fellow in Medical Genetics with Arno G. Motulsky at the University of Washington in Seattle. Motulsky was one of the creators of human genetics as a medical specialty. In Seattle, Goldstein initiated and completed a population genetic study to determine the frequency of the various hereditary lipid disorders in an unselected population of heart attack survivors. He and his colleagues discovered that 20% of all heart attack survivors have one of three single-gene determined types of hereditary hyperlipidemia. One of these disorders was the heterozygous form of familial hypercholesterolemia, which was found to affect 1 out of every 500 persons in the general population and 1 out of every 25 heart attack victims. During his fellowship in Seattle, he became conversant with tissue culture techniques, which proved to be invaluable in the subsequent studies with Michael Brown. In 1972, Goldstein returned to the University of Texas Health Science Center at Dallas, where he was appointed Assistant Professor in Seldin's Department of Internal Medicine and Head of the Medical School's first Division of Medical Genetics. He became Associate Professor of Internal Medicine in 1974 and Professor in 1976. In 1977 he was made Chairman of the Department of Molecular Genetics at the University of Texas Health Medical Center at Dallas and Paul J. Thomas Professor of Medicine and Genetics, a position that he currently holds. In 1985, he was named Regental Professor of the University of Texas.

Joseph Goldstein

Entry of cholesterol into a liver cell. After binding to the plasma membrane, the LDL particle is rapidly internalized and its content is delivered to acidic endosomes. While the receptors are returned to the plasma membrane, the LDL ends up in lysosomes. The lysosome enzymes hydrolyse the cholesteryl esters in the LDL particles and free cholesterol is released into the cytosol.

1986

Cohen, Stanley (Brooklyn, New York, November 17, 1922 -). American biochemist. Nobel Prize for Medicine, along with **Rita Levi-Montalcini**, for his research on factors produced in the body that influence the development of nerve and skin tissues. Stanley Cohen's father was a

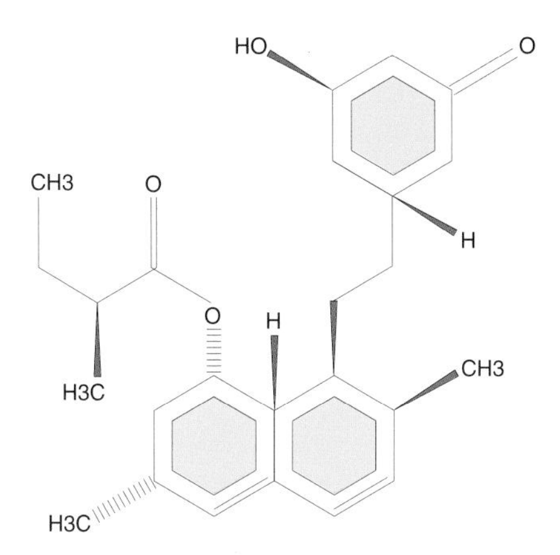

Lovastatine. As a potent inhibitor of enzyme involved in the metabolism of cholesterol, this substance is used clinically to induce an increase of LDL receptors on the hepatocyte membrane.

Russian immigrant and a tailor by trade. In his childhood he caught polio, becoming lame and handicapped for the rest of his life as a result. He started out in the laboratory of the famous zoologist Viktor Hamburger, a specialist in the biology of growth at the University of St Louis. He worked with Montalcini, an italian aristocrat who had discovered a nerve growth factor. Cohen succeeded in purifying this factor from a mouse tumour and raised an antibody against it. He demonstrated the existence of an equivalent of this substance in snake venom as well as in salivary glands from male mice. When he injected new-born mice with crude ex-tracts containing the factor, he found that they opened their eyes after seven days rather than at 14 days in control mice. Also, their teeth developed earlier. His work led to the discovery of a new growth factor called (EGF). He later identified and sequenced the receptor for this growth factor and discovered an important cell signalling pathway involved not only in cell growth, but also in cancer cell proliferation. The discovery of EGF soon led to some important therepeutic applications, notably for the healing and regeneration of skin damaged by burns. The application of a solution enriched in the growth factor accelerated wound

Stanley Cohen

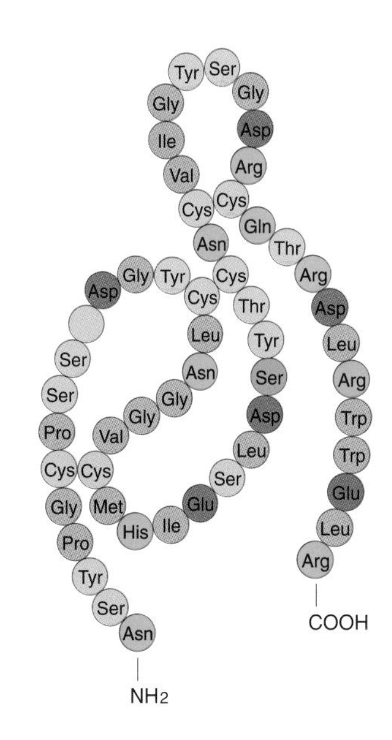

Epidermal growth factor. The amino acid sequence. EGF is widely used for treatment of severe burns.

healing endured by burns. Stanley Cohen earned his Ph.D. in biochemistry at the University of Michigan in 1948 and conducted research at Washington University under Levi-

Cell growth and cell division. These processes are under the control of a variety of growth factors. These substances most often act as signaling molecules.

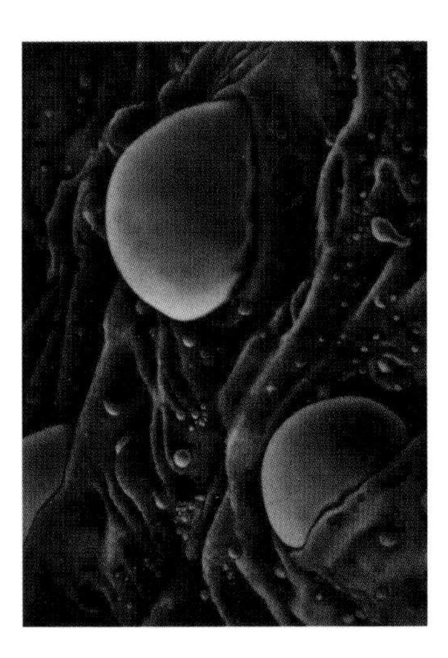

Bone marrow. A scanning electron micrograph. New blood cells are entering the circulation.

Montalcini until 1959. Then, he moved to Vanderbilt University, Nashville, Tennessee where he became professor of biochemistry in 1967.

Levi-Montalcini, Rita (Turin, Italy, April 22, 1909 -). Italian neurologist. 1986 Nobel Prize for Medicine, along with **Stanley Cohen**, for her discovery of nerve growth factor, a bodily substance that stimulates and influences the growth of nerve cells. Rita Levi-Montalcini finished her medical studies at the University of Turin just before the start of World War II. Because of her Jewish ancestry she went into hidding in Florence during the war. There she conducted research on the development of the chick embryo. The war forced her to continue her work in the Piedmont region, where she met another biologist named Levi. At the end of World War II, and after having taught at the University of Turin, she joined Washington University in St. Louis, Missouri, in 1947. There, she worked on chick embryos under the direc-tion of the physiologist Viktor Hamburger. In 1952 she succeeded in showing that the number of nerve fibres produced in chick embryos could be increased by an agent derived from a culture of mouse tumour cells. She further established that this agent could also act on embryonic nervous tissue cultured in vitro. The agent was in fact a growth factor, the first substance of its kind to have been discovered. From 1952 onward, Montalcini worked closely with Cohen to identify this substance, and in 1956, the latter managed to isolate it. It was named nerve growth factor (NGF).In 1980, it was established that the nerve growth factor stimulated the development of sensory and sympathetic neurons. Since the pioneering work of Rita Levi-Montalcini and Stanley Cohen many other growth factors have been discovered. Among them, the epidermal growth factor (discovered by Cohen), the platelet - derived growth factor, the fibroblast growth factor and the insulin-like growth factors.

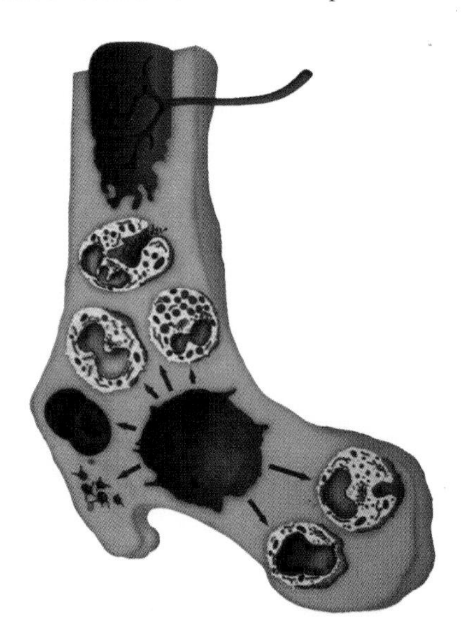

Bone marrow. It contains hematopoietic cells that differentiate into at least four types of specialized cells:erythrocytes, B - lymphocytes and T - lymphocytes, and various white blood cells including macrophages, neutrophiles, eosinophils and basophils. This process requires several growth factors, for example erythropoietin, a hormone that stimulates the production of erythroblasts from stem cells.

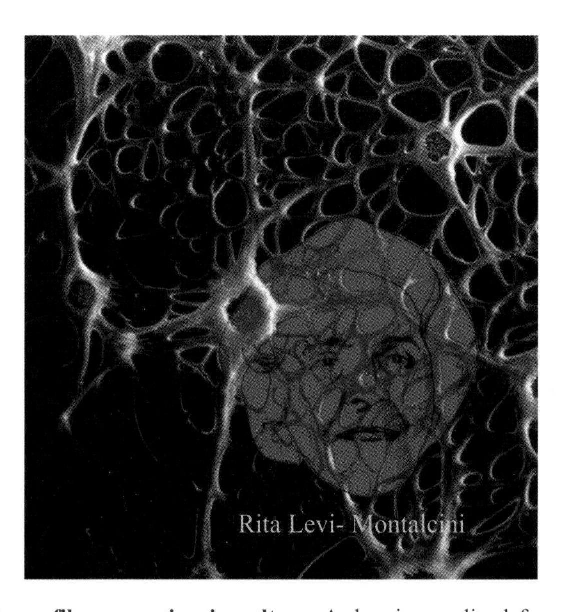

Nerve fibers growing in culture. A drawing realized from a scanning electron micrograph.

1987

Tonegawa, Susumu (Nagoya, Japan, September 5, 1939-). Japanese molecular biologist. Son of an engineer. Nobel Prize for Medicine for his recognition that the diversity of antibodies is a consequence of the huge numbers of lymphocytes present in the body, each with his own combination of functional antibody - producing genes. Susumu Tonegawa showed that the antibody diversity resulted from a process of somatic recombination. Such a process takes place during the development of B lymphocytes in the bone marrow. During this process the immunoglobulin genes rearrange themselves at random, thus generating billions of different antibodies. Incidentally, Tonegawa's work on immunoglobulin genes revealed, in an intron, a genetic regulatory element now known as an enhancer. This discovery contributed to our current understanding of cancer-forming mechanisms, notably in the case of blood cancers such as leukemias and lymphomas. Burkitt's lymphoma, for instance, is caused by the action of an enhancer sequence belonging to an immunoglobulin gene. Susumu Tonegawa was the first Japanese to win the Nobel Prize for Medicine. He became interested in chemistry while he was studying at the famous Hibyia High School. After presenting several papers on the genetics of transcription in bacteriophages at the University of California in San Diego, he took Renato Dulbecco's advice to go to the Institute of Immunology in Basel, where he began research that was to lead to explaining the origin of antibody diversity.

Susumu Tonegawa

1988

Black, James (Uddingston, Scotland, June 14, 1924 -). British pharmacologist. Nobel Prize for Medicine, along with **George H. Hitchings** and **Gertrude B. Elion**, for the development of two important drugs, Propranolol and Cimetidine. James Black and his colleagues designed novel drugs. They started from the known chemical structures of hormones. These molecules are targeted at cells in tissues and organs to coordinate various cellular functions. Cells can only respond to a hormone if they possess the appropriate receptor molecules at their cell surface. The hormone first recognizes its receptor and then binds to it specifically, thereby switching it on. Chemically, the two properties of hormone receptor binding and activation are separable. Black and his colleagues systematically modified the chemical structures of hormones in such a way that their binding properties were retained, whereas their activating properties were lost. In this way, they made hormone antagonists that bound to receptors, but blocked the action of the natural hormone. Using this principle, they designed drugs to reduce high blood pressure, heal stomach ulcers and treat epilepsy. James Black was Head of therepeutic research at the Wellcome pharmaceutical company from 1978 to 1984. He was also Professor of analytical pharmacology at the University of London. He was knighted in 1981.

Cross-section of the stomach. The stomach wall contains histamine receptors. The binding of cimetidine to these receptors does not cause allergical reactions. For this reason, a drug may be used for its anti-histaminic effects and for treatment of intestinal ulcers. Cimetidine was elaborated by James Black.

James Black

Elion, Gertrude (New York City, New York, January 23, 1918 - Chapel Hill, North Carolina, February 21, 1999). American pharmacologist. Nobel Prize for Medicine, along with **George H. Hitchings** and **James W. Black**, for her development of drugs used to treat several major diseases. Elion's father, Robert Elion, was a Lithuanian dentist. He emigrated to the United States along with his Russian wife Bertha Cohen. Her grandfather died of cancer and this motivated her to study medicine. Elion carried out much of her research in the Wellcome Pharmaceutical Laboratories in Great Britain. In 1967 she was appointed Head of the Experimental Therapy Department. Along with other duties she was president of the American Association for Cancer Research. At the outset, her scientific career was difficult. Although she had studied with success both chemistry and biochemistry, the fact that she was a woman was an obstacle to obtaining a research job in the chemical industry, although she coveted such a position. Most of Elion's work, conducted in close collaboration with Hitchings, involved developing ingenious methods for designing drugs that act specifically against diseased cells, without harming healthy ones.

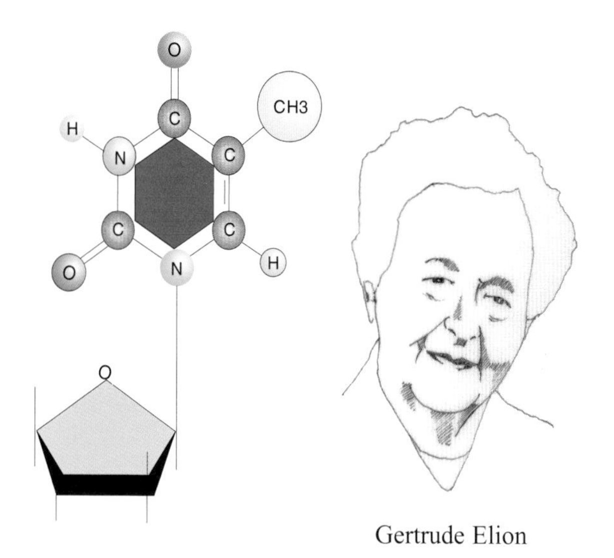

Gertrude Elion

Azido - thymidine. One of the most powerful drugs utilized at present for the treatment of acquired immunodeficiency.

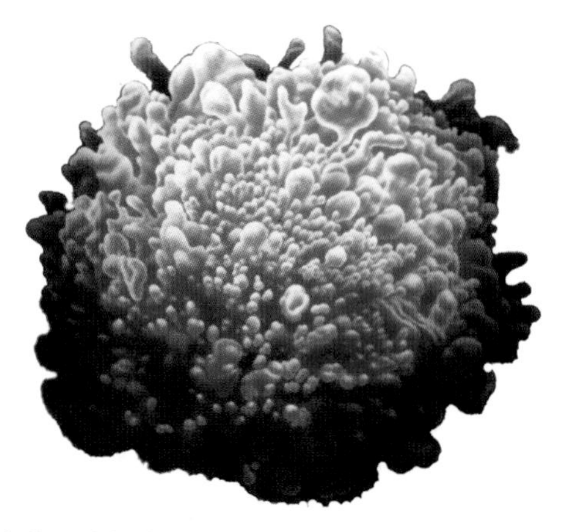

Cell infected by immunodeficiency virus. In this illustration, the cell is a macrophage but it could also be a T - lymphocyte, such as a T4 lymphocyte, which controls the functioning of the immune system.

Their research led, in 1951, to the synthesis of 6-mercaptopurine, also known as purinethol and sold under the name 6MP. Like thioguanine, this compound is a nucleic base analogue. In combination with other drugs, it proved truly effective against infantile leukaemia. Encouraged by this success to seek other applications, Elion and **Hitchings** went on to synthesise a derivative of 6MP: the immuno-suppressor azathioprine. Elion showed that this compound could prevent kidney transplant rejection in dogs and could also be used to treat auto-immune diseases such as rheumatoid arthritis. Another compound related to 6MP, allopurinol, was found to act against gout, a painful disease resulting from the accumulation of uric acid in the joints. Then came the discovery of pyrimethamine, an antimalarial compound, and thimethoprime, a drug effective against respiratory and urinary infections and notably against the *pneumococcus carinii pneumonia*, an opportunistic bacterium that infects AIDS patients. Although her laboratory was not equipped to conduct research on viruses, Elion developed and sent to England for testing a series of molecules, one of which was the drug to be known later as acyclovir. This compound is very effective against the herpes simplex and varicella-zoster viruses. It was in her laboratory that azidothymidine (AZT) was developped. AZT is used today in AIDS treatments.

Hitchings, George Herbert (Hoquiam, Washington, April 8, 1905 - Chapel Hill, North Carolina, February 27, 1998). American pharmacologist. Nobel Prize for Medicine, shared with **Gertrude Belle Elion** and **James Whyte Black**, for the development of drugs that became essential in the treatment of several major diseases. The major part of his contributions involved close collaboration with Elion and led to the discovery of numerous molecules of medical use, principally the antivirals. Several of these substances, as a result of their similarity to purines and pyrimidines, act by blocking the synthesis of nucleic acids, for example, 3' azido -2' 3' -dideoxythimidine, which is a major drug in the treatment of AIDS. This analog of deoxy thymidine triphosphate (dTTP), the right base, binds to the viral reverse transcriptase and blocks further chain elongation of DNA by this enzyme. Other strongly active compounds were the result of new pharmacological approaches introduced by George Hitchings and Elion. Among these were purethinol or 6MP, active against infantile leucemia, azthioprine (an immunodepressor) and allopurinol, used as a remedy for gout. Georges Herbert Hitchings

George Hitchings

studied at the University of Washington and received his Ph.D. from Harvard Medical School in 1933. He taught there before joining the Burroughs Wellcome Company and conducted research there until his retirement in 1975.

1989

Bishop, John Michael (York, Pennsylvania, February 22, 1936 -). American molecular biologist and virologist. Son of a Lutheran pastor. Nobel Prize for Medicine along with **Arnold Varmus** for their demonstration that cancer genes, or oncogenes, can arise from cellular genes called proto-oncogenes. Before the 1980s, many oncologists believed that retroviruses carried genes that caused certain cancers. When integrated into the host cell genomes during infection, these genes could be stimulated to induce transformation into cancer cells. The discovery of John Michael Bishop and Arnold Varmus showed that this hypothesis was incorrect. In essence, their discovery can be summarized as follows: genes, called proto - oncogenes, are present in normal cells and play a number of important roles, such as cell growth, cell division and cell differentiation. For example, the chicken gene src, whose existence had been postulated some 50 years earlier by **Francis Peyton Rous**, codes for an enzyme which is a protein tyrosine kinase. This particular enzyme catalyzes the phosphorylation of tyrosine residues in proteins. The gene normally participates in the regulation of cell growth. However, when a cell becomes infected by a retrovirus, the src gene can become incorporated in the viral genome at the end of the viral cycle and subsequently re-integrated into the DNA of a new cell following infection by the newly made viruses. If the gene is inserted incorrectly, for example at a site far from its normal chromosome site, it can escape the mechanisms regulating its expression and induce chaotic growth characteristic of cancer transformation. A similar imbalance can occur if the proto - oncogene is altered by mutagens, i.e. chemical substances found in the diet. Many cellular proto - oncogenes have since been discovered in animals and man. The Nobel Prize Jury insisted on the prime importance of this discovery by emphasizing the particular enthusiasm since then in searching for molecules that control cell growth. One controversy arose over the claim for the Nobel Prize by the French biologist, Dominique Stehlin, currently Head of the Pasteur Institute in Lille, France. Although he had proven the existence of proto - oncogenes, the Nobel Prize Jury considered that the fundamental ideas behind his experiments came from Bishop and Varmus. From 1968, Bishop was a lecturer at California University in San Francisco and Head of the G.W.Hooper Research Foundation.

Varmus, Harold Eliot (Oceanside, New York, December 18, 1939-). American microbiologist and virologist. Son of a farmer who left his native country before World War II. Nobel Prize for Medicine along with **Michael Bishop** for their demonstration that cancer genes, oncogenes, can arise from cellular genes, called proto - oncogenes. While studying a certain kind of viral gene that was responsible for causing tumors in chickens, they discovered that the cancer - causing gene was a normal gene. This discovery, which disproved the prevailing theory that cancer - causing

Chromosome 14
Chromosome 8
Translocation
myc gene
Immunoglobin gene
Immunoglobin gene
myc gene

Burkitt's lymphoma. Burkitt's lymphoma is a solid tumor of B lymphocytes. The genes for making antibodies are expressed only in B lymphocytes but in Burkitt's lymphoma, a reciprocal translocation has moved the proto-oncogene c-myc from its normal position on chromosome 8 to a location close to the enhancers of the antibody heavy chain genes on chromosome 14.

Harold Varmus

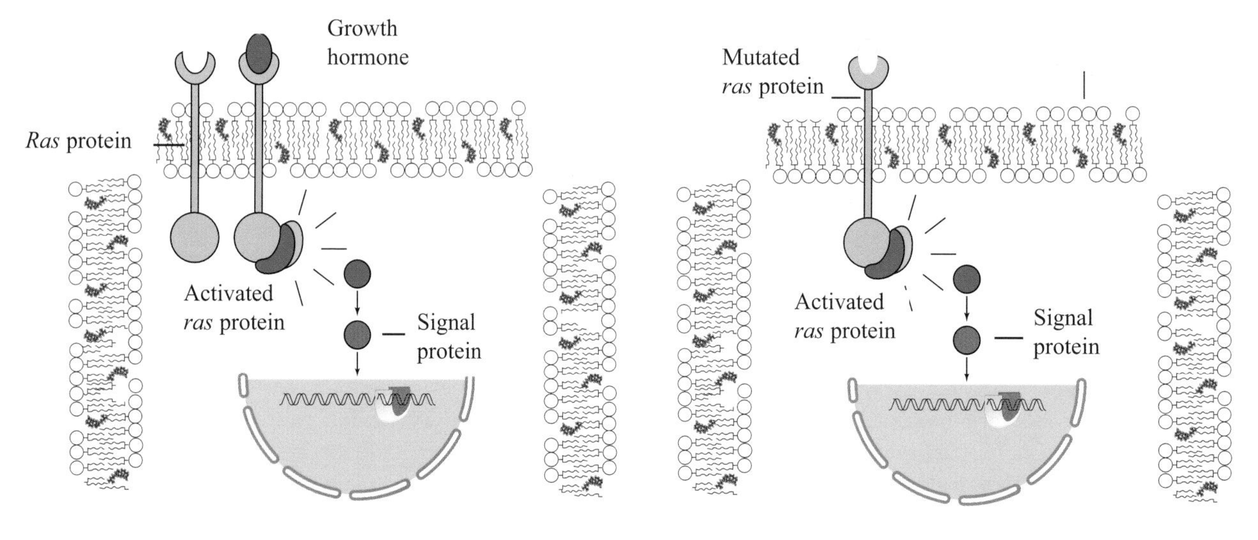

Proto - oncogenes and oncogenes. Several of these proteins are anchored in the cell membrane. Others are inside the cell. Many of them, called proto - oncogenes, normally appear to be involved in growth control. If a proto - oncogene coding for such a protein is mutated, it becomes an oncogene and it can produce an altered substance which stimulates a cell to divide continually. The altered protein can promote the transformation of a normal cell into a cancer cell.

genes had a viral origin, laid the cornestone for understanding the genetic origin of cancer. These normal cellular genes only caused problems when they or their control system were altered. Thus, these studies provided new insight into the complicated signal systems that govern the normal growth of cells and an unparalleled opportunity to discover how a normal cell becomes a cancer cell. A common mechanism for cancer is now widely accepted:cancer arises from previously normal cells, as a result of mutation in genes that control growth. Scientists have been able to identify which genes have been mutated in many cancers

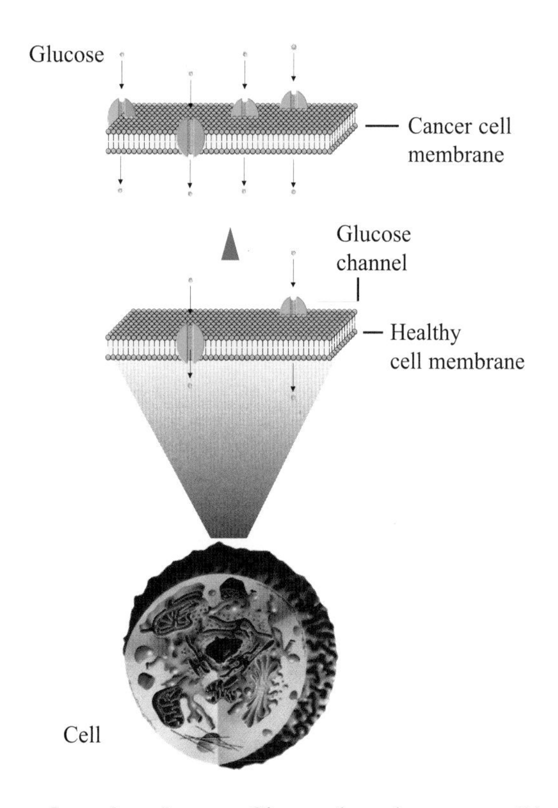

Glucose channels and cancer. Glucose channels are responsible for the uptake of glucose inside cells. In growing cancer cells, their number increases dramatically in order to provide the supplementary energy required.

Cultured cells that have been treated with certain viruses or chemical agents may be transformed. As a result, they lost normal control of cell division and continue to grow indefinitely.

and have been able in some cases to use this information in estimating the severity of a cancer or gauging a person's risk of developing cancer. The studies made possible by the finding of Varmus and Bishop may in time lead to the development of better means to diagnose, treat and prevent many kinds of cancer. Varmus is also widely recognized for his studies of the replication cycles of retroviruses and hepatitis B viruses. Harold Varmus is a graduate of Columbia University's

A cancer cell wandering in a blood vessel. An artistic drawing.

College of Physicians and Surgeons. His scientific training occurred first as a Public Health Service officer at the National Institutes of Health, where he studied bacterial gene expression with Dr. Pastan, and then as a post-doctoral fellow with Michael Bishop at the University of California, San Francisco. In 1993, he was named as the Director of the NIH, a position he held until the end of 1999.

1990

Murray, Joseph E. (Milford, Massachusetts, April 1, 1919 -). American immunologist. Nobel Prize for Medicine, along with **Edward Donnal Thomas**, for his work in developing fundamental techniques in transplantation surgery. Joseph E. Murray and his colleagues at Peter Bent Brigham Hospital in Boston performed the first truly successful kidney transplant from one twin to another. One of the main problems facing this surgeon was how to suppress, or at least weaken considerably, the immune response which a recipient usually develops against a graft from a donor who is not genetically identical. He finally succeeded in performing this operation by using the first immunosuppressant drugs such as azathioprine and later cyclosporine, two substances developed by Gertrude Elion. The successful transplantation of a kidney demonstrated conclusively that transplantation of solid organs in humans was possible. First performed on December 23, 1954 between identical twins with no genetic disparity, kidney transplantation was later performed between bro-

Joseph Murray

thers and, ultimately, from a cadaveric donnor. These all occured in the 1950s thus ushering in the era of organ transplantation. Subsequently, the transplantation of hearts, livers, lungs, pancreas and intestines have been successfully performed, initiating the single most important medical and surgical advance of the 20th century. We are also indebted to Murray for the first bone marrow transplantation in 1956. Joseph Murray served on the faculty at Harvard Medical School and its affiliated Peter Bent Brigham Hospital for several years. Murray graduated in medicine from Harvard Medical School in Cambridge, Massachusetts. After working in Boston (he began the work there for which he was awarded the

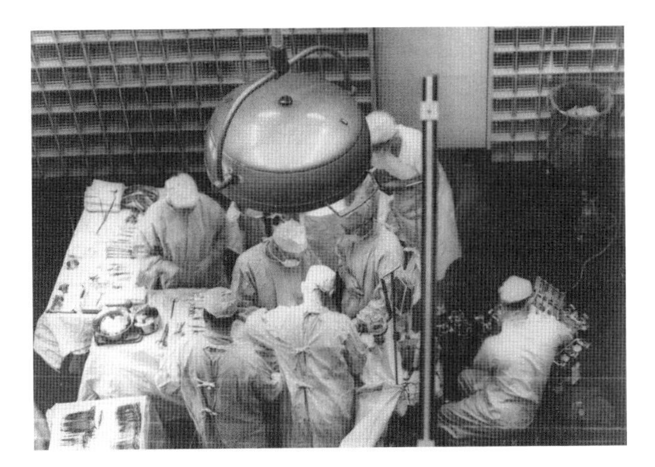

Photograph showing the surgical operation during which the first kidney transplant was performed. This operation was performed on a dog.

Nobel Prize), he moved to Brigham and then to the University of Harvard, Boston, where he taught plastic surgery.

Thomas, Edward Donnal (Mart, Texas, March 15, 1920 -). American immunologist. Nobel Prize for Medicine, along with **John Murray**, for his work in developing fundamental techniques in transplantation surgery. To this American physician we owe the development of an effective bone marrow transplantion technique in which blood cells and lymphocytes from a donor's bone marrow are trans-ferred to a recipient, the donor and recipient being as immunologically compatible as possible. It was parti- cularly hard to develop this technique. One danger to be avoided is rejection of the donated marrow by the reci -pient. Another is graft - versus - host disease, a potentially fatal reaction of the transplanted lymphocytes against the recipient's tissues. Developed first on laboratory dogs, this technique was applied with great success at the Medical School of the University of

Edward Thomas

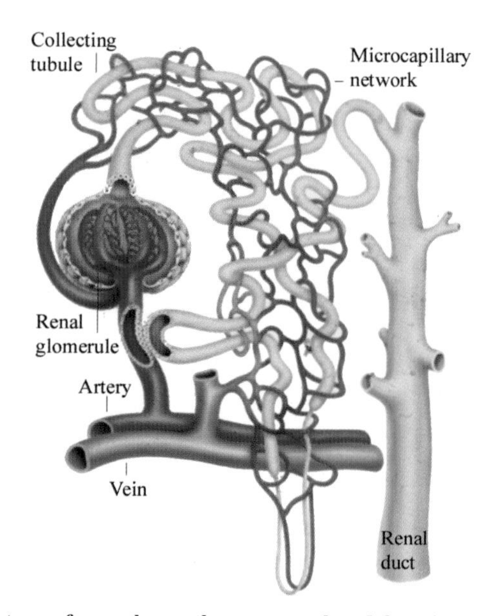

Structure of a nephron, the structural and functional urine-secreting unit of the kidney. Typically each nephron is made up of a glomerulus and a proximal convoluted tubule.

Washington where Thomas worked, a centre renowned worldwide for this type of technology. Thomas discovered an immunosuppressive drug called methotrexate. This substance is often given to prevent the graft - versus - host disease. He graduated from the University of Texas and Harvard Medical School and received his M.D. degree in 1946. From 1955 to 1963, he taught medicine at Columbia before moving to the University of Washington in Seattle at the invitation of Robert Williams, an endocrinologist who worked there. In

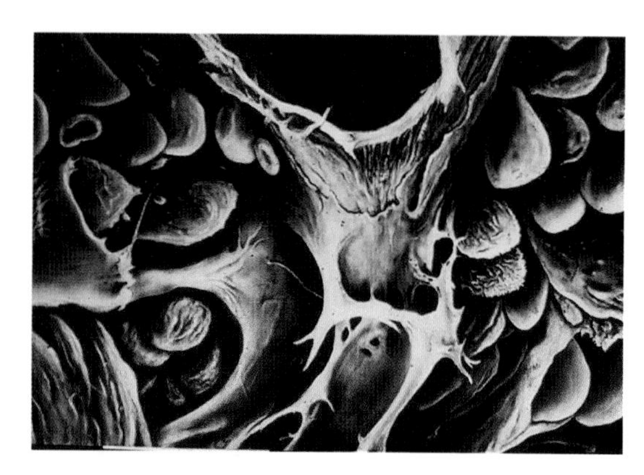

Bone marrow. A scanning electron micrograph. New blood cells are entering the circulation.

1975 he took a post at the Fred Hutchinson Cancer Research Center (now directed by LeeHartwell) in Seattle and carried out there more than 4,000 human marrow transplants.

1991

Neher, Erwin (Landsberg, Germany, March 20, 1944-). German biophysicist. Nobel Prize for Physiology or Medicine shared with **Bert Sackmann** for their research into basic cell function and for the development of the patch-clamp technique. This technique is remarkable in that it records how a single-channel molecule alters its shape and, in that way, controls the flow of current across the

Erwin Neher

membrane of a cell, within a time frame of only a few millionths of a second. Erwin Neher's interest lies in the basic signalling mechanisms of nerve cells and neuro-endocrine cells. In early years he developped, together with Sackmann, the patch clamp technique for recording single channel currents. He applied it to the transmitter activated currents at the neuromuscular junction and to voltage-acti-

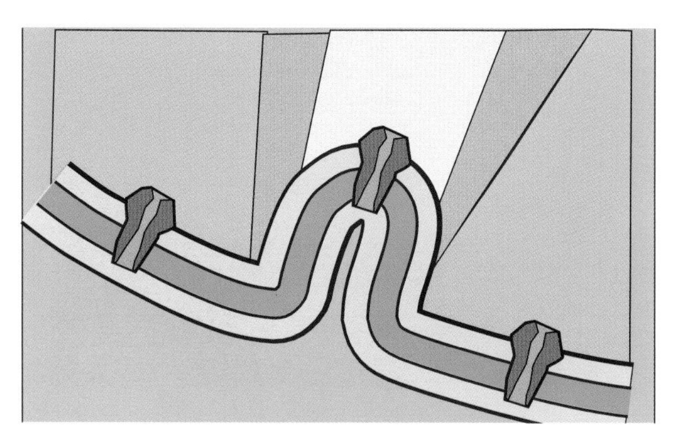

The patch clamp technique. It was developed by Erwin Neher and Bert Sackmann in 1976 to study the activity of individual ion channels. It allows measuring ion flow only through a single channel in the patch of plasma membrane as well as the effects of the membrane potential on the opening and closing of such a channel.

vated channels in neuroendocrine cells. More recently his interest shifted to the cellular action of channel - mediated signals, particularly to calcium concentration changes upon opening of calcium channels and to their role in triggering release of neurotransmitters and hormones. He studied the kinetics of the process of exocytosis and cellular calcium-buffering properties in order to establish quantitative relationships between calcium currents and neurotransmitter release. Erwin Neher graduated in physics from the Technical University of Munich and earned his Ph.D. there in science in 1970.He is now director of the Department of Membranbiophysik at the Max-Planck-Institut, Göttingen, Germany.

Bert Sackmann

Sackmann, Bert (Stuttgart, Germany, June 12, 1942 -). German molecular biologist. Nobel Prize for Medicine, along with **Erwin Neher**, for research into basic cell function and for their development of the patch-clamp technique. Bert Sackmann's interest in mechanisms of electrical excitability and synaptic transmission began in 1970, when he worked for three years as a postdoctorate in Bernard Katz's Department of Biophysics at University College, London. Dale Purves and Bert Sackmann found that the density and the molecular composition of acetylcholine receptor channels in muscle are regulated by electrical activity. This indicated that elucidation of function and structure of ion channels, in particular those associated with transmitter receptors, is essential for an understanding of "plastic" changes in synapses of the ner-

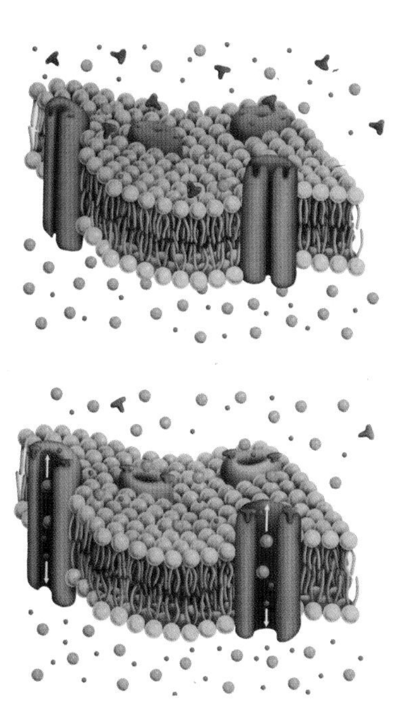

Ionic channels imbedded in a cell membrane. Thanks to the procedure developed by E. Neher and B. Sackman, the activity of such structures can now be individually measured.

vous system. Consequently, Sackmann characterized the elementary current events mediated by different classes of ion channels which presumably mediate those electrical signals that make our hearts beat and our brains think. With Neher, then working at the Max Planck Institute for Biophysical Chemistry in Göttingen, Germany, he succeeded in determining the conductance and gating properties of single ion channels in muscles and nervous cells. Since 1989, Sackmann has worked at the Max-Planck-Institute for Medical Research, Heidelberg. With his collaborators, he identified the basic excitatory circuitry of the rodents cortex and showed that action potentials also propagate backward into the dendritic arbor where the majority of synapses are located. This talking back is a trigger mechanism controlling the increase and decrease of synaptic efficacy in the cortex. In collaboration with Peter Seeburg, Sackmann showed, using genetic engineering techniques, that the abundance of one particular subunit of glutamate gated ion channels is one essential requirement for cortical synapses to undergo increases and decreases of synaptic efficacy evoked by their electrical activity. Bert Sackmann graduated from the universities of Tübingen and Munich. He also conducted research in Bernard Katz's Department at University College, London.

1992

Fischer, Edmond H. (Shanghaï, China, April 6, 1920 -). American biochemist. Son of an Austrian businessman. Nobel Prize for Medicine, along with **Edwin G. Krebs**, for his discoveries concerning reversible phosphorylation, a biochemical mechanism that governs the activities of a number of cell proteins. After having achieved his medical studies at the University of Geneva, where he earned his Ph.D in 1947, Edmond Fischer moved to the United States, in California first, then to the University of Seattle, Washington, where he joined Edwin Krebs. Most of his research was carried out in collaboration with him. It stemmed from their work in the mid - 1950s, that the activity of phosphorylase, the enzyme that catalyzes degradation of glycogen, was regulated by such a reaction brought about by the opposing actions of phosphorylase kinase and phosphatase. Phosphorylase kinase was purified and characterized and they further demonstrated that this enzyme was itself activated by another kinase (the cAMP - dependent protein kinase) thereby establishing the concept that signaling pathways consist of cascades of enzymatic reactions in which, more often than not, several kinases act successively on one another. Today, reversible protein phosphorylation is considered to be the most prevalent mechanism by which cellular events are regulated. It is involved in such diverse processes as signal transduction, gene transcription and translation, embryonic development, cell growth, differentiation and death, muscle contraction, nerve conduction, the immune response and diseases such as diabetes and oncogenic transformation. Edmond Fischer became a full professor at the University of Seattle in 1961.

Edmond Fischer

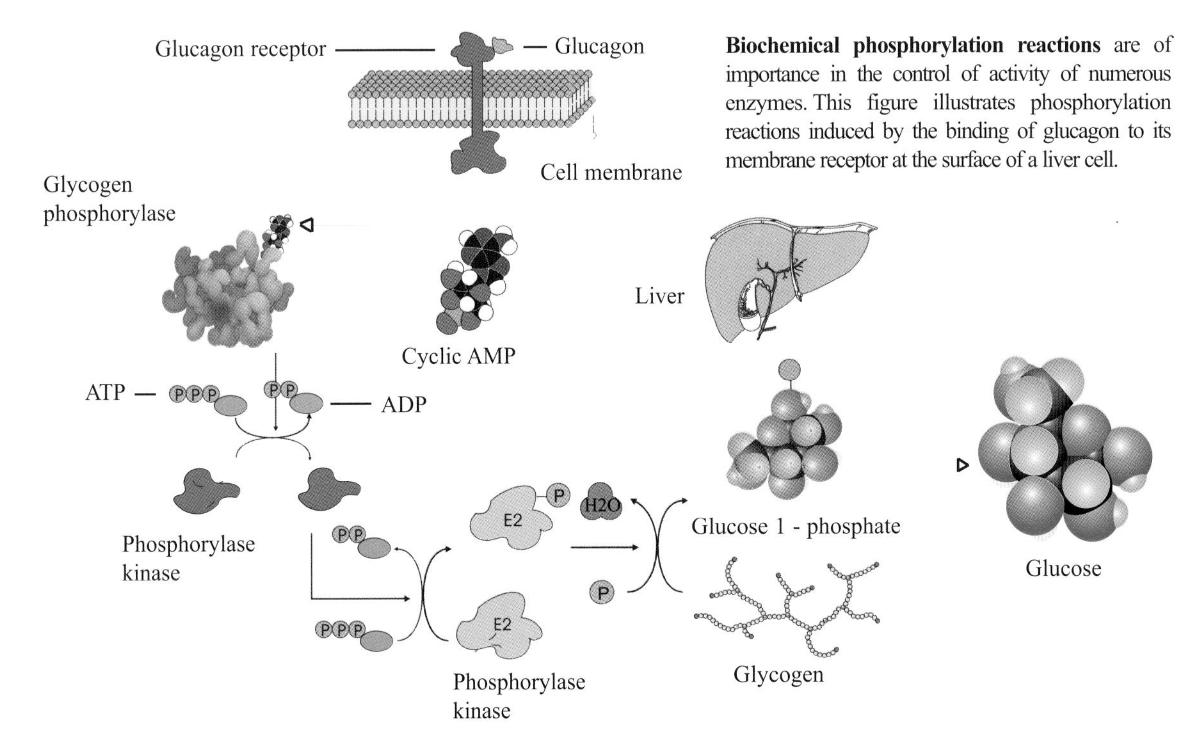

Biochemical phosphorylation reactions are of importance in the control of activity of numerous enzymes. This figure illustrates phosphorylation reactions induced by the binding of glucagon to its membrane receptor at the surface of a liver cell.

Krebs, Edwin (Landsing, Iowa, June 6, 1918 -). American biochemist. Son of a Presbyterian minister. Nobel Prize for Medicine, along with **Edmond Fischer**,

Edwin Krebs

for his contribution to the discovery of reversible protein phosphorylation. In the 1950s, the Coris had isolated a phosphorylase involved in glycogen metabolism. This enzyme could alternate between an active and an inactive form. Fischer and Krebs conducted research in order to elucidate the mechanism leading to this activation and the mechanism promoting the inactive state of the enzyme. The collaboration between the two researchers was so close that one can hardly attempt to dissociate their works. Their investigations essentially concerned cellular kinases and phosphatases. According to a commentary published in a recent issue of "New Scientist", the functions of kinases and phosphatases are so diverse that they could become one of the most important research focuses of the 21st century. Edwin Krebs earned a medical degree at Washington University in St. Louis, Missouri, in 1943. After working with the Cori's he moved to the University of Washington, Seattle, in 1968 before conducting research at the University of California. He finally settled down at Washington University in 1977 as Chairman of the Department of Pharmacology.

1993

Sharp, Phillip A. (Falmouth, Kentucky, June 6, 1944 -). American molecular biologist. Nobel Prize for medicine along with **Richard Roberts** for the discovery that genes in organisms more complex than bacteria have a split structure. Split genes contain nonsense segments that are removed by an "editing" process called RNA splicing in the course of expression of the genetic information. Before this discovery, a gene was conceived as a continuous segment within the long double - stranded DNA molecules. The discovery that informational sequences within a gene could be discontinuous, that is, present in the genetic material (DNA) as several well-separated segments was unanticipated and fundamental for understanding both normal and disease cellular functions. As an

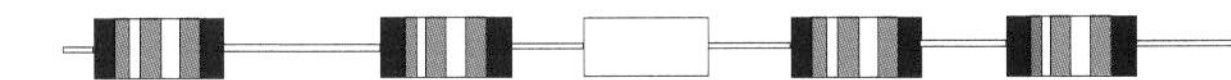

Introns and exons of the globin chain gene. The coding sequences (exons) are separated by non coding sequences (introns).

experimental model system, Phillip Sharp used a common cold-causing virus, called adenovirus, whose genes display important similarities to those in higher organisms. The discovery of split genes changed our view on how genes in higher organisms developed during evolution. The discovery of split genes has been of fundamental importance for research both in biology as well as in medicine. Phillip Sharp graduated from the University of Illinois where he earned his Ph.D. in 1969. That year, he worked at the California Institute of Technology before moving to the Cold Spring Harbor Laboratory in Long Island in 1971.

Roberts, Richard J (Derby, England, September 6, 1943 -). British molecular biologist. Nobel Prize for Medicine, along with **Philip Sharp** for his discovery that eukaryotic genes are often split. Richard Roberts's discovery of split genes showed that the coding information in the DNA of

Richard Roberts

higher organisms such as humans was arranged differently than had previously been known for bacteria. In bacteria the coding information is in a linear string with no gaps, whereas in higher organisms there are numerous gaps within the genes themselves. When the cell processes the information from the DNA it has to remove the sequences in the gaps and join together the meaningful coding information. The process is much like a film editor

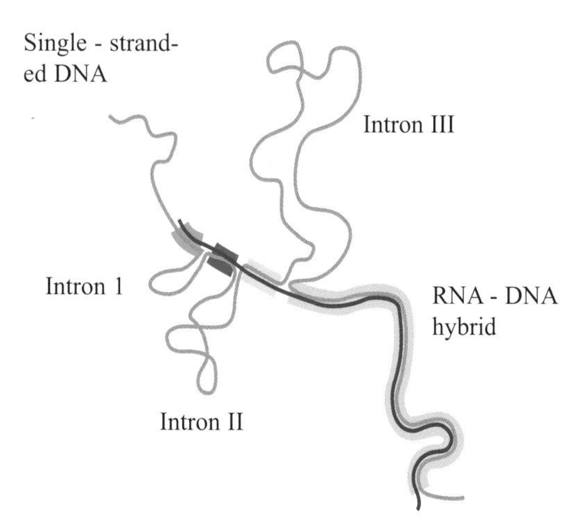

Figure showing that many eukaryotic genes are mosaics of coding and non-coding regions. An mRNA molecule was hybridized to genomic DNA containing the corresponding gene. Three single-stranded loops corresponding to introns separate regions of RNA-DNA hybrids. Such observations led to the discovery by Roberts and Sharp that much eukaryotic DNA was not continuous but was interrupted by introns, which were subsequently removed from primary transcripts by splicing.

who takes the raw footage and then cuts and splices it to produce the film that is shown in theaters. With Richard Gelinas, Roberts' work involved extensive biochemical and molecular biological studies of adenovirus-2 mRNA and the encoding DNA sequences, which indicated that an unusual mechanism of mRNA formation must be taking place. Finally, the structure of the genes encoding these mRNAs was revealed by electron microscopy experiments carried out in collaboration with Louise Chow and Tom Broker. Roberts graduated in organic chemistry from the University of Sheffield, England, in 1968. After conducting research at Harvard, he moved to Cold Spring Harbor Laboratory in New York in 1972. Richard Roberts is now Director of Research at New England Biolabs, United States. According to the Nobel Committee, Roberts' and Sharp's discovery has changed our view on how genes in higher organisms developed during evolution.

1994

Gilman, Alfred G. (New Haven, Connecticut, July 1, 1941-). American pharmacologist and molecular biologist. Nobel Prize for Medicine, along with American biochemist **Martin Rodbell** for discovering molecules called G proteins and their role in the transmission of cell signals. In working on mutant cells incapable of sending hormonal signals, Alfred Gilman more precisely defined the nature of membrane transductors whose role had already been established several years earlier by Rodbell. These transducers play a crucial role in the transmission of hormonal signals in metabolism and in many processes such as vision, sense of smell and cognition. They function as communicators between the membrane receptor and the enzyme adenylate cyclase. The binding of the hormone to the receptor activates a conformational change in this receptor. The activated receptor then binds to a component of protein G, called Gs alpha and the GDP bound to this subunit is replaced by GTP. This step is followed by the dissociation of Gs alpha-GTP from the rest of the protein G and its binding to adenylcyclase which in turn produces cyclic AMP. The hydrolysis of GTP causes Gs alpha to dissociate from adenylate cyclase and its re-association (as Gs-GDP) to the other part of protein G. Rise in intracellular cyclic AMP, which acts as a signal transducer, modifies the rates of different enzymatic reactions in specific tissues. There are multiple subtypes of G-protein-coupled receptors in tissues (for example the photoreceptor in retina or receptors involved in cell division or in cell differentiation), but all of them display common structural features, in particular seven sequences of hydrophobic amino acids forming seven transmembrane helical structures. For example studies on cholera (a desease

Alfred Gilman

caused by *vibrio cholerae*, a bacteria) showed that binding of the cholera toxin to its receptor in the membranes of sensitive cells inhibits GTPase (an enzyme that hydrolyses GTP to GDP and inorganic phosphate) activity and converts G protein to an irreversible activator of adenylcyclase. In the intestinal epithelial cell, the abnormally high level of cyclic AMP continously stimulates the active transport of ions and leads to a very large influx of sodium ions and water into the gut. The striking clinical feature of this disease is a massive diarrhea. Alfred G. Gilman received his B.S. (summa cum laude) in biochemistry in 1962 from Yale University, and his M.D. and Ph.D. in pharmacology in 1969 from Case Western Reserve University. He received further training as a

Pharmacology Research Associate in the Laboratory of Biochemical Genetics at the National Institutes of Health (1969-1971).

Rodbell, Martin (Baltimore, Maryland, December 1, 1925 - Chapel Hill, North Carolina, December 7, 1998). American biochemist. Nobel Prize for Medicine shared with **Alfred Gilman** for his discovery of cell signals transducers called G-proteins. Martin Rodbell was Director of the Laboratory of Nutrition and Endocrinology (1970 - 1985), then biochemist at the National Institute of Environmental Health in North Carolina. He showed hormone - induced cellular responses, at least in certain cases, are mediated by three independent elements that interact at the time of hormone stimulation. According to an hypothesis, these elements were localized in membranes and were composed of a receptor, a transducer, and an amplifier capable of large quantities of "second messengers" such as those described previously by **Earl Sutherland**. Rodbell was the first to realize that the receptor and the amplifier must be two independent units separated in space. In addition, he showed that the transducer was an intermediate element essential for signal transmission between the other two elements. Martin Rodbell first graduated from Johns Hopkins University and then from the University of Washington, where he earned his Ph.D. in biochemistry. He began his career as a biochemist at the National Institutes of Health in Bethesda, Md. Later, he worked at the National Institutes of Environmental Health Sciences in North Carolina.

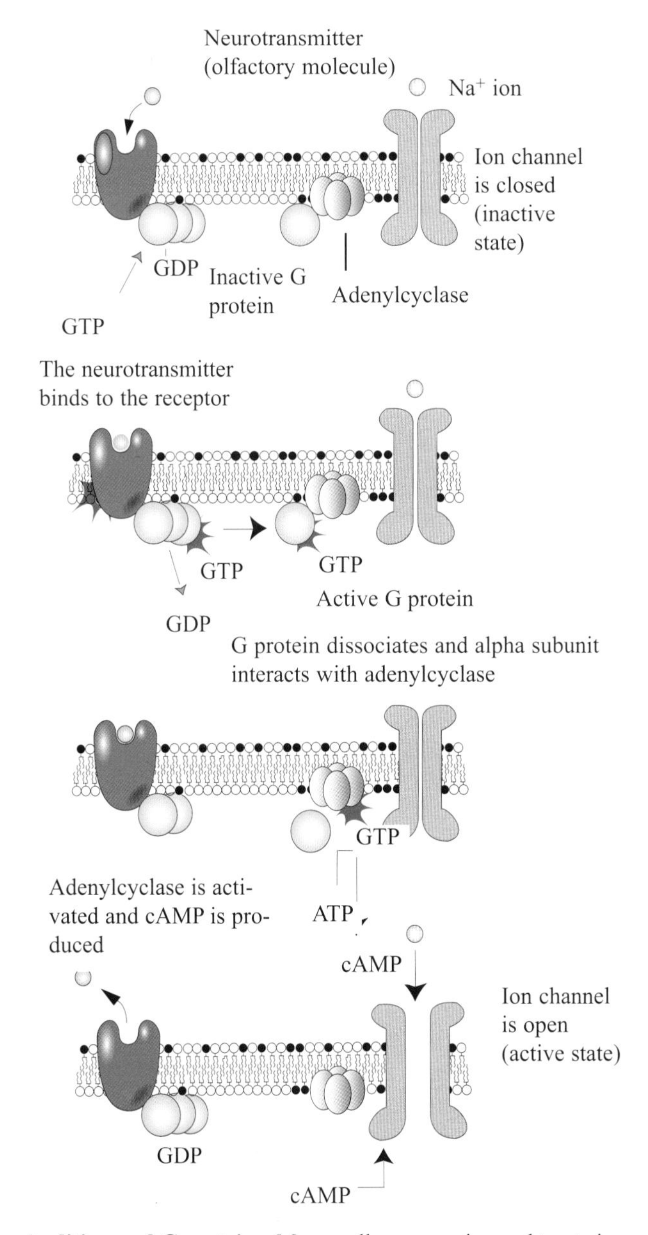

Audition and G proteins. Many cells can receive and treat signals. Such cells are equipped with a transduction system comprising a G protein. In this illustration, the G protein is involved in transmitting signals from auditory receptors anchored in the plasma membrane of a ciliated cell of the inner ear.

1995

Nüsslein-Volhard Christiane (Magdebourg, Germany, October 20, 1942 -). 1 Nobel Prize for Medicine with **Eric F. Wieschaus** and **Edward B. Lewis** for her research concerning the mechanisms of early embryonic development. While at the European Molecular Biology Laboratory (EMBL) in Heidelberg, Christiane Nüsslein-Volhard and Wieschaus were interested in discovering factors that regulate the first steps of embryogenesis. How does an animal develop from a single cell into a complex embryo with its many different structures and cell types? As a model organism for their studies they chose the fruit fly *Drosophila melanogaster*. In order to identify genes involved in this process, Nüsslein-Volhard and Wieshaus developed a powerful strategy: They treated fruit flies with chemicals that randomly caused mutations in the DNA of the flies. They reasoned that in some of these flies, mutations would be located in genes which play important roles in embryogenesis. In these cases, the damaged gene could no longer fulfil its normal function

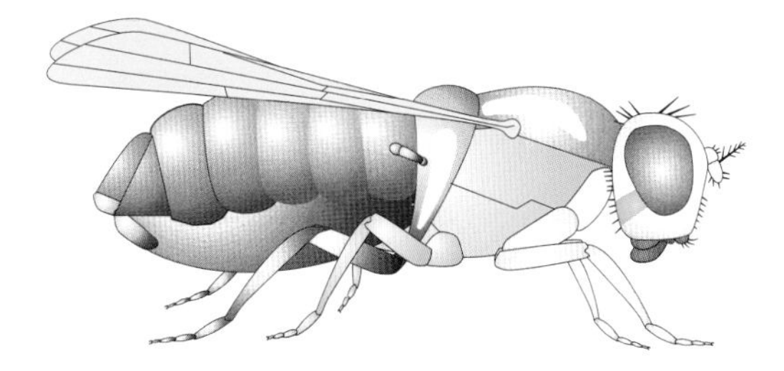

Homeotic genes were first discovered in *Drosophila melanogaster*. A mutation in such a gene could, e.g., result in a leg replacing an antenna. The vertebrate homeotic genes are expressed at particular stages in embryogenesis.

during development and as a result the fly would look different from normal flies. Having screened through around 40 000 mutations, Nüsslein-Volhard and Wieshaus had identified 120 genes with essential roles during embryogenesis.The first results of their studies were published in 1980. Although immediately recognised as a major breakthrough for embryology research by some scientists at the time, their discoveries were not fully appreciated until it was realised that in many cases related genes are present in most animals - including human beings - and often perform very similar tasks. More recently it has also been shown that they also play important roles in human disease, most notably in cancer. It would be difficult to overestimate the impact of Christiane Nüsslein-Volhard and Eric F. Wieshaus's contribution to the field of developmental biology. Their pioneering studies linking genetic techniques with embryogenesis form the basis for the work conducted in hundreds of laboratories throughout the world today and laid the ground for most of what we now know about how genes control development.

Wieschaus, Eric F. (South Bend, Indiana, June 8, 1947 -). Nobel Prize for Medicine, along with **Edward B. Lewis** and **Christiane Nüsslein-Volhard** for the genetic controls of early embryonic development. In 1980, Eric Wieschaus and Christiane Nusslein-Volhard published their work establishing the possibility to identify a number of genes whose expression controls the development of the fruit fly drosophila. Before their discovery it was clear that genes somehow controlled the process but such genes were not yet identified. To find an answer to this crucial question, Wieschaus and Nusslein-Volhardt decided to use drosophila, a current object of genetics because of its rapid reproduction. By inducing mutations in young fly embryos they succeded in recognizing genes whose expression precisely control the development of head, arms, eyes, muscles and other organs which normally appear in the adult fly. Moreover they showed that the genes that play essential roles in human development are not so different from those studied in this insect. Wieschaus earned his doctoral degree from Yale University in 1974. He then joined the University of Princeton in 1981 and is now currently teaching molecular biology.

Lewis, Edward B. (Wilkes-Barre, Pennsylvania, May 20, 1918 -). American developmental geneticist. Nobel Prize for Medicine, along with geneticists **Christiane Nüsslein-Volhard** and **Eric F. Wieschaus**, for discoveries concerning the genetic control embryonic development. Edward Lewis was educated at the California Institute of Technology. His is known for his studies of a group of genes affecting segmentation in *Drosophila.* Using a formal genetic approach, he discovered that the order of these genes in the chromosome reflected the order of the segments whose development they controlled. These so called "homeotic" genes, in controlling the expression of numerous subordinate genes, can direct the formation of organs such as the wings and legs of the fly.

1996

Doherty, Peter (Australia, October 15, 1940-). Australian immunologist and pathologist. Nobel Prize for Medicine, along with **Rolf Zinkernagel**, for his discovery of how the body's immune system distinguishes virus-infected cells from normal cells.A professor in the Department of Immunology of the Children's Hospital in Memphis, Tennessee, Peter Doherty showed the immune's system recognition of their targets by T cells, particularly by Tc cells, required the participation of histocompatibility antigens. Aside from their role in controlling the rejection of foreign grafts, these antigens are recognized by T - cell

Peter Doherty

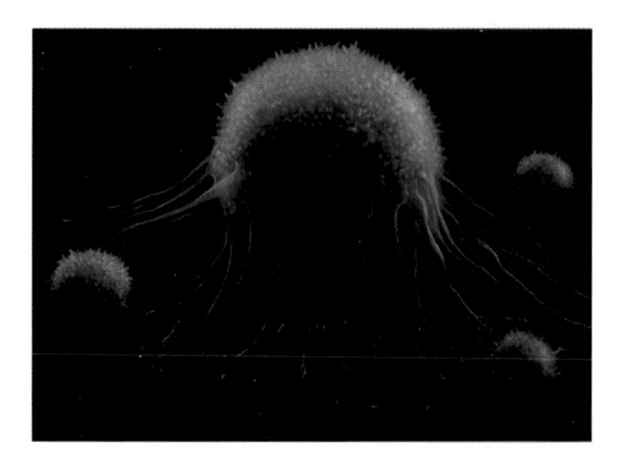

A coloured scanning electron micrograph showing a cancer cell targetted by three killer lymphocytes.

receptors. Doherty and Zinkernagel showed that the recognition of an antigen such as a viral component involved a simultaneous binding by a T-cell receptor of this component and of an HMC I molecule, to which it is tighly bound. After working from 1975 to 1982 as an Associate Professor at the Wistar Institute in Philadelphia, Pennsylvania, Peter Doherty headed the Department of Pathology at the Curtin School in Canberra from 1982 to 1988. He was then appointed Chairman of the Department of Immunology at St. Jude Children's Research Hospital in Memphis, Tennessee.

Rolf Zinkernagel

Zinkernagel, Rolf (Basel, Switzerland, January 6, 1944 -). Nobel Prize for Medicine, along with **Peter Doherty**, for his discovery of how the immune system distinguishes virus-infected cells from normal cells. Rolf Zinkernagel's discoveries developed how the immune system recognizes virus-infected cells and distinguishes it from normal cells. He found that white blood cells (cytotoxic T cells) simultaneously detect viral antigens and closely associated self-antigens (histocompatibility antigens) on the membranes of infected cells in the body. Recognition is followed by T-cell attack and destruction of the target cells. In other words, the immune system can distinguish between "self" and "non-self". Apart fom vaccines, advances in cancer research resulted from this work. These studies also led to new ways to suppress the immunune system's deleterious effects on the bodies' own tissues as occurs for example, in multiple sclerosis and in certain forms of diabetes. Rolf Zinkernagel graduated in medicine from the University of

T lymphocyte. A scanning electron micrograph.

Cross section of a T cell attacking a cancer cell, two steps. The binding of a T cell to its target induces its destruction. Both the viral antigen and the class I MHC molecule are involved in attachment of the T cell and killing results from the transfer of cytotoxins from the T cell to the target cell.

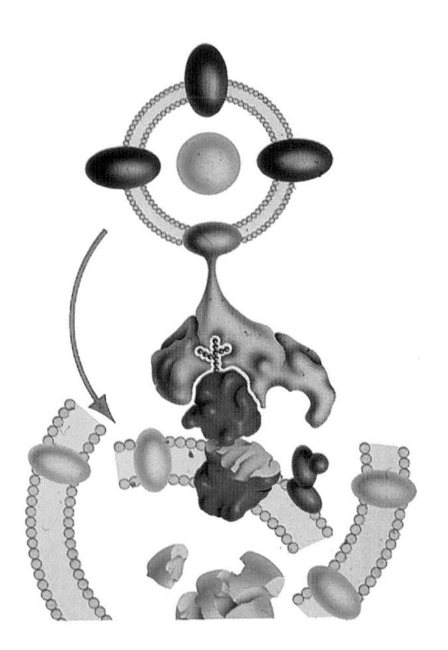

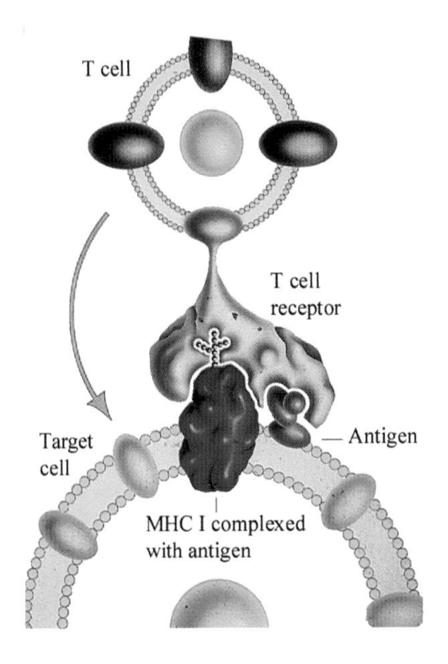

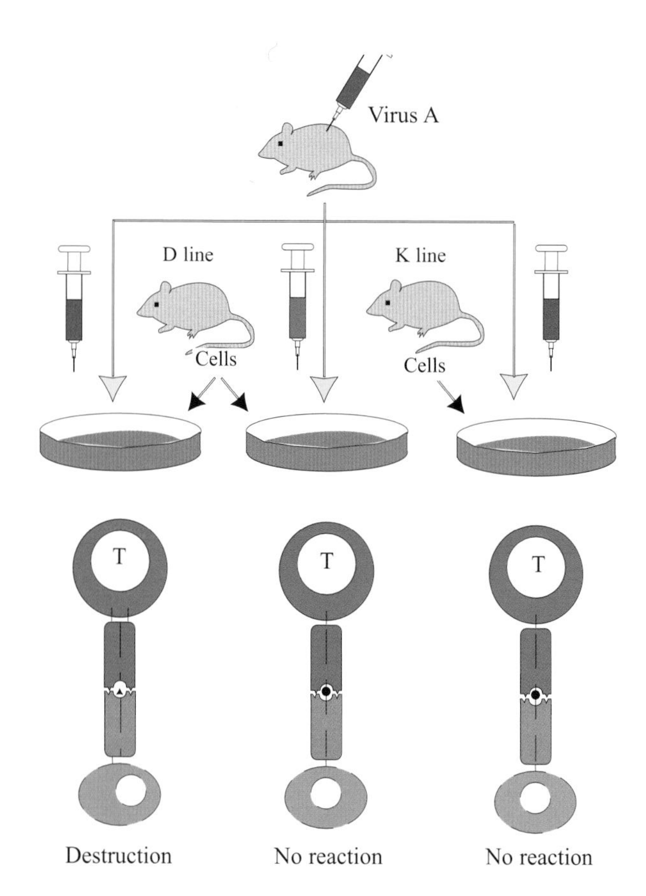

The Zinkernagel experiment. Tissue cells of a D line, cultivated in presence of a virus A, are lysed by lymphocytes derived from this cell line sensitised to the same virus (left). Cellular lysis does not occur if cells of the same line are cultivated in the presence of a virus of a different strain (middle), or inversely if cells of a different line are cultivated in the presence of virus A (right). This experiment demonstrates that lysis only occurs if the T lymphocytes recognise their targets, in addition to their specific antigen, an element belonging to themselves.

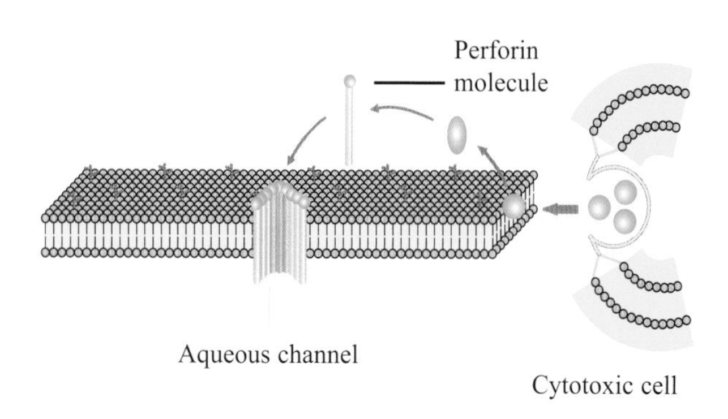

Perforin, a lymphocyte pore-forming protein. It is produced by cytotoxic T cells when these are in contact with target cells. With Ca^{2+}, perforin polymerizes into transmembrane tubules that form pores in the target cell membrane and is thus capable of lysing nonspecifically a variety of target cells.

Basel in 1970 and earned his Ph.D. from the Australian National University in Canberra, Australia, in 1975. He is

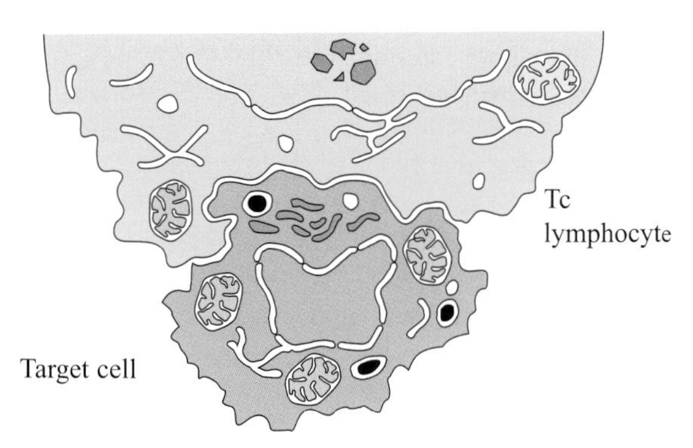

Cross-section of a cancer cell attacked by a Tc lymphocyte. The recognition of antigens on a tumor cell by a Tc lymphocyte is followed by the secretion of substances that destroy it.

now full Professor and Head of the Institute of Experimental Immunology at the University of Zurich, Switzerland.

1997

Prusiner, Stanley Ben (Des Moines, Iowa, May 28, 1942 -). American neurologist. Nobel Prize for Medicine for his discovery of the prion, a new kind of disease-causing agent that cause a number of degenerative brain disease in humans and animals. This discovery could explain some dementia-related disorders like Alzheimer's disease, spongiform encephalopathy, Creutzfeldt - Jakob disease or the famous "mad cow" disease, and promote research into new drugs for the treatment of these forms of progressive dementia. Prusiner and his colleagues hypothesized that

Stanley Prusiner

such illnesses could result from some infectious agent lacking nucleic acid, in fact a sort of yet unknown proteinaceous particle which they called prion. According to this hypothesis, a gene would be responsible for the synthesis of the prion, but this protein would be capable of folding into either a normal, unstable form or an abnormal, stable form, responsible for the observed symptoms. It is believed that this variant also induces the normal protein to fold abnormally. Prions, once dismissed as an impossibility, most scientists believed that these diseases were caused by unidentified viruses, have now gained wide recognition as extraordinary agents that cause a number of infectious, genetic and spontaneous disorders. Stanley Prusiner was educated at the University of Pennsylvania. In 1972, he did his residency at the University of California School of Medicine. He became a professor of neu-rology and biochemistry at the same institution in 1974. He is currently a member of the National Academy of Sciences.

1998

Furchgott, Robert F. (Charleston, South Carolina, June 4, 1916-). American pharmacologist. Nobel Prize for Medicine, along with **Louis J. Ignarro** and **Ferid Murad** for discovering that nitric oxide acts as a signaling molecule in

Robert Furchgott

the cardiovascular system. For more than four decades, Robert Furchgott has worked to unravel the mystery of how drugs interact with receptors in the blood vessels, achieving stunning results that form the basis for modern concepts of receptor pharmacology. In the 1950s, one of Furchgott's earliest contributions was the development of a laboratory method for studying how living blood vessels react to hormones, neurotransmitters and drugs. Using a helically cut, long strip of rabbit aorta in an organ chamber attached to a device that measured contractions and relaxations, he was the first to quantify accurately the effects of pharmacological agents on vascular smooth muscle. During his early studies on arteries, he also discovered and studied the phenomenon of hoto-relaxation, a relaxation of the vascular smooth muscle when exposed to ultraviolet light. As important as these advances were, they only laid the foundation for his later work which, despite its simplicity, has opened the door to a prodigious amount of critical research in cardiovascular disease, immune defense in degenerative diseases, and stroke, among others. This was the discovery in 1980 of a substance in the endothelium, a layer of cells at the innermost surface

Blood vessels in the brain. The maintenance of the opening of blood vessels and consequently the control of blood pressure appear to require the action of substances such as NO and EDRF.

of blood vessels, which causes the underlying smooth muscle to relax - a substance he called endothelium-derived relaxing factor, or EDRF. It was the first time attention had been called to this crucial role that endothelium plays in cardiovascular health and disease. The substance, though, was mysterious - it appeared only briefly and eluded attempts to capture and identify it. It took another six years for Furchgott and others to make the remarkable discovery that EDRF was in actuality nitric oxide, a simple but extremely labile chemical that has played a role in the management of cardiovascular disease for more than 100 years. A clue was found in nitroglycerin, used since the last century to control the chest pains associated with heart disease, which breaks down into nitric oxide. Furchgott observed that EDRF was released by a number of agents acting on endothelial receptors. The release of EDRF was shown by others to interfere with the aggregation of platelets and the adhesion of platelets to blood vessel walls, thereby preventing the formation of small blood clots. This effect combined with the vasodilating effect of EDRF to keep blood vessels open and facilitate blood flow. Later research has shown that nitric oxide or EDRF is a messenger molecule involved in a wide range of activities. It mediates the control of blood pressure and blood flow, airway tone, gastrointestinal motility, penile erection, and the fighting of cancer and infection. In the brain, nitric oxide is an important and unusual neurotransmitter that is revealing clues to unlock the mysteries of learning, memory, and emotion. To Robert Furchgott, for seminal research in drug receptor interactions that led to the discovery of EDRF, which has had profound implications for the treatment of cardiovascular conditions and other diseases, the Albert Lasker Basic Medical Research Award was given. Furchgott earned his Ph.D in 1940 from Nothwestern University at Evanston, Illinois. In 1956, he joined the Department of Pharmacology at the State University of New York Downstate Medical Center. He is presently distinguished professor emeritus.

Ignarro, Louis J. (Brooklyn, New York, May 31, 1941-). American pharmacologist. Nobel Prize for Medicine, along with **Ferid Murad** and **Robert F. Furchgott** for demonstrating that endothelium - derived relaxing factor was nitric oxide (NO). Louis Ignarro identified the nature of the endothelium - derived relaxing factor independently of Furchgott. With his collaborators, he hypothesized that nitric oxide could be responsible for the vascular smooth muscle relaxing action of nitroglycerine and they recognized that cyclic GMP was the second messenger which mediated the relaxant effect of nitric oxide. Guanylate cyclase, the enzyme which produces cyclic GMP from noncyclic GMP, intervenes in this process.

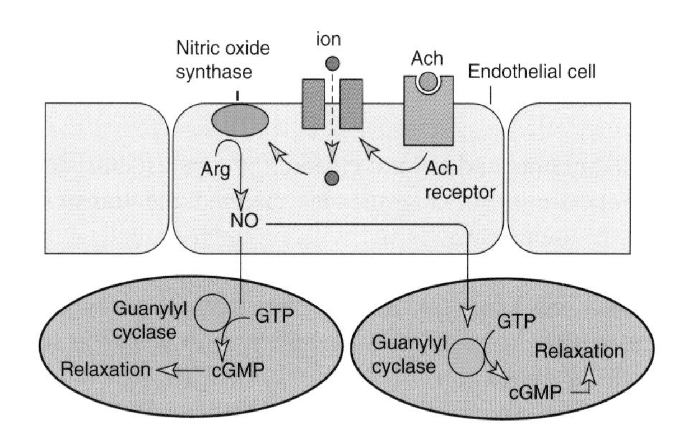

Action of nitric oxide and blood vessels. An acetylcholine receptor anchored on an endothelial cell membrane can activate intracellular signals such as nitric oxyde. In turn, this inorganic compound causes the diameter of small blood vessels to increase.

Ignarro showed that the relaxing effet of nitric oxide was abolished by haemoglobin, which oxidizes it and also by methylene blue, which inhibits guanylate cyclase and thus prevent the formation of nitric oxyde. An important part of Ignarro's work resulted in demonstrating another property of nitric oxide:its inhibitory effect on platelet aggregation, which is also mediated by cyclic GMP. Subsequent works focused to show how nitroglycerin was converted to nitric oxide by smooth vascular muscle and to elucidate the pathway by which nitric oxyde activated guanylate cyclase. The discovery of a heme group in this enzyme helped to explain the process. By reacting with heme iron of guanylate cyclase, nitric oxide stimulates the cyclisation of GMP. Louis Ignarro obtained his Ph.D. in pharmacology from the University of Minnesota. From 1979 to 1985 he was professor of pharmacology at Tulane University School of Medicine at New Orleans. An unexpected result of Ignarro's discovery was the development of a drug called viagra, which acts by increasing NO effect in penile blood vessels.

Ferid Murad

Murad, Ferid (Whiting, Indiana, September 14, 1936 -). Nobel Prize for Medicine along with **Robert F. Furchgott** and **Louis J. Ignarro** for his discoveries concerning the relaxing effect of nitroglycerin, and subsequently nitric oxide, on blood vessels. In 1965 Ferid Murad received his Ph.D. from Case Western Reserve University in Cleveland. His research focuses on the formation, meta-

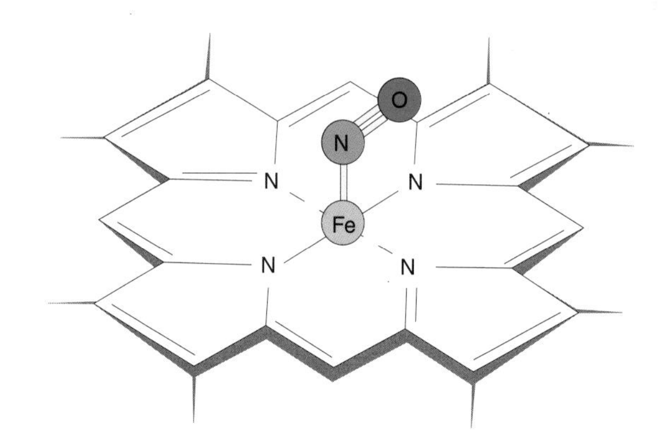

Nitrosoheme. The binding of a molecule of NO to the heme group of the enzyme adenylcyclase produces nitrosoheme.

bolism and function of nitric oxide and cyclic GMP as they participate in various cellular signaling processes. Nitric oxide is formed from L - arginine by one of several isoforms of nitric oxide synthase. Nitric oxide is also formed from a number of prodrugs or nitrovasodilators such as nitrogly- cerin and nitroprusside. There are few biological processes that are not regulated by nitric oxide. The formation of nitric oxide and/or cyclic GMP by many hormones, toxins, cytokines, growth factors and drugs can explain their mechanisms of action. Murad's laboratory has characterized and purified most of these nitric oxide synthases. Nitric oxide mediates some of its biological effects by increasing the synthesis of cyclic GMP by soluble guanylyl cyclase. Ferid Murad and his collaborators also found that the particulate isoforms of guanylyl cyclase can be activated by some bacterial enterotoxins or atriopeptins. They have also characterized, purified and cloned several of the isoforms of guanylyl cyclase. Their current research is focusing on the formation of peroxynitrite from nitric oxide and the nitration of various cellular proteins. They believe that protein nitration can represent another important cell signaling system to regulation of a number of important biological processes and explain some of the numerous effects of nitric oxide. Murad's laboratory utilizes intact cell cultures, tissues and cell-free preparations to characterize, purify and reconstitute these regulatory pathways using pharma-cological, biochemical and molecular biological approaches. It is viewed as one of the leading laboratories in the areas of nitric oxide and cyclic GMP research. Ferid Murad became Vice President of Abbott Laboratories but he left this position in 1993. He was later made Chairman of the Department of Integrative Biology, Pharmacology and Physiology at the University of Texas in Houston.

1999

Blobel Guenter (Waltersdorf, Germany, 1936 -) Nobel Prize for Medicine for the discovery that proteins contain a multitude of intrinsic signals that govern their intracellular transport and their asymmetric integration into cellular membranes. In 1954, Guenter Blobel graduated from the Geschwister Scholl Gymnasium in Freiberg, Saxony. He studied medicine in Frankfurt Am Main, Munich, Kiel and graduated with an M.D fromTuebingen University. After internships in various German hospitals, he moved to the University of Wisconsin, Madison, where he studied biochemistry with Van Rensselaer Potter. In 1966 he joined the laboratory of **George Emil Palade** at the Rockefeller University in New York City. Since 1976 he is Professor of Cell Biology at the Rockefeller University and since 1986 Investigator of the Howard Hughes Medical Institute. More precisely, he and his collaborator David D.Sabatini showed that ribosome attachment to the endoplasmic reticulum (ER) required a signal provided by an intrinsic amino acid sequence near the amino terminus of the growing polypeptide chain. A first version of the "signal hypothesis" was formulated in 1971. He also showed that the nascent polypeptide chain probably traversed the ER membrane through a protein-linked channel in the lipid bilayer. Most secretory proteins con-

Guenter Blobel

tain a signal sequence of 16 to 30 amino acids. Blobel, his collaborators and several research groups established that similar amino acid sequences targeted the transport of newly synthesized proteins to organelles such as mitochondria, peroxysomes or other bodies like chloroplasts and that this sequence is protein-specific. Peptide signals are thus topogenic signals. Blobel's discoveries have had an important impact on our knowledge of cell functionning as well as in our comprehension of several genetic diseases. Blobel is also engaged in the reconstruction of the city of Dresden. He saw the city shortly before its destruction in a devastating Allied bombing raid on

Artistic drawing showing the translocation of nascent polypeptide chains through the membrane of the endoplasmic reticulum.

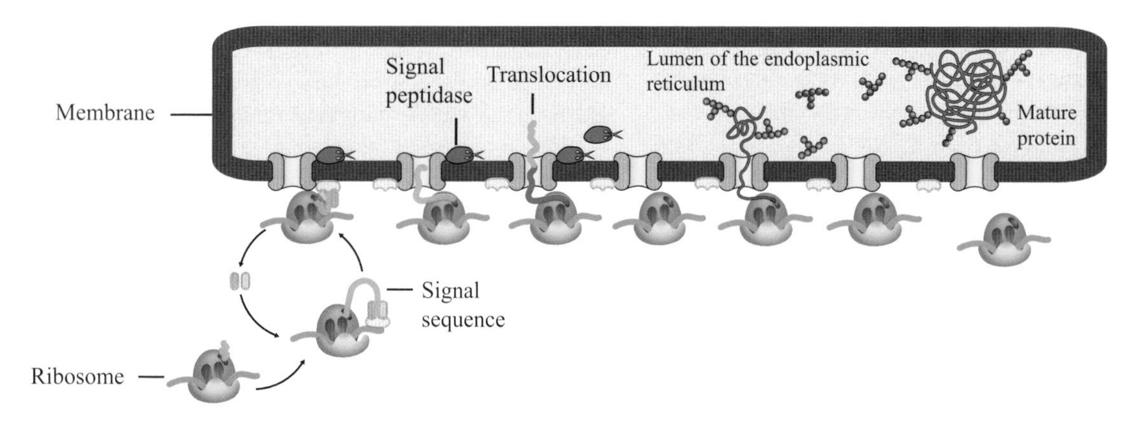

Peptide signal. It was proposed in the 1970s by G. Blobel and D. Sabatini. Peptide signal contains a sequence of amino acids that directs the ribosome and the nascent amino acid chain tothe membranes of the endoplasmic reticulum. There, the synthesis of the polypeptide encoded by the mRNA is resumed.

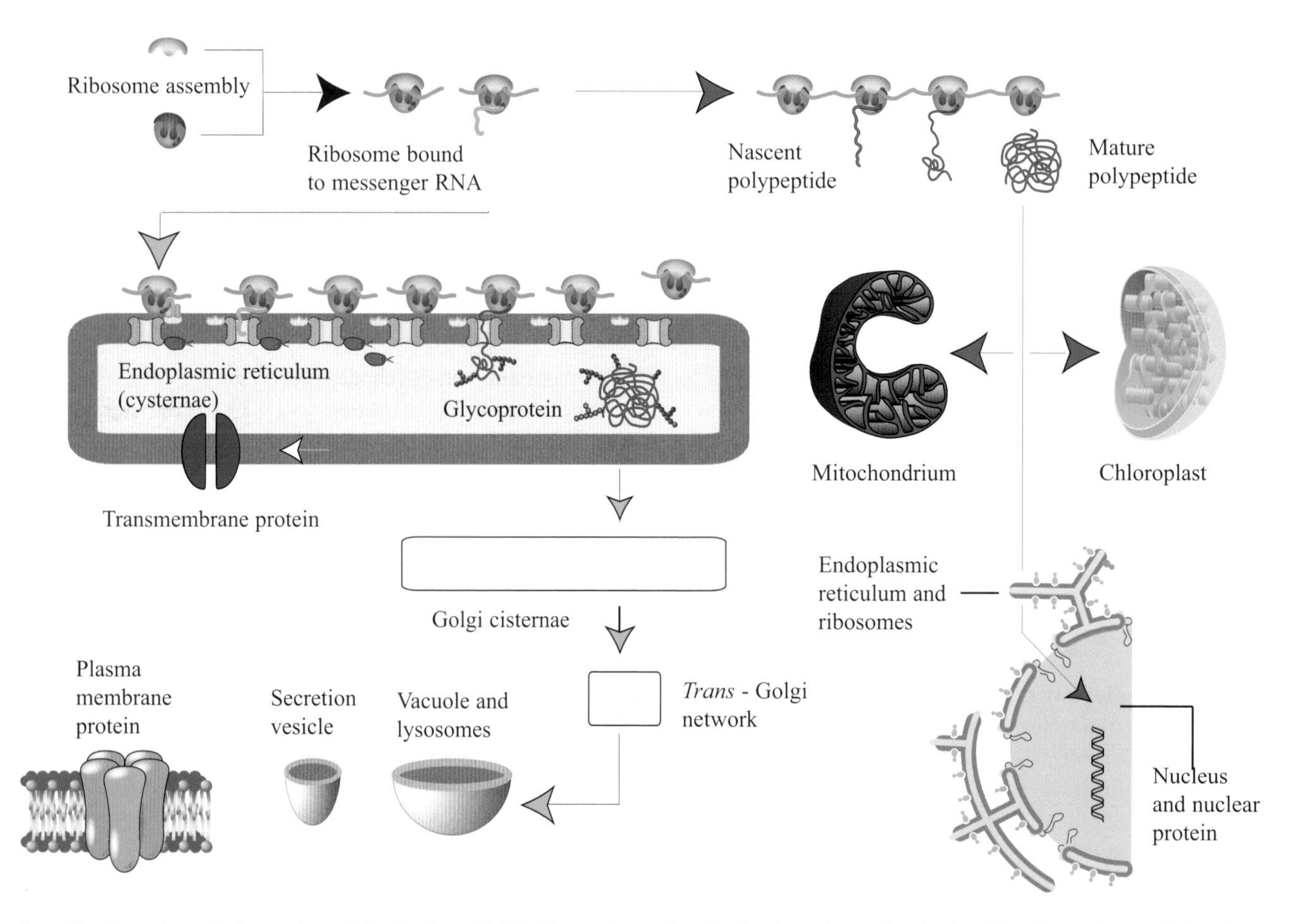

Localization of newly formed proteins in the cell. The figure shows the distribution of proteins destined for the nucleus and the plasma membrane, as well as to several organelles present in the cytosol. The localization of the proteins that are newly synthesized depends on specific signals present in their amino acid chain.

February 13/14, 1945. In 1994 he founded the "Friends of Dresden". Guenter Blobel donated the nearly one million dollar prize to the reconstruction of the famous Frauenkirche and the Synagogue in Dresden.

2000

Carlsson, Arvid (Upsala, Sweden, January 25 -). Nobel Prize for Medicine, jointly with **Paul Greengard** and **Eric Kandel** for their discoveries concerning signal transduction in the nervous system. Arvid Carlsson's main scientific contributions deal with neurotransmitters in the brain and their role in brain function as well as in neuropsychiatric disorders, such as Parkinson's disease, schizophrenia and depression. In the late 1950s he and his colleagues discovered the central stimulating and anti-

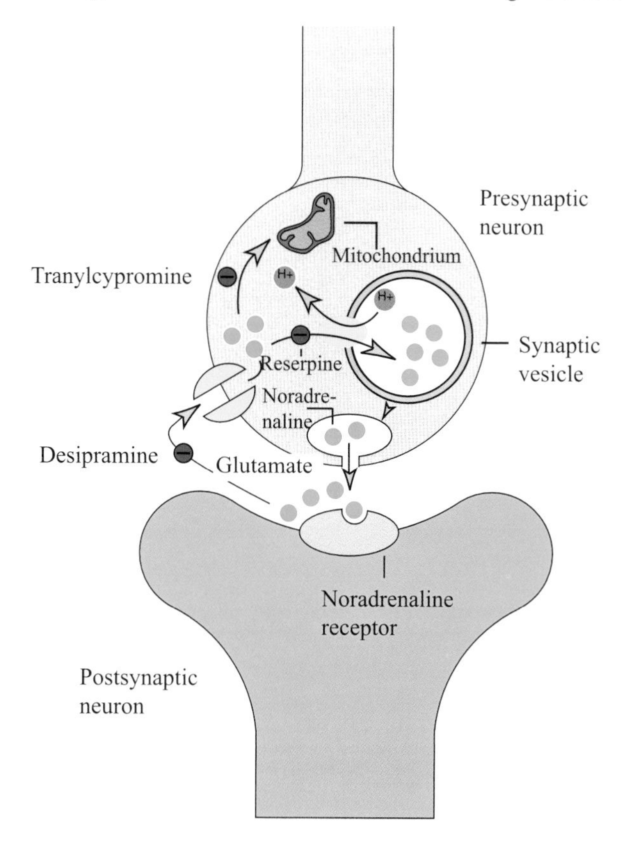

Dopamine. It was previously believed that dopamine was only a precursor of another transmitter, noradrenaline but A. Carlsson performed a series of pioneering studies during the late 1950's, which showed that dopamine is an important transmitter in itself and that it was concentrated in other areas of the brain than noradrenaline.

akinetic action of L-DOPA, identified dopamine as a normal brain constituent and proposed a role for it in mental and motor functions. In 1963 he and Margit Lindqvist

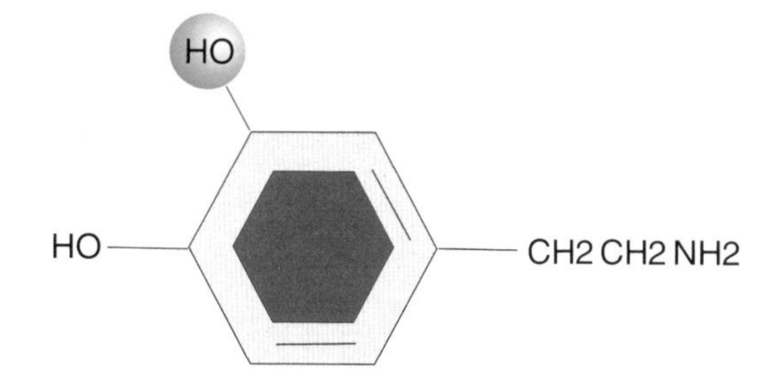

Dopamine. This substance is found in dopaminergic nerves and functions as a neurotrasmitter

discovered a specific action of major antipsychotic agents on dopamine metabolism and proposed that these agents act by blocking dopamine receptors. The current treatment of Parkinson's disease and the dopamine hypothesis of schizophrenia emanate from this work. In the late 1960s Carlsson and his collaborators discovered the serotonin-reuptake inhibiting action of major antidepressant drugs and subsequently developed the first SSRI, zimelidine. Carlsson is Emeritus Professor of Pharmacology, University of Göteborg, Sweden. He received an M.D. and Ph.D. in pharmacology at the University of Lund, Sweden, in 1951. He served as Assistant and Associate Professor of Pharmacology at the same university in 1951-1959 and thereafter as Professor of Pharmacology at the University of Göteborg, Sweden, until his retirement in 1989. Arvid Carlsson has been the leader of a research group consisting of medicinal chemists and pharmacologists, altogether about 20 people, for more than 20 years.

Greengard, Paul (New York City, New York, December 11, 1925 -). American neuroscientist. Nobel Prize for Medicine along with **Eric Kandel** and **David Carlsson** for his discoveries concerning signal transduction in the nervous system. Paul Greengard's discoveries have provided a conceptual framwork for understanding how the nervous system functions at the molecular level. He has also demonstrated that many effects, both therapeutic and toxic, of several classes of common anti-psychotic, hallucinogenic and antidepressant drugs can be explained in terms of distinct neurochemical actions which affect the transmission of nerve signals in the brain. Over the last 30 years, Greengard and his colleagues have developed a general model which provides a rational explanation, at the molecular and cellular levels, of the mechanism by which stimuli, -both electrical and chemical-, produce physiological responses in individual nerve cells. His research group has established that nerve cells respond to extracellular stimuli through an increase in the amount of a substance known as an intracellular ("second") messenger. Second messengers, in turn, produce many of their actions by regulation of the activity of protein kinases. A protein kinase is an enzyme that phosphorylates a protein. Greengard and his group found a large number of phosphorylated proteins that occur only in the brain. Of these, some are present in every nerve cell, and others in only one or a few cell types. These studies have demonstrated that various subclasses of neurons differ markedly from one to another in their chemical composition and suggest that it will be possible to develop highly specific therapeutic agents for the treatment of various neurological and psychiatric disorders.

Paul Greengard

Abnormalities in signaling by the neurotransmitter dopamine are associated with several neurological and psychiatric disorders, including Parkinson's disease, schizophrenia, attention deficit disorder, and substance abuse. His lab has shown that a protein called DARPP-32 is a major player in the mechanisms by which dopamine produces its effects in the brain. In 1998, Greengard's laboratory also offered the first physiological evidence that estrogen therapy may prevent the onset of Alzheimer's disease. In more recent research, the lab has shown that testosterone supplementation in elderly men may be protective in the treatment of Alzheimer's disease. He served as director of biochemical research at the Geigy Research Laboratories and as a professor of pharmacology at the **Albert Einstein** College of medicine and at the Vanderbilt University School of Medicine. In 1968 he was appointed Professor of Pharmacology at Yale University and was named Henry Bronson Professor in 1981. In 1983 he became a professor at The Rockefeller University, where he organized a new laboratory of molecular and cellular neuroscience. He was elected a member of the National Academy of Science in 1978. He is a member of the American Academy of Arts and Sciences.

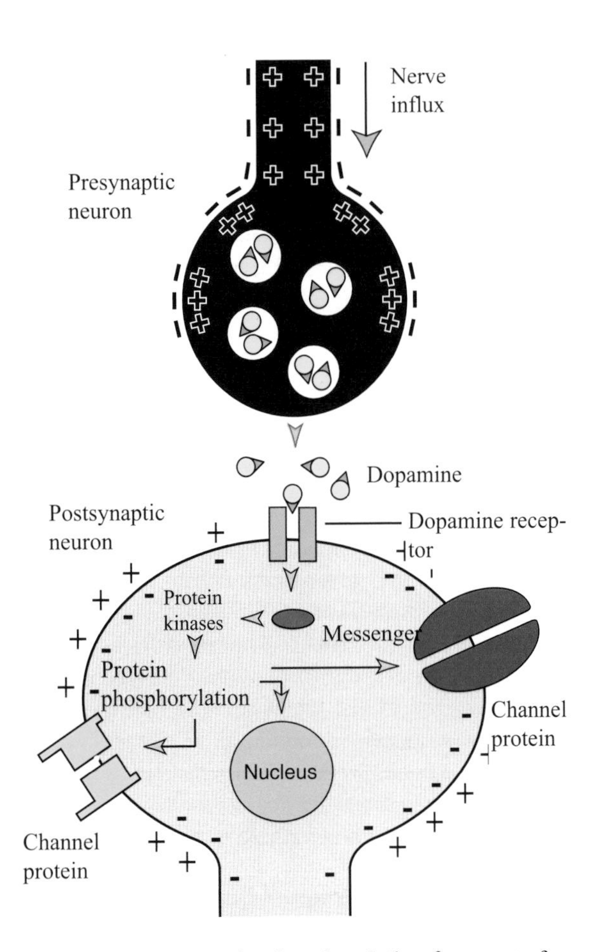

Protein kinases and protein phosphorylation form part of an important signal - transduction mechanism involved in dopamine mediated transmission of the nerve influx.

Kandel, Eric R. (Vienna, Austria, 1939 -). Nobel Prize for Medicine shared with **Paul Greengard** and **Arvid Carlsson** for their discoveries concerning signal transduction in the nervous system. Eric Kandel's research has focused on synaptic plasticity in the central nervous system and on the molecular basis of higher cognitive functions. His discoveries of cellular and molecular mechanisms contributing to simple forms of learning and memory storage provided the first biochemical insights into learned behaviour. To overcome the technical obstacles that previously kept the study of learning beyond the reach of a cell and molecular biological analysis, Kandel turned to a single invertebrate animal, the marine snail *Aplysia*. In this animal, he was able to delineate a simple behaviour, the gill-withdrawal reflex, to analyse its neural circuit in terms of its constituent nerve cells, and to discover elements in the circuit that were modifiable by learning. Using this new experimental system, Kandel analyzed three simple forms of learning:habituation, sensitization, and associative classical conditionning. In this simple system, Kandel's incisive combination of cell and molecular biology led to the first important links between events in individual neurons and behavior of the whole organism. Kandel has used this system to make several major discoveries, for example the first direct evidence that learning leads to the changes of synaptic strength at specific synaptic connections and that memory storage is associated with the persistence of these changes. Another result was the finding that the same synaptic connection can be modulated in opposite ways by different learning processes. By their work, Kandel and his colleagues have helped demystify the study of simple forms of learning and memory storage in both invertebrates and mice, and place them squarely within the context of modern cell and molecular biology. Eric R. Kandel, M.D., is University Professor at Columbia and a Senior Investigator at the Howard Hughes Medical Institute. A graduate of Harvard and New York University School of Medicine, he trained in Neurobiology at the National Institutes of Health, Bethesda, and in Psychiatry at Harvard Medical School, Cambridge. He joined the faculty of the College of

Eric Kandel

Physician and Surgeons at Columbia University, New York City, in 1974 as the Founding Director of the Center for Neurobiology and Behavior. Kandel is a member of the U.S. National Academy of Sciences as well as the National Science Academies of Germany and France. He has been recognized with the Albert Lasker Award, the Wolf Prize of Israel and the National Medal of Science.

2001

Hartwell, Lee (Los Angeles, California, October 30, 1939 -). American biologist. Nobel Prize for Medicine along with **Paul Nurse** and **R. Timothy Hunt** for pioneering the use of genetics to define the cell cycle and to understand its control and role in carcinogenesis. Lee Hartwell has been studying the yeast cell for more than 35 years. Using the budding yeast, *Saccharomyces cerevisiae*-, that is essential for brewing beer and baking bread, he identified many genes that control cell division. Yeast presented many advantages for this work, including its rapid division time, its facile genetic system and its budding shape. The genes that control cell division in yeast, called CDC genes (cell-division-cycle genes) subsequently have been found also to control cell division in humans and often to be the site of alteration in cancer. In addition to displaying alterations in cell division, cancer cells, unlike normal cells, also are unstable genetically. Hartwell also turned to yeast to investigate the basis for accurate cellular reproduction and discovered a new class of gene: "checkpoint" genes. These genes notice when mistakes have been made during cellular reproduction

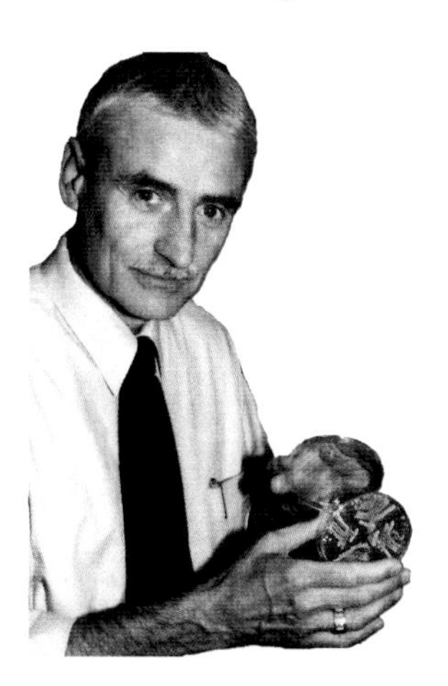

Lee Hartwell

and halt cell division so that repair can take place. In 1961 Hartwell earned a B.S. at the California Institute of Technology and in 1964 earned a Ph.D. from the Massachusetts Institute of Technology under the mentorship of Boris Magasanik. He engaged in postdoctoral work at the Salk Institute for Biological Studies from 1964 through 1965 with **Renato Dulbecco**. He joined the University of Washington faculty in 1968 and has been a professorof genetics there since 1973. In 1996 he joined the faculty of Seattle's Fred Hutchinson Cancer Research Center and in 1997 became its president and director. Hartwell is a member of the National Academy of Sciences.

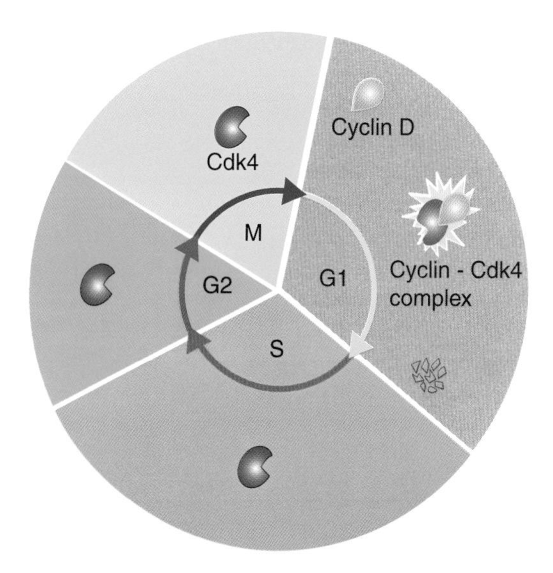

The cell cycle. It consists of a period of mitosis (**M**), a period called **G1**, a period of DNA replication and synthesis (**S**) and another period called **G2**, before mitosis. The length of a complete cycle varies from one cell type to another, and from one species to another.

Hunt,Timothy R. (Neston in the Wirral, Cheshire, England, February 19, 1943 -). Timothy Hunt and his colleagues identified proteins that showed repeated (cyclic) accumulation and degradation during the rapid cell division cycles that occur in fertilised sea urchin and clam eggs. These proteins (for which the term cyclins was coined) are made during interphase and disappear towards the end of mitosis, at about the time of sister chromatid separation. Hunt suggested that the oscillations in the levels of cyclins was connected with the entry and exit of the cell into and out of mitosis, and speculated that cyclins were intimately connected with 'maturation promoting factor', or MPF for short. MPF was discovered in frog oocytes as an enzyme that induced the cells to reinitiate meiosis, and was later shown to be present in mitotic cells as well. Not only that, but MPF can induce cells to enter mitosis as well as meiosis. Eventually, when cyclins and Cdc2 (see **Paul Nurse**) had been cloned and sequenced, it turned out that MPF was actually composed of a 1:1 complex between the Cdc2 protein kinase and cyclin B. At its simplest, the activity of MPF is controlled by the periodic rise and fall in the level of cyclins. Real life is more complicated than this, however, and the activity of MPF can be controlled by other means in response to signals, which may come from the outside or from within the cell. For example, chromosome damage can delay entry into mitosis, which requires retraint of the activation of MPF. In clam eggs, the arrival of a sperm at the surface of the egg is what activates MPF, whereas in frogs, the steroid hormone progesterone turns on MPF for the first time in the formation of the egg and then fertilization turns MPF off by triggering the destruction of cyclin. Thus, the signals that impinge on MPF can be thought of working through plug-in molecular modules that have evolved to turn this powerful entity on or off as appropriate. We now know that cyclins and CDKs control other aspects of cell growth and division besides mitosis. In fact, it would appear that all cell cycle transitions require these enzymes for their successful completion. Moreover, the highly regulated and specific proteolysis of the kind that degrades the mitotic cyclins turns out to be much more widespread and important than was appreciated at the time of its first discovery. R. Timothy Hunt studied biochemistry as an undergraduate in Cambridge. He stayed in Cambridge for his Ph.D. with Asher Korner, eventually studying haemoglobin synthesis in intact rabbit reticulocytes. After graduation in 1968, he went to the **Albert**

Timothy Hunt

Einstein College of Medicine, in New York, to work on the control of globin synthesis in Irving London's laboratory. He returned to England at the end of 1970, and until 1990 was a member of the Department of Biochemistry in Cambridge, continuing to work on translational control, in particular the phosphorylation of the protein synthesis initiation factor eIF1. He discovered cyclin in 1982, and at the end of 1990, moved from Cambridge to the ICRF Clare Hall Laboratories.

Nurse, Paul Maxime (Norwich, Norfolk, England, January 25, 1949-). 2001 Nobel Prize for Medicine shared with **Lee Hartwell** and **R. Timothy Hunt** for pioneering the use of genetics to define the cell cycle and to understand its control and role in carcinogenesis. In the mid 1970s Paul Nurse began working in Edinburgh with Murdoch Mitchison on the simple rod-shaped fission yeast, Schizosaccharomyces pombe. Before this, he spent time in Bern, Switzerland, with Urs Leopold learning S. pombe classical genetics. He wanted to understand the regulation of progress through the cell cycle and to do this

Paul Nurse

made a collection of mutants defective for this process. This collection included cdc (cell division cycle) mutants that are long because they grow but cannot divide and wee mutants which are advanced through the cell cycle and divide at a small size (hence the name wee, from Scottish origin). Paul Nurse showed that one of these genes, cdc2, was required at two stages of the cell cycle, regulating entry into S phase and also into mitosis. Cdc2 encoded a

Yeast budding. The study of this process produced a great deal of information about the mechanisms that control the cell cycle.

protein kinase regulated by phosphorylation on tyrosine 15 by the wee1 kinase and dephosphorylated by the cdc25 phosphatase to ensure timely entry into mitosis. Cdc2 was also found to interact with a cyclin B protein encoded by cdc13. The Cdc2 function is conserved in humans, as

Mitosis. A three-dimensional representation. Mitosis is the process by which cells divide. Cell division is not a hazardous process but is carefully coordinated with both cell growth and DNA replication.

demonstrated by cloning the human cdc2 homologue by rescue of the cdc2 mutant function. From other labo-ratories a parallel biochemical approach to understand cell cycle control, using eggs from xenopus and sea urchins, led to the identification of cyclins and purification of MPF (maturation promoting factor). The two areas of research came together when it was shown that the coponents of MPF were the homologues of cdc2 and cdc13. Thus the regulation of mitotic onset is conserved from yeast through humans. Cdc2 is now known to be a member of a family of protein kinases that are the catalytic components of a complex including a cyclin regulatory component known as CDKs (cyclin-dependent kinases). Theyare required at several different stages of the cell cycle. Paul Maxime Nurse graduated from the University of Birmingham. In 1973 he earned his Ph.D. degree from the University of East England in Cell Biology and Biochemistry. He is currently Head of the Laboratory Research at the Imperial Cancer Research Fund of London.

Table: The Nobel Prize Laureates

	Chemistry	Physics	Medicine
1901	**Hoff, Jacobus H, van 't.** (The Netherlands) Studies on rates of reaction, chemical equilibrium, and osmotic pressure	**Röntgen, Wilhelm** (Germany) Discovery of X - rays	**Von Behring, Emil A.** (Germany) Studies on serum therapy and its applications against diphtheria
1902	**Fischer, Emil Hermann** (Germany) Studies on carbohydrates and purines	**Lorentz, H.A.** (The Netherlands) **Zeeman, P** (The Netherlands) Theory of electromagnetic radiation	**Ross, Ronald** (Great Britain) Discovery of the life cycle of the malaria parasite
1903	**Arrhenius, Svante** (Sweden) Studies on electrolyte dissociation	**Becquerel, Henri** (France) **Curie, Pierre** (France) **Curie, Marie** (Poland - France) Discovery of natural radioactivity	**Finsen, Niels R.** (DenmarK) Use of ultraviolet light in the treatment of skin diseases
1904	**Ramsay, William** (Great Britain) Discovery of noble gases	**Strutt, Joseph** (Great Britain) Determination of gas density Discovery of argon	**Pavlov, I. Petrovich** (Russia) Discovery of the conditioned reflex
1905	**von Baeyer, Adolf** (Germany) Studies on organic dyes (indigo) and aromatic compounds	**Lenard, Philipp E.** (Hungary - Germany) Works on cathode rays	**Koch, Robert** (Germany) Discoveries in the field of tuberculosis
1906	**Moissan, Henri** (France) Isolation of fluorine	**Thomson, Joseph** (Great Britain) Electric conductivity in gases	**Golgi, Camillo** (Italy) **Ramon y Cajal, S.** (Spain) Discovery of the fine structure of the nervous system
1907	**Buchner, Edouard** (Germany) Discovery of enzyme action in yeast cell extracts	**Michelson, Albert** (USA) Precise experimental arrangements for metrology and spectroscopy	**Laveran, Charles - L.- A.** (France) Discovery of the malarial parasite
1908	**Rutherford, Ernest** (Great Britain) Studies on radioactive substances and development of nuclear physics	**Lippmann, Gabriel** (Luxembourg) Colour - photography process, utilizing interferometric methods	**Ehrlich, Paul** (Germany) Foundation of chemotherapy **Metchnikoff, I. Ilytch** (Russia) Discovery of phagocytosis and work on immunity
1909	**Ostwald, Wilhelm** (Germany) Studies on catalysis and chemical equilibrium	**Marconi, Guglielmo** (Italy) **Braun, Karl F.** (Germany) Development of wireless telegraphy	**Kocher E. Theodor** (Switzerland) Studies on the thyroid gland
1910	**Wallach, Otto** (Germany) Chemistry of vegetal oils and terpenes	**van der Waals, J.** (The Netherlands) Equation of state establishing the relation between volume, temperature and pressure	**Kossel, Albrecht** (Germany) Discovery of nuclein and study of nucleic acids
1911	**Curie, Marie** (France) Discovery of radium and polonium. Isolation of radium	**Wien, Wilhelm** (Germany) Law concerning the radiation emitted by the blackbody	**Gullstrand, Allvar** (Sweden) Research on the eye as a light - refracting apparatus
1912	**Grignard, Victor** (France) Discovery and development of the "Grignard reaction" **Sabatier, Paul** (France) Discovery of a new method for hydrogenation	**Dalen, Nils G.** (Sweden) Automatic light - controlled valves for navigation aid	**Carrel, Alexis** (France) Development of a method for suturing blood vessels
1913	**Werner, Alfred** (Switzerland) Studies on the structure of coordination compounds	**Kamerlingh - Onnes, H.** (The Netherlands) Low temperature physics. Production of liquid helium	**Richet, Charles Robert** (France) Discovery of anaphylaxis
1914	**Richards, Theodore W.** (USA) Precise determination of atomic weights of numerous elements	**von Laue, Max T.** (Germany) Discovery of the diffraction of X - rays by crystals	**Barany, Robert** (Austria) Studies on the physiology and pathology of the vestibular apparatus
1915	**Willstätter, R. Martin** (Germany) Research on plant pigments, especially chlorophylls	**Bragg, William H.** (Great Britain) **Bragg, William L.** (Australia - Great Britain) Crystal analysis using X - rays	Not awarded

1916	Not awarded	Not awarded	Not awarded
1917	Not awarded	**Barkla, Charles G.** (Great Britain) Identification of characteristic X - rays of elements	Not awarded
1918	**Haber, Fritz** (Germany) Development of a method of synthesizing ammonia	**Planck, Max** (Germany) Discovery of the quantization of energy	Not awarded
1919	Not awarded	**Stark, Johannes** (Germany) Influence of an electric field on the spectrum of light	**Bordet, Jules** (Belgium) Discovery of immunity factors in blood serum (complement system)
1920	**Nernst, Walter Hermann** (Germany) Research and studies in electrochemistry	**Guillaume, C. E.** (Switzerland) Discovery of special alloys, in particular invar	**Krogh, Schack A. S.** (Denmark) Discovery of a motor - regulating mechanism of capillaries
1921	**Soddy, Freddy** (Great Britain) Chemistry of radioactive substances and isotopes	**Einstein, Albert** (Germany) Contribution to mathematical physics and explanation of the photoelectric effect	Not awarded
1922	**Aston, Francis William** (Great Britain) Discovery of isotopes in non - radioactive elements	**Bohr, Niels** (Denmark) The structure of atoms. Study of the radiations emanating from them	**Hill, Archibald** (Great Britain) Studies and discoveries concerning heat production in muscles **Meyerhof, Otto Fritz** (Germany) Discovery of anaerobic glycolysis in muscles
1923	**Pregl, Fritz** (Austria) Developing new techniques in the microanalysis of organic compounds	**Millikan, Robert** (USA) Identification of the elementary electric charge	**Banting, Frederick** (Canada) Extraction of insulin from the pancreas
1924	Not awarded	**Siegbahn, Karl M.** (Sweden) X - ray spectroscopy	**Einthoven, Willem** (The Netherlands) Development of the electrocardiograph and advancements in the field of cardiac physiology
1925	**Zsigmondy, Richard** (Austria) Research on colloids and invention of the ultramicroscope	**Franck, James** (Germany) **Hertz, Gustav** (Germany) Laws governing the electron - atom interactions	Not awarded
1926	**Svedberg, Theodor** (Sweden) Chemistry of colloids	**Perrin, Jean - Baptiste** (France) Discontinuity in the structure of matter. The Brownian motion	**Fibiger, Johannes** (Denmark) Experimental induction of cancer
1927	**Wieland, Heinrich Otto** (Germany) Determination of the molecular structures of bile acids and related substances	**Compton, Arthur** (USA) Scattering of photons by electrons. The Compton effect **Wilson, Charles** (Scotland) Detection of charged particles using the cloud chamber	**Wagner, Julius Jaurreg** (Austria) Treatment of syphilis by shock therapy and discovery of vitamin B2
1928	**Windaus Adolf Otto R.** (Germany) Research on cholesterol and vitamin D	**Richardson, Owen** (Great Britain) Thermoionic phenomena	**Nicolle, Charles J. H.** (France) Discovery of how typhus is transmitted
1929	**Harden, Arthur** (Great Britain) **von Euler - Chelpin, H.** (Sweden) Studies on sugar fermentation and enzymes involved in this process	**de Broglie, Louis V.** (France) Wave properties of moving particles	**Eijkman, Christiaan** (The Netherlands) Discovery of vitamins **Hopkins, F. G.** (Great Britain) Studies on vitamins and other nutritional factors
1930	**Fischer, Hans** (Germany) Research on substances like hemin and chlorophylls	**Raman, C. Venkata** (India) Diffusion of light. The Raman effect	**Landsteiner, Karl** (Austria) Discovery of blood groups
1931	**Bosch, Carl** (Germany) **Bergius, Friedich** (Germany) Invention and development of analytical methods under high pressure	Not awarded	**Warburg, Otto H.** (Germany) Discovery of flavin enzymes and work on cytochromes
1932	**Langmuir, Irving** (USA) Studies of molecular films on solid and liquid surfaces that opened new fields in colloid research and biochemistry	**Heisenberg, W.** (Germany) Creation of quantum mechanics. Allotropic forms of hydrogen	**Adrian, Edgard** (Great Britain) Measurement of nerve influx **Sherrington, Charles** (Great Britain) Discovery of the reflex arc

1933	Not awarded	**Schrödinger, Erwin** (Austria) **Dirac, Paul A. M.** (Great Britain) New developments of the theory of atomic structure	**Morgan, Thomas H.** (USA) The chromosomal theory of heredity
1934	**Urey, Harold Clayton** (USA) Discovery of deuterium	Not awarded	**Minot, George R.** (USA) **Murphy, William P.** (USA) **Whipple, George H.** (USA) Role of liver diet in the treatment of pernicious anemia
1935	**Joliot, Frédéric** (France) **Joliot - Curie, Irène** (France) Synthesis of new radioactive elements	**Chadwick, James** (Great Britain) Discovery of the neutron	**Spemann, Hans** (Germany) Discovery of embryonic induction
1936	**Debye, Peter J. W.** (The Netherlands) Studies of dipole moments, X - rays, and light scattering in gases	**Hess, Victor F.** (Austria) Discovery and identification of cosmic rays **Anderson, Carl D.** (USA) Discovery of the positron	**Dale, Henri H.** (Great Britain) Isolation of acetylcholine **Loewi, Otto** (Germany) Discovery of neurotransmitters
1937	**Haworth, Walter N.** (Great Britain) Determination of the structure of carbohydrates and vitamin C **Karrer, Paul** (Switzerland) Studies on carotenoids, flavis, vitamins A and B2.	**Davisson, Clinton** (USA) **Thomson, George** (Great Britain) Diffraction of electrons by crystals	**Szent-Györghyi, Albert** (Hungary) Discovery of the role played by certain organic compounds in cell oxidations. Discovery of vitamin C
1938	**Kuhn, Paul** (Switzerland) Studies on carotenoids and vitamins	**Fermi, Enrico** (Italy) Nuclear reactions and production of artificial radioelements induced by slow neutrons	**Heymans, Corneille J. F.** (Belgium) Discovery of the relation between regulatory effects on respiration and specialized sensory organs
1939	**Butenandt, Adolf F. J.** (Germany) Discovery of sexual hormones **Ruzicka, Leopold** (Switzerland) Studies on terpenes and sex hormones	**Lawrence, Ernest** (USA) Invention of the cyclotron and its use to induce artificial radioactivity	**Domagk, Gerhard** (Germany) Discovery of the antibacterial effects of protonsil, a sulfonamide drug
1940-1942	Not awarded	Not awarded	Not awarded
1943	**de Hevesy, George** (Sweden) Development of isotopic tracer techniques	**Stern, Otto** (Poland) Measurement of the magnetic moment of the proton	**Dam, Carl Peter** (Denmark) **Doisy, Edward A.** (USA) Discovery of vitamin K
1944	**Hahn, Otto** (Germany) Discovery of nuclear fission	**Rabi, Isidor Isaac** (Poland - USA) Measurements of magnetic properties of atomic nuclei by resonant methods	**Erlanger, Joseph** (USA) **Gasser, Herbert S.** (USA) Discovery that fibers within the same nerve possess different functions
1945	**Virtanen, Artturi Ilmari** (Finland) Production and storage of protein - rich green fodder	**Pauli, Wolfgang** (Austria) Formulation of the Pauli exclusion principle	**Fleming, Alexander** (Great Britain) Discovery of penicillin Chain, Ernst B. (Great Britain) **Florey, Howard. W.** (Australia) Isolation, purification and production of penicillin
1946	**Sumner, James Batchelor** (USA) Discovery of the protein nature of enzymes **Northrop, John H.** (USA) **Stanley, Wendell M.**(USA) Purification and crystallization of enzymes and viruses	**Bridgman, Percy** (USA) Invention of apparatuses in order to produce high pressures	**Muller, Hermann J.** (USA) X - ray mutagenesis
1947	**Robinson, Robert** (Great Britain) Research on alkaloids and other substances of biological importance	**Appleton, Ed.** (Great Britain) Physics of upper atmosphere. Discovery of the so-called Appleton layer of the ionosphere	**Cori, Carl Ferdinand** (USA) **Cori, Gerty Theresa** (USA) Studies on glycogen metabolism **Houssay, Bernardo** (Argentina) Discovery of the role played by pituitary hormones in regulating the amount of blood

1948	**Tiselius, Arne Wilhelm** (Sweden) Development of electrophoresis and adsorption analysis	**Blackett, Patrick M.** (Great Britain) Development of the Wilson cloud chamber performances using peripheral detectors	**Müller, Paul. H.** (Switzerland) Discovery of the toxic effects of DDT on insects
1949	**Giauque, William Fr.** (USA) Studies of the properties of matter at temperatures close to absolute zero	**Yukawa, Ideki** (Japan) Theoritical prediction of the existence of pions to explain the nuclear interactions inside the atomic nucleus	**Egas Moniz, A.** (Portugal) Development of lobotomy **Hess, Walter R.** (Switzerland) Determination of the role played by the hypothalamus in coordinating the functions of the internal organs
1950	**Diels, Otto Paul H.** (Germany) **Alder, Kurt** (Germany) Discovery and development of diene synthesis	**Powell, Cecil F.** (Great Britain) Invention of a photographic method for the detection of particles. Identification of mesons	**Hench, Philip S.** (USA) Clinical use of cortisone **Kendall, Edward** (USA) Discovery of thyroxine and corticosteroids **Reichstein, Tadeus** (Switzerland) Discovery of corticosteroids
1951	**McMillan, Edwin M.** (USA) Seaborg, Glenn (USA) Chemistry of elements heavier than uranium	**Cockcroft, John D.** (Great Britain) **Walton, Ernest** (Ireland) Construction of a linear particle accelerator to transmute elements	**Theiler, Max** (South Africa) Development of a vaccine against yellow fever
1952	**Martin, Archer John P.** (Great Britain) **Synge, Richard L. M.** (Great Britain) Invention and development of paper partition chromatography	**Bloch, Félix** (Switzerland) **Purcell, Edouard** (USA) Precise measurements of magnetic moments	**Waksman, Selman A.** (USA) Discovery of streptomycin, the first specific agent effective in the treatment of tuberculosis
1953	**Staudinger, Hermann** (Germany) Chemistry of macromolecules, in particular cellulose, rubber and isoprenes	**Zernike, Frits** (The Netherlands) Invention of the phase - contrast microscope	**Krebs, Hans** (Great Britain) Discovery of the tricarboxylic acid cycle **Lipman, Fritz Albert** (USA) Discovery of coenzyme A
1954	**Pauling, Linus Carl** (USA) Studies on chemical bonds and protein structure	**Born, Max** (Poland - Germany) Statistical formulation of the wave function **Bothe, Walther** (Germany) Coincidence - counting method in nuclear physics	**Enders, John Franklin** (USA) **Robbins, Frederick C.** (USA) **Weller, Thomas Huckle** (USA) Development of a vaccine against poliomyelitis
1955	**du Vigneaud, Vincent** (USA) Discovery of vasopressin and oxytocin	**Lamb, Willis, Jr.** (USA) Discovery of the line shift in hydrogen **Kusch, Polykarp** (Germany - USA) Accurate determination of the magnetic moment of the electron	**Theorell Axel H.T.** (Sweden) Discoveries in the field of cellular oxidations
1956	**Hinshelwood, Cyril N.** (Great Britain) **Semenov, N.** (Soviet Union) Research on the mechanism of chemical reactions	**Bardeen, J.** (USA) **Brattain, W.** (USA) **Shockley, W.** (Great Britain - USA) Discovery of the transistor effect	**Forsmann, Werner** (Germany) Development of cardiac catheterization **Cournand, André F.** (USA) **Richards, Woodruff D.** (USA) Improvement of cardiac catheterization
1957	**Todd, Alexander** (Great Britain) Biochemistry of nucleic acids	**Yang, Chen N.** (China - USA) **Lee, Tsung D.** (China -USA) Demonstration of the parity violation in weak interactions	**Bovet, Daniel** (Switzerland) Discovery of antihistamine drugs and studies on curare
1958	**Sanger, Frederik** (Great Britain) Methods of protein and nucleic acids sequencing	**Cherenkov, P.** (Soviet Union) **Frank, Ilya M.** (Soviet Union) **Tamm, Igor Y.** (Soviet Union) Discovery and explanation of Cherenkov effect	**Beadle, George Wells** (USA) **Tatum, Edward Lawrie** (USA) Foundation of biochemical genetics **Lederberg, Joshua** (USA) Discovery of bacterial sexuality
1959	**Heyrovsky, J.** (Czechoslovakia) Modern analytical chemistry	**Chamberlain, Owen** (USA) **Segrè, Emilio G.** (Italy, USA) Discovery of antiproton	**Kornberg, Arthur** (USA) Discovery of DNA polymerase I **Ochoa, Severo** (Spain) Discovery of polynucleotide phosphorylase

1960	**Libby, Willard Frank** (USA) Use of carbon - 14 in archeology, geology and other scientific fields	**Glaser, Donald A.** (USA) Invention of the bubble chamber	**Burnet, Sir F. McF.** (Australia) **Medawar, Sir P. B.** (Great Britain) Discovery of acquired immunological tolerance to tissue transplants
1961	**Calvin, Melvin** (USA) Discovery of the reactions involved in photosynthesis	**Hofstadter, R.** (USA) Detemination of the structure of the atomic nucleus by electronic scattering **Mössbauer, R.** (Germany) Resonant absorption of gamma - rays : the Mössbauer effect	**Bekesy, Georg Von** (USA) Physiology of sound transmission in cochlea
1962	**Kendrew, John C.** (Great Britain) **Perutz, Max F.** (Great Britain) Determination of the structure of myoglobin and haemoglobin respectively	**Landau, Lev D.** (Soviet Union) Theories on condensed matter physics and explanation of the properties of liquid helium	**Crick, Francis H.** (Great Britain) **Watson, James D.** (USA) Discovery of the double helical structure of DNA **Wilkins, Maurice H.** (Great Britain) Study of DNA by X - ray diffraction
1963	**Natta, Giulio** (Italy) **Ziegler, Karl** (Germany) Development of Ziegler - Natta catalysis	**Wigner, Eugene** (Hungary - USA) Theory of the atomic nucleus through the discovery of symmetry principles **Goeppert-Mayer, M.** (Poland - USA) **Jensen, Johannes** (Germany) Development of the nuclear shell model	**Eccles, John Carrew** (USA) Studies on synaptic transmission **Hodgkin, Alan Lloyd** (USA) **Huxley, Andrew F. (Great Britain)** Studies dealing with the chemical phenomena involved in the transmission of nerve impulses
1964	**Hodgkin, Dorothy Cr.** (Great Britain) Determination of the structure of complex organic molecules, in particular vitamin B12	**Townes, Charles** (USA) **Basov, N.** (Australia - Soviet Union) **Prokhorov, A.** (Soviet Union) Discoveries in quantum electronics leading to the development of masers and lasers	**Bloch, Konrad Emil** (USA) Discoveries concerning cholesterol synthesis **Lynen, Feodor** (Germany) Cholesterol metabolism and works on fatty acid beta-oxidation
1965	**Woodward, R. Burns** (USA) Development of new methods in chemical synthesis	**Feynman, Richard** (USA) **Schwinger, Julian** (USA) **Tomonaga, Sin-Itiro** (Japan) Research in quantum electrodynamics and physics of high energy particles	**Lwoff, André Michael** (France) Discovery of lysogeny **Jacob, François** (France) **Monod, Jacques** (France) Genetic regulation in bacteria
1966	**Mulliken, Robert S.** (USA) Studies on chemical bonds and electronic structure of molecules	**Kastler, Alfred** (France) Double resonance technique (optical and radiofrequency) in atomic spectroscopy	**Huggins, Charles Brent** (USA) Cancer surgery **Rous, Francis Peyton** (USA) Discovery of a tumor virus
1967	**Eigen, Manfred** (Germany) **Norrish, Sir Ronald G.** (Great Britain) **Porter, Sir George** (Great Britain) Development of techniques for the study of fast chemical reactions	**Bethe, Hans** (Germany - USA) Theory of nuclear reactions and explanation of the production of energy in stars	**Granit, Ragnar** (Finland) **Hartline, Halden Keffer** (USA) Mechanism of vision **Wald, George** (USA) Biochemistry of vision. Role of vitamin A in vision
1968	**Onsager, Lars** (USA) Thermodynamics of irreversible processes	**Alvarez, Luis W.** (USA) New developments of bubble chambers	**Holley, Robert W.** (USA) Primary structure of tRNA **Khorana, Har Gobind** (USA) **Nirenberg, Marshall W.** (USA) Elucidation of the genetic code
1969	**Barton, Sir Derek H.R.** (Great Britain) **Hassel, Odd** (Norway) Studies on the three - dimensional structure of complex molecules	**Gell - Mann, Murray** (USA) Classification of elementary particles and of their interactions	**Delbrück, Max** (USA) **Hershey, Alfred** (USA) **Luria, Salvador** (USA) Bacterial genetics and bacterioph
1970	**Leloir, Luis F.** (Argentina) Discovery that sugar nucleotides are key elements in glycogen synthesis	**Alfvén, Hannes** (Sweden) Research in magneto - hydrodynamics **Néel, Louis** (France) Discovery of ferromagnetism and antiferromagnetism	**Axelrod, Julius** (USA) Discovery of antidepressants **Katz, Bernard** (Great Britain) Quantum theory of nerve transmission **Euler, Ulf Von** (Sweden) Identification of noradrenaline and discovery of prostaglandins
1971	**Herzberg, Gerhard** (Canada) Determination of the electronic structure and geometry of molecules	**Gabor, Dennis** (Hungary) Invention of holography	**Sutherland, E. Wilbur** (USA) Discovery of cyclic AMP and its role as a cell messenger

1972	**Anfinsen, Christian** (USA) Research on the relationship between the molecular structure of proteins and their biological functions **Moore, Stanford** (USA) **Stein, William** (USA) Research on the molecular structure of proteins	**Bardeen, John** (USA) **Cooper, Leon Neil** (USA) **Schrieffer, John** (USA) Theory of superconductivity	**Edelman, Gerald M.** (USA) **Porter, Rodney Robert** (Great Britain) Elucidation of antibody structure
1973	**Fischer, Ernst Otto** (Germany) **Wilkinson, Geoffrey** (Great Britain) Studies on organometallic compounds	**Esaki, Leo** (Japan) **Giaever, Ivar** (Norway) Tunneling of electrons in solid (quantum - mechanical effect) **Josephson, Brian** (Great Britain) Theory of superconductivity in junctions : the Josephson effect	**Lorenz, Konrad** (Austria) **Tinbergen, Nikolaas** (Austria) **Frisch, Karl Von** (Germany) Studies of animal behaviour
1974	**Flory, Paul J.** (USA) Theoretical and experimental studies of macromolecules	**Ryle, Martin** (Great Britain) Development of radiotelescopes **Hewish, Antony** (Great Britain) Discovery of pulsars	**Claude, Albert** (Belgium) **Palade, George** (USA) **de Duve, C. R.** (Belgium) Discovery of cell ultrastructures
1975	**Prelog, Vladimir** (Switzerland) Stereochemistry of organic molecules and reactions **Cornforth, J. W.** (Australia) Stereochemistry of enzyme - catalysed reactions	**Bohr, Aage Niels** (Denmark) **Mottelson, Benjamin** (USA) **Rainwater, L. James** (USA) Collective model of the atomic nucleus	**Baltimore, David** (USA) **Temin, Howard M.** (USA) Discovery of reverse transcriptase **Dulbecco, Renato** (USA) Works on oncogenic animal viruses
1976	**Lipscomb, William N.** (USA) Study of the structure and bonding of boron compounds	**Richter, Burton** (USA) **Ting, Samuel** (USA) Discovery of ψ/J, a subnuclear particle	**Gajdusek, Daniel C.** (USA) Research on the causal agents of various neurological disorders **Blumberg, Baruj S.** (USA) Discovery of an antigen that stimulates antibody response against hepatitis B
1977	**Prigogine Ylya** (Belgium) Contribution to the study of nonequilibrium thermodynamics	**van Vleck, John** (USA) Study of magnetic structures and disordered systems **Mott, Nevill** (Great Britain) **Anderson, Philip** (USA) Use of amorphous structures in electronics and in information processing	**Guillemin, Roger** (USA) **Schally, André** (USA) Discovery of hypothalamic hormones **Yalow, Rosalyn** (USA) Development of radioimmunoassay
1978	**Mitchell, Peter D.** (Great Britain) Explanation of ATP synthesis by the chemiosmotic mechanism	**Kapitza, Pyotr** (Soviet Union) Discoveries in the field of low-temperature physics **Penzias, Arno A.** (Germany - USA) **Wilson, Robert W.** (USA) Discovery of the cosmic microwave background radiation	**Arber, Werner** (Switzerland) Discovery of restriction enzymes **Nathans, Daniel** (USA) Genetic map of simian virus 40 **Smith, Hamilton** (USA) Discovery of the first specific restriction enzyme
1979	**Brown, Herbert C.** (USA) Study of boron compounds **Wittig, Georg** (Germany) Discovery of a new class of phosphorus compounds	**Weinberg, Steven** (USA) **Salam, Abdus** (Pakistan) **Glashow, Sheldon** (USA) Theory unifying electromagnetic and weak nuclear forces	**Cormak, A. Mcl.** (South Africa) **Hounsfield, Godfrey M.** (Great Britain) Development of computerized axial tomography (CAT)
1980	**Sanger, Frederik** (Great Britain) **Gilbert, Walter** (USA) Methods of sequencing DNA **Berg, Paul** (USA) Development of genetic engineering	**Cronin, James W.** (USA) **Fitch, V. Logsdon** (USA) Discovery of the violation of laws of symmetry	**Dausset, Jean** (France) **Snell, George Davis** (USA) Discovery of a major histocompatibility complex in mice and humans **Benacerraf, Baruj** (USA) Discovery of genes that control immunity

1980	**Sanger, Frederik** (Great Britain) **Gilbert, Walter** (USA) Methods of sequencing DNA **Berg, Paul** (USA) Development of genetic engineering	**Cronin, James W.** (USA) **Fitch, V. Logsdon** (USA) Discovery of the violation of laws of symmetry	**Dausset, Jean** (France) **Snell, George Davis** (USA) Discovery of a major histocompatibility complex in mice and humans **Benacerraf, Baruj** (USA) Discovery of genes that control immunity
1981	**Hoffmann, Roald** (USA) **Fukui, Kenichi** (Japan) Investigations of the mechanisms of chemical reactions	**Bloembergen, N.** (The Netherlands - USA) **Schawlow, Arthur** (USA) Development of laser spectroscopy **Siegbahn, Kai** (Sweden) Developments in high resolution electron spectroscopy	**Hubel, David H.** (USA) **Wiesel, Torsten M.** (USA) Information processing in visual systems **Sperry, Roger W.** (USA) Discovery of functional specialization in the cerebral hemispheres
1982	**Klug, Aaron** (Great Britain) Studies of the three - dimensional structure of viruses and development of crystallographic electron microscopy	**Wilson, Kenneth** (USA) Critical phenomena in connection with phase transition theory	**Bergström, Sune K.** (Sweden) **Samuelsson, Bengt I.** (Sweden) **Vane, John R.** (Great Britain) Discovery and function of prostaglandins
1983	**Taube, Henri** (USA) Studies of the reactional kinetics with dissolved inorganic substances	**Fowler, William** (USA) Studies of nuclear reactions leading to the formation of the elements in the universe **Chandrasekhar, S.** (India) Physical processes for the study of the structure and the evolution of stars	**McClintock, Barbara** (USA) Discovery of "jumping genes"
1984	**Merrifield, Robert B.** (USA) Development of a method for the automatic synthesis of amino acid chains	**Rubbia, Carlo** (Italy) **van der Meer, S.** (The Netherlands) Discovery of mesons (W+, W-, Z0) exchanged during weak nuclear interactions	**Jerne, Niels K.** (Denmark) Development of the immune system **Köhler, George J.F.** (Germany) **Milstein, Cesar** (Great Britain) Production of monoclonal antibodies
1985	**Hauptman, Herbert A.** (USA) **Karle, Jerome** (USA) Improvement of the X - ray diffraction method for the study of molecular structures	**Klitzing, Klaus von** (Poland - Germany) Discovery of quantized Hall effect	**Brown, Michael** (USA) **Goldstein, Joseph L.** (USA) Discovery and biology of cholesterol receptors
1986	**Herschbach, Dudley R.** (USA) **Polanyi, John C.** (Canada) **Lee, Yuan T.** (USA) Analysis of elementary chemical processes by using molecular beams	**Ruska, Ernest A.** (Germany) Design of the electron microscope **Binnig, Gerd** (Germany) **Röhrer H.** (Switzerland) Design of the scanning tunneling microscope (STM)	**Cohen, Stanley** (USA) Discovery of a epidermal growth factor **Levi-Montalcini, Rita** (Italy) Discovery of a nerve growth factor
1987	**Cram, Donald J.** (USA) **Lehn, Jean-Marie** (France) **Pedersen, Charles J.** (USA) Foundation of supramolecular chemistry	**Bednorz, Johannes** (Germany) **Müller, Karl A.** (Switzerland) Discovery of superconductivity in ceramic materials	**Tonegawa, Susumu** (Japan) Origin of antibody diversity
1988	**Deisenhofer, Johann** (Germany) **Huber, Robert** (Germany) **Michel, Hartmut** (Germany) Determination of the structure of a bacterial photosynthetic center	**Lederman, Leon** (USA) **Schwartz, Melvin** (USA) **Steinberger, Jack** (USA) Demonstration of the doublet structure of the leptons and discovery of the muon - neutrino	**Black, James W.** (Great Britain) **Elion, Gertrude B.** (USA) **Hitchings George H.** (USA) Synthesis of new drugs
1989	**Altman, Sidney** (Canada) **Cech, Thomas** (USA) Discovery of catalytic RNAs	**Dehmelt, Hans G.** (Germany) **Paul, Wolfgang** (Germany) Development of the ion trap technique **Ramsey, Norman** (USA) Development of atomic standards for accurate measurement of time	**Bishop, Michael J.** (USA) **Varmus, Harold E.** (USA) Discovery of oncogenes

1990	**Corey, E. James** (USA) Development of new methods for chemical synthesis	**Friedman, J.** (USA) **Kendall, Henry Way** (USA) **Taylor, Richard** (Canada) Innovations in electron - nucleon interaction in the field of the standard model	**Murray, Joseph E.** (USA) **Thomas, Edward D.** (USA) Development of fundamental techniques in transplantation surgery
1991	**Ernst, Richard** (Switzerland) Development of techniques for high - resolution nuclear magnetic resonance (NMR) spectroscopy	**de Gennes, P.G.** (France) Development of a simple model for the study of complex systems, in particular liquid crystals and polymers	**Neher, Erwin** (Germany) **Sakmann, Bert** (Germany) Development of "patch and clamp technique" and study of ionic channels
1992	**Marcus, Rudolf A.** (Canada) Studies on electron - transfer reactions in chemical systems	**Charpak, Georges** (Poland - France) New developments in subatomic particle detectors, in particular the multiwire proportional chamber	**Krebs, G. Edwin** (USA) **Fischer, Edmond H.** (USA) Discovery of cellular regulation by reversible phosphorylation
1993	**Mullis, Kary** (USA) Invention of the polymerase chain reaction (PCR) **Smith, Michael** (Great Britain) Development of site - directed mutagenesis	**Hulse, Russel** (USA) **Taylor, Joseph** (USA) Discovery of a new type of pulsar, which provided new possibilities for the study of gravitation	**Sharp, Phillip A.** (USA) **Roberts, Richard J.** (USA) Discovery of split genes
1994	**Olah, George A.** (USA) Isolation and study of carbocations	**Brockhouse, Bertram** (Canada) **Shull, Clifford** (USA) Development of neutron spectroscopy	**Gilman, Alfred G.** (USA) **Rodbell, Martin** (USA) Discovery of "G" proteins
1995	**Crutzen, Paul** (The Netherlands) **Molina, Mario** (USA) **Rowland, Sherwood F.** (USA) Studies of the chemistry of the atmosphere and in particular, of the formation and the destruction of the ozone layer	**Perl, Martin** (USA) Discovery of a subatomic particle called the tau lepton **Reines, Frederik** (USA) Detection of the neutrino	**Nüsslein-Volhard, Ch.** (Germany) **Wieschaus, Eric F.** (Germany) **Lewis, Edward B.** (USA) Genetic control of development. Discovery of homeotic genes
1996	**Kroto, Harold W.** (Great Britain) **Curl, Robert, Jr** (USA) **Smalley, Richard E.** (USA) Discovery of fullerenes	**Lee, David M.** (USA) **Osheroff, Douglas D.** (USA) **Richardson, Robert** (USA) Discovery of superfluidity in He^3	**Zinkernagel, Rolf M.** (Switzerland) **Doherty, Peter C.** (USA) How T - lymphocytes recognize virus - infected cells
1997	**Boyer, Paul** (USA) **Walker, John** (Great Britain) ATP - synthase. How the enzyme functions **Skou, Jens** (Denmark) Na^+/K^+ ATPase. Structure and function	**Cohen-Tannoudji, C.** (Algeria - France) **Chu, Steven** (USA) **Phillips, William** (USA) Development of techniques that use laser light for atom cooling	**Prusiner, Stanley B.** (USA) Discovery of prions
1998	**Kohn, Walter** (USA) **Pople, John A.** (Great Britain) Development of computational methods for use in quantum chemistry	**Laughlin, Robert B.** (USA) **Störmer, Horst L.** (Germany) **Tsui, Daniel C.** (China) Discovery of quantum fluid with fractionally charged excitation	**Furchgott, Robert F.** (USA) **Ignarro, Louis J.** (USA) **Murad, Ferid** (USA) Discovery that nitric oxide acts as a signaling molecule in the cardio - vascular system
1999	**Zewail, Ahmed H.** (Egypt – USA) Use of ultra - short laser flashes to study chemical reactions	**t'Hooft, Gerardus** (The Netherlands) **Veltman, Martinus** (The Netherlands) Elucidation of the quantum structure of electro-weak interactions	**Blobel, Guenter** (Germany) Discovery that proteins have intrinsic signals that govern their transport and localization in the cell
2000	**Heeger, Alan J.** (USA) **McDiarmid, Alan G.** (New Zealand) **Shirakawa, Hideki** (Japan) Electric conductivity of polyacetylene chains doped by halogens	**Alferov, Zhores** (Bielorussia) **Kroemer, Herbert** (Germany) Development of semiconductor heterostructures and materials for high speed optoelectronics **Kilby, Jack** (USA) Invention of integrated circuits	**Carlsson, Arvid** (USA) **Greengard, Paul** (USA) **Kandel, Eric R.** (USA) Signal transduction in the nervous system. Molecular basis of higher cognitive functions
2001	**Knowles, William S.** (United States) **Noyori, Ryoji** (Japan) **Sharpless, K. Barry** (United States) Studies on chirally catalysed hydrogenation reactions	**Cornell, Eric** (USA) **Ketterle, Wolfgang** (Germany) **Wieman, Carl E.** (USA) First experimental achievement of Bose - Einstein condensation	**Hartwell, Lee** (USA) **Hunt, R. Timothy** (Great Britain) **Nurse, Paul** (Great Britain) Control of the cell cycle.

Bibliography: Chemistry

Abbreviations used.
Dictionnary of Scientific Biography : (Dict. Sc. Biog.)
Biographical Dictionnary of Scientists : (Biog. Dict. of Sc.)
Biographical Memoirs, London : (Biog. Mem., London)
Biographical Memoirs, Academy of Sciences, Washington DC : Bioog. Mem. Ac. Sc., Washington DC
Current Biography : (Current Biog.)
Encyclopedia Universalis : (Enc. Univ.)
Encyclopedia Britannica : (Enc. Brit.)
Ouvrages généraux de biochimie et de biologie moléculaire : Ouv. Gen. Bioch & Biol. Mol.

Alder, K., (Enc. Brit.) Science, 1950. Nature, 1950.
Altman, S., Ouv. Gen. Bioch & Biol. Mol. Nature, 1989. Science, 1989.
Anfinssen, C., Ouv. Gen. Bioch & Biol. Mol. (Enc. Univ.)
Arrhénius, S.A., Dict. Sc. Biog., 1973, Vol 1.
Aston, F. W., Dict. Sc. Biog., 1970, Vol 1.
Baeyer, A von ,Dict. Sc. Biog., 1973, Vol 2. (Enc. Univ.).
Barton, D.H.R. ,(Enc. Brit.) Science, 1968. Nature, 1968.
Berg, P. ,Ouv. Gen. Bioch & Biol. Mol. (Enc. Univ.). Nature, 1980. Science, 1980.
Bergius, F., Dict. Sc. Biog., 1970, Vol 2.
Bosch, C., Dict. Sc. Biog., 1970, Vol 2. (Enc. Univ.).
Boyer, P., Ouv. Gen. Bioch & Biol. Mol. Nature, 1997. Science, 1997.
Brown, H.C., (Enc. Brit.) Science, 1979. Nature, 1979. (Enc. Univ.).
Buchner, E., Biog. Dict. of Sc., 1969. Dict. Sc. Biog., 1970, Vol 2. (Enc. Univ.).
Butenandt, A.F.J., Asimov's. (Enc. Univ.).
Calvin, M. ,Ouv. Gen. Bioch & Biol. Mol. (Enc. Univ.).
Cech, T., Ouv. Gen. Bioch & Biol. Mol. Nature, 1989. Science, 1989.
Corey, E.J., (Enc. Brit.) Science, 1968. Nature, 1968.
Cornforth, J., (Enc. Brit.) Science, 1975. Nature, 1975.
Cram, D.J., (Enc. Brit.) Science, 1987. Nature, 1987.
Crutzen, P., (Enc. Brit.) Science, 1993. Nature, 1993.
Curie, M., Dict. Sc. Biog., 1971, Vol 3. (Enc. Univ.). (Encyclop. Britannica).
Debye, P.J.W., Dict. Sc. Biog., 1971, Vol 3.
Deisenhofer, J., La Recherche, no 205. Pr la Science no 135. Nature, 1988. Science, 1988.
Diels, O., (Enc. Brit.) Science, 1950. Nature, 1950. Les Prix Nobel Allemands.
Eigen, M., (Enc. Brit.) Science, 1967. Nature, 1967. (Enc. Univ.).
Ernst, R. ,(Enc. Brit.) Science, 1991. Nature, 1991.
Euler-Chelpin, H. von , Dict. Sc. Biog., 1971, Vol 4. (Enc. Univ.).
Fischer, E., Dict. Sc. Biog., 1972, vol 5. (Enc. Univ.). (Encyclop. Britannica).
Fischer, E.O., (Enc. Brit.) Science, 1973. Nature, 1973.
Fischer, H. ,Dict. Sc. Biog., 1978, Vol 15. Biog. Dict. of Sc. 1981.
Flory, P.J. ,(Enc. Brit.) Science, 1974. Nature, 1974.
Fukui, K. , (Enc. Brit.) Science, 1987. Nature, 1987.
Giauque, W.F., (Enc. Brit.). (Enc. Univ.).
Gilbert, W., Ouv. Gen. Bioch & Biol. Mol.
Grignard, V., Dict. Sc. Biog., 1972, Vol 5. Enc. Univ., Enc. Brit.
Haber, F., Dict. Sc. Biog., 1972, Vol 5.
Hahn, O , Dict. Sc. Biog., 1972, Vol 6.
Harden, A., Dict. Sc. Biog., 1972, Vol 6. Biochemical Journal, 1941, 35.
Hassel, O., (Enc. Brit.) Science, 1969. Nature, 1969.
Hauptman, H., (Enc. Brit.) Science, 1985. Nature, 1985.
Haworth, W.N., Dict. Sc. Biog., 1972, Vol 6. (Enc. Univ.).
Herschbach, D.R., (Enc. Brit.) Science, 1986. Nature, 1986.
Herzberg, G., (Enc. Brit.) Science, 1971. Nature, 1971.
Hevesy, de Heves J., Dict. Sc. Biog., 1972, Vol 6. Current Biography Yearbook, 1959.
Heyrovsky, J. ,(Enc. Brit.) Science, 1959. Nature, 1959. (Enc. Univ.).
Hinshelwood, C.N., (Enc. Brit.) Science, 1956. Nature, 1956.
Hodgkin, D.M.C., (Enc. Brit.) Science, 1964. Nature, 1964. (Enc. Univ.).
Hoff, J.H. Van 't., Dict. Sc. Biog., 1976, vol 3. (Enc. Univ.).
Hoffmann, R., (Enc. Brit.) Science, 1981. Nature, 1981.
Huber, R. , La Recherche, no 205. Pr la Science no 135. Ouv. Gen. Bioch & Biol. Mol.
Joliot-Curie, Frédéric , Dict. Sc. Biog., 1973, Vol 7. Enc. Universalis. Enc. Britannica. Current Biography Yearbook, 1946.
Joliot-Curie, Irène , Current Biography Yearbook, 1940. Enc. Universalis.
Karle, J. , (Enc. Brit.) Science, 1985. Nature, 1985.
Karrer, P., Dict. Sc. Biog., 1978, Vol 15. Biog. Mem., London, 1978, Vol. 24.
Kendrew J.C., Ouv. Gen. Bioch & Biol. Mol.
Klug, A., Ouv. Gen. Bioch & Biol. Mol.
Knowles, W.S., Chem. Eng. News, 0ctober 15, 2001, p.5; Chem. Eng. News, November 5, 2001, p.37, Chimie Nouvelle, n°76, décembre 2001, p.3362; Pour la science, n°290, décembre 2001, p.7.
Kroto, H., (Enc. Brit.) Science, 1996. Nature, 1996.
Kuhn, R., Dict. Sc. Biog., 1973, Vol 7. Biog. Dict. of Sc. 1981.
Langmuir, I., Dict. Sc. Biog., 1980, Suppl.6.
Lee, Yuan Tseh ,(Enc. Brit.) ; Science, 1986. Nature, 1986
Lehn, J.M., (Enc. Brit.) Science, 1987. Nature, 1987.
Leloir, L.F., Ouv. Gen. Bioch & Biol. Mol.
Libby, W.F., (Enc. Brit.) Science, 1960. Nature, 1960.
Lipscomb, W.N., (Enc. Brit.) Science, 1976. Nature, 1976.
Marcus, R., (Enc. Brit.) Science, 1992. Nature, 1992.
Martin, A.J.P., (Enc. Brit.) Science, 1952. Nature, 1952.
McMillan, E.M., (Enc. Brit.) Science, 1951. Nature, 1951.
Merrifield, B., (Enc. Brit.) Science, 1984. Nature, 1984.
Michel, H., La Recherche, no 205. Pr la Science, no 135
Mitchell,P., Ouvrages généraux Bioch.&Biol. Mol.
Moissan, H., Dict. Sc. Biogr, 1973, Vol 9.
Molina, M., (Enc. Brit.) ; Science, 1993. Nature, 1993.
Moore, S., Ouv. Gén. Bioch.& Biol. Mol.
Mullikan, R.S., (Enc. Brit.) ; Science, 1966. Nature, 1966
Mullis, K., La Recherche no 260.
Natta, G., (Enc. Bri.) ; Science, 1963. Nature, 1963.
Nernst, W., Dict. Sc. Biog., 1978, Vol 15.
Norrish, R.G.W., (Enc. Brit.) Science, 1967. Nature, 1967.
Northrop, J.H, Biog. Dict. of Sc., 1981. Current Biography Yearbook, 1947.
Noyori, R., Chem. Eng. News, 0ctober 15, 2001, p.5; Chem. Eng. News, November 5, 2001, p.37. Chimie Nouvelle, n°76, décembre 2001, p.3362; Pour la science, n°290, décembre 2001, p.7.
Olah,G. ,(Enc. Brit.) Science, 1994. Nature, 1994.
Onsager, L., (Enc. Brit.) Science, 1967. Nature, 1967.
Ostwald, W., Dict. Sc. Biog., 1978, Vol 15.
Pauling, L.C., Ouv. Gen. Bioch & Biol. Mol.
Pedersen, D.J., (Enc. Brit.) Science, 1987. Nature, 1987.
Perutz, M.F., Ouv. Gen. Bioch & Biol. Mol.
Polanyi, J.Ch., (Enc. Brit.) Science, 1986. Nature, 1986.
Porter, G. ,(Enc. Brit.) Science, 1967. Nature, 1967.
Pregl, F., Dict. Sc. Biog., 1975, Vol 11.
Prelog, V., (Enc. Brit.) Science, 1975. Nature, 1975.
Prigogine, I., La Recherche, 1977. Nature, 1977. Science, 1977. (Enc. Univ.). (Encyclop. Britannica).
Ramsay, W., Dict. Sc. Biog., 1973, Vol 1.
Richards, W., Dict. Sc. Biog., 1975, Vol 11.
Robert, F. Curl ,(Enc. Brit.) Science, 1996. Nature, 1996.
Robinson, R., Biog. Mem., London ; 1976.
Rowland, Frank S., Science 1995. (Enc. Brit.). Nature, 1995.
Rutheford, E., Dict. Sc. Biog., 1975, Vol 12. Ouvrages généraux de physique.
Ruicka, L., Asimov's. Biog. Mem., London ; 1980, Vol. 26. New-York Times ; Sept, 27 ; 1976.
Sabatier, F., Biog. Mem., London, 1942-1944 Vol. 4.
Sanger, F., Ouv. Gen. Bioch & Biol. Mol.
Seaborg, G.T., (Enc. Brit.) Science, 1951. Nature, 1951. La Recherche.
Semenov, N.N., (Enc. Brit.) Science, 1956. Nature, 1956.
Sharpless, K. B , Chem. Eng. News, 0ctober 15, 2001, p.5; Chem. Eng. News, November 5, 2001, p.37. Chimie Nouvelle, n°76, décembre 2001, p.3362; Pour la science, n°290, décembre 2001, p.7
Skou, J., Science, 1997. Nature, 1997.
Smalley, R., Scienc 1997.Nature, 1997.
Smith, M., Voir : ADN recombinant p. 200 fig 11-8 / Pr la Science / Oligonuc. synthétiques.
Soddy, F., Biog. Mem., London ; 1957, Vol. 3.
Stanley, W.M., Current Biography Yearbook, 1947. Asimov's. National Cyclopedia of American Biogr. 1977, Vol 57.
Staudinger, H., (Enc. Brit.) Science, 1953. Nature, 1953.
Stein, W., (Enc. Brit.) Science, 1972. Nature, 1972.
Sumner, J.B., Biog. Mem., Ac. Sciences, Washington DC 1958, Vol. 31.
Svedberg, T., Dict. Sc. Biog., 1976, Vol 13. Biog. Mem., London 1972, Vol.18.
Synge, R.L.M., Ouv. Gen. Bioch & Biol. Mol.
Taube, H., (Enc. Brit.) Science, 1983. Nature, 1983.
Tiselius, A.W.K. Biog. Mem., London ; 1974, Vol 20.
Todd, A.R. Ouv. Gen. Bioch & Biol. Mol. The eight day of creation ; Horace F. Judson. Penguin Science.
Urey, H.C., Current Biography Yearbook, 1960. New-York Times Magazine January, 7 ; 1981.
Vigneaud, V. Du (Enc. Brit.) Science, 1955. Nature, 1955.
Virtanen, A.I., Dict. Sc. Biog., 1976, Vol 14. Biog. Dict. of Sc. 1981.
Walker, J. ,(Enc. Brit.) Science, 1997. Nature, 1997.
Wallach, O., Dict. Sc. Biog., Vol 14.
Wieland, H., Dict. Sc. Biog., 1976, Vol 14.
Wilkinson, G., (Enc. Brit.) Science, 1973. Nature, 1973.
Willstätter, R.M.., Dict. Sc. Biog., 1976, Vol 14. Les Prix Nobel allemands.
Windaus, A., Dict. Sc. Biog., 1976, Vol 14.
Wittig, G., (Enc. Brit.) Science, 1979. Nature, 1979.
Woodwaard, R.B , (Enc. Brit.) Science, 1965. Nature, 1965.
Ziegler, K. , (Enc. Brit.) Science, 1963. Nature, 1963.
Zsigmondy, R., Dict. Sc. Biog., 1976, Vol 14. Asimov's.

Bibliography: Physics

Abbreviations used.

BES - Biographical Encyclopedia of Scientists (2ème édition) 2 volumes. John Dainith, Sarah Mitchell, Elisabeth Tootill, Derek Gjertsen Institute of Physics Publishing, Bristol and Philadelphia, 1993 / BHL - A Brief History of Light. Richard J.Weiss, World Scientific, 1996 / Charpak 1 - La vie à fil tendu, Georges Charpak, Dominique Saudinos, Editions Odile Jacob, 1993 / Charpak 2 - Feux follets et champignons nucléaires, Georges Charpak, Richard L.Garwin, Editions Odile Jacob, 1997 / CSB - Les Conseils Solvay et les débuts de la physique moderne. Pierre Marage et Grégoire Wallenborn eds. Université Libre de Bruxelles, 1995 / DOM - La découverte des ondes de matière. Colloque organisé à l'occasion du centenaire de la naissance de Louis de Broglie, 16-17 juin 1992. Technique et documentation, Lavoisier Paris, 1994 / EPN - Europhysics News, / EU - Einstein's Universe. Nigel Calder, Wings Books, New York 1979 / JP - Le jardin des particules : l'univers tel que le voient les physiciens. Gordon Kane. Masson Paris, 1996 / LaL - Lasers and Light. Readings from Scientific American, W.H.Freeman & company, San Francisco, 1969 / Lederman - Une sacrée particule. L.Lederman Ed. Editions Odile Jacob / LPN - Les Prix Nobel - Imprimerie Royale, P.A.Norstedt et Söner, Stockholm, numéro annuel / LOS - Los Alamos, beginning of an era 1943-1945 - Los Alamos Historical Society, 1993, 64 pp / MP - Le Monde des Particules, Steven Weinberg, Univers des Sciences, Paris, 1985 / PIR - Physics, imagination and reality. P.R.Wallace, World Scientific, Singapore, 1991 / PLS - Pour la Science, / PMD - Les Physiciens Modernes et leurs Découvertes. E.Segrè, Fayard, 1984 / PS - Pioneers of Science ; Nobel Prize Winners in Physics. Robert L.Weber, second edition, Adam Hilger, Bristol, 1988 (2 volumes) / Rech - La Recherche, Revue Mensuelle / SAP - The discovery of Subatomic Particles. Steven Weinberg. W.H.Freeman & company, Oxford, 1983. / Sc.Am., Scientific American, / Silk - The Big Bang. The creation and evolution of the universe. Joseph Silk. W.H.Freeman & company, San Francisco, 1980 / TCP - Twentieth Century Physics, IOP Bristol 1995. 3 volumes. Editors : L.M.Brown, A.Pais, B.Pippard / UN - Universalia, complément annuel de " Encyclopedia Universalis " / XR - X-rays : the first hundred years. Alan Michette and Slawka Pfauntsch eds. John Wiley and Sons, Chichester 1996.

Alferov, Z., Physics today, Jan. 2000, Dec. 2000, Dec. 2001.
Alfven, H., BES 11 / LPN 1970 / PS 217 / UN 1996 p.473
Alvarez, L., Physics Today, vol. 90-8, 1987 / BES 15 / PMD / PS 212 / TCP ch.9.11. ; ch.15 / UN 1989 p.549.
Anderson, C., BES 20 / CSB / JP p.15 / PMD / PS 106 / SAP 163 / TCP p.386 / UN 1992 p.518
Anderson, Ph., BES 20 / PS 253 / Sc.Am., décembre 1992
Appleton, E. (Sir), BES 25 / PS 129
Bardeen, J., BES 52 / LPN 1972 / PLS, n° hors série, septembre 1995 / PS 159 / TCP p.939 / UN 1992 p.522 ; 1995 p.265
Barkla, Ch., BES 53 / CSB / PMD / PS 56
Basov, N., BES 58 / BHL ch.13 / EPN, vol.12, juillet 1981/ LaL / PS 197
Becquerel, A., Clefs CEA-Saclay-France n°spécial Radioactivité 34, hiver 96-97 / BES 63 / CSB / EPN 27, 1996 / JP p.31 / PMD / PS 15 / SAP
Bednorz, J.G., BES 62 / PLS, novembre 1987 ; août 1988/ Rech n°187, avril 1987 ; n°194, décembre 1987 ; n°195, janvier 1988 ; n°201, juillet 1988 ; n°240, février 1992/ PS 302 / UN 1995, p.265
Bethe, H., BES 81 / BHL ch.12 / Charpak 2 / JP 32 / LOS / PMD / PS 210 / Sc.Am., octobre 1992
Binnig, G., BES 83 / EPN, n°25, 1994 no 33, 2002. / PLS octobre 1985 / PS 297 / Rech n°138, novembre 1982 ; n°181, octobre 1986 ; n°183, décembre 1986 / UN 1991 p.241
Blackett, P,. (Baron), BES 87 / CSB / PMD / PS 131 / SAP 107 / TCP p.400 ; Ch.9.11 / UN 1975 p.447
Bloch, F., BES 90 / Charpak 1 / PS 142 / UN 1984 p.547
Bloembergen, N., BES 90 / BHL ch.13 / EPN, juillet 1981 ; février 1982 / LaL / PS 273 / Rech n°128, décembre 1981 / TCP p.1430
Bohr, A., BES 94 / LPN 1975 / PLS n° 69, juillet 1983 / PS 239
Bohr, N., B.Hoffman " L'étrange histoire des quanta ", Seuil, 1967 / BES 95 / BHL ch.10 / CSB / DOM / EU / LOS / PIR / PLS n°213, Juillet 93 / PMD / PS 67 / Rech n°125, mai 1981 ; n°171, novembre 1985 / TCP p.79-1188
Born, M.., BES 102 / BHL ch.8 / CSB / PMD / PS 150 / TCP p.985
Bothe, W., BES 105 / CSB / PMD / PS 153 / SAP 142 / TCP ch.9.11
Bragg, W.H.(Sir), BES 112 / CSB / LaL / PS 51 / XR
Bragg, W.L.(Sir), BES 112 / CSB / LaL / PMD / PS 54 / XR
Brattain, W., BES 115 / PS 161 / UN 1995 p.265
Braun, K., BES 116 / PS 38
Bridgman, P., BES 118 / LOS / PS 127
Brockhouse, B., EPN 1994, p.176 / TCP p.993
Chadwick, J. (Sir), Jules Six " La découverte du neutron ", Editions du CNRS 1987 / BES 152 / Charpak 1/ CSB / PMD / PS 101 / SAP/ TCP p.120 / UN 1975, p.453
Chamberlain, O., BES 153 / EPN 1993, p.24 / PLS, mai 1982 / PMD / PS 175 / Rech n°108, février 1980 / SAP 160
Chandrasekhar, S., BES 156 / EPN, décembre 1983 / PIR 209/ PS 281 / Rech n°150, décembre 1983 / Sc.Am. n°237, octobre 1977 / UN 1996 p.484
Charpak G., BES 161 / Charpak 1 / Charpak 2 / EPN août 1979 ; 1992, p.184 / Lederman / Rech n°128, décembre 1981 ; n°138, novembre 1982 ; n°249, décembre 1992 / TCP ch.9.11 / UN 1993, p.490
Cherenkov, P., BES 162 / Charpak 1 / PS 169 / TCP ch.9.11
Chu, S., Physics Today, octobre 1990 ; janvier 1996 / Sc.Am., mars 1987 ; février 1992 / Science, juillet 1991
Cockroft, J. (Sir), BES 170 / CSB / EPN vol.13, n°8/9, août 1992/ EU / LaL / PMD / PS 137 / TCP 1638
Cohen-Tannoudji Cl., CNRS info., janvier 1996 / Physics Today, octobre 1990, janvier 1996 / PLS, n° hors série, septembre 1995 / Rech., vol 25, janvier 1994
Compton, A., BES 173 / BHL ch.5 ; Ch.7 ; Ch.12 / CSB / LaL 8 / LOS / PMD / PS 82
Cooper, L., BES 177 / LPN 1972 / PLS, n° hors série, septembre 1995 / PS 224 / Sc.Am., décembre 1992
Cornell, E., (Science 269, (1995), 285 (1999), 292 (2001); Nature 412 (2001), 417 (2002); Physics today Dec. 1999.
Cronin, J., Les Particules Elémentaires, Bibliothèque Pour la Science / BES 189 / Charpak 1 / EPN 23, 1992 / PS 267 / Rech n°117, décembre 1980
Curie, P., Clefs CEA n° spécial (34) radiochimie, 1996/ BES 193 / CSB / JP 32 / PMD / PS 17 / SAP
Curie-Sklodowska, M., Eve Curie: " Marie Curie ", Gallimard, Paris 1938 / " Marie Curie et la Belgique " éditions de l'Université Libre de Bruxelles, 1990/ Revue du Palais de la Découverte, vol.16, n°157, avril 1988 / Clefs CEA-Saclay-France n°spécial Radioactivité 34, hiver 96-97 / BES 191 / BHL ch.10 / CSB / JP 32 / PLS, janvier 1983/ PMD / PS20 / SAP / TCP p.57
Dalen, N., BES 198 / PS 45
Davisson, C., BES 207 / CSB 178 / DOM / PS 109
de Broglie, L.-V., (Prince), B.Hoffman " L'étrange histoire des quanta ", Seuil, 1967 / BES 211 / CSB / DOM / PIR / PMD / PS 91/ Rech n°178, juin 1986, p.841 / UN 1988, p.536
de Gennes, P.-G., BES 341 / EPN 1992, n°23 / PLS , n°79, mai 1984 ; n°111, janvier 1987 ; n° hors série, septembre 1995 / Rech n°191, septembre 1987 ; n°238, décembre 1991 / UN 1992 p.504
Dehmelt, H., Bulletin de l'American Institute of Physics, n° 241, 22 septembre 1995 / BES 214 / PLS n°36 , octobre 1980 / Rech n°209, avril 1989 ; n°214, octobre 1989 ; n°216, décembre 1989
Dirac, P., B.Hoffman " L'étrange histoire des quanta " Seuil, 1967 / Hideki Yukawa " Tahibito " (the traveler). Translated by L.Brown and R.Yoshida. World Scientific, Singapore, 1982 / BES 228 / CSB / JP / PMD / PIR / PS 97 / UN 1981 p.185 ; 1985 p.561
Einstein, A., Carl Seeling (Editeur) " Ideas and opinions by Albert Einstein " Wings Books, NY / Kenji Sugimoto " Einstein ", Edition Belin 1990 (avec 500 photos) / BES 254 / BHL ch.8, Ch.10 / CSB / EU/ JP / LaL / PIR / PMD / PS 64 / Rech n°171, novembre 1985 ; n°220, avril 1990 / SAP / Silk / TCP p.250]
Esaki, L., BES 268 / LPN 1973 / PS 229
Fermi, E., Pierre de Latil " Enrico Fermi " collection Savants du Monde Entier, Editions Seghers 1963 / BES 283 / CSB / JP p.36 / LOS / PLS février 1982 / PMD / PS 114 / SAP / TCP p.1187-1197]
Feynman, R., Feynman Lectures on Physics, 3 volumes, Cal Tech / R.Feynman " La Nature de la Physique ", éditions du Seuil, Paris, 1980 / R.Feynman : Six easy pieces (1995) Helix Books, Addison Wesley et " Vous rêvez Monsieur Feynman ", Interéditions Paris, 1985 / James Gleick " Le génial Professeur Feynman ", Editions Odile Jacob 1994 / Les Particules Elémentaires. Bibliothèque Pour la Science / BES 287 / JP / LOS / PIR / PMD / PS 20/ / TCP 648 / UN 1989, p.573
Fitch, V., Les Particules Elémentaires. Bibliothèque Pour la Science / BES 294 / PS 261 / Rech n°117, décembre 1980
Fowler, W., BES 306 / EPN, décembre 1983 / PLS juillet 1987 / PS 279 / Rech n°150, décembre 1983 / Sc.Am. 237, octobre 1977 / Silk / UN 1996 p.496
Franck, J., BES 308 / CSB 158 ;180 / Lal / LOS / PS 75
Frank, I., BES 309 / PS 171 / TCP ch.9.11
Friedman, J., BES 316 / Lederman / PLS n°31, mai 1980 / Rech n°227, décembre 1990 / UN 1995, p.262
Gabor, D., BES 323 / LaL / LPN 1971 / PS 221/ TCP 1415 ; 1567 / UN 1980 p.549
Gell-Mann, M., Les Particules Elémentaires. Bibliothèque Pour la Science / BES 340 / Charpak 1 / JP 28 / LPN 1969 / PIR / PLS, octobre 1977 ; août 1988/ PMD / PS 215 / Sc.Am. juillet 1979 ; mars 1980 ; mars 1992
Giaever, I., BES 344 / LPN 1973 / PS 231.
Glaser, D., BES 348 / PS 179 / TCP ch.9.11
Glashow, S., BES 348 / Charpak 1 / EPN n° 11-2, janvier-février 1980 / JP 32 / PLS, n°2, décembre 1977 / PS 285 / Rech n°106, décembre 1979 / TCP 705 / UN 1977 p.284
Goeppert-Mayer, M.., BES 35 2/ PS 190 / TCP 1211
Guillaume, Ch., Advanced Materials and Processes, vol.151, mai 1997 / BES 372 / PS 62
Heisenberg, W., Elisabeth Heisenberg : " Heisenberg " (Collection Un savant, une époque) Editions Belin, Paris 1990) / Hideki Yukawa " Tahibito " (the traveler). Translated by L.Brown and R.Yoshida. World Scientific, Singapore, 1982 / "What is Life" / BES 399 / BHL ch.12 / CSB / JP 25 / PLS n°213, juillet 1993 / PMD / PS 95/ Rech n°130, février 1982 / TCP p.373 / UN 1977, p.497
Hertz, G., BES 409 / CSB / LaL / PS 78 / UN 1975, p.492
Hess, V., BES 413 / PS 104 / TCP p.369
Hewish, A., BES 415 / LPN 1974 / PS 235 / Rech, n°201, octobre 1988 / Sc.Am., novembre 1993
Hofstadter, R., BES 427 / JP 32 / Lederman / PS 182.